Principles of Neurocomputing for Science and Engineering

Principles of Neurocomputing for Science and Engineering

Fredric M. Ham, Ph.D.
Harris Professor of Electrical Engineering
Florida Institute of Technology

Ivica Kostanic
Senior RF Engineer
Agilent Technologies, Inc.

McGRAW-HILL, INC.

New York St. Louis San Francisco Auckland Bogotá
Caracas Lisbon London Madrid Mexico City Milan
Montreal New Delhi San Juan Singapore
Sydney Tokyo Toronto

McGraw-Hill Higher Education

A Division of The ***McGraw-Hill*** *Companies*

PRINCIPLES OF NEUROCOMPUTING FOR SCIENCE AND ENGINEERING

Published by McGraw-Hill, an imprint of The McGraw-Hill Companies, Inc. 1221 Avenue of the Americas, New York, NY, 10020.

This book is printed on acid-free paper.

1 2 3 4 5 6 7 8 9 0 DOC/DOC 0 9 8 7 6 5 4 3 2 1 0

ISBN 0070259666

Vice president/Editor-in-chief: *Kevin Kane*
Publisher: *Thomas Casson*
Executive editor: *Elizabeth A. Jones*
Sponsoring editor: *Catherine Fields*
Developmental editor: *Michelle L. Flomenhoft*
Senior marketing manager: *John T. Wannemacher*
Project manager: *Scott Scheidt*
Production supervisor: *Gina Hangos*
Senior supplement coordinator: *Rose M. Range*
Media technology producer: *Christopher Styles*
Cover design: *Craig E. Jordan*
Interior design: *Matthew Baldwin*
Compositor: *Keyword Publishing Services*
Typeface: *10/12 Times Roman*
Printer: *R. R. Donnelley & Sons Company*

Library of Congress Cataloging-in-Publication Data

Ham, Fredric M.
Principles of neurocomputing for science and engineering / Fredric M. Ham, Ivica Kostanic.
p. cm.
Includes bibliographical references and index.
ISBN 0-07-025966-6
1. Neural computers. 2. Neural networks (Computer science) I. Kostanic, Ivica. II. Title.

QA76.87.H352 2001
006.3′2—dc21 00-042710

www.mhhe.com

At McGraw-Hill Higher Education, the authors would like to thank Catherine Fields Shultz, Sponsoring Editor; Scott Scheidt, Project Manager; Craig E. Jordan for the cover design; and last but certainly not least, Michelle L. Flomenhoft, Developmental Editor. The authors truly appreciate Michelle's constant and meticulous attention to the development of this book.

Finally, the first author would like to thank his wife Emily and son Matt for their love and patience. The second author would like to thank Jasmina for her love and support.

To Emily and Matt (FMH)
To Nikola and Radojka (IK)

ABOUT THE AUTHORS

FREDRIC M. HAM, Ph.D., is the Harris Professor of Electrical Engineering at the Florida Institute of Technology, Melbourne. He received his B.S., M.S., and Ph.D. degrees in electrical engineering from Iowa State University in 1976, 1979, and 1980, respectively. Professor Ham has written numerous research papers in the areas of signal processing, biomedical engineering, neural networks, and control systems. He is head of the Information Processing Laboratory at Florida Tech and a consultant to industry. Professor Ham is a Senior Member of the Institute of Electrical and Electronics Engineers (IEEE). His society affiliations in IEEE include control systems, biomedical engineering, signal processing, and neural networks. He is also a member of Eta Kappa Nu, Tau Beta Pi, Phi Kappa Phi, and Sigma Xi.

IVICA KOSTANIC received the B.S. degree in electrical engineering from the University of Belgrade, Yugoslavia, in 1993, and the M.S. degree in electrical engineering from the Florida Institute of Technology in 1995. He is currently pursuing his Ph.D. degree in electrical engineering at the University of Central Florida, Orlando. He is a Senior RF Engineer and a researcher at Agilent Technologies, Inc., Melbourne, Florida. His principal research interests are in the areas of wireless communication, signal processing, and practical applications of neural networks. He is a Member of the Institute of Electrical and Electronics Engineers (IEEE), and he has published several papers in the areas of wireless communication, neural networks, and nonlinear controls.

PREFACE

> *"One cannot learn anything so well as by experiencing it oneself."*
>
> —Albert Einstein

There are four major questions that should be answered in the preface to any textbook: (1) What is the book about? (2) Why publish another book on the particular subject, in this case artificial neural networks (or neural networks)? (3) Who is the intended audience for the book, and what are the prerequisites? (4) What is specifically contained in the book?

Question: What is the book about?

Answer: This textbook is about artificial neural networks (or neural networks). More specifically, the book covers *neurocomputing*. So the question really is, What is neurocomputing? Neurocomputing is concerned with processing information. Unlike its *programmed computing* counterpart, a neurocomputing approach to information processing first involves a *learning* process within an artificial neural network (neural network) architecture. This neural architecture *learns* or adaptively responds to inputs according to a defined learning rule. After the neural network has *learned* what it needs to *know*, the trained network can be used to perform certain tasks depending on the particular application. Neural networks have the ability to learn from their environment and adapt to it in an interactive manner similar to their biological counterparts. Neurocomputing can play an important role in solving certain problems in science and engineering that would otherwise be difficult to solve, problems such as pattern recognition, optimization, event classification, control and identification of nonlinear systems, and statistical analysis, to name a few. Therefore, this book is primarily intended for those individuals who want to understand the underlying principles of artificial neural networks for neurocomputing and want to be able to apply various neurocomputing techniques to solve problems in science and engineering.

Question: Why another book on neural networks?

Answer: The field of neural networks is vast and interdisciplinary. It has drawn interest from researchers from many different areas, and the contributions have been enormous. There are many very good books available in the area of neural networks. However, the authors felt that there was a need for a book on neural networks that scientists and engineers could specifically identify with, that is, a book for scientists and engineers who want to apply neural networks to solve complex problems. This statement should not be misconstrued to mean this book is a smorgasbord of neural architectures with their associated training algorithms. Instead, this textbook on neural networks presents a variety of neurocomputing approaches that can be used to solve a vast number of problems in science and engineering. In almost every case, a solid mathematical foundation is presented for each neurocomputing concept along with illustrative examples to accompany the particular architecture and associated training algorithm.

Question: Who is the intended audience for the book, and what are the prerequisites?

Answer: This book is primarily for courses offered in neural networks at the graduate level; however, advanced undergraduate students could manage the material in this textbook with the proper background (the second part to this answer). Moreover, practicing engineers and scientists can also learn the material through self-study. The prerequisites for the successful study of neurocomputing by using this textbook are primarily a background in linear algebra and differential equations. It would be desirable to have knowledge in the areas of random variables and stochastic processes but it is not necessary because this material is briefly (though sufficiently) covered in Appendix A.

Question: What is specifically contained in the book?

Answer: The book is divided into two major parts, detailed below. Appendix A covers the mathematical foundation for neurocomputing.

PART I: Fundamental Neurocomputing Concepts and Selected Neural Network Architectures and Learning Rules

Overview: Part I consists of Chapters 1 through 5.

Chapter 1 introduces the reader to the basic idea of neural networks and neurocomputing. Also included is a brief history of neural networks.

Chapter 2 begins with a discussion of basic models of artificial neurons that are the building blocks of neural networks. Next is a discussion of different types of activation functions followed by a presentation of the adaptive linear element (Adaline) and the multiple Adaline (Madaline). The least mean-square (LMS) algorithm is presented next; then the simple perceptron is detailed followed by a brief discussion of feedforward multilayer perceptrons. Some basic learning rules are covered next. Many of these learning rules are the basis for training more sophisticated neural network architectures. Chapter 2 concludes with an overview of selected data preprocessing methods. All the material in Chapter 2 should be thoroughly covered if the reader is not familiar with artificial neural networks. This chapter sets the stage for the in-depth coverage of selected neural network architectures and associated learning rules presented in Chapters 3 through 5.

Chapter 3 presents a variety of mapping neural networks, beginning with associative memories, followed by backpropagation used to train feedforward multilayer perceptrons. Next, more advanced training methods are given for backpropagation. Counterpropagation is then presented, and the chapter concludes with a presentation of radial basis function neural networks.

Chapter 4 discusses selected self-organizing neural networks. This includes the Kohonen self-organizing map (SOM) and learning vector quantization (LVQ). The chapter concludes with adaptive resonance theory (ART) neural networks; the ART1 network is presented in detail.

Chapter 5 presents recurrent neural networks along with temporal feedforward networks (which are also recurrent networks). The distinction is made between these temporal feedforward neural networks and those that are not multilayer feedforward networks. Included in this chapter are the Hopfield network, simulated annealing, the Boltzmann machine, the simple recurrent network (SRN), time-delay networks, and distributed time-lagged feedforward neural networks.

PART II: Applications of Neurocomputing

Overview: Part II consists of Chapters 6 through 10.

Chapter 6 presents selected neurocomputing approaches for solving constrained optimization problems. Neural networks for linear programming and quadratic programming problems are presented. The chapter concludes with a discussion of neural networks for nonlinear continuous constrained optimization problems. This section includes neural networks for nonlinear programming penalty and barrier function methods. Also included are neural networks for ordinary and augmented Lagrange multiplier methods.

Chapter 7 covers structured neural network architectures and associated learning rules for solving various matrix algebra problems. A wide variety of important matrix decompositions (or factorizations) are presented along with a neurocomputing solution for each method. Neurocomputing approaches are also given for computing the pseudoinverse of a matrix, solving the algebraic Lyapunov equation, and solving the algebraic Riccati equation.

Chapter 8 covers neurocomputing approaches for solving linear algebraic equations. The various methods include a least-squares neurocomputing approach, a conjugate gradient learning rule, a generalized robust neurocomputing approach, regularization methods for ill-posed problems with ill-determined numerical rank, matrix splittings for iterative discrete-time methods, and the total least-squares problem. Also, L_∞-norm and L_1-norm neural network approaches are presented for solving linear algebraic equations.

Chapter 9 contains many neural network architectures for statistical analysis of data, including neural networks for principal-component analysis (PCA), principal-component regression (PCR), classical least squares (CLS), neural networks for nonlinear PCA and robust PCA, a neural network approach for partial least-squares regression (PLSR), and a neural network approach for robust PLSR.

Chapter 10 covers neural networks for signal processing applications, linear and nonlinear system identification, nonlinear control, and estimation. Many different examples are explained in detail. Also included is independent-component analysis (ICA) using neural networks for blind source separation. In addition, a fast ICA algorithm is presented along with an example that applies the fast ICA algorithm to separation of digital images.

Key Features of the Book

- Most training algorithms are highlighted in the book so they can be easily referenced.
- Some of these training algorithms have an accompanying MATLAB® function implementation given in the text (also highlighted). The code is relatively short and takes only a few minutes to enter into MATLAB.
- In addition, the MATLAB neural network toolbox is used extensively to experimentally illustrate certain neurocomputing concepts.

- Some of the problems have data that reside on the McGraw-Hill Higher Education website for this book and can be easily accessed. The URL for this book is: http://www.mhhe.com/engcs/electrical/ham.
- Detailed examples are presented in most sections to illustrate the neurocomputing concept covered.
- An extensive set of problems is given at the end of each chapter (except Chap. 1). Some problems require the use of MATLAB and the MATLAB neural network toolbox. In some cases the code is provided for the MATLAB function.
- Appendix A contains a comprehensive mathematical foundation for neurocomputing.

F. M. Ham
I. Kostanic

ACKNOWLEDGMENTS

The authors would like to express their gratitude to the reviewers of this book: Okan K. Ersoy, Purdue University; Sylvian R. Ray, University of Illinois; Yu Hen Hu, University of Wisconsin-Madison; Bruce MacLennan, University of Tennessee; Simon Y. Foo, Florida State University; and Risto Miikkulainen, The University of Texas at Austin. Their comments and suggestions were invaluable and essential in preparing the final manuscript of this book.

We would also like to express our gratitude to: Glenn Cohen for contributing to Section 1.3, "Neurocomputing and Neuroscience"; Sungjin Park for his assistance in the final preparation of Chapter 10; Tom McDowall for his contribution to the material in Section 9.7, "Robust PLSR: A Neural Network Approach"; the many students of ECE 5268 and CSE 5294 ("Theory and Applications of Neural Networks") taught by Professor Ham at Florida Tech; Joseph C. Wheeler with The Boeing Company for his support; Matt Ham for helping draw some of the figures; Edwin Sherman for his assistance in the Information Processing Laboratory; and The MathWorks, especially Naomi Fernandes, Brian Bostek, and Peter Trogos, for their support.

Contact Information for The MathWorks:

The MathWorks, Inc.
3 Apple Hill Drive
Natick, MA 01760
USA
Tel: 508-647-7000
Fax: 508-647-7001
E-mail: *info@mathworks.com*
Web:www.mathworks.com

LIST OF IMPORTANT SYMBOLS AND OPERATORS

$\boldsymbol{A} > 0$	positive definite $\boldsymbol{A}$ matrix
$\boldsymbol{A} \geq 0$	positive semi-definite (non-negative definite) $\boldsymbol{A}$ matrix
$\boldsymbol{A} < 0$	negative definite $\boldsymbol{A}$ matrix
$\boldsymbol{A} \leq 0$	negative semi-definite (non-positive definite) $\boldsymbol{A}$ matrix
$\boldsymbol{A}_{:k}$	denotes the kth column of the matrix $\boldsymbol{A}$
$\boldsymbol{A}_{k:}$	denotes the kth row of the matrix $\boldsymbol{A}$
$\tilde{A}$	complement of the set A
a_{ij}	i (row) j (column) element of the matrix $\boldsymbol{A}$
arg	argument of a complex quantity
adj	adjoint operator
$\hat{\boldsymbol{b}}_f$	calibration model
β	bias
$\mathscr{C}$	complex numbers
$\mathscr{C}^{n \times m}$	complex $n \times m$ matrices
$\mathscr{C}^{n \times n}$	complex $n \times n$ (square) matrices
$\mathscr{C}^{n \times 1}$	complex n-dimensional column vector
$\mathscr{C}^{1 \times n}$	complex n-dimensional row vector
$\mathscr{C}^{n}$	complex n-dimensional vector (row or column)
$(\mathscr{C}^{n}, \mathscr{C})$	n-dimensional complex vector space
C_X	autocovariance function
$\boldsymbol{C}_X$	covariance matrix
cof	cofactor operator
$\mathrm{cond}_p(\boldsymbol{A})$	condition number of the matrix $\boldsymbol{A}$
$\Delta\mathscr{E}$	energy change
diag	square matrix operator that selects the diagonal elements
δ_{ij}	Kronecker delta
det	determinant
exp	exponential
E	expectation operator
$\mathscr{E}$	energy function
E	energy function
e	exponential
$\mathscr{F}$	field
f_{bs}	binary sigmoid activation function
f_{lin}	linear activation function
f_L	logistic function

f_{hl}	hard limiter activation function
f_H	Huber's function
f_{hts}	hyperbolic tangent sigmoid activation function
f_M	M-estimator function
f_T	Talwar's function
f_s	sampling frequency, Hz
f_{shl}	symmetric hard limiter activation function
f_{sl}	saturating linear activation function
f_{ssl}	symmetric saturating linear activation function
g_{bs}	derivative of the binary sigmoid activation function
g_L	derivative of the logistic function
g_{hts}	derivative of the hyperbolic tangent sigmoid activation function
g_H	derivative of Huber's function
g_M	derivative of the M-estimator function
g_T	derivative of Talwar's function
gof	composition of g with f
γ	forgetting factor or leakage factor
$\boldsymbol{H}$	Hessian matrix
h^O	optimal number of factors
$\boldsymbol{I}_n$	$n \times n$ identity matrix
$\boldsymbol{I}$	identity matrix of appropriate dimension
inf	infimum
J	cost function
$\boldsymbol{J}$	Jacobian matrix
k	discrete time index
kurt	kurtosis operator
L	Lyapunov function
LT	lower triangular matrix operator
$\mathscr{L}$	Lagrangian function, Laplace transform operator, or successive relaxation matrix
ℓ	learning signal
λ	eigenvalue or Lagrange multiplier
m_x	mean of a stochastic process
$\boldsymbol{m}_x$	mean vector of a vector stochastic process
MIN	fuzzy intersection
MAX	fuzzy union
min	minimum value operator
max	maximum value operator
μ	learning rate parameter
ν	nullity

$\boldsymbol{\Omega}$	nonlinear mapping
$O(\gamma^2)$	higher-order effects in γ
ω_S	sampling frequency, $\frac{\text{rad}}{\text{sec}}$
Pr	probability
p_X	probability density function
$\mathscr{P}_X$	probability distribution function
Ψ	loss function
$\boldsymbol{\Phi}$	regression matrix
$\boldsymbol{\phi}$	state transition matrix
$\boldsymbol{\phi}_\mathbf{h}$	prototype memory
$\Re$	real numbers
$\Re^{n\times m}$	real $n \times m$ matrices
$\Re^{n\times n}$	real $n \times n$ (square) matrices
$\Re^{n\times 1}$	real n-dimensional column vector
$\Re^{1\times n}$	real n-dimensional row vector
$\Re^{n}$	real n-dimensional vector (row or column)
$(\Re^n, \Re)$	n-dimensional real vector space
$(\Re^n(s), \Re(s))$	n-dimensional rational vector space
$\mathscr{R}$	Rayleigh quotient
$\boldsymbol{R}_X$	correlation matrix
R_x	autocorrelation function
$\mathscr{R}_x$	time autocorrelation function
ρ	matrix rank or vigilance parameter
ρ_X	correlation coefficient
$\boldsymbol{S}_x$	power spectral density matrix
$\mathscr{S}$	probability space or sample space
sup	supremum
sgn	signum function
σ	standard deviation, singular value, spread parameter, or regularization parameter
σ^2	variance
$\boldsymbol{\sigma}_h$	fixed stable point
t	continuous time
tr	trace of a matrix
trace	trace of a matrix
T	temperature
T_{binary}	threshold logic operator
T_s	sampling period
Tr	sum over all possible configurations
θ	threshold
$\boldsymbol{\theta}$	parameter vector
$V(\boldsymbol{x})$	Lyapunov function of the state vector $\boldsymbol{x}$
var	variance operator

vec	operator to form a vector from a matrix by "stacking" the columns
vecd	operator to select the major diagonal elements from a square matrix
$\mathscr{X}$	set of vectors
$(\mathscr{X}, \mathscr{F})$	vector (linear) space
$\overline{X}$	mean of the random variable X
$\overline{x}$	complement or NOT of x
$\boldsymbol{x}^*$	complex conjugate transpose of the vector $\boldsymbol{x}$ or optimal solution
$\boldsymbol{x}^T\boldsymbol{y}$	inner product of two vectors $\boldsymbol{x}$ and $\boldsymbol{y}$
$\langle \boldsymbol{x}, \boldsymbol{y} \rangle$	inner product of two vectors $\boldsymbol{x}$ and $\boldsymbol{y}$
$\boldsymbol{x}\boldsymbol{y}^T$	outer product of two vectors $\boldsymbol{x}$ and $\boldsymbol{y}$
z^{-1}	unit delay operator
$(\boldsymbol{A})^T$	transpose of the matrix $\boldsymbol{A}$
$(\boldsymbol{A})^{-1}$	inverse of the matrix $\boldsymbol{A}$
$(\boldsymbol{A})^{1/2}$	square root of the matrix $\boldsymbol{A}$
$(\boldsymbol{A})^{T/2}$	transpose of the square root of the matrix $\boldsymbol{A}$
$(\boldsymbol{A})^+$	pseudo-inverse of the matrix $\boldsymbol{A}$
$(\boldsymbol{A})^*$	complex conjugate transpose of the matrix $\boldsymbol{A}$
$(\boldsymbol{A})^H$	Hermitian transpose of the matrix $\boldsymbol{A}$
$\lvert\boldsymbol{A}\rvert$	determinant of the matrix $\boldsymbol{A}$
$\lvert\alpha + j\beta\rvert$	magnitude or absolute value of a complex quantity
$\angle(\alpha + j\beta)$	angle or argument of a complex quantity
$\det(\boldsymbol{A})$	determinant of the matrix $\boldsymbol{A}$
$\Delta\boldsymbol{w}$	weight vector change
∇	gradient operator
$\nabla_x\mathscr{E}(\boldsymbol{x})$	gradient of the function $\mathcal{E}$ with respect to the vector $\boldsymbol{x}$
$\boldsymbol{\nabla}^T\boldsymbol{f}(\boldsymbol{x})$	Jacobian matrix of the vector function $\boldsymbol{f}$ with respect to the vector $\boldsymbol{x}$
∇^2	Laplacian operator
$\nabla_x^2 f(\boldsymbol{x})$	Hessian matrix of the function f with respect to the vector $\boldsymbol{x}$
$\oplus$	Kronecker sum or OR logic operator
$\otimes$	Kronecker product
$\odot$	Khatri-Rao product
$\cup$	union
$\cap$	intersection
$\subset$	subset of
$\in$	contained in
$\notin$	does not belong to
$\forall$	for all
$\ni$	such that
$\wedge$	and (also the MIN operator)
$\vee$	or (also the MAX operator)
$\varnothing$	empty set
$\Rightarrow$	implies
$\rightarrow$	maps to
$\lVert\boldsymbol{x}\rVert_p$	L_p norm of the vector $\boldsymbol{x}$
$\lVert\boldsymbol{x}\rVert_1$	L_1 norm (absolute value norm) of the vector $\boldsymbol{x}$
$\lVert\boldsymbol{x}\rVert_2$	L_2 norm (Euclidean norm) of the vector $\boldsymbol{x}$
$\lVert\boldsymbol{x}\rVert_\infty$	L_∞ norm (Chebyshev norm) of the vector $\boldsymbol{x}$

$\|\boldsymbol{x}\|_{-\infty}$	$L_{-\infty}$ norm (negative infinity norm) of the vector $\boldsymbol{x}$
$\|\boldsymbol{x}\|_W$	inner-product-generated norm of the vector $\boldsymbol{x}$
$\|\boldsymbol{x}\|_{2-Q}$	weighted Euclidean norm of the vector $\boldsymbol{x}$
$\|\boldsymbol{A}\|_p$	L_p norm of the matrix $\boldsymbol{A}$
$\|\boldsymbol{A}\|_1$	L_1 norm of the matrix $\boldsymbol{A}$ (largest column sum of absolute values)
$\|\boldsymbol{A}\|_2$	spectral norm of the matrix $\boldsymbol{A}$
$\|\boldsymbol{A}\|_\infty$	L_∞ norm of the matrix $\boldsymbol{A}$ (largest row sum of absolute values)
$\|\boldsymbol{A}\|_F$	Frobenius norm of the matrix $\boldsymbol{A}$
$\sigma_r(\boldsymbol{A})$	spectral radius of the matrix $\boldsymbol{A}$

LIST OF IMPORTANT ABBREVIATIONS

Adaline	adaptive linear element
AIC	Akaike's information theoretic criterion
AND	AND logic function
ANN	artificial neural network
APEX	adaptive principal component extraction
ARMA	autoregressive moving average
ARMAX	autoregressive moving average with exogenous inputs
BER	bit error rate
BFGS	Broyden-Fletcher-Goldfarb-Shanno
BIBO	bounded-input bounded-output
BP	backpropagation
CAM	content addressable memory
CLS	classical least-squares
CPCA	constrained PCA
DOA	direction of arrival
DPC	discrete Picard condition
DTLFNN	distributed time lagged feedforward neural network
EVD	eigenvalue decomposition
FFPA	fast fixed-point algorithm
FFT	fast Fourier transform
FIR	finite impulse response
FMMC	fuzzy min-max classifier
GHA	generalized Hebbian algorithm
GSVD	generalized SVD
IC	independent component
ICA	independent component analysis
ILS	inverse least squares
isL	in the sense of Lyapunov
KO	Karhunen-Oja
LAPART	laterally primed adaptive resonance theory
LDU	lower diagonal upper matrix decomposition
LMBP	Levenberg-Marquardt backpropagation
LMS	least mean-square

LP	linear programming
LVQ	learning vector quantization
LSR	linear shift register
LU	lower upper matrix decomposition
Madaline	multiple Adaline
MAJ	majority logic function
MAW	mountain associated wave
MDL	minimum description length
NGE	nested generalized exemplar
MIMO	multiple-input multiple-output
MLP	multilayer perceptron
MLP NN	multilayer perceptron neural network
MRAC	model reference adaptive control
MRI	Madaline rule I
MRII	Madaline rule II
MSE	mean square error
MUSIC	multiple signal classification
NARMA	nonlinear autoregressive moving average
NARMAX	nonlinear autoregressive moving average with exogenous inputs
NARX	nonlinear autoregressive with exogenous inputs
NIPALS	nonlinear iterative partial least squares
NIR	near-infrared
NLPCA	nonlinear PCA
NN_c	neural network controller
NN_I	neural network to perform system identification
NOR	NOT-OR logic function
NP	nonlinear programming or nondeterministic polynomial (time complete)
OLS	orthogonal least-squares
OR	OR logic function
PCA	principal component analysis
PCR	principal component regression
PLSNET	partial least-squares regression neural network
PLSNET-C	PLSNET-calibration
PLSNET-P	PLSNET-prediction
PLSR	partial least-squares regression
PN	pseudo noise
PRESS	predicted residual error sum of squares
QP	quadratic programming
RBF	radial basis function
RBF NN	radial basis function neural network
RLS	recursive least-squares

RMS	root-mean-square
ROB	robust
RWLS	recursive weighted least-squares
SEC	standard error of calibration
SEE	standard error of estimation
SEP	standard error of prediction
SISO	single-input single-output
SGA	stochastic gradient ascent
SNR	signal-to-noise ratio
SOM	self-organizing map
SOR	successive overrelaxation
SRN	simple recurrent network
SSE	sum-squared error
SVD	singular value decomposition
TDNN	time delay neural network
TLS	total least-squares
TLU	threshold logic unit
TSVD	truncated SVD
VOL	volcano
wss	wide-sense stationary
XNOR	exclusive NOR logic function
XOR	exclusive OR logic function

CONTENTS IN BRIEF

CONTENTS

PART II Applications of Neurocomputing

PART I

Fundamental Neurocomputing Concepts and Selected Neural Network Architectures and Learning Rules

CHAPTER 1

Introduction to Neurocomputing

1.1 WHAT IS NEUROCOMPUTING?

Neurocomputing is concerned with processing information. Unlike its *programmed computing* counterpart, a neurocomputing approach to information processing first involves a learning process within an artificial neural network† (or neural network) architecture that adaptively responds to inputs according to a learning rule. After the neural network has *learned* what it needs to *know*, the trained network can be used to perform certain tasks depending on the particular application. Neural networks have the ability to learn from their environment and to adapt to it in an interactive manner similar to their biological counterparts. Indeed, this is an exciting prospect because of the vast possibilities that exist for performing certain functions with *artificial* neural networks that can emulate (to a limited degree) the comparable *biological* function.

As an example, individuals can perform the task of *pattern recognition* fairly well. We "see" an automobile that passes by us on the street, and it "catches our eye." Maybe the car was traveling fast, but we were able to recognize some distinctive feature(s) of the body styling that we *associated* with our stored image of a 1984 Ferrari Testarossa. Now it is apparent why we were able to capture only a glimpse of the car, but nonetheless, that was enough for us to make the correct identification of the type of vehicle. We take this process for granted; however, it is very complex. It was not necessary for the car to be parked and every minute detail of it scrutinized in order for us

†Artificial neural networks are also referred to as connectionist networks (or systems), parallel distributed processing systems, or neurocomputers.

to make the proper identification. On the contrary, a quick glance at the moving car, observing, for example, the distinctive side louvers *and* maybe the horse standing up on its hind legs on the logo, was enough to make a positive identification. This type of pattern recognition can be performed with an *artificial neural network*, specifically, a Hopfield network (cf. Sect. 5.3) which is a recurrent neural network. One of the powerful features of such a network is its ability to recall a stored memory given only partial input data, for example, the side louvers and standing horse on the logo of the Ferrari Testarossa.

Even though people are not as fast or precise as a digital computer (e.g., a digital computer is much faster at multiplying two seven-digit numbers), they are typically much better at perceiving and identifying an object of interest in a natural scene, or interpreting natural language, and many other natural cognitive tasks, than a digital computer. Why can we perform certain tasks much better than a digital computer? The answer to this question is still not totally understood; however, enough is known that allows certain functions that we as humans perform very well to be emulated by artificial neural networks. One reason why we are much better at recognizing objects in a complex scene, for example, is due to the way that our brain is organized. Our brain employs a computational architecture that is well suited to solve very complex problems that a digital computer would have a difficult time with. The basic processing unit in the brain is the *neuron* (nerve cell), whereas in a digital computer we can think in terms of silicon-based logic gates. Neurons are approximately six orders of magnitude slower than silicon logic gate events. However, the brain can compensate for the relatively slow operational speed of the neuron by processing data in a highly parallel architecture that is massively interconnected. It is estimated that the human brain must contain on the order of 10^{11} neurons and approximately three orders of magnitude more connections or synapses. Therefore, the brain is an *adaptive, nonlinear, parallel* computer that is capable of organizing neurons to perform certain tasks. Neurocomputing systems very crudely model the human brain. Although research in the area of modeling the human brain is an interesting and stimulating field, this is not the thrust of this book. Section 1.3 presents an overview of biological neural networks; however, this is done for the sake of completeness and to present certain neurobiological concepts that are directly relevant to artificial neural networks for neurocomputing. Our presentation of artificial neural networks is made from the perspective of the engineer; however, the material has relevance to the scientist as well as the engineer. Therefore, this book is primarily intended for those individuals who want to understand the underlying principles of artificial neural networks for neurocomputing, and to be able to apply various neurocomputing techniques to solve problems in science and engineering.

The ability to *learn* by example and *generalize* are the principal characteristics of artificial neural networks. A neural network is trained by presenting several patterns that the network must learn according to a learning rule (algorithm). In the above example, being able to recognize a Ferrari Testarossa in the first place means that the individual had observed this type of automobile at least once before and was told what it was (the training process). The learning process is what sets a neural network apart from processing information on a digital computer that must be *programmed.* If a neural network has *generalized*, this means that it can classify input patterns to an acceptable level of accuracy even if they were never used during the training process. The Ferrari Testarossa that the person identified in the example above probably had never been seen before; that is, this *exact* car probably had never been seen before, but others like it had been observed.

An artificial neural network stores the knowledge that has been learned during the training process in the synaptic weights of the neurons. The artificial neuron† is the building block for any artificial neural network. How the network of neurons is organized, the learning rule that prescribes the adjustment of the synaptic weights, and the criterion that dictates when the training process is complete all characterize the particular type of neural network. There are many different types of neural networks with varying degrees of complexity; however, most of them possess similar characteristics. For example, the basic parallel computational architecture that allows for fast computations is probably the commonality among most all types of neural networks. Most artificial neural architectures have many neurons that are highly interconnected (many synapses), like their biological counterparts. The nonlinearity at the neuron output is also a distinctive component of a neural network, although several types of neural networks are "linear" architectures. For example, the structured networks presented in Chapter 7 are all linear networks.

The field of neural networks is now extremely vast and interdisciplinary, drawing interest from researchers in many different areas such as engineering (including biomedical engineering), physics, neurology, psychology, medicine, mathematics, computer science, chemistry, and economics. Artificial neural networks provide a neurocomputing approach for solving complex problems that might otherwise not have a tractable solution. Applications of neural networks include (but are not limited to) prediction and forecasting, associative memory, function approximation, clustering, data compression, speech recognition and synthesis, nonlinear system modeling, nonlinear control, pattern classification, feature extraction, combinatorial optimization, solution of matrix algebra problems, blind source separation, and solution of differential equations. Taking a neurocomputing approach to solving certain problems has many advantages, for example, fault tolerance. The implementation of a neural network in hardware tends to be inherently fault-tolerant. If the neural network architecture has a neuron that is damaged, or if the connections are damaged, the performance of the entire network is typically only slightly affected due to this failure. This is a contributing factor to the *robust* nature of neural networks. Because the information is *distributed* in a neural network, the damage would have to be very serious (i.e., many damaged neurons and/or connections) for the neural network to completely fail. As stated above, arti-

---

†Artificial neurons are also referred to as (information) processing elements, units, cells, nodes, and neurodes.

ficial neurons are typically nonlinear; therefore, the network itself is a nonlinear system. Many times a nonlinear system is viewed as being sinister; however, the nonlinear nature of a neural network is a very important property. This is especially true when the physical process of interest is absolutely nonlinear and the measurements made from this system are used for training a neural network. The adaptive nature of a neural network is also an important quality. Because of the adaptability of the network's synaptic weights, it can interact and respond to its environment. Therefore, if the environment changes, the network can respond to these changes and basically retrain itself after an initial training session. Moreover, for a nonstationary environment, a neural network can be designed to perform real-time adaptation of its weights. Neural networks are good at performing functions such as adaptive control, adaptive pattern classification, and adaptive signal processing. Therefore, one way to describe the presentation of much of the material in this book is from the viewpoint of adaptive signal processing.

Now that we have defined what an artificial neural network is, and the advantages of taking a neurocomputing approach to solving certain problems, we give some of the historical background about the field of neural networks. This is not meant to be a comprehensive historical discourse on neural networks, but hopefully it serves to place the material in this book into proper perspective.

1.2 HISTORICAL NOTES

It is not the intention to present a complete history of neural networks, but to introduce some of the major research accomplishments that carved the way for others. An overview of early significant achievements is presented that should provide the reader with an appreciation of how certain contributions have led to the field's development over the years. Summaries of early artificial neural network research can be found in [1–4].

Historical overview

- McCulloch and Pitts, 1943
  It is said that the modern age of neural networks began with the work of Warren S. McCulloch and Walter Pitts in 1943 [5]. They presented five assumptions governing the operation of neurons. These describe what is now known as the McCulloch–Pitts neuron. There is no training of these neurons; however, they can act as certain logic functions. The McCulloch–Pitts neuron model laid the foundation for future developments in neural networks.
- Hebb, 1949
  In the paper by Donald Hebb in 1949 [6], he describes a *learning process* that was postulated from a neurobiological viewpoint. Hebb stated that information is stored in the connections of the neurons and postulated a learning strategy for adjustment of the connection weights. This was the first time a

learning rule was presented that allowed for the adjustment of the synaptic weights, which has had a major impact on later work in this field.

- von Neumann, 1958
  John von Neumann played an important role in the development of the digital computer and was one of the great figures in science in the first part of the 20th century. He was also an active member of the Manhattan Project. John von Neumann saw the potential analogy between digital computers and the brain from the earliest days of the computer [7]. For example, he mentions the importance of memory in biological nervous systems as in the electronic one.
- Rosenblatt, 1958
  In 1958, the original concept of the *perceptron* was developed by Frank Rosenblatt [8]. It was the first precisely defined, computationally oriented neural network that attracted much attention by engineers because of its complex adaptive behavior. It was a machine that could be trained to classify certain patterns.
- Widrow and Hoff, 1960
  The Adaline (*ada*ptive *lin*ear *e*lement) of Bernard Widrow, trained by the least-mean-square (LMS) learning rule [9], closely resembles Rosenblatt's perceptron. The Adaline was extended to many Adalines (Madaline). There were many applications of the Adaline and Madaline, for example, in adaptive control and pattern recognition.
- Minsky and Papert, 1969
  Marvin Minsky and Seymour Papert slowed down neural network research in 1969 (and for several years thereafter) when they pointed out in their book, *Perceptrons* [10], that single-layer neural networks have limited capabilities. One such example is the exclusive-OR (XOR) problem. Rosenblatt had studied architectures with more than one layer of neurons and believed that they could overcome the limitations of the simple perceptron; however, there was no learning rule known at the time [11]. Minsky and Papert doubted that one could be found.
- Kohonen, 1972; and Anderson, 1972
  In spite of the slowdown of neural network research in the 1970s, several people continued research. One of the main areas was content-addressable associative memories. In 1972, Teuvo Kohonen published his paper on correlation matrix memories [12]. In the same year, James Anderson independently proposed the same model as Kohonen [13].
- von der Malsburg, 1973
  The pioneering work of Christoph von der Malsburg studied networks possessing synaptic modification rules that could produce a model cortex that displayed the ability to modify and organize itself [14]. His work was inspired by animal experiments in the early 1970s.
- Werbos, 1974
  The first description of the *backpropagation* algorithm for training multilayer feedforward perceptrons was given by Werbos in 1974 [15].
- Little and Shaw, 1975
  In their paper [16], Little and Shaw describe a neural network that uses a probabilistic model of a neuron instead of a deterministic one.

- Lee and Lee, 1975
  Lee and Lee presented the fuzzy McCulloch–Pitts neuron model [17].
- Grossberg, 1976
  In 1976, Grossberg published a paper that presents a theoretical analysis inspired by the developmental physiology of cortical organization [18]. He stated that there is strong evidence for the development and modification of feature detectors in the visual cortex in response to an environment.
- Amari, 1977
  In his 1977 paper [19], Amari discusses pattern associators. In one type of associator, the input pattern evokes an appropriate, but different, output pattern. In the other type, using what he calls concept-forming networks (these are recurrent networks), the input and output patterns are the same pattern, and the output pattern can be fed back to the input of the network.
- Hopfield, 1982
  It is said that the *modern* age of neural networks began with the publication of the paper by John Hopfield [20] (Nobel laureate in physics). Hopfield presented a sophisticated and comprehensive description of the workings of a *recurrent* neural network and what it could actually do. The network can store information (e.g., patterns) in a dynamically stable environment and is able to perform the function of data storage and retrieval. Given a noisy input to the network, the associated pattern stored in memory can be properly retrieved in spite of the incomplete (corrupted) version that was presented to the network.
- Kohonen, 1982
  Teuvo Kohonen presented the self-organizing feature map in 1982 [21]. It is an unsupervised, competitive learning, clustering network in which only one neuron (or only one neuron in a group) is "on" at a time.
- Oja, 1982
  Erkki Oja presented a single linear neuron trained by a normalized Hebbian learning rule that acts as a principal-component analyzer [22]. The neuron is capable of adaptively extracting the first principal eigenvector from the input data. Extension of his work has led to many different neural network approaches for adaptively estimating multiple principal eigenvectors.
- Fukushima, Miyake, and Ito, 1983
  The neocognitron was developed by Fukushima, Miyake, and Ito for character recognition [23].
- Kirkpatrick, Gelatt, and Vecchi, 1983
  Even though the paper by Kirkpatrick, Gelatt, and Vecchi [24] was not a neural network paper, it set the stage for the Boltzmann machine.
- Kampfner and Conrad, 1983
  Kampfner and Conrad investigated evolutionary computation methods for training neural networks [25]. The history of evolutionary computation is detailed in the book edited by D. B. Fogel [26].
- Ackley, Hinton, and Sejnowski, 1985
  A learning algorithm for Boltzmann machines was presented by Ackley, Hinton, and Sejnowski in 1985 [27].
- Parker, 1985; and LeCun, 1985

Independent discoveries of the backpropagation training algorithm were made by Parker [28] and LeCun [29].

- Herault, Jutten, and Ans, 1985
  Herault, Jutten, and Ans used neural networks for blind separation of independent source signals [30], that is, neural network realization of independent-component analysis (ICA) for blind source separation [31].
- Rumelhart, Hinton, and Williams, 1986
  Independent discovery of the backpropagation training algorithm was made by Rumelhart, Hinton, and Williams [32, 33].
- Carpenter and Grossberg, 1987
  Carpenter and Grossberg developed self-organizing neural networks based on adaptive resonance theory (ART) [34].
- Sivilotti, Mahowald, and Mead, 1987
  The first VLSI realization of neural networks is due to Sivilotti, Mahowald, and Mead [35].
- Broomhead and Lowe, 1988
  First exploitation of radial basis functions in designing neural networks is due to Broomhead and Lowe [36].

McCulloch–Pitts neuron

As previously mentioned, at the beginning of the chronological overview of neural network research, the work by Warren S. McCulloch and Walter Pitts in 1943 essentially started the modern age of neural networks. Therefore, it seems appropriate to begin with their concept of a neuron.

The McCulloch–Pitts neuron is a very simple *two-state* device. It is either on or off; that is, its output is binary. The output of a particular neuron cannot coalesce with output of another neuron; however, it can branch to another neuron and terminate as an input to that neuron, or terminate on itself. Two types of terminations are allowed: an *excitatory* input or an *inhibitory* input. There can be any number of inputs to a neuron. Whether the neuron will be on (firing) or off (quiet) depends on the *threshold* of the neuron. The following are the physical assumptions that they made regarding their simple neuron [5]:

1. The neuron activity is an all-or-nothing process; that is, the activation of the neuron is *binary*. At any discrete time step when the neuron "fires," the activation is 1; and when it is "quiet," the activation is 0. These are the two possible states of the neuron.
2. A certain number of fixed (weighted) neuron synapses must be excited within a discrete time step to excite the neuron, and this number is independent of any previous activity.
3. The only significant delay within the "nervous system" is the synaptic delay, that is, the time it takes to broadcast the information on the synapses.
4. The activity of *any nonzero* inhibitory synapse will absolutely prevent excitation of the neuron at that discrete time step.
5. The structure of the network does not change with time.

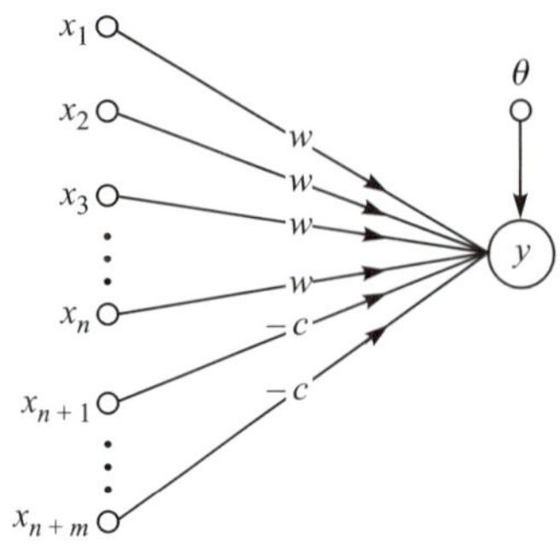

Figure 1.1 Architecture of a McCulloch–Pitts neuron y with threshold θ.

Figure 1.1 shows an example of a McCulloch–Pitts neuron. The connections from x_1 to x_n are excitatory inputs because the synaptic weights w are positive. The connections from x_{n+1} to x_{n+m} are inhibitory because the synaptic weights $(-c)$ are negative. Therefore, there are n excitatory inputs and m inhibitory inputs to neuron y. In one discrete time step, the signals pass from the neuron inputs (x_1 to x_{n+m}) to y. The state of neuron y at the discrete time instant k is determined by the states of the inputs x_1 to x_{n+m} at time $k-1$. The total input signal received is u and if $u \geq \theta$, then the neuron output is $y = 1$ (neuron output is excitatory, or firing); however, if $u < \theta$, then $y = 0$ (neuron output is inhibitory, or quiet), that is,

$$y = \begin{cases} 1 & \text{for } u \geq \theta \\ 0 & \text{for } u < \theta \end{cases} \tag{1.1}$$

The absolute inhibition state of the neuron requires the following condition to hold:

$$nw - c < \theta \tag{1.2}$$

This is a direct result of assumption 4 stated above; that is, *any* nonzero inhibitory synapse will absolutely prevent excitation of the neuron. (It only takes one!) If the threshold of the neuron is set to $\theta = hw$ (h is an integer), then neuron y will fire if $hw \geq \theta$, that is, if h or more (up to n) excitatory inputs are received with no inhibitory input. By setting the weights and the threshold of the McCulloch–Pitts neuron, some simple logic (Boolean) functions can be realized.

EXAMPLE 1.1. The AND logic function can be realized using the McCulloch–Pitts neuron having two synaptic connections with equal weights of 1 and a threshold set to $\theta = 2$ (see Fig. 1.2). The AND logic function yields a *true* response only if both inputs are on; otherwise the response of the neuron is *false* (one or both of the inputs are off, i.e., inhibitory). Therefore, a true response is represented by a 1 and a false response is a 0. The "truth" table associated with the AND logic function is given by the following:

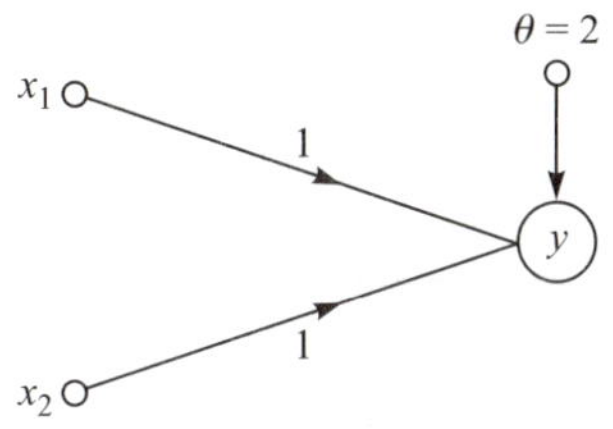

Figure 1.2 McCulloch–Pitts neuron used to perform the AND logic function.

AND logic function

x_1	x_2	→	y
1	1		1
1	0		0
0	1		0
0	0		0

These "patterns" in the AND logic function table above could be training patterns for a different neural network (remember, here with the McCulloch–Pitts neuron, all weights and thresholds are set a priori). Specifically, the x's would be the training input patterns, and the associated y's would be the target (desired output) values. We will address this in depth later when discussing supervised training of neural networks.

EXAMPLE 1.2. The OR logic function can be realized with the McCulloch–Pitts neuron using again two synaptic connections; however, now the weights are set to 2, and the threshold is still set to $\theta = 2$ (see Fig. 1.3). The neuron response is always 1 (true) if either of the two inputs (or both) is on; however, if both inputs are off, the neuron response is 0 (false). More specifically, this is referred to as the *inclusive* OR logic function because the neuron output is on (true) if *both* inputs are on (true). The truth table associated with the OR logic function is given by the following:

OR logic function

x_1	x_2	→	y
1	1		1
1	0		1
0	1		1
0	0		0

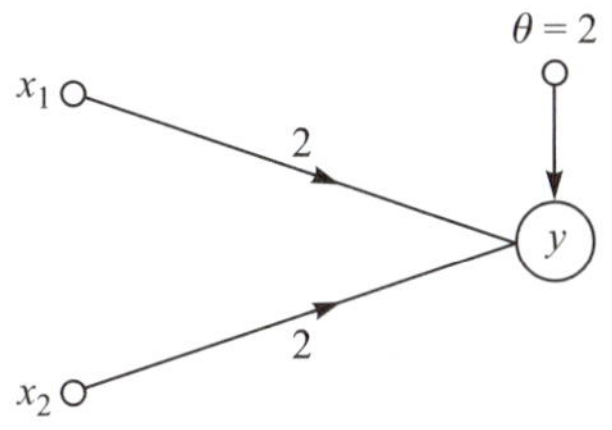

Figure 1.3 McCulloch–Pitts neuron used to perform the OR logic function.

EXAMPLE 1.3. The realization of the XOR (exclusive-OR) logic function is more involved using McCulloch–Pitts neurons. It actually takes three neurons to realize this logic function. The XOR logic function is the same as the previous (inclusive) OR logic function with the exception that the neuron fires (i.e., the neuron gives a true response, a 1) only if *exactly* one of the input values is on (or true); otherwise the response is false (or off). Therefore, the truth table for this function is as follows:

XOR logic function

x_1	x_2	$\rightarrow$	y
1	1		0
1	0		1
0	1		1
0	0		0

The neural network in Figure 1.4 is a two-layer network. Let's observe the network for the case when both x_1 and x_2 are 1. At the middle layer of the neural network $x_3 = x_4 = 0$ because both of the total input signals received at neurons x_3 and x_4 are less than the threshold: $u_3 < \theta$ and $u_4 < \theta$ (that is, $1 < 2$). Therefore, since both x_3 and x_4 are inhibitory, the output $y = 0$, which is precisely what is shown above in the XOR logic function table. If both inputs are off, that is, $x_1 = x_2 = 0$, it is obvious that again $y = 0$ (as shown above in the XOR logic function table). Let's see what happens in Figure 1.4 if $x_1 = 1$ and $x_2 = 0$. At x_3 the neuron response will be on because the total input signal received at the neuron equals the neuron threshold; therefore, $x_3 = 1$. However, at x_4 the opposite is true, that is, $x_4 = 0$, because total input signal received at the neuron is less than the neuron threshold. Therefore, the network output (response) will be $y = 1$ (firing, or *true*) because the total input signal received at the neuron equals the neuron threshold. For the inputs $x_1 = 0$ and $x_2 = 1$, this is the dual of the previous scenario, and the outcome is the same, that is, $y = 1$, which completes the XOR logic function truth table shown above.

As we will see in Chapter 2, the fundamental concepts that lay the foundation for neurocomputing have their roots in the simple McCulloch–Pitts neuron. There have been tremendous strides in neural network research since the earlier efforts of McCulloch and Pitts. This is truly an interdisciplinary

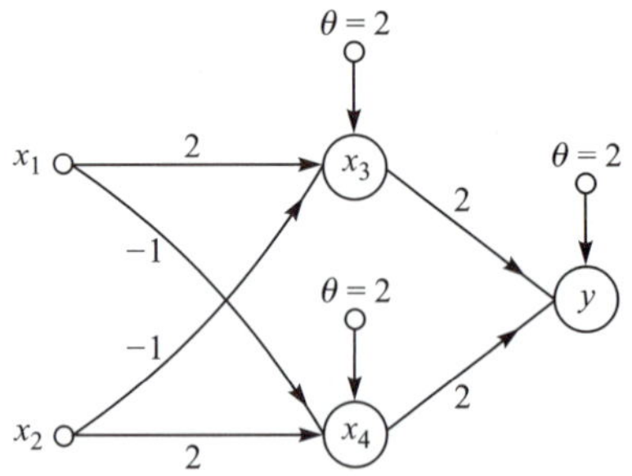

Figure 1.4 McCulloch–Pitts neural network used to perform the XOR logic function.

subject, and there is every reason to believe that the trend will continue, resulting in new theories and applications for neural networks.

1.3 NEUROCOMPUTING AND NEUROSCIENCE

As we previously mentioned, neurocomputing systems very crudely model the human brain. In fact, many researchers in the neural networks field do not acknowledge any connection between neuroscience and neurocomputing. However, many developments in neurocomputing have been inspired by neuroscience research. Therefore, it is appropriate to present an overview of biological neural networks. There are many excellent sources of comprehensive accounts of research in neuroscience. It is not our intention to present an exhaustive or extensive chronicle of the field of neuroscience. However, we do want to give some details of pertinent material on biological neural networks so that certain artificial neural network concepts that are presented have a linkage to their biological counterpart.

Biological neural networks

The nervous system is a vast and complex neural network. The brain is the central element of the nervous system. It is connected to receptors that shuttle sensory information to it, and it delivers action commands to effectors. The brain itself consists of a network of about 10^{11} neurons that are interconnected through subnetworks called nuclei. The nuclei consist of clusters (assemblies) of neurons with specific and defined functions. The subnetworks usually divide up and modify the incoming sensory information before sending the information to other subnetworks. The final form of the processed signals is delivered to effectors to initiate an action.

The sensory system and its subnetworks in the brain are extremely good at decomposing complex sensory information into those fundamental components that are the essential features of the sense. These decompositions are distinct for each sense. For example, the eyes and brain divide up the visual image according to colors, intensity, directionality, movements, scale, binocularity, etc. These components are not reconstructed to the original visual form but instead are sent to other subnetworks for selected evaluations and partial reconstructions.

The biological neuron has served as the inspiration for the artificial neuron (processing element). For this reason, the two bear a close structural analogy to each other. Fausett [37] has addressed the importance of biological plausibility in order to model biological and artificial neurons most effectively. The brain and the other parts of the nervous system consist of a wide variety of different types of neurons, which differ in their electrical properties and their numbers, sizes, and patterns of connectivity (Fig. 1.5 and Table 1.1). These differences are often quite striking.

A biological neuron consists of three main components: dendrites, cell body, and axon; see Figure 1.6(*a*). The dendrites receive signals from other neurons. The axons of other neurons connect to the dendrite and cell body

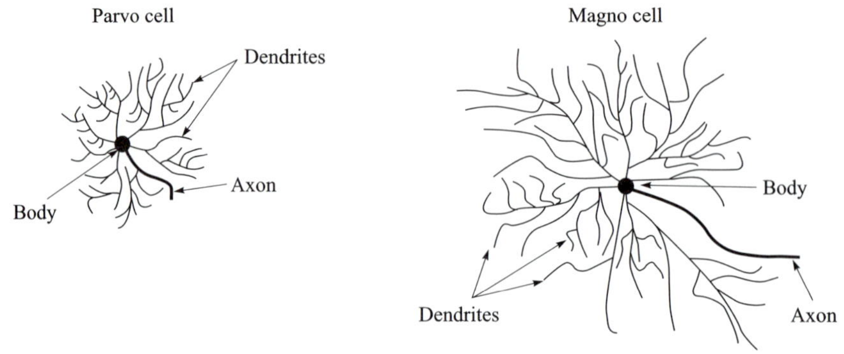

Figure 1.5 Examples of parvo (small) and magno (large) cells.

surfaces by means of connectors called synapses. Depending on the type of neuron, the number of synaptic connections from other neurons may range from a few hundred to 10,000 [39]. Because of the electrical properties of the neuronal membranes, the signals that reach the dendrite quickly decay in strength in time (temporal) and over distance (spatial), and thereby lose the ability to stimulate the neuron unless they are strengthened by other signals occurring at almost the same time and/or at nearby locations; see Figure 1.7.

The cell body (soma) sums the incoming signals from the dendrites and sums the signals from the numerous synapses on its surface. When sufficient input is received to stimulate the neuron to its threshold, the neuron generates an action potential, that is, fires, and transmits an action potential along its axon to other neurons or target cells outside the nervous system, such as a muscle. However, if the inputs do not reach the threshold, the inputs will quickly decay and do not generate an action potential, see Figure 1.8(*a*). Thus, the generation of the action potential is considered to be all-or-nothing

TABLE 1.1
Selected anatomical and physiological differences between parvo and magno ganglion cells, along with some possible consequences for behavior

	Parvo ganglion cells	**Magno ganglion cells**
Anatomical differences	Small cell body	Large cell body
	Dense branching	Sparse branching
	Short branching	Long branches
	Majority of cells	Minority of cells
Physiological differences	Slow conduction rate	Rapid conduction rate
	Sustained response	Transient response
	Small receptive field	Large receptive field
	Low-contrast sensitivity	High-contrast sensitivity
	Color-sensitive	Color-blind
Possible behavioral consequences	Detailed form analysis	Motion detection
	Spatial analysis	Temporal analysis
	Color vision	Depth perception

Source: Adapted with permission from Coren et al. [38], p. 83.

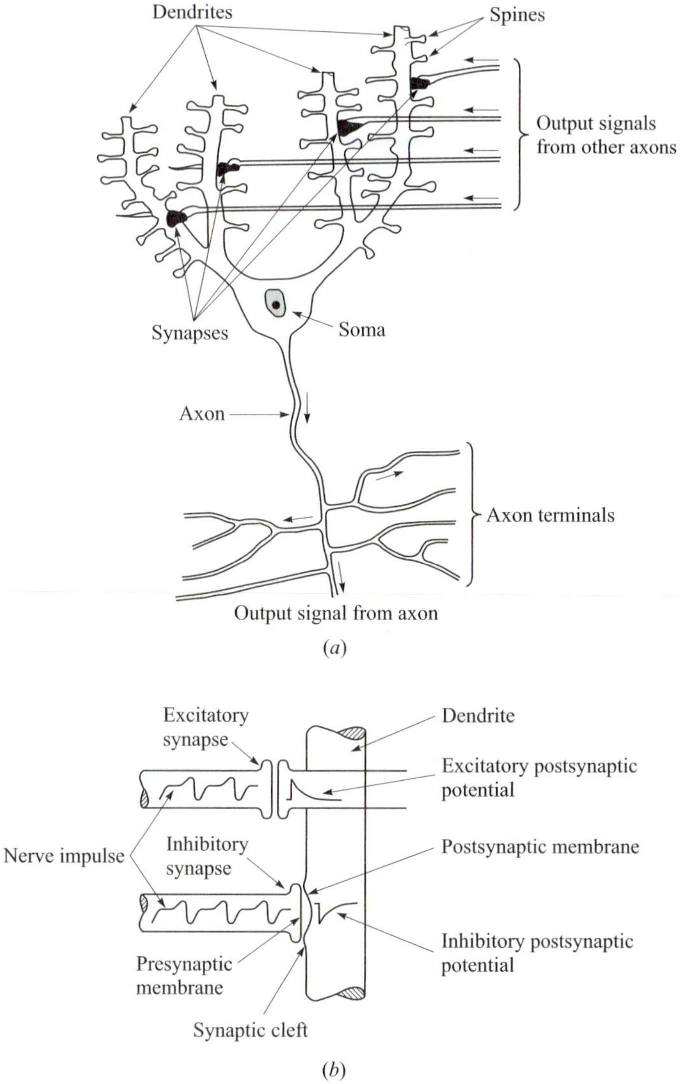

Figure 1.6 (*a*) Schematic structure of a biological neuron. (*b*) Simplified drawing of the synapses. (*Source*: Reproduced with permission from Cichocki and Unbehauen [40], p. 39.)

because the axon either generates it or does not. Moreover, the strength of inputs is expressed by the number of action potentials generated per second rather than by their size. For example, the size and shape of the action potentials from a strong input will be the same as those from a weak input. However, a strong input will generate more action potentials per second (unit time) than a weak input.

Synapses are the points of contact that connect the axon terminals to their targets. These specialized structures, which, for example, connect the axons to dendrites, cell bodies, nerve terminals, muscles, or glands, are characterized by

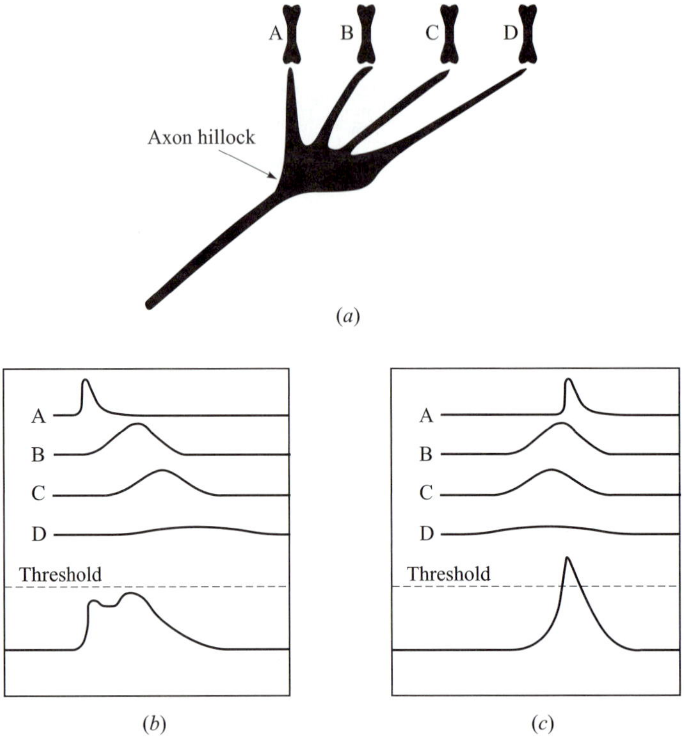

Figure 1.7 (*a*) Neuron used as a motion detector because of the geometry of its dendrites. (*b*) Exciting *A*, *B*, *C*, and *D* simultaneously does not excite the neuron; (*c*) however, if the synapses are activated in the order *D*–*C*–*B*–*A* so the four peaks reach the axon hillock nearly simulaneously, the *combined* effect will exceed the threshold and the neuron fires. (*Source*: Reproduced with permisson from Bose and Liang [41], p. 20.)

specific structural, chemical, and electrical properties. The synapse consists of three primary elements: (1) the nerve terminal, (2) the synaptic cleft or gap, and (3) postsynaptic membrane; see Figure 1.9. When the action potential reaches ("invades") the nerve terminal, the nerve terminal converts the electric signal to a chemical signal through a series of biochemical events. The nerve terminal releases a chemical substance called a neurotransmitter as the final step in this conversion. The neurotransmitter acts on the postsynaptic membrane. The nervous system releases dozens of different neurotransmitters. However, an individual nerve terminal releases only one type of neurotransmitter. The neurotransmitter will diffuse through the synaptic cleft, which is a relatively open space, in about 2 milliseconds (ms). Each neurotransmitter attaches to specific binding sites called receptors that are embedded in the postsynaptic membrane. When the neurotransmitter attaches to the receptor, it triggers electrical and biochemical responses that lead to a change in electrical potential at the postsynaptic membrane. There are two major classes of neurotransmitters: excitatory and inhibitory. Excitatory neurotransmitters depolarize the membrane, but a single synapse is too small to generate an ean-square (LMS) algorithm. The simple perceptron is detailed, fol- py a brief discussion of feedforward multilayer perceptrons. Some basic g rules are covered next that are in some cases extensions of those ted earlier in the chapter. Many of these learning rules are the basis aining more sophisticated neural network architectures. Chapter 2 con- s with an overview of selected data preprocessing methods. This is a very rtant subject that, in the opinion of the authors, is not emphasized ugh in the study of neurocomputing. All the material in Chapter 2 should thoroughly covered if the reader is not already familiar with artificial neural works. This chapter sets the stage for the in-depth coverage of selected ural network architectures and associated learning rules presented in hapters 3 through 5. The intention of the neural networks presented in hese chapters is to give a comprehensive presentation of the selected networks and learning rules that are historically significant and have tremendous application in the real world. Moreover, a lot of the information in these chapters has relevance to material covered in Part II of the book. In addition, Part I serves the purpose of introducing the more advanced readers to various neural architectures and learning rules that might allow readers to develop their own robust neural approaches for solving real-world problems.

Chapter 3 presents a variety of mapping neural networks. It begins with associative memories, followed by backpropagation used to train feedforward multilayer perceptrons. Then more advanced training methods are given for backpropagation. Counterpropagation is presented next, and the chapter concludes with a presentation of radial basis function neural networks. Selected self-organizing neural networks are presented in Chapter 4. These include the Kohonen self-organizing map (SOM), learning vector quantization (LVQ), and adaptive resonance theory (ART) neural networks; the ART1 network is presented in detail. Part I concludes with a presentation of recurrent neural networks along with temporal feedforward networks (which are also recurrent neural networks) in Chapter 5. We make a distinction between temporal feedforward networks and those that are not multilayer feedforward networks. Included in this chapter are the Hopfield network, simulated annealing, the Boltzmann machine, the simple recurrent network (SRN), time-delay networks, and distributed time-lagged feedforward neural networks. The objectives for studying neurocomputing using this book will dictate how much of the material in Chapters 3 through 5 should be covered.

What is in Part II?

Part II consists of Chapters 6 through 10. These chapters contain many different applications of neurocomputing to solve myriad problems in engineering and science. Chapter 6 presents selected neurocomputing approaches for solving optimization problems. Included are neural networks for linear programming and quadratic programming and neural networks for nonlinear continuous constrained optimization problems. Chapter 7 presents a series of *structured* neural networks for solving selected matrix algebra problems. A wide variety of important matrix decompositions (or factorizations) are presented along with a neurocomputing solution for each method. Neurocomputing approaches are also given for computing the pseudoinverse

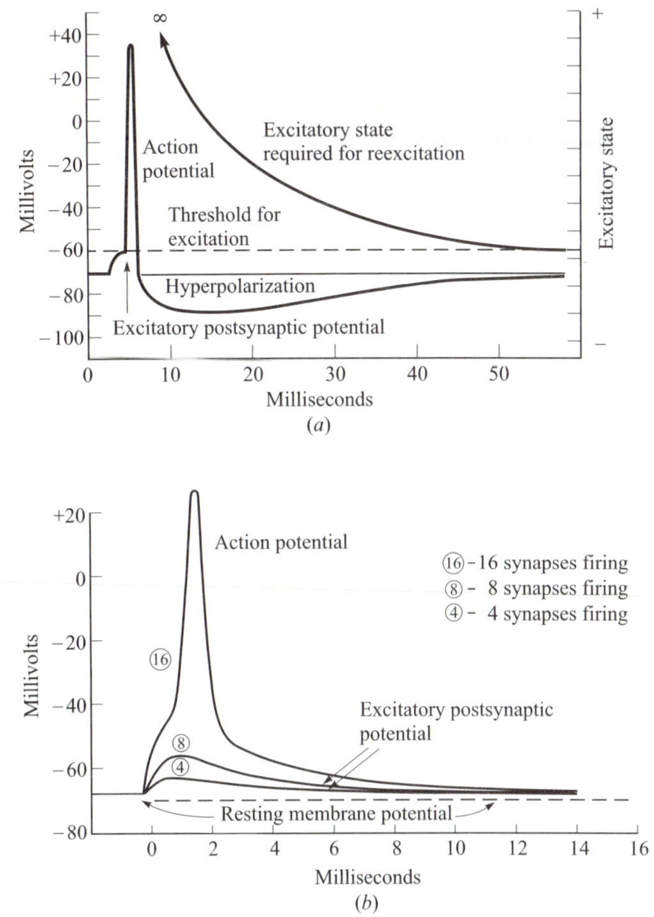

Figure 1.8 (*a*) An example of action potential. (*b*) Excitatory postsynaptic potentials, showing that simultaneous firing of only a few synapses will not cause sufficient summated potential. However, the simultaneous firing of many synapses will elevate the summated potential to the threshold for excitation and cause a superimposed action potential. (*Source*: Reproduced with permisson from Guyton [42], (*a*) p. 492 and (*b*) p. 490.)

action potential by itself. When summed with concurrent depolarizations at hundreds of other synapses, they can collectively generate an action potential. Inhibitory neurotransmitters cause the opposite effect and hyperpolarize the postsynaptic membrane, thereby nullifying the actions of excitatory neurotransmitters and in some cases preventing the generation of action potentials. Thus, the actions represent the sum of excitatory activities on the neurons. In turn, the action potentials travel to the ends of the axons where neurons communicate with other neurons by way of synapses. Synapses serve as the direct connections to establish neuronal circuits among neurons.

As an example of artificial and biological-type neural network similarities, Table 1.2 compares three different neural architectures.

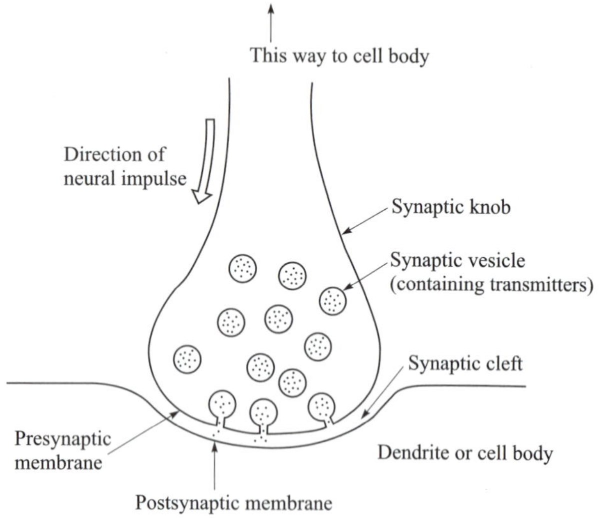

Figure 1.9 A typical synapse. The synaptic vesicles store transmitter substances that are released across the synaptic cleft. (*Source*: Reproduced with permission from Coren et al. [38], p. 642.)

1.4 CLASSIFICATION OF NEURAL NETWORKS

There are many different approaches used to classify artificial neural networks. For example, neural networks can be classified according to how they learn or the type of training that is required, the various applications they can perform, those that use activation functions versus basis functions, whether they are recurrent or nonrecurrent, the type of training inputs [43], etc. Therefore, the taxonomy of neural networks is not necessarily a straightforward exercise. As an example, when neural networks are classified according to unsupervised learning versus supervised learning, there can be overlap for certain networks. Figure 1.10 shows some examples of different types of neural networks and how they would be categorized according to the type of

TABLE 1.2
Examples of neural network architectures and their nervous system counterparts

Neural network	Nervous system
Single-layer feedforward networks	Rare except in the simplest of reflex arc pathways.
Multilayer feedforward networks	Common and complex; typically have several levels of hidden layers. Partially connected are more common than fully connected.
Recurrent networks	Negative feedback; quite a bit more complicated than neural network analogs. An essential and dominant feature of the nervous system.

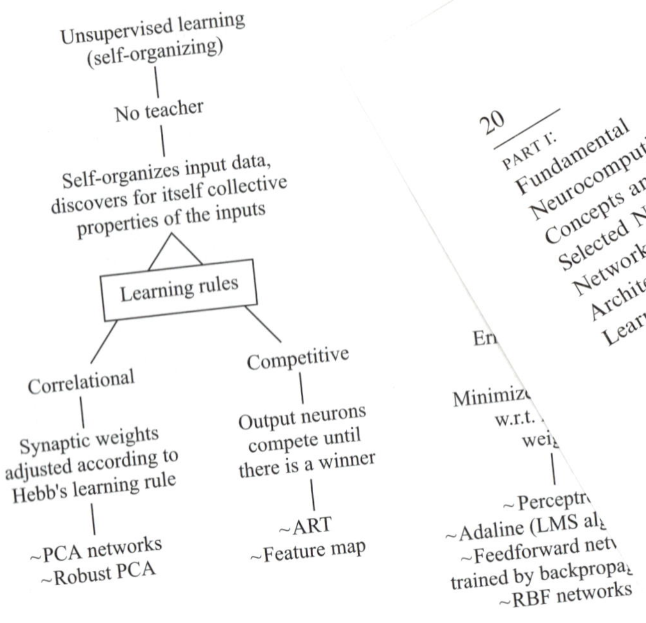

Figure 1.10 Classification of selected neural networks acco[rding ...] learn, that is, supervised or unsupervised.

learning, that is, supervised learning versus unsupervised [...] wanted to place counterpropagation (cf. Sect. 3.5) in one of [...] it is not necessarily clear how it would be classified. Another e[...] Hopfield network (cf. Sect. 5.3), which is a recurrent neural ne[...] network is actually considered to be a fixed-weight network.

1.5 GUIDE TO THE BOOK

This book is divided into two major parts:

- Part I: Fundamental Neurocomputing Concepts and Selected Neural Network Architectures and Learning Rules
- Part II: Applications of Neurocomputing

What is in Part I?

Part I consists of Chapters 1 through 5. Chapter 1 is an introductory chapter to present to the reader the basic idea of neural networks and neurocomputing. Chapter 2 starts with a presentation of basic models of artificial neurons that are the building blocks of neural networks, followed by a discussion of different types of activation functions. The adaptive linear element (Adaline) and multiple Adaline (Madaline) are presented next along with the

of a matrix, solving the algebraic Lyapunov equation, and solving the algebraic Riccati equation. Chapter 8 addresses how to solve linear algebraic equations using neural networks. The various methods include a least-squares neurocomputing approach, a conjugate gradient learning rule, a generalized robust neurocomputing approach, regularization methods for ill-posed problems with ill-determined numerical rank, matrix splittings for iterative discrete-time methods, and the total least-squares problem. Also the L_∞-norm and L_1-norm neural network approaches are presented for solving linear algebraic equations. Chapter 9 gives an in-depth discussion of different neurocomputing methods for statistical analysis. Included are the following: neural networks for principal-component analysis (PCA), principal-component regression (PCR), classical least squares (CLS), neural networks for nonlinear PCA and robust PCA, a neural network approach for partial least-squares regression (PLSR), and a neural network approach for robust PLSR. Chapter 10 covers applications of neural networks for signal processing problems, linear and nonlinear system identification, nonlinear control, and estimation. Many different examples are covered in detail. Also included is independent-component analysis (ICA) using neural networks for blind source separation. In addition, a fast ICA algorithm is presented along with an example that applies the fast ICA algorithm to separation of digital images. An appendix is included that presents some mathematical fundamentals for studying and applying neurocomputing techniques.

REFERENCES

1. J. A. Anderson and E. Rosenfeld, eds., *Neurocomputing: Foundations of Research*, Cambridge, MA: M.I.T. Press, 1988.
2. G. Nagy, "Neural Networks — Then and Now," *IEEE Transactions on Neural Networks*, vol. 2, 1991, pp. 316–18.
3. B. Widrow and M. A. Lehr, "30 Years of Neural Networks: Perceptron, Madaline and Backpropagation," *Proceedings of the IEEE*, vol. 78, 1990, pp. 1415–42.
4. S. Saarinen, R. B. Bramley, and G. Cybenko, "Neural Networks, Backpropagation, and Automatic Differentiation," in *Automatic Differentiation of Algorithms: Theory, Implementation, and Application*, eds.: A. Griewank and G. F. Corliss, Philadelphia, PA: Society for Industrial and Applied Mathematics, 1992, pp. 31–42.
5. W. S. McCulloch and W. Pitts, "A Logical Calculus of the Ideas Immanent in Nervous Activity," *Bulletin of Mathematical Biophysics*, vol. 5, 1943, pp. 115–33. Reprinted in 1988, Anderson and Rosenfeld [1], pp. 18–27.
6. D. O. Hebb, *The Organization of Behavior*, New York: Wiley, 1949, introduction and chapter 4, "The First Stage of Perception: Growth of the Assembly," pp. xi–xix, 60–78. Reprinted in 1988, Anderson and Rosenfeld [1], pp. 484–507.
7. J. von Neumann, *The Computer and the Brain*, New Haven, CT: Yale University Press, 1958.
8. F. Rosenblatt, "The Perceptron: A Probabilistic Model for Information Storage and Organization in the Brain," *Psychological Review*, vol. 65, 1958, pp. 386–408. Reprinted in 1988, Anderson and Rosenfeld [1], pp. 92–114.

9. B. Widrow and M. E. Hoff, Jr., "Adaptive Switching Circuits," *IRE WESCON Convention Record*, part 4, New York: IRE, 1960, pp. 96–104. Reprinted in 1988, Anderson and Rosenfeld [1], pp. 126–34.
10. M. L. Minsky and S. A. Papert, *Perceptrons*, Cambridge, MA: M.I.T. Press, 1969.
11. J. Hertz, A. Krogh, and R. G. Palmer, *Introduction to the Theory of Neural Computation*, Redwood City, CA: Addison-Wesley, 1991.
12. T. Kohonen, "Correlation Matrix Memories," *IEEE Transactions on Computers*, vol. C-21, 1972, pp. 353–59. Reprinted in 1988, Anderson and Rosenfeld [1], pp. 174–80.
13. J. A. Anderson, "A Simple Neural Network Generating an Interactive Memory," *Mathematical Bioscience*, vol. 14, 1972, pp. 197–220. Reprinted in 1988, Anderson and Rosenfeld [1], pp. 181–92.
14. C. von der Malsburg, "Self-Organization of Orientation Sensitive Cells in the Striata Cortex," *Kybernetik*, vol. 14, 1973, pp. 85–100. Reprinted in 1988, Anderson and Rosenfeld [1], pp. 212–27.
15. P. J. Werbos, "Beyond Regression: New Tools for Prediction and Analysis in the Behavioral Sciences," Ph.D. thesis, Cambridge, MA: Harvard University, 1974.
16. W. A. Little and G. L. Shaw, "A Statistical Theory of Short and Long Term Memory," *Behavioral Biology*, vol. 14, 1975, pp. 115–33. Reprinted in 1988, Anderson and Rosenfeld [1], pp. 231–41.
17. S. C. Lee and E. T. Lee, "Fuzzy Neural Networks," *Mathematical Biosciences*, vol. 23, 1975, pp. 155–77.
18. S. Grossberg, "Adaptive Pattern Classification and Universal Recoding: I. Parallel Development and Coding of Neural Feature Detectors," *Biological Cybernetics*, vol. 23, 1976, pp. 121–34. Reprinted in 1988, Anderson and Rosenfeld [1], pp. 245–58.
19. S.-I. Amari, "Neural Theory of Association and Concept-Formation," *Biological Cybernetics*, vol. 26, 1977, pp. 175–85. Reprinted in 1988, Anderson and Rosenfeld [1], pp. 271–81.
20. J. J. Hopfield, "Neural Networks and Physical Systems with Emergent Collective Computational Abilities," *Proceedings of the National Academy of Sciences*, vol. 79, 1982, pp. 2554–58. Reprinted in 1988, Anderson and Rosenfeld [1], pp. 460–4.
21. T. Kohonen, "Self-Organizing Formation of Topologically Correct Feature Maps," *Biological Cybernetics*, vol. 43, 1982, pp. 59–69. Reprinted in 1988, Anderson and Rosenfeld [1], pp. 511–21.
22. E. Oja, "A Simplified Neuron Model as a Principal Component Analyzer," *Journal of Mathematical Biology*, vol. 15, 1982, pp. 267–73.
23. K. Fukushima, S. Miyake, and T. Ito, "Neocognitron: A Neural Network Model for a Mechanism of Visual Pattern Recognition," *IEEE Transactions on Systems, Man, and Cybernetics*, vol. SMC-13, 1983, pp. 826–34. Reprinted in 1988, Anderson and Rosenfeld [1], pp. 526–34.
24. S. Kirkpatrick, C. D. Gelatt, Jr., and M. P. Vecchi, "Optimization by Simulated Annealing," *Science*, vol. 220, 1983, pp. 671–80. Reprinted in 1988, Anderson and Rosenfeld [1], pp. 554–67.
25. R. R. Kampfner and M. Conrad, "Computational Modeling of the Evolutionary Learning Processes in the Brain," *Bulletin of Mathematical Biology*, vol. 45, 1983, pp. 931–68.
26. D. B. Fogel, ed., *Evolutionary Computation: The Fossil Record*, Piscataway, NJ: IEEE Press, 1998.
27. D. H. Ackley, G. E. Hinton, and T. J. Sejnowski, "A Learning Algorithm for Boltzmann Machines," *Cognitive Science*, vol. 9, 1985, pp. 147–69. Reprinted in 1988, Anderson and Rosenfeld [1], pp. 638–49.

28. D. Parker, "Learning Logic," Technical Report TR-87, Cambridge, MA: Center for Computational Research in Economics and Management, M.I.T., 1985.
29. Y. LeCun, "Une procedure d'apprentissage pour reseau a seuil asymmetrique," in *In Cognitiva 85: A la Frontiere de l'Intelligence Artificielle des Sciences de la Connaissance des Neurosciences*, CESTA, vol. 85, Paris, 1985, pp. 599–604.
30. J. Herault, C. Jutten, and B. Ans, "Détection de grandeur primitives dans un message composite par une architecture de calcul neuromimétique en apprentissage nonsupervisé," in *Proceedings GRETSI Conference*, Nice, France, May 1985, pp. 1017–22 (in French).
31. C. Jutten and J. Herault, "Blind Separation of Sources, Part I: An Adaptive Algorithm Based on Neuromimetic Architecture," *Signal Processing*, vol. 24. 1991, pp. 1–10.
32. D. E. Rumelhart, G. E. Hinton, and R. J. Williams, "Learning Internal Representations by Error Propagation," in *Parallel Distributed Processing: Explorations in the Microstructures of Cognition*, vol. 1, *Foundations*, eds. D. E. Rumelhart and J. L. McClelland, Cambridge, MA: M.I.T. Press, 1986, pp. 318–62. Reprinted in 1988, Anderson and Rosenfeld [1], pp. 675–95.
33. D. E. Rumelhart, G. E. Hinton, and R. J. Williams, "Learning Representations by Back-Propagating Errors," *Nature*, vol. 323, 1986, pp. 533–36. Reprinted in 1988, Anderson and Rosenfeld [1], pp. 696–9.
34. G. A. Carpenter and S. Grossberg, "A Massively Parallel Architecture for a Self-Organizing Neural Pattern Recognition Machine," *Computer Vision, Graphics, and Image Processing*, vol. 37, 1987, pp. 54–115.
35. M. A. Sivilotti, M. A. Mahowald, and C. A. Mead, "Real-Time Visual Computations Using Analog CMOS Processing Arrays," *Advanced Research in VLSI: Proceedings of the 1987 Stanford Conference*, ed. P. Losleben, Cambridge, MA: M.I.T. Press, 1987, pp. 295–312. Reprinted in 1988, Anderson and Rosenfeld [1], pp. 703–11.
36. D. S. Broomhead and D. Lowe, "Multivariable Functional Interpolation and Adaptive Networks," *Complex Systems*, vol. 2, 1988, pp. 321–55.
37. D. W. Fausett, "Strictly Local Backpropagation," in *Proceedings of the International Joint Conference on Neural Networks*, vol. 3. San Diego, CA, 1990, pp. 125–30.
38. S. Coren, L. M. Ward, and J. T. Enns, *Sensation and Perception,* 4th ed., Orlando, FL: Harcourt Brace, 1994.
39. G. M. Shepherd, *Neurobiology,* 3rd ed., New York: Oxford University Press, 1994.
40. A. Cichocki and R. Unbehauen, *Neural Networks for Optimization and Signal Processing*, New York: Wiley, 1993.
41. N. K. Bose and P. Liang, *Neural Network Fundamentals with Graphs, Algorithms, and Applications,* New York: McGraw-Hill, 1996.
42. A.C. Guyton, *Textbook of Medical Physiology*, 8th ed., Philadelphia, PA: W.B. Saunders, 1991.
43. *DARPA Neural Networks Study*, AFCEA International Press, November 1988, Chapter 8, "Classification and Clustering Models."

CHAPTER 2

Fundamental Neurocomputing Concepts

2.1 INTRODUCTION

The material presented in this chapter is the foundation for many neural architectures that can perform various functions. As stated in Chapter 1, artificial neural networks consist of many interconnected processing elements (artificial neurons or nodes); therefore, this chapter begins with the presentation of models of artificial neurons. These models are the building blocks for many different types of neural networks that can be used for myriad applications. The principal interest here is the presentation of fundamental concepts from a science and engineering perspective to solve problems with artificial neural networks. We are not concerned with addressing issues relating to biological neural networks, such as cognition, neural modeling, neurophysiological considerations, and details of the human brain.

A number of basic learning rules for a single neuron are presented in Section 2.8. These concepts can be extended to networks with multiple neurons. Therefore, many of the learning rules presented here are used in subsequent chapters for building more sophisticated neural structures. The last part of this chapter deals with the very important subject of data preprocessing. Several preprocessing methods are presented. The performance of a particular neural network is dependent on the training phase (specifically, the training data used). Many times it is necessary to preprocess the training data to extract important features from the data that can be used to train the network instead of the "raw" data. The preprocessing of the training data can, therefore, improve the performance of the neural network.

2.2 BASIC MODELS OF ARTIFICIAL NEURONS

In Chapter 1 the basic concept of an artificial neural network (ANN) was presented where it was stated that an ANN is typically made up of many single neurons. An artificial neuron can also be referred to as a processing element, node, or a threshold logic unit; however, we typically refer to it as a *neuron*. A neuron is an information processing unit that roughly resembles its biological counterpart (cf. Sect. 1.3). Figure 2.1 shows a model of an artificial neuron. There are four basic components of the model: (1) There is a set of synapses with associated synaptic weights. As shown in Figure 2.1, the continuous-valued input to the synapses is a vector signal $\boldsymbol{x} \subset \Re^{n\times 1}$, with the individual vector components given as x_j, for $j = 1, 2, \ldots, n$, that is, $\boldsymbol{x} = [x_1, x_2, \ldots, x_n]^T$. Therefore, each vector component x_j is input to the jth synapse and connected to neuron q through a synaptic weight w_{qj}; that is, x_j is multiplied by the synaptic weight w_{qj}. Our convention for subscripting the synaptic weights is that the first subscript is associated with the particular neuron, and the second subscript relates to the element of the input vector which the synaptic weight multiplies. This convention is arbitrary (the opposite notation could be adopted); however, the convention must be adhered to for consistency in the network architecture. (2) The summing device acts to add all the signals that are broadcast into the adder; that is, each input is multiplied by the associated synaptic weight and then summed. All the operations up to and including the output of the adder u_q constitute a linear combiner, because u_q is a linear combination of the inputs to the synapses. (3) The activation function (or squashing function) $f(\cdot)$, shown in Figure 2.1, serves to limit the amplitude of the neuron output y_q when $f(\cdot)$ is a nonlinear function. It can be continuous-valued, binary, or bipolar; or it can even be a linear function in certain cases. When the activation function is nonlinear, its finite limits are typically normalized in the range of either [0, 1] (binary) or [−1, 1] (bipolar). In a highly interconnected, massively parallel artificial neural architecture, the nonlinearities can serve to enhance the network's classification, approximation, and noise-immunity capabilities. (4) The threshold θ_q is usually externally applied and lowers the cumulative input to the activation

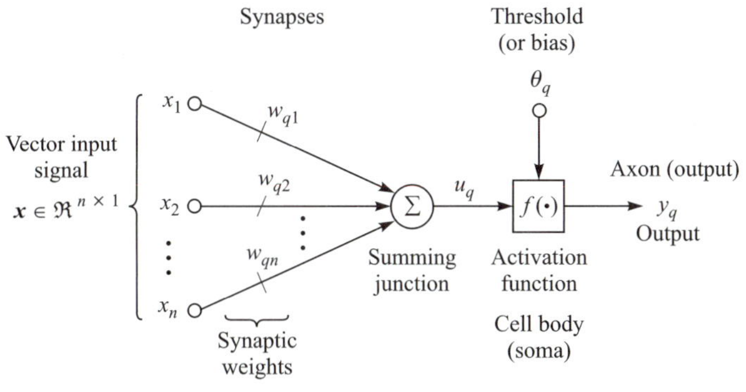

Figure 2.1 Nonlinear model of an artificial neuron.

function. Therefore, θ_q is subtracted from the output of the linear combiner u_q before the activation is applied. The input to the activation function could be raised by adding a term to u_q, that is, a *bias*. In this case, θ_q would be added to u_q; thus, bias is the negative of the threshold. Specifically, the relationship between the effective internal *activation potential* or *activity level*

$$v_q = u_q - \theta_q \tag{2.1}$$

and the linear combiner output u_q of neuron q depends on whether the threshold θ_q is positive or negative. The threshold can be thought of as applying an *affine transformation* to the output of the linear combiner u_q [1].

Mathematically, we can describe the operation of the artificial neuron in Figure 2.1 by writing the following equations. The output of the linear combiner is given by

$$u_q = \sum_{j=1}^{n} w_{qj} x_j = \boldsymbol{w}_q^T \boldsymbol{x} = \boldsymbol{x}^T \boldsymbol{w}_q \tag{2.2}$$

where $\boldsymbol{x}$ is described above, $\boldsymbol{w}_q = [w_{q1}, w_{q2}, \ldots, w_{qn}]^T \in \Re^{n \times 1}$, and the output of the activation function is

$$y_q = f(v_q) = f(u_q - \theta_q) \tag{2.3}$$

Therefore, by using (2.2) and (2.3), the output of the neuron is given by

$$y_q = f\left(\sum_{j=1}^{n} w_{qj} x_j - \theta_q\right) \tag{2.4}$$

Figure 2.2 shows an alternate model for the artificial neuron. In this model the threshold (or bias) is incorporated into the synaptic weight vector $\boldsymbol{w}_q$ for neuron q, and the input vector is augmented with x_0; therefore, $\boldsymbol{w}_q \in \Re^{n+1 \times 1}$ and $\boldsymbol{x} \in \Re^{n+1 \times 1}$. The effective internal activation potential is now written as

$$v_q = \sum_{j=0}^{n} w_{qj} x_j \tag{2.5}$$

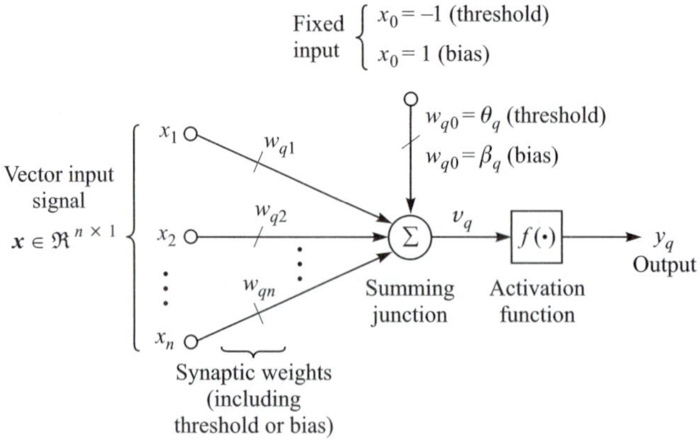

Figure 2.2 Alternate nonlinear model of an artificial neuron.

and the output of neuron q is written as

$$y_q = f(v_q) \tag{2.6}$$

Therefore, in (2.5) a new synapse is added, and from Figure 2.2, depending on whether a threshold or a bias is externally applied, the input to the synapse will be $x_0 = -1$ or $x_0 = 1$, respectively. Also the associated synaptic weight will be $w_{q0} = \theta_q$ for an externally applied threshold, or $w_{q0} = \beta_q$ for an externally applied bias.

2.3 BASIC ACTIVATION FUNCTIONS

The *activation function* (sometimes called a *transfer function*) shown in Figure 2.1, or Figure 2.2, can be a linear or nonlinear function. There are many different types of activation functions. Selection of one type over another depends on the particular problem that the neuron (or neural network) is to solve. Here we present four of the most common types of activation functions. In subsequent chapters more sophisticated functions are presented for the type of problem to be solved. Our basic reference model for the presentation of the activation functions is shown in Figure 2.2. It is assumed that the threshold (or bias) is set to zero, unless otherwise indicated.

The first type is the *linear* (or *identity*) *function*, which is continuous-valued. Mathematically, the output of the linear activation function at the q neuron is written as

$$y_q = f_{\text{lin}}(v_q) = v_q \tag{2.7}$$

where v_q (the effective internal activation potential), as shown in Figure 2.2, is the output of the linear combiner and is input to the activation function $f(\bullet)$. The output of the identity function, that is, the output of neuron y_q, is simply v_q (or the output of the linear combiner); this is shown in Figure 2.3. This might appear to be a trivial activation function; however, later we will see that some linear networks are very useful.

The second type of activation function is a *hard limiter*. This is a *binary* (or *bipolar*) function that hard-limits the input to the function to either a 0 or a 1 for the binary type, and a -1 or 1 for the bipolar type. The binary hard limiter is sometimes called the *threshold function* (we refer to it as just a hard limiter), and the bipolar hard limiter is referred to as the *symmetric hard limiter*, that is, symmetric about the abscissa for the function values plotted versus the input values v_q. The output of the binary hard limiter (or the q neuron output) can be written as

$$y_q = f_{\text{hl}}(v_q) = \begin{cases} 0 & \text{if } v_q < 0 \\ 1 & \text{if } v_q \geq 0 \end{cases} \tag{2.8}$$

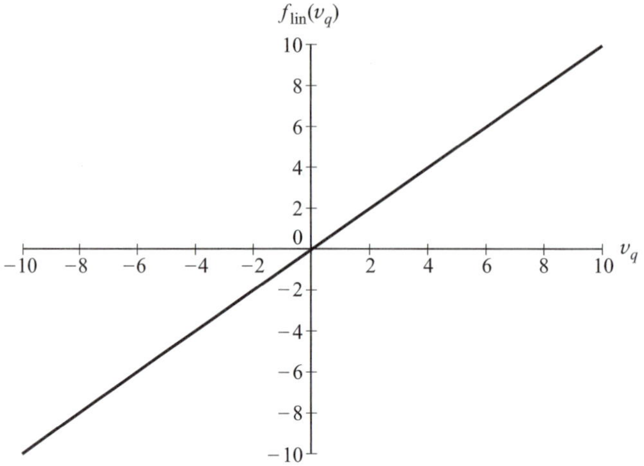

Figure 2.3 Linear (identity) activation function.

For the symmetric hard limiter (suscript "shl"), the q neuron output is written as

$$y_q = f_{\text{shl}}(v_q) = \begin{cases} -1 & \text{if } v_q < 0 \\ 0 & \text{if } v_q = 0 \\ 1 & \text{if } v_q > 0 \end{cases} \tag{2.9}$$

This function is sometimes referred to as the *signum* (or *sign*) function, that is, $f(\bullet) = \text{sgn}\,(\bullet)$. Figures 2.4 and 2.5 show the hard limiter and symmetric hard limiter characteristics, respectively. The artificial neuron with the hard limiter activation function is referred to as the McCulloch–Pitts model (or the *threshold unit*) [2], as discussed in Chapter 1. In their original work, the weights of

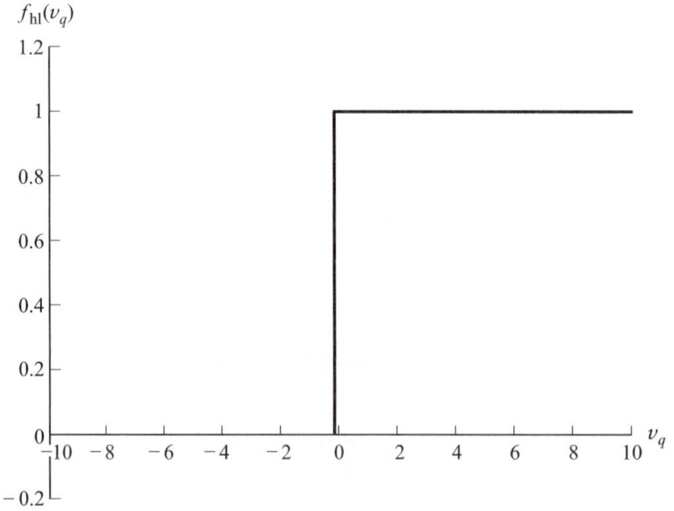

Figure 2.4 Hard limiter activation function.

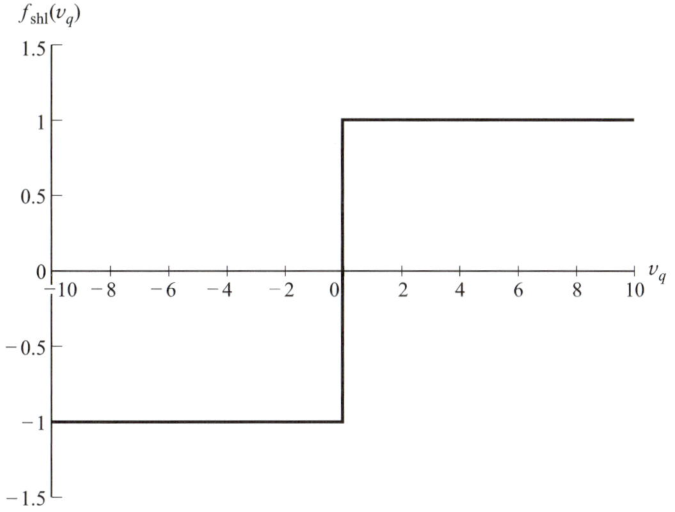

Figure 2.5 Symmetric hard limiter activation function.

the neuron, along with the threshold level, were preset. Thus there was no *training* associated with the neuron; however, with the preset weights derived from analysis, the neuron could perform simple logic functions. Three of these simple logic functions were presented in Chapter 1. Many individuals feel that the origin of the field of artificial neural networks came from their pioneering work.

The third type of basic activation function is the *saturating linear function*, or *piecewise linear function*. This type of function can have either a binary or bipolar range for the saturation limits of the output. The bipolar saturating linear function will be referred to as the *symmetric saturating linear function*. The q neuron output for the saturating linear function (i.e., binary outputs) is given by

$$y_q = f_{sl}(v_q) = \begin{cases} 0 & \text{if } v_q < -\frac{1}{2} \\ v_q + \frac{1}{2} & \text{if} -\frac{1}{2} \leq v_q \leq \frac{1}{2} \\ 1 & \text{if} v_q > \frac{1}{2} \end{cases} \tag{2.10}$$

and for the symmetric saturating linear function the output is

$$y_q = f_{ssl}(v_q) = \begin{cases} -1 & \text{if } v_q < -1 \\ v_q & \text{if } -1 \leq v_q \leq 1 \\ 1 & \text{if } v_q > 1 \end{cases} \tag{2.11}$$

Figures 2.6 and 2.7 show the saturating linear and symmetric saturating linear function characteristics, respectively.

The fourth basic activation function is commonly referred to as a sigmoid (S-shaped) function, and here we present two types. This nonlinear sigmoid function is the most common type of activation used to construct artificial neural networks. It is a mathematically well-behaved, strictly increasing func-

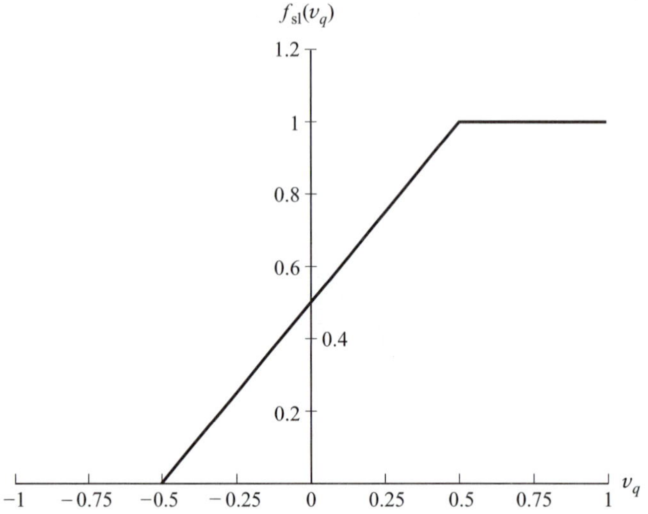

Figure 2.6 Saturating linear activation function.

tion. The first type of sigmoid function is the *binary sigmoid function*. The saturating output values of this function have a binary range, and mathematically the q neuron output is written as

$$y_q = f_{bs}(v_q) = \frac{1}{1 + e^{-\alpha v_q}} \tag{2.12}$$

where α is the *slope parameter* of the binary sigmoid function. By varying this parameter, different shapes of the function can be obtained, as illustrated in

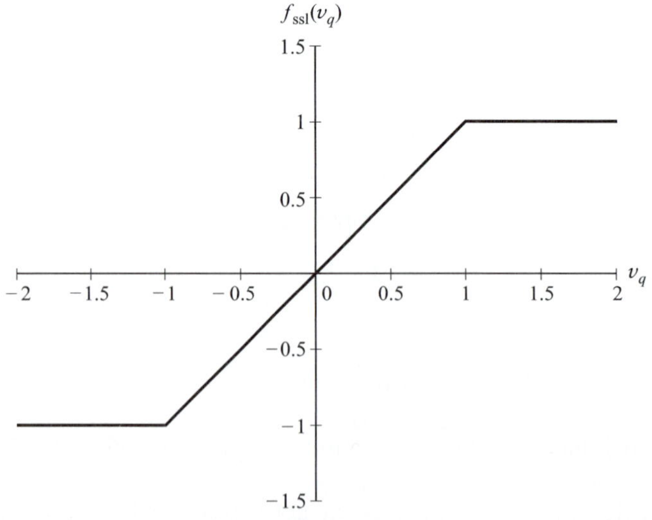

Figure 2.7 Symmetric saturating linear activation function.

Figure 2.8. Unlike the hard limiter which has no derivative at the origin, the binary sigmoid is a continuous and differentiable function. We will see later that differentiability of an activation function plays an important role in neurocomputing. The derivative of the binary sigmoid function with respect to the output of the linear combiner can be written as

$$g_{\text{bs}}(v_q) = \frac{\mathrm{d}f_{\text{bs}}(v_q)}{\mathrm{d}v_q} = \frac{\alpha e^{-\alpha v_q}}{(1+e^{-\alpha v_q})^2} = \alpha f_{\text{bs}}(v_q)[1 - f_{\text{bs}}(v_q)] \tag{2.13}$$

At the origin, the slope of the binary sigmoid function is given by $\alpha/4$. This is obvious from (2.13) by setting $v_q = 0$. Therefore, as α increases, the binary sigmoid approaches the hard limiter. Figure 2.9 shows the derivative of the binary sigmoid function for two different values of the slope parameter, $\alpha = 1$ and $\alpha = 0.5$.

The bipolar form of a sigmoid function can be a *hyperbolic tangent sigmoid*. Thus, the saturating limits of this function have a bipolar range, and the output of neuron q can be written as

$$y_q = f_{\text{hts}}(v_q) = \tanh(\alpha v_q) = \frac{e^{\alpha v_q} - e^{-\alpha v_q}}{e^{\alpha v_q} + e^{-\alpha v_q}} = \frac{1 - e^{-2\alpha v_q}}{1 + e^{-2\alpha v_q}} \tag{2.14}$$

where α is the slope parameter. The hyperbolic tangent sigmoid function is shown plotted in Figure 2.10 for three different values of the slope parameter. The derivative of the hyperbolic tangent sigmoid with respect to v_q is given by

$$g_{\text{hts}}(v_q) = \frac{\mathrm{d}f_{\text{hts}}(v_q)}{\mathrm{d}v_q} = \alpha[1 + f_{\text{hts}}(v_q)][1 - f_{\text{hts}}(v_q)] \tag{2.15}$$

It is interesting to note that the derivatives of the two sigmoid activation functions in (2.12) and (2.14), given in (2.13) and (2.15), respectively, can be expressed in terms of the individual activation function itself. This will become

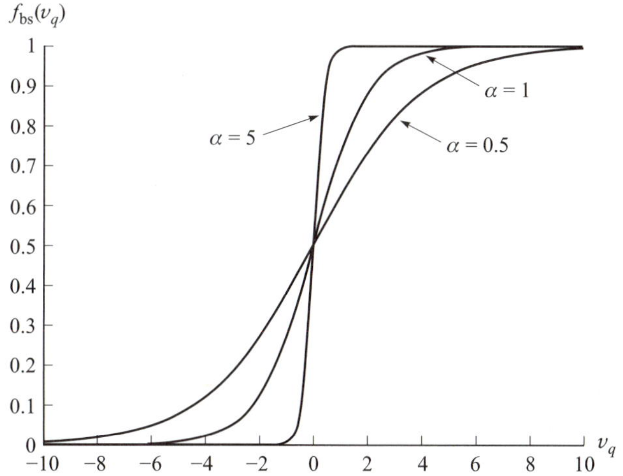

Figure 2.8 Binary sigmoid activation function for three different values of the slope parameter.

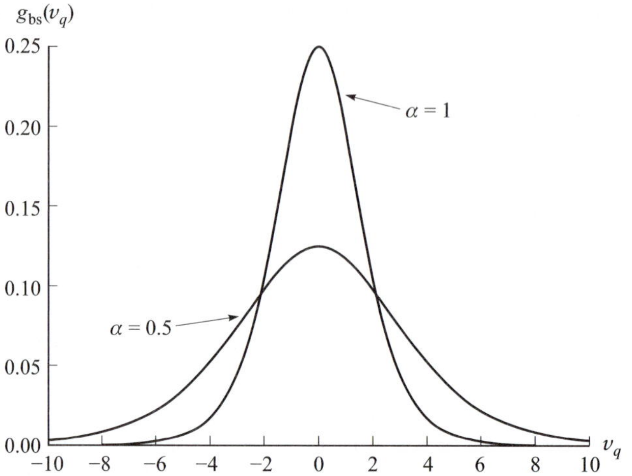

Figure 2.9 Derivative of the binary sigmoid function for two different values of the slope parameter.

important later, for example, when we are developing learning rules to train simple perceptrons (cf. Sect. 2.6), and feedforward multilayer perceptrons (cf. Sect. 3.3).

The activation functions shown thus far that have both negative as well as positive values can be analytically beneficial, as we will see in the next chapter. Moreover, there is neurophysiological experimental evidence to support the use of such activation functions [3]. However, rarely does there exist exact antisymmetry about the origin as in the hyperbolic tangent sigmoid activation function.

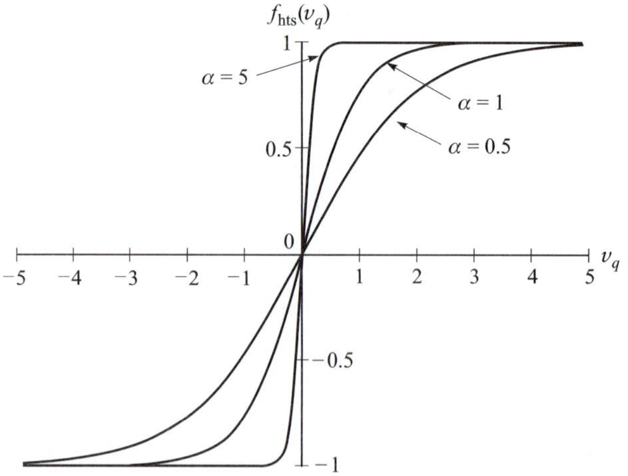

Figure 2.10 Hyperbolic tangent sigmoid activation function for three different values of the slope parameter.

The effect of the threshold θ_q and the bias β_q, shown in Figure 2.2, can be illustrated by observing the binary sigmoid function in (2.12). Figure 2.11 shows three plots of the binary sigmoid function, with a threshold ($\theta_q = 2$), with a bias ($\beta_q = 2$), and the nominal case ($\theta_q = \beta_q = 0$). For all three cases the slope parameter was set to unity (that is, $\alpha = 1$). From Figure 2.11 it can be seen that applying a threshold is analogous to *delaying* a time-domain signal, and adding a bias is analogous to an *advance* of a signal.

2.4 THE HOPFIELD MODEL OF THE ARTIFICIAL NEURON

As stated in [4], regarding the paper by John Hopfield [5], "As far as public visibility goes, the *modern era* in neural networks dates from the publication of this paper by John Hopfield." In this 1982 paper, Hopfield presented a neural architecture made up of simple processing units based on the formal neuron of McCulloch and Pitts [2]. Even though considerable research in the field of neural networks had been previously carried out, Hopfield's paper was clearly written and brought together several seemingly unrelated concepts in the literature. Its impact was mainly due to his presentation of these concepts in a highly coherent fashion and the relationship of the theoretical ideas to practical issues of how a neural network works and its possible applications.

The basic model is shown in Figure 2.2, with the activation function $f(\bullet)$ being the hard limiter shown in Figure 2.4. The neuron has two states which are determined by the activation potential of the neuron. The on state (or firing state) is given by the output of the neuron $y_q = 1$, and the off state (or no-firing state) is represented by the neuron output $y_q = 0$. The Hopfield neural network based on these neurons is an asynchronous parallel proces-

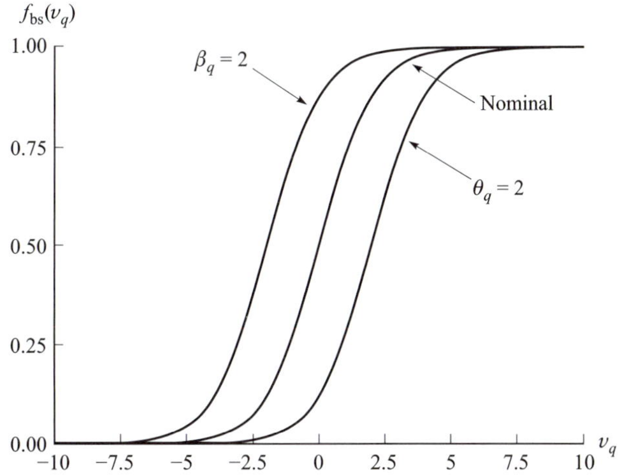

Figure 2.11 Binary sigmoid activation function ($\alpha = 1$) with a threshold ($\theta_q = 2$) and bias ($\beta_q = 2$).

sing, fully interconnected, *content-addressable memory* (or *associative memory*) that has a primary function of retrieving a stored pattern in response to the presentation of a noisy or incomplete version of that pattern. Typically, the activation function is taken as the symmetric hard limiter (shown in Fig. 2.5) for the discrete-time model, and the hyperbolic tangent sigmoid (shown in Fig. 2.10) for the continuous-time model to take advantage of the bipolar outputs. Unlike feedforward perceptrons, the Hopfield network has *feedback,* and as such it is considered a *recurrent* neural network. We discuss the details of the Hopfield network in Section 5.3. Here we want to set the stage by presenting the discrete-time and continuous-time Hopfield models of the neuron.

The discrete-time Hopfield model of the artificial neuron is shown in Figure 2.12. From Figure 2.12 we can write the output of the neuron before the unit delay z^{-1} as

$$y_q(k+1) = f_{\text{shl}}[v_q(k+1)] \tag{2.16}$$

where

$$v_q(k+1) = \sum_{j=1}^{n} w_{qj} x_j(k) - \theta_q \tag{2.17}$$

θ_q is an externally applied threshold, and $v_q(k) = v_q(kT_s)$, where $k = 0, 1, 2, \ldots$ is the discrete-time index and T_s is the sampling period. Without loss of generality, we assume the sampling period is normalized to unity (that is, $T_s = 1$). Using (2.16) and (2.17), we can write the output of neuron $y_q(k+1)$ as

$$y_q(k+1) = f_{\text{shl}}\left[\sum_{j=1}^{n} w_{qj} x_j(k) - \theta_q\right] \tag{2.18}$$

The unit delay z^{-1}, where z is a complex variable [6, 7], delays the output of the activation function by one sample period to give $y_q(k)$. Because we are formulating the neuron response in discrete time, (2.18) is a *difference equation.* As we will see in Section 5.3 for the Hopfield network, this quantity is fed back to the inputs of all *other* neurons (i.e., all neurons except the one that the

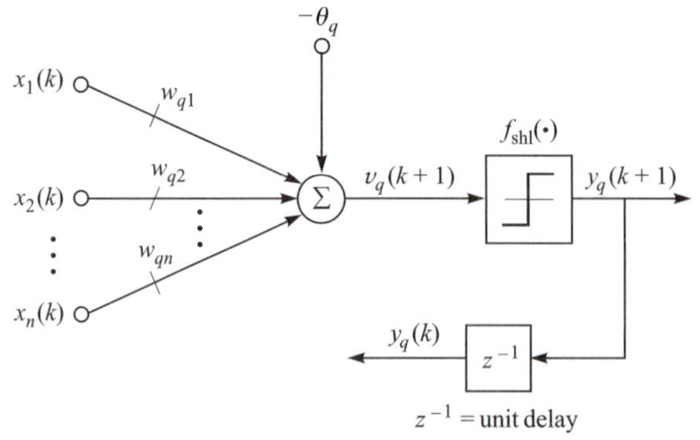

Figure 2.12 Discrete-time model of the Hopfield artificial neuron.

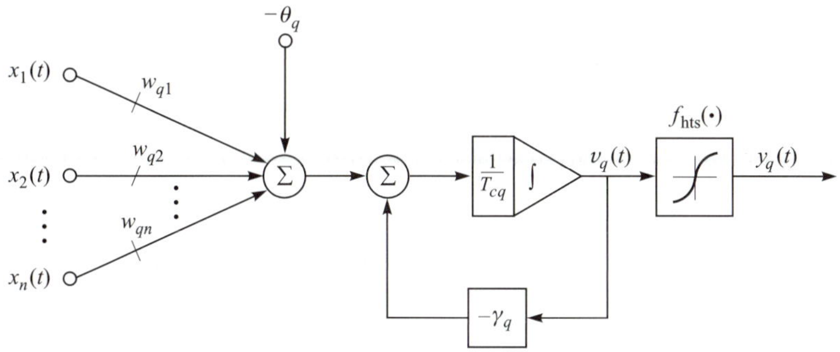

Figure 2.13 Continuous-time model of the Hopfield artificial neuron.

delayed output is taken from) that make up the discrete-time Hopfield neural network. Therefore, the inputs to each neuron are bipolar values, and the synaptic weight matrix for the discrete-time Hopfield neural network $\boldsymbol{W} = [w_{qj}]$, for $q, j = 1, 2, \ldots, n$, is real, symmetric, with zeros down the diagonal.

The continuous-time Hopfield model of the artificial neuron is shown in Figure 2.13. In Figure 2.13, $T_{cq} = R_q C_q$ is the integration *time constant* of the qth neuron, and θ_q is an externally applied threshold. The (leaky) integrator can be realized with an operational amplifier [8], capacitor C_q, and resistor R_q, and $\gamma_q > 0$ is called the *leakage* (or *forgetting*) *factor* of the integrator shown in the feedback loop in Figure 2.13. The leakage factor forces the internal signal v_q to zero for a zero input. From Figure 2.13 we can write a *differential equation* for the activation potential $v_q(t)$ as

$$T_{cq}\frac{\mathrm{d}v_q}{\mathrm{d}t} = -\gamma_q v_q + \left(\sum_{j=1}^{n} w_{qj}x_j - \theta_q\right) \tag{2.19}$$

The output of the neuron shown in Figure 2.13 is given by

$$y_q = f_{\mathrm{hts}}(v_q) \tag{2.20}$$

where the activation function $f_{\mathrm{hts}}(\bullet)$ is the hyperbolic tangent sigmoid function shown in Figure 2.10 and defined in (2.14). The discrete-time model difference equation in (2.17) can be derived from (2.19) for $\mathrm{d}v_q/\mathrm{d}t = 0$ and $\gamma_q = 1$, and for the hyperbolic tangent sigmoid function slope parameter set $\alpha >> 1$ [9]. This is actually a classic exercise in developing a discrete-time state-space model from the state-space model of the continuous-time system [10].

2.5 ADALINE AND MADALINE

The *least-mean-square* (LMS) *algorithm* was originally proposed by Bernard Widrow and M. E. (Ted) Hoff, Jr. (Widrow's student), in 1960 [11] at Stanford University, Stanford, California. This learning rule is sometimes referred to as the Widrow–Hoff learning rule or the delta rule [12, 13]. The

Adaline (*ada*ptive *lin*ear *e*lement) is a single neuron whose synaptic weights are updated according to the LMS algorithm [14]. At Stanford University, Widrow and his students developed one of the first trainable layered neural networks with multiple adaptive elements, known as the Madaline (*m*ultiple *Adaline*) [14–16]. The LMS algorithm for training a single-layer network is the predecessor to the *backpropagation* learning rule for feedforward multilayer perceptrons (cf. Sect. 3.3). The Adaline and Madaline were the basis for the first commercial neurocomputers [15] from Memistor Corporation (founded by Bernard Widrow). The LMS learning rule (delta rule) is an adaptive algorithm that computes adjustments of the neuron synaptic weights. The algorithm is based on the method of *steepest descent* [17, 18], and it adjusts the neuron weights to minimize the mean square error between the inner product of the weight vector with the input vector and the *desired* output of the neuron. We begin with an explanation of a simple adaptive linear combiner and then proceed to the details of the LMS algorithm, the Adaline, and finally the Madaline.

2.5.1 Simple Adaptive Linear Combiner and the LMS Algorithm

The architecture of a simple adaptive linear combiner is presented in Figure 2.14 [11]. The basic structure of the linear neuron is derived from Figure 2.2 with $f(\bullet)$ taken as a linear activation function, $x_0 = 1$ and $w_{q0} = \beta_q$ (bias), and $q = 1$ (because we are only working with a single neuron; however, we drop this subscript). During the process of supervised training of the neuron, the adaptive linear combiner is presented with training input vectors $\boldsymbol{x}(k) \in \Re^{n \times 1}$ (that is, $\boldsymbol{x}(k) = [x_1(k)\ x_2(k) \ldots\ x_n(k)]^T$) and the associated desired responses $d(k) \in \Re$,

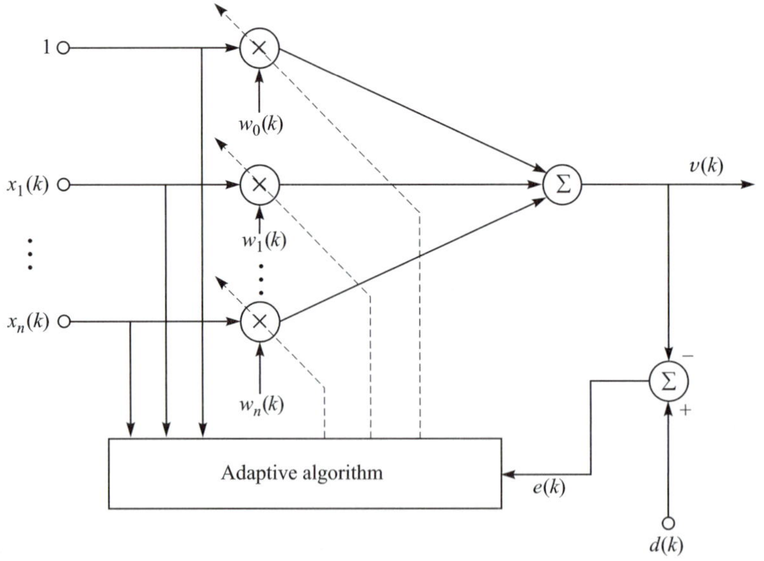

Figure 2.14 Simple adaptive linear combiner.

for $k = 1, 2, 3, \ldots$ (the discrete-time index); and the sampling period is assumed to be $T_s = 1$. The components of $\boldsymbol{x}$ can be either continuous analog values [possibly sampled as shown above, that is, $\boldsymbol{x}(k)$ with $T_s = 1$] or binary (or bipolar) values. We assume that the input $\boldsymbol{x}(k)$ is zero-mean; that is, $\mathrm{E}\{\boldsymbol{x}\} = \mathbf{0}$ where $\mathrm{E}(\bullet)$ is the expectation operator (cf. Sect. A.7.4), wide-sense stationary vector stochastic process. It is straightforward to show that the assumption $\mathrm{E}\{\boldsymbol{x}\} = \mathbf{0}$ is not a restriction on the LMS algorithm. The output of the network at any time step k is computed as the inner product of the input vector $\boldsymbol{x}(k)$ with the vector of weights $\boldsymbol{w}(k) = [w_1(k), w_2(k), \ldots, w_n(k)]^T \in \Re^{n\times 1}$, with $x_0 = 1$ and $w_0(k) = \beta$. We could incorporate the bias into the weight vector and $x_0 = 1$ into $\boldsymbol{x}(k)$ as we did previously. However, without loss of generality we assume the bias $\beta = 0$; thus the output of the network is written as

$$v(k) = \boldsymbol{x}^T(k)\boldsymbol{w}(k) = \boldsymbol{w}^T(k)\boldsymbol{x}(k) \tag{2.21}$$

The output $v(k)$ is then compared with the desired response $d(k)$, and the difference [or the error $e(k)$] of the two quantities is taken. The error along with the input vector $\boldsymbol{x}(k)$ is then fed to the adaptation or learning algorithm, as shown in Figure 2.14. The learning algorithm determines how the network synaptic weights are to be modified so that the difference between the output of the network $v(k) = \boldsymbol{w}^T(k)\boldsymbol{x}(\mathrm{k})$ (network response), and the desired response $d(k)$ is made as small as possible. The desired response is an auxiliary input to the network that is used only during the training phase. Commonly, the learning algorithms are derived so that they optimize some defined error criterion, which is explained in detail below, leading to the LMS algorithm.

A reader with a digital signal processing background would easily recognize the linear combiner as being a *linear transversal filter* [19] or a *finite impulse response* (FIR) *filter* [20]. Indeed, the linear combiner is very popular and widely used in adaptive signal processing [19]. The most commonly used performance criterion for deriving the learning rule for a linear combiner is minimization of the square of the error between the output of the network and the desired response. This is referred to as the mean square error (MSE) criterion. The simplest learning rule derived from the MSE criterion is called the least mean-square (LMS) learning rule, or the Widrow–Hoff learning rule, or the delta rule [19]. Before we develop the LMS learning rule, let us look at the classical approach to determining the optimal weight vector $\boldsymbol{w}^*$, known as the Wiener–Hopf solution [19, 21].

Let $\boldsymbol{x}(k)$ and $d(k)$ be the training input and the desired response, respectively, that are presented to the network at the kth iteration; and let $\boldsymbol{w}(k)$ denote the current weight vector of the network. The response of the linear combiner is given in (2.21); and the *error,* that is, the difference between the desired response and the network response, can be written as

$$e(k) = d(k) - v(k) = d(k) - \boldsymbol{w}^T(k)\boldsymbol{x}(k) \tag{2.22}$$

Assuming that both $\boldsymbol{x}(k)$ and $d(k)$ are from a statistically wide-sense stationary population, the MSE criterion can be formed as an ensemble average given as

$$J(\boldsymbol{w}) = \frac{1}{2}\mathrm{E}\{e^2(k)\} = \frac{1}{2}\mathrm{E}\left\{\left[d(k) - \boldsymbol{w}^T(k)\boldsymbol{x}(k)\right]^2\right\} \tag{2.23}$$

Recall that the objective of the learning rule is to adaptively modify the weights of the network so that (2.23) is minimized. Expanding (2.23), we can write

$$J(\boldsymbol{w}) = \frac{1}{2}\mathrm{E}\{d^2(k)\} - \mathrm{E}\{d(k)\boldsymbol{x}^T(k)\}\boldsymbol{w}(k) + \frac{1}{2}\boldsymbol{w}^T(k)\mathrm{E}\{\boldsymbol{x}(k)\boldsymbol{x}^T(k)\}\boldsymbol{w}(k) \tag{2.24}$$

Because $\boldsymbol{x}(k)$ and $d(k)$ are wide-sense stationary stochastic processes, (2.24) can be written as

$$J(\boldsymbol{w}) = \frac{1}{2}\mathrm{E}\{d^2(k)\} - \boldsymbol{p}^T\boldsymbol{w}(k) + \frac{1}{2}\boldsymbol{w}^T(k)\boldsymbol{C}_x\boldsymbol{w}(k) \tag{2.25}$$

where $\boldsymbol{p} = \mathrm{E}\{d(k)\boldsymbol{x}(k)\}$ represents the cross correlation vector between the desired response and the input patterns, and $\boldsymbol{C}_x = \mathrm{E}\{\boldsymbol{x}(k)\boldsymbol{x}^T(k)\}$ represents the covariance matrix for the input patterns [recall that input vector $\boldsymbol{x}(t)$ is zero-mean]. Equation (2.25) is a quadratic function of the synaptic weights; thus, it is positive for all possible network weights. In the vector space of the weights, the MSE surface for $J(\boldsymbol{w})$ has a unique minimum [19]. Therefore, we can mathematically compute the gradient of the performance measure in (2.25), with respect to the weight vector $\boldsymbol{w}$, and set this result equal to zero for the optimum condition, that is,

$$\nabla_w J(\boldsymbol{w}) = \frac{\partial J(\boldsymbol{w})}{\partial \boldsymbol{w}} = -\boldsymbol{p} + \boldsymbol{C}_x\boldsymbol{w}(k) = \boldsymbol{0} \tag{2.26}$$

See Section A.3.4.1 for details of differentiating a scalar with respect to a vector. Also note that the derivative of the MSE criterion with respect to the weight vector in (2.26) is a continuous derivative (i.e., the weights are continuous in amplitude). From (2.26) the optimal weights $\boldsymbol{w}^*$ are obtained as

$$\boldsymbol{w}^* = \boldsymbol{C}_x^{-1}\boldsymbol{p} \tag{2.27}$$

Equation (2.27) is the vector-matrix form of the well-known Wiener–Hopf solution for the optimal weights of a linear combiner [19]. Practical use of (2.27) is limited in neural networks and signal processing applications for two reasons: (1) Evaluation of the inverse of the covariance matrix is very computationally costly; (2) Equation (2.27) is not suitable for online modifications of the weights because in most cases the covariance matrix and the cross-correlation vector are not known a priori.

To circumvent these problems, Widrow and Hoff [11] developed the LMS algorithm. The idea of the LMS algorithm can be easily explained by examining the nature of the MSE surface shown in Figure 2.15, which depicts a typical error surface defined in (2.25) for a linear combiner with two weights, that is, $\boldsymbol{w} \in \Re^{2\times 1}$. To obtain the optimal values of the synaptic weights when $J(\boldsymbol{w})$ is a minimum, we can search the error surface using a gradient descent method and find the minimum value (i.e., when the gradient is zero). This was the same idea when we set the gradient of $J(\boldsymbol{w})$ equal to zero in (2.26). It is obvious that we can reach the bottom of the error surface in Figure 2.15 by changing the weights in the direction of the *negative gradient* on the surface. Because the gradient on the surface cannot be computed without knowledge of the input covariance matrix and the cross-correlation vector, these must be

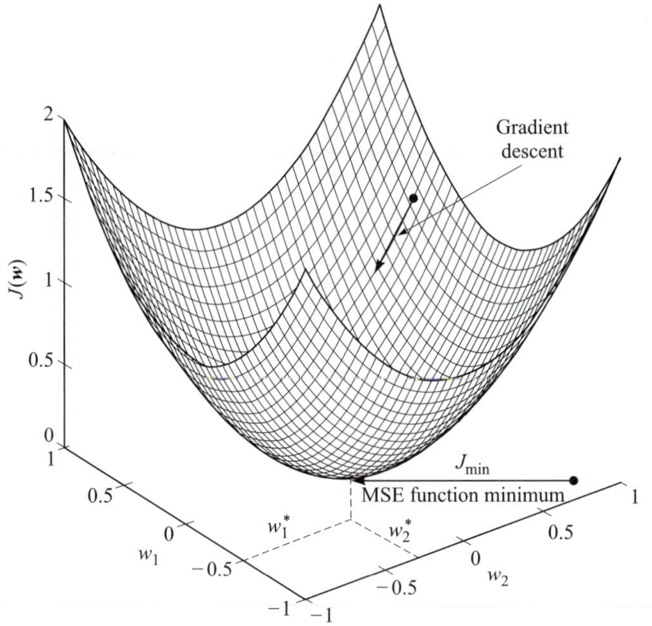

Figure 2.15 Typical MSE surface of an adaptive linear combiner.

estimated during an iterative training procedure. The simplest, and because of that, the crudest, estimate of the MSE gradient surface can be obtained by taking the gradient of the *instantaneous* error surface. That is, in (2.24) the expectation operator E(•) is dropped, and the gradient of $J(\boldsymbol{w})$ approximated as

$$\begin{aligned}
\nabla_{w} J(\boldsymbol{w}) &\approx \frac{1}{2}\frac{\partial e^2(k)}{\partial \boldsymbol{w}}\bigg|_{\boldsymbol{w}=\boldsymbol{w}(k)} \\
&= \frac{1}{2}\frac{\partial}{\partial \boldsymbol{w}(k)}\left[d^2(k) - 2d(k)\boldsymbol{x}^T(k)\boldsymbol{w}(k) + \boldsymbol{w}^T(k)\boldsymbol{x}(k)\boldsymbol{x}^T(k)\boldsymbol{w}(k)\right] \\
&= -d(k)\boldsymbol{x}(k) + \boldsymbol{x}(k)\boldsymbol{x}^T(k)\boldsymbol{w}(k) = -d(k)\boldsymbol{x}(k) + \boldsymbol{w}^T(k)\boldsymbol{x}(k)\boldsymbol{x}(k) \\
&= -\underbrace{[d(k) - \boldsymbol{w}^T(k)\boldsymbol{x}(k)]}_{e(k)}\boldsymbol{x}(k) = -\,e(k)\boldsymbol{x}(k)
\end{aligned} \tag{2.28}$$

Using our result in (2.28) for the estimate of the gradient of the MSE criterion, we can write the learning rule for updating the weights using the steepest-descent gradient method as

$$\boldsymbol{w}(k+1) = \boldsymbol{w}(k) + \mu[-\nabla_{w} J(\boldsymbol{w})] = \boldsymbol{w}(k) + \mu e(k)\boldsymbol{x}(k) \tag{2.29}$$

where the error $e(k)$ is defined in (2.22). Equations (2.22) and (2.29) together are known as the LMS algorithm for updating the weights of the adaptive linear combiner. The LMS algorithm has diverse applications, for example, adaptive equalization and noise cancellation. In (2.29) the real parameter

$\mu > 0$ is commonly referred to as the learning rate parameter. It specifies the magnitude of the *update step* for the weights in the negative gradient direction. If the value of μ is chosen to be too small, the learning algorithm will modify the weights sluggishly and a relatively large number of iterations will be required before the bottom of the error surface is reached. On the other hand, if the value of the learning parameter is set too large, the learning rule can become numerically unstable. The reason for this is the *approximation* that was used for the gradient evaluation in (2.28). Therefore, setting the learning rate parameter too large will result in repeated multiples of any errors leading to the weights not converging, that is, possible divergence problems. The scalar form of the LMS algorithm can be written directly from (2.22) and (2.29) as

$$e(k) = d(k) - \sum_{h=1}^{n} w_h(k)x_h(k) \tag{2.30}$$

and

$$w_i(k+1) = w_i(k) + \mu e(k)x_i(k) \tag{2.31}$$

respectively, for $i = 1, 2, \ldots, n$ (i.e., for all synapses).

The continuous-time form of the LMS algorithm can have an arbitrarily large learning rate parameter without affecting the numerical stability of the algorithm. However, the discrete-time form in (2.22) and (2.29) must have an upper bound established for the learning rate parameter to ensure stability. Rigorous treatment of the convergence properties of the discrete-time form of the LMS algorithm [19, 22] establishes bounds on the learning rate parameter given as

$$0 < \mu < \frac{2}{\lambda_{\max}} \tag{2.32}$$

where $\lambda_{\max}$ represents the largest eigenvalue of the input covariance matrix $\boldsymbol{C}_x$. Simply stated, the LMS algorithm is *convergent in the mean* if the learning rate parameter is positive with an upper bound of $2/\lambda_{\max}$. In order to have convergence of the LMS algorithm be less susceptible to stability problems, the acceptable values for the learning rate parameter are commonly bounded by

$$0 < \mu < \frac{2}{\mathrm{trace}\{\boldsymbol{C}_x\}} \tag{2.33}$$

This inequality can be derived from a convergence analysis of the *LMS algorithm in the mean square*. The bound on the learning rate parameter in (2.33) is more conservative than (2.32). This is true because

$$\mathrm{trace}\{\boldsymbol{C}_x\} = \sum_{h=1}^{n} \lambda_h = \sum_{h=1}^{n} c_{x_{hh}} \geq \lambda_{\max} \tag{2.34}$$

[23], where $c_{x_{hh}}$, for $h = 1, 2, \ldots, n$, are the diagonal elements of the covariance matrix $\boldsymbol{C}_x$ and λ_h, for $h = 1, 2, \ldots n$, are the eigenvalues of $\boldsymbol{C}_x$ which have to be nonnegative. Therefore, if the LMS algorithm is convergent in a mean

square sense, then it is also convergent in the mean; however, the converse of this statement is not true. Moreover, the trace of the covariance matrix is equal to the total input power. Thus, in (2.33) the upper bound could be understood as being twice the reciprocal of the total input power.

Both (2.32) and (2.33) assume that we at least have an estimate of the input covariance matrix. In most practical cases such an estimate is difficult to obtain. However, even if some estimate of the covariance matrix is available, the learning rate parameter is frequently set to a fixed value. Although a fixed value for the learning rate parameter produces the simplest form of the steepest descent algorithm, it is perhaps more appropriate to have the learning rate parameter change with time. One of the major problems with a fixed learning rate parameter (even if it is selected to result in LMS algorithm convergence) is the accuracy of the results. That is, if the fixed learning rate parameter is set relatively large, the numerical accuracy of the synaptic weights is directly dependent on this value. The smaller the value of the parameter, the more accurate the results. However, if the fixed learning rate parameter is set relatively small, convergence of the LMS algorithm is painfully slow. In stochastic approximation theory, which can be traced back to Robbins and Monro's root-finding algorithm in 1951 [24], the learning rate parameter varies with time. The most commonly used form in the stochastic approximation literature is given by

$$\mu(k) = \frac{\kappa}{k} \tag{2.35}$$

where κ is a constant. Equation (2.35) will guarantee convergence of the stochastic approximation algorithm if κ is chosen relatively small [25, 26]. As we can see from (2.35), as training progresses, the value of μ decreases. On the other hand, for κ set too large, for example, outside the range of values given in (2.32), divergence of the algorithm will occur for small values of the time step k.

The basic problem with the stochastic approximation schedule in (2.35) is the rapid initial decrease of the learning rate parameter. It seems reasonable that during the learning process μ should be large at the beginning of training and then gradually decrease as the network converges. This is precisely what the *search-then-converge* algorithm of Darken and Moody [27] accomplishes. In the first phase of the search-then-converge strategy (called the search phase), μ is relatively large and almost constant (i.e., μ decreases very slowly). In the second phase (called the converge phase), μ decreases exponentially to zero. Equation (2.36) gives the simplest form of this type of adaptive adjustment of the learning rate parameter.

$$\mu(k) = \frac{\mu_0}{1 + k/\tau} \tag{2.36}$$

In (2.36), $\mu_0 > 0$ and $\tau >> 1$ (called the *search time constant*); however, typically $100 \leq \tau \leq 500$. For suitably chosen parameters μ_0 and τ, considerable improvement in the speed of convergence of the LMS algorithm can be achieved for training of the simple adaptive linear combiner (see Example 2.1). These methods of adjusting the learning rate parameter are commonly called *learning rate schedules*. From (2.36) we can see that for small k (i.e., the beginning of the adaptation of μ) and k small with respect to the search time constant τ, $\mu(k) \approx \mu_0$. Therefore, the algorithm essentially behaves as the classical LMS algorithm with a fixed learning rate parameter μ_0. However, when k becomes relatively large with respect to the search time constant τ, then the adaptation of μ is essentially a stochastic approximation schedule given in (2.35), with $\kappa = \tau\mu_0$. Ideally, μ_0 in (2.36) should be set at a relatively large value, but within the permissible range given in (2.33). Figure 2.16 shows a comparison of the stochastic approximation and search-then-converge schedules for the LMS learning rate parameter. There are variations of the simple scheduling strategy for μ given in (2.36); for example, see [28, 29].

Another approach that can be used to adjust the learning rate parameter is not a scheduled-type adjustment, which continually decreases the value every time step as the stochastic approximation and search-then-converge schedules. Instead, the third method is an adaptive normalization approach in the sense that μ is adjusted according to the input data every time step as

$$\mu(k) = \frac{\mu_0}{\|\boldsymbol{x}(k)\|_2^2} \tag{2.37}$$

where μ_0 is a fixed constant. Stability is guaranteed if $0 < \mu_0 < 2$; however the practical range is $0.1 \leq \mu_0 \leq 1$ [16, 22]. Table 2.1 summarizes the LMS algorithm for the simple adaptive linear combiner using the search-then-converge strategy for adjusting the learning rate parameter given in (2.36). The stochastic approximation schedule in (2.35), or the adaptive normalization approach in (2.37), or even a constant learning rate parameter (for the classical LMS algorithm) could replace the search-then-converge scheduling presented in Table 2.1 for adjusting the learning rate parameter.

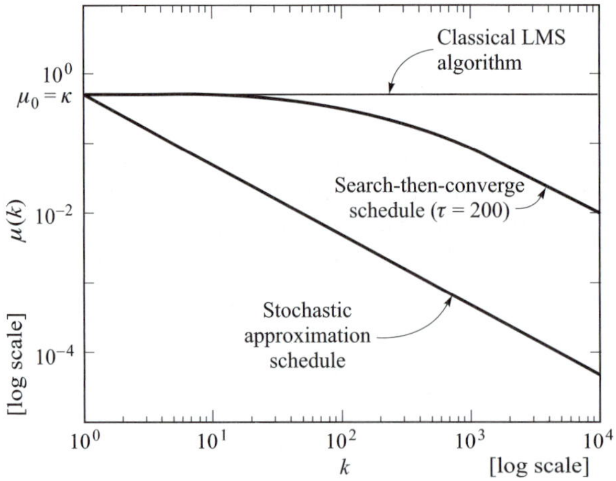

Figure 2.16 Comparison of two learning rate schedules: stochastic approximation schedule and the search-then-converge schedule.

TABLE 2.1
Summary of the LMS algorithm

Step 1. Set $k = 1$, initialize the synaptic weight vector $\boldsymbol{w}(k = 1)$, and select values for μ_0 and τ.

Step 2. Compute the learning rate parameter as

$$\mu(k) = \frac{\mu_0}{1 + k/\tau}$$

Step 3. Compute the error

$$e(k) = d(k) - \sum_{h=1}^{n} w_h(k)x_h(k)$$

Step 4. Update the synaptic weights $w_i(k+1) = w_i(k) + \mu(k)e(k)x_i(k)$, for $i = 1, 2, \ldots, n$.

Step 5. If convergence is achieved, *stop*; else set $k \leftarrow k + 1$, then go to step 2.

EXAMPLE 2.1. This example illustrates the ability of the adaptive linear combiner, trained with the LMS algorithm, to estimate the parameters of a linear model. The input data consist of 1,000 zero-mean Gaussian random vectors with three components, that is, $\boldsymbol{x} \in \Re^{3\times 1}$; and the bias is set to zero, or $\beta = 0$. The variances of the components of $\boldsymbol{x}$ are 5, 1, and 0.5, respectively. The assumed linear model is given by $\boldsymbol{b} = [1, 0.8, -1]^T$. To generate the target values (desired outputs), the 1,000 input vectors are used to form a matrix $\boldsymbol{X} = [\boldsymbol{x}_1 \boldsymbol{x}_2 \cdots \boldsymbol{x}_{1,000}]$, and the desired outputs are computed according to $\boldsymbol{d} = \boldsymbol{b}^T \boldsymbol{X}$. The covariance matrix of the vector input signals can be estimated as

$$\boldsymbol{C}_x \approx \frac{1}{1{,}000} \sum_{h=1}^{1{,}000} \boldsymbol{x}\boldsymbol{x}^T = \frac{\boldsymbol{X}\boldsymbol{X}^T}{1{,}000}$$

[30]. Using the LMS algorithm in Table 2.1, with a value of $\mu_0 = 0.9/\lambda_{\max} = 0.1936$, where $\lambda_{\max}$ is the largest eigenvalue of the covariance matrix $\boldsymbol{C}_x$, and $\tau = 200$ (the search time constant), the input vectors along with the associated desired output values are presented to the linear combiner. The criterion used to terminate the learning process involved monitoring the square root of the MSE values every time step k. The learning process was terminated when $\sqrt{J} = \sqrt{{}^1\!/\!{}_2 e^2(k)} \le 10^{-8}$, where $e(k) = d(k) - \boldsymbol{w}^T(k)\boldsymbol{x}(k)$. The initial values of the synaptic weight vector were selected as zero-mean Gaussian random numbers with a variance of 0.25, $\boldsymbol{w}_{\text{initial}} = [-0.3043, -0.8195, 0.3855]^T$. The LMS learning process was terminated after only 204 iterations (training epochs). In other words, after the first 204 input vectors along with the associated desired output values were presented, the network converged. The final synaptic weight matrix was given as $\boldsymbol{w}_{\text{final}} = [1.000000, 0.800000, -1.000000]^T$, which is exactly the assumed linear model $\boldsymbol{b}$ (out to six decimal places). In fact, the L_2-norm of the difference between the linear model $\boldsymbol{b}$ and the final weight vector $\boldsymbol{w}_{\text{final}}$ is $\|\boldsymbol{b} - \boldsymbol{w}_{\text{final}}\|_2 = 1.505404 \times 10^{-7}$. Figure 2.17 shows the progress of the learning rate parameter as it is adjusted according to the search-then-converge schedule. As we can see from the plot, at the beginning of training μ does not change much; then toward the end of training it becomes much smaller. Figure 2.18 shows the root mean square (RMS) value of the performance measure, that is, $\sqrt{J}$, as training of the network progresses. This exercise is similar to problems in system identification of estimating a parameter vector associated with a dynamic model of a system given only input/output data from the system, that is, *parametric system identification* [31].

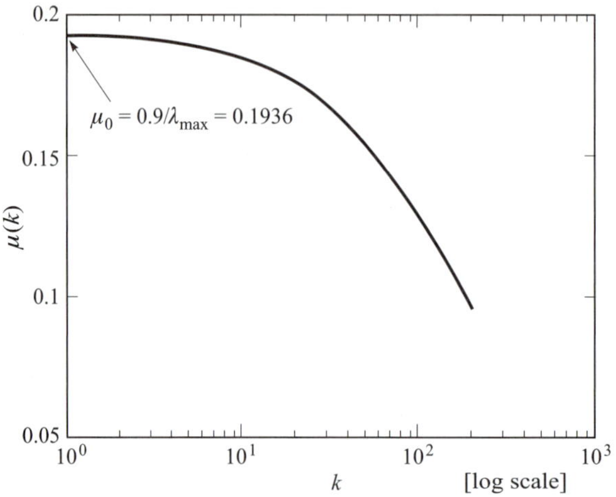

Figure 2.17 Learning rate for the search-then-converge schedule.

2.5.2 Adaptive Linear Element (Adaline)

The Adaline (adaptive linear element) is an adaptive pattern classification network that is trained by the LMS algorithm. The Adaline is the basic building block used in many neural networks. The architecture of the Adaline [11, 16] is presented in Figure 2.19. We can see that the network consists of a linear combiner cascaded with a symmetric hard limiter (or symmetric hard-limiting quantizer), that is, a *signum* function. The symmetric-hard-limiting quantizer produces a bipolar (± 1) output, that is, $y(k) = \text{sgn}[v(k)]$, although this is not a restriction; for example, a hard limiter producing binary outputs $\{0, 1\}$ could be used as well. The adjustable bias

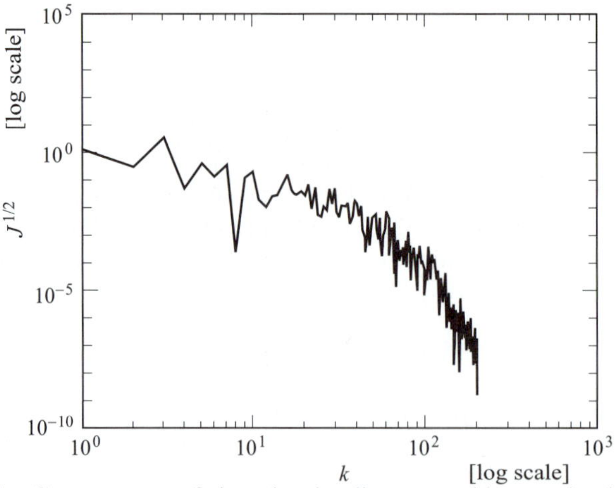

Figure 2.18 Convergence of the adaptive linear combiner trained by the LMS algorithm.

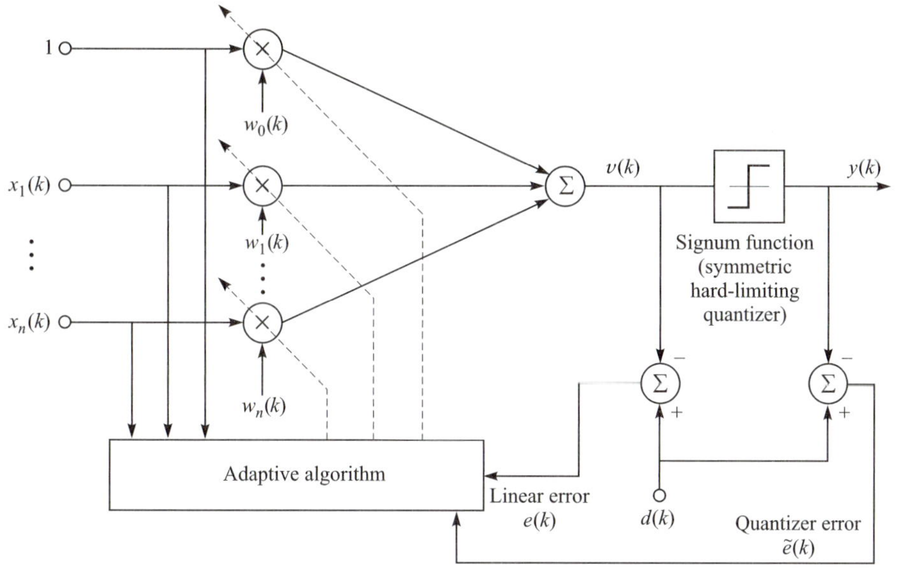

Figure 2.19 Adaptive linear element (Adaline).

weight $w_0(k) = \beta$ connected to the constant input $x_0(k) = 1$ effectively serves to control the *threshold* level of the quantizer. During the training process of the network, the error is generated as the difference between the desired output and the output of the *linear combiner*, and the resulting error is called the *linear error*. The linear error is the basis for the LMS learning algorithm that we discussed in previous sections. An alternative method of describing the error is to take the difference between the desired output and the output from the symmetric hard limiter. This error is called the *quantizer error*, as shown in Figure 2.19, and is the basis for the perceptron learning rule (cf. Sect. 2.6). Therefore, the Adaline and the perceptron are very similar; however, their differences are discussed in Section 2.6.

During training of the Adaline, the input vectors $\boldsymbol{x} = [1, x_1, x_2, \cdots, x_n] \in \Re^{n+1\times 1}$, are presented to the network along with the associated target (or desired output) values ($d \in \Re$). The synaptic weights $\boldsymbol{w} = [\beta, w_1, w_2, \cdots, w_n] \in \Re^{n+1\times 1}$ are adaptively changed according to the *linear* LMS algorithm. After the Adaline is trained, an input vector presented to the network with fixed weights will result in a scalar output. Therefore, the network performs a mapping of an n-dimensional vector space to a scalar ($\Re^n \to \Re$). The activation function is not used during the training of the Adaline; therefore, the training process is identical to that of the adaptive linear combiner. The symmetric hard limiter (quantizer) is used only after the Adaline is trained. Once the weights are properly adjusted, the response of the trained neuron can be tested by applying various inputs that were not used during the training phase. If the Adaline produces responses (outputs) that are consistent to a high degree of probability with the test inputs, it can be said that *generalization* has occurred. The process of *training* and *generalization* are two very important attributes of the Adaline, and of neural networks in general. One common application of the Adaline is for the realization of a small class of logic functions, for example, the AND, NOT, OR, and MAJ (majority) logic functions [9, 32]. Only those logic functions that are *linearly separable* can be realized by the (single) Adaline. Three of these logic functions can be written mathematically as

$$y = \operatorname{sgn}\left[\left(\sum_{j=1}^{n} x_j\right) + 1 - n\right] = \mathsf{AND}(x_1, x_2, \ldots, x_n) = \begin{cases} +1 & \text{if all } x_j = +1 \\ -1 & \text{otherwise} \end{cases} \tag{2.38}$$

$$y = \operatorname{sgn}\left[\left(\sum_{j=1}^{n} x_j\right) + n - 1\right] = \mathsf{OR}(x_1, x_2, \ldots, x_n) = \begin{cases} +1, & \text{if some } x_j = +1 \\ -1 & \text{otherwise} \end{cases} \tag{2.39}$$

$$y = \operatorname{sgn}\left[\sum_{j=1}^{n} x_j\right] = \mathsf{MAJ}(x_1, x_2, \ldots, x_n) = \begin{cases} +1 & \text{if the majority of } x_j = +1 \\ -1 & \text{otherwise} \end{cases} \tag{2.40}$$

Figure 2.20 shows a single-neuron (Adaline) realization of the AND, OR, and MAJ logic functions given in (2.38), (2.39), and (2.40), respectively.

Linear separability

When presented with an input pattern, the Adaline produces an output which is either -1 or $+1$ (assuming a symmetric hard limiter for the activation function). Consequently, the Adaline acts as a classifier which separates all possible input patterns into two categories. Consider a simple example of an Adaline having just two inputs, as shown in Figure 2.21. There are actually three inputs; however, we always assume a bias (or threshold). The output of the linear combiner is given as

$$v(k) = w_1(k)x_1(k) + w_2(k)x_2(k) + w_0(k) \tag{2.41}$$

The output of the hard limiter will be determined by the sign of $v(k)$, and therefore the borderline for the classification is defined by

$$v(k) = 0 \tag{2.42}$$

That is,

$$w_1(k)x_1(k) + w_2(k)x_2(k) + w_0(k) = 0 \tag{2.43}$$

or

$$x_2(k) = -\frac{w_1(k)}{w_2(k)}x_1(k) - \frac{w_0(k)}{w_2(k)} \tag{2.44}$$

Equation (2.44) represents a straight line in a two-dimensional vector space of the input vectors, as illustrated in Figure 2.22.

As seen in Figure 2.22, the straight line effectively separates the input space into two domains, that is, $v(k) > 0$ and $v(k) < 0$. All the input vectors

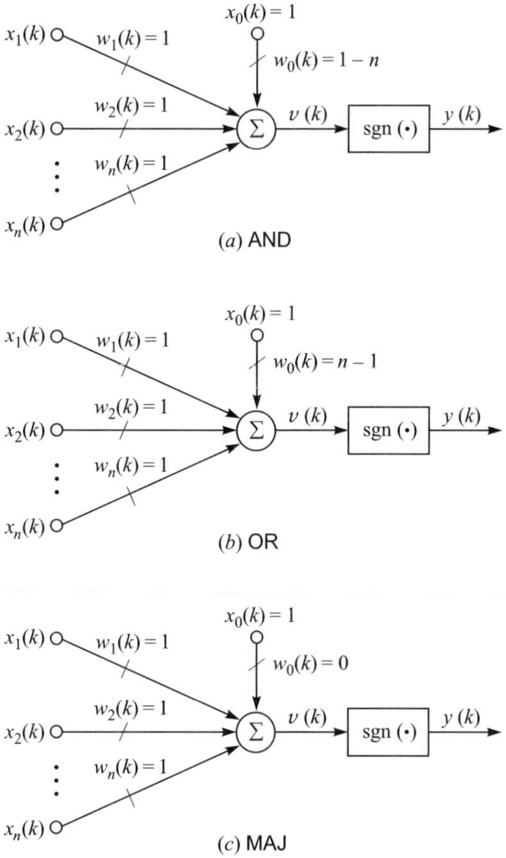

Figure 2.20 Single-neuron (Adaline) realization of the (*a*) AND, (*b*) OR, and (*c*) MAJ logic functions. For the MAJ (majority) logic function in (*c*), n is always taken to be an odd integer.

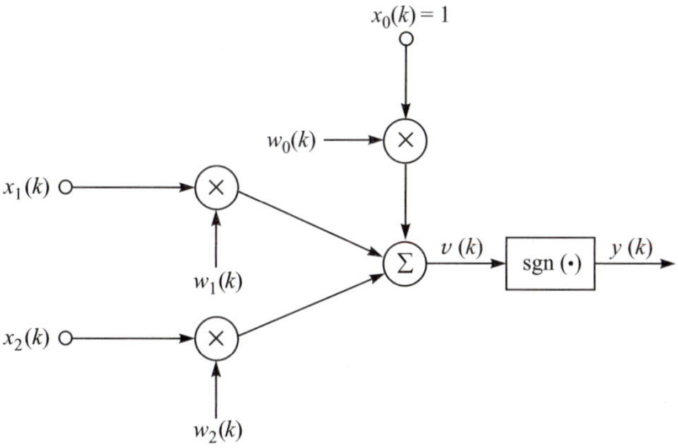

Figure 2.21 Adaline with two inputs.

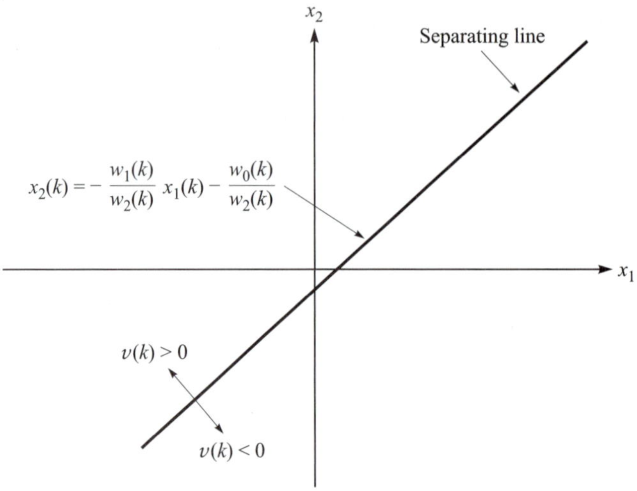

Figure 2.22 Linear separability of the Adaline.

that belong to the same domain will be classified in the same way, either 1 or −1. If the dimension of an input vector is 3 (excluding the bias), the domains are separated by a plane. If the dimension of the input vector is greater than 3, the border represents a hyperplane. In general, the Adaline represents a linear classifier, and this limits its application to cases in which input patterns are linearly separable. To illustrate the point, consider an example shown in Figure 2.23. Suppose that a network is to be trained to perform the separation of the input space as indicated by the separation boundary in Figure 2.23. Since the boundary is not a straight line and linear separability does not exist for this case, the Adaline cannot be used to accomplish this task.

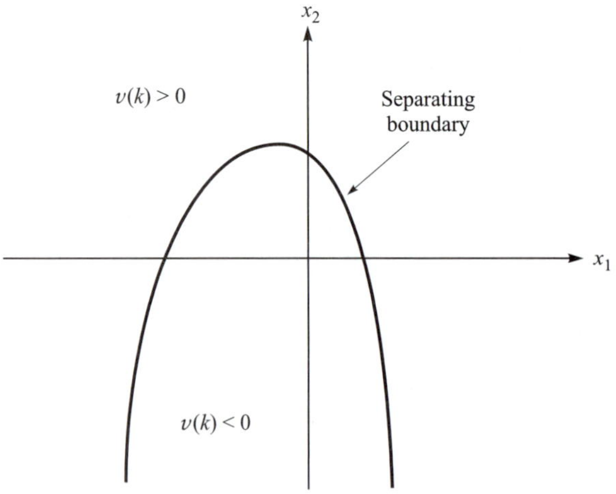

Figure 2.23 Nonlinear separation problem.

Adaline with nonlinearly transformed inputs (polynomial discriminant functions)

To solve the classification problem for patterns that are not linearly separable, the inputs to the Adaline can be preprocessed with fixed nonlinearities. Useful nonlinearities include the polynomial functions to preprocess the network inputs [16]. Consider an example of the network shown with two inputs in Figure 2.24. In this case, the signal at the input to the symmetric hard limiter is given as

$$v(k) = w_0(k) + w_1(k)x_1^2 + w_2(k)x_1 + w_3(k)x_1x_2 + w_4(k)x_2 + w_5(k)x_2^2 \quad (2.45)$$

The critical thresholding condition for this Adaline with nonlinearly transformed inputs occurs when $v(k)$ in (2.45) is set to zero. This condition represents an ellipse in the two-dimensional input vector space, as shown in Figure 2.25, and is a solution to the exclusive-NOR (XNOR) problem. By introducing the nonlinearities in the input layer, a separation boundary is generated which is not a straight line (i.e., an elliptical separating boundary). Therefore, if the appropriate nonlinearities are chosen, the network can be trained to separate the input space into two subspaces which are not linearly separable. In general, the Adaline with nonlinearly transformed inputs can be trained in the same manner as the linear Adaline network [16].

The use of nonlinearities, as illustrated in the previous example, can be generalized for more than two network inputs and for many other types of nonlinearities. The obvious advantage of transforming the inputs to the network with nonlinear functions is a possibility of forming a nonlinear separation boundary. Thus, a wide variety of adaptive nonlinear discriminant

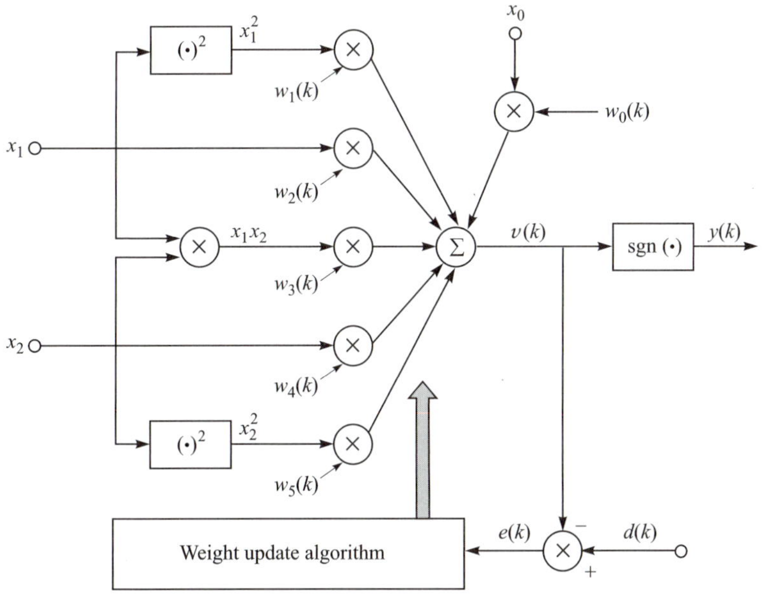

Figure 2.24 Adaline with nonlinearly transformed inputs.

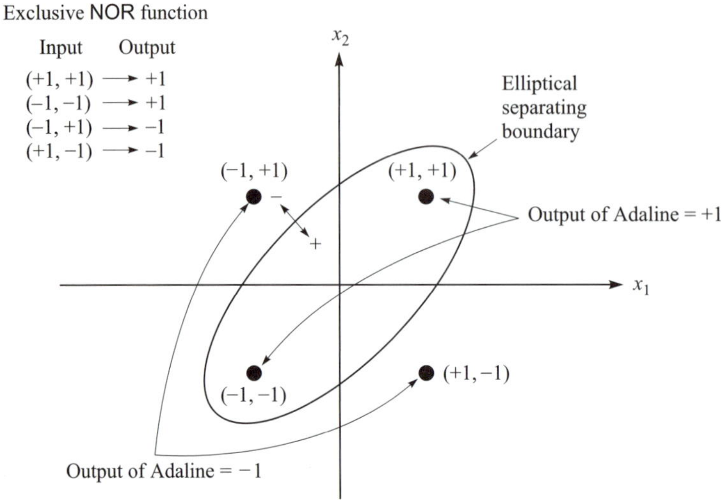

Figure 2.25 Realizing a nonlinearly separable function (*exclusive*-NOR function, i.e., XNOR) with an elliptical separating boundary. The algebraic expression for this type of logic gate is given by $y = x_1x_2 + \bar{x}_1\bar{x}_2 = \overline{x_1 \oplus x_2}$, where $\bar{x}_j$ denotes the NOT (or the *complement* operation) of x_j.

functions can be realized by adapting a single Adaline [16]. Polynomial discriminant functions have been used by Specht [33, 34] for classification and analysis of electrocardiogram (ECG) data, and other work in this area can be found in [35–38].

Linear error correction rules

For the Adaline there are two basic linear correction rules that can be used to adaptively adjust the synaptic weights of the network. They are called μ-LMS and α-LMS. For these linear correction rules, the change of the network weights is proportional to the difference between the output of the linear combiner and the desired output signal. The μ-LMS learning rule is the same as the LMS learning rule given in (2.22) and (2.29) for the simple adaptive linear combiner. The α-LMS learning rule can be thought of as a self-normalizing version of the μ-LMS learning rule. The weights of the network are updated according to the α-LMS learning rule

$$\boldsymbol{w}(k+1) = \boldsymbol{w}(k) + \alpha \frac{e(k)\boldsymbol{x}(k)}{\|\boldsymbol{x}(k)\|_2^2} \tag{2.46}$$

where $e(k)$ is given by (2.22). Comparing the adaptive normalization method in (2.37) for adjusting the learning rate parameter in the LMS algorithm in (2.22) and (2.29) to the α-LMS learning rule in (2.46), we see that they are identical for $\mu_o = \alpha$. Moreover, comparing the classical LMS algorithm (with a fixed learning rate parameter) in Equations (2.22) and (2.29) to the α-LMS learning rule in (2.46), we see that the only difference between the two learning rules is the normalization term in the α-LMS algorithm. However, the interpretation of the mechanisms for the two learning rules is entirely different. Specifically, the α-LMS algorithm is designed in accordance with the *minimal-disturbance principle* [16]; that is, when adapting to respond properly to a new input pattern, the responses to previous training patterns are minimally disturbed (on the average). Therefore, the μ-LMS is based on minimization of the MSE surface, while α-LMS updates the weights so that the current error is reduced. To clarify this point, consider the change in the error for α-LMS written as [16]

$$\begin{aligned}\Delta e(k) &= e(k+1) - e(k) = \left[d(k) - \boldsymbol{w}(k+1)^T \boldsymbol{x}(k)\right] - e(k) \\ &= \left\{d(k) - \left[\boldsymbol{w}^T(k) + \alpha \frac{e(k)\boldsymbol{x}^T(k)}{\|\boldsymbol{x}(k)\|_2^2}\right]\boldsymbol{x}(k)\right\} - e(k) \\ &= \left[e(k) - \alpha \frac{e(k)\boldsymbol{x}^T(k)\boldsymbol{x}(k)}{\|\boldsymbol{x}(k)\|_2^2}\right] - e(k) = -\alpha e(k)\end{aligned} \tag{2.47}$$

From (2.47) we see that $\alpha = -\Delta e(k)/e(k)$, which represents the relative change in the error or the part of the current error that is corrected at each step of the training process. The choice of α controls stability and speed of convergence [22], and is typically set in the range

$$0.1 < \alpha < 1 \tag{2.48}$$

The α-LMS algorithm is self-normalizing in the sense that the choice of α does not depend on the magnitude of the inputs to the network. When the inputs are bipolar [−1, 1], the normalization term $\|\boldsymbol{x}(k)\|_2^2$ in (2.46) is equal to the number of weights and does not vary from input pattern to input pattern. However, if the inputs are binary [0, 1], no adaptation will occur for those weights with a zero input; but for bipolar inputs all network weights are adapted each cycle, and convergence tends to be faster. Therefore, bipolar input patterns are generally preferred.

Detailed comparison of the μ-LMS and α-LMS learning rules

The relationship between the μ-LMS and α-LMS learning rules can be demonstrated in the following manner. We start with the α-LMS algorithm given in (2.46)

$$\begin{aligned}\boldsymbol{w}(k+1) &= \boldsymbol{w}(k) + \alpha \frac{e(k)\boldsymbol{x}(k)}{\|\boldsymbol{x}(k)\|_2^2} \\ &= \boldsymbol{w}(k) + \alpha \frac{\left[d(k) - \boldsymbol{w}^T(k)\boldsymbol{x}(k)\right]\boldsymbol{x}(k)}{\|\boldsymbol{x}(k)\|_2^2} \\ &= \boldsymbol{w}(k) + \alpha \left[\frac{d(k)}{\|\boldsymbol{x}(k)\|_2} - \boldsymbol{w}^T(k)\frac{\boldsymbol{x}(k)}{\|\boldsymbol{x}(k)\|_2}\right]\frac{\boldsymbol{x}(k)}{\|\boldsymbol{x}(k)\|_2}\end{aligned} \tag{2.49}$$

and then define

$$\widehat{d}(k) \triangleq \frac{d(k)}{\|\boldsymbol{x}(k)\|_2} \tag{2.50}$$

and also define

$$\widehat{\boldsymbol{x}}(k) \triangleq \frac{\boldsymbol{x}(k)}{\|\boldsymbol{x}(k)\|_2} \tag{2.51}$$

as the *normalized* desired response and *normalized* training vector, respectively. Using the definitions in (2.50) and (2.51), we can write (2.49) as

$$\boldsymbol{w}(k+1) = \boldsymbol{w}(k) + \alpha\left[\widehat{d}(k) - \boldsymbol{w}^T(k)\widehat{\boldsymbol{x}}(k)\right]\widehat{\boldsymbol{x}}(k) \tag{2.52}$$

which has the same form as the μ-LMS learning rule. Therefore, we can conclude that the α-LMS algorithm represents a μ-LMS learning strategy for normalized input patterns.

Although the α-LMS algorithm is slightly more complicated to analyze and implement than the μ-LMS algorithm, it has been shown to converge faster, for a given level of gradient noise propagated into the network weights, in the cases when there is a relatively large eigenvalue spread associated with the covariance matrix $\boldsymbol{C}_x = \mathrm{E}\{\boldsymbol{x}\boldsymbol{x}^T\}$ [16]. Gradient noise is the difference between the estimate of the gradient and the actual (or true) gradient. However, it can also be shown that the μ-LMS algorithm has the advantage of converging to the minimum of the MSE surface while α-LMS converges in the mean to a slightly biased version of the optimal least-squares solution for nonbipolar inputs [16]. Moreover, for cases when the input patterns have the same norm (e.g., for bipolar inputs), the two learning rules yield identical results.

Nonlinear weight correction rules

It has been shown in certain situations that the linear weight correction rules previously given may fail to separate certain training patterns that are linearly separable [39]. When this situation arises, it may be beneficial to use a nonlinear learning rule for the adaptation of the network weights. We defer discussion of these types of learning rules until Section 2.6.

2.5.3 Multiple Adaline (Madaline)

We have already demonstrated in Section 2.5.2 that a single Adaline cannot resolve input space separation problems that require a separation boundary that is nonlinear, unless a nonlinear transform of the inputs is applied. One approach to the nonlinear separation problem (without a nonlinear transform of the inputs to the network) is to use Madaline (multiple Adaline) networks. The basic architecture of a Madaline network consists of combining several Adalines into a single feedforward structure. There are two basic types of Madaline networks, termed Madaline I and Madaline II. Madaline I is a single-layer network that was first introduced by Widrow [15] and Hoff [42]. An example of a Madaline I architecture, which consists of three Adalines, is shown in Figure 2.26. To generate the output $y(k)$ of the Madaline I network, a fixed logic function (e.g., an OR, AND, or MAJ logic element) is applied to the outputs of the Adalines. The Madaline II architecture is a multilayer network with multiple outputs [32, 43, 44]. An example of a Madaline II architecture is shown in Figure 2.27. At this point it is important to discuss the convention for counting the number of layers in a multilayer network. The number of layers in a multilayer network is defined to be the total number of layers that have processing elements. Thus, the *input layer* is not considered in counting the total number of layers in a multilayer network. For example, in Figure 2.27 there are two layers of processing elements; therefore, the Madaline II is considered a two-layer network.

Before we discuss methods to adapt the weights of a Madaline, let us again address the exclusive-NOR (XNOR) problem and see how we can realize this logic function with a Madaline architecture. Figure 2.28 shows a two-input Madaline I architecture consisting of two Adalines. The two outputs of the Adalines are presented to a logic AND gate whose output y gives the logic decision of either 1 or -1. The outputs of the two linear combiners set equal to zero, that is, $v_1(k) = 0$ and $v_2(k) = 0$, will form two borderlines in the two-dimensional input (pattern) space

$$x_2(k) = -\frac{w_{11}(k)}{w_{12}(k)}x_1(k) - \frac{w_{10}(k)}{w_{12}(k)} \tag{2.53}$$

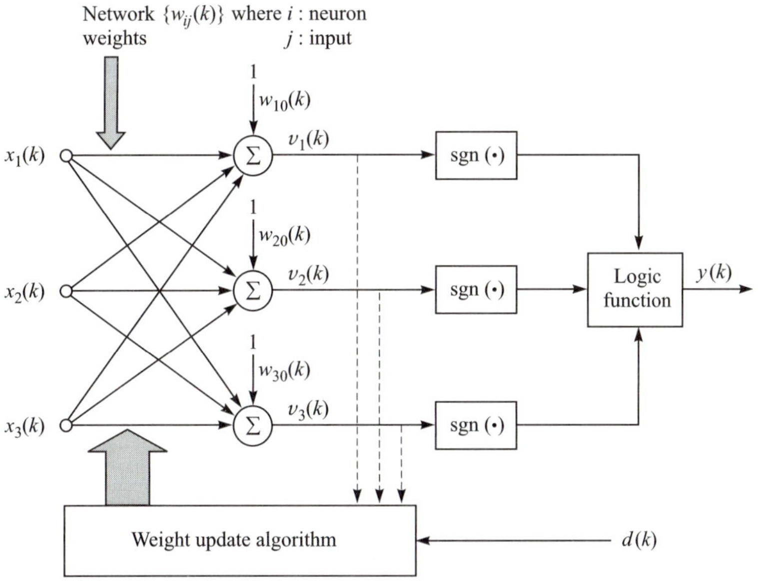

Figure 2.26 Example of a Madaline I network consisting of three Adalines.

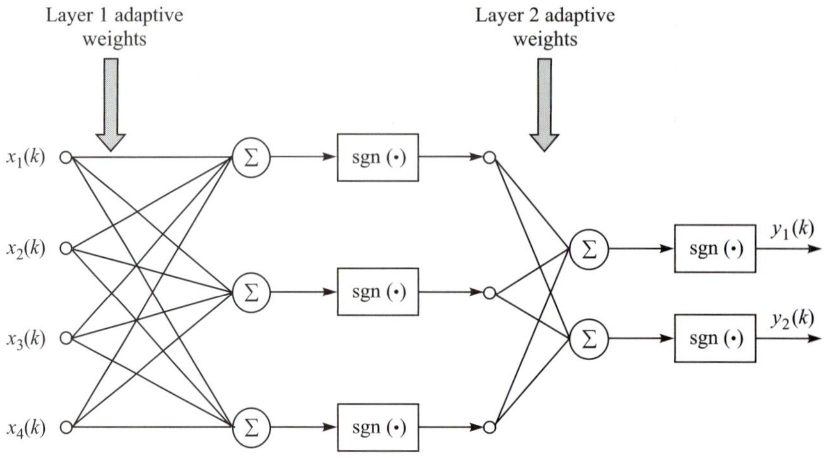

Figure 2.27 Two-layer Madaline II architecture.

and

$$x_2(k) = -\frac{w_{21}(k)}{w_{22}(k)}x_1(k) - \frac{w_{20}(k)}{w_{22}(k)} \tag{2.54}$$

With the network weights properly set, the borderlines together form a separation boundary that implements the XNOR logic function as illustrated in Figure 2.29.

Madaline learning strategies

There are two basic learning strategies for adjusting the weights of Madalines [16]. The first is Madaline rule I (MRI), which is a learning strategy

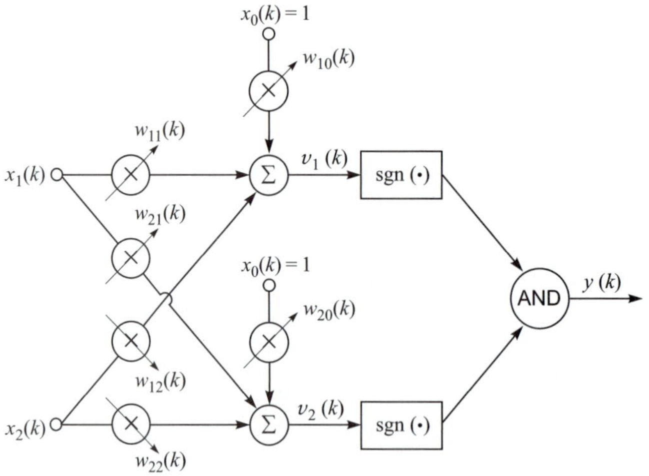

Figure 2.28 Madaline I realization of an XNOR logic function.

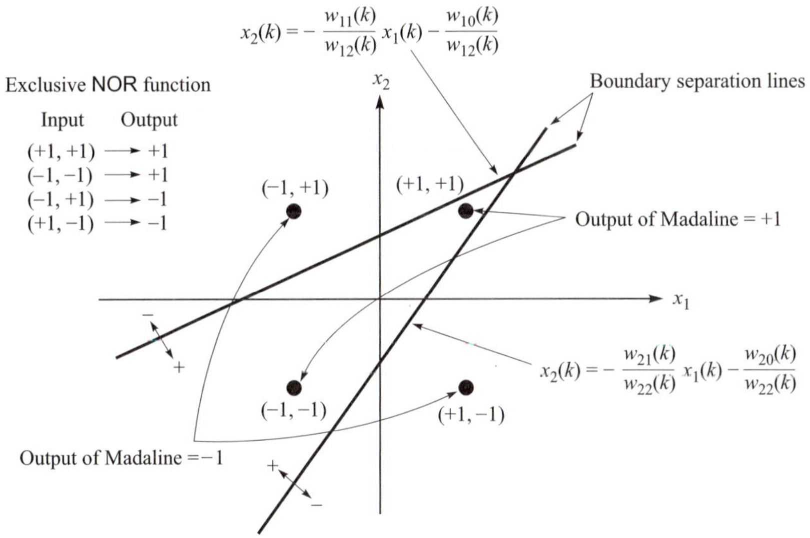

Figure 2.29 Madaline I separation properties for the XNOR problem.

for adaptation of Madaline I weights [32, 43]. In the Madaline I architecture shown in Figure 2.26, assuming the logic function is the MAJ function, the weights update algorithm would adjust the appropriate weight(s) to perform the correction of the output y relative to the desired response d. The weights of the network are initially set to small random values. There are many variations of this learning strategy, for example, *absolute correction* (or "fast" learning) or using the α-LMS algorithm for *statistical* (or "slow") learning. The basic idea is to adjust those weights for the neuron whose linear output $v_j(k)$ is closest to zero (i.e., the neuron whose analog response is closest to the desired response) because they require the smallest weight changes to reverse their output responses. In general, MRI follows the *minimal-disturbance principle* in the sense that no more neurons are adapted than necessary to correct the output decision and any "dead-zone" constraint. It is possible using the MRI algorithm to "hang up" on local minima [16]. The second basic learning strategy, Madaline rule II (MRII), is an extension of MRI [44]. For example, the MRII would be used to adjust the weights in both layers of the Madaline II architecture shown in Figure 2.27. The weights are initially set to small random values. The training patterns are presented in a random fashion with the objective of minimizing the average Hamming error over the training set. Like the MRI algorithm, MRII can also "hang up" on local minima [16]. The important ingredient that is missing in these learning strategies is the ability to backpropagate the errors through the architecture that can be used to adapt the network weights. This is the backpropagation learning rule that can be applied to multilayer feedforward networks. Details concerning backpropagation are addressed in Chapter 3.

2.6 SIMPLE PERCEPTRON

The simple *perceptron* (single-layer perceptron) has probably had the most significant impact on neural networks as we know them today, and as mentioned in Section 2.5.2, it is very similar to the Adaline. There are several different types of perceptrons; however, the original concept was developed by Frank Rosenblatt in the late 1950s [40, 41, 45] along with a learning procedure to adjust the network weights. His perceptron was based on the McCulloch–Pitts model of the neuron [2]. Other perceptron concepts can be found in the work by Block [46] and Minsky and Papert [47, 48]. Minsky and Papert [47] also discuss limitations of perceptrons in terms of the types of problems that cannot be solved by using perceptrons. One significant limitation that was demonstrated was the perceptron's inability to solve the *exclusive*-OR (XOR) problem. However, it was shown later that with the correct processing layer, a perceptron can solve the XOR problem, or even its more general form, the parity function [49]. In spite of Minsky and Papert's demonstrated limitations of the simple perceptron, significant neural network research continued in the 1970s. The simple (single-layer) perceptron is closely related to a classical pattern classifier known as the *maximum-likelihood Gaussian classifier* [1, 50] in that both can be considered as linear classifiers [51].

Most perceptrons are trained according to a supervised learning rule; however, some perceptrons are self-organizing. In Rosenblatt's early work, his perceptron had three layers. The first layer was called the sensory surface (the "retina") that projected to the next layer, which he referred to as the *association area* and had localized random connectivity. The association area was also referred to as the A unit (i.e., association unit), and this A unit was reciprocally connected to the third and final layer consisting of the R units (i.e., response units). The appropriate R unit was activated for a given input pattern (or class of input patterns), and it was not allowed to have more than one R unit on at a time. A set of reciprocal inhibitory connections was used to accomplish this, so that when an R unit was activated, it indirectly suppressed any competitors. The winner-take-all system, which is part of many neural network models, behaves similarly [52].

The simple perceptron can be viewed by referring to Figure 2.19. In Rosenblatt's original perceptron the inputs were binary and no bias was included. Here we consider bipolar inputs and a bias term associated with the neuron. According to Figure 2.19, the output of the *neuron* $y \in \{-1, 1\}$, which is the output of the symmetric hard limiter, that is, the *quantizer*, is subtracted from the desired output $d \in \{-1, 1\}$, forming the quantizer error $\tilde{e}$. The quantizer error is used to adjust the synaptic weights of the neuron. The adaptive algorithm for adjusting the neuron weights (the perceptron learning rule [16]) is given as

$$\boldsymbol{w}(k+1) = \boldsymbol{w}(k) + \alpha \frac{\tilde{e}(k)}{2} \boldsymbol{x}(k) \tag{2.55}$$

where the quantizer error is

$$\tilde{e}(k) = d(k) - \text{sgn}\left[\boldsymbol{w}^T(k)\boldsymbol{x}(k)\right] = d(k) - y(k) \tag{2.56}$$

and Rosenblatt normally set α to unity in (2.55). Unlike with the α-LMS algorithm, the choice of the learning rate parameter α does not affect the numerical stability of the perceptron learning rule; however, α can affect the speed of convergence. The perceptron learning rule is considered a nonlinear algorithm because the quantizer error (which is nonlinear) is used instead of the linear error, as in the Adaline. The perceptron learning rule performs the update of the weights until all the input patterns are classified correctly. Thereafter, the quantizer error will be zero for all training pattern inputs, and no weight adjustments will occur. Since this learning rule is not based on a defined optimization criterion, the weights are not guaranteed to be optimal in any sense. The learning rule for the perceptron is similar to the Widrow–Hoff delta rule; however, its behavior is quite different [16]. Comparing the α-LMS algorithm in (2.46) with the perceptron learning rule in (2.55), we see that if the normalized linear error $e(k)/\|x(k)\|_2^2$ in (2.46) is replaced by $\tilde{e}(k)/2$ (one-half of the quantizer error), then we obtain the perceptron learning rule in (2.55). Unlike learning in the Adaline, the perceptron learning rule has been shown to be capable of separating any linearly separable set of training patterns [39, 41, 46, 53]. Detailed convergence analysis of the error correction learning algorithm for the simple perceptron will not be presented here. However, this can be found in many excellent sources [1, 48, 50, 54, 55].

2.6.1 Mays's Perceptron Learning Rules

Mays proposed two modifications to the standard perceptron learning rule [16]. Both of these modifications use the so called dead zone, $\pm\gamma$ about zero. The output of the linear combiner $v(k)$ (see Figure 2.19) is considered to be in the dead zone if its magnitude is less than γ, that is, if $|v(k)| < \gamma$. Mays's weight adaptation algorithms are summarized as follows.

Mays's *increment adaptation algorithm*:

$$\boldsymbol{w}(k+1) = \begin{cases} \boldsymbol{w}(k) + \alpha\,\tilde{e}(k)\dfrac{\boldsymbol{x}(k)}{2\|\boldsymbol{x}(k)\|_2^2} & \text{if}\,|v(k)| \geq \gamma \\ \boldsymbol{w}(k) + \alpha\,d(k)\dfrac{\boldsymbol{x}(k)}{\|\boldsymbol{x}(k)\|_2^2} & \text{if}\,|v(k)| < \gamma \end{cases} \tag{2.57}$$

where $\tilde{e}(k)$ (the quantizer error) is given in (2.56). If the dead zone is set to zero, Mays's increment adaptation algorithm reduces to a normalized version of the perceptron learning rule given in (2.55). If the training patterns are linearly separable, Mays proved that his increment adaptation algorithm will always converge and is able to separate the patterns in a finite number of steps. For cases of nonlinearly separable training patterns, Mays's increment adaptation algorithm will also typically outperform the standard perceptron learning rule because of the dead zone. A sufficiently large dead zone will cause the adaptation of the weight vector away from zero, for existence of a reasonably good solution, and remain in a region that is associated with a relatively low average error. For the standard perceptron learning rule, input patterns that are not linearly separable result in training ad infinitum and often do not yield a low-error solution (even if one exists). Typically, the weight vector for this case tends toward zero. Mays also showed that the dead zone can reduce sensitivity to errors in the weights. The increment adaptation algorithm was proposed by others before Mays, but from a different perspective [46].

Mays's *modified relaxation algorithm*:

$$\boldsymbol{w}(k+1) = \begin{cases} \boldsymbol{w}(k) & \text{if} \left| v(k) \right| \geq \gamma \text{ and } \tilde{e}(k) = 0 \\ \boldsymbol{w}(k) + \alpha\, e(k) \dfrac{\boldsymbol{x}(k)}{\left\| \boldsymbol{x}(k) \right\|_2^2} & \text{otherwise} \end{cases} \tag{2.58}$$

where $\tilde{e}(k)$ is the quantizer error given in (2.56) and $e(k)$ is the linear error given in (2.22) for the simple adaptive linear combiner. This learning rule for the perceptron is more like the α-LMS algorithm previously shown for the Adaline, and as the dead zone approaches infinity ($\gamma \to \infty$), the modified relaxation algorithm approaches the standard perceptron learning rule. For the dead zone $0 < \gamma < 1$ and the learning rate $0 < \alpha \leq 2$, the modified relaxation algorithm is guaranteed to converge and separate any linearly separable set of input patterns in a finite number of steps. However, if the input patterns are not linearly separable, the algorithm will perform similar to Mays's increment adaptation algorithm. Both of Mays's perceptron learning algorithms achieve similar pattern separation results, and as with the standard perceptron learning rule, numerical stability is not affected by the choice of learning rate α.

All previous discussions for training a simple perceptron did not involve a learning strategy based on a defined performance measure. John Shynk in 1990 [56] viewed a perceptron learning algorithm as a steepest-descent method based on an instantaneous performance function that is iteratively minimized, and alternative perceptron learning rules were also derived.

2.6.2 Simple Perceptron with a Sigmoid Activation Function

We now consider the simple perceptron with a sigmoid activation function, as shown in Figure 2.30, instead of the previously used symmetric hard limiter, as shown in Figure 2.19. The perceptron learning rule in this case, like the LMS algorithm of Widrow and Hoff, is based on the method of steepest descent and attempts to minimize an instantaneous performance function. A learning rule for adjusting the network weights can be derived from a performance measure that is based on the MSE, that is,

$$J(\boldsymbol{w}_q) = \frac{1}{2} \mathrm{E}\{\tilde{e}_q^2(k)\} \tag{2.59}$$

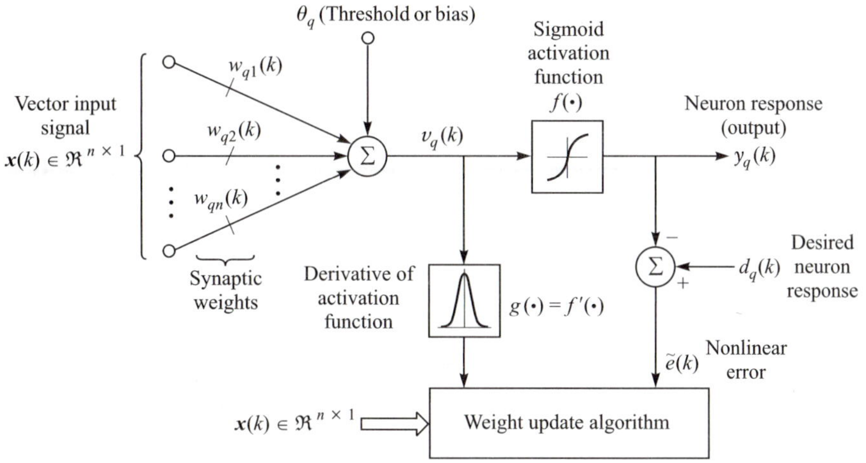

Figure 2.30 Simple perceptron with a sigmoid activation function. This could be the qth neuron in a multilayer feedforward perceptron (cf. Sect. 2.7).

where $\tilde{e}_q(k) = d_q(k) - y_q(k)$ is the nonlinear error shown in Figure 2.30, and E[•] is the expectation operator. However, we are restricted to *instantaneous* values; therefore, the instantaneous performance function to be minimized is given as

$$J(\mathbf{w}_q) = \frac{1}{2}\tilde{e}_q^2(k) = \frac{1}{2}\left[d_q(k) - y_q(k)\right]^2 = \frac{1}{2}\left[d_q^2(k) - 2d_q(k)y_q(k) + y_q^2(k)\right] \quad (2.60)$$

where

$$y_q(k) = f\left[v_q(k)\right] = f\left[\mathbf{x}^T(k)\mathbf{w}_q(k) + \theta_q\right] \quad (2.61)$$

The sigmoid activation function that we assume here is the hyperbolic tangent sigmoid; see (2.14). Therefore, the output of the neuron in Figure 2.30 can be written as

$$y_q(k) = f_{\text{hts}}\left[v_q(k)\right] = \tanh\left[\alpha v_q(k)\right] \quad (2.62)$$

where α is the slope parameter of the function. We will drop the subscript "hts" for the hyperbolic tangent sigmoid in (2.62) for convenience. According to (2.15), the derivative of the hyperbolic tangent sigmoid with respect to the neuron activity level v_q is given as

$$g\left[v_q(k)\right] = f'\left[v_q(k)\right] = \alpha\left\{1 - f^2\left[v_q(k)\right]\right\} \quad (2.63)$$

Note that the derivative of the activation function in (2.63) is a *continuous* derivative because we consider the synaptic weights to be continuous in amplitude but discretized in time. The discrete-time learning rule for the perceptron with a sigmoid activation function in Figure 2.30, taking the steepest descent approach, will have the form

$$\mathbf{w}_q(k+1) = \mathbf{w}_q(k) - \mu\nabla_{\mathbf{w}} J(\mathbf{w}_q) \quad (2.64)$$

where k is the discrete-time index and $J(\boldsymbol{w}_q)$ is the instantaneous performance function given in (2.60). Computing the gradient in (2.64) gives

$$\begin{aligned}\nabla_w J(\boldsymbol{w}_q) &= -d_q(k) f'[v_q(k)]\boldsymbol{x}(k) + f[v_q(k)] f'[v_q(k)]\boldsymbol{x}(k) \\ &= \underbrace{\{-d_q(k) + f[v_q(k)]\}}_{-\tilde{e}_q(k)} f'[v_q(k)]\boldsymbol{x}(k) \\ &= -\tilde{e}_q(k) f'[v_q(k)]\boldsymbol{x}(k)\end{aligned} \tag{2.65}$$

From (2.63), we can substitute $f'[v_q(k)] = \alpha\{1 - f^2[v_q(k)]\}$ in (2.65) to give

$$\nabla_w J(\boldsymbol{w}_q) = -\alpha\tilde{e}_q(k)\{1 - \underbrace{f^2[v_q(k)]}_{y_q^2(k)}\}\boldsymbol{x}(k) = -\alpha\tilde{e}_q(k)[1 - y_q^2(k)]\boldsymbol{x}(k) \tag{2.66}$$

Therefore, using the gradient result in (2.66), we can write the discrete-time learning rule (in vector form) from (2.64) for the simple perceptron as

$$\boldsymbol{w}_q(k+1) = \boldsymbol{w}_q(k) + \mu\alpha\tilde{e}_q(k)[1 - y_q^2(k)]\boldsymbol{x}(k) \tag{2.67}$$

Equation (2.67) can be written in scalar form as

$$w_{qj}(k+1) = w_{qj}(k) + \mu\alpha\tilde{e}_q(k)[1 - y_q^2(k)]x_j(k) \tag{2.68}$$

where $j = 1, 2, \ldots, n$. From Figure 2.30

$$\tilde{e}_q(k) = d_q(k) - y_q(k) \tag{2.69}$$

and

$$y_q(k) = f[v_q(k)] = f\left[\sum_{j=1}^{n} x_j(k) w_{qj}(k) + \theta_q\right] \tag{2.70}$$

Equations (2.68), (2.69), and (2.70) together are considered the standard form of backpropagation training algorithm for training the simple perceptron shown in Figure 2.30 [57]. The backpropagation training algorithm is extended to feedforward multilayer perceptrons in Chapter 3. As discussed in Section 2.3, the binary sigmoid activation function and the hyperbolic tangent sigmoid have derivatives that can be expressed in terms of their original functions. This can be seen in (2.63) for the hyperbolic tangent sigmoid. This is a highly desirable characteristic for the activation function for obvious reasons.

EXAMPLE 2.2. In this example we use the simple perceptron with the sigmoid activation function shown in Figure 2.30 to learn one character, namely, the E character; this is shown in Figure 2.31(a). This character (image) consists of 25 pixels, that is, a 5×5 array of pixels. The dark pixels in the image are given the numerical value of 1, and the "off" (white) pixels are given a numerical value of 0. Therefore, the array of binary numbers is square, and the vec operation (cf. Sect. A.2.17), operating on this array, will yield a 25×1 binary vector used for training the perceptron

$$\boldsymbol{x} = [1\ 1\ 1\ 1\ 1\ 1\ 0\ 1\ 0\ 1\ 1\ 0\ 1\ 0\ 1\ 1\ 0\ 0\ 0\ 1\ 0\ 0\ 0\ 0\ 0]^T$$

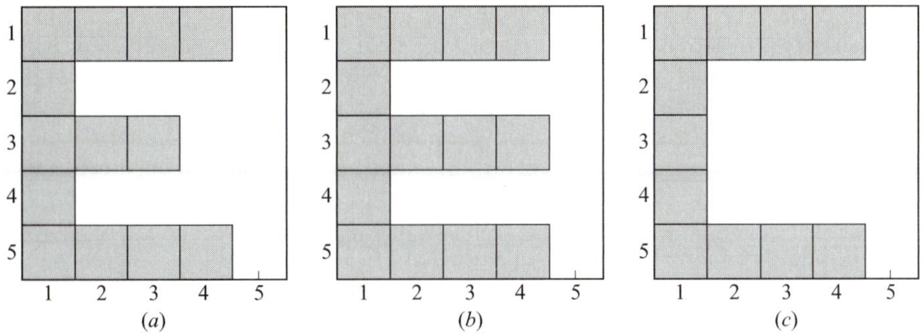

Figure 2.31 (*a*) Original E character. (*b*) Modified version of the original character used for "testing." (*c*) Another modified version of the original character used for testing.

The perceptron learning rule given in (2.67) is used with $\alpha = 1$ and the learning rate parameter set to $\mu = 0.25$. The desired neuron response is set to $d = 0.5$; that is, this is a numerical value that identifies the character E after training is completed, and the input pattern, the vector $\boldsymbol{x}$ shown above, is presented to the neuron. An error goal of 10^{-8} was established to terminate training. In other words, when the square of the difference between the desired response d and the actual response y is less than 10^{-8}, training of the neuron stops. The initial weights of the neuron were randomized, and after 39 presentations of the input pattern, the actual neuron output was $y = 0.50009$. Figure 2.32 shows the progression of the neuron output error during training. One attractive feature of taking a neurocomputing approach for character recognition is the ability of the network to "correct" or compensate for inputs that can be corrupted by

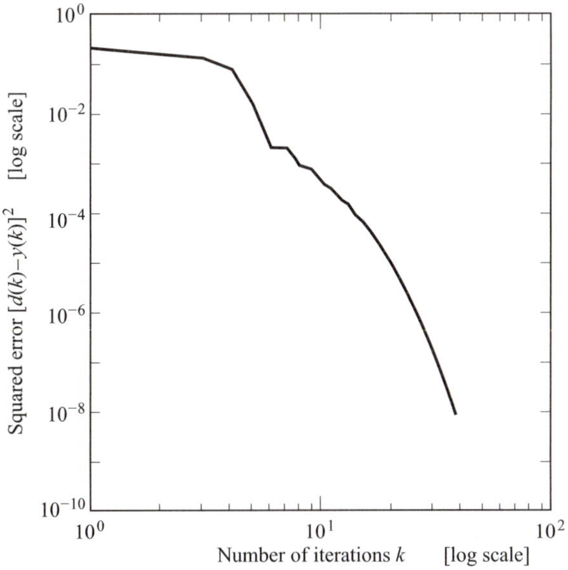

Figure 2.32 Square of the error at the neuron output versus the iteration number during training.

noise. In this example it is shown that the *single neuron* can*not* correct for a noisy input; that is, alone it does not possess an error correction capability. For example, if the E is modified as shown in Figure 2.31(*b*), the resulting input to the single neuron yields an output of $y = 0.5204$. This is relatively close to the actual value that identifies a "true" E because the modified image looks similar to the E. However, there was no error correction carried out by the simple perceptron. If the E is now modified as shown in Figure 2.31(*c*), the resulting neuron output is $y = 0.6805$, which is even further from the actual target value that identifies an E. The image in Figure 2.31(*c*) looks more like a C now; however, the neuron trained to only recognize an E has no idea what a C is supposed to "look like." Obviously more neurons are needed not only to allow for compensation of noisy inputs but also to recognize more characters. This can be accomplished with multilayer perceptrons (cf. Sect. 3.3). Also when we investigate the Hopfield associative memory (cf. Sect. 5.3), it will be shown that this recurrent network has the capability to compensate for noisy or incomplete inputs presented to it and to recognize many different characters.

2.7 FEEDFORWARD MULTILAYER PERCEPTRON

Figure 2.33 shows a standard feedforward multilayer perceptron (MLP) with three layers. This type of architecture is part of a large class of feedforward neural networks with the neurons arranged in cascaded layers. The neural network architectures in this class share a common feature that all neurons in a layer (or sometimes called a *slab*) are connected to all neurons in adjacent layers through unidirectional branches. That is, the branches or links can only broadcast information in one direction, that is, the "forward direction." The branches have associated transmittances, that is, synaptic weights, that can be adjusted according to a defined learning rule. Feedforward networks do not allow connections between neurons within any layer of the architecture. At every neuron the output of the linear combiner, that is, the neuron activity level v_q, is input to a nonlinear activation function $f(\bullet)$, whose output is the response of the neuron. The neurons in the network typically have activity levels in the range $[-1, 1]$, and in some applications the range $[0, 1]$ is used. In Figure 2.33, there are actually four layers; however, in Section 2.5.3 the number of layers in a multilayer network was defined to be the total number of layers that only have processing elements. Therefore, we see that the "zero" layer (or input layer) in Figure 2.33 does not perform any computations, but only serves to feed the input signal to the neurons of the "first" layer (called the first *hidden* layer). The outputs of the first hidden layer are presented the neurons of the second hidden layer (the "second" layer) whose outputs are then input to the "third" layer (or the *output* layer). The output of the output layer is the network response vector. We can refer to this architecture as an *h-p-m* feedforward MLP neural network, that is, *h* neurons (nodes) in the first layer (first hidden layer), *p* neurons in the second layer (second hidden layer), and *m* nodes in the third layer (output layer). Therefore, the network can perform the *nonlinear input/output mapping* $\mathbf{\Omega} : \Re^{n\times 1} \rightarrow \Re^{m\times 1}$. In general, there can be any number of hidden layers in the architecture; however, from

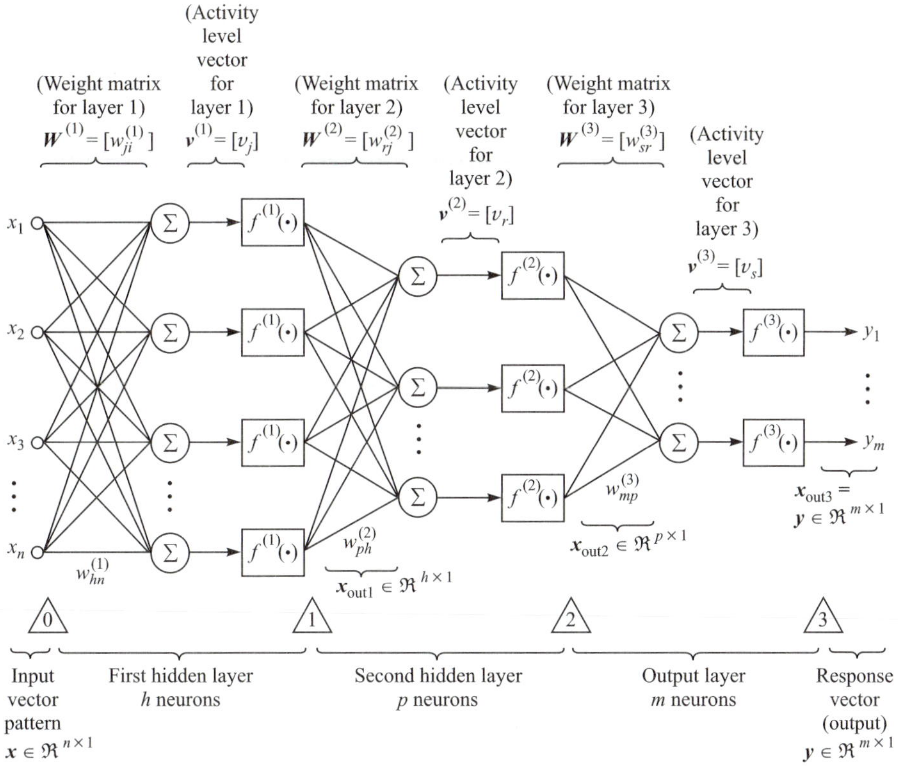

Figure 2.33 Feedforward three-layer perceptron architecture, where $i = 1, 2, \ldots, n$; $j = 1, 2, \ldots, h$; $r = 1, 2, \ldots, p$; $s = 1, 2, \ldots, m$; and $f^{(1)}(\bullet)$ is the nonlinear activation function at each neuron in the first layer, $f^{(2)}(\bullet)$ is the nonlinear activation function at each neuron in the second layer, and $f^{(3)}(\bullet)$ is the nonlinear activation function at each neuron in the third layer. A bias (or threshold) is assumed at each of the neurons.

a practical perspective, only one or two hidden layers are typically used. In fact, it can be shown that an MLP that has only one hidden layer, with a sufficient number of neurons, acts as a universal approximator of nonlinear mappings (cf. Sect. 3.3.2).

In Figure 2.33, each layer has a synaptic weight matrix associated with all the connections made from the previous layer to the next layer, that is, $\boldsymbol{W}^{(\ell)}$, for $\ell = 1, 2, 3$. The first layer has the weight matrix $\boldsymbol{W}^{(1)} = \left[w_{ji}^{(1)}\right] \in \Re^{h\times n}$, the second layer's weight matrix is $\boldsymbol{W}^{(2)} = \left[w_{rj}^{(2)}\right] \in \Re^{p\times h}$, and the third layer's weight matrix is given by $\boldsymbol{W}^{(3)} = \left[w_{sr}^{(3)}\right] \in \Re^{m\times p}$, for $i = 1, 2, \ldots, n$; $j = 1, 2, \ldots, h$; $r = 1, 2, \ldots, p$; and $s = 1, 2, \ldots, m$. The nonlinear input-output mapping $\boldsymbol{\Omega} : \Re^{n\times 1} \to \Re^{m\times 1}$ can be determined directly from Figure 2.33. First we define a diagonal nonlinear operator matrix

$$\boldsymbol{f}^{(\ell)}[\bullet] \triangleq \mathrm{diag}\left[f^{(\ell)}[\bullet], f^{(\ell)}[\bullet], \cdots, f^{(\ell)}[\bullet]\right] \tag{2.71}$$

whose dimension depends on ℓ. When $\ell = 1, \boldsymbol{f}^{(1)}[\bullet]$ is an $h \times h$ diagonal matrix; for $\ell = 2, \boldsymbol{f}^{(2)}[\bullet]$ is $p \times p$ in dimension; and for $\ell = 3, \boldsymbol{f}^{(3)}[\bullet]$ is $m \times m$ in dimension. Given the network input vector $\boldsymbol{x} \in \Re^{n \times 1}$, the output of the first layer $\boldsymbol{x}_{\text{out1}} \in \Re^{h \times 1}$ can be written as

$$\boldsymbol{x}_{\text{out1}} = \boldsymbol{f}^{(1)}[\boldsymbol{v}^{(1)}] = \boldsymbol{f}^{(1)}[\boldsymbol{W}^{(1)}\boldsymbol{x}] \tag{2.72}$$

which is the input to the second layer. The output of the second layer $\boldsymbol{x}_{\text{out2}} \in \Re^{p \times 1}$ can be written as

$$\boldsymbol{x}_{\text{out2}} = \boldsymbol{f}^{(2)}[\boldsymbol{v}^{(2)}] = \boldsymbol{f}^{(2)}[\boldsymbol{W}^{(2)}\boldsymbol{x}_{\text{out1}}] \tag{2.73}$$

which is the input to the third layer. The output of the third layer, which is the response of the network $\boldsymbol{y} = \boldsymbol{x}_{\text{out3}} \in \Re^{m \times 1}$, can be written as

$$\boldsymbol{y} = \boldsymbol{x}_{\text{out3}} = \boldsymbol{f}^{(3)}[\boldsymbol{v}^{(3)}] = \boldsymbol{f}^{(3)}[\boldsymbol{W}^{(3)}\boldsymbol{x}_{\text{out2}}] \tag{2.74}$$

Substituting (2.72) into (2.73) for $\boldsymbol{x}_{\text{out1}}$, and then substituting this result into (2.74) for $\boldsymbol{x}_{\text{out2}}$ gives the final response of the network as

$$\boldsymbol{y} = \boldsymbol{f}^{(3)}[\boldsymbol{W}^{(3)}\boldsymbol{f}^{(2)}[\boldsymbol{W}^{(2)}\boldsymbol{f}^{(1)}[\boldsymbol{W}^{(1)}\boldsymbol{x}]]] = \boldsymbol{\Omega}[\boldsymbol{x}] \tag{2.75}$$

In the nonlinear mapping of (2.75) the synaptic weights are assumed to be fixed. However, a training process must be carried out a priori to properly adjust the weights to perform a desired mapping, for example, to solve a pattern classification problem. The details concerning training of MLPs using backpropagation are discussed in Chapter 3.

2.8 OVERVIEW OF BASIC LEARNING RULES FOR A SINGLE NEURON

Here we present several basic learning rules for a single neuron. These concepts can be extended to multidimensional networks (i.e., networks with more than one neuron). Therefore, many of the learning rules presented here will be used in subsequent chapters for building more sophisticated neural structures.

2.8.1 Generalized LMS Learning Rule

In Section 2.5.1 the LMS algorithm was presented which was originally developed for adaptation of the weights of a simple linear combiner. It was then shown that this algorithm could be applied to training an Adaline using the linear error term (cf. Sect. 2.5.2). Next, extensions of the LMS algorithm led to training algorithms for the simple (single-layer) perceptron (cf. Sect. 2.6). Finally, in Section 2.6.2 a learning rule was derived for a simple perceptron with a sigmoid activation function. Here we want to establish a generalized LMS learning rule for a single neuron, and from this general form we can derive several important variations. To begin, we can define a performance function (or energy function), to be minimized, as

$$\mathscr{E}(\boldsymbol{w}) = \psi(e) + \frac{\alpha}{2}\|\boldsymbol{w}\|_2^2 \tag{2.76}$$

where $\|\boldsymbol{w}\|_2$ is the Euclidean norm of the vector $\boldsymbol{w}$ (cf. Sect. A.2.13). In (2.76) $\psi(\bullet)$ can be any differentiable function, $e \in \Re$ is the linear error (see Fig. 2.19), that is,

$$e = d - \boldsymbol{w}^T\boldsymbol{x} \tag{2.77}$$

where $d \in \Re$ is the desired (linear) output, $\boldsymbol{x} \in \Re^{n+1\times 1}$ is the input vector, and $\boldsymbol{w} \in \Re^{n+1\times 1}$ is the weight vector. Taking the steepest descent approach, we can develop a continuous-time learning rule as a set of vector differential equations in the form

$$\frac{d\boldsymbol{w}}{dt} = -\mu\nabla_{\boldsymbol{w}}\mathscr{E}(\boldsymbol{w}) \tag{2.78}$$

The discrete-time form of (2.78) can be written as

$$\boldsymbol{w}(k+1) = \boldsymbol{w}(k) - \mu\nabla_{\boldsymbol{w}}\mathscr{E}(\boldsymbol{w}) \tag{2.79}$$

where $\boldsymbol{w}(k) = \boldsymbol{w}(kT_s)$, k is the discrete-time index, and T_s is the sampling period (however, we will assume T_s is normalized to unity without loss of generality, that is, $T_s = 1$). The sampling period T_s can be normalized because the Euler approximation to the derivative in (2.78) can be written as $[\boldsymbol{w}(k+1) - \boldsymbol{w}(k)]/T_s$, and the discrete-time form of the learning rule is

$$\frac{\boldsymbol{w}(k+1) - \boldsymbol{w}(k)}{T_s} = -\mu\nabla_{\boldsymbol{w}}\mathscr{E}(\boldsymbol{w})$$

However, multiplying both sides by the sampling period, adding $\boldsymbol{w}(k)$ to both sides, and incorporating the sampling period T_s into the learning rate parameter μ give (2.79). In Section 2.5.1 a range of values for the learning rate parameter was given in (2.32) for the LMS algorithm. Therefore, regardless of the actual value of the sampling period, the inequality in (2.32) must be adhered to for ensuring convergence (in the mean) of the LMS algorithm. Computing the gradient of the energy function in (2.76) gives

$$\begin{aligned}\nabla_{\boldsymbol{w}}\mathscr{E}(\boldsymbol{w}) &= \frac{\partial\mathscr{E}(\boldsymbol{w})}{\partial\boldsymbol{w}} = \frac{\partial\psi(e)}{\partial\boldsymbol{w}} + \frac{\alpha}{2}\frac{\partial\boldsymbol{w}^T\boldsymbol{w}}{\partial\boldsymbol{w}} = \underbrace{\psi'(e)}_{g(e)}(-\boldsymbol{x}) + \alpha\boldsymbol{w} \\ &= -g(e)\boldsymbol{x} + \alpha\boldsymbol{w}\end{aligned} \tag{2.80}$$

Using the continuous-time form of the learning rule in (2.78) and our gradient result in (2.80), we can write the *general LMS algorithm* as

$$\frac{d\boldsymbol{w}}{dt} = \mu[g(e)\boldsymbol{x} - \alpha\boldsymbol{w}] \tag{2.81}$$

and from (2.79), the discrete-time form is

$$\boldsymbol{w}(k+1) = \boldsymbol{w}(k) + \mu[g(e)\boldsymbol{x}(k) - \alpha\boldsymbol{w}(k)] \tag{2.82}$$

where $\mu > 0$ is the learning rate parameter and $\alpha \geq 0$ is the leakage factor. If the function $\psi(t) = {}^1\!/_2 t^2$ (quadratic weighting) in (2.76), and $\psi'(t) = g(t) = t$, where t is a dummy variable, then (2.81) is written as

$$\frac{d\boldsymbol{w}}{dt} = \mu(e\boldsymbol{x} - \alpha\boldsymbol{w}) = \mu e\boldsymbol{x} - \underbrace{\mu\alpha}_{\gamma}\,\boldsymbol{w} = \mu e\boldsymbol{x} - \gamma\boldsymbol{w} \tag{2.83}$$

and the discrete-time form of (2.83) is given as

$$\boldsymbol{w}(k+1) = \boldsymbol{w}(k) + \mu e(k)\boldsymbol{x}(k) - \gamma\boldsymbol{w}(k) = (1-\gamma)\boldsymbol{w}(k) + \mu e(k)\boldsymbol{x}(k) \tag{2.84}$$

where $0 \leq \gamma < 1$ is the leakage factor. This learning rule is called the *leaky LMS algorithm*. Now, if the leakage factor in (2.84) is set to zero, that is, $\gamma = 0$, the learning rule becomes

$$\boldsymbol{w}(k+1) = \boldsymbol{w}(k) + \mu e(k)\boldsymbol{x}(k) \tag{2.85}$$

which is the *standard LMS algorithm* [see (2.29)]. This can be written in scalar form as

$$w_j(k+1) = w_j(k) + \mu e(k)x_j(k) \tag{2.86}$$

for $j = 0, 1, 2, \ldots, n$, where $e(k) = d(k) - \sum_{j=1}^{n} w_j(k)x_j(k)$.

There are three important variations of the standard LMS algorithm. The first modification involves adding a *momentum* term to the right side of (2.85). The momentum term is intended to provide a specified inertia (or momentum) to change the weight vector in the direction of the average downhill "force" that it feels instead of sustaining oscillations during the training process. The momentum term can be formed as a weighted difference between the current and previous weight vectors, that is,

$$\alpha\Delta\boldsymbol{w}(k) = \alpha[\boldsymbol{w}(k) - \boldsymbol{w}(k-1)] \tag{2.87}$$

Therefore, (2.85) is rewritten as

$$\boldsymbol{w}(k+1) = \boldsymbol{w}(k) + \mu e(k)\boldsymbol{x}(k) + \alpha[\boldsymbol{w}(k) - \boldsymbol{w}(k-1)] \tag{2.88}$$

where $0 < \alpha < 1$ is the *momentum parameter* and (2.88) is referred to as the *standard LMS algorithm with momentum*.

The second variation of the standard LMS algorithm is *recursive weighted least-squares*. This is a standard algorithm used in parametric system identification based on autoregressive moving-average (ARMA) models [7, 31] and adaptive filtering [19]. The system identification problem typically involves estimating the elements of a parameter vector associated with a strictly proper rational transfer function or a canonical-form state-space realization of a linear time-invariant system [7, 31] (cf. Chap. 10). In [7] the recursive weighted least-squares algorithm involves an update expression for the parameter vector $\boldsymbol{w} \in \Re^{n+1\times 1}$, an update expression for the *gain* vector, $\boldsymbol{L} \in \Re^{n+1\times 1}$, and another update expression for the *weighting matrix* $\boldsymbol{P} \in \Re^{n+1\times n+1}$ and $\boldsymbol{P}^T = \boldsymbol{P}$. The update expression for the parameter vector is given as

$$\boldsymbol{w}(k+1) = \boldsymbol{w}(k) + \boldsymbol{L}(k+1)e(k) \tag{2.89}$$

the update expression for the gain vector is

$$L(k+1) = P(k)x(k)\left[\lambda + x^T(k)P(k)x(k)\right]^{-1} \tag{2.90}$$

and the update expression for the weighting matrix is given by

$$P(k+1) = \frac{1}{\lambda}\left[P(k) - L(k+1)x^T(k)P(k)\right] \tag{2.91}$$

where the error term $e(k)$ is given by

$$e(k) = d(k) - w^T(k)x(k) \tag{2.92}$$

and λ determines the type of weighting, for example, if $0 < \lambda < 1$, this leads to exponentially weighted recursive least squares. Substituting (2.90) into (2.89) gives

$$w(k+1) = w(k) + \underbrace{\frac{e(k)P(k)x(k)}{\lambda + x^T(k)P(k)x(k)}}_{\mu^{-1}(k)} \tag{2.93}$$

Therefore, the modified update expression for the synaptic weight vector is given as

$$w(k+1) = w(k) + \mu(k)e(k)P(k)x(k) \tag{2.94}$$

where

$$\mu(k) = \frac{1}{\lambda + x^T(k)P(k)x(k)} \tag{2.95}$$

is the adaptive learning rate parameter. The update expression for the weighting matrix can be written in terms of the gain vector by substituting (2.90) into (2.91)

$$\begin{aligned} P(k+1) &= \frac{1}{\lambda}\left[P(k) - \frac{P(k)x(k)x^T(k)P(k)}{\lambda + x^T(k)P(k)x(k)}\right] \\ &= \frac{1}{\lambda}\left[P(k) - \mu(k)P(k)x(k)x^T(k)P(k)\right] \\ &= \frac{1}{\lambda}\left[I - \mu(k)P(k)x(k)x^T(k)\right]P(k) \end{aligned} \tag{2.96}$$

From (2.96), the update expression for the weighting matrix is written as

$$P(k+1) = \frac{1}{\lambda}\left[I - \mu(k)P(k)x(k)x^T(k)\right]P(k) \tag{2.97}$$

where $\mu(k)$ is given in (2.95). Therefore, (2.94), (2.95), and (2.97) constitute the *recursive weighted least-squares (RWLS) algorithm.*

The third variation of the standard LMS algorithm involves the *minimal-disturbance principle* (cf. Sect. 2.5.2). We can further modify (2.46) by introducing a positive constant to the denominator. This ensures that the update to the weight vector does not become unbounded. Thus, the *modified normalized LMS algorithm* is given by

TABLE 2.2
Modified LMS algorithms (discrete-time, vector-matrix form)

Algorithm name	Algorithm form
General LMS algorithm	$\mathbf{w}(k+1) = \mathbf{w}(k) + \mu[g(e)\mathbf{x}(k) - \alpha \mathbf{w}(k)]$, $\mu > 0, \alpha \geq 0$
Leaky LMS algorithm	$\mathbf{w}(k+1) = (1-\gamma)\mathbf{w}(k) + \mu e(k)\mathbf{x}(k)$, $\mu > 0, 0 \leq \gamma < 1$
Standard LMS algorithm	$\mathbf{w}(k+1) = \mathbf{w}(k) + \mu e(k)\mathbf{x}(k)$, $\mu > 0$
Standard LMS algorithm with momentum	$\mathbf{w}(k+1) = \mathbf{w}(k) + \mu e(k)\mathbf{x}(k) + \alpha[\mathbf{w}(k) - \mathbf{w}(k-1)]$, $\mu > 0, 0 < \alpha < 1$
Recursive weighted least-squares (RWLS) algorithm	$\mathbf{w}(k+1) = \mathbf{w}(k) + \mu(k)e(k)\mathbf{P}(k)\mathbf{x}(k)$ $\quad \mu(k) = \dfrac{1}{\lambda + \mathbf{x}^T(k)\mathbf{P}(k)\mathbf{x}(k)} \qquad 0 < \lambda < 1$ $\quad \mathbf{P}(k+1) = \dfrac{1}{\lambda}[\mathbf{I} - \mu(k)\mathbf{P}(k)\mathbf{x}(k)\mathbf{x}^T(k)]\mathbf{P}(k)$
Modified normalized LMS algorithm	$\mathbf{w}(k+1) = \mathbf{w}(k) + \mu\left[\dfrac{e(k)\mathbf{x}(k)}{\alpha + \|\mathbf{x}(k)\|_2^2}\right]$, $\alpha \geq 0,\ 0 < \mu < 2$, typically $0.1 < \mu < 1$

$$\mathbf{w}(k+1) = \mathbf{w}(k) + \mu\left\{\frac{e(k)\mathbf{x}(k)}{\alpha + \|\mathbf{x}(k)\|_2^2}\right\} \tag{2.98}$$

where $\alpha \geq 0$, and the learning rate parameter is set according to $0 < \mu < 2$, typically $0.1 < \mu < 1$; see (2.48). In the paper by Douglas [58], a derivation of the normalized LMS algorithm is generalized. This results in a family of projectionlike algorithms that is based on an L_p-minimized filter coefficient change. There are other variations of the standard LMS algorithm, and a summary of these can be found in Cichocki and Unbehauen [9]. Table 2.2 summarizes the various discrete-time forms of the LMS algorithms presented in this section.

EXAMPLE 2.3. The problem addressed in this example is the same as that in Example 2.1. A comparison is made between the various modified LMS algorithms shown in Table 2.2, except the general LMS algorithm, and the standard LMS algorithm with the search-then-converge scheduling of the learning rate parameter summarized in Table 2.1. The general LMS algorithm is excluded from this comparative analysis because it is actually in another class of learning algorithms that are considered to be *robust*. These types of learning rules are discussed in Chapters 8 and 9. In all cases run in the simulations, the same initial weight vector was used, as indicated in Example 2.1. When appropriate, the same initial learning rate was used, as also indicated in Example 2.1; that is, $\mu_0 = 0.9/\lambda_{\max} = 0.1936$. The obvious exception is for the RWLS algorithm. Another exception is for the modified normalized LMS algorithm. For this learning rule it was necessary to set the learning rate parameter much larger than $\mu_0 = 0.9/\lambda_{\max} = 0.1936$ to achieve convergence in a reasonable number of training steps. In addition, for all six cases, the same learning process *termination criterion*

TABLE 2.3
Simulation results for Example 2.3

Algorithm used	Relevant parameters	Number of training steps required for convergence	$\|\|b - w_{final}\|\|_2$ Truth model: $b = [1, 0.8, -1]^T$
Leaky LMS algorithm	$\mu_0 = 0.1936$, $\gamma = 10^{-8}$	217	1.940037×10^{-6}
Standard LMS algorithm	$\mu_0 = 0.1936$	234	5.803031×10^{-7}
Standard LMS algorithm with momentum	$\mu_0 = 0.1936$, $\alpha = 0.01$	192	1.077056×10^{-5}
Recursive weighted least-squares (RWLS) algorithm	$\lambda = 0.1$, $P(0) = I_{3\times 3}$	12	3.105226×10^{-10}
Modified normalized LMS algorithm	$\mu_0 = 0.999$, $\alpha = 0.1$	34	3.394817×10^{-7}
Standard LMS algorithm with search-then-converge scheduling	$\mu_0 = 0.1936$, $\tau = 200$	204	1.505354×10^{-7}

was used, as discussed in Example 2.1; that is, the RMS value $\sqrt{J} = \sqrt{\frac{1}{2}e^2(k)} \leq 10^{-8}$, where $e(k) = d(k) - w^T(k)x(k)$. When appropriate, for each LMS algorithm the particular parameter(s) was (were) empirically optimized to yield the "best" performance results. Table 2.3 shows the simulation results. Figure 2.34 shows the convergence profile for each of the LMS algorithms. By observing Table 2.3 and Figure 2.34 it is obvious that the RWLS algorithm is the superior learning rule.

2.8.2 Hebbian Learning

In the original work by Donald Hebb [4, 59], a learning process was postulated from a neurobiological viewpoint. Based on the concept of *cell assemblies*, Hebb suggested that certain subsets of cells within cortical networks tend to behave as functional units with coordinated activity patterns in accordance with changing synaptic strength throughout the network [60]. The subsets of cells were referred to as assemblies. Hebb suggested that the strength of a synapse between cells A and B increased slightly for the situation when the firing in A was followed by firing in B with a very small time delay. That is, for two neurons on either side of a synapse that are *synchronously* activated, then the strength of the synapse is increased. However, this is not a mathematical statement, and thus he did not provide a quantitative learning rule that could be followed. Therefore, several mathematical learning rules can be defined that are justifiably called *Hebb synapses*. Extensions of Hebb's concepts were presented by Stent [61] and by Changeux and Danchin [62]. These extensions essentially expanded Hebb's original statement to include the case when two neurons on either side of a synapse are asynchronously activated, leading to a weakened synapse, or a synapse that is eliminated. Also, Rumelhart and McClelland [57] point out that Hebb's original proposal was not sufficient to quantitatively develop an explict model. They state an extension of Hebb's original rule to account for positive and negative activation values as follows:

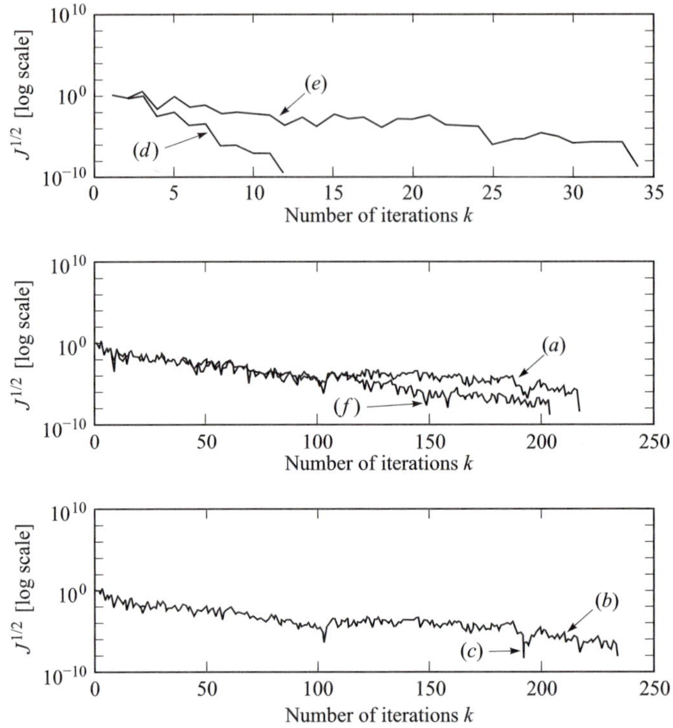

Figure 2.34 Convergence profiles for each of the LMS algorithms: (*a*) leaky LMS algorithm (217 iterations), (*b*) standard LMS algorithm (234 iterations), (*c*) standard LMS algorithm with momentum (192 iterations), (*d*) RWLS algorithm (12 iterations), (*e*) modified normalized LMS algorithm (34 iterations), and (*f*) standard LMS algorithm with search-then-converge scheduling (204 iterations).

> Adjust the strength of the connection between units A and B in proportion to the product of their simultaneous activation.[†]

Simply put, this statement implies that if the product of the activations is positive, the modification to the synaptic connection is more *excitatory*; however, if the product is negative, the modification to the synaptic connection is more *inhibitory*.

We can now be more precise in our definition of a Hebbian synapse. A *Hebbian synapse* is defined as a synapse that uses a highly local, time-dependent, and strongly interactive mechanism to increase synaptic efficiency as a function of the correlation between the presynaptic and postsynaptic activity levels [1]. From this definition, four key properties of a Hebbian synapse can be stated [63]: (1) *Time-dependent mechanism.* This refers to changes in a Hebbian synapse that depend on the precise time of occurrence of the presynaptic and postsynaptic activity levels. (2) *Local mechanism.* Within a synapse, ongoing activity levels in the presynaptic and postsynaptic units

[†]Quoted with permission from Rumelhart and McClelland [57], p. 36.

(locally available information) are used by a Hebbian synapse to produce an input-dependent, local synaptic modification. This local mechanism within a neural network consisting of Hebbian synapses provides the means for unsupervised learning. (3) *Interactive mechanism*: Any form of Hebbian learning depends on the (deterministic or statistical) interaction between presynaptic and postsynaptic activities. (4) *Conjunctional (correlational) mechanism*. The "co-occurrence" of presynaptic and postsynaptic activities within a relatively short time interval is sufficient to produce a synaptic modification. Therefore, a Hebbian synapse can be referred to as a *conjunctional synapse*. Another viewpoint of Hebb's postulate of learning is based on a statistical characterization of the interactive mechanism within a Hebbian synapse. That is, the correlation between presynaptic and postsynaptic activities over time accounts for a synaptic modification; therefore, a Hebbian synapse can also be referred to as a *correlational synapse*.

We can think of the activity in a Hebbian synapse as being enhanced or depressed. That is, positively correlated activity connecting a pair of neurons results in synaptic strengthening (or enhancement), whereas either uncorrelated or negatively correlated activity produces synaptic weakening (or synaptic depression). Synaptic depression may also occur as a result of noncoincident presynaptic or postsynaptic activity. Synaptic activities can be categorized as *Hebbian*, *anti-Hebbian*, or *non-Hebbian* [64]. Accordingly, for positively correlated presynaptic or postsynaptic activities, a Hebbian synapse increases its strength, and its strength is decreased when the activities are either uncorrelated or negatively correlated. An anti-Hebbian synapse strengthens negatively correlated presynaptic or postsynaptic activities and weakens positively correlated activities. A non-Hebbian synapse does not involve the strongly interactive, highly local, time-dependent mechanism that Hebbian and anti-Hebbian synapses possess.

With the ideas stated above in mind, we can derive the standard Hebbian learning rule for a single neuron from an energy function defined as

$$\mathscr{E}(\boldsymbol{w}) = -\psi(\boldsymbol{w}^T\boldsymbol{x}) + \frac{\alpha}{2}\|\boldsymbol{w}\|_2^2 \tag{2.99}$$

where $\boldsymbol{w} \in \Re^{n+1\times 1}$ is the synaptic weight vector (including a bias or threshold), $\boldsymbol{x} \in \Re^{n+1\times 1}$ is the input to the neuron, $\psi(\bullet)$ is a differentiable function, and $\alpha \geq 0$ is the forgetting factor. Also,

$$y = \frac{\mathrm{d}\psi(v)}{\mathrm{d}v} = f(v) \tag{2.100}$$

is the output of the neuron, where $v = \boldsymbol{w}^T\boldsymbol{x} \in \Re$ is the activity level of the neuron. Taking the steepest descent approach to derive the continuous-time learning rule

$$\frac{\mathrm{d}\boldsymbol{w}}{\mathrm{d}t} = -\mu\nabla_{\boldsymbol{w}}\mathscr{E}(\boldsymbol{w}) \tag{2.101}$$

where $\mu > 0$ is the learning rate parameter, we see that the gradient of the energy function in (2.99) must be computed with respect to the synaptic weight vector, that is, $\nabla_{\boldsymbol{w}}\mathscr{E}(\boldsymbol{w}) = \partial\mathscr{E}(\boldsymbol{w})/\partial\boldsymbol{w}$. The gradient of (2.99) is given as

$$\nabla_w \mathscr{E}(w) = -\underbrace{f(v)}_{y}\underbrace{\frac{\partial v}{\partial w}}_{x} + \alpha w = -yx + \alpha w \tag{2.102}$$

Therefore, by using the result in (2.102) along with (2.101), the continuous-time standard Hebbian learning rule for a single neuron is given as

$$\frac{dw}{dt} = \mu[yx - \alpha w] \tag{2.103}$$

The discrete-time standard Hebbian learning rule (in vector form) is given by

$$w(k+1) = w(k) + \mu[y(k)x(k) - \alpha w(k)] \tag{2.104}$$

and in scalar discrete-time form by

$$w_j(k+1) = w_j(k) + \mu\big[y(k)x_j(k) - \alpha w_j(k)\big] \tag{2.105}$$

for $j = 0, 1, \ldots, n$.

The results obtained above can be derived from a more general setting. Amari [65] showed that the *generalized* Hebbian learning rule can be considered as a gradient optimization process when an appropriate energy or Lyapunov function $\mathscr{E}(w)$ is selected, that is,

$$\frac{dw}{dt} = -\mu \frac{\partial \mathscr{E}(w)}{\partial w} \tag{2.106}$$

The resulting generalized Hebbian learning rule is given as

$$\frac{dw}{dt} = \mu(\ell x - \alpha w) \tag{2.107}$$

where $\ell \triangleq \ell(w, x, v, y, d)$ is the *learning signal* and $d \in \Re$ is the desired signal. Equation (2.107) can be written in discrete-time form as

$$\begin{aligned} w(k+1) &= w(k) + \mu[\ell(k)x(k) - \alpha w(k)] \\ &= w(k) + -\underbrace{\mu\alpha}_{\gamma} w(k) + \mu\ell(k)x(k) = (1-\gamma)w(k) + \mu\ell(k)x(k) \end{aligned} \tag{2.108}$$

where $\mu > 0$ and $0 \leq \gamma < 1$ (forgetting factor). Many types of learning algorithms can be derived from the expressions in (2.107) and (2.108). Assuming that the synaptic weights are adjusted by only local signals, a local learning algorithm can be immediately derived from (2.107) if we assume the learning signal is the output of the neuron, that is, from (2.100)

$$\ell = y = \frac{d\psi(v)}{dv} = f(v) \tag{2.109}$$

Therefore, (2.107) becomes

$$\frac{dw}{dt} = \mu(yx - \alpha w) \tag{2.110}$$

which is identical to (2.103). The typical Hebbian co-occurrence term appears in (2.110) as yx. Figure 2.35 shows the continuous-time standard Hebbian

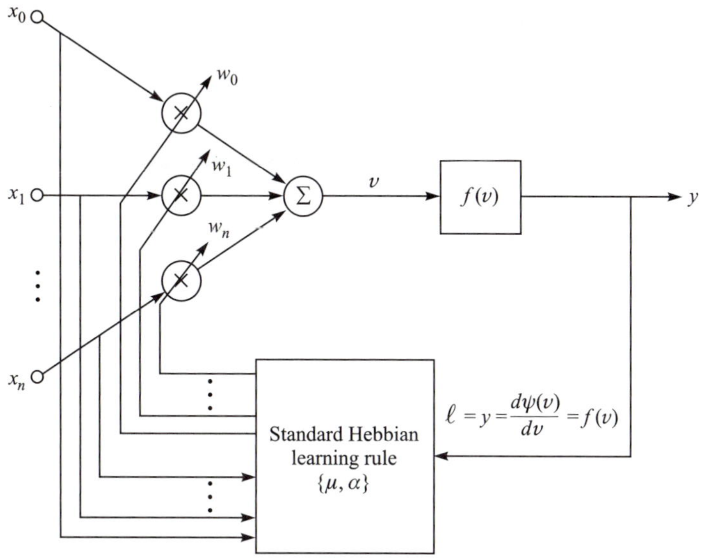

Figure 2.35 Standard Hebbian learning for a single neuron.

learning (local learning rule) for a single neuron. There are many variations of Hebb's learning rule. One very important extension is Oja's learning rule.

2.8.3 Oja's Learning Rule

Oja's learning rule [66] can be derived by minimizing the energy function

$$\mathscr{E}(\boldsymbol{w}) = \frac{1}{2}\|\boldsymbol{e}\|_2^2 \tag{2.111}$$

where

$$\boldsymbol{e} = \boldsymbol{x} - \hat{\boldsymbol{x}} \tag{2.112}$$

is the representation error, given the input to the neuron and its estimate $\hat{\boldsymbol{x}}$. Two basic assumptions are now made: (1) the neuron synaptic weight vector is considered to be normalized, that is, $\|\boldsymbol{w}\|_2 = 1$; and (2) a linear activation function is assumed, that is, $y = v = \boldsymbol{w}^T\boldsymbol{x}$. The estimate of the input is assumed to be the synaptic weight vector weighted by the neuron output, that is,

$$\hat{\boldsymbol{x}} = \boldsymbol{w}y \tag{2.113}$$

Therefore, using (2.112) and (2.113), we can write the energy function in (2.111) as

$$\begin{aligned}\mathscr{E}(\boldsymbol{w}) &= \frac{1}{2}\|\boldsymbol{x} - \boldsymbol{w}y\|_2^2 = \frac{1}{2}(\boldsymbol{x}^T - \boldsymbol{w}^T y)(\boldsymbol{x} - \boldsymbol{w}y) \\ &= \frac{1}{2}(\boldsymbol{x}^T\boldsymbol{x} - 2\boldsymbol{w}^T\boldsymbol{x}y + \boldsymbol{w}^T\boldsymbol{w}y^2)\end{aligned} \tag{2.114}$$

By using the steepest descent method, Oja's continuous-time learning rule can be written as a vector differential equation, given as

$$\frac{d\boldsymbol{w}}{dt} = -\mu \nabla_{\boldsymbol{w}} \mathscr{E}(\boldsymbol{w}) \tag{2.115}$$

where $\mu > 0$ is the learning rate parameter. Computing the necessary gradient of the energy function in (2.114) gives

$$\nabla_{\boldsymbol{w}} \mathscr{E}(\boldsymbol{w}) = -\boldsymbol{x}y + \boldsymbol{w}y^2 \tag{2.116}$$

Substituting (2.116) into (2.115) gives Oja's continuous-time learning rule

$$\frac{d\boldsymbol{w}}{dt} = \mu(\boldsymbol{x}y - \boldsymbol{w}y^2) \tag{2.117}$$

where

$$y = v = \boldsymbol{w}^T \boldsymbol{x} \tag{2.118}$$

The first term on the right-hand side of (2.117) is the typical Hebbian co-occurrence term. The second term is regarded as an active decay (or stabilization) term that prevents the synaptic weight vector from becoming unbounded by normalizing the vector to unit length, that is, $\|\boldsymbol{w}\|_2 = 1$.

The learning rule in (2.117) can be written in discrete-time form as

$$\boldsymbol{w}(k+1) = \boldsymbol{w}(k) + \mu y(k)[\boldsymbol{x}(k) - \boldsymbol{w}(k)y(k)] \tag{2.119}$$

The scalar form of (2.119) can be written as

$$w_j(k+1) = w_j(k) + \mu y(k)\left[x_j(k) - w_j(k)y(k)\right] \tag{2.120}$$

for $j = 0, 1, \ldots, n$.

Oja's learning rule can also be derived from the typical (simple) form of Hebbian learning [66]. We will restrict our discussions to the discrete-time scalar form of the learning rule. In this simple form of Hebbian learning, the learning rule only includes the co-occurrence term, that is,

$$w_j(k+1) = w_j(k) + \mu y(k)x_j(k) \tag{2.121}$$

for $j = 0, 1, \ldots, n$. This simple form of Hebbian learning typifies the statement made by Rumelhart and McClelland [57] as an extension to Hebb's original proposal, which is quoted in Section 2.8.2. However, the learning rule in (2.121) can lead to unlimited growth of the synaptic weights without some form of normalization (or saturation) incorporated into the learning rule. Oja [66] invoked normalization in the learning rule by dividing the entire right-hand side of (2.121) by the L_2-norm of $w_j(k) + \mu y(k)x_j(k)$ (with the inclusion of a *plasticity coefficient*) over the entire set of synapses associated with the neuron. With the appropriate assumptions, the resulting normalized learning rule is identical to the learning rule given in (2.120). The details pertaining to the normalization of (2.121) leading to (2.120) are presented in Section 9.3.1. The normalized Hebbian learning rule of Oja will become important in Chapter 9 (cf. Sect. 9.3) when we study methods to adaptively extract principal components. The following example illustrates this.

EXAMPLE 2.4. In this example we want to analyze zero-mean random data, specifically, 5,000 stochastic vectors, where the respective components of the vectors are from a normal (Gaussian) distribution. The first component has a variance of 10, and the other two have equal variances of 0.002. Oja's discrete-time learning rule given in (2.119) is used to process each of the consecutive vectors presented to the single neuron learning rule, and the weight vector is adjusted accordingly. A fixed learning rate parameter of $\mu = 0.001$ is used in (2.119). A criterion is used to decide whether enough training has occurred. This simply involves monitoring the progress of the weight vector convergence by the following rule:

$$\text{if} \left\| \boldsymbol{w}(k) - \boldsymbol{w}(k-1) \right\|_2 10^{-6} \rightarrow \text{quit}$$
$$\text{else} \rightarrow \text{continue}$$

The initial weight vector was randomized to $\boldsymbol{w}(0) = \left[0.5949, \ -0.5585, \ 0.4811 \right]^T$. After 1,182 iterations (i.e., only the first 1,182 vectors were presented to the single neuron), convergence was achieved. The final synaptic weight vector is

$$\boldsymbol{w} = \left[1.0000, \quad -0.0007, \quad 0.0002 \right]^T$$

This is very close to the first principal eigenvector of the estimated covariance matrix using the entire data set (i.e., all 5,000 vectors). This vector can be easily computed as

$$\boldsymbol{w}_1 = \left[1.0000, \quad 0.0001, \quad 0.0001 \right]^T$$

The associated (largest) eigenvalue for this eigenvector is $\lambda_1 = 9.9772$. Appropriate MATLAB functions can be used for this analysis. The eigenvalue that can be computed from the weight vector above ($\boldsymbol{w}$), found by using Oja's learning rule, is

$$\lambda = \text{var}(\boldsymbol{w}^T \boldsymbol{X}) = 9.9792$$

where $\boldsymbol{X} \in \Re^{3 \times 5,000}$ (all 5,000 stochastic vectors) and corresponds closely to the largest eigenvalue computed from the estimated covariance matrix. The `var` function is a standard MATLAB function used to compute the variance of the value in a vector (or the columns of a matrix). In Figure 2.36(*a*), elements of the weight vector are shown as the training of the neuron progresses. The final values are those shown above in the weight vector $\boldsymbol{w}$. Note that the norm of this vector will be essentially 1 (which is one of the constraints for Oja's learning rule). Figure 2.36(*b*) shows the progression of the weight vector residual in a norm sense.

2.8.4 Potential Learning Rule

Potential learning does not rely on a desired signal; therefore, it is in the class of unsupervised learning. However, the learning is performed exclusively on the basis of an *internal potential* [65], that is, v, the activity level of the neuron. The *potential learning rule* can be derived by minimizing the energy function

$$\mathscr{E}(\boldsymbol{w}) = -\psi(\boldsymbol{w}^T \boldsymbol{x}) + \frac{\alpha}{2} \|\boldsymbol{w}\|_2^2 \tag{2.122}$$

where $\alpha \geq 0$, $v = \boldsymbol{w}^T \boldsymbol{x}$, and $\psi(\bullet)$ is the *loss* function. The gradient of (2.122) with respect to the synaptic weight vector is given as

$$\nabla_{\boldsymbol{w}} \mathscr{E}(\boldsymbol{w}) = -\ell(v) \boldsymbol{x} + \alpha \boldsymbol{w} \tag{2.123}$$

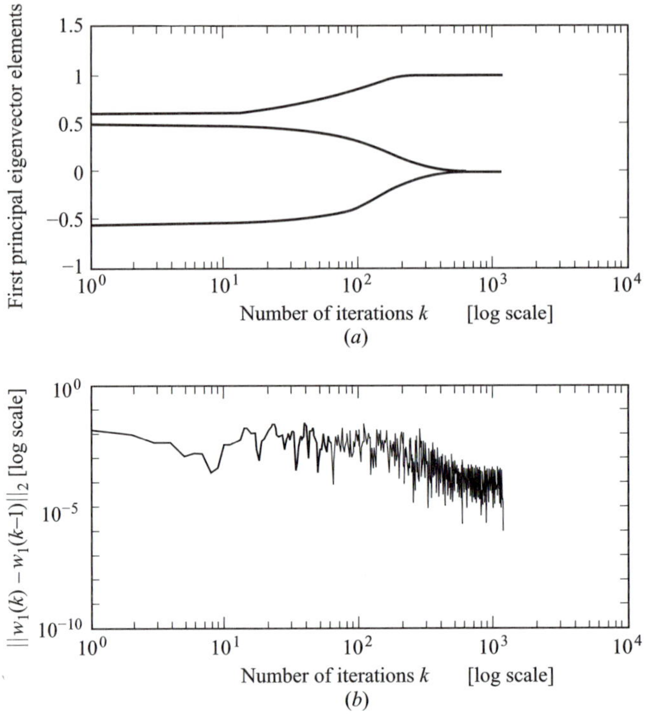

Figure 2.36 (*a*) Convergence of the elements of the weight vector during training using Oja's learning rule. (*b*) The L_2-norm of the consecutive weight vector differences during training.

where the learning signal is

$$\ell(v) = \frac{d\psi(v)}{dv}$$

Using Amari's result in (2.106) and the gradient in (2.123), we obtain the continuous-time potential learning rule

$$\frac{d\boldsymbol{w}}{dt} = \mu[\ell(v)\boldsymbol{x} - \alpha\boldsymbol{w}] \tag{2.124}$$

The discrete-time form of (2.124) can be written as

$$\boldsymbol{w}(k+1) = \boldsymbol{w}(k) + \mu[\ell(v)\boldsymbol{x}(k) - \alpha\boldsymbol{w}(k)] \tag{2.125}$$

and the discrete-time scalar form is

$$w_j(k+1) = w_j(k) + \mu\big[\ell(v)x_j(k) - \alpha w_j(k)\big] \tag{2.126}$$

where $j = 0, 1, \ldots, n$.

2.8.5 Correlation Learning Rule

We can derive the *correlation learning rule* by minimizing the energy function

$$\mathcal{E}(w) = -dw^T x + \frac{\alpha}{2}\|w\|_2^2 \tag{2.127}$$

where the gradient with respect to the synaptic weight matrix is given as

$$\nabla_w \mathcal{E}(w) = -dx + \alpha w \tag{2.128}$$

where $\ell = d$ is the learning signal, and is the desired response due to x. Therefore, correlation learning is supervised. Again, using Amari's result in (2.106) and the gradient in (2.128), we obtain the continuous-time correlation learning rule

$$\frac{dw}{dt} = \mu(dx - \alpha w) \tag{2.129}$$

The discrete-time form of (2.129) can be written as

$$w(k+1) = w(k) + \mu[d(k)x(k) - \alpha w(k)] \tag{2.130}$$

and the discrete-time scalar form is

$$w_j(k+1) = w_j(k) + \mu\left[d(k)x_j(k) - \alpha w_j(k)\right] \tag{2.131}$$

where $j = 0, 1, \ldots, n$. The correlation learning rule is typically applied to recording data in memory networks with binary response-type neurons. Interestingly enough, if d in (2.129) is replaced by y (the neuron output), we obtain the Hebbian learning rule in (2.103) (unsupervised learning).

2.8.6 Standard Perceptron Learning Rule

In Section 2.6.2 we studied the simple perceptron with a sigmoid activation function. Now we want to develop a more general learning rule based on any differentiable activation function. The resulting learning rule will be called the *standard perceptron learning rule* and can be derived by minimizing the MSE criterion (i.e., the *instantaneous* energy function)

$$\mathcal{E}(w) = \frac{1}{2}e^2 \tag{2.132}$$

where $e = d - y$. The output of the neuron is written as

$$y = f\left(w^T x\right) = f(v) \tag{2.133}$$

and $f(\bullet)$ is the neuron activation function. We now assume that the threshold (or bias) term is included in the synaptic weight vector; therefore, $w \in \Re^{n+1\times 1}$. Taking the steepest descent approach, the continuous-time learning rule is given by

$$\frac{dw}{dt} = -\mu \nabla_w \mathcal{E}(w) \tag{2.134}$$

where the gradient of (2.132) is given by

$$\nabla_w \mathscr{E}(\boldsymbol{w}) = -d\frac{\mathrm{d}f(v)}{\mathrm{d}v}\boldsymbol{x} + \psi(v)\frac{\mathrm{d}f(v)}{\mathrm{d}v}\boldsymbol{x} = -\underbrace{[d - \psi(v)]}_{e}\frac{\mathrm{d}f(v)}{\mathrm{d}v}\boldsymbol{x}$$
$$= \underbrace{-e\frac{\mathrm{d}f(v)}{\mathrm{d}v}}_{\ell}\boldsymbol{x} = -\ell\boldsymbol{x} \tag{2.135}$$

where

$$\ell = e\frac{\mathrm{d}f(v)}{\mathrm{d}v} = ef'(v) = eg(v) \tag{2.136}$$

is the learning signal. Therefore, using (2.134), (2.135), and (2.136), we can write the continuous-time standard perceptron learning rule for a single neuron as

$$\frac{\mathrm{d}\boldsymbol{w}}{\mathrm{d}t} = \mu\ell\boldsymbol{x} \tag{2.137}$$

Equation (2.137) can be written in discrete-time form as

$$\boldsymbol{w}(k+1) = \boldsymbol{w}(k) + \mu\ell(k)\boldsymbol{x}(k) \tag{2.138}$$

where $\ell(k)$ is the discrete-time form of (2.136). The scalar form of the discrete-time learning rule in (2.138) can be written as

$$w_j(k+1) = w_j(k) + \mu\ell(k)x_j(k) \tag{2.139}$$

for $j = 0, 1, \ldots, n$.

2.8.7 Generalized Perceptron Learning Rule

The standard perceptron learning rule in Section 2.8.6 can be generalized for the case when the energy function is not necessarily defined to be the MSE criterion; that is, we can define a general energy function as

$$\mathscr{E}(\boldsymbol{w}) = \psi(e) = \psi(d - y) \tag{2.140}$$

where $\psi(\bullet)$ is a differentiable (weighting or loss) function. If $\psi(e) = {}^1\!/_2 e^2$, this would lead to the standard perceptron learning rule. However, for any suitable function $\psi(\bullet)$, the *general perceptron learning rule* can be derived by minimizing (2.140) with respect to the synaptic weight vector $\boldsymbol{w}$. The generalized energy function in (2.140) is frequently used to derive learning rules for the *robust* case (cf. Chaps. 8 and 9). The generalized energy function is robust in the sense that the errors will be *weighted* less than quadratically, resulting in outlier rejection. The MSE performance criterion, as in (2.132), places a quadratic weighting on the errors. Using the steepest descent method, the form of the continuous-time general perceptron learning rule is given in (2.134). Therefore, the gradient of (2.140) must be computed, and it can be determined by applying the chain rule as

$$\nabla_w \mathscr{E}(w) = \frac{\partial \psi}{\partial e} \underbrace{\frac{\partial e}{\partial y}}_{-1} \frac{\partial y}{\partial v} \underbrace{\frac{\partial v}{\partial w}}_{x} \tag{2.141}$$

where

$$\frac{d\psi(e)}{de} = \psi'(e) \triangleq \delta(e) \tag{2.142}$$

and

$$y = f(w^T x) = f(v) \tag{2.143}$$

where $f(\bullet)$ is a differentiable function, and

$$\frac{dy(v)}{dv} - \frac{df(v)}{dv} = f'(v) \triangleq g(v) \tag{2.144}$$

Therefore, (2.141) can be written as

$$\nabla_w \mathscr{E}(w) = -\delta(e)g(v)x \tag{2.145}$$

and the continuous-time general perceptron learning rule is given as

$$\frac{dw}{dt} = \mu\delta(e)g(v)x \tag{2.146}$$

where $\mu > 0$ is the learning rate parameter. If we define the learning signal as

$$\ell \triangleq \delta(e)g(v) \tag{2.147}$$

Equation (2.146) can be written as

$$\frac{dw}{dt} = \mu\ell x \tag{2.148}$$

The generalized perceptron learning rule can be written in discrete-time form as

$$w(k+1) = w(k) + \mu\ell(k)x(k) \tag{2.149}$$

where $w(k) = w(kT_s)$ (T_s is the sampling period), $\ell(k)$ is given in (2.147), and the scalar form of (2.149) is given by

$$w_j(k+1) = w_j(k) + \mu\ell(k)x_j(k) \tag{2.150}$$

for $j = 0, 1, \ldots, n$. Figure 2.37 shows the generalized perceptron learning rule.

2.9 DATA PREPROCESSING

In general, the performance of a neural network is strongly dependent on the preprocessing that is performed on the training data [67, 68]. The neural network training process can also be made more efficient if certain preprocessing steps are carried out on the input patterns and target values. That is, many times the "raw" data are not the best data to use for training a neural network. For example, in the backpropagation algorithm used to train a feedfor-

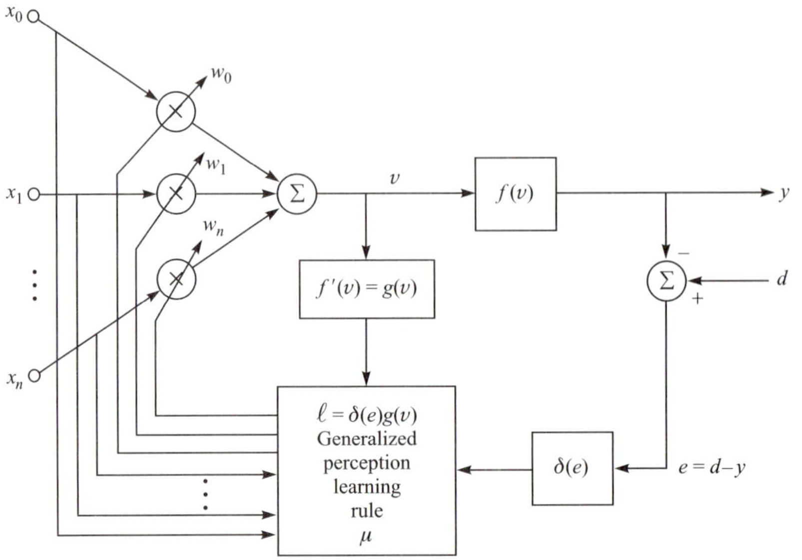

Figure 2.37 Generalized perceptron learning rule.

ward perceptron, if a binary sigmoid function (cf. Sect. 2.3) is used as the nonlinear activation function for the neurons in the network, the saturation limits are 0 and 1. If the training patterns have large values compared to these limits, the nonlinear activation functions could be operating almost exclusively in a saturated mode and not allow the network to train. Therefore, the training data (input patterns and target values) should be *range-scaled* to avoid this problem. This type of preconditioning of the training data is what we define to be *scaling preprocessing*. This will be discussed first, followed by what we term *transformation preprocessing*. The material presented here is not intended to exhaustively cover all forms and methods of data preprocessing. However, the methods that are presented have been used extensively to preprocess data for training neural networks.

2.9.1 Scaling

The training data can be amplitude-scaled in basically two ways: so that the values of the patterns lie between −1 and 1, or so that the values of the patterns lie between 0 and 1. These two types of amplitude scaling are usually referred to as min/max scaling. In the case of fuzzy neural networks, it is necessary to scale the input data in the [0, 1] range [69–71]. There is a function **`premnmx`** available in the MATLAB Neural Network Toolbox [68] to scale in the input data in the range [−1, 1]. The **`premnmx`** function has an option to scale the input data or both the input and target data.

There is another important scaling process called *mean centering* and *variance scaling* [68, 72–76] that can prove useful for training a neural network. We will assume that the input patterns (vectors) are arranged as columns in a matrix $A \in \Re^{n \times m}$ and that the target vectors are arranged as columns in a matrix $C \in \Re^{p \times m}$. The mean centering process involves computing a mean value for each *row* of A and C. Therefore, A would have n computed mean values, and C would have p computed mean values. The associated mean value is then subtracted from each element in the particular row for all rows in both A and C. Variance scaling involves computing the n *standard deviations* for the n rows of the input matrix A, and the p standard deviations for the p rows of the target matrix C. The associated standard deviation is then divided into each element in the particular row for all rows in both A and C. Mean centering and variance can be carried out separately or together. In the MATLAB neural network toolbox [68], the function **`prestd`** will mean-center and variance-scale both the input data and the target data or just the input data. Mean centering can be important if the data contain biases; if A is mean-centered, then C should also be mean-centered [72, 73]. Variance scaling is advisable if the collected data used for training are measured with different units. Again, if A is variance-scaled, C should also be variance-scaled [72, 73]. Many individuals insist that the data should always be mean-centered and variance-scaled, and others maintain that the data should never be preconditioned in this manner [77]. We feel that mean-centering and variance-scaling the data should only be preprocessed in this manner if there is (are) a compelling reason(s) to do so; otherwise, these processes should not be performed arbitrarily.

2.9.2 Transformations

Many times the *features* of certain "raw" signals are used for training inputs to a neural network and provide better results than the raw signals themselves. Therefore, a front-end feature extractor can be used to discern salient or distinguishing characteristics of the data, and these signal features can then be used as inputs for training the network. It is also highly desirable to reduce the input vector length of the training patterns, which will typically decrease the size of the overall network architecture. Four transform methods are discussed that can be used as feature extractors and have these highly desirable features as preconditioning signal processing methods for preprocessing the training data.

2.9.3 Fourier Transform

The Fourier transform can be useful for preconditioning data for training neural networks. From a practical viewpoint, the fast Fourier transform (FFT) [20] of temporal signals is an efficient method to carry out the discrete Fourier transform (DFT). The FFT is a built-in function **`fft`** in MATLAB. A major advantage of using the magnitude spectrum of the FFT is that it is insensitive to the phase of the signal. Therefore, if the phase of the signal is not important, the FFT magnitude samples can be used as a feature vector for each training pattern (signal). Typically, the number of Fourier frequencies required to properly represent the signal is significantly less than the length of the raw signal itself. Figure 2.38 illustrates these two points. First, in Figure 2.38(*a*) we observe three identical signals (each having 1,024 time samples), and their associated FFT magnitude response (the first 16 samples) is shown in Figure 2.38(*b*). Observing Figure 2.38(*b*), we see that the FFT *magnitude* responses are identical. Second, it appears again from Figure 2.38(*b*) that the first 16 magnitude samples in all likelihood would be sufficient to represent the 1,024 sample length signals, as a set of unique features of the data. Therefore, the 16 magnitude samples would form a training vector that is substantially smaller than the input pattern represented by the raw time-domain signal. As

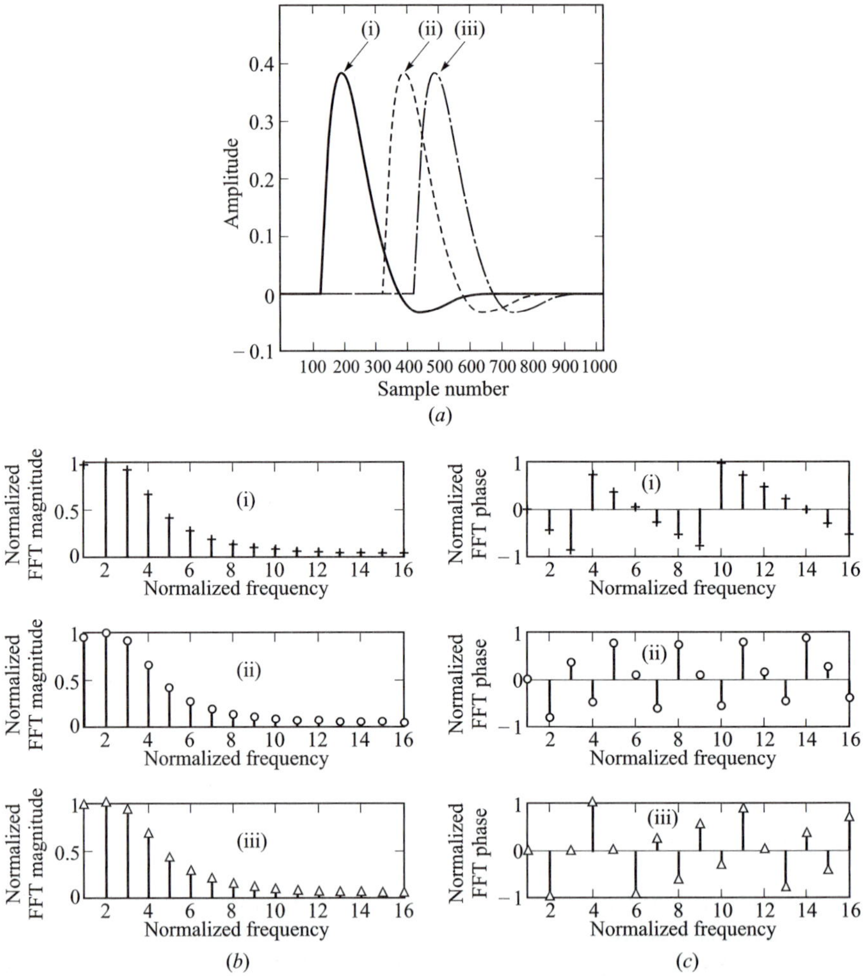

Figure 2.38 (*a*) Three identical signals shifted in time. The signals each have 1,024 time samples. (*b*) The first 16 FFT magnitude samples of each signal. (*c*) The first 16 FFT phase samples of each signal. (*d*) The first 16 FFT real-part samples of each signal. (*e*) The first 16 FFT imaginary-part samples of each signal.

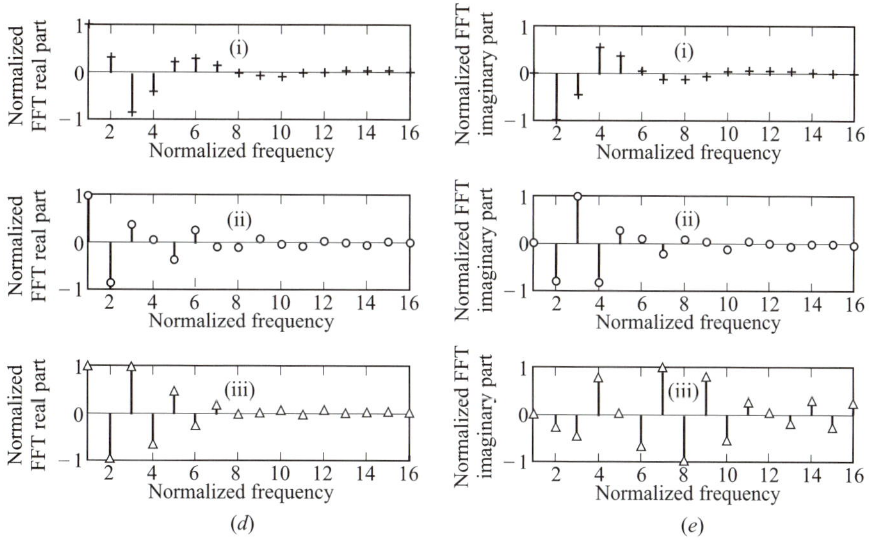

Figure 2.38 (continued)

shown in Figure 2.38(*c*) through (*e*), the respective phase data, real-part features, and imaginary-part features do not possess the phase-invariant feature that the magnitude response holds.

A related approach to the FFT method involves using the *complex cepstrum* [20, 78] of the training patterns. The complex cepstrum is used many times to separate signals that are convolved; this is called *homomorphic deconvolution* [20, 78]. As with the FFT, the complex cepstrum of the input patterns can be used to extract the important features of the data, and then these predominant characteristic features can be used to train the neural network. Cepstrum processing of speech signals for robust speaker recognition is detailed in the paper by Mammone et al. [79].

2.9.4 Principal-Component Analysis

Principal-component analysis (PCA) (cf. Sect. 9.2) can be used to "compress" the input training data set (or reduce the dimension of the inputs). Using PCA allows the dimension of the input vectors to be reduced by determining the important features of the data according to an assessment of the variance of the data. The resulting "compressed" input vectors will have elements that are uncorrelated. This is evidenced by the fact that the estimated covariance matrix of the transformed inputs is diagonal. The data should be mean-centered before PCA is applied. In the MATLAB neural network toolbox [68], a function **`prepca`** is provided to perform PCA on the training data.

Given a set of training data $\boldsymbol{A} \in \Re^{n \times m}$, where it is assumed that $m \gg n$. However, n (the dimension of the input training patterns) is assumed to be relatively large. Using PCA, an "optimal" orthogonal transformation matrix

$W_{pca} \in \Re^{h \times n}$ can be determined (cf. Sect. 9.2), where typically $h \ll n$ (the degree of dimension reduction) . Using this transformation matrix, the dimension of the input vectors (i.e., the columns of A) can be reduced according to the transformation

$$A_r = W_{pca} A \tag{2.151}$$

where $A_r \in \Re^{h \times m}$ is the reduced-dimension set of training patterns. The columns of A_r are the principal components for each of the inputs from A (i.e., the associated columns in A). Note in Section 9.2 that we define the *rows* of the orthogonal transformation matrix W, W_{pca} in (2.151), to be the orthonormal principal eigenvectors.

2.9.5 Partial Least-Squares Regression

Partial least-squares regression (PLSR) (cf. Sect. 9.5) can also be used to compress the input training data set. PLSR is restricted for use with supervised trained neural networks because both the input and target training data are required. Furthermore, only scalar target values (i.e., scalar response variables) are allowed.† The factor analysis in PLSR (cf. Sect. 9.5) can determine the degree of compression of the input data. That is, after the optimal number of PLSR factors h has been determined, the weight loading vectors can be used to transform the data similar to the PCA approach given in Section 2.9.4. Therefore, the optimal number weight loading vectors can form an orthogonal transformation matrix as the *columns* of the matrix $W_{plsr} \in \Re^{n \times h}$. Using this transformation matrix, the dimension of the input vectors (i.e., the columns of A) can be reduced according to the transformation

$$A_r = W_{plsr}^T A \tag{2.152}$$

The compressed vectors in (2.152) are actually the *scores* that are generated in the partial least-squares algorithm (cf. Sect. 9.5). This data compression approach has been used with much success to train a multilayer feedforward perceptron by backpropagation [80]. Figure 2.39 shows a comparison of the PCA and PLSR orthogonal transformation vectors used for data compression. The difference in the two sets of vectors is the fact that PLSR uses both the input data and the target data to generate the weight loading vectors in the orthogonal transformation W_{plsr}.

2.9.6 Wavelets and Wavelet Transforms

In basic terms, a *wave* is an oscillating function of time (or space), say, a sinusoid with a particular amplitude and frequency. Fourier analysis is used for analyzing waves; that is, certain functions (or signals) can be *expanded* in

†This is a restriction only because the *single-component* case is presented in this book (cf. Sect. 9.5). However, in general the target values (response variables) in PLSR can be vector quantities [76].

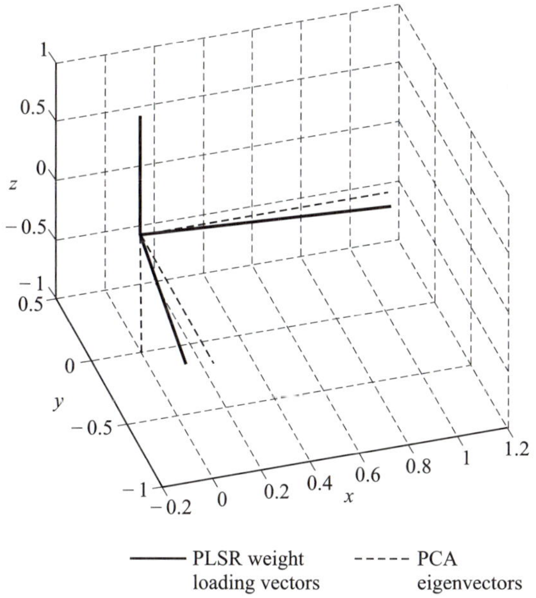

Figure 2.39 PCA and PLSR orthogonal transformation vectors used for data compression.

terms of sinusoidal waves (or complex exponentials). This method of signal analysis is very useful when we deal with physical data that are periodic, time-invariant, or stationary. Therefore, in the case of Fourier series, sinusoids are chosen as basis functions; then properties of the resulting expansion are analyzed. The Fourier transform provides an analytical tool that transforms a signal to the frequency domain and gives an explicit representation of the frequency content of the signal with respect to the basis functions (i.e., how much of each frequency component is required to synthesize the signal).

In the case of *wavelet* analysis, the desired properties are defined; then the resulting basis functions are derived. A wavelet can be considered as a *small wave* [81], whose energy is concentrated (in time). A wavelet can provide a useful tool for analyzing signals that are time-varying, transient, or nonstationary. Thus many types of signals that cannot be properly analyzed using Fourier methods can be investigated using a wavelet approach. Moreover, the wavelet has the ability to allow for simultaneous time and frequency analysis.

The wavelet transform [81–84] is much more local than say the classic Fourier transform. That is, wavelets are local waves; instead of oscillating ad infinitum, they eventually fall to zero. This is unlike the Fourier transform that is based on infinite sinusoidal functions. Wavelets are useful for signal and image compression, detection, and denoising, to name a few. The wavelet transform can provide a *time-frequency* description of signals and can be used to compress data for training neural networks [85, 86]. In the MATLAB wavelet toolbox [87], there are several functions that can be used for denoising and compression for signals and images.

PROBLEMS

2.1. The derivative of a scalar function with a vector argument $f(\boldsymbol{x}): \ \Re^{n \times 1} \rightarrow \Re$ is defined as

$$\frac{\partial f(\boldsymbol{x})}{\partial \boldsymbol{x}} = \begin{bmatrix} \dfrac{\partial f(\boldsymbol{x})}{\partial x_1} \\ \dfrac{\partial f(\boldsymbol{x})}{\partial x_2} \\ \vdots \\ \dfrac{\partial f(\boldsymbol{x})}{\partial x_n} \end{bmatrix}$$

Prove the following:

(*a*) $\dfrac{\partial}{\partial \boldsymbol{x}}\left(\boldsymbol{h}^T \boldsymbol{x}\right) = \dfrac{\partial}{\partial \boldsymbol{x}}\left(\boldsymbol{x}^T \boldsymbol{h}\right) = \boldsymbol{h}$, where $\boldsymbol{x}, \boldsymbol{h} \in \Re^{n \times 1}$

(*b*) $\dfrac{\partial}{\partial \boldsymbol{x}}\left(\boldsymbol{x}^T \boldsymbol{A} \boldsymbol{x}\right) = 2\boldsymbol{A}\boldsymbol{x}$, where $\boldsymbol{x} \in \Re^{n \times 1}$, $\boldsymbol{A} \in \Re^{n \times n}$, and $\boldsymbol{A} = \boldsymbol{A}^T$

2.2. Consider the Adaline shown in Figure 2.19. Use the μ-LMS algorithm to train the network to perform the OR logic function. As training inputs use (*a*) bipolar vectors and (*b*) binary vectors. (*c*) Compare the rate of convergence for cases (*a*) and (*b*). Explain the difference.

2.3. Repeat Problem 2.2, using the α-LMS algorithm.

2.4. Write a computer program for training the Adaline with nonlinearly transformed inputs given in Figure 2.24 to perform the logic function XOR. As training inputs use bipolar vectors. In your own words, explain why this network structure has superior separability properties compared to the perceptron.

2.5. The XOR function can be expressed as

A XOR B = (NOT A AND B) OR (A AND NOT B)

(*a*) Design a Madaline network to implement the above logic function.
(*b*) Train the network using the MRI learning rule.
(*c*) Draw the boundary separation lines.

2.6. Consider a separation problem shown in Figure 2.40. It should be obvious that the circles and squares are not linearly separable.
(*a*) Write a computer program implementing an Adaline with nonlinearly transformed inputs, trained with an LMS algorithm.
(*b*) Use your program to separate the circles and squares given in Figure 2.40.
(*c*) Modify your code so that the training is performed using α-LMS learning.

2.7. Consider a set of two-dimensional vectors defined as $\{\boldsymbol{x} \in \Re^{n \times 1} \mid -2 \leq x_1 \leq 2, \text{and} -2 \leq x_2 \leq 2\}$. Train the Adaline neural network with nonlinearly transformed inputs given in Figure 2.24 to perform the following classification:

$\boldsymbol{x}$ is classified as 1 if $x_1^2 + x_2^2 < 1$; otherwise, $\boldsymbol{x}$ is classified as 0.

2.8. Write a computer program using a perceptron (see Figure 2.30) to classify the digits given in Figure 2.41. The number of neurons in the output layer should be

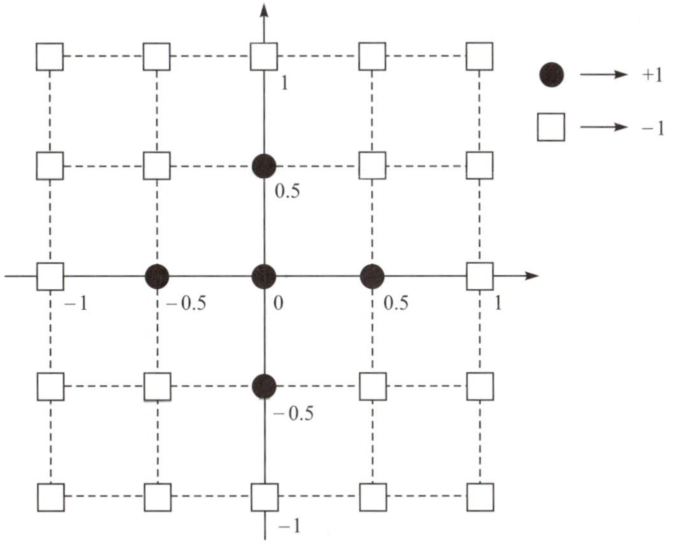

Figure 2.40 Separation of patterns for Problem 2.6.

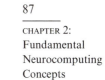

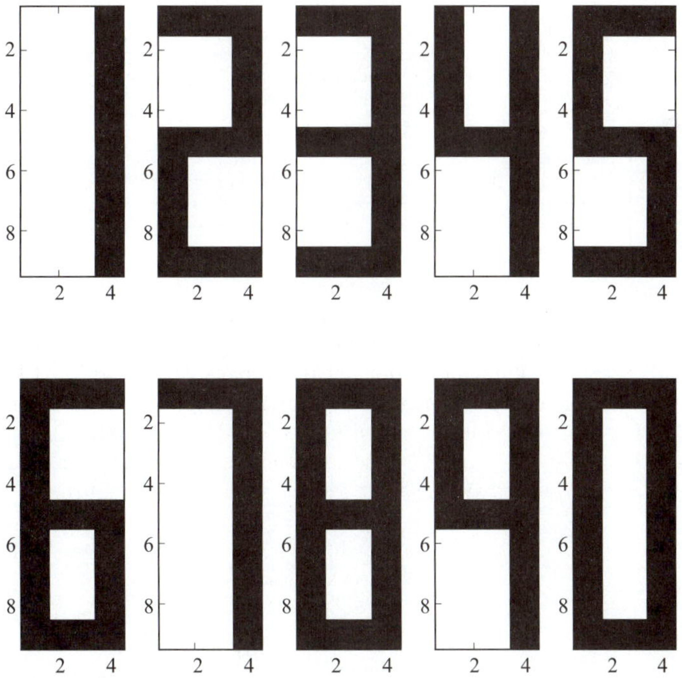

Figure 2.41 Digits for Problem 2.8.

equal to the number of digits. Each of the digits is represented as a 9×4 matrix of binary (or bipolar) numbers. The input training patterns can be generated from each digit as a vector resulting from applying the vec operator (cf. Sect. A.2.17) to each matrix representing the digit. After the network is trained, introduce random noise into the digit representations, and test the performance of the neural network. Experiment with different activation functions in the output layer.

2.9. (*a*) By substituting (2.27) into (2.25) prove that

$$J_{\min}(\boldsymbol{w}) = \frac{1}{2}\mathrm{E}\{d^2(k)\} - \frac{1}{2}\boldsymbol{p}^T\boldsymbol{w}^*$$

(*b*) Using the equation derived in part (a), prove that (2.25) can be expressed as

$$J(\boldsymbol{w}) = J_{\min}(\boldsymbol{w}) + (\boldsymbol{w} - \boldsymbol{w}^*)^T \boldsymbol{C}_x(\boldsymbol{w} - \boldsymbol{w}^*)$$

2.10. Suppose that the weights of a simple linear combiner are set in accordance with (2.27).

(*a*) Show that

$$\mathrm{E}\{e(k)\boldsymbol{x}(k)\} = \boldsymbol{0}$$

(*b*) What is a physical significance of the result in part (*a*)?

M

2.11. One of the most popular uses of the adaptive linear combiner is for adaptive equalization of a communication channel. Consider the problem of the data transfer over a channel corrupted with an unknown interference, as shown in Figure 2.42. For this problem, assume that the interference is a sinusoid of

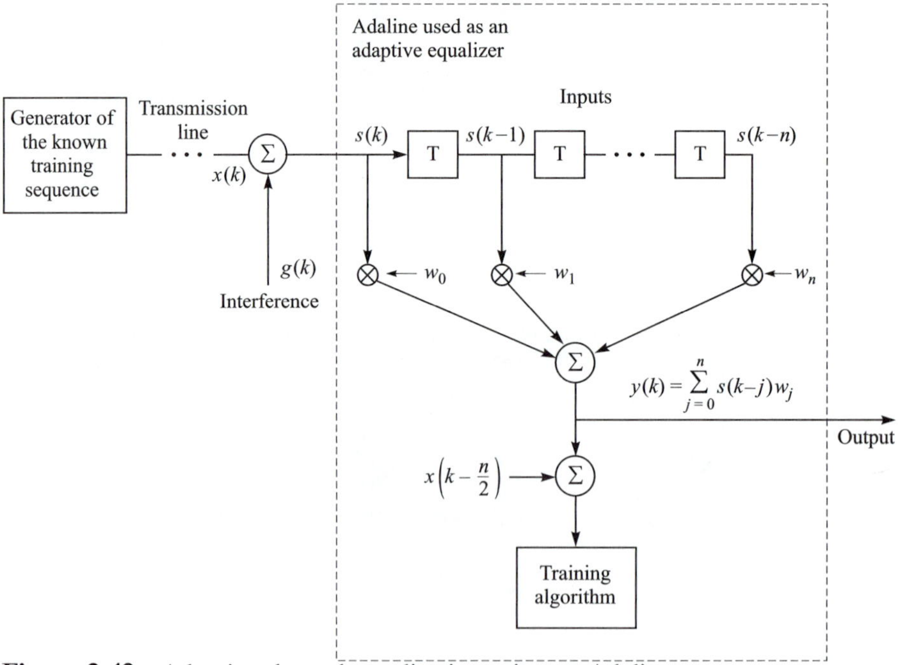

Figure 2.42 Adaptive channel equalization using an Adaline.

frequency f. To achieve an interference-free link, an Adaline network is placed at the end of the transmission line. Before the transmission of the information data bit stream, a known sequence is sent over the line, as shown in the figure. At the end of the link the Adaline compares the received signal with the known sequence and adjusts its weights so that difference between the two is minimized.

Perform a computer simulation to carry out the following:

(*a*) Generate a random bit sequence 1,024 bits long. (The bit sequence can be generated by using a random number generator.) Let each bit be represented with at least four samples (either bipolar or binary).

(*b*) Add an interference signal to the bit sequence (see Figure 2.43).

```
N = 1024;                    % size of the sequence
x0 = round(rand(N,1));       % generate a random binary sequence
x = zeros(N*4,1);            % let each bit be represented with
x(1:4:4*N,1) = x0;           % four samples
x(2:4:4*N,1) = x0;
x(3:4:4*N,1) = x0;
x(4:4:4*N,1) = x0;

t = 1:4*N;                   % generate the interference
g = sin(2*pi*t/10)';

s = x+g;                     % signal at the end of the channel
```

Figure 2.43 Example of MATLAB code to generate the training data.

(*c*) Use the LMS algorithm to train the Adaline with n neurons in the input and one neuron in the output layer. Let the input vector be a delayed version of the signal $s(k)$, and let the desired network output be the corresponding bit of the known sequence, as shown in Figure 2.42.

(*d*) Test the output of the Adaline after it has been trained, and compare its output to the undistorted known sequence.

(*e*) Experiment with sequences of different lengths, different numbers of input neurons, and different learning rules (i.e., α-LMS and μ-LMS).

2.12. In our derivation of the LMS algorithm, we have estimated the direction of the weight updates by taking the derivative of the instantaneous error surface. In essence, the gradient of the MSE surface given as

$$\nabla_w J(\boldsymbol{w}) = -\mathrm{E}\{d(k)\boldsymbol{x}(k)\} + \mathrm{E}\{\boldsymbol{x}(k)\boldsymbol{x}^T(k)\}\boldsymbol{w}(k)$$

was approximated by

$$\nabla_w J(\boldsymbol{w}) \approx -d(k)\boldsymbol{x}(k) + \boldsymbol{w}^T(k)\boldsymbol{x}(k)\boldsymbol{x}(k) = -[d(k) - \boldsymbol{w}^T(k)\boldsymbol{x}(k)]\boldsymbol{x}(k)$$
$$= -e(k)\boldsymbol{x}(k)$$

The instantaneous error is the crudest approximation of the MSE surface. To ensure stable behavior of the algorithm, the learning rate must be kept relatively small.

(*a*) Write a computer program using the following approximations

$$\mathrm{E}\{d(k)\boldsymbol{x}(k)\} \cong \sum_{m=k-M+1}^{k} d(m)\boldsymbol{x}(m)$$

and

$$\mathrm{E}\{\boldsymbol{x}(k)\boldsymbol{x}^T(k)\} \cong \sum_{m=k-M+1}^{k} \boldsymbol{x}(m)\boldsymbol{x}^T(m)$$

(*b*) Test your program for the adaptive equalization example described in Problem 2.11. Experiment with different values of M and discuss the results. For example, compare the speed, memory requirements, and stability of the learning algorithm developed in part (*a*) with the original LMS algorithm.

2.13. (*a*) Let $\boldsymbol{e} \in \Re^{n\times 1}$. Prove that

$$\frac{\partial\|\boldsymbol{e}\|_2}{\partial \boldsymbol{e}} = \frac{\boldsymbol{e}}{\|\boldsymbol{e}\|_2}$$

(*b*) In the derivation of the leaky LMS learning rule in (2.84) (cf. Sect. 2.8.1) we have assumed $\psi(t) = {}^1\!/_2 t^2$. Frequently, to make the learning rule more robust relative to the outliers and noise in the training samples, the function $\psi(\bullet)$ is defined as $\psi(t) = |t|$. Show that in this case the general (discrete-time) LMS learning rule in (2.82) assumes the form

$$\boldsymbol{w}(k+1) = \boldsymbol{w}(k) + \mu\left[\frac{e}{\|e\|_2}\boldsymbol{x}(k) - \alpha\boldsymbol{w}(k)\right]$$

2.14. A problem frequently encountered in the application of the standard LMS algorithm is its poor performance when the network operates in a flat plateau of the error surface. One way to overcome this problem is to use the *leaky* LMS algorithm (see Table 2.2) of the form

$$\boldsymbol{w}(k+1) = \mu e(k)\boldsymbol{x}(k) + (1-\gamma)\boldsymbol{w}(k)$$

Show that if the error surface is relatively flat, the leaky LMS can be approximated by

$$\boldsymbol{w}(k+1) = \frac{\mu}{\gamma}e(k)\boldsymbol{x}(k) + (1-\gamma)^{n+1}\boldsymbol{w}(k-n)$$

We can see that the learning rate parameter has the effective value of $\mu_{eff} = \mu/\gamma$.

2.15. The LMS algorithm with momentum given by

$$\boldsymbol{w}(k+1) = \boldsymbol{w}(k) + \mu e(k)\boldsymbol{x}(k) + \alpha[\boldsymbol{w}(k) - \boldsymbol{w}(k-1)]$$

outperforms the standard LMS algorithm in the case when the error surface is relatively flat. Show that for flat error surfaces updating with momentum can be approximated by

$$\boldsymbol{w}(k+1) \approx \boldsymbol{w}(k) + \frac{1}{1-\alpha}\mu e(k)\boldsymbol{x}(k)$$

Note that the effective learning rate is now $\mu_{eff} = [1/(1-\alpha)]\mu$.

REFERENCES

1. S. Haykin, *Neural Networks—A Comprehensive Foundation*, 2nd ed., Upper Saddle River, NJ: Prentice Hall, 1999.
2. W. S. McCulloch and W. Pitts, "A Logical Calculus of the Ideas Immanent in Nervous Activity," *Bulletin of Mathematical Biophysics*, vol. 5, 1943, pp. 115–33. Reprinted in 1988, Anderson and Rosenfeld [4], pp. 18–27.
3. F. H. Eeckman and W. J. Freeman, "The Sigmoid Nonlinearity in Neural Computation: An Experimental Approach," in *Neural Networks for Computing*, ed. J. S. Denker, New York: American Institute of Physics, 1986, pp. 135–45.
4. J. A. Anderson and E. Rosenfeld, eds., *Neurocomputing: Foundations of Research*, Cambridge, MA: M.I.T. Press, 1988.
5. J. J. Hopfield, "Neural Networks and Physical Systems with Emergent Collective Computational Abilities," *Proceedings of the National Academy of Sciences*, vol. 79, 1982, pp. 2554–8. Reprinted in 1988, Anderson and Rosenfeld [4], pp. 460–4.
6. R. V. Churchill, J. W. Brown, and R. F. Verhey, *Complex Variables and Applications*, 3rd ed., New York: McGraw-Hill, 1976.
7. G. F. Franklin, J. D. Powell, and M. L. Workman, *Digital Control of Dynamical Systems*, 2nd ed., Reading, MA: Addison-Wesley, 1990.
8. D. A. Bell, *Operational Amplifiers: Applications, Troubleshooting, and Design*, Englewood Cliffs, NJ: Prentice-Hall, 1990.
9. A. Cichocki and R. Unbehauen, *Neural Networks for Optimization and Signal Processing*, New York: Wiley, 1993.
10. B. C. Kuo, *Digital Control Systems*, 2nd ed., New York: Saunders College Publ., 1992.
11. B. Widrow and M. E. Hoff, Jr., "Adaptive Switching Circuits," *IRE WESCON Convention Record*, part 4, New York, IRE, 1960, pp. 96–104. Reprinted in 1988, Anderson and Rosenfeld [4], pp. 126–34.
12. M. Caudill, "Neural Networks Primer: Part II," *AI Expert*, 1988, pp. 55–61.
13. D. E. Rumelhart, G. E. Hinton, and R. J. Williams, "Learning Internal Representations by Error Propagation," in *Parallel Distributed Processing*, vol. 1, chap. 8, eds. D. E. Rumelhart and J. L. McClelland, Cambridge, MA: M.I.T. Press, 1986.
14. B. Widrow, "ADALINE and MADALINE—1963, Plenary Speech," *Proceedings of First IEEE International Conference on Neural Networks*, vol. 1, San Diego, CA, June 23, 1987, pp. 145–58.
15. B. Widrow, "Generalization and Information Storage in Networks of Adaline 'Neurons,'" in *Self-Organizing Systems 1962*, eds. M. Yovitz, G. Jacobi, and G. Goldstein, Washington: Spartan Books, 1962, pp. 435–61.
16. B. Widrow and M. A. Lehr, "30 Years of Neural Networks: Perceptron, Madaline and Backpropagation," *Proceedings of the IEEE*, vol. 78, 1990, pp. 1415–42.
17. D. G. Luenberger, *Linear and Nonlinear Programming*, 2nd ed., Reading, MA: Addison-Wesley, 1984.
18. S. G. Nash and A. Sofer, *Linear and Nonlinear Programming*, New York: McGraw-Hill, 1996.
19. S. Haykin, *Adaptive Filter Theory*, 3rd ed., Upper Saddle River, NJ: Prentice-Hall, 1996.
20. J. G. Proakis and D. G. Manolakis, *Digital Signal Processing: Principles, Algorithms, and Applications*, 3rd ed., Upper Saddle River, NJ: Prentice-Hall, 1996.
21. N. Wiener, *The Extrapolation, Interpolation, and Smoothing of Stationary Time Series*, New York: Wiley, 1949.

22. B. Widrow and S. D. Stearns, *Adaptive Signal Processing*, Englewood Cliffs, NJ: Prentice-Hall, 1985.
23. G.H. Golub and C.F. Van Loan, *Matrix Computations*, 3rd ed., Baltimore, MD: Johns Hopkins University Press, 1996.
24. H. Robbins and S. Monro, "A Stochastic Approximation Method," *Annals of Mathematical Statistics*, vol. 22, 1951, pp. 400–7.
25. L. Ljung, "Analysis of Recursive Stochastic Algorithms," *IEEE Transactions on Automatic Control*, vol. AC-22, 1977, pp. 551–75.
26. H. J. Kushner and D. S. Clark, *Stochastic Approximation Methods for Constrained and Unconstrained Systems*, New York: Springer-Verlag, 1978.
27. C. Darken and J. Moody, "Towards Faster Stochastic Gradient Search," in *Advances in Neural Information Processing Systems 4*, eds. J. E. Moody, S. J. Hanson, and R.P. Lippmann, San Mateo, CA: Morgan Kaufmann, 1992, pp. 1009–16.
28. C. Darken and J. Moody, "Learning Rate Schedules for Faster Stochastic Gradient Search," in *Proceedings of the 1992 IEEE Workshop on Neural Networks for Signal Processing 2*, Piscataway, NJ: IEEE Press, 1992.
29. R. S. Sutton, "Gain Adaptation Beats Least Squares?" in *Proceedings of the Seventh Yale Workshop on Adaptive and Learning Systems*, New Haven, CT: Yale University, 1992, pp. 161–6.
30. G. E. P. Box, G. M. Jenkins, and G. C. Reinsel, *Time Series Analysis—Forecasting and Control*, 3rd ed., Englewood Cliffs, NJ: Prentice-Hall, 1994.
31. L. Ljung, *System Identification: Theory for the User*, Englewood Cliffs, NJ: Prentice-Hall, 1987.
32. B. Widrow and R. G. Winter, "Neural Nets for Adaptive Filtering and Adaptive Pattern Recognition," *IEEE Computer*, vol. 31, 1988, pp. 25–39.
33. D. F. Specht, "Vector Cardiographic Diagnosis Using the Polynomial Discriminant Method of Pattern Recognition," *IEEE Transactions on Biomedical Engineering*, vol. BME-14, 1967, pp. 90–95.
34. D. F. Specht, "Generation of Polynomial Discriminant Functions for Pattern Recognition," *IEEE Transactions on Electronic Computation*, vol. EC-16, 1967, pp. 308–19.
35. A. R. Barron, "Adaptive Learning Networks: Development and Application in the United States of Algorithms Related to GMDH," in *Self-Organizing Methods in Modeling*, ed. S. J. Farlow, New York: Marcel Dekker, 1984, pp. 25–65.
36. A. R. Barron, "Predicted Squared Error: A Criterion for Automatic Model Selection," in *Self-Organizing Methods in Modeling*, ed. S. J. Farlow, New York: Marcel Dekker, 1984, pp. 87–103.
37. A. R. Barron and R. L. Barron, "Statistical Learning Networks: A Unifying View," *1988 Symposium on the Interface: Statistics and Computing Science*, Reston, VA, April 21–23, 1988, pp. 192–203.
38. A. G. Ivakhnenko, "Polynomial Theory of Complex Systems," *IEEE Transactions on Systems, Man, and Cybernetics*, vol. SMC-1, 1971, pp. 364–78.
39. C. H. Mays, "Adaptive Threshold Logic," Ph.D. thesis, Tech. Report 1557-1, Stanford Electronic Labs., Stanford, CA, April 1963.
40. F. Rosenblatt, "On the Convergence of Reinforcement Procedures in Simple Perceptrons," *Cornell Aeronautical Lab. Report VG-1196-G-4*, Buffalo, NY, February 1960.
41. F. Rosenblatt, *Principles of Neurodynamics: Perceptrons and the Theory of Brain Mechanisms*, Washington: Spartan Books, 1962.

42. M. E. Hoff, Jr., "Learning Phenomena in Networks of Adaptive Switching Circuits," Ph.D. thesis, Tech. Report 1554-1, Stanford Electronic Labs., Stanford, CA, July 1962.
43. B. Widrow, R. G. Winter, and R. Baxter, "Learning Phenomena in Layered Neural Networks," in *Proceedings of the First IEEE International Conference on Neural Networks*, vol. 2, San Diego, CA, June 1987, pp. 411–29.
44. R. G. Winter, "Madaline Rule II: A New Method for Training Networks of Adalines," Ph.D. thesis, Stanford University, Stanford, CA, January 1989.
45. F. Rosenblatt, "The Perceptron: A Probabilistic Model for Information Storage and Organization in the Brain, *Psychological Review*, vol. 65, 1958, pp. 386–408. Reprinted in 1988, Anderson and Rosenfeld [4], pp. 92–114.
46. H. D. Block, "The Perceptron: A Model for Brain Functioning. I," *Reviews of Modern Physics*, vol. 34, 1962, pp. 123–35. Reprinted in 1988, Anderson and Rosenfeld [4], pp. 138–50.
47. M. L. Minsky and S. A. Papert, *Perceptrons*, Cambridge, MA: M.I.T. Press, 1969. Introduction, pp. 1–20, and p. 73 (Figure 5.1) reprinted in 1988, Anderson and Rosenfeld [4], pp. 161–9.
48. M. L. Minsky and S. A. Papert, *Perceptrons*, expanded ed., Cambridge, MA: M.I.T. Press, 1988.
49. N. K. Bose and P. Liang, *Neural Network Fundamentals with Graphs, Algorithms, and Applications,* New York, McGraw-Hill, 1996.
50. R. P. Lippmann, "An Introduction to Computing with Neural Nets," *IEEE ASSP Magazine*, April 1987, pp. 4–22.
51. R. O. Duda and P. E. Hart, *Pattern Classification and Scene Analysis*, New York: Wiley, 1973.
52. J. A. Feldman and D. H. Ballard, "Connectionist Models and Their Properties," *Cognitive Science*, vol. 6, 1982, pp. 205–54. Reprinted in 1988, Anderson and Rosenfeld [4], pp. 484–507.
53. N. Nilsson, *Learning Machines*, New York: McGraw-Hill, 1965.
54. J. Hertz, A. Krogh, and R. G. Palmer, *Introduction to the Theory of Neural Computation*, Redwood City, CA: Addison-Wesley, 1991.
55. M. A. Arbib, *Brains, Machines, and Mathematics*, 2nd ed., New York: Springer-Verlag, 1987.
56. J. J . Shynk, "Performance Surfaces of a Single-Layer Perceptron," *IEEE Transactions on Neural Networks*, vol. 1, 1990, pp. 268–74.
57. D. E. Rumelhart and J. L. McClelland, eds., *Parallel Distributed Processing*, vol. 1, Cambridge, MA: M.I.T. Press, 1986.
58. S. C. Douglas, "A Family of Normalized LMS Algorithms," *IEEE Signal Processing Letters*, vol. 1, 1994, pp. 49–51.
59. D. O. Hebb, *The Organization of Behavior*, New York, Wiley, 1949. Introduction and chapter 4, "The First Stage of Perception: Growth of the Assembly," pp. xi–xix, 60–78. Reprinted in 1988, Anderson and Rosenfeld [4], pp. 484–507.
60. R. J. MacGregor and E. R. Lewis, *Neural Modeling: Electrical Signal Processing in the Nervous System*, New York: Plenum Press, 1977.
61. G. S. Stent, "A Physiological Mechanism for Hebb's Postulate of Learning," in *Proceedings of the National Academy of Sciences of the U.S.A.*, vol. 70, 1973, pp. 997–1001.
62. J. P. Changeux and A. Danchin, "Selective Stabilization of Developing Synapses as a Mechanism for the Specification of Neural Networks," *Nature* (London), vol. 264, 1976, pp. 705–12.

63. T. H. Brown, E. W. Kairiss, and C. L. Keenan, "Hebbian Synapses: Biophysical Mechanisms and Algorithms," *Annual Review of Neuroscience*, vol. 13, 1990, pp. 475–511.
64. G. Palm, *Neural Assemblies: An Alternative Approach*, New York: Springer-Verlag, 1982.
65. S. Amari, "Mathematical Theory of Neural Learning," *New Generation Computing*, vol. 8, 1991, pp. 281–94.
66. E. Oja, "A Simplified Neuron Model as a Principal Component Analyzer," *Journal of Mathematical Biology*, vol. 15, 1982, pp. 267–73.
67. K. Yale, "Preparing the Right Data Diet for Training Neural Networks," *IEEE Spectrum*, vol. 34, March 1997, pp. 64–6.
68. H. Demuth and M. Beale, *Neural Network Toolbox—For Use with MATLAB*, version 3, User's Guide, The Mathworks, Inc., Natick, MA: 1998.
69. G. A. Carpenter, S. Grossberg, and D. B. Rosen, "Fuzzy ART: Fast Stable Learning and Categorization of Analog Patterns by an Adaptive Resonance System," *Neural Networks*, vol. 4, 1991, pp. 759–71.
70. G. A. Carpenter, S. Grossberg, N. Markuzon, J. H. Reynolds, and D. B. Rosen, "Fuzzy ARTMAP: A Neural Network Architecture for Incremental Supervised Learning of Analog Multidimensional Maps," *IEEE Transactions on Neural Networks*, vol. 3, 1992, pp. 698–713.
71. F. M. Ham, G. Han, and L. V. Fausett, "Fuzzy LAPART: A Neural Architecture for Supervised Learning through Inferencing for Stable Category Recognition," *Journal of Artificial Neural Networks*, vol. 2, 1995, pp. 241–64.
72. F. M. Ham and I. Kostanic, "Partial Least-Squares: Theoretical Issues and Engineering Applications in Signal Processing," *Mathematical Problems in Engineering,* vol. 2, 1995, pp. 63–93.
73. D. M. Haaland and E. V. Thomas, "Partial Least-Squares Methods for Spectral Analysis. 1. Relation to Other Quantitative Calibration Methods and the Extraction of Qualitative Information," *Analytical Chemistry*, vol. 60, 1988, pp. 1193–1202.
74. P. Geladi and B. R. Kowalski, "Partial Least-Squares: A Tutorial," *Analytica Chimica Acta*, vol. 185, 1986, pp. 1–17.
75. K. R. Beebe and B. R. Kowalski, "An Introduction to Multivariate Calibration and Analysis," *Analytical Chemistry*, vol. 59, 1987, pp. 1007A–17A.
76. H. Martens and T. Naes, *Multivariate Calibration*, New York: Wiley, 1989.
77. E. R. Malinowski, *Factor Analysis in Chemistry*, 2nd ed., New York: Wiley, 1991.
78. A. V. Oppenheim and R. W. Schafer, *Discrete-Time Signal Processing*, Englewood Cliffs, NJ: Prentice-Hall, 1989.
79. R. J. Mammone, X. Zhang, and R. P. Ramachandran, "Robust Speaker Recognition: A Feature-Based Approach," *IEEE Signal Processing Magazine*, vol. 13, 1996, pp. 58–71.
80. P. Bhandare, Y. Mendelson, R. Peura, G. Janatsch, J. D. Kruse-Jarres, R. Marbach, and H. M. Heise, "Multivariate Determination of Glucose in Whole Blood Using Partial Least-Squares and Artificial Neural Networks Based on Mid-infrared Spectroscopy," *Applied Spectroscopy*, vol. 47, 1993, pp. 1214–21.
81. C. S. Burrus, R. A. Gopinath, and H. Guo, *Introduction to Wavelets and Wavelet Transforms: A Primer*, Upper Saddle Creek, NJ: Prentice-Hall, 1998.
82. O. Rioul and M. Vetterli, "Wavelets and Signal Processing," *IEEE Signal Processing Magazine*, October 1991, pp. 14–38.
83. G. Strang and T. Nguyen, *Wavelets and Filter Banks*, Wellesley, MA: Wellesley-Cambridge Press, 1996.

84. I. Daubechies, *Ten Lectures on Wavelets*, Philadelphia, PA: Society for Industrial and Applied Mathematics (CBMS-61), 1992.
85. H. H. Szu, X. Y. Yang, B. A. Telfer, and Y. Sheng, "Optical Wedge-Filter on Wavelet Transform Domain for Scale-Invariant Signal Classification," in *Proceedings of the World Congress on Neural Networks*, Portland, OR, vol. 4, 1993, pp. 803–7.
86. H. Szu, Y. Zhang, M. Sun, and C. C. Li, "Neural Network Adaptive Digital Image Screen Halftoning (DISH) Based Wavelet Transform Preprocessing," in *Proceedings of the International Joint Conference on Neural Networks*, Nagoya, Japan, vol. 2, 1993, pp. 1215–18.
87. M. Misiti, Y. Misiti, G. Oppenheim, and J.-M. Poggi, *Wavelet Toolbox—For Use with MATLAB*, version 1, User's Guide, Natick, MA: The Mathworks, 1997.

CHAPTER 3

Mapping Networks

3.1 INTRODUCTION

In this chapter several *mapping* neural network architectures and associated leaning rules are presented. As mentioned in Section 1.4, there are several approaches that can be taken to classify neural networks, and thus the taxonomy of neural networks is not necessarily straightforward. One classification approach was given in Figure 1.10 where selected neural networks were categorized according to how the network learns, that is, supervised learning versus unsupervised learning. In this chapter four types of important mapping networks are presented. This grouping is chosen because it represents networks that are trained in a supervised manner and have many different capabilities, for example, pattern association and classification, function approximation, and estimation, to name a few. The four different types of mapping networks presented are (1) associative memory networks; (2) feedforward multilayer perceptrons trained by the backpropagation algorithm, with some variants of the backpropagation training algorithm; (3) counterpropagation networks; and (4) radial basis function networks. The commonality among these four types of mapping neural networks can be seen by observing Figure 3.1. The mapping function shown in Figure 3.1, that is, $\mathbf{\Omega}(\bullet)$, can be linear or nonlinear; typically it is nonlinear. The mapping $\mathbf{\Omega} : \Re^{n\times 1} \rightarrow \Re^{m\times 1}$, could provide a means of storing and retrieving different patterns, for example, in a character recognition system.

In the case of associative memory networks, the operation is relatively simple. During the learning process, key patterns are presented to the network (that starts off with a "clean slate"), and the memory transforms these patterns to *memorized* (or *stored*) patterns. The synaptic weights of the network are adjusted during the learning process. After training, the recall (or retrieval)

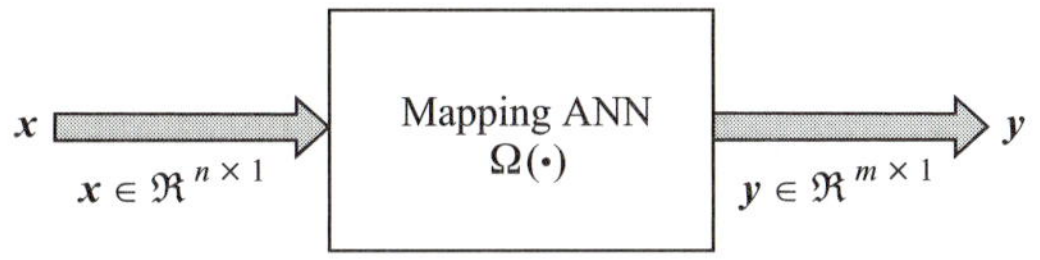

Figure 3.1 General structure for mapping ANNs.

phase involves presenting a stimulus (input) to the network. This input could be incomplete or noisy; however, in spite of the corrupted input the memory network has the capability to properly recall the "associated" correct pattern. Even Aristotle realized association is a prominent feature of human memory. The feedforward multilayer perceptron trained by the backpropagation algorithm is probably the most well-known and often used neural network today. The standard backpropagation algorithm is based on the steepest descent (cf. Sect. A.5.2) approach, and the synaptic weights in the network are updated in proportion to the computed error at the output of the network, that is, the difference between the actual output and the desired output. The result after training is a specific nonlinear mapping from the network input to the output. Variants of the standard backpropagation algorithm are given that can increase the speed of convergence of the network and its performance. Counterpropagation neural networks operate as statistically optimal self-programming lookup tables. These networks provide a bidirectional mapping between the input and output training patterns. A radial basis function network is another powerful supervised-trained network that can be used for pattern classification and function approximation. The basis functions (i.e., the nonlinearites) in the hidden layer of the network produce a significant nonzero response when the input to them falls within a relatively small, localized region of the input space. In many cases, radial basis function networks will train much more quickly than feedforward multilayer perceptrons trained by backpropagation.

3.2 ASSOCIATIVE MEMORY NETWORKS

The ability to store and retrieve information is critical in any type of neural network. This *memory* capability is crucial for the information processing system to remember and deduce information that is stored. The stored information must be properly addressed in the network's memory and output to the outside world. That is, given a *key* input (the stimulus), the appropriate *memorized pattern* is retrieved from the (associative) memory and is output as the appropriate response to the stimulus. This is done to the best of the network's ability to respond to a possibly incomplete (noisy) input key or network crosstalk.

In a neurobiological system, the concept of memory refers to neural alterations that are induced by the interaction of the organic structure with its environment [1]. If this change did not occur, memory would not exist.

Also, if the memory is to be useful, it must be accessible to the nervous system so that learning can take place and information retrieval can be carried out. This is basically the general description given previously for any information processing system, whether artificial or biological. An activity pattern is stored in memory through a learning process; thus, memory and learning are abstrusely related. After a particular activity pattern has been learned, it is stored in memory, after which it can be retrieved (or recalled) at a later time when it is required to obtain that particular information. There are two types of memory: *long-term memory* and *short-time memory*. The distinction made between the two types of memory depends on the retention time.

We will study the memory dynamics of information processing systems that operate by *association*, which is an essential feature of human memory. Most all models of cognition use association in some form, and we refer to any memory system that uses association as an *associative memory* [1–11]. Associative memories are distributed in the sense that many neurons are grouped together, so activity patterns that are to be stored in memory can form a large spatial pattern inside the memory that will contain information about the stimuli. Therefore, the memory performs a *distributed mapping* that transforms an activity pattern in the input space to another activity pattern in the output space. There are, in general, interactions between the individual patterns that are stored in memory. This must be the case; otherwise, the memory would have to be prohibitively large to offer total isolation of individual patterns from one another. Because of the interactions, there is the possibility of having memory errors made during the pattern retrieval process. There are two basic types of associative memories: *autoassociative memory* and *heteroassociative memory*. In the case of an autoassociative memory, a key input vector is associated (or mapped) to itself, whereas in a heteroassociative memory, key input vectors are associated with (or mapped to) arbitrary memorized vectors. In this case, the output space dimension may be different from the input space dimension (for the autoassociative memory, the input and output space dimensions are the same).

3.2.1 General Linear Distributed Associative Memories

In the case of a general linear distributed associative memory, the learning process involves presenting a key input pattern (vector) to the network; then the memory transforms this vector to a stored (or memorized) pattern. The linear neural network architecture, shown in Figure 3.2, can act as an associative memory model, using simple linear combiners as the neurons. This single-layer linear neural network has an input (vector)

$$\boldsymbol{x}_k = \begin{bmatrix} x_{k1}, & x_{k2}, & \cdots, & x_{kn} \end{bmatrix}^T \tag{3.1}$$

called the *key input pattern*, and an output (vector)

$$\boldsymbol{y}_k = \begin{bmatrix} y_{k1}, & y_{k2}, & \cdots, & y_{kn} \end{bmatrix}^T \tag{3.2}$$

called the *memorized pattern*. The elements in both $\boldsymbol{x}_k$ and $\boldsymbol{y}_k$ can be either positive or negative values, which is neurobiologically plausible. For a speci-

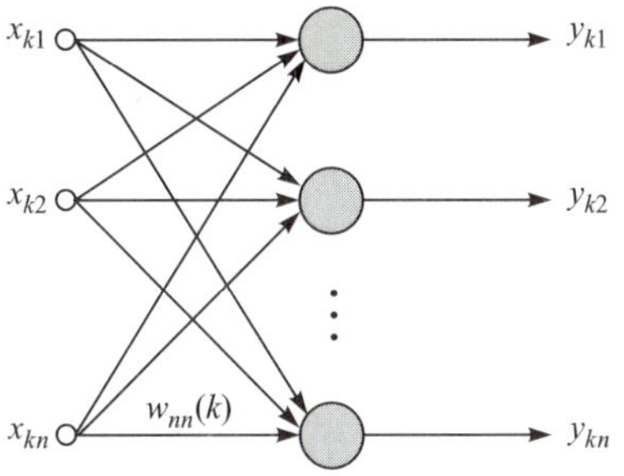

Figure 3.2 Single-layer linear neural network associative memory. It is assumed that the input and output dimensions are the same, that is, n.

fied dimension n, the neural architecture in Figure 3.2 can associate h patterns; however, $h \leq n$ (where n is the maximum storage capacity of the network). Realistically, the working capacity of the network is $h < n$. The linear association mapping between the key vector $\boldsymbol{x}_k$ and the memorized vector $\boldsymbol{y}_k$ can be written in matrix form as

$$\boldsymbol{y}_k = \boldsymbol{W}(k)\boldsymbol{x}_k \tag{3.3}$$

where the weight matrix $\boldsymbol{W}(k) \in \Re^{n \times n}$ is determined by the input/output pairs, that is, $\{\boldsymbol{x}_k, \boldsymbol{y}_k\}$, for $k = 1, 2, \ldots, h$. For each input/output pair, there is a corresponding weight matrix $\boldsymbol{W}(1), \boldsymbol{W}(2), \ldots, \boldsymbol{W}(h)$. From this set of weight matrices we can construct a *memory matrix* $\boldsymbol{M} \in \Re^{n \times n}$, that describes the *sum* of the weight matrices for every input/output pair, or the entire set of pattern associations. This can be written as

$$\boldsymbol{M} = \sum_{k=1}^{h} \boldsymbol{W}(k) \tag{3.4}$$

This memory matrix can then define the overall linkage between any input pattern (key pattern) and the associated output pattern (memorized pattern). Moreover, the memory matrix can be thought of as representing the collective experience that is gained by presenting the h input/output patterns to the network. Equation (3.4) can be written in the form of a recursion given by

$$\boldsymbol{M}_k = \boldsymbol{M}_{k-1} + \boldsymbol{W}(k), \quad \text{for} \quad k = 1, 2, \ldots, h \tag{3.5}$$

where $\boldsymbol{M}_0 = \boldsymbol{0}$. The final result from (3.5) gives the identical result shown in (3.4). As the memory matrix is "built" from each weight matrix increment $\boldsymbol{W}(k)$, produced by the kth $\{\boldsymbol{x}_k, \boldsymbol{y}_k\}$ association, the current weight matrix loses its distinct identity among the blend of other weight matrices. However, information pertaining to the current association may not be completely lost in the synaptic mixture of other associations. Presented in the next section is a method to estimate the memory matrix, given a key pattern and an associated memorized pattern.

3.2.2 Correlation Matrix Memory

An estimate of the memory matrix $\boldsymbol{M}$ shown in (3.4) can be formed from pairs of key and memorized patterns as

$$\hat{\boldsymbol{M}} = \sum_{k=1}^{h} \boldsymbol{y}_k \boldsymbol{x}_k^T \tag{3.6}$$

From (3.6) we see that the basis of the estimate of the memory matrix $\boldsymbol{M}$ is the sum of h *outer product* matrices $\boldsymbol{y}_k \boldsymbol{x}_k^T$, for $k = 1, 2, \ldots, h$. Hence, each outer product matrix is an estimate of the weight matrix $\boldsymbol{W}(k)$ that maps the output pattern $\boldsymbol{y}_k$ onto the input pattern $\boldsymbol{x}_k$. For each input/output pair $\{\boldsymbol{x}_k, \boldsymbol{y}_k\}$, the elements in the estimate of the weight matrix are $y_{ki}x_{kj}$. The associated weight matrix element $w_{ij}(k)$ for the kth association has as an input to the network (x_{kj}) a presynaptic node j, and the ith neuron in the output layer of the network is a postsynaptic node (y_{ki}). The learning rule in (3.6) is actually a localized learning process, and it can be viewed as a form of Hebbian learning (cf. Sect. 2.8.2). Each term in the weight matrix estimates $(y_{ki}x_{kj})$ are co-occurrence terms that are typical in Hebbian learning.

The learning process shown in (3.6) is referred to as the *outer product rule*, because the memory matrix estimate $\hat{\boldsymbol{M}}$ is built by summing h outer product matrices $\boldsymbol{y}_k \boldsymbol{x}_k^T$, $k = 1, 2, \ldots, h$. Therefore, an associative memory designed in this manner is called a *correlation matrix memory* [12]. We can also write the correlation learning process given in (3.6) in the form of a recursion given by

$$\hat{\boldsymbol{M}}_k = \hat{\boldsymbol{M}}_{k-1} + \boldsymbol{y}_k \boldsymbol{x}_k^T, \quad \text{for} \quad k = 1, 2, \ldots, h \tag{3.7}$$

As in the general recursion given in (3.5), for $k = 1$ in (3.7), $\hat{\boldsymbol{M}}_0 = \boldsymbol{0}$. The final value $\hat{\boldsymbol{M}}_k\Big|_{k=h} = \hat{\boldsymbol{M}}_h$ in (3.7) is identical to the result $\hat{\boldsymbol{M}}$ obtained from (3.6).

Given an estimate of the memory matrix $\hat{\boldsymbol{M}}$, it is desired to address and recall the proper *memorized pattern* stored in the associative memory when the *key pattern* is introduced to the network. It is assumed that the estimate of the memory matrix for the associative memory has been built from the learning process in (3.6), given h pattern associations. We will let the key pattern $\boldsymbol{x}_q$ be selected randomly and applied to the network as a stimulus to the memory, and the corresponding response is

$$\boldsymbol{y} = \hat{\boldsymbol{M}}\boldsymbol{x}_q = \sum_{k=1}^{h} \boldsymbol{y}_k \boldsymbol{x}_k^T \boldsymbol{x}_q = \sum_{k=1}^{h} (\boldsymbol{x}_k^T \boldsymbol{x}_q)\boldsymbol{y}_k \tag{3.8}$$

Equation (3.8) can be rewritten as

$$\boldsymbol{y} = (\boldsymbol{x}_q^T \boldsymbol{x}_q)\boldsymbol{y}_q + \sum_{\substack{k=1 \\ k \neq q}}^{h} (\boldsymbol{x}_k^T \boldsymbol{x}_q)\boldsymbol{y}_k \tag{3.9}$$

We assume that each key vector, $\boldsymbol{x}_1, \boldsymbol{x}_2, \ldots, \boldsymbol{x}_h$, has normalized unit length, that is,

$$\boldsymbol{x}_k^T \boldsymbol{x}_k = 1 \quad \text{for} \quad k = 1, 2, \ldots, h \tag{3.10}$$

This can be carried out by dividing each key vector by its Euclidean norm, that is,

$$\boldsymbol{x}_k \leftarrow \frac{\boldsymbol{x}_k}{\|\boldsymbol{x}_k\|_2} \tag{3.11}$$

Then (3.9) can be written as

$$\boldsymbol{y} = \boldsymbol{y}_q + \sum_{\substack{k=1 \\ k\neq q}}^{h} (\boldsymbol{x}_k^T \boldsymbol{x}_q)\boldsymbol{y}_k = \boldsymbol{y}_q + \boldsymbol{z}_q \tag{3.12}$$

where

$$\boldsymbol{z}_q = \sum_{\substack{k=1 \\ k\neq q}}^{h} (\boldsymbol{x}_k^T \boldsymbol{x}_q)\boldsymbol{y}_k \tag{3.13}$$

From (3.12) we see that if the associative memory was able to perfectly reproduce the memorized pattern when the associated key pattern was introduced to the network, $\boldsymbol{y} = \boldsymbol{y}_q$ and $\boldsymbol{z}_q = \boldsymbol{0}$. Therefore, $\boldsymbol{y}_q$ in (3.12) is the desired response (*signal*), and $\boldsymbol{z}_q$ is considered to be *noise*, or *crosstalk*. From (3.13) it can be seen that the crosstalk occurs because of the interaction of the key vector $\boldsymbol{x}_q$, (i.e., the stimulus) with the other key vectors stored in memory. If the individual patterns are statistically independent, from the central limit theorem [13] it can be deduced that the noise vector $\boldsymbol{z}_q$ is a stochastic vector with elements from a Gaussian distribution. This is a classical problem of separating a signal of interest from additive noise [13]. The level of the "noise" will dictate the degree of reconstruction fidelity for the memorized patterns stored in the associative memory.

From (3.13) it is obvious that if the various key vectors formed an orthogonal set (actually an orthonormal set because we assume each key vector is normalized to unit length), the crosstalk would be zero and the memorized pattern would be perfectly reproduced. Given a set of key vectors that are linearly independent but not necessarily orthonormal, the Gram–Schmidt orthogonalization [14] procedure could be performed on the key vectors prior to development of the memory matrix $\hat{\boldsymbol{M}}$. That is, given a set of key vectors $\{\boldsymbol{x}_1, \boldsymbol{x}_2, \ldots, \boldsymbol{x}_k\}$, it is desired to create a new set of orthonormal vectors $\{\boldsymbol{g}_1, \boldsymbol{g}_2, \ldots, \boldsymbol{g}_k\}$ with a linear one-to-one correspondence to the original set. The Gram–Schmidt orthogonalization is carried out by first letting $\boldsymbol{g}_1 = \boldsymbol{x}_1$, and the remaining $\boldsymbol{g}_k$ vectors for $k = 2, 3, \ldots, h$ can be generated from

$$\boldsymbol{g}_k = \boldsymbol{x}_k - \sum_{i=1}^{k-1} \left(\frac{\boldsymbol{g}_i^T \boldsymbol{x}_k}{\boldsymbol{g}_i^T \boldsymbol{g}_i}\right)\boldsymbol{g}_i \tag{3.14}$$

After the Gram–Schmidt orthogonalization is carried out, the associations are performed on the $\{\boldsymbol{g}_k, \boldsymbol{y}_k\}$ pairs, for $k = 1, 2, \ldots, h$. If the key vectors form a set of orthonormal vectors, the upper limit on the storage capacity of the associative memory is n (the dimension of the network, or the dimension of the input space). In general, the storage capacity of the associative memory is

$\rho(\hat{M}) \leq n$. In other words, the storage limit depends on the rank of the memory matrix.

EXAMPLE 3.1. In this example we want to develop an *autoassociative memory*. The memory is trained with the three key vectors

$$x_1 = \begin{bmatrix} -0.3333 \\ 0.7778 \\ 0.5329 \end{bmatrix} \quad x_2 = \begin{bmatrix} 0.4444 \\ -0.5556 \\ 0.7027 \end{bmatrix} \quad x_3 = \begin{bmatrix} 0.4969 \\ 0.6667 \\ 0.5556 \end{bmatrix} \tag{3.15}$$

Each of these vectors has unit length. As discussed above, if the training vectors were mutually orthogonal, no crosstalk would exist, and each memorized pattern stored in the associative memory would be perfectly reproduced. Therefore, it would be beneficial to compute the angles between these vectors, and thus determine how well the memory recall would be. The angles between the vectors in (3.15) are computed as

$$\theta_{12} = \cos^{-1}\frac{x_1 x_2^T}{\|x_1\|_2 \|x_2\|_2} = 101.9^{\circ} \tag{3.16}$$

$$\theta_{13} = \cos^{-1}\frac{x_1 x_3^T}{\|x_1\|_2 \|x_3\|_2} = 49.5^{\circ} \tag{3.17}$$

$$\theta_{23} = \cos^{-1}\frac{x_2 x_3^T}{\|x_2\|_2 \|x_3\|_2} = 76.1^{\circ} \tag{3.18}$$

From (3.16) through (3.18), we see that these three vectors are far from being mutually orthogonal. The memory matrix of the autoassociative network can be found from (3.6) as

$$\hat{M} = x_1 x_1^T + x_2 x_2^T + x_3 x_3^T = \begin{bmatrix} 0.5555 & -0.1749 & 0.4107 \\ -0.1749 & 1.3582 & 0.3945 \\ 0.4107 & 0.3945 & 1.0865 \end{bmatrix} \tag{3.19}$$

Using the memory matrix in (3.19), we can obtain an estimate of the input key patterns by presenting the key vectors given in (3.15). The results are

$$\hat{x}_1 = \hat{M}x_1 = \begin{bmatrix} -0.1023 \\ 1.3249 \\ 0.7489 \end{bmatrix} \quad \hat{x}_2 = \hat{M}x_2 = \begin{bmatrix} 0.6326 \\ -0.5551 \\ 0.7268 \end{bmatrix}$$
$$\hat{x}_3 = \hat{M}x_3 = \begin{bmatrix} 0.3876 \\ 1.0378 \\ 1.0707 \end{bmatrix} \tag{3.20}$$

Comparing these results to the original key vectors in (3.15), we see that the estimates are not perfect replicas. This was expected because of the nonorthogonality of the vectors, as evidenced from the angles between the vectors given in (3.16) through (3.18), that is, not all 90°. However, if we compute the Euclidean distances of each of the response (output) vectors from the normalized key inputs in (3.15), we can obtain a quantitative measure of how "close" the response is to the original key vectors.

Let's first determine the Euclidean distance of the response vector $\hat{x}_1$ from each of the key vectors. The results are given by

$$\delta_{11} = \|x_1 - \hat{x}_1\|_2 = 0.6319 \qquad \delta_{21} = \|x_2 - \hat{x}_1\|_2 = 1.9589$$
$$\delta_{31} = \|x_3 - \hat{x}_1\|_2 = 0.9108 \tag{3.21}$$

From (3.21), clearly the response vector $\hat{x}_1$ is much closer to x_1 than either x_2 or x_3 because δ_{11} is minimal. Carrying out this same exercise with the response vector $\hat{x}_2$, we obtain

$$\delta_{12} = \|x_1 - \hat{x}_2\|_2 = 1.6575 \qquad \delta_{22} = \|x_2 - \hat{x}_2\|_2 = 0.1898$$
$$\delta_{32} = \|x_3 - \hat{x}_2\|_2 = 1.2412 \tag{3.22}$$

As expected, we see that the response vector $\hat{x}_2$ is closest to the key vector x_2. For the response vector $\hat{x}_3$, the Euclidean distances are computed as

$$\delta_{13} = \|x_1 - \hat{x}_3\|_2 = 0.9363 \qquad \delta_{23} = \|x_2 - \hat{x}_3\|_2 = 1.6363$$
$$\delta_{33} = \|x_3 - \hat{x}_3\|_2 = 0.6442 \tag{3.23}$$

and as expected, the response vector $\hat{x}_3$ is closest to the key vector x_3.

Let's try a different set of (unit length) key vectors, given by

$$x_1 = \begin{bmatrix} 0.1309 \\ -0.9779 \\ -0.1629 \end{bmatrix} \qquad x_2 = \begin{bmatrix} -0.7548 \\ 0.0587 \\ -0.6533 \end{bmatrix} \qquad x_3 = \begin{bmatrix} -0.6354 \\ -0.2370 \\ 0.7349 \end{bmatrix} \tag{3.24}$$

The angles between these vectors are now given by

$$\theta_{12} = \cos^{-1} \frac{x_1 x_2^T}{\|x_1\|_2 \|x_2\|_2} = 92.9^\circ \tag{3.25}$$

$$\theta_{13} = \cos^{-1} \frac{x_1 x_3^T}{\|x_1\|_2 \|x_3\|_2} = 88.3^\circ \tag{3.26}$$

$$\theta_{23} = \cos^{-1} \frac{x_2 x_3^T}{\|x_2\|_2 \|x_3\|_2} = 90.8^\circ \tag{3.27}$$

From (3.25) through (3.27), we would expect the autoassociative recall process to be better than before, because the unit-length vectors in (3.24) are "closer" to being mutually orthogonal than those in (3.15). This is indicated in Figure 3.3. The memory matrix is now given by

$$\hat{M} = x_1 x_1^T + x_2 x_2^T + x_3 x_3^T = \begin{bmatrix} 0.9906 & -0.0217 & 0.0048 \\ -0.0217 & 1.0159 & -0.0532 \\ 0.0048 & -0.0532 & 0.9934 \end{bmatrix} \tag{3.28}$$

Using (3.28), the responses to the key vectors in (3.24) are given by

$$\hat{x}_1 = \hat{M} x_1 = \begin{bmatrix} 0.1501 \\ -0.9876 \\ -0.1092 \end{bmatrix} \qquad \hat{x}_2 = \hat{M} x_2 = \begin{bmatrix} -0.7521 \\ 0.1108 \\ -0.6558 \end{bmatrix}$$
$$\hat{x}_3 = \hat{M} x_3 = \begin{bmatrix} -0.6207 \\ -0.2661 \\ 0.7396 \end{bmatrix} \tag{3.29}$$

The Euclidean distance of the response vector $\hat{x}_1$ from each of the key vectors is given by

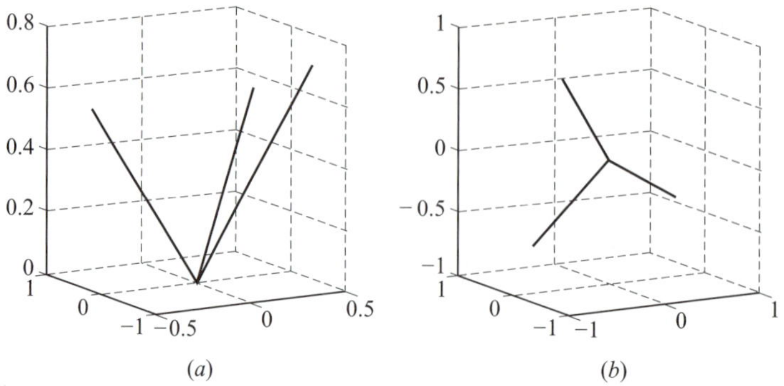

Figure 3.3 (*a*) Key vectors shown in (3.15), with the angles between the vectors given in (3.16) through (3.18). (*b*) Key vectors shown in (3.24), with the angles between the vectors given in (3.25) through (3.27).

$$\begin{aligned} \delta_{11} &= \|\boldsymbol{x}_1 - \hat{\boldsymbol{x}}_1\|_2 = 0.0579 \qquad \delta_{21} = \|\boldsymbol{x}_2 - \hat{\boldsymbol{x}}_1\|_2 = 1.4865 \\ \delta_{31} &= \|\boldsymbol{x}_3 - \hat{\boldsymbol{x}}_1\|_2 = 1.3758 \end{aligned} \tag{3.30}$$

From (3.30), the response vector $\hat{\boldsymbol{x}}_1$ is much closer to $\boldsymbol{x}_1$ than either $\boldsymbol{x}_2$ or $\boldsymbol{x}_3$. Comparing these results to the equivalent results in (3.21) for the previous set of key vectors, we see that the second set of key vectors yields better results because they are more orthogonal. The results for response vector $\hat{\boldsymbol{x}}_2$ are given by

$$\begin{aligned} \delta_{12} &= \|\boldsymbol{x}_1 - \hat{\boldsymbol{x}}_2\|_2 = 1.4859 \qquad \delta_{22} = \|\boldsymbol{x}_2 - \hat{\boldsymbol{x}}_2\|_2 = 0.0522 \\ \delta_{32} &= \|\boldsymbol{x}_3 - \hat{\boldsymbol{x}}_2\|_2 = 1.4382 \end{aligned} \tag{3.31}$$

and as expected, $\hat{\boldsymbol{x}}_2$ is closest to $\boldsymbol{x}_2$. Finally, for the response vector $\hat{\boldsymbol{x}}_3$, the results are

$$\begin{aligned} \delta_{13} &= \|\boldsymbol{x}_1 - \hat{\boldsymbol{x}}_3\|_2 = 1.3734 \qquad \delta_{23} = \|\boldsymbol{x}_2 - \hat{\boldsymbol{x}}_3\|_2 = 1.4365 \\ \delta_{33} &= \|\boldsymbol{x}_3 - \hat{\boldsymbol{x}}_3\|_2 = 0.0329 \end{aligned} \tag{3.32}$$

and $\hat{\boldsymbol{x}}_3$ is closest to $\boldsymbol{x}_3$.

3.2.3 An Error Correction Approach for Correlation Matrix Memories

Even though a correlation matrix memory is a fairly simple design, a major drawback to this approach for an associative memory is the relatively large number of errors that can occur during recall. In its simplest form, a correlation matrix memory has no provision for correcting errors that may occur during the recall of a memorized pattern, given a key input stimulus. The underlying reason for the network's inability to correct for errors is the lack of *feedback* from the output to the input. For a memory matrix $\hat{\boldsymbol{M}}$ that has been built from the associations $\boldsymbol{x}_k \rightarrow \boldsymbol{y}_k$ for $k = 1, 2, \ldots, h$, according to (3.6), the actual output response $\boldsymbol{y}$ that is produced from the stimulus $\boldsymbol{x}_q$

may not be "close" enough (in a Euclidean sense) to the desired (or true) response $\boldsymbol{y}_q$ for the memory to perfectly associate.

We want to incorporate an error correction mechanism into the recursion formula to force the design of the memory to be such that it can associate perfectly [15, 16]. Therefore, the main objective here is to have the associative memory reconstruct the memorized patterns in an *optimal* sense, thus improving the quality of the memory responses $\boldsymbol{y}_k$. We will let $\hat{\boldsymbol{M}}(\tau)$ be the memory matrix learned up through iteration τ. A randomly selected key vector pattern $\boldsymbol{x}_k$ is applied to the memory at iteration τ, and the resulting actual response $\hat{\boldsymbol{M}}(\tau)\boldsymbol{x}_k$ is used to form the *error vector*

$$\boldsymbol{e}_k(\tau) = \boldsymbol{y}_k - \hat{\boldsymbol{M}}(\tau)\boldsymbol{x}_k \tag{3.33}$$

where $\boldsymbol{y}_k$ is the desired activity pattern associated with the key input pattern $\boldsymbol{x}_k$. It is desired to use this error term in some way to compute an adjustment to the memory matrix at iteration τ as it is being built, to reduce reconstruction errors. We will take a *steepest descent* (cf. Sect. A.5.2) approach for the development of a discrete-time learning rule which is a modification of the recursion in (3.6). This discrete-time learning rule based upon steepest descent will have the form

$$\hat{\boldsymbol{M}}(\tau+1) = \hat{\boldsymbol{M}}(\tau) - \mu\nabla_{\hat{\boldsymbol{M}}}\mathscr{E}(\hat{\boldsymbol{M}}) \tag{3.34}$$

where $\mathscr{E}(\hat{\boldsymbol{M}})$ is an energy (or Lyapunov) function, given as

$$\mathscr{E}(\hat{\boldsymbol{M}}) = \frac{1}{2}\|\boldsymbol{e}_k\|_2^2 \tag{3.35}$$

and $\boldsymbol{e}_k = \boldsymbol{y}_k - \hat{\boldsymbol{M}}\boldsymbol{x}_k$ from (3.33), and $\mu > 0$ in (3.34) is the learning rate parameter. Computing the gradient of (3.35) with respect to the memory matrix $\hat{\boldsymbol{M}}$ gives

$$\nabla_{\hat{\boldsymbol{M}}}\mathscr{E}(\hat{\boldsymbol{M}}) = -\boldsymbol{y}_k\boldsymbol{x}_k^T + \hat{\boldsymbol{M}}\boldsymbol{x}_k\boldsymbol{x}_k^T \tag{3.36}$$

using the results from Section A.3.4.2 (i.e., differentiation of a scalar with respect to a matrix). Substituting the result from (3.36) into (3.34) gives

$$\hat{\boldsymbol{M}}(\tau+1) = \hat{\boldsymbol{M}}(\tau) + \mu[\boldsymbol{y}_k - \hat{\boldsymbol{M}}(\tau)\boldsymbol{x}_k]\boldsymbol{x}_k^T \tag{3.37}$$

Note that inside the square brackets in (3.37) is the error vector $\boldsymbol{e}_k$ defined in (3.33). Therefore, this learning strategy has an error correction mechanism built into the algorithm to correct for errors as the memory matrix $\hat{\boldsymbol{M}}$ is being learned. Rewriting (3.37) as

$$\hat{\boldsymbol{M}}(\tau+1) = \hat{\boldsymbol{M}}(\tau) + \mu\boldsymbol{y}_k\boldsymbol{x}_k^T - \mu\hat{\boldsymbol{M}}(\tau)\boldsymbol{x}_k\boldsymbol{x}_k^T \tag{3.38}$$

and comparing this result to the original recursion in (3.7) show the additional term $\hat{\boldsymbol{M}}(\tau)\boldsymbol{x}_k\boldsymbol{x}_k^T$ that is responsible for allowing corrections to be made to the memory matrix as it is being built. The supervised learning algorithm in (3.37), based on error correction, is repeated for each of the h associations

$$\boldsymbol{x}_k \rightarrow \boldsymbol{y}_k, \quad \text{for} \quad k = 1, 2, \ldots, h \tag{3.39}$$

selected at random.

Care must be taken when one is selecting the learning rate parameter μ to ensure stability of the feedback during learning. A fixed learning rate parameter can be used, or an adjustable one with respect to time can also be used (cf. Sect. 2.5.1). For each association, the iterative adjustments to the memory matrix in (3.37) continue until the error vector $\boldsymbol{e}_k(\tau)$ in (3.33) becomes negligibly small. In other words, learning for the kth association $\boldsymbol{x}_k \to \boldsymbol{y}_k$ can be stopped when the actual response $\hat{\boldsymbol{M}}(\tau)\boldsymbol{x}_k$ is "close" to the desired response $\boldsymbol{y}_k$, that is, $\boldsymbol{y}_k - \hat{\boldsymbol{M}}(\tau)\boldsymbol{x}_k \cong \boldsymbol{0}$. This results in minimizing the performance criterion in (3.35), and allowing the associative memory to reconstruct the memorized patterns in an *optimal* sense. To initialize the algorithm for learning the memory matrix $\hat{\boldsymbol{M}}$ from the input/output pairs $\{\boldsymbol{x}_k, \boldsymbol{y}_k\}$, using the error correction mechanism in (3.37), we set $\hat{\boldsymbol{M}}(0) = \boldsymbol{0}$. Comparing these results to the least mean-square (LMS) algorithm in Section 2.5.1, we see that (3.37) is in the form of the LMS algorithm, or delta rule.

3.3 BACKPROPAGATION LEARNING ALGORITHMS

We now consider *supervised* learning in a feedforward multilayer perceptron (MLP). Specifically, we want to study the backpropagation algorithm, or the generalized delta rule, for training MLPs. Backpropagation is the most widely used learning process in neural networks today, and it was first developed by Werbos in 1974 [17]; however, this work remained unknown for many years [18, 19]. This method has been rediscovered several times, in 1982 by Parker [20] (also see [21, 22]), in 1985 by LeCun [23], and by Rumelhart et al. in 1986 [24, 25]. The presentation of backpropagation by Rumelhart et al. is probably responsible for the popularization of the algorithm in the areas of science and engineering. Training MLPs with backpropagation algorithms results in a nonlinear mapping or an association task. Thus, given two sets of data, that is, input/output pairs, the MLP can have its synaptic weights adjusted by the backpropagation algorithm to develop a specific nonlinear mapping (cf. Sect. 2.7). The MLP, with fixed weights after the training process, can provide an association task for classification, pattern recognition, diagnosis, etc. During the training phase of the MLP, the synaptic weights are adjusted to minimize the disparity between the actual and desired outputs of the MLP, averaged over all input patterns (or learning examples).

3.3.1 Basic Backpropagation Algorithm for the Feedforward Multilayer Perceptron

In this section we present a derivation of the standard backpropagation learning algorithm. For the sake of simplicity we will derive the learning rule for a multilayer perceptron neural network (MLP NN) having three layers of weights, namely, one output layer and two hidden layers, since it is the most frequently used MLP NN architecture. An example of this type of neural network is shown in Figure 3.4. The extension of the derivation to the general case when the network has more than two hidden layers is straightforward.

The standard backpropagation algorithm for training of the MLP NN is based on the steepest descent gradient approach applied to the minimization of an energy function representing the instantaneous error.

In other words, we desire to minimize a function defined as

$$E_q = \frac{1}{2}\left(\boldsymbol{d}_q - \boldsymbol{x}_{\text{out}}^{(3)}\right)^T \left(\boldsymbol{d}_q - \boldsymbol{x}_{\text{out}}^{(3)}\right) = \frac{1}{2}\sum_{h=1}^{n_3}\left(d_{qh} - x_{\text{out},h}^{(3)}\right)^2 \tag{3.40}$$

where $\boldsymbol{d}_q$ represents the desired network output for the qth input pattern and $\boldsymbol{x}_{\text{out}}^{(3)} = \boldsymbol{y}_q$ is the actual output of the MLP network shown in Figure 3.4. Very often the method for the weight updates derived from minimizing (3.40) is called the *online* method, emphasizing the fact that it has minimum memory storage requirements.

Using the steepest-descent gradient approach, the learning rule for a network weight in any one of the network layers is given by

$$\Delta w_{ji}^{(s)} = -\mu^{(s)} \frac{\partial E_q}{\partial w_{ji}^{(s)}} \tag{3.41}$$

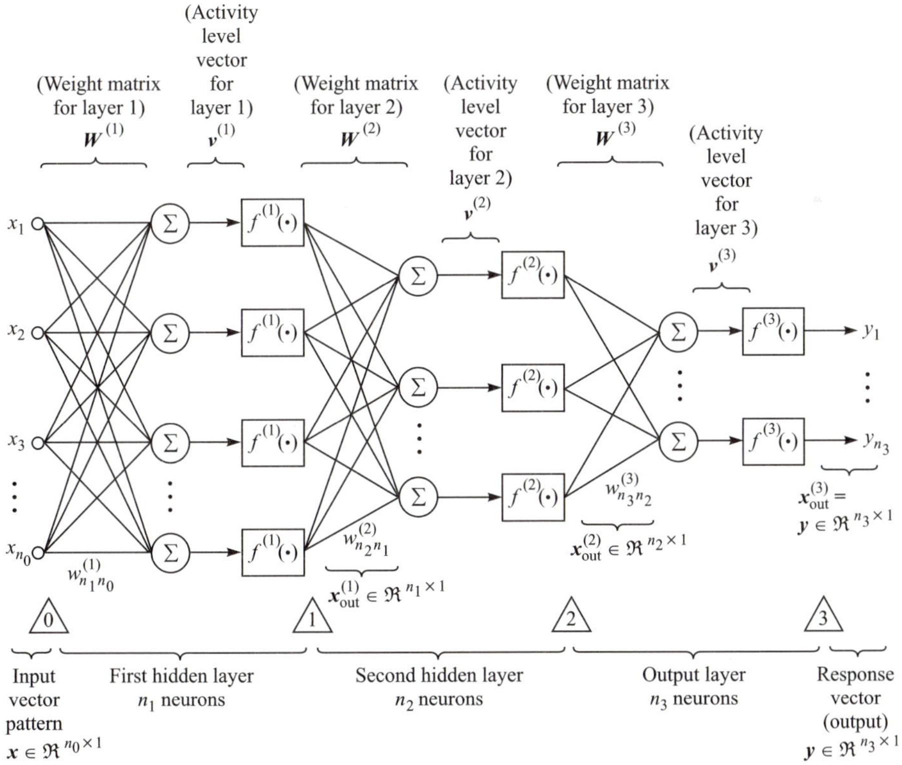

Figure 3.4 An example of a three-layer feedforward MLP NN architecture.

where $s = 1, 2, 3$ designates the appropriate network layer and $\mu^{(s)} > 0$ is the corresponding learning rate parameter. For reasons which will become apparent shortly, we will derive separate learning rules for weights in the output and in the hidden layers of the MLP NN. Let us first consider the output layer of the network. The weights in the output layer can be updated according to

$$\Delta w_{ji}^{(3)} = -\mu^{(3)} \frac{\partial E_q}{\partial w_{ji}^{(3)}} \tag{3.42}$$

Using the chain rule for the partial derivatives, (3.42) can be rewritten as

$$\Delta w_{ji}^{(3)} = -\mu^{(3)} \frac{\partial E_q}{\partial v_j^{(3)}} \frac{\partial v_j^{(3)}}{\partial w_{ji}^{(3)}} \tag{3.43}$$

Separated terms in (3.43) can be evaluated as

$$\frac{\partial v_j^{(3)}}{\partial w_{ji}^{(3)}} = \frac{\partial}{\partial w_{ji}^{(3)}} \left(\sum_{h=1}^{n_2} w_{jh}^{(3)} x_{\text{out},h}^{(2)} \right) = x_{\text{out},i}^{(2)} \tag{3.44}$$

and

$$\frac{\partial E_q}{\partial v_j^{(3)}} = \frac{\partial}{\partial v_j^{(3)}} \left\{ \frac{1}{2} \sum_{h=1}^{n_3} \left[d_{qh} - f\left(v_h^{(3)}\right) \right]^2 \right\} = -\left[d_{qj} - f\left(v_j^{(3)}\right) \right] g\left(v_j^{(3)}\right) \tag{3.45}$$

or

$$\frac{\partial E_q}{\partial v_j^{(3)}} = -\left(d_{qj} - x_{\text{out},j}^{(3)} \right) g\left(v_j^{(3)}\right) \hat{=} -\delta_j^{(3)} \tag{3.46}$$

where $g(\bullet)$ represents the first derivative of the nonlinear activation function $f(\bullet)$. The term defined in (3.46) is commonly referred to as *local error*, or *delta*.

Combining (3.43), (3.44), and (3.46) we can write the learning rule equation for the weights in the output layer of the network as

$$\Delta w_{ji}^{(3)} = \mu^{(3)} \delta_j^{(3)} x_{\text{out},i}^{(2)} \tag{3.47}$$

or

$$w_{ji}^{(3)}(k+1) = w_{ji}^{(3)}(k) + \mu^{(3)} \delta_j^{(3)} x_{\text{out},i}^{(2)} \tag{3.48}$$

The update equations for the weights in the hidden layers of the network can be derived in essentially the same way. Applying the steepest descent gradient approach, we have

$$\Delta w_{ji}^{(2)} = -\mu^{(2)} \frac{\partial E_q}{\partial w_{ji}^{(2)}} = -\mu^{(2)} \frac{\partial E_q}{\partial v_j^{(2)}} \frac{\partial v_j^{(2)}}{\partial w_{ji}^{(2)}} \tag{3.49}$$

The second partial derivative on the right-hand side in (3.49) can be evaluated as

$$\frac{\partial v_j^{(2)}}{\partial w_{ji}^{(2)}} = \frac{\partial}{\partial w_{ji}^{(2)}} \left(\sum_{h=1}^{n_1} w_{jh}^{(2)} x_{\text{out},h}^{(1)} \right) = x_{\text{out},i}^{(1)} \tag{3.50}$$

Evaluation of the first partial derivative in (3.49) is more complex since the change in $v_j^{(2)}$ propagates through the output layer of the network and affects all the network outputs. We can pursue the derivation by expressing this quantity as a function of quantities that are already known and of other terms which are easily evaluated. To proceed, we can write

$$\frac{\partial E_q}{\partial v_j^{(2)}} = \frac{\partial}{\partial x_{\text{out},j}^{(2)}} \left\{ \frac{1}{2} \sum_{h=1}^{n_3} \left[d_{qh} - f\left(\sum_{p=1}^{n_2} w_{hp}^{(3)} x_{\text{out},p}^{(2)} \right) \right]^2 \right\} \frac{\partial x_{\text{out},j}^{(2)}}{\partial v_j^{(2)}} \tag{3.51}$$

or

$$\begin{aligned} \frac{\partial E_q}{\partial v_j^{(2)}} &= -\left[\sum_{h=1}^{n_3} \left(d_{qh} - x_{\text{out},h}^{(3)} \right) g\left(v_h^{(3)} \right) w_{hj}^{(3)} \right] g\left(v_j^{(2)} \right) \\ &= -\left(\sum_{h=1}^{n_3} \delta_h^{(3)} w_{hj}^{(3)} \right) g\left(v_j^{(2)} \right) \triangleq -\delta_j^{(2)} \end{aligned} \tag{3.52}$$

Combining equations (3.49), (3.50), and (3.52) yields

$$\Delta w_{ji}^{(2)} = \mu^{(2)} \delta_j^{(2)} x_{\text{out},i}^{(2)} \tag{3.53}$$

or

$$w_{ji}^{(2)}(k+1) = w_{ji}^{(2)}(k) + \mu^{(2)} \delta_j^{(2)} x_{\text{out},i}^{(2)} \tag{3.54}$$

Comparing (3.48) and (3.54), we see that the update equations for the weights in the output layer and the hidden layer have the same form. The only difference lies in how we compute the local error. For the output layer, the local error is proportional to the difference between the desired output and the actual network output. By extending the same concept to the "outputs" of the hidden layers, the local error for a neuron in a hidden layer can be viewed as being proportional to the difference between the desired output and actual output of the particular neuron. Of course, during the training process, the desired outputs of the neurons in the hidden layer are not known, and therefore the local errors need to be recursively estimated in terms of the error signals of all connected neurons. Equation (3.54) can be generalized for the MLP NN having an arbitrary number of hidden layers. For such a network we can write

$$w_{ji}^{(s)}(k+1) = w_{ji}^{(s)}(k) + \mu^{(s)} \delta_j^{(s)} x_{\text{out},i}^{(s)} \tag{3.55}$$

where

$$\delta_j^{(s)} = \left(d_{qh} - x_{\text{out},j}^{(s)} \right) g\left(v_j^{(s)} \right) \tag{3.56}$$

for the output layer, and

$$\delta_j^{(s)} = \left(\sum_{h=1}^{n_{s+1}} \delta_h^{(s+1)} w_{hj}^{(s+1)} \right) g\left(v_j^{(s)}\right) \tag{3.57}$$

for the hidden layers.

Summary of the standard backpropagation algorithm

Training the MLP NN by using the standard backpropagation algorithm can be performed according to the following algorithm.

Standard backpropagation algorithm

Step 1. Initialize the network synaptic weights to small random values.

Step 2. From the set of training input/output pairs, present an input pattern and calculate the network response.

Step 3. The desired network response is compared with the actual output of the network, and by using (3.56) and (3.57) all the local errors can be computed.

Step 4. The weights of the network are updated according to (3.55).

Step 5. Until the network reaches a predetermined level of accuracy in producing the adequate response for all the training patterns, continue steps 2 through 4.

From the above algorithm we see that the classical backpropagation can be interpreted as performing two independent tasks. The first is backpropagation of the errors from the nodes in the output layer to the nodes in the hidden layers, and the second is using the LMS algorithm to update the weights in every layer.

3.3.2 Some Practical Issues in Using Standard Backpropagation

Standard backpropagation and its derivatives are by far the most widely used learning algorithm for training of the MLP NN. In this section we address some of the practical problems involved in its effective application.

Initialization of synaptic weights

The weights of the MLP NN are initially set to small random values. They have to be sufficiently small so that network training does not start from a point in the error space that corresponds to some of the nodes being saturated. When the network operates in saturation, it may take a lot of iterations for the learning to converge. One commonly used heuristic algorithm for weight initialization is to set the weights as uniformly distributed random numbers in the interval from $-0.5/fan\_in$ to $0.5/fan\_in$, where *fan_in* represents the total number of the neurons in the layer that the weights are fed into [26]. For the case of MLP NN with one hidden layer, an alternate approach was suggested by Nguyen and Widrow [27]. The authors demonstrate that this approach can significantly improve the speed of the network training. Nguyen and Widrow's initialization of MLP NNs can be summarized in the following algorithm.

Define:

n_0 = number of components in input layer
n_1 = number of neurons in hidden layer
γ = scaling factor

Step 1. Compute the scaling factor according to

$$\gamma = 0.7 \sqrt[n_0]{n_1} \tag{3.58}$$

Step 2. Initialize the weights w_{ij} of a layer as random numbers between −0.5 and 0.5.
Step 3. Reinitialize the weights according to

$$w_{ij} = \gamma \frac{w_{ij}}{\sqrt{\sum_{i=1}^{n_1} w_{ij}^2}} \tag{3.59}$$

Step 4. For the ith neuron in the hidden layer, set the bias to be a random number between $-w_{ij}$ and w_{ij}.

Network configuration and ability of the network to generalize

The configuration of the MLP NN is determined by the number of hidden layers, number of the neurons in each of the hidden layers, as well as the type of the activation functions used for the neurons. While it has been proved that the performance of the network does not depend much on the type of the activation function (as long as it is nonlinear), the choice of the number of hidden layers and the number of units in each of the hidden layers is critical.

Hornik et al. [28] established that an MLP NN that has only one hidden layer, with a sufficient number of neurons, acts as a universal approximator of nonlinear mappings. In practice, it is very difficult to determine a *sufficient number* of neurons necessary to achieve the desired degree of approximation accuracy. Frequently, the number of units in the hidden layer is determined by trial and error. Furthermore, if the network has only one hidden layer, the neurons seem to "interact" with one another [29]. In such a situation, it is difficult to improve the approximation for one point in the mapping without degrading it at some other point. For the above reason, MLP NNs are commonly designed with two hidden layers.

Typically, to solve a real-world problem using MLP NNs, we need to train a relatively large neural network architecture. Having a large number of units in the hidden layers guarantees good network performance when it is presented with input patterns that belong to the training set. However, an "overdesigned" architecture will tend to "overfit" the training data [30–32], which results in the loss of the generalization property of the network. To clarify this point, consider the following example.

EXAMPLE 3.2. An MLP NN is to be trained to approximate the nonlinear function

$$y = e^{-x} \sin(3x) \tag{3.60}$$

in the interval [0, 4]. This is a fairly simple problem for a neural network with one hidden layer consisting of 50 neurons. For this example, the interval [0, 4] was

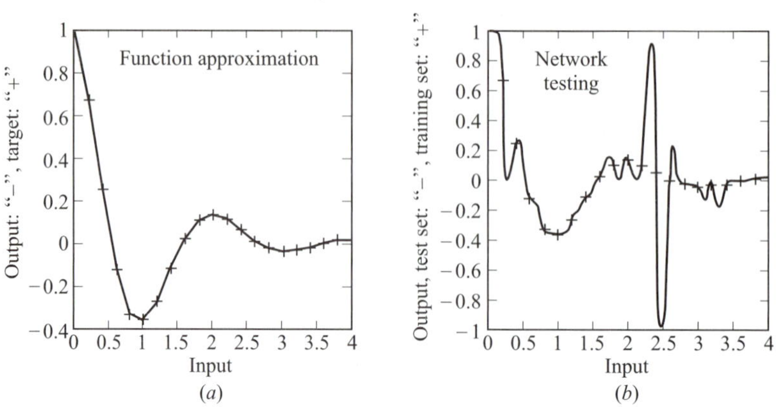

Figure 3.5 Illustration of data overfitting. (*a*) Response of the network to the training inputs. (*b*) Response of the network to the test data set.

sampled with 21 points separated by 0.2. MATLAB routine `trainlm` was used to perform the network training with hyperbolic tangent nonlinearities and a target mean square error of 0.01 over the entire data set. The network converged in only 5 epochs, and Figure 3.5(*a*) illustrates the agreement between the desired and actual network outputs for the training data set. To test the network's generalization capability, the same interval was sampled with 401 points separated by 0.01, and the network response is presented in Figure 3.5(*b*). As seen in Figure 3.5(*b*), the response of the network does not show good agreement with the function we are trying to approximate. The reason for this is *overfitting* of the training data. In this case, the network with a considerably smaller number of neurons would perform the approximation task in a better way.

Independent validation

It is never a good idea to assess the generalization properties of a neural network based on the training data alone. Using the training data to assess the final performance quality of the network can lead to overfitting. This can be avoided by using a standard method in statistics called *independent validation* (cf. Sect. 9.4). The method involves dividing the available data into a training set and a test set. The entire data set is usually randomized first. The training data are next split into two partitions; the first partition is used to update the weights in the network, and the second partition is used to assess (or validate) the training performance (i.e., it is used to decide when to *stop training*). The test data are then used to assess how well the network has generalized.

Speed of convergence

As seen from the derivation of the standard backpropagation algorithm, it is a generalization of the LMS algorithm presented in Section 2.5.1. Conversely, the LMS algorithm for training a single-layer perceptron can be seen as a special case of the standard backpropagation algorithm. In Section 2.5.1 it was demonstrated that the convergence properties of the LMS algorithm (in particular its speed and stability) depend critically on the magnitude of the learning rate parameter. To guarantee network convergence, and avoid the oscillations during the training, the learning rate parameter must be set to a relatively small value. This clearly affects the speed of the algorithm since a small value of the learning rate parameter restricts the change in the weights of the network. Furthermore, if the starting point of the network training is far from the global minimum, some of the neurons can be operating in saturation. When that happens, the derivative of the activation function is small. Since the magnitude of the weight changes depends directly on the magnitude of the activation function derivative, the network can get stuck on a flat plateau of the error surface, and it might require many iterations before convergence is achieved. It is not uncommon for even moderately complex real-world problems to require hours, even days, of network training.

Slow convergence of the backpropagation algorithm has encouraged research in alternate (faster) algorithms for MLP NN training. Research on faster algorithms can be roughly divided into two main categories. The first category consists of various heuristic improvements to the standard backpropagation algorithm. Although useful, and in many cases easily understandable, the heuristic algorithms are very much case-by-case specific, and their performance characteristics cannot be easily established. The second category involves use of standard numerical optimization techniques. Most of the algorithms in this category give significant improvement in the network convergence speed at the expense of increased network computational complexity. Some representative and popular algorithms from both categories are presented in the following sections.

In addition to modification of backpropagation learning, preprocessing and reduction in the input data can result in improved performance and faster learning. That is, reduction in the network size reduces its complexity, and this significantly improves the convergence speed. Some of the methods for the data preprocessing are covered in Section 2.9.

3.3.3 Backpropagation Learning Algorithm with Momentum Updating

Backpropagation with momentum updating is one of the most popular modifications to the standard algorithm presented in Section 3.3.1. The idea of the algorithm is to update the weights in the direction which is a linear combination of the current gradient of the instantaneous error surface and the one obtained in the previous step of the training. Namely, the weights are updated according to

$$\Delta w_{ji}^{(s)}(k+1) = \mu^{(s)}\delta_j^{(s)}(k)x_{\text{out},i}^{(s)}(k) + \alpha\Delta w_{ji}^{(s)}(k-1) \tag{3.61}$$

or

$$w_{ji}^{(s)}(k+1) = w_{ji}^{(s)}(k) + \mu^{(s)}\left[\delta_j^{(s)}(k)x_{\text{out},i}^{(s)}(k) + \alpha\delta_j^{(s)}(k-1)x_{\text{out},i}^{(s)}(k-1)\right] \tag{3.62}$$

where α is commonly referred to as forgetting factor and is typically chosen in interval (0, 1). The second term in (3.61) is called the momentum term, and it improves the convergence speed of the standard backpropagation algorithm by introducing stabilization in weight changes. Intuitively, according to (3.61), if the weights are to be changed in the same direction as in the previous step, the rate of change is increased. Alternatively, if the change in the current step is not in the same direction as that in the previous step, the rate of change is decreased. This type of learning significantly improves convergence in some very important cases which are handled poorly by the standard backpropagation algorithm. First, if the training patterns contain some element of uncertainty, for example, noise, then updating with momentum provides a sort of low-pass filtering by preventing rapid changes in the direction of the weight updates. Second, this kind of behavior renders the training relatively immune to the presence of outliers or erroneous training pairs. Also, if the network is operating on a flat plateau of the error surface, the presence of momentum will increase the rate of weight change, and the speed of convergence is increased. This can be conveniently illustrated by considering the weight update equation [26]

$$\Delta w_{ji}^{(s)}(k+1) = -\mu^{(s)} \frac{\partial E_q}{\partial w_{ji}^{(s)}} + \alpha \Delta w_{ji}^{(s)}(k-1) \tag{3.63}$$

If the network is operating on a flat area of the error surface, the value of the gradient is not changing substantially from step to step; therefore (3.63) can be approximated as

$$\begin{aligned}\Delta w_{ji}^{(s)}(k+1) &\approx -\mu^{(s)} \frac{\partial E_q}{\partial w_{ji}^{(s)}} - \alpha\mu^{(s)} \frac{\partial E_q}{\partial w_{ji}^{(s)}} - \alpha^2\mu^{(s)} \frac{\partial E_q}{\partial w_{ji}^{(s)}} + \dots \\ &= -\mu^{(s)}(1+\alpha+\alpha^2+\cdots)\frac{\partial E_q}{\partial w_{ji}^{(s)}} \\ &\approx -\frac{\mu^{(s)}}{1-\alpha}\frac{\partial E_q}{\partial w_{ji}^{(s)}}\end{aligned} \tag{3.64}$$

Because the forgetting factor α is always smaller than unity, updating with momentum increases the *effective* learning rate to

$$\mu_{\text{eff}}^{(s)} = \frac{\mu^{(s)}}{1-\alpha} \tag{3.65}$$

3.3.4 Batch Updating

The standard backpropagation algorithm assumes that the weights are updated for every input/output training pair. A batch-updating approach accumulates the weight corrections over several training patterns (possibly one entire epoch) before actually performing the update. The update is commonly formed as an average of the corrections for each individual input/output pair.

The batch-updating approach has the following advantages:

1. Using several (possibly all) training pairs gives a much better estimate of the error surface than the instantaneous one used for standard backpropagation.
2. Through the process of correction averaging, the batch update procedure provides some inherent low-pass filtering of the training data. This can be advantageous in the cases when the training data are corrupted with noise.
3. The batch algorithm is suitable for more sophisticated optimization procedures such as the conjugate gradient method or Newton's method.

The advantages listed above are achieved at the expense of these factors:

1. Batch updating is more demanding from the standpoint of memory requirements. It is obvious that we need extra storage space for the weight corrections before the weights are updated. The memory storage requirement becomes critical in cases of networks with a large number of weights.
2. Averaging of the weight corrections adds extra computational complexity to the algorithm.
3. For the no-batch-training mode (i.e., standard backpropagation), the presence of the noise in the training pairs helps the network training to escape local minima on the error surface. The smoothing effect of the batch updating makes the learning algorithm more prone to converging to a local minimum.

In general, the performance of the batch updating backpropagation algorithm is very *case-dependent*. A good compromise between the batch update, where the averaging is performed over the entire training set, and standard backpropagation is to accumulate the changes over several training pairs before the weights are updated. This allows the learning algorithm to yield a better estimate of the error surface without significantly increasing the possibility of being trapped in a local minimum.

3.3.5 Search-Then-Converge Method

The search-then-converge method is a relatively simple heuristic strategy for speeding up backpropagation learning, proposed by Darken et al. [33, 34] (cf. Sect. 2.5.1). According to this strategy, the backpropagation learning in an MLP NN can be divided into two phases. In the first phase, the network is relatively far from the global minimum. This phase is called the *search phase*. During the search phase the learning rate is kept sufficiently large and relatively constant, so that the network can perform a fast decrease toward the minimum of the performance surface. The second phase is called the *converge phase*, and it commences when the network is approaching the global minimum. In the converge phase the learning rate is decreased at each iteration, allowing the network to perform a fine-tuning of the weights. In practice, it is impossible to tell *how far* the network is from the global minimum, and therefore a strategy for the learning rate decrease must be adopted in advance, that is, before the beginning of network training. Two common learning rate decrease strategies have been suggested [33–35]

$$\mu(k) = \mu_0 \frac{1}{1 + k/k_0} \tag{3.66}$$

and

$$\mu(k) = \mu_0 \frac{1 + (c/\mu_0)(k/k_0)}{1 + (c/\mu_0)(k/k_0) + k_0(k/k_0)^2} \tag{3.67}$$

where $\mu_0 > 0$ represents the initial learning rate parameter and c and k_0 are suitably chosen constants. Typically, $1 < c/\mu_0 < 100$ and $100 < k_0 < 500$. In both (3.66) and (3.67), when $k << k_0$, the learning rate is approximately constant and equal to μ_0. This corresponds to the search phase. When $k >> k_0$, the learning rate decreases in proportion to $1/k$ in (3.66), and to $1/k^2$ in (3.67). It has been demonstrated that for suitably chosen parameters c and k_0, the search-then-convergence strategy can significantly improve the speed of the backpropagation algorithm [35].

3.3.6 Batch Updating with Variable Learning Rate

Batch updating with a variable learning rate represents a simple heuristic strategy to increase the convergence speed of the backpropagation algorithm with a batch update. The idea behind the approach is to increase the magnitude of the learning rate if the learning in the previous step has decreased the total error function. Conversely, if the error function has increased, the learning rate needs to be decreased. The algorithm can be summarized as follows [32]:

1. If the error function over the entire training set has decreased, increase the learning rate by multiplying it by a number $\eta > 1$ (typically $\eta = 1.05$).
2. If the error function has increased more than some set percentage ξ (typically a few percent), decrease the learning rate by multiplying it by a number $\chi < 1$ (typically $\chi = 0.7$).
3. If the error function is increased less than the percentage ξ, the learning rate remains unchanged.

Applying the variable learning rate to batch updating can significantly speed up the convergence in the cases of smooth and slowly decreasing error functions. However, the algorithm can easily be trapped in a local minimum of the error surface. To avoid this, the learning rate is not allowed to fall below a certain value μ_{min}.

3.3.7 Vector-Matrix Form of the Backpropagation Algorithm

Here we present the vector-matrix form of the backpropagation algorithm. From the practical implementation standpoint, the vector-matrix form of the algorithm has limited application since most of the contemporary hardware and software is not capable of parallel processing. However, the matrix-vector form gives considerable insight into the backpropagation and, more importantly, allows a more straightforward application of some of the advanced numerical optimization techniques for speeding up the convergence of learning.

Referring to Figure 3.4, we can specify the energy function as

$$E_q = \frac{1}{2}\left(\boldsymbol{d}_q - \boldsymbol{x}_{\text{out},q}^{(3)}\right)^T\left(\boldsymbol{d}_q - \boldsymbol{x}_{\text{out},q}^{(3)}\right) \tag{3.68}$$

where $\boldsymbol{d}_q$ represents the desired network output for the qth input pattern and $\boldsymbol{x}_{\text{out}}^{(3)} = \boldsymbol{y}_q$ is the actual output of the MLP network shown in Figure 3.4. Therefore (3.68) represents an instantaneous estimate of the error surface. Using the steepest descent approach, the synaptic weight update equations can be written as

$$w_{ij}^{(s)}(k+1) = w_{ij}^{(s)}(k) - \alpha^{(s)}\frac{\partial E_q}{\partial w_{ij}^{(s)}(k)} \tag{3.69}$$

where $s = 1, 2, 3$ indicates the number of the network layer and $\alpha^{(s)}$ is the learning rate parameter associated with the particular layer. Applying the chain rule to the partial derivative of the energy in the (3.68) function, we can write

$$\frac{\partial E_q}{\partial w_{ij}^{(s)}} = \frac{\partial E_q}{\partial v_i^{(s)}} \cdot \frac{\partial v_i^{(s)}}{\partial w_{ij}^{(s)}} \tag{3.70}$$

where the discrete-time index k is omitted for the sake of simplicity.

The second term on the right-hand side of (3.70) can be evaluated as

$$\frac{\partial v_i^{(s)}}{\partial w_{ij}^{(s)}} = \frac{\partial}{\partial w_{ij}^{(s)}}\left(\sum_{h=1}^{n_{s-1}} w_{ih}^{(s)} x_{\text{out},h}^{(s-1)}\right) = x_{\text{out},j}^{(s-1)} \tag{3.71}$$

The first term on the right side of (3.70) is commonly referred to as the *sensitivity* term. Physically, it quantifies the change in the energy function given in (3.68) as a result of the weight change of the sth layer. Therefore, we can define

$$\delta_i^{(s)} \triangleq -\frac{\partial E_q}{\partial v_i^{(s)}} \tag{3.72}$$

and use (3.70) and (3.71) to rewrite (3.69) as

$$w_{ij}^{(s)}(k+1) = w_{ij}^{(s)}(k) + \alpha^{(s)}\delta_i^{(s)} x_{\text{out},j}^{(s-1)} \tag{3.73}$$

or in vector-matrix form

$$W^{(s)}(k+1) = W^{(s)}(k) + \alpha^{(s)} D^{(s)} x_{\text{out}}^{(s-1)} \tag{3.74}$$

where $D^{(s)}$ represents the sensitivity vector of the sth layer, defined as

$$D^{(s)} = -\left[\frac{\partial E_q}{\partial v_1^{(s)}}, \frac{\partial E_q}{\partial v_2^{(s)}}, \ldots, \frac{\partial E_q}{\partial v_{n_s}^{(s)}}\right]^T \tag{3.75}$$

Calculation of the sensitivities

Let us first consider the output layer of the network given in Figure 3.4. A single term in the sensitivity vector can be evaluated as

$$\begin{aligned}\frac{\partial E_q}{\partial v_i^{(3)}} &= \frac{\partial}{\partial v_i^{(3)}}\left[\frac{1}{2}\left(d_q - x_{\text{out}}^{(3)}\right)^T\left(d_q - x_{\text{out}}^{(3)}\right)\right] \\ &= \frac{\partial}{\partial v_i^{(3)}}\left\{\frac{1}{2}\sum_{h=1}^{n_3}\left[d_{qh} - f\left(v_h^{(3)}\right)\right]^2\right\} \\ &= -\left[d_{qi} - f\left(v_i^{(3)}\right)\right]f'\left(v_i^{(3)}\right) = -\left(d_{qi} - x_{\text{out},i}^{(3)}\right)g\left(v_i^{(3)}\right)\end{aligned} \tag{3.76}$$

where $g(x) = df(x)/dx$ represents the first derivative of the activation function.

Substituting (3.76) into (3.75), we can express the sensitivity vector for the output layer of the network as

$$D^{(3)} = G\left(v^{(3)}\right)\left(d_q - x_{\text{out}}^{(3)}\right) \tag{3.77}$$

where $G\left(v^{(3)}\right) = \text{diag}\left[g\left(v_1^{(3)}\right), g\left(v_2^{(3)}\right), \ldots, g\left(v_{n_3}^{(3)}\right)\right]$. Now, let us find the sensitivity vector for the second and first layers of the network. Using the definition of the sensitivity vector in (3.75) and the chain rule, we can write the sensitivity of the second layer as

$$D^{(2)} = -\frac{\partial E_q}{\partial v^{(3)}} = -\left(\frac{\partial v^{(3)}}{\partial v^{(2)}}\right)^T \frac{\partial E_q}{\partial v^{(3)}} = \left(\frac{\partial v^{(3)}}{\partial v^{(2)}}\right)^T D^{(3)} \tag{3.78}$$

Equation (3.78) shows that the sensitivity vector of the second layer can be expressed as a function of the output layer sensitivity. The linear transformation between the two sensitivities in (3.78) is given by the following Jacobian matrix:

$$\left(\frac{\partial v^{(3)}}{\partial v^{(2)}}\right)^T = \begin{bmatrix} \frac{\partial v_1^{(3)}}{\partial v_1^{(2)}} & \frac{\partial v_1^{(3)}}{\partial v_2^{(2)}} & \cdots & \frac{\partial v_1^{(3)}}{\partial v_{n_2}^{(2)}} \\ \frac{\partial v_2^{(3)}}{\partial v_1^{(2)}} & \frac{\partial v_2^{(3)}}{\partial v_2^{(2)}} & \cdots & \frac{\partial v_2^{(3)}}{\partial v_{n_2}^{(2)}} \\ \cdots & \cdots & \cdots & \cdots \\ \frac{\partial v_{n_3}^{(3)}}{\partial v_1^{(2)}} & \frac{\partial v_{n_3}^{(3)}}{\partial v_2^{(2)}} & \cdots & \frac{\partial v_{n_3}^{(3)}}{\partial v_{n_2}^{(2)}} \end{bmatrix}^T \tag{3.79}$$

In a similar manner, we can show that the sensitivity vector of the first layer can be expressed as

$$\boldsymbol{D}^{(1)} = \left(\frac{\partial \boldsymbol{v}^{(2)}}{\partial \boldsymbol{v}^{(1)}}\right)^T \boldsymbol{D}^{(2)} \tag{3.80}$$

Consider a single entry in the transformation Jacobian matrix given in Equation (3.79)

$$\begin{aligned}\frac{\partial v_i^{(3)}}{\partial v_j^{(2)}} &= \frac{\partial}{\partial v_j^{(2)}}\left[\sum_{h=1}^{n_2} w_{ih}^{(3)} x_{\text{out},h}^{(2)}\right] = w_{ij}^{(3)} \frac{\partial x_{\text{out},j}^{(2)}}{\partial v_j^{(2)}} \\ &= w_{ij}^{(3)} \frac{\partial f\left(v_j^{(2)}\right)}{\partial v_j^{(2)}} = w_{ij}^{(3)} f'\left(v_j^{(2)}\right) = w_{ij}^{(3)} g\left(v_j^{(2)}\right)\end{aligned} \tag{3.81}$$

Substituting (3.81) into (3.79), we see that the Jacobian matrix can be rewritten as

$$\left(\frac{\partial \boldsymbol{v}^{(3)}}{\partial \boldsymbol{v}^{(2)}}\right)^T = \left[\boldsymbol{W}^{(3)} \boldsymbol{G}\left(\boldsymbol{v}^{(2)}\right)\right]^T = \boldsymbol{G}\left(\boldsymbol{v}^{(2)}\right) \boldsymbol{W}^{(3)T} \tag{3.82}$$

where $\boldsymbol{G}\left(\boldsymbol{v}^{(2)}\right) = \text{diag}\left[g\left(v_1^{(2)}\right), g\left(v_2^{(2)}\right), \ldots, g\left(v_{n_2}^{(2)}\right)\right]$.

Finally, combining (3.82) and (3.80) gives

$$\boldsymbol{D}^{(2)} = \boldsymbol{G}\left(\boldsymbol{v}^{(2)}\right) \boldsymbol{W}^{(3)T} \boldsymbol{D}^{(3)} \tag{3.83}$$

Using the same approach, we can express the sensitivity of the first layer as

$$\boldsymbol{D}^{(1)} = \boldsymbol{G}\left(\boldsymbol{v}^{(1)}\right) \boldsymbol{W}^{(2)T} \boldsymbol{D}^{(2)} \tag{3.84}$$

Equations (3.74), (3.77), (3.83), and (3.84) form the vector-matrix representation of the backpropagation learning rule.

Vector-matrix form of the backpropagation algorithm

Step 1. Present an input pattern and calculate the outputs of the network and at all the internal layers.

Step 2. For each of the layers, calculate the sensitivity vector according to

$$\boldsymbol{D}^{(s)} = \boldsymbol{G}\left(\boldsymbol{v}^{(s)}\right)\left(\boldsymbol{d}_q - \boldsymbol{x}_{\text{out}}^{(s)}\right) \qquad \text{for output layer}$$

$$\boldsymbol{D}^{(s-1)} = \boldsymbol{G}\left(\boldsymbol{v}^{(s-1)}\right) \boldsymbol{W}^{(s)T} \boldsymbol{D}^{(s)} \qquad \text{for all hidden layers}$$

Step 3. Update the synaptic weights of the network according to

$$\boldsymbol{W}^{(s)}(k+1) = \boldsymbol{W}^{(s)} + \alpha^{(s)} \boldsymbol{D}^{(s)} \boldsymbol{x}_{\text{out}}^{(s-1)T}$$

Step 4. Continue steps 1 through 3 until the network reaches the desired mapping accuracy.

3.4 ACCELERATED LEARNING BACKPROPAGATION ALGORITHMS

In this section we present several modifications of the standard backpropagation algorithm. As we have pointed out previously, learning in the MLP NN is in effect minimization of the mean square error (MSE) between the actual and desired network outputs. The problems associated with using the MSE criterion are well studied in numerical analysis. The modifications of the backpropagation algorithm presented here are based on the application of advanced numerical techniques to the problem of MLP NN training.

3.4.1 Conjugate Gradient Backpropagation for the Feedforward Multilayer Perceptron

The conjugate gradient method (cf. Sect. A.5.5) is a well-known numerical technique used for solving various optimization problems. It is widely used since it represents a good compromise between simplicity of the steepest descent algorithm and the fast quadratic convergence of Newton's method (cf. Sect. A.5.3). Several methods for training MLP NNs based on the conjugate gradient method have been developed, and a comprehensive survey is given in [36]. Most of these algorithms are based on the assumption that in the vicinity of the solution, the error function of all the weights in the network can be accurately approximated by a quadratic function. That is,

$$J(\boldsymbol{w}) = \frac{1}{2P}\sum_{p=1}^{P}\sum_{h=1}^{n_s}\left(d_{ph} - y_{ph}\right)^2 = \frac{1}{2}\boldsymbol{w}^T\boldsymbol{Q}\boldsymbol{w} - \boldsymbol{b}^T\boldsymbol{w} \tag{3.85}$$

where $\boldsymbol{w}$ represents all the weights in the network, P is the total number of training patterns, n_s is the number of neurons in the output layer, d_{ph} is the desired output of the hth neuron in the output layer to the pth training input, and y_{ph} is the actual output of the hth neuron in the output layer to the pth training input. The matrix $\boldsymbol{Q}$ in (3.85) is a square matrix of second partial derivatives, the Hessian matrix. The dimension of the Hessian matrix is equal to the total number of weights in the network. Conjugate gradient algorithms attempt to find a system of conjugate directions on the error surface and to perform the update of the weights along these directions. Since the Hessian matrix $\boldsymbol{Q}$ is typically large, its computations are impractical, and most of the conjugate gradient algorithms for training MLP NNs seek to find conjugate gradient directions without explicitly computing the Hessian. Some practical implementations of the conjugate gradient method for training MLP NNs can be found in [35, 37, 38].

The algorithm presented in this section adopts a different approach to conjugate gradient-based training of the MLP NNs. Instead of considering an error surface, this algorithm formulates a set of *normal equations* at each of the neurons which are then solved iteratively by using the conjugate gradient approach. In the presentation of this algorithm we closely follow the approach given in [39].

In deriving the standard backpropagation algorithm, we have already pointed out that it can be viewed as being comprised of two distinct processes working together over the course of the network training. The first process is the estimation of the local error at each node of the MLP NN. This is accomplished through backpropagation of the error from the output layer, where it can be explicitly computed as a difference between the actual and the desired network response to the neurons in the hidden layers. The second process is the update of the network weights. In Section 3.3.1 we see that the standard backpropagation algorithm uses the LMS algorithm to perform the weight updates. It is very important to understand that these two processes are separate and in essence independent of each other. It allows development of different weight update algorithms while still maintaining the error backpropagation process. Using more sophisticated gradient descent techniques for updating the network weights will speed up the training process.

We have seen that the MLP NN is considered to be trained when the output of the network is within a certain specified tolerance of the desired output for each of the training patterns. In order for this to be accomplished, each neuron of the network has to be adequately trained. In other words, training the MLP NN implies training of each of the nodes in the network for the desired response. Referring to Figure 3.4, we see that each of the nodes consists of adaptive linear elements (commonly referred to as linear combiners) followed by sigmoidal nonlinearities. Existence of the nonlinearities is the source of the powerful mapping capabilities of the MLP NN. On the other hand, it is the introduction of these nonlinearities that increases the complexity of the network training. We can observe that the output of the nonlinear activation function will be the desired response if the linear combiner produces an appropriate input to the activation function. Therefore, we can conclude that training the MLP NN essentially involves adjusting the weights so that each of the network's linear combiners produces the desired output.

Normal equations for the linear combiner

Consider the ith linear combiner in the sth layer of the MLP NN given in Figure 3.4. When the qth input pattern is presented to the network, the output of the combiner is calculated as an inner product between the combiner weights $\boldsymbol{w}_i^{(s)} \in \Re^{n_{s-1} \times 1}$ and the input vector to the particular layer $\boldsymbol{x}_{\text{out},q}^{(s-1)} \in \Re^{n_{s-1} \times 1}$. That is,

$$v_i^{(s)} = \boldsymbol{w}_i^{(s)T} \boldsymbol{x}_{\text{out},q}^{(s-1)} \tag{3.86}$$

Suppose for a moment that the desired output for the particular combiner is $d_{i,q}^{(s)}$ and is known for every pattern in the training set. Training the MLP NN effectively assumes training of all its linear combiners; therefore, the goal of the learning algorithm is to minimize the squared error cost function, given by

$$J_i^{(s)} = \frac{1}{2}\sum_{q=1}^{M}\left(d_{i,q}^{(s)} - v_{i,q}^{(s)}\right)^2 \tag{3.87}$$

where M represents the total number of vectors in the training set. Substituting (3.86) into (3.87), we can write

$$J_i^{(s)} = \frac{1}{2}\sum_{q=1}^{M}\left(d_{i,q}^{(s)} - \boldsymbol{w}_i^{(s)T}\boldsymbol{x}_{\text{out},q}^{(s-1)}\right)^2 \tag{3.88}$$

To find the weight vector which minimizes the cost function given in (3.88), we can take its partial derivative with respect to $\boldsymbol{w}_i^{(s)}$ and equate it to zero, that is,

$$\frac{\partial J_i^{(s)}}{\partial \boldsymbol{w}_i^{(s)}} = \sum_{q=1}^{M}\left(-d_{i,q}^{(s)}\boldsymbol{x}_{\text{out},q}^{(s-1)} + \boldsymbol{x}_{\text{out},q}^{(s-1)}\boldsymbol{x}_{\text{out},q}^{(s-1)T}\boldsymbol{w}_i^{(s)}\right) = 0 \tag{3.89}$$

Defining

$$\boldsymbol{C}_i^{(s)} \triangleq \sum_{q=1}^{M}\boldsymbol{x}_{\text{out},q}^{(s-1)}\boldsymbol{x}_{\text{out},q}^{(s-1)T} \tag{3.90}$$

and

$$\boldsymbol{p}_i^{(s)} \triangleq \sum_{q=1}^{M} d_{i,q}^{(s)}\boldsymbol{x}_{\text{out},q}^{(s-1)} \tag{3.91}$$

Equation (3.89) can be rearranged in vector-matrix form as

$$\boldsymbol{C}_i^{(s)}\boldsymbol{w}_i^{(s)} = \boldsymbol{p}_i^{(s)} \tag{3.92}$$

The matrix $\boldsymbol{C}_i^{(s)}$ can be interpreted as an estimate of the covariance matrix of the vector inputs to the sth layer, and vector $\boldsymbol{p}_i^{(s)}$ is an estimate of the cross-correlation vector between the inputs to the sth layer and the desired outputs of the linear combiner. Notice that matrix $\boldsymbol{C}_i^{(s)}$ does not depend on the position of the linear combiner in the layer, and therefore the subscript i can be omitted. The vector-matrix equation in (3.92) is well known in the context of adaptive filtering under the name of *deterministic normal equation* [29]. The solution of the normal equation minimizes the quadratic error function given in (3.87). In summary, equations of the form given in (3.92) can be written for every linear combiner output in the MLP NN, and the network training can be conveniently viewed as a process involving their solutions.

Conjugate gradient method applied to solving the normal equations

There are several numerical techniques that can be used to solve the system of linear equations given in (3.92). One such technique is the conjugate gradient method. In this section, we only present the algorithm. The reader is strongly encouraged to read the details presented in Section A.5.5.

Step 1. Initialize the components of the weight vector $\boldsymbol{w}_i^{(s)}(0) \in \Re^{n_s \times 1}$ to some small arbitrary values.
Step 2. Set $k = 0$. Calculate the initial conjugate direction $\boldsymbol{d}_0$ and the gain vector $\boldsymbol{g}_0$ as

$$\boldsymbol{d}_0 = -\boldsymbol{g}_0 = \boldsymbol{p}_i^{(s)} - \boldsymbol{C}^{(s)}\boldsymbol{w}_i^{(s)}(0)$$

Step 3. Determine the conjugate vector coefficient

$$\alpha_k = -\frac{\boldsymbol{g}_k^T \boldsymbol{d}_k}{\boldsymbol{d}_k^T \boldsymbol{C}^{(s)} \boldsymbol{d}_k} \qquad \text{where} \qquad \boldsymbol{g}_k = \boldsymbol{C}^{(s)}\boldsymbol{w}_i^{(s)}(k) - \boldsymbol{p}_i^{(s)}$$

Step 4. Update the weight vector

$$\boldsymbol{w}_i^{(s)}(k+1) = \boldsymbol{w}_i^{(s)}(k) + \alpha_k \boldsymbol{d}_k$$

Step 5. Determine new gain vector

$$\boldsymbol{g}_{k+1} = \boldsymbol{C}^{(s)}\boldsymbol{w}_i^{(s)}(k+1) - \boldsymbol{p}_i^{(s)}$$

Step 6. Determine the new conjugate gradient direction

$$\boldsymbol{d}_{k+1} = -\boldsymbol{g}_{k+1} + \beta_k \boldsymbol{d}_k \qquad \text{where} \qquad \beta_k = \frac{\boldsymbol{g}_{k+1}^T \boldsymbol{C}^{(s)} \boldsymbol{d}_k}{\boldsymbol{g}_k^T \boldsymbol{g}_k}$$

Step 7. Set $k = k + 1$ and test the exit condition. If $k < n$, go to step 3; otherwise, stop.

The vector $w_i^{(s)*}$ which results from the completion of the above steps solves the normal equations for the particular linear combiner.

Training algorithm

There are some principal difficulties associated with applying the conjugate gradient algorithm as it was presented in the previous section. The desired output of the particular nodes is not known except for the nodes in the output layer. This means that they have to be estimated for all the nodes in the hidden layers. To accomplish this, we can refer to Section 3.3.1 and the physical interpretation of the local error. As we have already mentioned, the local errors represent estimates of the errors between the actual outputs of the neurons and their desired outputs. This estimate is based on only one training input/output pair and the current values of the weights in the network; and therefore, we do not expect it to be accurate at the beginning of network training. However, as the training progresses, the estimate of the error becomes increasingly more accurate. Knowing the actual output of a particular node, and knowing the local error, the desired output of the ith neuron in the sth layer to the qth training pattern can be calculated as

$$\bar{d}_{i,q}^{(s)} = x_{\text{out},i,q}^{(s)} + \mu \delta_{i,q}^{(s)} \tag{3.93}$$

where μ is some positive number commonly taken in the range from 10 to 400 [39]. Equation (3.93) gives an estimate of the desired output of each neuron of the MLP NN. When deriving the normal equations, we assume knowledge of the desired outputs of the linear combiners. Since the activation function is commonly chosen as a monotonically increasing sigmoid function, there is a one-to-one correspondence between the output of the neuron and the output of the linear combiner. Given the output of the neuron, the output of the combiner can be calculated as

$$\hat{v}_{i,q}^{(s)} = f^{-1}\left(\bar{d}_{i,q}^{(s)}\right) \tag{3.94}$$

By selecting a suitable activation function, we can easily express the inverse in (3.94). For example, if the activation function is selected to be

$$y = f(t) = \frac{1 - e^{-\sigma t}}{1 + e^{\sigma t}} \tag{3.95}$$

where σ is the parameter controlling the slope, the inverse function can be expressed as

$$t = f^{-1}(y) = \frac{1}{\sigma}\ln\frac{1 + y}{1 - y} \tag{3.96}$$

Applying (3.96) to (3.95), we have the desired output of the linear combiner as

$$\hat{v}_{i,q}^{(s)} = \frac{1}{\sigma}\ln\frac{1 + \bar{d}_{i,q}^{(s)}}{1 - \bar{d}_{i,q}^{(s)}} \tag{3.97}$$

The conjugate gradient algorithm described in the previous section assumes an explicit knowledge of the covariance matrices $\boldsymbol{C}^{(s)}$ and the cross-correlation vectors $\boldsymbol{p}_i^{(s)}$. Of course, they are not known in advance and have to be estimated during the training process. A convenient way to do this is to update their estimates with each presentation of the input/output training pair. The estimate of the correlation matrix for the sth layer can be written as

$$\boldsymbol{C}^{(s)}(k+1) = b\boldsymbol{C}^{(s)}(k) + \boldsymbol{x}_{\text{out},q}^{(s-1)}\boldsymbol{x}_{\text{out},q}^{(s-1)T} \tag{3.98}$$

Similarly, the cross-correlation vector for each of the linear combiners can be estimated as

$$\boldsymbol{p}_i^{(s)}(k+1) = b\boldsymbol{p}_i^{(s)}(k) + \hat{v}_i^{(s)}\boldsymbol{x}_{\text{out},q}^{(s-1)} \tag{3.99}$$

The b coefficient in (3.98) and (3.99) is called the *forgetting factor* and determines the weighting of the previous instantaneous estimates of the covariance matrix and cross-correlation vector [first terms on the right-hand side in (3.98) and (3.99)]. Typically b it is set to be in the range from 0.9 to 0.99.

Based on the above comments, we can provide a summary of the algorithm.

Conjugate-gradient-based algorithm for training an MLP NN

Step 1. Initialize the network weights to some small random values. Any of the weight initialization techniques described in Section 3.3.2 can be used.

Step 2. Propagate the qth training pattern throughout the network, calculating the output of every node.

Step 3. Calculate the local error at every node in the network. For the output nodes the local error is calculated as

$$\delta_{i,q}^{(s)} = \left(d_{i,q} - x_{\text{out},i,q}^{(s)}\right)g\left(v_{i,q}^{(s)}\right)$$

where $g(\bullet)$ is the derivative of activation function $f(\bullet)$. For each of the hidden layer nodes, the local error is calculated as

$$\delta_{i,q}^{(s)} = \left(\sum_{h=1}^{n_{s+1}} \delta_{h,q}^{(s+1)} w_{h,i}^{(s+1)} \right) g\left(v_{i,q}^{(s)}\right)$$

Step 4. For each of the linear combiner estimates, the desired output value is given by

$$\hat{v}_{i,q}^{(s)} = f^{-1}\left(\bar{d}_{i,q}^{(s)}\right) \qquad \text{where} \qquad \bar{d}_{i,q}^{(s)} = x_{\text{out},i,q}^{(s)} + \mu \delta_{i,q}^{(s)}$$

Step 5. Update the estimate of the covariance matrix in each layer

$$\boldsymbol{C}^{(s)}(k) = b\boldsymbol{C}^{(s)}(k-1) + \boldsymbol{x}_{\text{out},q}^{(s-1)} \boldsymbol{x}_{\text{out},q}^{(s-1)T}$$

Update the estimate of the cross-correlation vector for each node

$$\boldsymbol{p}_i^{(s)}(k) = b\boldsymbol{p}_i^{(s)}(k-1) + \hat{v}_i^{(s)} \boldsymbol{x}_{\text{out},q}^{(s-1)}$$

where k is the pattern presentation index.

Step 6. Update the weight vector for every node in the network as follows.

(*a*) At every node calculate

$$\boldsymbol{g}_i^{(s)}(k) = \boldsymbol{C}^{(s)}(k)\boldsymbol{w}_i^{(s)}(k) - \boldsymbol{p}_i^{(s)}(k), \text{ else}$$

If $\boldsymbol{g}_i^{(s)} = 0$, do not update the weight vector for the node and go to step 7; else perform the following steps:

(*b*) Find the direction $\boldsymbol{d}(k)$. If the iteration number is an integer multiple of the number of weights in the node, then

$$\boldsymbol{d}_i^{(s)}(k) = -\boldsymbol{g}_i^{(s)}(k),$$

else

$$\boldsymbol{d}_i^{(s)}(k) = -\boldsymbol{g}_i^{(s)}(k) + \beta_i^{(s)} \boldsymbol{d}_i^{(s)}(k-1)$$

where

$$\beta_i^{(s)} = -\boldsymbol{g}_i^{(s)T}(k) \frac{\boldsymbol{C}^{(s)}(k)\boldsymbol{d}_i^{(s)}(k-1)}{\boldsymbol{d}_i^{(s)T}(k-1)\boldsymbol{C}^{(s)}(k)\boldsymbol{d}_i^{(s)}(k-1)}$$

(*c*) Compute the step size

$$\alpha_i^{(s)}(k) = -\frac{\boldsymbol{g}_i^{(s)T}(k)\boldsymbol{d}_i^{(s)}(k)}{\boldsymbol{d}_i^{(s)T}(k)\boldsymbol{C}^{(s)}(k)\boldsymbol{d}_i^{(s)}(k)}$$

(*d*) Modify the weight vector according to

$$\boldsymbol{w}_i^{(s)}(k) = \boldsymbol{w}_i^{(s)}(k-1) + \alpha_i^{(s)}(k)\boldsymbol{d}_i^{(s)}(k)$$

Step 7. If the network has not converged, go back to step 2.

Several comments are in order regarding the algorithm presented above. The conjugate gradient algorithm for solving the normal equations requires at least n steps to converge. At first glance, it would appear that n steps have to be taken in the process of the linear combiner weight vector update every time a new training input/output pair is presented. However, this is not done for the following reasons. During training, only the estimates of the covariance matrices and cross-correlation vectors are known; therefore, an exact solution of the corresponding normal equations does not yield the required weight vector. This is especially true at the beginning of network training. As training progresses, the estimates of cross-correlation vectors and covariance matrices stabilize, and because they remain approximately constant for several training input/output pairs, it is sufficient to perform *one* conjugate gradient step for each input/output pair. Of course, since the entire process is inherently nonlinear, the training of the network will not be completed in n steps. From step

6 we see that after n conjugate gradient steps, the conjugate gradient update gets reinitialized (i.e., starts with a steepest descent step).

3.4.2 Recursive Least-Squares-Based Backpropagation Algorithm

We have already seen in the previous section that training of the MLP NN can be understood as a process of solving a set of deterministic normal equations. Normal equations can be written for every linear combiner in the network and are given as

$$\boldsymbol{C}^{(s)}\boldsymbol{w}_i^{(s)} = \boldsymbol{p}_i^{(s)} \tag{3.100}$$

where $\boldsymbol{C}^{(s)}$ indicates the covariance matrix of the inputs to the sth layer, $\boldsymbol{p}_i^{(s)}$ is the cross-correlation vector between inputs to the sth layer and desired outputs of the ith linear combiner in the sth layer, and $\boldsymbol{w}_i^{(s)}$ represents the vector of synaptic weights to the ith linear combiner in the sth layer. If we assume that the covariance matrix $\boldsymbol{C}^{(s)}$ and cross-correlation vector $\boldsymbol{p}_i^{(s)}$ are known, the appropriate weight vector can be calculated as

$$\boldsymbol{w}_i^{(s)} = \left[\boldsymbol{C}^{(s)}\right]^{-1}\boldsymbol{p}_i^{(s)} \tag{3.101}$$

However, we do not have explicit knowledge of either the covariance matrices or the cross-correlation vectors, and over the course of training of the network they have to be estimated. By using (3.98) and (3.99), these estimates can be made, and the system of normal equations given in (3.100) has to be solved in an iterative fashion. In the previous section we applied the conjugate gradient method to accomplish the task. Here we explore an alternate method for solving the equations in (3.100) based on the recursive least-squares (RLS) approach [29]. The RLS algorithm can solve for the weight vector in (3.101) by directly performing an adaptive recursive estimate of the inverse of the correlation matrix $[\boldsymbol{C}^{(s)}]^{-1}$ with each presentation of the input/output training pairs. The RLS algorithm can be viewed as a special case of Kalman filtering [29], and for that reason the method presented here is sometimes referred to as Kalman filtering-based backpropagation [40]. To pave the way for development of the RLS backpropagation algorithm, we start with a very important result from matrix algebra known as the *matrix inversion lemma.*

Matrix inversion lemma (Woodbury's identity)

Let $\boldsymbol{A} \in \Re^{m\times m}$ and $\boldsymbol{B} \in \Re^{m\times m}$ be two positive definite matrices related by (cf. Sect. A.2.7)

$$\boldsymbol{A} = \boldsymbol{B}^{-1} + \boldsymbol{C}\boldsymbol{D}^{-1}\boldsymbol{C}^T \tag{3.102}$$

where $\boldsymbol{D} \in \Re^{n\times n}$ is another positive definite matrix, and $\boldsymbol{C} \in \Re^{m\times n}$. Then the inverse of matrix $\boldsymbol{A}$ can be written as

$$\boldsymbol{A}^{-1} = \boldsymbol{B} - \boldsymbol{B}\boldsymbol{C}\left(\boldsymbol{D} + \boldsymbol{C}^T\boldsymbol{B}\boldsymbol{C}\right)^{-1}\boldsymbol{C}^T\boldsymbol{B} \tag{3.103}$$

Derivation of the RLS backpropagation algorithm

Consider the update equation for the covariance matrix given in (3.98), and repeated in (3.104) for convenience,

$$C^{(s)}(k) = bC^{(s)}(k-1) + x_{\text{out},q}^{(s-1)}(k)x_{\text{out},q}^{(s-1)T}(k) \tag{3.104}$$

Comparing (3.104) and (3.102), we can identify $A = C^{(s)}(k)$, $B^{-1} = bC^{(s)}(k-1)$, $C = x_{\text{out},q}^{(s-1)}(k)$, and $D = 1$. To simplify the notation, let us define

$$x_{\text{out},q}^{(s)}(k) \triangleq X^{(s)} \tag{3.105}$$

Using the result stated in (3.103), we can write

$$\begin{aligned}\left[C^{(s)}(k)\right]^{-1} &= b^{-1}\left[C^{(s)}(k-1)\right]^{-1} \\ &\quad - \frac{b^{-1}\left[C^{(s)}(k-1)\right]^{-1} X^{(s-1)} X^{(s-1)T} b^{-1}\left[C^{(s)}(k-1)\right]^{-1}}{1 + X^{(s-1)T} b^{-1}\left[C^{(s)}(k-1)\right]^{-1} X^{(s-1)}}\end{aligned} \tag{3.106}$$

Defining

$$K(k) \triangleq \frac{\left[C^{(s-1)}(k-1)\right]^{-1} X^{(s-1)}}{b + X^{(s-1)T}\left[C^{(s-1)}(k-1)\right]^{-1} X^{(s-1)}} \tag{3.107}$$

we can substitute (3.107) into (3.106) and write

$$\left[C^{(s)}(k)\right]^{-1} = b^{-1}\left\{\left[C^{(s)}(k-1)\right]^{-1} - K(k)X^{(s-1)T}\left[C^{(s)}(k-1)\right]^{-1}\right\} \tag{3.108}$$

Equations (3.107) and (3.108) give a mechanism for recursive estimation of the inverse of the covariance matrix at every hidden layer of the MLP NN. Combining Equations (3.99) and (3.101) yields

$$w_i^{(s)}(k) = \left[C^{(s)}(k)\right]^{-1} p^{(s)}(k) = \left[C^{(s)}(k)\right]^{-1}\left[bp_i^{(s)}(k-1) + \hat{v}_i^{(s)}(k)X^{(s-1)}\right] \tag{3.109}$$

Substituting the expression for the inverse of the covariance matrix given in (3.108) into (3.109) gives

$$\begin{aligned} w_i^{(s)}(k) &= b^{-1}\left\{\left[C^{(s)}(k-1)\right]^{-1} - K(k)X^{(s-1)T}\left[C^{(s)}(k-1)\right]^{-1}\right\} \\ &\quad \cdot \left[bp_i^{(s)}(k-1) + \hat{v}_i^{(s)}(k)X^{(s-1)}\right] \\ &= \left[C^{(s)}(k-1)\right]^{-1} p_i^{(s)}(k-1) - K(k)X^{(s-1)T}\left[C^{(s)}(k-1)\right]^{-1} p_i^{(s)}(k-1) \\ &\quad + b^{-1}\left[C^{(s)}(k-1)\right]^{-1} \hat{v}_i^{(s)}(k)X^{(s-1)} - b^{-1}K(k)X^{(s-1)T} \\ &\quad \cdot \left[C^{(s)}(k-1)\right]^{-1} \hat{v}_i^{(s)}(k)X^{(s-1)} \end{aligned} \tag{3.110}$$

or

$$
\begin{aligned}
\boldsymbol{w}_i^{(s)}(k) = {} & \boldsymbol{w}_i^{(s)}(k-1) - \boldsymbol{K}(k)\boldsymbol{X}^{(s-1)T}\boldsymbol{w}_i^{(s)}(k-1) \\
& + \hat{v}_i^{(s)}(k)b^{-1}\left[\boldsymbol{C}^{(s)}(k-1)\right]^{-1}\boldsymbol{X}^{(s-1)} \\
& - \hat{v}_i^{(s)}(k)b^{-1}\frac{\left[\boldsymbol{C}^{(s)}(k-1)\right]^{-1}\boldsymbol{X}^{(s-1)}\boldsymbol{X}^{(s-1)T}\left[\boldsymbol{C}^{(s)}(k-1)\right]^{-1}\boldsymbol{X}^{(s-1)}}{b+\boldsymbol{X}^{(s-1)T}\left[\boldsymbol{C}^{(s)}(k-1)\right]^{-1}\boldsymbol{X}^{(s-1)}}
\end{aligned}
\tag{3.111}
$$

By combining the last two terms in (3.111), the weight update equation simplifies to

$$
\begin{aligned}
\boldsymbol{w}_i^{(s)}(k) &= \boldsymbol{w}_i^{(s)}(k-1) - \boldsymbol{K}(k)\boldsymbol{X}^{(s-1)T}\boldsymbol{w}_i^{(s)}(k-1) + \hat{v}_i^{(s)}(k)\boldsymbol{K}(k) \\
&= \boldsymbol{w}_i^{(s)}(k-1) + \boldsymbol{K}(k)\left[\hat{v}_i^{(s)}(k) - \boldsymbol{X}^{(s-1)T}\boldsymbol{w}_i^{(s)}(k-1)\right] \\
&= \boldsymbol{w}_i^{(s)}(k-1) + \boldsymbol{K}(k)\left[\hat{v}_i^{(s)}(k) - v_i^{(s)}(k)\right]
\end{aligned}
\tag{3.112}
$$

Equation (3.112) is the weight update equation for the ith linear combiner in the sth layer of the network. The update is proportional to the difference between the desired output of the linear combiner and the actual one. This is the case in all learning algorithms of the Hebbian type, and the RLS is just one special case. The key property of the RLS algorithm is that learning rates vary dynamically for every neuron in the network and at every step of network training. The learning rates are calculated as entries in the Kalman gain matrix in (3.112).

Based on the update equation (3.112), the RLS algorithm for training an MLP NN can be summarized in the following steps [40]:

Recursive least-squares version of the backpropagation algorithm

Step 1. Initialize the weights in the network according to any of the standard initialization procedures discussed in Section 3.3.2.

Step 2. Present an input pattern, and calculate the responses of all linear combiners $v_i^{(s)}$ and all neuron outputs $x_{\text{out},i}^{(s)}$ in the network.

Step 3. For each layer of the network, calculate the Kalman gain matrix and update the covariance matrix estimate according to the following equations:

$$\boldsymbol{A}^{(s)}(k) = \left[\boldsymbol{C}^{(s)}(k-1)\right]^{-1}\boldsymbol{x}_{\text{out}}^{(s-1)}(k)$$

Update the Kalman gain matrix for the sth layer

$$\boldsymbol{K}^{(s)}(k) = \frac{\boldsymbol{A}^{(s)}(k)}{b + \boldsymbol{x}_{\text{out}}^{(s-1)T}(k)\boldsymbol{A}^{(s)}(k)}$$

Update the covariance matrix for the sth layer according to

$$\left[\boldsymbol{C}^{(s)}(k)\right]^{-1} = b^{-1}\left\{\left[\boldsymbol{C}^{(s)}(k-1)\right]^{-1} - \boldsymbol{K}^{(s)}(k)\boldsymbol{A}^{(s)T}(k)\right\}$$

Step 4. Calculate and backpropagate the local errors for the output layer according to the equations

$$\delta_i^{(s)} = \left(d_{qh} - x_{\text{out},i}^{(s)}\right)g\left(v_i^{(s)}\right)$$

and for hidden layers

$$\delta_i^{(s)} = \left(\sum_{h=1}^{n_2} \delta_h^{(s+1)} w_{hi}^{(s+1)}\right) g\left(v_i^{(s)}\right)$$

where $g(z) = df(z)/dz$, and $f(z)$ is the activation function of the neuron.

Step 5. For each of the linear combiners, estimate the desired output according to

$$\hat{v}_i^{(s)} = f^{-1}\left(x_{\text{out},i}^{(s)} + \mu\delta_i^{(s)}\right)$$

where $f^{-1}(z)$ denotes the inverse of the neuron activation function.

Step 6. Update the weights in each layer of the network according to

$$\boldsymbol{w}_i^{(s)}(k) = \boldsymbol{w}_i^{(s)}(k-1) + \boldsymbol{K}(k)\left[\hat{v}_i^{(s)}(k) - v_i^{(s)}(k)\right]$$

Step 7. Stop if the network has converged; else go back to step 2.

From the above algorithm summary, it is obvious that RLS algorithm involves added computations to the training of the MLP NN compared to the standard backpropagation algorithm. However, detailed analysis presented in [40] and [41] demonstrates that the RLS algorithm needs a considerably smaller number of iterations to converge, and the overall training time is more than an order of magnitude less when compared to standard backpropagation. The RLS algorithm has been shown to be less sensitive to the initial choice of weights, and the adaptive nature of the learning rate makes the algorithm less likely to get stuck in a local minimum. Significant improvement in the algorithm performance can be gained by its parallel implementation described in [40].

3.4.3 Backpropagation with Adaptive Slopes of Activation Functions

From the weight update equations for backpropagation we have seen thus far, we observe that the rate of the update is proportional to the derivative of the nonlinear activation function. As discussed earlier, a typical activation function for neurons in an MLP NN is of a sigmoidal type with a bell-shaped derivative, as shown in Figure 3.6. During training of the network, the output of the linear combiner may fall in the saturation region of the activation function. The derivative of the activation function in that region is very small, and since the weight updates depend directly on the magnitude of the derivative, the rate of learning becomes extremely slow. It may take many iterations before the output of the linear combiner moves out of the saturation region. A straightforward approach to prevent this saturation would involve increasing the size of the nonsaturated part of the activation function by decreasing its slope. However, decreasing the slope makes the network behave more as a linear network, which in effect diminishes the advantages of having a multilayer network (since any number of layers having linear activation functions can be replaced by a single layer). Hence, there is an optimum value for the activation function slope that balances the speed of the network training and its mapping capabilities. This value is not necessarily the same for every neuron in the network. Due to the complexity of the MLP NN structure,

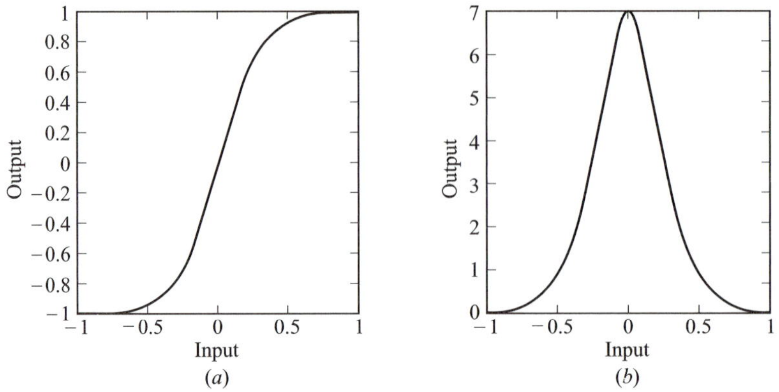

Figure 3.6 (*a*) Typical MLP NN activation function. (*b*) Derivative of the activation function in (*a*).

it is impossible to determine the optimal slope of the activation function for every neuron in the network before the training commences; therefore, their estimation in an adaptive fashion offers a viable alternative. This is the approach that we take in this section.

The procedure for adaptive estimation of the slopes can be derived by using the same optimization criterion used in the process of deriving weight update equations. Specifically, the slopes are to be chosen so as to minimize the performance criterion

$$E_q = \frac{1}{2}\left(\boldsymbol{d}_q - \boldsymbol{x}_{\text{out}}^{(s)}\right)^T \left(\boldsymbol{d}_q - \boldsymbol{x}_{\text{out}}^{(s)}\right) = \frac{1}{2}\sum_{h=1}^{n_s}\left(d_{qh} - x_{\text{out},h}^{(s)}\right)^2 \tag{3.113}$$

where s denotes the number of layers in the network and $\boldsymbol{d}_q \in \Re^{n_s \times 1}$ and $\boldsymbol{x}_{\text{out}}^{(s)}$ are desired and actual outputs, respectively, of the network due to the qth training pattern. Consider an activation function of the sigmoidal type given by

$$f(v, \gamma) = \frac{1 - \exp(-\gamma v)}{1 + \exp(-\gamma v)} \tag{3.114}$$

where v is the input to the nonlinearity (output of the linear combiner) and γ is the slope parameter to be adjusted so that (3.113) is minimized. Considering the nonlinearity of the ith neuron in the sth layer of the network, we apply the same approach used in the derivation of the weight update equations in standard backpropagation, obtaining

$$\gamma_i^{(s)}(k+1) = \gamma_i^{(s)}(k) - \beta \frac{\partial E_q}{\partial \gamma_i^{(s)}} \tag{3.115}$$

Using the chain rule, the second term on the right side in (3.115) can be rewritten as

$$\frac{\partial E_q}{\partial \gamma_i^{(s)}} = \frac{\partial E_q}{\partial v_i^{(s)}} \frac{\partial v_i^{(s)}}{\partial x_{\mathrm{out},i}^{(s)}} \frac{\partial x_{\mathrm{out},i}^{(s)}}{\partial \gamma_i^{(s)}}$$

$$= -\delta_i^{(s)} \frac{1}{\partial x_{\mathrm{out},i}^{(s)} / \partial v_i^{(s)}} \frac{\partial x_{\mathrm{out},i}^{(s)}}{\partial \gamma_i^{(s)}} = -\delta_i^{(s)} \frac{f_\gamma(v, \gamma)}{f_v(v, \gamma)} \tag{3.116}$$

where $\delta_i^{(s)}$ is the local error for the ith neuron of the sth layer, and $f_\gamma(v, \gamma)$ and $f_v(v, \gamma)$ denote partial derivatives of the activation function with respect to γ and v, respectively. If the activation function given in (3.114) is used, we have

$$f_\gamma(v, \gamma) = \frac{1}{2}[1 - f(v, \gamma)^2] \tag{3.117}$$

and

$$f_v(v, \gamma) = \frac{1}{2}[1 - f(v, \gamma)^2] \tag{3.118}$$

By substituting (3.116), (3.117), and (3.118) into (3.115), the update equation for the slope of the activation function becomes

$$\gamma_i^{(s)}(k+1) = \gamma_i^{(s)}(k) + \beta\delta_i^{(s)} \tag{3.119}$$

Frequently, a momentum term is added to this update equation (3.119) to improve stability. In addition, to avoid possible linearization of the neural network mapping, the value of the slope is prevented from becoming smaller than some predetermined value $\gamma_{\min}$. The entire backpropagation algorithm with adaptive slopes can be summarized in the following six steps:

Backpropagation algorithm with adaptive activation function slopes

Step 1. Initialize the weights in the network according to the standard initialization procedure discussed in Section 3.3.2.

Step 2. From the set of the training input/output pairs, present the input pattern and calculate the network response.

Step 3. Compare the desired network response with the actual output of the network, and the local errors are computed using

$$\delta_i^{(s)} = \left(d_{qh} - x_{\mathrm{out},i}^{(s)}\right) g\left(v_i^{(s)}\right) \qquad \text{for output layer}$$

$$\delta_i^{(s)} = \left(\sum_{h=1}^{n_2} \delta_h^{(s+1)} w_{hi}^{(s+1)}\right) g\left(v_i^{(s)}\right) \qquad \text{for hidden layers}$$

Step 4. The weights of the network are updated according to

$$w_{ij}^{(s)}(k+1) = w_{ij}^{(s)}(k) + \mu^{(s)} \delta_i^{(s)} x_{\mathrm{out},j}^{(s)}$$

Step 5. The slopes of the activation functions are updated according to

$$\gamma_i^{(s)}(k+1) = \gamma_i^{(s)}(k) + \beta\delta_i^{(s)} + \rho\left[\gamma_i^{(s)}(k) - \gamma_i^{(s)}(k-1)\right]$$

If $\gamma_i^{(s)}(k+1) < \gamma_{\min}$, then $\gamma_i^{(s)}(k+1) = \gamma_{\min}$.

Step 6. Stop if the network has converged; else go back to step 2.

Comparing the algorithm above with standard backpropagation, we see that the only difference lies in step 5 where the update of the slopes is performed. Since the local error is already computed as a necessary part of the weight update equation, the update of the slopes does not impose any significant computational burden.

Although this method was presented as a modification of the standard backpropagation algorithm, adaptation of the slopes can be performed in conjunction with other versions of backpropagation-based algorithms. In addition, an interesting version of the adaptive slope method, when the activation function is piecewise linear or quantized, is presented in [42, 43]. Piecewise linear and quantized activation functions are of extreme importance from the standpoint of practical hardware implementation of MLP NNs.

3.4.4 Levenberg-Marquardt Algorithm

The Levenberg–Marquardt backpropagation (LMBP) algorithm represents a simplified version of Newton's method applied to the problem of training MLP NNs. Newton's method is a well-established numerical optimization technique with quadratic speed of convergence. Here, we present only the algorithm. A detailed explanation of Newton's method is provided in Section A.5.3, and readers are strongly encouraged to review that section.

Summary of Newton's optimization algorithm

Consider the numerical optimization problem of finding a vector $\boldsymbol{w} \in \Re^{N\times 1}$ that minimizes a given energy function $E(\boldsymbol{w}) : \Re^{N\times 1} \rightarrow \Re^{+}$. According to Newton's method, an iterative procedure that will accomplish this minimization task can be formulated as follows:

Step 1. Initialize the components of the vector $\boldsymbol{w} \in \Re^{N\times 1}$ to some random values.

Step 2. Update the vector $\boldsymbol{w}$ according to

$$\boldsymbol{w}(k+1) = \boldsymbol{w}(k) - \boldsymbol{H}_k^{-1}\boldsymbol{g}_k \tag{3.120}$$

where matrix $\boldsymbol{H}_k^{-1} \in \Re^{N\times N}$ represents the inverse of the Hessian matrix. The Hessian matrix is given by

$$\boldsymbol{H} = \begin{bmatrix} \dfrac{\partial^2 E(\boldsymbol{w})}{\partial w_1^2} & \dfrac{\partial^2 E(\boldsymbol{w})}{\partial w_1 \partial w_2} & \cdots & \dfrac{\partial^2 E(\boldsymbol{w})}{\partial w_1 \partial w_N} \\ \dfrac{\partial^2 E(\boldsymbol{w})}{\partial w_2 \partial w_1} & \dfrac{\partial^2 E(\boldsymbol{w})}{\partial w_2^2} & \cdots & \dfrac{\partial^2 E(\boldsymbol{w})}{\partial w_2 \partial w_N} \\ \cdots & \cdots & \cdots & \cdots \\ \dfrac{\partial^2 E(\boldsymbol{w})}{\partial w_N \partial w_1} & \dfrac{\partial^2 E(\boldsymbol{w})}{\partial w_N \partial w_2} & \cdots & \dfrac{\partial^2 E(\boldsymbol{w})}{\partial w_N^2} \end{bmatrix} \tag{3.121}$$

and evaluated at the point $w = w(k)$. The vector $g_k \in \Re^{N\times 1}$ represents the gradient of the energy function, computed as

$$g = \left[\frac{\partial E(w)}{\partial w_1} \quad \frac{\partial E(w)}{\partial w_2} \quad \cdots \quad \frac{\partial E(w)}{\partial w_N}\right]^T \tag{3.122}$$

and evaluated at the point $w = w(k)$.

An obvious problem with Newton's method is the computational requirements involved in calculating the inverse of the Hessian matrix. Even for moderate-sized neural networks, the complexity of the algorithm limits its practical use. The LMBP algorithm offers a viable alternative to Newton's method with approximately the same convergence speed and significantly less complexity. To apply the LMBP algorithm, the problem of training the MLP NN has to be formulated as a nonlinear optimization problem.

Consider the MLP NN shown in Figure 3.4. The task of the neural network training can be viewed as finding a set of network weights that minimizes the error between the desired and actual network output for all the patterns in the training set. If the number of patterns is finite, the energy function can be written as

$$E(w) = \frac{1}{2}\sum_{q=1}^{Q}\left(d_q - x_{\text{out},q}^{(3)}\right)^T\left(d_q - x_{\text{out},q}^{(3)}\right) = \frac{1}{2}\sum_{q=1}^{Q}\sum_{h=1}^{n_3}\left(d_{qh} - x_{\text{out},qh}^{(3)}\right)^2 \tag{3.123}$$

where Q is the total number of training patterns, w represents the vector containing all the weights in the network, d_q is the desired output, and $x_{\text{out},q}^{(3)}$ is the actual network output due to the qth training pattern. According to Newton's method, the set of optimal weights that minimizes the energy function in (3.123) can be found by applying

$$w(k+1) = w(k) - H_k^{-1}g_k \tag{3.124}$$

where

$$H_k = \nabla^2 E(w)\big|_{w=w(k)} \tag{3.125}$$

and

$$g_k = \nabla E(w)\big|_{w=w(k)} \tag{3.126}$$

By defining $P = n_3 Q$, (3.123) can be rewritten as

$$E(w) = \frac{1}{2}\sum_{p=1}^{P}\left(d_p - x_{\text{out},p}^{(3)}\right)^2 = \frac{1}{2}\sum_{p=1}^{P} e_p^2 \tag{3.127}$$

where

$$e_p = d_p - x_{\text{out},p}^{(3)} \tag{3.128}$$

The gradient of the energy function in (3.126) can be computed as

$$g = \frac{\partial E(\boldsymbol{w})}{\partial \boldsymbol{w}} = \frac{1}{2}\begin{bmatrix} \dfrac{\partial \sum_{p=1}^{P} e_p^2}{\partial w_1} \\ \dfrac{\partial \sum_{p=1}^{P} e_p^2}{\partial w_2} \\ \vdots \\ \dfrac{\partial \sum_{p=1}^{P} e_p^2}{\partial w_N} \end{bmatrix} = \begin{bmatrix} \sum_{p=1}^{P} e_p \dfrac{\partial e_p}{\partial w_1} \\ \sum_{p=1}^{P} e_p \dfrac{\partial e_p}{\partial w_2} \\ \vdots \\ \sum_{p=1}^{P} e_p \dfrac{\partial e_p}{\partial w_N} \end{bmatrix} = \boldsymbol{J}^T \boldsymbol{e} \tag{3.129}$$

where $\boldsymbol{J} \in \Re^{P\times N}$ is the Jacobian matrix defined by

$$\boldsymbol{J} = \begin{bmatrix} \dfrac{\partial e_1}{\partial w_1} & \dfrac{\partial e_1}{\partial w_2} & \cdots & \dfrac{\partial e_1}{\partial w_N} \\ \dfrac{\partial e_2}{\partial w_1} & \dfrac{\partial e_2}{\partial w_2} & \cdots & \dfrac{\partial e_2}{\partial w_N} \\ \cdots & \cdots & \cdots & \cdots \\ \dfrac{\partial e_P}{\partial w_1} & \dfrac{\partial e_P}{\partial w_2} & \cdots & \dfrac{\partial e_P}{\partial w_N} \end{bmatrix} \tag{3.130}$$

Next we need to find an expression for the Hessian matrix. The k, j element of the Hessian matrix can be expressed as

$$[\nabla^2 E(\boldsymbol{w})]_{k,j} = \frac{\partial^2 E(\boldsymbol{w})}{\partial w_k \partial w_j} = \sum_{p=1}^{P}\left(\frac{\partial e_p}{\partial w_k}\frac{\partial e_p}{\partial w_j} + e_p \frac{\partial^2 e_p}{\partial w_k \partial w_j}\right) \tag{3.131}$$

By using the expression for the Jacobian matrix in (3.130), the Hessian can be expressed as

$$\nabla^2 E(\boldsymbol{w}) = \boldsymbol{J}^T \boldsymbol{J} + \boldsymbol{S} \tag{3.132}$$

where matrix $\boldsymbol{S} \in \Re^{N\times N}$ is the matrix of second-order derivatives given by

$$\boldsymbol{S} = \sum_{p=1}^{P} e_p \nabla^2 e_p \tag{3.133}$$

When approaching the minimum of the energy function, the elements of matrix $\boldsymbol{S}$ become small, and the Hessian can be closely approximated by

$$\boldsymbol{H} \approx \boldsymbol{J}^T \boldsymbol{J} \tag{3.134}$$

Substituting (3.129) and (3.134) into the expression for Newton's method given in (3.124), we obtain

$$\boldsymbol{w}(k+1) = \boldsymbol{w}(k) - \left[\boldsymbol{J}_k^T \boldsymbol{J}_k\right]^{-1} \boldsymbol{J}_k^T \boldsymbol{e}_k \tag{3.135}$$

where subscript k indicates the evaluation of the appropriate matrices at $\boldsymbol{w} = \boldsymbol{w}(k)$.

One problem with the iterative update given in (3.135) is that it requires the inversion of matrix $\boldsymbol{H} = \boldsymbol{J}^T\boldsymbol{J}$ which may be ill conditioned or even singular. This problem can be easily resolved by the following modification of (3.134)

$$\boldsymbol{H} \approx \boldsymbol{J}^T\boldsymbol{J} + \mu\boldsymbol{I} \tag{3.136}$$

where μ is a small number and $\boldsymbol{I} \in \Re^{N\times N}$ is the identity matrix. Substituting (3.136) into (3.135) results in the Levenberg–Marquardt algorithm [35] for updating the network weights

$$\boldsymbol{w}(k+1) = \boldsymbol{w}(k) - \left[\boldsymbol{J}_k^T\boldsymbol{J}_k + \mu_k\boldsymbol{I}\right]^{-1}\boldsymbol{J}_k^T\boldsymbol{e}_k \tag{3.137}$$

Before we show how (3.137) can be implemented in the neural network environment, we need to realize that it represents a transition between the steepest descent method and Newton's method. For a small value of μ_k in (3.137), it approaches the algorithm given in (3.135) which approximates Newton's algorithm. When the value of μ_k is increased, the second term inside the square brackets in (3.137) becomes dominant and the update equation can be written as

$$\begin{aligned}\boldsymbol{w}(k+1) &= \boldsymbol{w}(k) - \left[\boldsymbol{J}_k^T\boldsymbol{J}_k + \mu_k\boldsymbol{I}\right]^{-1}\boldsymbol{J}_k^T\boldsymbol{e}_k \\ &\approx \boldsymbol{w}(k) - [\mu_k\boldsymbol{I}]^{-1}\boldsymbol{J}_k^T\boldsymbol{e}_k = \boldsymbol{w}(k) - \frac{1}{\mu_k}\boldsymbol{J}_k^T\boldsymbol{e}_k\end{aligned} \tag{3.138}$$

Defining $\alpha_k = 1/\mu_k$ and using (3.129), we see that (3.138) can be rewritten as

$$\boldsymbol{w}(k+1) = \boldsymbol{w}(k) - \alpha_k\boldsymbol{g}_k \tag{3.139}$$

which is the steepest descent gradient method.

The biggest problem in implementing the LMBP algorithm is the calculation of the Jacobian matrix $\boldsymbol{J}(\boldsymbol{w})$. Each term in the matrix has the form

$$J_{i,j} = \frac{\partial e_i}{\partial w_j} \tag{3.140}$$

The simplest approach to compute the derivatives in (3.140) is to use the approximation

$$J_{i,j} \approx \frac{\Delta e_i}{\Delta w_j} \tag{3.141}$$

where Δe_i represents the change in the output error due to the small perturbation of the weight Δw_j. This method is relatively straightforward and simple to implement. The perturbations of the weights are kept small, at least an order of magnitude smaller than current learning rate parameter μ_k. After the Jacobian matrix is computed, (3.137) can be used to perform the weight updates.

Levenberg–Marquardt version of backpropagation algorithm

Step 1. Initialize the network weights to small random values. One of the initialization procedures suggested in Section 3.3.2 can be used. Set the learning rate parameter.

Step 2. Present an input pattern, and calculate the output of the network.

Step 3. Use (3.141) to calculate the elements of the Jacobian matrix associated with the input/output pairs.
Step 4. When the last input/output pair is presented, use (3.137) to perform the update of the weights.
Step 5. Stop if the network has converged; else, go back to step 2.

Additional remarks:

1. The weight update approach presented here is a batch version of the LMBP algorithm. A scalar version of the algorithm can be found in [35].
2. The learning rate parameter μ_k in (3.137) can be modified dynamically during the training process. At early stages of network training it should be kept relatively small, and the algorithm approaches Newton's method. To prevent oscillations, and to perform fine-tuning of the network weights, at the later stages of network training the learning rate parameter is increased, and the algorithm approaches the steepest descent gradient method.
3. The Jacobian matrix used in the update equation need not be computed for the entire set of input/output training pairs. In the interest of reducing memory requirements, the updates can be performed after a subset of the training patterns is presented to the network.

3.5 COUNTERPROPAGATION

The counterpropagation network was developed by Hecht-Nielsen [44–46]. It functions as a self-programming optimal lookup table, providing a bidirectional mapping between input and output training patterns. It can be used as an alternative to the MLP NN trained by backpropagation in cases when the speed of the network training is of prime interest. Counterpropagation networks typically converge much more quickly than MLP NNs. However, the number of neurons required to achieve a desired accuracy is typically much greater than that needed for an MLP NN. Therefore, the most common use of counterpropagation is for the prototype stage of developing a neurocomputing system. After the neurocomputing system is developed using counterpropagation, this network is replaced by an MLP NN. In this section we will present two versions of the counterpropagation network, *forward-only* counterpropagation and *full* counterpropagation.

Forward-only counterpropagation neural network

The architecture for the forward-only counterpropagation network is shown in Figure 3.7. As we can see, the network consists of an input, an output, and one hidden layer. There are two sets of weights that are adjusted with two different learning algorithms. Weights connecting the input and the hidden layer are trained using Kohonen's self-organizing learning rule, and the weights between the hidden and the output layers are trained using Grossberg's learning algorithm.

During the training process the network is presented with examples of the desired mapping, that is, with input vector $\boldsymbol{x} \in \Re^{n \times 1}$ and output vector

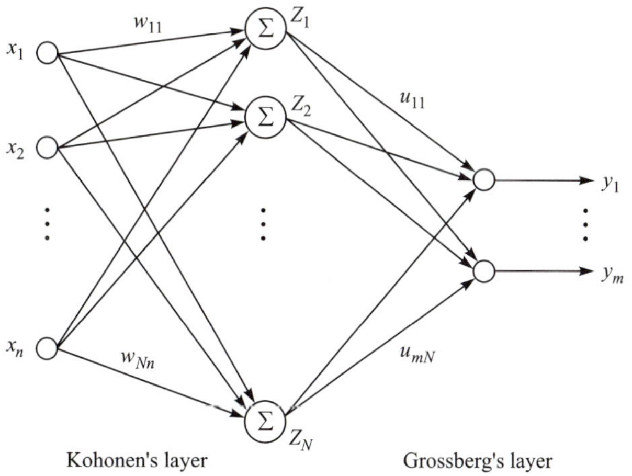

Figure 3.7 Forward-only counterpropagation network architecture.

$y \in \Re^{m\times 1}$. The weights in the Kohonen and Grossberg layers are trained independently. First, the network computes the distance between the input vector and the weights for the connections between the input layer and each of the neurons in the hidden layer. The distance can be computed according to

$$z_j = \text{dist}(\boldsymbol{x}, \boldsymbol{w}_j) = \|\boldsymbol{x} - \boldsymbol{w}_j\|_2 = \sqrt{\sum_{k=1}^{n} (x_k - w_{jk})^2} \tag{3.142}$$

After the distances are computed, the neurons in the hidden layer are allowed to compete and their outputs are set according to

$$z_i = \begin{cases} 1 & \text{if } i \text{ is the smallest integer for which } z_i \le z_j, \text{ for all } j \\ 0 & \text{otherwise} \end{cases} \tag{3.143}$$

In other words, the neuron with weights "closest" to the input pattern wins, and its output is set to 1. All other neurons have a 0 output. Finally, the weights on the connections to the winning neuron are updated according to Kohonen's self-organizing learning rule

$$\boldsymbol{w}_j(k+1) = [1 - \alpha(k)]\boldsymbol{w}_j(k) + \alpha(k)\boldsymbol{x} \tag{3.144}$$

Other processing elements in the hidden layer do not adjust their weights. The learning rate α commonly starts at a relatively large value, say, $\alpha = 0.9$, and decreases gradually over the course of network training. The weights in the Kohonen layer distribute themselves accordingly to the statistical properties of the input vectors. Thus, this stage of training essentially performs an optimal sampling of the input vector space.

The weights in the output layer of the counterpropagation network are updated according to the Grossberg learning rule

$$u_{ji}(k+1) = u_{ji}(k) + \beta(k)\left[-u_{ji}(k) + y_j\right]z_i \tag{3.145}$$

where β represents the learning rate parameter for the output layer. Only the weights connecting the winning neuron in the hidden layer with the neurons in the output layer will be updated.

Algorithm for training counterpropagation networks

Step 1. Select the number of neurons in the clustering layer to be N, and initialize the weights to random values within the interval bounded by the variations of the input vector components.

Step 2. Present an input pattern $\boldsymbol{x} \in \Re^{n\times 1}$ and corresponding output target vector $\boldsymbol{y} \in \Re^{m\times 1}$.

Step 3. Calculate the distance between the input vector and the weights of the connections to the cluster units according to

$$z_j = \left[\sum_{k=1}^{n} (x_k - w_{jk})^2\right]^{1/2} \qquad j = 1, 2, \ldots, N$$

Step 4. Calculate the output of the cluster units according to

$$z_i = \begin{cases} 1 & \text{if } i \text{ is the smallest integer for which } z_i \leq z_j, \text{ for all } j \\ 0 & \text{otherwise} \end{cases}$$

Step 5. Update the weights in the Kohonen layer according to

$$\boldsymbol{w}_j(k+1) = [1 - \alpha(k)]\boldsymbol{w}_j(k) + \alpha(k)\boldsymbol{x}$$

where the learning rate α is decreased gradually during the training. One possibility is to gradually decrease α according to

$$\alpha(k) = \alpha(0)\exp\left(-\frac{k}{k_0}\right)$$

where k_0 is a suitably chosen time constant.

Step 6. Update the weights in the output layer according to

$$u_{ji}(k+1) = u_{ji}(k) + \beta(k)[-u_{ji}(k) + y_j]z_i \qquad j = 1, 2, \ldots, m;\ i = 1, 2, \ldots, N$$

where z_i is the *winning* cluster unit.

Step 7. Stop if network has converged; else, go back to step 2.

EXAMPLE 3.3. Consider the problem of designing a neural network to approximate the function

$$y = \frac{1}{x+1}$$

in the interval [0, 4]. To accomplish the task, we can use a forward-only counterpropagation network that has 20 neurons in the hidden layer. During network training the learning rate for the Kohonen layer is set to $\alpha_0 = 0.95$ and decreased exponentially according to

$$\alpha(k) = \alpha_0 \exp\left(-\frac{k}{10}\right)$$

The network trained for 50 epochs, and the approximation results are presented in Figure 3.8.

Full counterpropagation

While *forward-only* counterpropagation networks are trained to provide the mapping in only one direction, *full* counterpropagation networks are designed to learn bidirectional mappings. Through the process of supervised training, the network adaptively constructs a lookup table approximating the

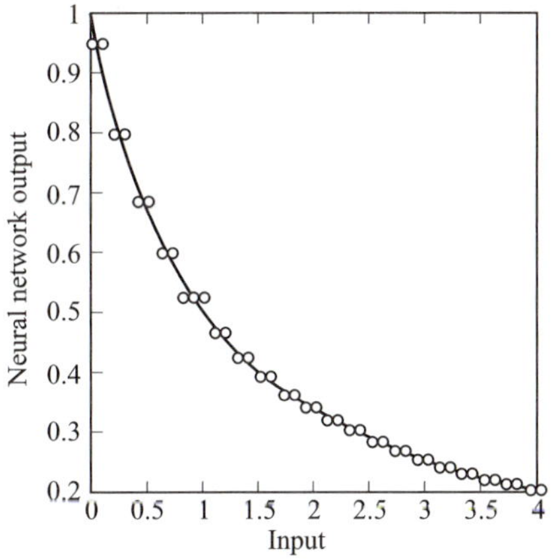

Figure 3.8 Approximation of the function $y = 1/(x + 1)$ using a forward-only counterpropagation neural network.

mapping between the presented input/output training pairs: $\boldsymbol{x} \in \Re^{n \times 1}$, $\boldsymbol{y} \in \Re^{m \times 1}$. This represents an extension of the forward-only counterpropagation network. After training of the *full* counterpropagation network, it can be used to reconstruct the corresponding $\boldsymbol{y}^*$ vector if $\boldsymbol{x}$ is known, and vice versa. The architecture for the full counterpropagation network is presented in Figure 3.9. For the sake of simplicity, individual weights are omitted from the figure.

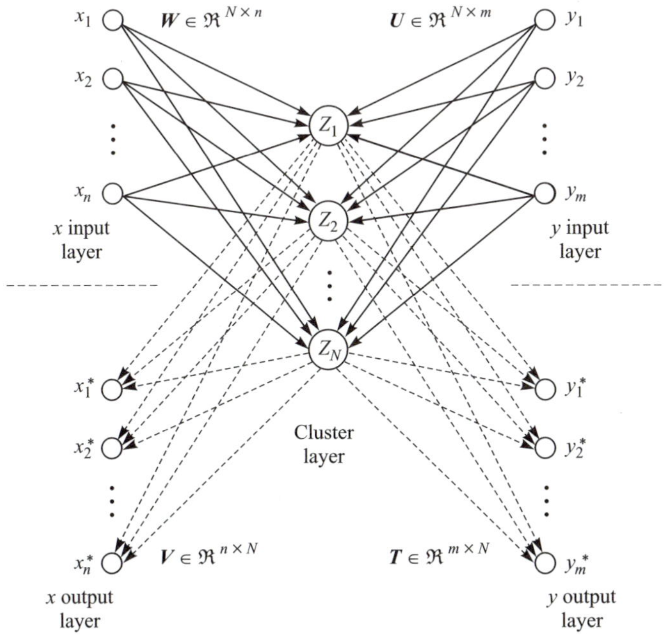

Figure 3.9 Architecture for the full counterpropagation network.

Full counterpropagation networks have four sets of weights. The weights connecting the x and y input layers with the clustering layer are trained using Kohonen's self-organizing learning rule, and the weights connecting the clustering layer with the two output layers are trained using the Grossberg learning rule. In the basic version of the algorithm, only the weights associated with the winning neuron are allowed to learn.

Algorithm for training full counterpropagation networks

Step 1. Select the number of neurons N for the clustering layer, and initialize the weights of the network. Initialize the weights of the network to random values when
  (*a*) $\boldsymbol{W}$ and $\boldsymbol{V}$ are initialized to random values bounded by the maximum and minimum for the components of $\boldsymbol{x}$ inputs.
  (*b*) $\boldsymbol{U}$ and $\boldsymbol{T}$ are initialized to random values bounded by the maximum and minimum for the components of $\boldsymbol{y}$ inputs.

Step 2. Calculate the *distance* between the input $\boldsymbol{x}, \boldsymbol{y}$ pair and the units in the clustering layer according to

$$z_i = \sum_{k=1}^{n} (x_k - w_{i,k})^2 + \sum_{k=1}^{m} (y_k - u_{i,k})^2$$

Step 3. Set the activation function for the neurons in the clustering layer according to

$$z_i = \begin{cases} 1 & \text{if } i \text{ is smallest integer for which } z_i \leq z_j, \text{ for all } j \\ 0 & \text{otherwise} \end{cases}$$

Step 4. Update the weights in the Kohonen layers according to

$$\boldsymbol{w}_i(k+1) = [1 - \alpha_x(k)]\boldsymbol{w}_i(k) + \alpha_x(k)\boldsymbol{x}$$

and

$$\boldsymbol{u}_i(k+1) = [1 - \alpha_y(k)]\boldsymbol{u}_i(k) + \alpha_y(k)\boldsymbol{y}$$

where α_x and α_y are learning rate parameters commonly decreased over the course of training, and i is the index of the winning neuron, that is, the one whose activation function is set to 1.

Step 5. Update the weights in the Grossberg layers according to

$$v_{ji}(k+1) = v_{ji}(k) + \beta_x(k)[-v_{ji}(k) + x_j]z_i \qquad j = 1, 2, \ldots, n;\ i = 1, 2, \ldots, N$$

and

$$t_{ji}(k+1) = t_{ji}(k) + \beta_y(k)[-t_{ji}(k) + y_j]z_i \qquad j = 1, 2, \ldots, m;\ i = 1, 2, \ldots, N$$

Step 6. Stop if the network has converged; else, go back to step 2.

3.6 RADIAL BASIS FUNCTION NEURAL NETWORKS

In practice, the supervised training of the neural networks can be viewed as a curve-fitting process. The network is presented with training pairs, each consisting of a vector from an input space and a desired network response. Through a defined learning algorithm, the network performs the adjustments of its weights so that the error between the actual and desired responses is minimized relative to some optimization criterion. Once trained, the network performs the interpolation in the output vector space, and this is referred to as the generalization property. In previous sections we saw that backpropagation and counterpropagation networks can be trained to perform a nonlinear mapping between the input and output vector spaces. In this section we present an alternative network that is capable of accomplishing the same task. This is the radial basis function neural network (RBF NN).

The architecture of the RBF NN is presented in Figure 3.10. The network consists of three layers: an input layer, a single layer of nonlinear processing neurons, and an output layer. The output of the RBF NN is calculated according to

$$y_i = f_i(\boldsymbol{x}) = \sum_{k=1}^{N} w_{ik}\phi_k(\boldsymbol{x}, \boldsymbol{c}_k) = \sum_{k=1}^{N} w_{ik}\phi_k(\|\boldsymbol{x} - \boldsymbol{c}_k\|_2), \quad i = 1, 2, \ldots, m \quad (3.146)$$

where $\boldsymbol{x} \in \Re^{n\times 1}$ is an input vector, $\phi_k(\cdot)$ is a function from $\Re^+$ (set of all positive real numbers) to $\Re$, $\|\cdot\|_2$ denotes the Euclidean norm, w_{ik} are the weights in the output layer, N is the number of neurons in the hidden layer, and $\boldsymbol{c}_k \in \Re^{n\times 1}$ are the RBF centers in the input vector space. For each neuron in the hidden layer, the Euclidean distance between its associated center and the input to the network is computed. The output of the neuron in a hidden layer is a nonlinear function of the distance. Finally, the output of the network is computed as a weighted sum of the hidden layer outputs. The functional form of $\phi_k(\cdot)$ is assumed to have been given, and some typical choices are [26, 47] as follows:

1. $\phi(x) = x$	linear function
2. $\phi(x) = x^3$	cubic approximation
3. $\phi(x) = x^2 \ln x$	thin-plate-spline function
4. $\phi(x) = \exp(-x^2/\sigma^2)$	Gaussian function
5. $\phi(x) = \sqrt{x^2 + \sigma^2}$	multiquadratic function
6. $\phi(x) = \dfrac{1}{\sqrt{x^2 + \sigma^2}}$	inverse multiquadratic function

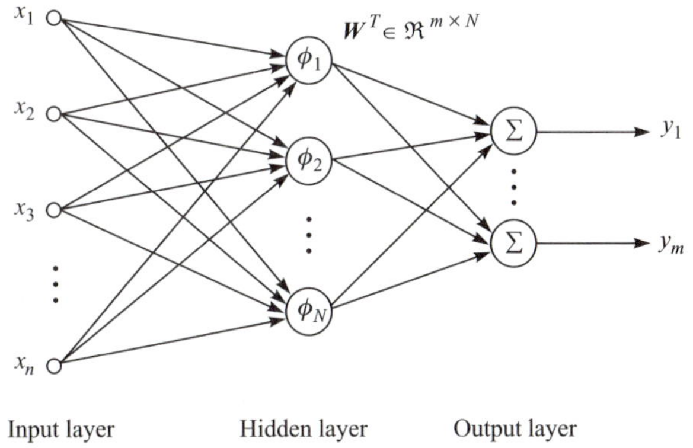

Figure 3.10 The RBF NN architecture.

where parameter σ controls the "width" of the RBF and is commonly referred to as the spread parameter. In practical applications, the most widely used RBF is the Gaussian RBF.

The centers $\mathbf{c}_k$ are defined points that are assumed to perform an adequate sampling of the input vector space. They are usually chosen as a subset of the input data. We see from Figure 3.10 that there is no interaction between the neurons in the output layer. For this reason, without loss of generality, we can consider a single-output RBF NN. The network having more than one output can be considered as a superposition of several single-output networks sharing a common hidden layer.

3.6.1 Training the RBF NN with Fixed Centers

It follows from (3.146) that there are two set of parameters governing the mapping properties of the RBF NN: the weights w_{ik} in the output layer and the centers $\mathbf{c}_k$ of the radial basis functions. The simplest form of RBF NN training is with fixed centers. In particular, they are commonly chosen in a random manner as a subset of the input data set. This approach was first suggested by Broomhead and Lowe [48]. The reasoning behind their method can be summarized as follows. A sufficient number of centers randomly selected from the input data set would distribute according to the probability density function of the training data, thus providing an adequate sampling of the input space. Qualitatively, this method can be considered as a "sensible" approach [29]; however, it is very difficult to quantify what a sufficient number of centers should be so that an adequate sampling of the input space is achieved. Common practice is to select a relatively large number of input vectors as the centers. Thus we can ensure an adequate input space sampling. After the network has trained, some of the centers may be removed in a systematic manner without significant degradation of the network mapping performance.

Once the centers are chosen, the output of the network to the input vectors in the training data set can be computed as

$$\hat{y}(q) = \sum_{k=1}^{N} w_{ik}\phi(\mathbf{x}(q), \mathbf{c}_k), \quad q = 1, 2, \ldots, Q \tag{3.147}$$

where Q is the total number of the training pairs. Arranging (3.147) in vector-matrix form, we have

$$\begin{bmatrix} \hat{y}(1) \\ \hat{y}(2) \\ \vdots \\ \hat{y}(Q) \end{bmatrix} = \begin{bmatrix} \phi(\mathbf{x}(1), \mathbf{c}_1) & \phi(\mathbf{x}(1), \mathbf{c}_2) & \cdots & \phi(\mathbf{x}(1), \mathbf{c}_N) \\ \phi(\mathbf{x}(2), \mathbf{c}_1) & \phi(\mathbf{x}(2), \mathbf{c}_2) & \cdots & \phi(\mathbf{x}(2), \mathbf{c}_N) \\ \cdots & \cdots & \cdots & \cdots \\ \phi(\mathbf{x}(Q), \mathbf{c}_1) & \phi(\mathbf{x}(Q), \mathbf{c}_2) & \cdots & \phi(\mathbf{x}(Q), \mathbf{c}_N) \end{bmatrix} \begin{bmatrix} w_1 \\ w_2 \\ \vdots \\ w_N \end{bmatrix} \tag{3.148}$$

or

$$\hat{\mathbf{y}} = \mathbf{\Phi}\mathbf{w} \tag{3.149}$$

where $\hat{\boldsymbol{y}} \in \Re^{Q\times 1}$ is the actual network output vector, $\boldsymbol{w} \in \Re^{N\times 1}$ is the vector of the weights in the output layer, and $\Phi \in \Re^{Q\times N}$ is the matrix of RBF nonlinear mappings performed by the hidden layer. Because the centers are fixed, the mapping performed by the hidden layer is fixed as well. Therefore, the network training task is to determine the appropriate settings of the weights in the network output layer so that the performance of the network mapping is optimized in some sense. A common optimization criterion to use is the mean-squared error between the actual and desired network outputs. In other words, the optimal set of weights minimizes the performance measure

$$J(\boldsymbol{w}) = \frac{1}{2}\sum_{q=1}^{Q}[y_d(q) - \hat{y}(q)]^2 = \frac{1}{2}(\boldsymbol{y}_d - \hat{\boldsymbol{y}})^T(\boldsymbol{y}_d - \hat{\boldsymbol{y}}) \tag{3.150}$$

where $\boldsymbol{y}_d \in \Re^{Q\times 1}$ denotes the vector of desired network outputs. Substituting (3.149) into (3.150) gives

$$J(\boldsymbol{w}) = \frac{1}{2}(\boldsymbol{y}_d - \Phi\boldsymbol{w})^T(\boldsymbol{y}_d - \Phi\boldsymbol{w}) = \frac{1}{2}(\boldsymbol{y}_d^T\boldsymbol{y}_d - 2\boldsymbol{y}_d^T\Phi\boldsymbol{w} + \boldsymbol{w}^T\Phi^T\Phi\boldsymbol{w}) \tag{3.151}$$

Minimizing the performance measure $J(\boldsymbol{w})$ is achieved by

$$\frac{\partial J(\boldsymbol{w})}{\partial \boldsymbol{w}} = 0 \tag{3.152}$$

or

$$-\Phi^T\boldsymbol{y}_d + \Phi^T\Phi\boldsymbol{w} = 0 \tag{3.153}$$

Solving for $\boldsymbol{w}$, we have

$$\boldsymbol{w} = (\Phi^T\Phi)^{-1}\Phi^T\boldsymbol{y}_d = \Phi^+\boldsymbol{y}_d \tag{3.154}$$

where Φ^+ denotes the pseudoinverse (cf. Sect. A.2.7) of the nonlinear mapping matrix Φ.

From (3.154) we see that in the case of fixed network centers, the problem of network training has a "closed-form" solution. This effectively means that the RBF NN can be trained very fast in comparison to backpropagation, and even counterpropagation, networks. The fact that it performs a nonlinear multidimensional interpolation, and yet it can be trained using a linear least-squares algorithm, makes the RBF NN very attractive for various types of signal processing applications [29, 47, 49–52].

There is another important point to be made. Depending on the size of the RBF NN system of equations, (3.153) can be underdetermined, can be overdetermined, or can have a unique solution. If the number of centers is greater than or equal to the number of training patterns, the error between the desired and actual network outputs can be made arbitrarily small. In fact, if (3.154) is used, the error will always be equal to 0. Consider the following example.

EXAMPLE 3.4. An RBF NN is to be trained to approximate the nonlinear function

$$y = e^{-x}\sin(3x)$$

in the interval [0, 4]. This is the same task that we examined in Example 3.2 using an MLP NN. To ensure a valid comparison, the interval [0, 4] was again sampled with 21 points separated by 0.2. The MATLAB function **trainrbfe** was used to train the network. The number of hidden units is 21, and the centers correspond to the input training vectors. Therefore, the system of equations in (3.153) has a unique solution. The MATLAB function **trainrbfe** uses a Gaussian RBF, and the spread parameter σ was set to 0.2. After training, the response of the network to the training patterns was plotted, and is shown in Figure 3.11(*a*). To test the generalization capability of the network, the same interval was sampled with 401 points, separated by 0.01. This generated the test data set. The network response to this test data set is presented in Figure 3.11(*b*).

The results presented in Example 3.4 represent a case of data overfitting using a RBF NN. In Example 3.2 we saw that overfitting entirely destroys the performance of the MLP NN. Figure 3.11(*b*) shows that the RBF NN at least maintains acceptable performance even in the case when it overfits the training data.

Setting the spread parameter

In the case of the Gaussian RBF, the spread parameter σ is commonly set according to the following simple heuristic relationship [29]

$$\sigma = \frac{d_{\text{max}}}{\sqrt{K}} \tag{3.155}$$

where d_{max} is the maximum Euclidean distance between the selected center and K is the number of the centers. Using (3.155), the RBF of a neuron in the hidden layer of the network is given as

$$\phi(\boldsymbol{x}, \boldsymbol{c}_k) = \exp\left(-\frac{K}{d_{\text{max}}^2}\left\|\boldsymbol{x} - \boldsymbol{c}_k\right\|^2\right) \tag{3.156}$$

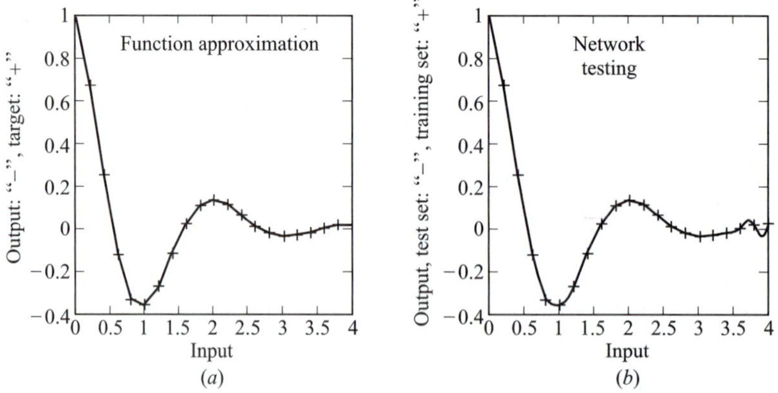

Figure 3.11 RBF neural network response (*a*) to the training data and (*b*) to the testing data.

Step 1. Choose the centers for the RBF functions. The centers are chosen from the set of input vectors. A sufficient number of centers have to be selected in order to ensure adequate sampling of the input vector space.
Step 2. Calculate the spread parameter σ for the RBF functions according to (3.155).
Step 3. Initialize the weights in the output layer of the network to some small random values.
Step 4 Calculate the output of the neural network according to (3.149).
Step 5. Solve for the network weights, using (3.154).

The algorithm described above is commonly referred to as a *batch* training algorithm. The pseudoinverse of matrix Φ is obtained in one step, which means that all the training data have to be available in advance. For cases of real-time processing, the pseudoinverse can be computed using one of the iterative numerical procedures such as steepest descent, recursive least-squares, conjugate gradient, Newton's method, and methods presented in Section 7.2.

3.6.2 Training the RBF NN Using the Stochastic Gradient Approach

In the previous section we described a process for training RBF NNs in which the only adjustable parameters of the network were the weights in the output layer. As we saw, this approach leads to an extremely simple training algorithm. However, to perform adequate sampling of the input, a large number of centers must be selected from the input data set. This produces a relatively large network, even for relatively simple problems.

The stochastic gradient approach for training RBF NNs allows the adjustments of all three sets of the network parameters (that is, the weights, position of the RBF centers, and the width of the RBFs). Therefore, along with the weights in the output layer, both the position of the centers as well as the spread parameter for every processing unit in the hidden layer undergo the process of supervised training. The first step in the development of the stochastic gradient-based supervised learning algorithm is to define the instantaneous error cost function as

$$J(n) = \frac{1}{2}|e(n)|^2 = \frac{1}{2}\left[y_d(n) - \sum_{k=1}^{N} w_k(n)\phi\{\boldsymbol{x}(n), \boldsymbol{c}_k(n)\}\right]^2 \tag{3.157}$$

If the RBF is chosen to be Gaussian, (3.157) becomes

$$J(n) = \frac{1}{2}\left[y_d(n) - \sum_{k=1}^{N} w_k(n)\exp\left(-\frac{\|\boldsymbol{x}(n) - \boldsymbol{c}_k(n)\|_2^2}{\sigma_k^2(n)}\right)\right]^2 \tag{3.158}$$

The update equations for the network parameters are given by

$$\boldsymbol{w}(n+1) = \boldsymbol{w}(n) - \mu_w \frac{\partial}{\partial \boldsymbol{w}} J(n)\bigg|_{\boldsymbol{w}=\boldsymbol{w}(n)} \tag{3.159}$$

$$\boldsymbol{c}_k(n+1) = \boldsymbol{c}_k(n) - \mu_c \frac{\partial}{\partial \boldsymbol{c}_k} J(n)\bigg|_{\boldsymbol{c}_k=\boldsymbol{c}_k(n)} \tag{3.160}$$

$$\sigma_k(n+1) = \sigma_k(n) - \mu_\sigma \frac{\partial}{\partial \sigma_k} J(n)\bigg|_{\sigma_k=\sigma_k(n)} \tag{3.161}$$

Stochastic gradient-based method for training an RBF NN

Step 1. Choose the centers for the RBF functions. The centers are chosen from the set of input vectors.
Step 2. Calculate the initial value of the spread parameter for the RBF functions according to (3.155).
Step 3. Initialize the weights in the output layer of the network to some small random values.
Step 4. Present an input vector, and compute the network output according to

$$\hat{y}(n) = \sum_{k=1}^{N} w_k \phi\{\boldsymbol{x}(n), \boldsymbol{c}_k, \sigma_k\}$$

Step 5. Update the network parameters according to

$$\boldsymbol{w}(n+1) = \boldsymbol{w}(n) + \mu_w e(n)\Psi(n)$$

$$\boldsymbol{c}_k(n+1) = \boldsymbol{c}_k(n) + \mu_c \frac{e(n)w_k(n)}{\sigma_k^2(n)} \phi\{\boldsymbol{x}(n), \boldsymbol{c}_k(n), \sigma_k\}[\boldsymbol{x}(n) - \boldsymbol{c}_k(n)]$$

$$\sigma_k(n+1) = \sigma_k(n) + \mu_\sigma \frac{e(n)w_k(n)}{\sigma_k^3(n)} \phi\{\boldsymbol{x}(n), \boldsymbol{c}_k(n), \sigma_k\}\|\boldsymbol{x}(n) - \boldsymbol{c}_k(n)\|^2$$

where

$$\Psi(n) = \left[\phi\{\boldsymbol{x}(n), \boldsymbol{c}_1, \sigma_1\}, \quad \phi\{\boldsymbol{x}(n), \boldsymbol{c}_2, \sigma_2\}, \quad \cdots, \quad \phi\{\boldsymbol{x}(n), \boldsymbol{c}_N, \sigma_N\}\right]^T$$
$$e(n) = \hat{y}(n) - y_d(n)$$

$y_d(n)$ is the desired network output, and μ_w, μ_c, and μ_σ are appropriate learning rate parameters.
Step 6. Stop if the network has converged; else, go back to step 4.

Having the ability to update the center positions and spread parameters of the processing units in the hidden layer greatly improves the performance of the RBF NN. For a given size of the hidden layer, the RBF NN that is trained with the stochastic gradient approach outperforms a network with fixed centers. However, the price paid for this is the increased complexity of the training algorithm, which increases the time needed for the network to be trained. Observing the update equations given above, we note the following [29]:

1. The instantaneous error cost function $J(n)$ is convex relative to the weights in the output layer. However, this is not necessarily true for the position of the centers and associated spread parameters for the RBF of the hidden-layer units. This makes the training algorithm prone to getting stuck in a local minimum.
2. In general, learning rate parameters μ_w, μ_c, and μ_σ are set to different values.

3. The learning rules are still less complex than backpropagation. Because an RBF NN has only one set of adjustable weights (in the output layer), backpropagation of errors is not required.

3.6.3 Orthogonal Least Squares

We have seen that a major challenge in the design of the RBF NN is the selection of the centers. Their selection in a random fashion, and even their modification using the stochastic gradient algorithm, commonly leads to a relatively large network. The orthogonal least-squares (OLS) method offers a systematic method for center selection, and it significantly reduces the size of the RBF NN. Before we present the OLS method for selecting the centers in RBF NN, we review the basics of the Gram–Schmidt orthogonalization procedure.

Gram–Schmidt orthogonalization

The Gram–Schmidt orthogonalization [29, 50] is a process for decomposition of a matrix $\boldsymbol{M} \in \Re^{n \times m}$ into the product of two matrices according to

$$\boldsymbol{M} = \boldsymbol{W}\boldsymbol{A} \tag{3.162}$$

where $\boldsymbol{A} \in \Re^{m \times m}$ is an upper triangular matrix of the form

$$\boldsymbol{A} = \begin{bmatrix} 1 & \alpha_{12} & \cdots & \alpha_{1m} \\ 0 & 1 & \cdots & \alpha_{2m} \\ \cdots & \cdots & \cdots & \cdots \\ 0 & 0 & \cdots & 1 \end{bmatrix} \tag{3.163}$$

and $\boldsymbol{W} \in \Re^{n \times m}$ is a matrix with m mutually orthogonal vectors given by

$$\boldsymbol{W}^T \boldsymbol{W} = \text{diag}(h_1, h_2, \cdots, h_m) \tag{3.164}$$

Gram–Schmidt orthogonalization of a matrix is based on one of the fundamental results from linear vector space theory, the *orthogonal decomposition theorem*. This theorem can be stated as follows.

THEOREM 3.1. Any vector $\boldsymbol{m} \in \Re^{n \times 1}$ can be decomposed uniquely with respect to the subspace $\boldsymbol{Y} \in \Re^{n \times 1}$ into two mutually orthogonal parts. One part is parallel to the subspace $\boldsymbol{Y}$ (i.e., it lies within it), and the other is perpendicular to it. That is,

$$\boldsymbol{m} = \hat{\boldsymbol{m}} + \boldsymbol{e} \tag{3.165}$$

with $\hat{\boldsymbol{m}} \in \boldsymbol{Y}$ and $\boldsymbol{e} \perp \boldsymbol{Y}$. The component $\hat{\boldsymbol{m}}$ is called the orthogonal projection of $\boldsymbol{m}$ onto the subspace $\boldsymbol{Y}$.

In our case, the columns of matrix $\boldsymbol{M}$ can be considered as vectors in the n-dimensional vector space. The orthogonalization is used to obtain an orthonormal set of basis vectors in the subspace spanned by the columns of matrix $\boldsymbol{M}$.

Step 1. Set the first basis vector equal to one of the columns of matrix $\boldsymbol{M}$.

$$\boldsymbol{w}_1 = \boldsymbol{m}_1$$

Step k. Extract the kth basis vector so that it is orthogonal to the previous $k-1$ vectors

$$\alpha_{ik} = \boldsymbol{m}_k^T \boldsymbol{w}_i \qquad 1 \le i \le k-1$$

$$\boldsymbol{w}_k = \boldsymbol{m}_k - \sum_{i=1}^{k-1} \alpha_{ik} \boldsymbol{w}_i$$

Repeat step k until $k = m$.

At every step k the subspace $\boldsymbol{Y}^{(k-1)} = [\boldsymbol{w}_1 \quad \boldsymbol{w}_2 \quad \cdots \quad \boldsymbol{w}_{k-1}]$ is expanded with vector $\boldsymbol{w}_k$ which represents the portion of $\boldsymbol{m}_k$ orthogonal to subspace $\boldsymbol{Y}^{(k-1)}$. Obviously, $\boldsymbol{W} = \boldsymbol{Y}^{(m)}$. Elements of matrix $\boldsymbol{A}$ represent the coordinates of the vectors $\boldsymbol{m}_k$, $1 \le k \le m$, in the coordinate system formed by the columns of $\boldsymbol{W}$.

Orthogonal least-squares regression

The OLS method has its origin in linear regression models. The mapping performed by the RBF NN can be viewed as a regression model of the form

$$\begin{bmatrix} y_d(1) \\ y_d(2) \\ \vdots \\ y_d(Q) \end{bmatrix} = \begin{bmatrix} \phi(\boldsymbol{x}_1, \boldsymbol{c}_1, \sigma_1) & \phi(\boldsymbol{x}_1, \boldsymbol{c}_2, \sigma_2) & \cdots & \phi(\boldsymbol{x}_1, \boldsymbol{c}_N, \sigma_N) \\ \phi(\boldsymbol{x}_2, \boldsymbol{c}_1, \sigma_1) & \phi(\boldsymbol{x}_2, \boldsymbol{c}_2, \sigma_2) & \cdots & \phi(\boldsymbol{x}_2, \boldsymbol{c}_N, \sigma_N) \\ \cdots & \cdots & \cdots & \cdots \\ \phi(\boldsymbol{x}_Q, \boldsymbol{c}_1, \sigma_1) & \phi(\boldsymbol{x}_Q, \boldsymbol{c}_2, \sigma_2) & \cdots & \phi(\boldsymbol{x}_Q, \boldsymbol{c}_N, \sigma_N) \end{bmatrix} \begin{bmatrix} w_1 \\ w_2 \\ \vdots \\ w_N \end{bmatrix} + \begin{bmatrix} e_1 \\ e_2 \\ \vdots \\ e_Q \end{bmatrix} \tag{3.166}$$

or

$$\boldsymbol{y}_d = \boldsymbol{\Phi}\boldsymbol{w} + \boldsymbol{e} \tag{3.167}$$

where $\boldsymbol{y}_d \in \Re^{Q\times 1}$ is the vector of desired network outputs, $\boldsymbol{\Phi} \in \Re^{Q\times N}$ can be regarded as a linear regression matrix with each column vector $\boldsymbol{\phi} \in \Re^{Q\times 1}$ being a regression vector or regressor, $\boldsymbol{w} \in \Re^{N\times 1}$ is the vector of weights, and $\boldsymbol{e} \in \Re^{N\times 1}$ is the vector of errors between the desired and actual network outputs.

The centers of the RBF NN are chosen from the set of input patterns. As we can see from (3.166), there are a total of Q candidates. Using all the Q input patterns as centers will lead to a network performing an error-free mapping. However, in most cases, the network using all Q candidates from the input vector space is of a prohibitively large size. The task of OLS regression is to perform a systematic selection of $N < Q$ centers so that the size of the network is significantly reduced with minimal degradation in network performance.

From (3.166) we see that there is a one-to-one correspondence between the centers of the RBF NN and regressors in the regression matrix Φ. At each step of OLS regression, a new center can be selected in such a manner that the variance increment of the desired output is maximized. Suppose that we have selected $N < Q$ centers. The least-squares solution yielding the weights is given by

$$\hat{\boldsymbol{w}} = \Phi^{+}\boldsymbol{y}_d \tag{3.168}$$

The actual output of the network is given by

$$\hat{\boldsymbol{y}} = \Phi\hat{\boldsymbol{w}} = [\phi_1, \phi_2, \ldots, \phi_N]\hat{\boldsymbol{w}} \tag{3.169}$$

where $\hat{\boldsymbol{y}}$ represents the portion of $\boldsymbol{y}_d$ that resides within the vector space spanned by the columns ϕ_i of the regression matrix Φ.

By using the Gram–Schmidt orthogonalization, the regression matrix can be decomposed as

$$\boldsymbol{\Phi} = \boldsymbol{BA} = [\boldsymbol{b}_1, \boldsymbol{b}_2, \cdots, \boldsymbol{b}_N]\begin{bmatrix} 1 & a_{11} & a_{12} & \cdots & a_{1N} \\ 0 & 1 & a_{23} & \cdots & a_{2N} \\ \cdots & \cdots & \cdots & \cdots & \cdots \\ 0 & 0 & 0 & \cdots & 1 \end{bmatrix} \tag{3.170}$$

where $\boldsymbol{A} \in \Re^{N\times N}$ is an upper triangular matrix with 1s on the main diagonal and $\boldsymbol{B} \in \Re^{Q\times N}$ is a matrix with mutually orthogonal columns $\boldsymbol{b}_i$ such that

$$\boldsymbol{B}^T\boldsymbol{B} = \boldsymbol{H} = \text{diag}(h_1, h_2, \cdots, h_N) \tag{3.171}$$

The matrix $\boldsymbol{H} \in \Re^{N\times N}$ is diagonal with elements h_i given by

$$h_i = \boldsymbol{b}_i^T\boldsymbol{b}_i = \sum_{k=1}^{N} b_{ik}^2 \tag{3.172}$$

The vector space spanned by the set of orthogonal basis vectors $\boldsymbol{b}_i$ is the same vector space spanned by the columns of the regression matrix Φ. Substituting (3.170) into (3.167) we obtain

$$\boldsymbol{y}_d = \boldsymbol{BAw} + \boldsymbol{e} = \boldsymbol{Bg} + \boldsymbol{e} \tag{3.173}$$

where $\boldsymbol{g} = \boldsymbol{Aw}$. In (3.173) the desired output vector $\boldsymbol{y}_d$ is expressed as a linear combination of the mutually orthogonal columns of matrix $\boldsymbol{B}$. A least-squares solution for the coordinate vector $\boldsymbol{g}$ is given by

$$\hat{\boldsymbol{g}} = \left(\boldsymbol{B}^T\boldsymbol{B}\right)^{-1}\boldsymbol{B}^T\boldsymbol{y}_d = \boldsymbol{B}^{+}\boldsymbol{y}_d = \boldsymbol{H}^{-1}\boldsymbol{B}^T\boldsymbol{y}_d \tag{3.174}$$

The ith coordinate of vector $\hat{\boldsymbol{g}}$ is given by

$$g_i = \frac{\boldsymbol{b}_i^T\boldsymbol{y}_d}{\boldsymbol{b}_i^T\boldsymbol{b}_i} \tag{3.175}$$

and, as we would expect, it is a normalized projection of vector $\boldsymbol{y}_d$ in the direction of column $\boldsymbol{b}_i$. Because the Gram–Schmidt orthogonalization ensures the orthogonality between the approximation error and $\boldsymbol{Bg}$ in (3.173), we have

$$y_d^T y_d = g^T B^T B g + e^T e \;= g^T H g + e^T e = \sum_{i=1}^{N} h_i g_i^2 + e^T e \tag{3.176}$$

It is relatively easy to give a physical interpretation of (3.176). The term $y_d^T y_d$ represents the total energy of the desired output vector. The term $\sum_{i=1}^{N} h_i g_i^2$ represents the portion of the energy explained by the regression, and $e^T e$ is the energy of the regression error. Each term in the summation on the right side of (3.176) represents an increment in the energy due to the inclusion of the ith regression vector. Since there is a one-to-one correspondence between the elements of the regression vector g_i and the RBF centers c_i, each term in the summation reflects the contribution of each of the RBF centers. We can define the error reduction ratio (err) due to the inclusion of the pth RBF center as

$$[\mathrm{err}]_p = \frac{h_p g_p^2}{y_d^T y_d} \tag{3.177}$$

The error reduction ratio in (3.177) offers a simple and effective criterion for selection of the RBF centers in a forward regression manner. At every step of the forward regression, an RBF center is selected so that the error reduction ratio is maximal. The classical Gram–Schmidt orthogonalization procedure can be incorporated into the forward regression scheme, and the entire procedure can be summarized as in the following algorithm.

Orthogonal least-squares algorithm for training an RBF network

Step 1. $k = 1$. For $1 \le i \le Q$, set

$$b_1^{(i)} = \phi_i$$

Compute the error reduction ratio of the ith center as

$$[\mathrm{err}]_1^i = \frac{\left(b_1^{(i)T} y_d\right)^2}{b_1^{(i)T} b_1^{(i)} \cdot y_d^T y_d}$$

Find

$$[\mathrm{err}]_1^{i_1} = \max\{[err]_1^i, 1 \le i \le Q\}$$

and select

$$b_1 = \phi_{i_1}$$

and the center $c_1 = c_{i_1}$.

Step k. $k \ge 2$. For $1 \le i \le Q, i \ne i_1, i \ne i_2, \cdots, i \ne i_{k-1}$ compute

$$a_{jk}^{(i)} = \frac{b_j^T \phi_i}{b_j^T b_j} \qquad 1 \le j \le k-1$$

Set

$$b_k^{(i)} = \phi_i - \sum_{j=1}^{k-1} a_{jk}^{(i)} b_j$$

Compute

$$[\text{err}]_k^{(i)} = \frac{\left(\boldsymbol{b}_k^{(i)T}\boldsymbol{y}_d\right)^2}{\boldsymbol{b}_k^{(i)T}\boldsymbol{b}_k^{(i)} \cdot \boldsymbol{y}_d^T\boldsymbol{y}_d}$$

Find

$$[\text{err}]_k^{i_k} = \max\left\{[\text{err}]_k^i,\ 1 \le i \le Q,\ i \ne i_1,\ i \ne i_2, \cdots, i \ne i_{k-1}\right\}$$

and select

$$\boldsymbol{b}_k = \boldsymbol{b}_k^{(i_k)}$$

and the center $\boldsymbol{c}_k = \boldsymbol{c}_{i_k}$.

Step k + 1. Repeat step k. The regression is stopped at step N_1 when

$$1 - \sum_{j=1}^{N_1} [\text{err}]_j < \rho$$

where $0 < \rho < 1$ is a selected tolerance value.

The geometric interpretation of the above procedure can be stated as follows. At the kth step, the dimension of the space spanned by the selected regressors increases from $k-1$ to k by introducing one more basis vector. The newly added vector maximizes the amount of energy of the desired network output (i.e., it maximizes the error reduction ratio). Each vector that is included in the regression corresponds to a center in the set of input data points. The addition of the new vectors (and selection of the new centers) stops when the regression has accounted for a sufficient portion of the energy of the desired network output. Consequently, the OLS approach produces a relatively small network.

The tolerance parameter ρ is important for balancing the accuracy and complexity of the network. If ρ is set too high (i.e., close to 1), the resulting network will approximate the mapping with a high degree of accuracy, but it will result in a large number of centers. Moreover, the obtained accuracy can most likely be attributed to overfitting. On the other hand, setting ρ to be small will result in relatively poor modeling properties of the network; however, the network size will be significantly reduced. One reasonable way to set ρ is $\rho = 1 - \sigma_n^2/\sigma_d^2$, where σ_n^2 is some estimate of measurement noise power and σ_d^2 is the total power of the target signal.

The version of the OLS algorithm presented in this section applies to RBF NNs having a single output. Extension to the multioutput networks is relatively straightforward.

EXAMPLE 3.5. Consider the task of designing an RBF NN that approximates the mapping $z = f(x, y)$ given by

$$z = \cos(3x)\sin(2y)$$

over the area determined by $-1 \le x \le 1$ and $-1 \le y \le 1$. To accomplish this, we use an RBF NN having 121 centers distributed over the defined input space, as indicated by the + symbol in Figure 3.12. Using a Gaussian RBF function with the spread parameter set to 0.3, the network is trained to perform the exact mapping of all 121 centers. The OLS algorithm was used to perform the reduction

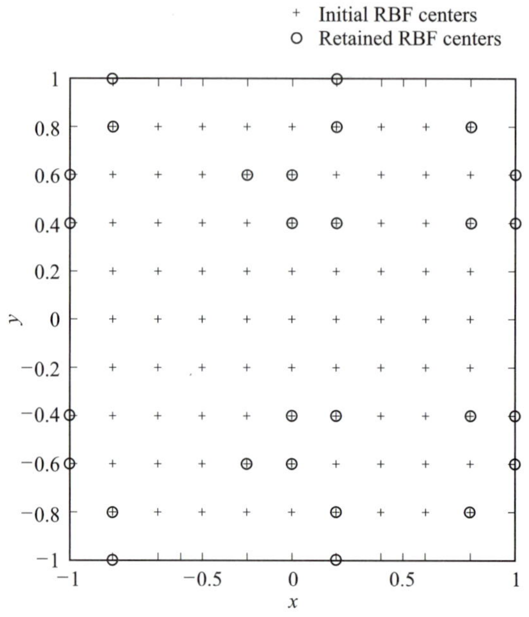

Figure 3.12 RBF centers used to design the mapping $z = \cos(3x)\sin(2y)$ over the area bounded by $-1 \le x \le 1$ and $-1 \le y \le 1$. After application of OLS, the reduced network retained only 28 centers.

of the network size with a required accuracy of $\rho = 0.99$. The network retained 28 centers, indicated by the "o" symbols in Figure 3.12. For a minimal reduction in performance of the RBF NN mapping, the size has been reduced by $1 - {}^{28}/_{121} \approx$ 77 percent.

PROBLEMS

3.1. Consider an associative memory network described in Section 3.2. In general, the dimensions of the *key* vectors and the *memorized* vectors do not have to be the same. Consider the following input key vectors

$$\boldsymbol{x}_1 = \begin{bmatrix} 1 & 0 & 0 & 0 \end{bmatrix}^T \quad \boldsymbol{x}_2 = \begin{bmatrix} 0 & 1 & 0 & 0 \end{bmatrix}^T \quad \boldsymbol{x}_3 = \begin{bmatrix} 0 & 0 & 0 & 1 \end{bmatrix}^T$$

and the output memorized vectors

$$\boldsymbol{y}_1 = \begin{bmatrix} 2 & -2 & 3 \end{bmatrix}^T \quad \boldsymbol{y}_2 = \begin{bmatrix} 1 & 1 & -2 \end{bmatrix}^T \quad \boldsymbol{y}_3 = \begin{bmatrix} -2 & 4 & 2 \end{bmatrix}^T$$

(*a*) Compute the (rectangular) memory matrix $\hat{\boldsymbol{M}}$, using the outer product rule.

(*b*) Does the resulting memory associate perfectly? This can be determined by using each of the key patterns as inputs to the network, that is,

$$\hat{\boldsymbol{y}}_i = \hat{\boldsymbol{M}}\boldsymbol{x}_i$$

and comparing the estimates $\hat{\boldsymbol{y}}_i$ to the memorized vectors.

3.2. In some situations the input patterns to a network can become distorted or *masked*. For example, in Example 3.1, the first key vector given in (3.15) could have the first element zeroed out for some reason, resulting in a masked version $\boldsymbol{x}_{1m}$ of the original key vector $\boldsymbol{x}_1$, that is,

$$\boldsymbol{x}_{1m} = \begin{bmatrix} 0 & 0.7778 & 0.5329 \end{bmatrix}^T$$

Compute the response of the memory to $\boldsymbol{x}_{1m}$, using the memory matrix given in (3.19); then compute the Euclidean distances between this response and the original key vectors in (3.15). What can you conclude from this? Carry out the same analysis for the case when the first element in the first key vector given in (3.24) has been zeroed out; that is, the masked vector is

$$\boldsymbol{x}_{1m} = \begin{bmatrix} 0 & -0.9779 & -0.1629 \end{bmatrix}^T$$

and now use the memory matrix given in (3.28). Again, what can be concluded from the results? Furthermore, by comparing these results with those from the first part of this problem, what conclusions, if any, can be drawn?

3.3. Consider again Problem 3.1. Using the error correction recursive algorithm given in (3.37), compute the memory matrix. Write a MATLAB function that implements the recursive algorithm and includes in the argument list the learning rate parameter μ, the total number of iterations N, and a tolerance value tol that will define when convergence has been achieved. Specifically, the "stopping criterion" in your program would occur mathematically when

$$\left\| \boldsymbol{Y} - \hat{\boldsymbol{M}}\boldsymbol{X} \right\|_2 \leq tol$$

where $\|\bullet\|_2$ is the largest singular value of the matrix $\boldsymbol{Y} - \hat{\boldsymbol{M}}\boldsymbol{X}$ (cf. Sect. A.2.13), $\boldsymbol{Y}$ contains all the memorized patterns as column vectors, $\boldsymbol{X}$ contains all the key patterns as column vectors, and $\hat{\boldsymbol{M}}$ is a *rectangular* memory matrix. The MATLAB function as an "m-file" would begin like this:

```
function M=corrmm(X,Y,mu,N,tol)
```

Compare your results to those obtained in Problem 3.1. Experiment with different values of the learning rate parameter μ and the tolerance value tol. A reasonable value to use for tol is 10^{-8}.

3.4. Heteroassociative memory neural networks can be used for detection of known sequences in the presence of noise. As an illustration, design a heteroassociative network to store the following pairs of vectors:

$$\boldsymbol{x}(1) = \begin{bmatrix} 1 & 1 & 1 & 1 & 1 & 1 & 1 & 1 \end{bmatrix}^T \qquad \boldsymbol{y}(1) = \begin{bmatrix} 1 & -1 \end{bmatrix}^T$$
$$\boldsymbol{x}(2) = \begin{bmatrix} 1 & -1 & -1 & 1 & 1 & -1 & -1 & 1 \end{bmatrix}^T \qquad \boldsymbol{y}(2) = \begin{bmatrix} 1 & 1 \end{bmatrix}^T$$
$$\boldsymbol{x}(3) = \begin{bmatrix} 1 & -1 & 1 & -1 & -1 & 1 & -1 & 1 \end{bmatrix}^T \qquad \boldsymbol{y}(2) = \begin{bmatrix} -1 & 1 \end{bmatrix}^T$$

Use the architecture shown in Figure 3.13 and the learning rule given in (3.7).

(*a*) Test the performance of the network when the input data are corrupted with zero-mean Gaussian noise. Experiment with different noise powers.

(*b*) Store an additional pair of associative vectors, given by

$$\boldsymbol{x}(4) = \begin{bmatrix} -1 & 1 & 1 & -1 & 1 & 1 & 1 & 1 \end{bmatrix}^T \qquad \boldsymbol{y}(4) = \begin{bmatrix} -1 & -1 \end{bmatrix}^T$$

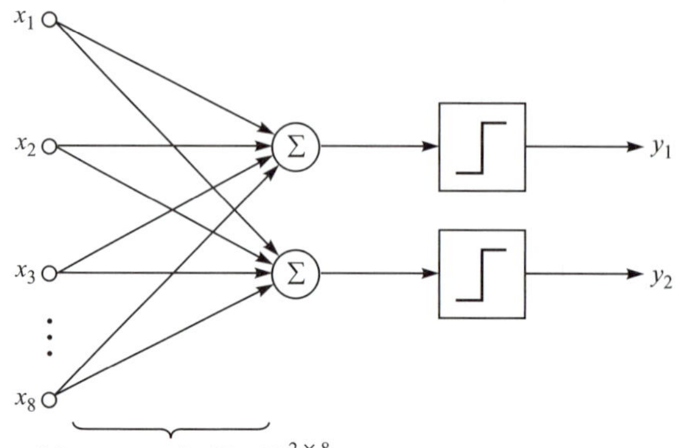

Figure 3.13 Heteroassociative memory neural network for Problem 3.4.

Test the performance of the network when the input data are corrupted with zero-mean white noise with a standard deviation of 0.3. Explain the performance degradation relative to part (*a*).

3.5. One of the most common applications of the MLP NN trained with backpropagation is the approximation of nonlinear functional mappings. Write a computer program and design an MLP NN with one hidden layer, and train by backpropagation to perform the following mappings:

1. $f(x) = \dfrac{1}{x} \qquad x \in (0.1, 1)$

2. $f(x, y) = x^2 + y^2 \qquad x \in (-1, 1)$ and $y \in (-1, 1)$

3. $f(x, y) = \sin(\pi x)\cos(\pi x) \qquad x \in (-2, 2)$ and $y \in (-2, 2)$

4. $f(x, y, z) = \dfrac{x^2}{2} + \dfrac{y^2}{3} + \dfrac{z^2}{4} \qquad x \in (-2, 2), y \in (-2, 2),$ and $z \in (-2, 2)$

For each of the above examples, do the following:

(*a*) Generate three independent sets of input patterns:
Training set: 200 patterns
Testing set: 100 patterns
Validation set: 50 patterns
For each of the sets generate the target values, using the analytical expression for the function to be approximated.

(*b*) In the process of network training, use the training data set to modify the weights. Use the testing data set at the end of each training epoch to monitor the ability of the network to generalize and prevent overfitting. Finally, use the validation set to verify the overall performance of the network after it has been trained.

(*c*) Experiment with different numbers of neurons in the hidden layer.

(*d*) Compare the speed of network convergence when the weights are initialized as small random numbers versus the Nguyen–Widrow initialization procedure (cf. Sect. 3.3.2).

3.6. Write a computer program implementing the backpropagation learning algorithm with momentum updating. Use your program to train an MLP NN having 10 neurons in the hidden layer to perform the nonlinear mapping defined by

$$f(x_1, x_2) = \begin{cases} +1 & \text{if } x_1^2 + x_2^2 \leq 1 \\ -1 & \text{if } x_1^2 + x_2^2 > 1 \end{cases} \quad \text{over the region}$$

$$-2 < x_1 < 2 \qquad \text{and} \qquad -2 < x_2 < 2$$

(*a*) As a training set use 441 data points, defined by

$$\boldsymbol{x} = (x_i, x_j)$$

where $x_i = -2 + i \cdot 0.2 \qquad i = 0, 1, \ldots, 20$

$x_j = -2 + j \cdot 0.2 \qquad j = 0, 1, \ldots, 20$

(*b*) Experiment with following values for the forgetting factor:

$$\alpha \in \{0,\ 0.2,\ 0.4,\ 0.6,\ 0.8,\ 0.9,\ 0.99\}$$

For each value of the forgetting factor, train the network 50 times and record the average number of epochs needed for the network to converge. Perform the nonlinear mapping with the accuracy specified by

$$\frac{1}{441} \sum_{i=0}^{40} \sum_{j=0}^{40} [f_{NN}(x_i, x_j) - f(x_i, x_j)]^2 \leq 0.1$$

Initialize the weights to *small* random numbers.

(*c*) Repeat the experiment, using Nguyen–Widrow's initialization procedure (cf. Sect. 3.3.2).

3.7. The MLP NN trained by backpropagation can be successfully used for equalization of a memoryless communication channel. Consider the situation depicted in Figure 3.14(*a*). Nonlinear distortions introduced by the channel may cause the degradation in the quality of the signal transmission. As shown in Figure 3.14(*b*), the MLP NN can be used as an adaptive filter that eliminates the nonlinearities. Assume that the communication channel nonlinear input/output relation can be approximated by

$$y(nT) = Ax(nT) + Bx^2(nT)$$

where $A = 1$ and $B = 0.2$.

(*a*) Design and train an MLP NN having one input, seven hidden neurons, and one output neuron to perform the equalization of the communication channel. Set $T = 1$, and as a training sequence use the signal

$$s(nT) = s(n) = 2\sin\left(\frac{2\pi n}{20}\right) \qquad n = 1, 2, \ldots$$

(*b*) Test the performance of the equalizer, using the following test signals

$s_1(nT) = s_1(n) = 0.8\sin\left(\frac{2\pi n}{10}\right) + 0.25\cos\left(\frac{2\pi n}{25}\right)$ and

$s_2(nT) = s_2(n)$ is a sequence of random numbers following a normal distribution with a mean of zero and unity variance.

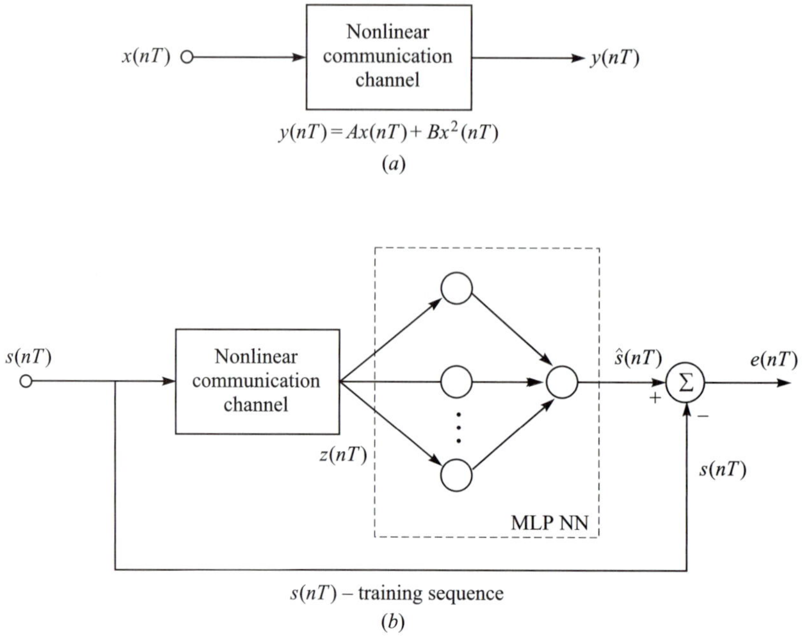

Figure 3.14 Using the MLP NN for equalization of the nonlinear channels.

(*c*) Repeat the experiment, assuming that the channel input/output relation can be approximated as

$$y(nT) = Ax(nT) + Bx^2(nT) + Cx^3(nT)$$

where $A = 1$, $B = 0.3$, and $C = -0.2$. Experiment with the number of the neurons in the MLP NN hidden layer.

3.8. The multilayer perceptron neural network can be used to perform lossy data compression. Consider the neural network architecture shown in Figure 3.15. The network consists of two layers: a hidden layer and an output layer. The number of neurons in the hidden layer is smaller than the dimension of the input vectors, and it performs the compression of the input data to a lower-dimensional vector space. The vectors used as input patterns are used as target patterns as well. Therefore, the output layer of the network is trained to reconstruct the input data from their lower-dimensional representation.

After the network is trained, the compression is performed by storing the outputs of the hidden layer instead of the input data. Also, to be able to reconstruct the data, the weights of the output layer need to be stored as well. Let $\Psi_1(\bullet)$ and $\Psi_2(\bullet)$ be the nonlinear mappings performed by neurons in the hidden and output layers, respectively. After the network is trained, the compression step is accomplished as

$$\boldsymbol{y} = \Psi_1(\boldsymbol{W}\boldsymbol{x})$$

The reconstruction of the original signal is performed as

$$\hat{\boldsymbol{x}} = \Psi_2(\boldsymbol{V}\boldsymbol{y})$$

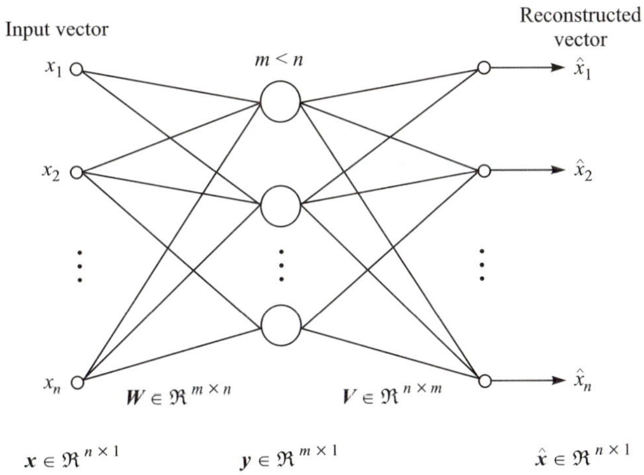

Figure 3.15 Use of an MLP NN for data compression.

Write a computer program that uses the MLP NN trained by backpropagation to perform image compression. To generate the input/target vectors $\boldsymbol{x}$, divide the image into 8-by-8 pixel blocks and arrange the elements of each block into a 64-dimensional vector. Experiment with different numbers of neurons in the hidden layer.

3.9. Pseudorandom sequences, sometimes referred to as pseudonoise (PN) sequences because they have noiselike properties, are used extensively in communication systems, for example, in spread spectrum systems. These binary sequences possess many interesting properties and can be generated using a linear shift register (LSR). Using an LSR of length $n = 5$ (e.g., see Fig. 3.16), there are six different *maximal-length* PN sequences that can be generated. In general, a maximal-length sequence has length $N = 2^n - 1$ with $2^{n-1} - 1$ zeros, and the remaining bits are 1. A PN sequence will be maximal length if the tap locations on the LSR used to generate the sequence are associated with a *primitive polynomial.* The six primitive polynomials are

$$g_1(x) = x^5 + x^3 + 1 \qquad g_4(x) = x^5 + x^4 + x^3 + x^2 + 1$$
$$g_2(x) = x^5 + x^2 + 1 \qquad g_5(x) = x^5 + x^4 + x^3 + x + 1$$
$$g_3(x) = x^5 + x^3 + x^2 + x + 1 \qquad g_6(x) = x^5 + x^4 + x^2 + x + 1$$

The vector representations of the six primitive polynomials are given by

$$g_1 \rightarrow [0 \quad 1 \quad 0 \quad 0 \quad 1] \qquad g_4 \rightarrow [1 \quad 1 \quad 1 \quad 0 \quad 1]$$
$$g_2 \rightarrow [0 \quad 0 \quad 1 \quad 0 \quad 1] \qquad g_5 \rightarrow [1 \quad 1 \quad 0 \quad 1 \quad 1]$$
$$g_3 \rightarrow [0 \quad 1 \quad 1 \quad 1 \quad 1] \qquad g_6 \rightarrow [1 \quad 0 \quad 1 \quad 1 \quad 1]$$

Figure 3.17 shows a MATLAB function **lprs** that will generate a maximal-length pseudorandom sequence given a primitive polynomial coefficient vector and an initial-condition vector. The initial-condition vector can be any proper length sequence of 1s and 0s but not all 0s (typically all values are selected to be 1).

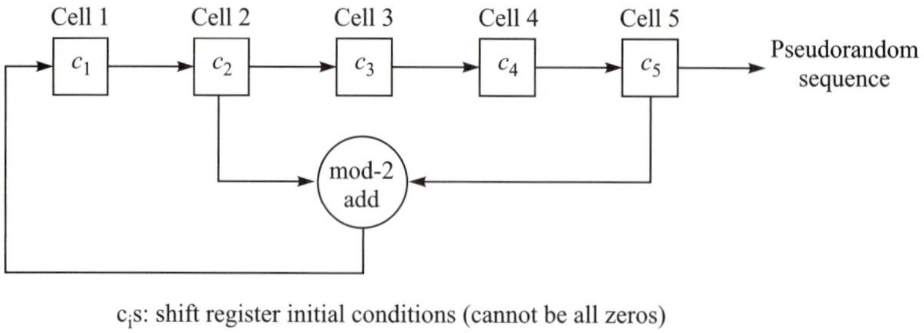

Figure 3.16 Linear shift register configured to generate a maximal-length pseudorandom sequence using the primitive polynomial $g_1(x) = x^5 + x^3 + 1$.

(*a*) Generate the six PN sequences (length: $N = 2^n - 1|_{n=5} = 31$), using the vector representations of the six primitive polynomials given above and the MATLAB function in Figure 3.17.

(*b*) Use the six vectors from part (a) as training inputs to a feedforward multilayer perceptron, and let the target vectors be the associated vector representations of the primitive polynomial coefficients. Use **trainbp** in the MATLAB Neural Network Toolbox. A single hidden layer is sufficient to perform the proper mapping. Experiment with different numbers of neurons in the hidden layer and activation functions. After the network has

```
function z=lprs(n,ppc,b)
% z:      output binary sequence
% n:      number of tubes in shift register
%         (or length of shift register)
% ppc:    primitive polynomial coefficients
%         (n-dimensional row vector)
% b:      initialization vector
%         (n-dimensional row vector)
M=zeros(n,n);
M(:,1)=ppc';
for i=1:n-1
     M(i,i+1)=1;
end
z(1)=b(1);
for k=1:2^n-2
     b=b*M;
     for i=1:n
        if b(i)/2 == fix(b(i)/2)
            b(i)=0;
        else
          b(i)=1;
        end
     end
z(k+1)=b(1);
end
```

Figure 3.17 MATLAB function to generate a maximal-length pseudorandom sequence given a primitive polynomial coefficient vector and an initial-condition vector.

been trained, present the six inputs and observe the outputs. Did the neural network correctly classify the inputs?

(*c*) Add random noise to the six input vectors used to train the network; that is, randomly "toggle" some of the bits in the vector. Use these as inputs to the network that you trained in part (*b*) of this problem. Observe the network's error-correcting capability. An example of how the input vectors can be corrupted with noise is shown below (Fig. 3.18) in the MATLAB function `errors`. Comment on the error-correcting capability of the neural network for different bit error rates (BERs) and different numbers of neurons in the hidden layer. In the paper by Ham et al. [53], two unsupervised neural networks are used to classify pseudorandom sequences.

(*d*) Also experiment with different initial conditions to generate the PN sequences. Remember that the sequences are periodic; that is, they will repeat their binary pattern every $N = 2^n - 1$ bits. Try different initial conditions for each of the six different maximal-length sequences, and train the network with the resulting patterns. Comment on the results.

3.10. Consider a problem of classifying the points on the circle

$$x^2 + y^2 = 1$$

relative to the four quadrants of the rectangular coordinate system. In other words,

$$\text{Point } z = (x, y) \text{ is classified as } \begin{cases} 1 & \text{if } x > 0 \text{ and } y > 0 \\ 2 & \text{if } x < 0 \text{ and } y > 0 \\ 3 & \text{if } x < 0 \text{ and } y > 0 \\ 4 & \text{if } x > 0 \text{ and } y < 0 \end{cases}$$

```
function ZC=errors (Z,BER)
% Z:      matrix containing input
%         patterns in columns
% BER:    bit-error-rate given in a percentage,
%         for example, BER=30 means that on the
%         average 30% of the bits will be
%         ''toggled''
% ZC:     Corrupted output matrix
[nr,nc]=size(Z);
RN=rand(size(Z));
ZC=Z;
T=1-BER/100;
for i=1:nr
     for j=1:nc
        if RN(i, j)>=T
           if ZC(i, j)==1
              ZC(i, j)=0;
           else
               ZC(i, j)=+1;
           end
        end
     end
end
```

Figure 3.18 MATLAB function **errors** for generating bit errors.

(*a*) Use a *forward-only* counterpropagation neural network with 36 neurons in the hidden layer to accomplish the classification.

(*b*) In Section 3.5 it was shown how the Euclidean distance can be used for computation of the distance in Kohonen's layer of the counterpropagation. Repeat the above-stated classification problem, using the dot-product measure of distance

$$\text{dist}(z, \boldsymbol{w}_j) = 1 - \frac{xw_{j1} + yw_{j2}}{\sqrt{x^2 + y^2}\sqrt{w_{j1}^2 + w_{j2}^2}}$$

(*c*) Test the classification performance of the network for the inputs from the set

$$\{z = (x, y) \; : \; x^2 + y^2 < 1\}$$

using both Euclidean and the dot-product measures of distance. Explain any difference in the results.

3.11. Use (3.158) through (3.161) to derive the learning rules for the stochastic gradient method for training an RBF neural network.

3.12. Design an RBF NN to approximate the mapping defined by

$$f(x_1, x_2) = \begin{cases} +1 & \text{if } x_1^2 + x_2^2 \le 1 \\ -1 & \text{if } x_1^2 + x_2^2 > 1 \end{cases}$$

$$\text{over region} - 2 < x_1 < 2 \text{ and} - 2 < x_2 < 2$$

As a training set, use 441 data points defined by

$$\boldsymbol{x} = (x_i, x_j)$$

where

$$x_i = -2 + i \cdot 0.2 \qquad i = 0, 1, \ldots, 20$$
$$x_j = -2 + j \cdot 0.2 \qquad j = 0, 1, \ldots, 20$$

(*a*) Carry out the design of the RBF NN, using all the points in the training set as centers of the RB functions.

(*b*) Perform the design of the RBF NN, using 150 centers randomly selected from the input data. Compare the performance of this network to the one designed in part (*a*).

(*c*) Randomly choose 150 centers and use stochastic gradient approach to perform the RBF NN design. Compare the performance of this network relative to networks designed in parts (*a*) and (*b*).

(*d*) Design an RBF NN, using the entire data set as centers for the RB functions. Apply the orthogonal least-squares procedure to reduce the size of the network. Plot the centers that are selected by OLS regression, and compare them with the centers obtained in part (*c*).

3.13. Repeat Problem 3.7, using an RBF neural network. Discuss the pros and cons of using the two neural network architectures.

3.14. RBF neural networks are often used for identification of the nonlinear dynamic systems. Consider the situation depicted in Figure 3.19. The nonlinear plant

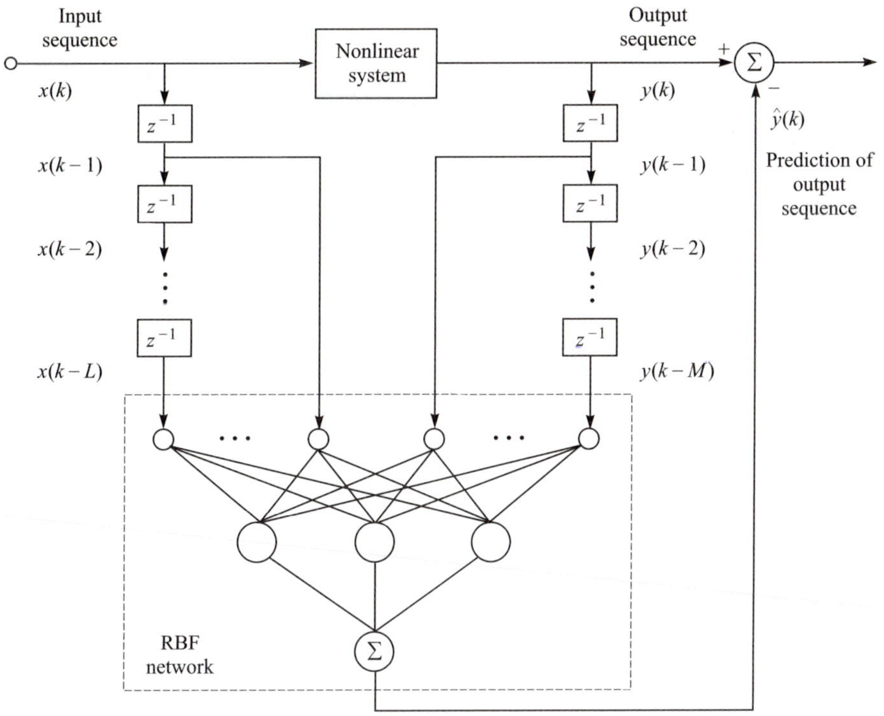

Figure 3.19 Use of the RBF neural network for nonlinear system identification.

performs the mapping between the input sequence $x(k)$ and output sequence $y(k)$. The neural network is trained to predict the output of the nonlinear system based on the previous values of the input and output sequences. After training, the neural network identifies the nonlinear system.

Let us assume that the nonlinear plant in Figure 3.19 can be characterized with the input/output difference equation given as

$$y(k) = \frac{1}{1 + y(k-1)^2} + 0.25x(k) - 0.3x(k-1)$$

(*a*) Design an RBF neural network to perform the identification of the nonlinear plant.

(*b*) Implement the batch learning approach. For an input sequence, use white noise with zero mean and standard deviation of unity.

(*c*) Test the results of identification, using the following test signals:

$$x_1(k) = \sin\left(\frac{2\pi k}{10}\right)$$

$$x_2(k) = 4\sin\left(\frac{2\pi k}{10}\right)$$

Explain the results.

(*d*) Rework this problem, using the learning rules given in (3.159) to (3.161).

REFERENCES

1. T. J. Teyler, "Memory: Electrophysiological Analogs," in *Learning and Memory: A Biological View*, eds. J. L. Martinez, Jr., and R. S. Kesner, Orlando, FL: Academic Press, 1986, pp. 237–65.
2. T. Kohonen, *Self-Organization and Associative Memory*, 3rd ed., Berlin: Springer-Verlag, 1989.
3. T. Kohonen, *Associative Memory: A System-Theoretical Approach*, Berlin: Springer-Verlag, 1977.
4. K. Nakano, "Association—A Model of Associative Memory," *IEEE Transactions on Systems, Man, and Cybernetics*, vol. SMC-2, 1972, pp. 380–8.
5. J. A. Anderson, "A Simple Neural Network Generating an Interactive Memory," *Mathematical Bioscience*, vol. 14, 1972, pp. 197–220. Reprinted in 1988, Anderson and Rosenfeld, pp. 181–92. J. A. Anderson and E. Rosenfeld, eds., *Neurocomputing: Foundations of Research*, Cambridge, MA: M.I.T. Press, 1988.
6. S. Amari, "Learning Patterns and Pattern Sequences by Self-Organizing Nets of Threshold Elements," *IEEE Transactions on Computers*, vol. C-21, 1972, pp. 1197–1206.
7. S. Amari and K. Maginu, "Statistical Neurodynamics of Associative Memory," *Neural Networks*, vol. 2, 1988, pp. 63–73.
8. B. Kosko, "Bidirectional Associative Memories," *IEEE Transactions on Systems, Man, and Cybernetics*, vol. SMC-18, 1988, pp. 49–60.
9. S. Amari, "Characteristics of Sparsely Encoded Associative Memories," *Neural Networks*, vol. 3, 1989, pp. 451–7.
10. M. Hassoun, ed., *Associative Neural Memories: Theory and Implementation*, Oxford, England: Oxford University Press, 1993.
11. J. J. Hopfield, "Neural Networks and Physical Systems with Emergent Collective Computational Abilities," in *Proceedings of National Academy of Science, USA*, vol. 79, 1982, pp. 2554–8.
12. T. Kohonen, "Correlation Matrix Memories," *IEEE Transactions on Computers*, vol. C-21, 1972, pp. 353–9. Reprinted in 1988, Anderson and Rosenfeld, pp. 174–80.
13. A. D. Whalen, *Detection of Signals in Noise*, New York, Academic Press, 1971.
14. C. W. Therrien, *Discrete Random Signals and Statistical Signal Processing*, Englewood Cliffs, NJ: Prentice-Hall, 1992.
15. J. A. Anderson, "Cognitive and Psychological Computation with Neural Models," *IEEE Transactions on Systems, Man, and Cybernetics*, vol. SMC-13, 1983, pp. 799–815.
16. J. A. Anderson and G. L. Murphy, "Concepts in Connectionist Models," *Neural Networks for Computing*, ed. J. S. Denker, New York: American Institute of Physics, 1986, pp. 17–22.
17. P. J. Werbos, "Beyond Regression: New Tools for Prediction and Analysis in the Behavioral Sciences," Ph.D. thesis, Cambridge, MA: Harvard University, 1974.
18. P. J. Werbos, "Backpropagation through Time: What It Does and How to Do It," *Proceedings of the IEEE*, vol. 78, 1990, pp. 1550–60.
19. P. J. Werbos, *The Roots of Backpropagation*, New York: Wiley, 1994.
20. D. B. Parker, *Learning-Logic*, Invention Report: S81-64, File 1, Stanford, CA: Office of Technology Licensing, Stanford University, October 1982.
21. D. B. Parker, "Optimal Algorithms for Adaptive Networks: Second Order Back Propagation, Second Order Direct Propagation, and Second Order Hebbian Learning," in *Proceedings of the First IEEE International Conference On Neural Networks*, June 1987, vol. 2, San Diego, CA, pp. 593–600.

22. D. B. Parker, "Learning-Logic: Casting the Cortex of the Human Brain in Silicon," Technical Report TR-47, Cambridge, MA: Center for Computational Research in Economics and Management Sciences, M.I.T., 1985.
23. Y. LeCun, "Une procedure d'apprentissage pour reseau à seuil asymmetrique," in *In Cognitiva 85: A la Frontiere de l'Intelligence Artificielle des Sciences de la Connaissance des Neurosciences*, CESTA, vol. 85, Paris, 1985, pp. 599–604.
24. D. E. Rumelhart, G. E. Hinton, and R. J. Williams, "Learning Internal Representations by Error Propagation," in *Parallel Distributed Processing*, vol. 1, chap. 8, eds. D. E. Rumelhart and J. L. McClelland, Cambridge, MA: M.I.T. Press, 1986, pp. 318–62.
25. D. E. Rumelhart, G. E. Hinton, and R. J. Williams, "Learning Representations by Back-Propagating Errors," *Nature*, vol. 323, 1986, pp. 533–6.
26. A. Cichocki and R. Unbehauen, *Neural Networks for Optimization and Signal Processing*, New York: Wiley, 1993.
27. D. Nguyen and B. Widrow, "Improving the Learning Speed of the 2-Layer Neural Networks by Choosing Initial Values of Adaptive Weights," in *Proceedings of the International Joint Conference on Neural Networks*, vol. 3, San Diego, CA, 1990, pp. 21–26.
28. K. Hornik, M. Stinchcombe, and H. White, "Multilayer Feedforward Networks Are Universal Approximators," *Neural Networks*, vol. 2, no. 5, 1989, pp. 359–66.
29. S. Haykin, *Adaptive Filter Theory*, 3rd ed., Upper Saddle River, NJ: Prentice-Hall, 1996.
30. F. M. Ham, "Detection and Classification of Biological Substances Using Infrared Absorption Spectroscopy and Hybrid Artificial Network," *Journal of Artificial Neural Networks*, vol. 1, no. 1, 1994, pp. 101–4.
31. K. S. Narendra, "Neural Networks for Identification and Control," *Series of Lectures Presented at Yale University*, December 1994.
32. T. P. Vogl, J. K. Mangis, A. K. Zigler, W. T. Zink, and D. L. Alkon, "Accelerating the Convergence of the Backpropagation Method," *Biological Cybernetics*, vol. 59, 1988, pp. 257–63.
33. C. Darken and J. Moody, "Towards Faster Stochastic Gradient Search," in *Advances in Neural Information Processing Systems*, vol. 4, San Mateo, CA: Morgan Kaufmann, 1991, pp. 1009–16.
34. C. Darken, J. Chang, and J. Moody, "Learning Rate Schedules for Faster Stochastic Gradient Search," in *Proceedings of Neural Network Signal Processing, 1992 IEE Workshop*, New York: IEEE Press, 1992.
35. M. T. Hagan, H. B. Demuth, and M. Beale, *Neural Network Design,* Boston, MA: PWS Publishing Company, 1996.
36. R. Battit, "First and Second Order Methods of Learning: Between the Steepest Descent and Newton's Method," *Neural Computation*, vol. 4, no. 2, 1991, pp. 141–66.
37. M. F. Moller, "A Scaled Conjugate Gradient Algorithm for Fast Supervised Learning," PB-339 Reprint, Computer Science Department, University of Aaurhus, Denmark, 1990.
38. E. Barnard and R. Cole, "A Neural-Net Training Program Based on Conjugate Gradient Optimization," Beaverton, OR: Oregon Graduate Institute, Report CSE 89-014, 1988.
39. R. Rosario, "A Conjugate-Gradient Based Algorithm for Training of the Feed-Forward Neural Networks," Ph.D. dissertation, Melbourne: Florida Institute of Technology, September 1989.

40. R. S. Scalero, "A Fast New Algorithm for Training of the Feed Forward Neural Networks," Ph.D. dissertation, Melbourne: Florida Institute of Technology, September 1989.
41. R. S. Scalero and N. Tepedelenlioglu, "A Fast Training Algorithm for Neural Networks," in *Proceedings of International Joint Conference on Neural Networks*, Washington, 1990.
42. A. Rezgui, "The Effect of an Adaptive Slope of the Activation Function on the Performance of the Backpropagation Algorithm," Ph.D. dissertation, Florida Institute of Technology, Melbourne, August 1990.
43. A. Rezgui and N. Tepedelenlioglu, "The Effect of the Slope of the Activation Function on the Performance of the Backpropagation Algorithm," in *Proceedings of International Joint Conference on Neural Networks,* vol. 1, Washington, January 1990, pp. 707–710.
44. R. Hecht-Nielsen, "Counterpropagation Networks," in *Proceedings of International Conference on Neural Network*, vol. 2, New York: IEEE, 1987, pp. 19–32.
45. R. Hecht-Nielsen, "Counterpropagation Networks," *Applied Optics*, vol. 26, 1987, pp. 4979–84.
46. R. Hecht-Nielsen, *Neurocomputing*, Reading, MA: Addison-Wesley, 1990.
47. S. Chen, S. A. Billings, C. F. N. Cowan, and P. M. Grant, "Practical Identification of NARMAX Models Using Radial Basis Functions," *International Journal of Control*, vol. 52, 1990, pp. 1327–50.
48. D. S. Broomhead and D. Lowe, "Multivariable Functional Interpolation and Adaptive Networks," *Complex Systems*, vol. 2, 1988, pp. 269–303.
49. E. S. Chen, S. Chen, and B. Mulgrew, "Efficient Computational Schemes for the Orthogonal Least Squares Algorithm," *IEEE Transactions on Signal Processing*, vol. 43, 1995, pp. 373–6.
50. S. Chen, C. F. Cowan, and P. M. Grant, "Orthogonal Least Squares Algorithm for Radial Basis Function Networks," *IEEE Transactions on Neural Networks*, vol. 2, pp. 302–9, 1991.
51. S. Chen, B. Mulgrew, and P. M. Grant, "A Clustering Technique for Digital Communications Channel Equalization Using Radial Basis Function Neural Networks," *IEEE Transactions on Neural Networks*, vol. 4, July 1993, pp. 570–9.
52. S. Chen, S. A. Billings, C. F. Cowan, and P. M. Grant, "Non-linear Systems Identification Using Radial Basis Functions," *International Journal of Control,* vol. 21, 1990, pp. 2513–39.
53. F. M. Ham, L. V. Fausett, and Byoungho Cho, "Classification of Maximal Length Pseudo-random Sequences Using ART and SOM," in *Proceedings of Artificial Neural Networks in Engineering 1993*, November 14–17, 1993, pp. 851–6.

CHAPTER 4

Self-Organizing Networks

4.1 INTRODUCTION

There are many different types of self-organizing neural networks; however, they all share a common characteristic. This is the ability to assess the input patterns presented to the network, organize itself to learn, on its own, similarities among the collective set of inputs, and categorize (or cluster) them into groups of similar patterns. Therefore, these types of neural networks learn *without* a "teacher," that is, through *unsupervised learning*. We will revisit this idea of self-organization in Chapter 9 when a discussion of principal-component analysis (PCA) (cf. Sect. 9.2) neural networks is given. The same idea of learning without a teacher is put into play for (adaptively) extracting principal-component information from the input data.

In general, self-organized learning (or unsupervised learning) involves the frequent modification of the network's synaptic weights in response to a set of input patterns. The weight modifications are carried out in accordance with a set of learning rules. After repeated applications of these patterns to the network, a configuration emerges that is of some significance. Basically, from numerous originally random local interactions within a network, in response to the input patterns, there emerges global order. This global order can ultimately lead to some form of congruous behavior. However, in order for self-organized learning to perform a meaningful information processing function, there has to exist redundancy in the input patterns presented to the network. From this redundancy in the input patterns order and structure emerge, and thus information that the neural network can assimilate as knowledge. For example, in the case of PCA, from a large amount of random input patterns presented to the network, the neuron weights converge to a mapping operator that maps the inputs to the principal components of the input data. In other words, from large amounts of (redundant) input data, the network learns the underlying latent characteristics of the input patterns, and the mapping (matrix), that is, the synaptic weights of the network, contains this information.

A special class of self-organizing neural networks is based on competitive learning. In competitive learning networks the output neurons compete among themselves to determine a winner. These are the networks that we want to study in this chapter. There are three basic types of neural networks presented: the Kohonen self-organizing map (SOM), learning vector quantization (LVQ), and adaptive resonance theory (ART) networks. LVQ may not fall exactly into this class of networks because it does not learn unsupervised. However, in a sense it does belong in this class of neural networks because it is so closely related to the Kohonen SOM.

4.2 KOHONEN SELF-ORGANIZING MAP

The self-organizing map developed by Kohonen is an unsupervised, competitive learning, clustering network, in which only one neuron (or only one neuron in a group) is "on" at a time [1]. The SOM is an artificial system that emulates certain mappings that occur in the brain. In the visual system, for example, there exist several topographic mappings of visual space onto the surface of the visual cortex. The basic idea of this type of self-organizing neural network is that the inputs (from a primary event space) are received by a simple network of adaptive elements. The signal representations are mapped (automatically) onto a set of outputs in such a manner that the responses attain the same topological order as that of the primary events. Therefore, the network can achieve an automatic formation of topological correct maps of features of observable events. In other words, the SOM transforms input patterns (of arbitrary dimension) to a one-dimensional or two-dimensional map of features in a topological ordered fashion. Figure 4.1 shows a conventional feature-mapping architecture (two-dimensional mapping from a set of input vectors). Figure 4.2 shows a biologically motivated mapping network that could be emulating a mapping from the retina to the cortex (this type of architecture is studied less often than the conventional one shown in Figure 4.1).

We will now study Kohonen's training algorithm as applied to the architecture in Figure 4.1. Even though there are *no* lateral connections between output neurons, neurons in the neighborhood of the neuron with the best match to the input (i.e., the *winning neuron*) are modified so that they respond more like the winning unit than they did previously.

It is critical to the formation of an ordered map that the neurons doing the learning do so nonindependently of one another, but in a *topologically* related fashion. In physiologically inspired neural network models, lateral feedback connections and other lateral interactions are used to achieve correlated learning by spatially neighboring neurons. An earlier feature-mapping architecture using lateral connections was developed by Willshaw and von der Malsburg

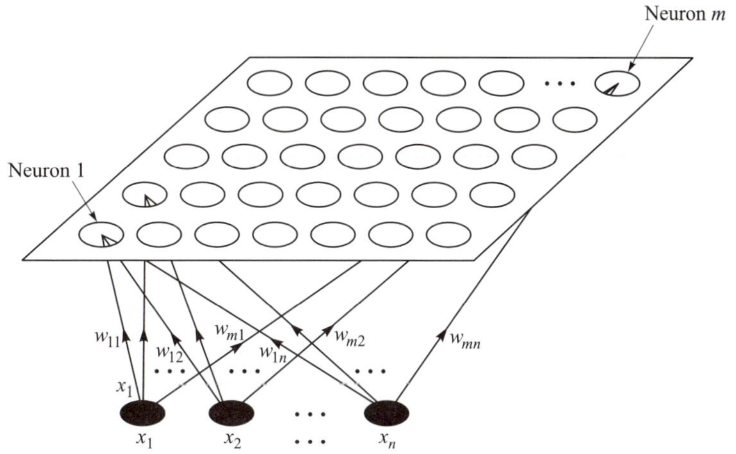

Figure 4.1 Self-organizing map—conventional feature-mapping architecture. The inputs are fully connected to each output neuron; however, only a few connections are shown.

[3] for the retinotopic map problem. They used the architecture shown in Figure 4.2 with lateral connections (of the Mexican hat form) between the output neurons.

The inputs to the network shown in Figure 4.1 can be written in vector form as

$$\boldsymbol{x} = [x_1, x_2, \cdots, x_n]^T \tag{4.1}$$

and the synaptic weight vector of neuron i in the two-dimensional (2-D) array is given by

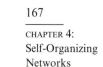

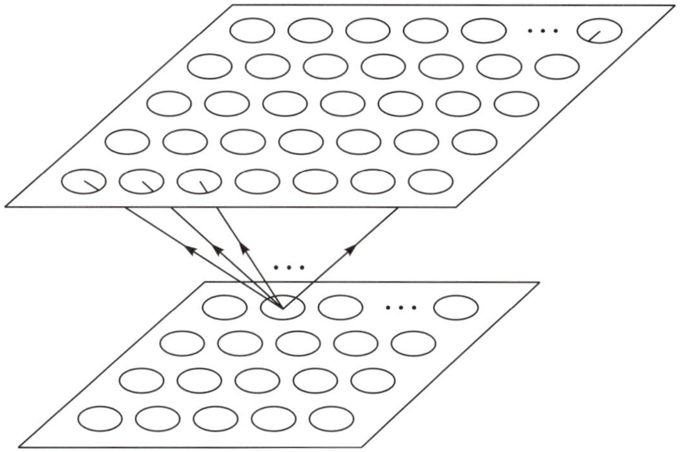

Figure 4.2 Self-organizing map—a biologically inspired mapping (could be emulating the mapping that occurs from the retina to the cortex). Each input is fully connected to each output neuron; however, only a few connections are shown.

$$\boldsymbol{w}_i = [w_{i1}, w_{i2}, \cdots, w_{in}]^T \qquad i = 1, 2, \ldots, m \tag{4.2}$$

where m is the total number of output neurons in the 2-D array. The *best match* of the input vector $\boldsymbol{x}$ with the synaptic weight vector $\boldsymbol{w}_i$ is determined from

$$q(\boldsymbol{x}) = \min_{\forall i} \|\boldsymbol{x} - \boldsymbol{w}_i\|_2 \qquad i = 1, 2, \ldots, m \tag{4.3}$$

where $q(\boldsymbol{x})$ is the index into the output neuron array that specifically identifies the *winning neuron*, and $\|\bullet\|_2$ is the L_2 or Euclidean norm. Therefore, by using (4.3), a continuous input space is mapped onto a discrete array of neurons. The response of the network could also be the synaptic weight vector that is closest to the input instead of the index position in the output neuron array. The next step in Kohonen's algorithm is to update the synaptic weight vector associated with the winning neuron and the neurons within a defined neighborhood of the winning neuron. The learning rule is given by [4, 5]

$$\boldsymbol{w}_i(k+1) = \boldsymbol{w}_i(k) + \eta_{qi}(k)[\boldsymbol{x}(k) - \boldsymbol{w}_i(k)] \tag{4.4}$$

where

$$\eta_{qi}(k) = \begin{cases} \mu(k) & \text{within } N_q \text{ (neighborhood set for winning neuron } q) \\ & \text{where } 0 < \mu(k) < 1 \text{, the learning rate parameter that} \\ & \text{should decrease with time} \\ 0 & \text{outside } N_q \end{cases} \tag{4.5}$$

is a scalar *kernel* function (or neighborhood function). More specifically, from (4.4) and (4.5) we can write the learning rule as

$$\boldsymbol{w}_i(k+1) = \begin{cases} \boldsymbol{w}_i(k) + \mu(k)[\boldsymbol{x}(k) - \boldsymbol{w}_i(k)] & \text{if } i \in N_q(k) \\ \boldsymbol{w}_i(k) & \text{if } i \notin N_q(k) \end{cases} \tag{4.6}$$

where $0 < \mu(k) < 1$ (the learning rate parameter). Note that N_q in (4.6) is considered to be in general a function of the discrete time index k, that is, $N_q(k)$. It has been shown to be advantageous to let the neighborhood set $N_q(k)$ be relatively wide in the beginning of training and then shrink monotonically with time [4] (this is illustrated in Figure 4.3).

The definition of the neighborhood function η_{qi} can be more general in the sense that a biological lateral interaction can have a *bell-shaped curve* [4]. To incorporate this into the kernel function, we denote the coordinates of the neurons q and i by the vectors $\boldsymbol{r}_q$ and $\boldsymbol{r}_i$, respectively. Therefore, a typical choice for η_{qi} in (4.4) is

$$\eta_{qi} = \eta_0 \exp\left(-\|\boldsymbol{r}_i - \boldsymbol{r}_q\|^2/\sigma^2\right) \tag{4.7}$$

with $\eta_0 = \eta_0(k)$ and $\sigma = \sigma(k)$ chosen as suitable decreasing functions of time. The learning rule in (4.4), according to Hertz et al. [6], drags the weight vector $\boldsymbol{w}_q$ associated with the winning unit toward $\boldsymbol{x}$, and it also drags the $\boldsymbol{w}_i$'s of the closest units along with it. An elastic net can be thought of in the input space that wants to come as close as possible to the inputs to the network. The

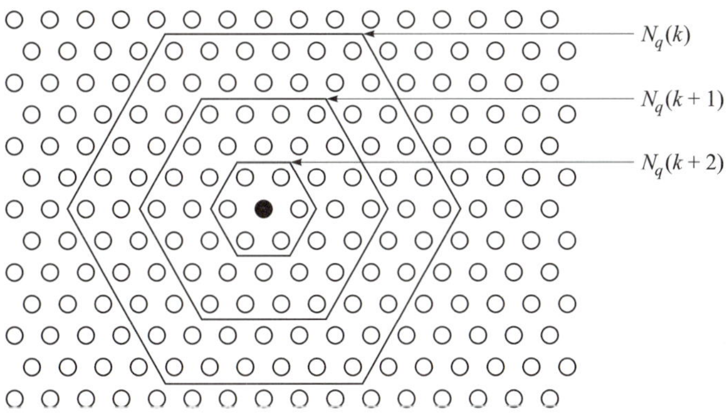

Figure 4.3 Topological neighborhood examples, showing a monotonic decrease in the neighborhood.

elastic net has the topology of the output array, and the points of the net can be thought of as having the weights as coordinates.

During the learning process there are two separate but related phases, that is, the ordering phase and the convergence phase. During the initial learning process, that is, the *ordering phase*, the learning rate parameter should be set close to unity and then gradually decreased (but not allowed to go below 0.1). It is during this part of the learning process that the topological ordering of the weight vectors is carried out. The *convergence phase* is the second phase of

Summary of the SOM algorithm

Step 1. Initialize network weight vectors† $\boldsymbol{w}_i$, initialize learning rate parameter, define topological neighborhood function, initialize parameter(s), set $k = 0$.

Step 2. Check stopping condition.‡ If false, continue; if true, quit.

Step 3. For each training vector $\boldsymbol{x}$, perform steps 4 through 7.

Step 4. Compute the best match of a weight vector with the input

$$q(\boldsymbol{x}) = \min_{\forall i} \|\boldsymbol{x} - \boldsymbol{w}_i\|_2$$

Step 5. For all units in the specified neighborhood $i \in N_q(k)$ (q is the winning neuron), update the weight vectors according to

$$\boldsymbol{w}_i(k+1) = \begin{cases} \boldsymbol{w}_i(k) + \mu(k)[\boldsymbol{x}(k) - \boldsymbol{w}_i(k)] & \text{if } i \in N_q(k) \\ \boldsymbol{w}_i(k) & \text{if } i \notin N_q(k) \end{cases}$$

where $0 < \mu(k) < 1$ (the learning rate parameter).

Step 6. Adjust the learning rate parameter.

Step 7. Appropriately reduce the topological neighborhood $N_q(k)$.

Step 8. Set $k \leftarrow k + 1$; then go to step 2.

---

†Initialization of the network weights can be carried out by either randomly initializing them, or selecting a set of weights that reflect some a priori knowledge about the input data, that is, information regarding the possible distribution of the output clusters.

‡Stopping conditions can be, for example, based on the total number of specified iterations or based on monitoring the weight changes.

the learning process that is generally the longest part of the network learning. The remaining iterations are necessary during this phase for carrying out fine adjustments of the map. The learning rate parameter should attain relatively small values for a long time. For example, the learning rate parameter should be on the order of (or less than) 0.01 [4].

EXAMPLE 4.1. In this example, 1,000 two-dimensional vectors in the unit square are generated from a uniform distribution; see Figure 4.4. These vectors are mapped onto a 5 × 5 planar array of neurons. The function **trainsm** in the MATLAB neural network toolbox is used to carry out the simulation. The neighborhoods and learning rate parameter are both gradually reduced during training. Figure 4.5 shows the results of the neural network training as the network learns to represent the distribution of the input. In Figure 4.5(*b*) the map begins to "unfold" after 800 iterations. In frame (*c*), after 1,600 iterations, the map continues to unfold. Finally after 10,000 iterations, frame (*d*) shows the map essentially unfolded. The statistical distribution of the neurons in the map is approaching that of the input vectors, which is a result of the uniform distribution of the input vectors. This will not be the case for other input distributions. Figure 4.6 shows an anomaly that can occasionally occur, that is, the twisting of the map. This can occur when different parts of the map fit the topology of separate parts of the input space. The only difference between this example and the previous one is a different initial weight matrix. After 4,000 iterations the map has still not unfolded, and it is unlikely that it will; therefore, it is best to simply restart the training process with a different set of initial weights.

EXAMPLE 4.2. A very important feature of the Kohonen SOM is referred to as the *topological ordering property*. This property accounts for the ability of the SOM to form an abstract two-dimensional representation from a higher-dimensional input space. The example presented here illustrates this property. Two sets of 1,000 three-dimensional Gaussian vectors are generated. Both sets of data have a variance of $\sigma^2 = 0.1$; however, the first set of data is centered at (0, 0, 0) (class

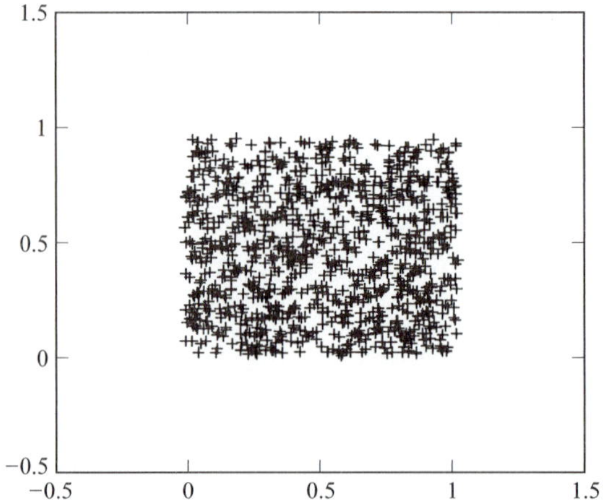

Figure 4.4 Distribution of 1,000 two-dimensional random vectors used in Example 4.1. The vectors are drawn from a uniform distribution in the range [0, 1] in both dimensions.

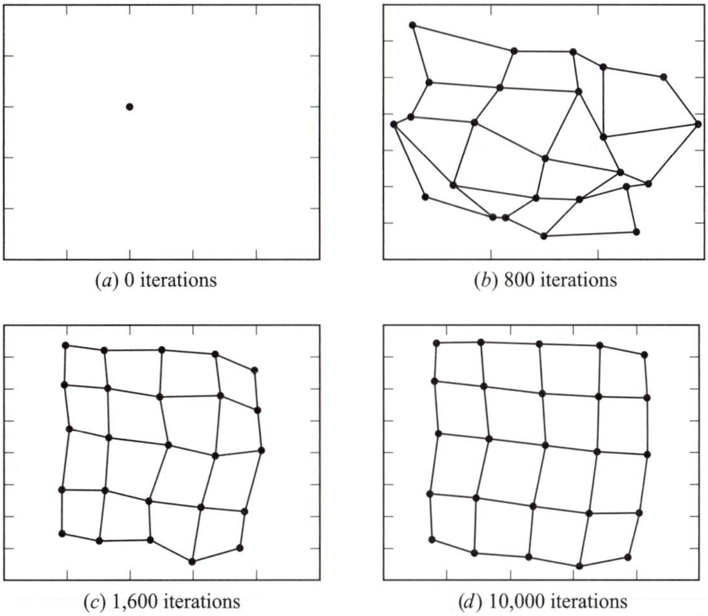

Figure 4.5 Mapping of uniform random vectors from a unit square onto a 5×5 array of neurons. The initial learning rate parameter was set to $\mu(0) = 1$.

1), and the second set of data is centered at (5, 5, 5) (class 2). Figure 4.7 shows the two Gaussian clouds. We train a SOM map with 25 neurons in a square. The two sets of three-dimensional vectors are presented to the network 5,000 times, that is, 5,000 training epochs. As in the previous example, the function **`trainsm`** in the MATLAB neural network toolbox is used to carry out the simulation, and the neighborhoods and learning rate parameter (initially set to one) are both gradually reduced during training. The two-dimensional map after 5,000 training epochs is shown in Figure 4.8(*a*) (this is the topological grid that MATLAB

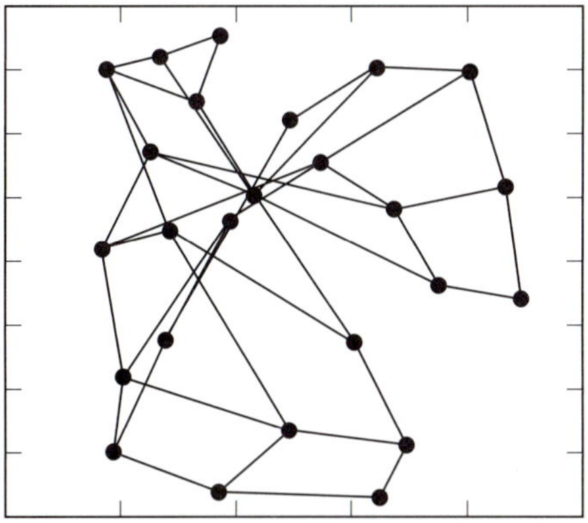

Figure 4.6 Twisted feature map after 4,000 iterations.

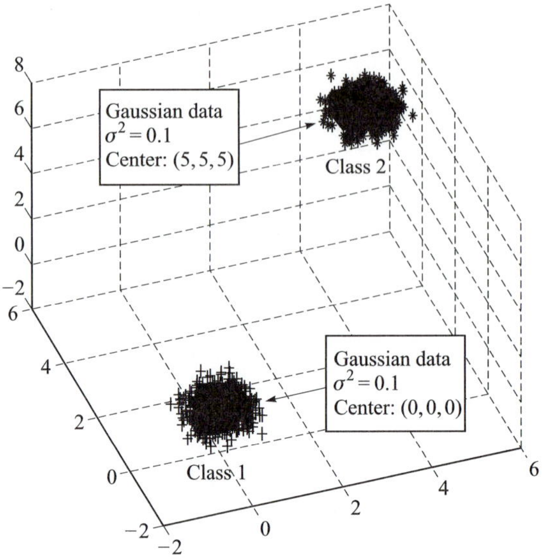

Figure 4.7 Two sets of three-dimensional Gaussian vectors. Both have a variance of $\sigma^2 = 0.1$ and are centered as shown.

generates during training). In Figure 4.8(b), the three-dimensional weight vectors are plotted, and the two classes shown outlined are determined by computing the mean of each weight vector. From both Figure 4.8(a)and Figure 4.8(b), we see that the SOM is able to properly categorize the three-dimensional Gaussian vectors.

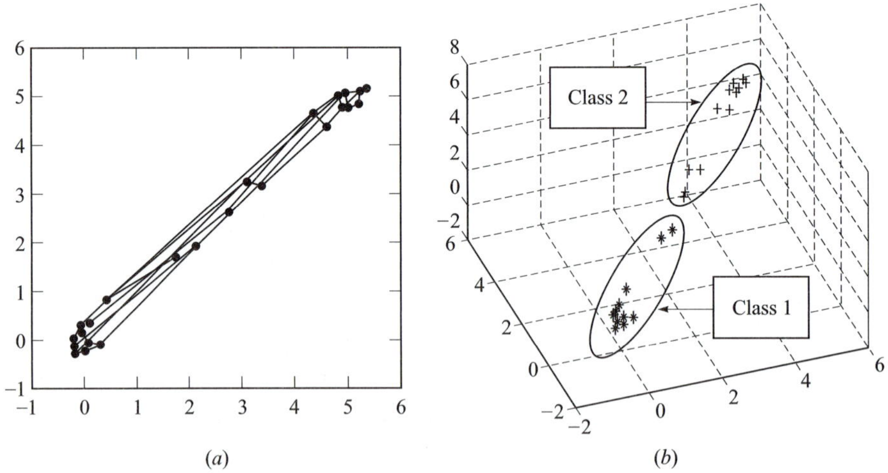

Figure 4.8 (a) Kohonen SOM network mapping of two sets of three-dimensional Gaussian vectors onto a 5 × 5 array of neurons. (b) SOM three-dimensional weights. The two classes were determined by computing the mean of each weight vector.

4.3 LEARNING VECTOR QUANTIZATION

Vector quantization [7–10] is a technique used for compression of speech and image data. The basic idea is to represent the input vectors with a smaller set of *prototypes* that provide a *good* approximation to the input space $\mathscr{X}$, where the vectors $\boldsymbol{x}_i$, for $i = 1, 2, \ldots, N$, constitute the input space. Given no a priori knowledge of a probability model for the input space, it is assumed that a long training sequence of data is available (or a relatively large set of input training vectors, that is, $N \gg 1$). The objective is to develop a "codebook" of quantization vectors and then use these vectors to encode any input vector. To develop a *dependable* codebook, the large set of training vectors is used to form groups according to a predetermined number of clusters, and each cluster j is represented by its particular centroid $\hat{\boldsymbol{x}}_j$ (i.e., the reproduction or reconstruction vector). The centroid clustering is based on a given distortion measure. Several different distortion measures can be used; however, the distortion measure should be tractable, computable from sampled data, and subjectively meaningful [11]. One such distortion measure is based on the L_2 (Euclidean) norm of a vector, that is,

$$d(\boldsymbol{x}, \hat{\boldsymbol{x}}) = \|\boldsymbol{x} - \hat{\boldsymbol{x}}^2\|_2^2 = (\boldsymbol{x} - \hat{\boldsymbol{x}})^T(\boldsymbol{x} - \hat{\boldsymbol{x}}) \tag{4.8}$$

referred to as the squared-error distortion. Once the codebook is developed, it is stored at both the "transmitter" and the "receiver." The quantization of an input vector $\boldsymbol{x}_i$ is then performed as follows: (1) The input is presented to the vector quantizer and is compared to the codebook vectors $\hat{\boldsymbol{x}}_j$ (for $j = 1, 2, \ldots, m$), and the codebook vector that yields the minimum distortion, that is, the minimum distance according to (4.8), is selected. (2) This selected vector $\hat{\boldsymbol{x}}_q$ then represents $\boldsymbol{x}_i$, and the index q (associated with the appropriate *class* to which the input vector belongs) is "transmitted" to the receiver where the appropriate reproduction (reconstruction) vector $\hat{\boldsymbol{x}}_q$ is selected as the representation of the input vector $\boldsymbol{x}_i$. Figure 4.9 illustrates the vector quantization process.

An alternative distortion measure is the Mahalanobis distortion [12], given by

$$d_M(\boldsymbol{x}, \hat{\boldsymbol{x}}) = (\boldsymbol{x} - \hat{\boldsymbol{x}})^T \boldsymbol{C}_x^{-1}(\boldsymbol{x} - \hat{\boldsymbol{x}}) \tag{4.9}$$

where the weighting matrix $\boldsymbol{C}_x$ is the covariance matrix of the input, that is, $\boldsymbol{C}_x = \mathrm{E}\{(\boldsymbol{x} - \bar{\boldsymbol{x}})(\boldsymbol{x} - \bar{\boldsymbol{x}})^T\}$, where $\bar{\boldsymbol{x}} = \mathrm{E}\{\boldsymbol{x}\}$ (cf. Sect. A.7.6). Another more sophisticated distortion measure of Itakura and Saito [13, 14] is given by

$$d_{\mathrm{IS}}(\boldsymbol{x}, \hat{\boldsymbol{x}}) = (\boldsymbol{x} - \hat{\boldsymbol{x}})^T \boldsymbol{R}(\boldsymbol{x})(\boldsymbol{x} - \hat{\boldsymbol{x}}) \tag{4.10}$$

where for each $\boldsymbol{x}$, $\boldsymbol{R}(\boldsymbol{x})$ is a positive definite symmetric matrix.

When the Euclidean distance (distortion) measure is used as in (4.8) to decide to which region (class, group, category, or cluster) the input vector $\boldsymbol{x}_i$ belongs, the quantizer is referred to as a *Voronoi quantizer* [15]. The Voronoi quantizer performs a partitioning of the input space into various Voronoi cells [8], where each cell is represented by one of the reproduction vectors $\hat{\boldsymbol{x}}_j$. The

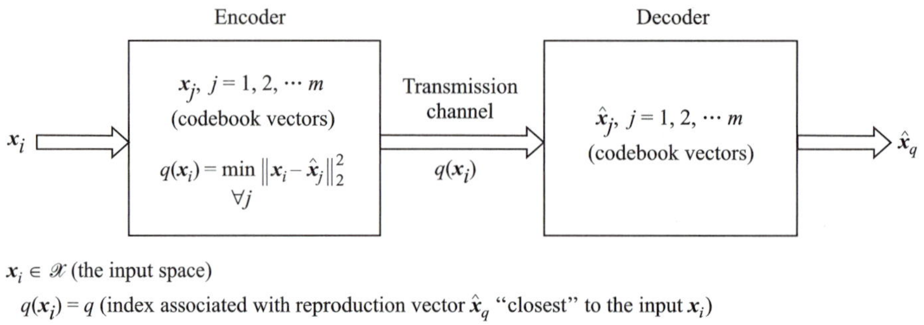

Figure 4.9 Vector quantizer.

qth Voronoi cell then contains those points of the input space $\mathcal{X}$ that are closer to the reproduction vector $\hat{\boldsymbol{x}}_q$, in the Euclidean sense, than to any other reproduction vector $\hat{\boldsymbol{x}}_j$ $(j \neq q)$. Figure 4.10 shows an example of partitioning the input space $\mathcal{X}$ into four cells with four associated reproduction vectors.

Learning vector quantization was developed by Kohonen [16], and in [17] he summarizes three versions of the algorithm. This is a *supervised learning* technique that can classify input vectors based on vector quantization. The version of LVQ presented here is LVQ1, which was Kohonen's first version of learning vector quantization [16]. The LVQ1 training process proceeds with an input vector being randomly selected (along with the correct class for that vector, thus the supervised learning) from the "labeled" training set. There can actually exist several reproduction (prototype) vectors per class or category. LVQ1 is very similar to the Kohonen SOM, even though LVQ1 is a supervised network and the Kohonen SOM is unsupervised.

Given an input vector $\boldsymbol{x}_i$ to the network, the "output neuron" (i.e., the class or category) in LVQ1 is deemed to be a "winner" according to

$$\min_{\forall j} d(\boldsymbol{x}_i, \boldsymbol{w}_j) = \min_{\forall j} \|\boldsymbol{x}_i - \boldsymbol{w}_j\|_2^2 \tag{4.11}$$

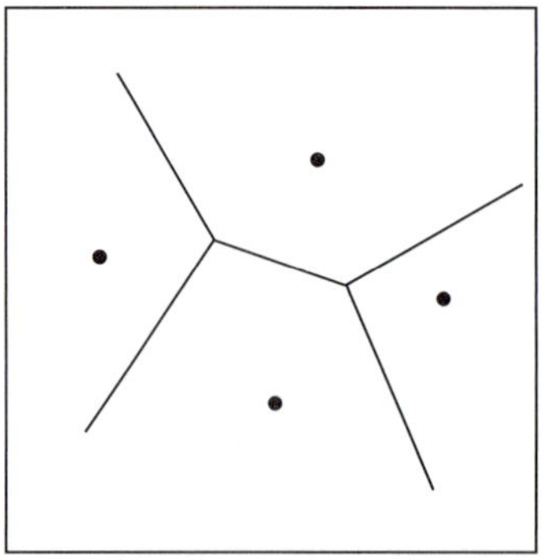

Figure 4.10 Partitioning of the input space $\mathcal{X}$ as realized by a Voronoi quantizer with four reproduction vectors representing the four cells.

which is essentially the same "winning rule" we established for the Kohonen SOM. The only difference is that in (4.11) we take the square of the Euclidean norm. In (4.11) the synaptic weight vector $\mathbf{w}_j$ has replaced the reproduction vector $\hat{\mathbf{x}}_j$ shown in (4.8) and Figure 4.9. The major difference between LVQ1 and the Kohonen SOM lies in how the weight vectors are updated. We let the set of input vectors be denoted as $\{\mathbf{x}_i\}$, for $i = 1, 2, \ldots, N$, and the network synaptic weight vectors (Voronoi vectors) are denoted as $\{\mathbf{w}_j\}$, for $j = 1, 2, \ldots, m$. We also let $C_{\mathbf{w}_j}$ be the class (or category) that is associated with the (weight) Voronoi vector $\mathbf{w}_j$, and $C_{\mathbf{x}_i}$ is the *class label* of the input vector $\mathbf{x}_i$ to the network. The weight vector $\mathbf{w}_j$ is adjusted in the following manner:

1. If the class associated with the weight vector and the class label of the input are the same, that is, $C_{\mathbf{w}_j} = C_{\mathbf{x}_i}$, then

$$\mathbf{w}_j(k+1) = \mathbf{w}_j(k) + \mu(k)[\mathbf{x}_i - \mathbf{w}_j(k)] \tag{4.12}$$

   where $0 < \mu(k) < 1$ (the learning rate parameter).
2. But if $C_{\mathbf{w}_j} \neq C_{\mathbf{x}_j}$, then

$$\mathbf{w}_j(k+1) = \mathbf{w}_j(k) - \mu(k)[\mathbf{x}_i - \mathbf{w}_j(k)] \tag{4.13}$$

   and the other weight vectors are *not* adapted.

Therefore, the update rule for modifying a weight vector in (4.12) is the standard one if the class is correct. In other words, according to the learning rule in (4.12), the weight vector $\mathbf{w}_j$ is moved in the direction of the input $\mathbf{x}_i$ if the class labels of the input vector and of the weight vector agree. However, if the class is not correct, the weight vector is moved in the opposite direction away from the input vector according to (4.13). The learning rate parameter $\mu(k)$ is typically monotonically decreased in accordance with the discrete-time index k (e.g., linearly decreased in time, starting at 0.01 or 0.02 [17]; however, many times 0.1 is used as the initial value). The convergence properties of LVQ have been studied by Baras and LaVigna [18], and their approach is based on stochastic approximation theory. The weights can be initialized by using several methods; for example, the first m (total number of classes) vectors from the set of training vectors can be used to initialize the m weight vectors, that is, $\mathbf{w}_j(0)$ for $j = 1, 2, \ldots, m$. Another approach is to randomly initialize the weight vectors (within the dynamic range of the input vectors). The stopping condition can be based on the total number of desired training epochs, or on monitoring the convergence of the weight vectors. Another stopping condition can be based on monitoring the learning rate parameter directly, and when it is sufficiently small, training can be terminated. In the example shown below we will establish the stopping condition to be the total number (predefined) of iterations. The basic LVQ1 algorithm can be summarized as follows:

LVQ1 algorithm

Step 1. Initialize all weight vectors $\mathbf{w}_j(0)$, initialize the learning rate parameter $\mu(0)$, and set $k = 0$.
Step 2. Check the stopping condition. If false, continue; if true, quit.
Step 3. For each training vector $\mathbf{x}_i$ perform steps 4 and 5:
Step 4. Determine the weight vector index $(j = q)$ such that $\min_{\forall j} \|\mathbf{x}_i - \mathbf{w}_j(k)\|_2^2$ [using the square of the Euclidean distance as given in (4.11)]. [*Note*: $\mathbf{w}_q(k)$ will be the weight vector that minimizes the square of the norm.]
Step 5. Update the appropriate weight vector $\mathbf{w}_q(k)$ as follows:

$$\text{If } C_{\mathbf{w}_q} = C_{\mathbf{x}_i} \quad \text{then} \quad \mathbf{w}_q(k+1) = \mathbf{w}_q(k) + \mu(k)[\mathbf{x}_i - \mathbf{w}_q(k)]$$

$$\text{If } C_{\mathbf{w}_q} \neq C_{\mathbf{x}_i} \quad \text{then} \quad \mathbf{w}_q(k+1) = \mathbf{w}_q(k) - \mu(k)[\mathbf{x}_i - \mathbf{w}_q(k)]$$

Step 6. Set $k \leftarrow k + 1$, reduce the learning rate parameter,† then go to step 2.

The neural architecture for LVQ1 is shown in Figure 4.11. Actually the architecture is basically the same as that of the Kohonen SOM map but without the topological structure.

Kohonen improved on LVQ1 and called the newer version LVQ2 [17]. The LVQ2 algorithm was based on differentially shifting the decision boundaries toward the Bayes limits. The LVQ2 version was next modified, resulting in LVQ2.1 [17], and finally LVQ3 was developed [4]. These later LVQ versions have in common that the weight vector of the winning neuron and the "runner up" neuron weight vector are both updated.

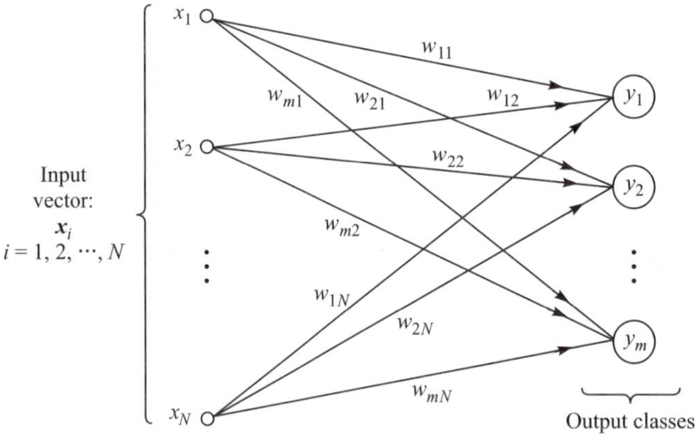

Figure 4.11 Neural architecture for learning vector quantization. The output classes y_j $(j = 1, 2, \ldots, m)$ have class labels associated with the synaptic weight vector for that output class, that is, $C_{\mathbf{w}_j}$ $(j = 1, 2, \ldots, m)$.‡

†The learning rate parameter μ can be reduced in accordance with k (discrete-time index) using $\mu(k) = \mu(k-1)/(k+1)$, for $k > 0$.

‡Note that the convention established in Chapter 2 (cf. Sect. 2.2) to subscript the weights in a network is followed. However, in the LVQ1 algorithm presented above, the transpose of the weight vectors is considered.

EXAMPLE 4.3. We present a very simple example of seven four-dimensional vectors assigned to two classes. The seven vectors and associated classes are as follows:

Vector	**Class** (C_{x_i})
$x_1 = [1, 0, 0, 1]^T$	$\rightarrow 1$
$x_2 = [0, 1, 1, 0]^T$	$\rightarrow 2$
$x_3 = [0, 0, 0, 1]^T$	$\rightarrow 2$
$x_4 = [1, 0, 0, 0]^T$	$\rightarrow 1$
$x_5 = [1, 1, 1, 0]^T$	$\rightarrow 1$
$x_6 = [0, 1, 1, 1]^T$	$\rightarrow 2$
$x_7 = [1, 1, 1, 1]^T$	$\rightarrow 1$

As described above, one method for initialization of the network is to use the first two vectors from the above set and initialize w_1 and w_2 with x_1 and x_2, respectively. The associated classes for w_1 and w_2 are then $C_{w_1} = 1$ and $C_{w_2} = 2$, respectively. The remaining five vectors can be used for training (for the first training epoch). Let's go through the details of one training epoch. We initialize the learning rate parameter to $\mu(k = 1) = \mu(1) = 0.1$ and decrease it by k every training epoch, for example, $\mu(2) = \mu(1)/2$, $\mu(3) = \mu(2)/3$, etc.[†]

Training epoch 1 ($k = 1$):

1. Initialize weights:

$$w_1 = [1, 0, 0, 1]^T \quad (\text{class } C_{w_1} = 1) \quad \text{and}$$
$$w_2 = [0, 1, 1, 0]^T \quad (\text{class } C_{w_2} = 2)$$

2. For input vector $x_3 = [0, 0, 0, 1]^T$ with class $C_{x_3} = 2$, check:

$$\|x_3 - w_2\|_2^2 = 3$$
$$\|x_3 - w_1\|_2^2 = 1 \Rightarrow \text{minimum} \Rightarrow q = 1$$

   and since $C_{x_3} \neq C_{w_1}$, move the weight vector w_1 away from x_3, that is, use (4.13):

$$w_1 = [1, 0, 0, 1]^T - 0.1([0, 0, 0, 1]^T - [1, 0, 0, 1]^T)$$
$$= [1.1, 0, 0, 1]^T$$

3. For input vector $x_4 = [1, 0, 0, 0]^T$ with class $C_{x_4} = 1$, check:

$$\|x_4 - w_2\|_2^2 = 3$$
$$\|x_4 - w_1\|_2^2 = 1.01 \Rightarrow \text{minimum} \Rightarrow q = 1$$

   and since $C_{x_4} = C_{w_1}$, move the weight vector w_1 in the direction of x_4, that is, use (4.12):

$$w_1 = [1.1, 0, 0, 1]^T + 0.1([1, 0, 0, 0]^T - [1.1, 0, 0, 1]^T)$$
$$= [1.09, 0, 0, 0.9]^T$$

[†]Note that we take the initial discrete-time index in this example to be $k = 1$ instead of $k = 0$; therefore, the learning rate parameter is now adjusted according to $\mu(k) = \mu(k-1)/k$, for $k > 1$.

4. For the input vector $x_5 = [1, 1, 1, 0]^T$ with class $C_{x_5} = 1$, check:

$$\|x_5 - w_1\|_2^2 = 2.8181$$
$$\|x_5 - w_2\|_2^2 = 1 \Rightarrow \text{minimum} \Rightarrow q = 2$$

and since $C_{x_5} \neq C_{w_2}$, move the weight vector w_2 away from x_5, that is, use (4.13):

$$\begin{aligned} w_2 &= [0, 1, 1, 0]^T - 0.1([1, 1, 1, 0]^T - [0, 1, 1, 0]^T) \\ &= [-0.1, 1, 1, 0]^T \end{aligned}$$

5. For the input vector $x_6 = [0, 1, 1, 1]^T$ with class $C_{x_6} = 2$, check:

$$\|x_6 - w_1\|_2^2 = 3.1981$$
$$\|x_6 - w_2\|_2^2 = 1.01 \Rightarrow \text{minimum} \Rightarrow q = 2$$

and since $C_{x_6} = C_{w_2}$, move the weight vector w_2 in the direction of x_6, that is, use (4.12):

$$\begin{aligned} w_2 &= [-0.1, 1, 1, 0]^T + 0.1([0, 1, 1, 1]^T - [-0.1, 1, 1, 0]^T) \\ &= [-0.09, 1, 1, 0.1]^T \end{aligned}$$

6. For the input vector $x_7 = [1, 1, 1, 1]^T$ with class $C_{x_6} = 1$, check:

$$\|x_7 - w_1\|_2^2 = 2.0181$$
$$\|x_7 - w_2\|_2^2 = 1.9981 \Rightarrow \text{minimum} \Rightarrow q = 2$$

and since $C_{x_7} \neq C_{w_2}$, move the weight vector w_2 away from x_7, that is, use (4.13):

$$\begin{aligned} w_2 &= [-0.09, 1, 1, 0.1]^T - 0.1([1, 1, 1, 1]^T - [-0.09, 1, 1, 0.1]^T) \\ &= [-0.199, 1, 1, 0.01]^T \end{aligned}$$

End of Training Epoch 1

Now according to the algorithm given above, we set $k = 2$, and $\mu(2) = \mu(1)/2 = 0.05$, then go to step 2 and check the stopping condition. The next training epoch we start with the first training vector x_1 and proceed. In this example, the MATLAB function given in Table 4.1 was used with the number of training epochs set to 500.

After 500 training epochs the final weights are

$$w_1 = \begin{bmatrix} 1.2996 \\ 0.4952 \\ 0.4952 \\ 0.4848 \end{bmatrix} \quad \text{and} \quad w_2 \begin{bmatrix} -0.1881 \\ 1 \\ 1 \\ 0.3603 \end{bmatrix}$$

The convergence profile for the elements of the two weight vectors as a function of the training epoch is shown in Figure 4.12. Finally, if we compute the minimum distance between each of the input vectors and the computed weight vectors w_1 and w_2 shown above, this will indicate the class to which each input vector x_i belongs. The results are as follows:

TABLE 4.1
MATLAB function for LVQ1

```
function W = lvq1 (X,CX,m,mu,maxiter)
%
% W =       LVQ1 (X,CX,mu,maxiter) computes the weight
%           matrix for learning vector quantization 1 (LVQ1)
%
% X:        is the matrix of inputs, i.e., each column
%           vector is an input
% CX:       is a row vector of scalar ''classes'' associated
%           with the column vectors of X
% m:        number of different classes
% mu:       initial learning rate parameter
% maxiter: maximum number of training epochs (iterations)
%
N=size (X,2);
% Initialize the weight vectors with the first nc vectors
% from the training set (Note: must have training vectors
% arranged so first nc vectors have the full set of classes
% to be represented) .
W=X(:,1:m) ;
CW=CX(1:m) : % classes for weight vectors
snorm=zeros (1,m);
niter=1;
while niter <= maxiter
   if niter == 1
      for i=m+1:N
         for j=1:m
            snorm (1,j)=norm(X(:,i)-W(:,j))^2;
         end
         [mind,index]=min(snorm) ;
         if CX(i)==CW(index)
            W(:,index)=W(:,index)+mu*(X(:,i)-W(:,index)) ;
         else
            W(:,index)=W(:,index)-mu*(X(:,i)-W(:,index)) ;
         end
      end
   else
      for i=1:N
         for j=1:m
            snorm(1,j)=norm(X(:,i)-W(:,j))^2;
         end
         [mind,index]=min(snorm) ;
         if CX(i)==CW(index)
            W(:,index)=W(:,index)+(mu/niter)*(X(:,i)-W(:,index));
         else
            W(:,index)=W(:,index)-(mu/niter)*(X(:,i)-W(:,index));
         end
      end
   end
   niter=niter+1;
end
```

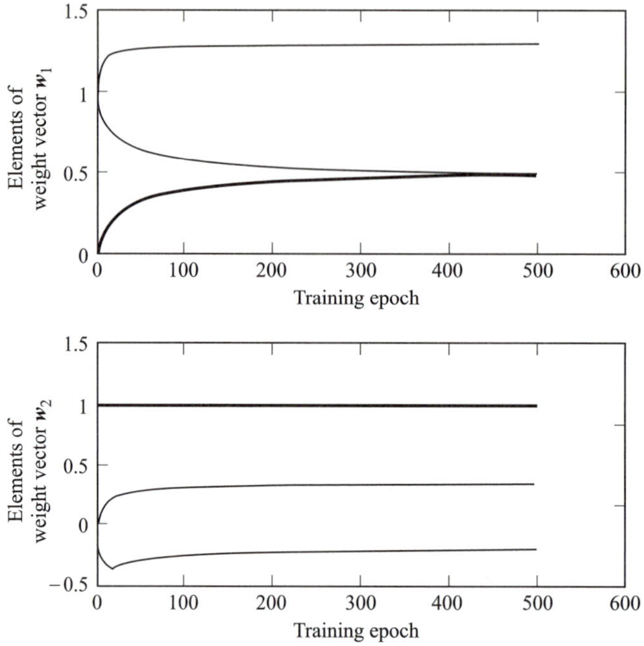

Figure 4.12 Convergence profiles for the elements of the two weight vectors as a function of the training epoch.

$$\min\{\|x_1 - w_1\|_2^2, \|x_1 - w_2\|_2^2\} \Rightarrow x_1 \text{ closest to } w_1 \Rightarrow \text{class 1}$$
$$\min\{\|x_2 - w_1\|_2^2, \|x_2 - w_2\|_2^2\} \Rightarrow x_2 \text{ closest to } w_2 \Rightarrow \text{class 2}$$
$$\min\{\|x_3 - w_1\|_2^2, \|x_3 - w_2\|_2^2\} \Rightarrow x_3 \text{ closest to } w_2 \Rightarrow \text{class 2}$$
$$\min\{\|x_4 - w_1\|_2^2, \|x_4 - w_2\|_2^2\} \Rightarrow x_4 \text{ closest to } w_1 \Rightarrow \text{class 1}$$
$$\min\{\|x_5 - w_1\|_2^2, \|x_5 - w_2\|_2^2\} \Rightarrow x_5 \text{ closest to } w_1 \Rightarrow \text{class 1}$$
$$\min\{\|x_6 - w_1\|_2^2, \|x_6 - w_2\|_2^2\} \Rightarrow x_6 \text{ closest to } w_2 \Rightarrow \text{class 2}$$
$$\min\{\|x_7 - w_1\|_2^2, \|x_7 - w_2\|_2^2\} \Rightarrow x_7 \text{ closest to } w_1 \Rightarrow \text{class 1}$$

These results correspond exactly to the classes established for each of the input vectors.

Self-organizing map and LVQ

By combining the Kohonen SOM with LVQ, an adaptive pattern classification system can be developed. And *K*-means clustering [19], used in pattern recognition, could be used in lieu of the Kohonen SOM; however, we will consider only the latter. The first step in the process is to use the SOM to select a relatively small set of features that will contain the pertinent information relating to the input data, which can be classified by the LVQ network in the second step. Therefore, the LVQ network would function as the actual classifier, using the selected features (generated by the SOM) from the input data and assigning them to individual classes or categories. This hybrid adaptive

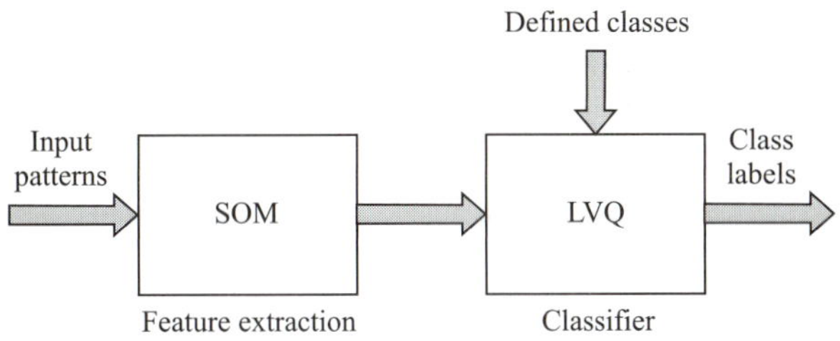

Figure 4.13 Adaptive pattern classification system using the Kohonen SOM and LVQ.

pattern classification system thus combines an unsupervised network (SOM) with a supervised network (LVQ), and it is shown in Figure 4.13.

EXAMPLE 4.4. This example illustrates the use of the hybrid adaptive pattern classification network shown in Figure 4.13. We use the results from Example 4.2, that is, the three-dimensional weight vectors that were generated using the Kohonen SOM [shown in Figure 4.8(*b*), and repeated in Figure 4.14(*a*)]. The 25 SOM weight vectors are inputs to the LVQ network with two classes established, that is, class 1 and class 2, as shown in Figure 4.14(*a*). The MATLAB LVQ1 program shown in Table 4.1 was modified to randomly initialize the two three-dimensional weights. The initial learning rate parameter is set to $\mu(1) = 0.1$, and the total number of training epochs is 1,000. The final two, three-dimensional, weight vectors are shown in Figure 4.14(*b*). The two weight vectors are given as

$$w_1 = \begin{bmatrix} 0.3232 \\ 0.3770 \\ 0.3113 \end{bmatrix} \quad \text{and} \quad w_2 = \begin{bmatrix} 4.2837 \\ 4.2825 \\ 4.2575 \end{bmatrix}$$

These are relatively close to the centroids of the Gaussian clouds shown in Figure 4.7. Comparing the SOM results shown in Figure 4.14(*a*) with those shown in

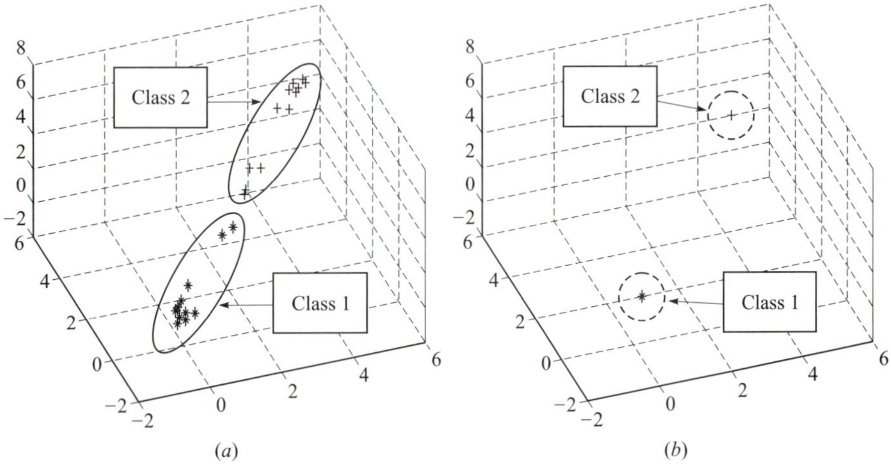

Figure 4.14 (*a*) SOM three-dimensional weight vectors from Example 4.2 with the two defined classes. (*b*) Final weight vectors from the LVQ1 network after 1,000 training epochs.

Figure 4.14(*b*) using LVQ, we can readily see the improvement obtained by using LVQ to classify the patterns and assign them to two distinct categories.

4.4 ADAPTIVE RESONANCE THEORY (ART) NEURAL NETWORKS

Clusters (categories) formed by a competitive learning network are not guaranteed to be stable. For example, even if the same set of input vectors is continuously presented to the network, the winning unit may continue to change. One way to prevent this is to gradually reduce the learning rate to zero, thus freezing the learned categories. However, when this is carried out, *stability* is gained at the expense of losing *plasticity*, or the ability of the network to react to new data (that is, the network will not be able to learn new categories). This problem is commonly referred to as Grossberg's *stability/plasticity dilemma.* Adaptive resonance theory (ART), developed by Carpenter and Grossberg [20], overcomes this stability/plasticity dilemma by accepting and adapting the stored prototype of a category only when the input is sufficiently *similar* to it. When an input pattern is not sufficiently similar to any existing prototype, a new category is formed with the input pattern as the prototype using a previously uncommitted output (cluster) unit. Therefore, this type of network can essentially produce new clusters by itself. If there are no such uncommitted units remaining, then a novel input yields no response (i.e., a form of outlier rejection). The meaning of being sufficiently *similar* is dependent on a *vigilance parameter*. If the vigilance parameter is large, the similarity condition becomes very stringent and many finely divided categories are formed. On the other hand, a small vigilance parameter gives a coarser categorization. As the network is trained, each training pattern may be presented to the network several times. The first time a pattern is presented to the network, it may be placed on one cluster unit and then placed on a different cluster when it is presented later. This is due to changes in the network weights for the first cluster if it has learned other input patterns. A *stable* network will not return a pattern to a previous cluster.

Figure 4.15 shows the basic features of an ART network. Stability and plasticity can be achieved by using the gain control units G_1 and G_2 and the orienting subsystem M unit vigilance parameter ρ. There is a layer of processing units in the attentional subsystem called the *feature representation field* F_1 as well as a layer of output units called the *category representation field* F_2. In these fields reside short-term memory (STM) traces, because they exist only in association with a single application of an input vector. The weights associated with the bottom-up and top-down connections between F_1 and F_2 are called long-term memory traces because they encode information that remains as a part of a network for an extended period. Excitatory signals are indicated with a + sign and inhibitory signals a − sign. In the process of learning, once a cluster unit is selected, the bottom-up and top-down signals are sustained for an extensive period. During this period the weight changes occur. This is the *resonance* condition. Each unit in an ART network has three sources from which it can receive a signal. Input units can receive signals from an input

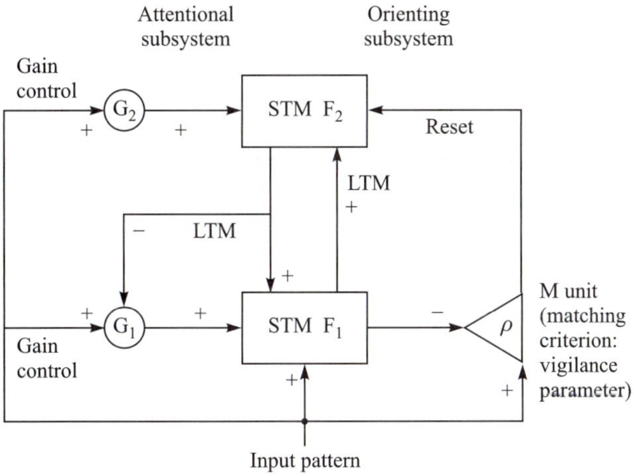

Figure 4.15 Basic features of the ART architecture.

vector, a top-down signal from an F_2 neuron, or the gain control unit G_1. Similarly, an F_2 unit can receive a signal from an F_1 unit, the M unit, or the gain control unit G_2. An F_1 or F_2 unit must receive two excitatory signals in order to be "on". Since there are three possible sources of signals, this is called the $^2\!/_3$ rule. The supplemental unit M (the orienting subsystem) controls the vigilance matching.

The changes in the activity levels of the network neurons, and the changes in the network weights, are actually governed by coupled differential equations. However, in practice, it can be assumed that the activity level of a neuron is typically changing more rapidly than the weights; therefore, this process can be viewed in a simplified manner. Specifically, this applies to the manner in which ART networks are trained. There are basically two types of simplified learning approaches in ART networks. They differ in the assumptions made and their performance characteristics. In the case of *fast learning*, it is assumed that during *resonance* the weight updates are carried out relatively fast (with respect to the time duration that an input pattern is presented to the network). As a result, the weights of the network reach equilibrium on each trial. Conversely, in the case of *slow learning*, the network weights change more slowly than in the fast learning mode, and thus, the weights do not reach equilibrium on any particular trial. There are tradeoffs between these two basic learning modes. In the case of slow learning, more training epochs are required (compared to fast learning) for the weights to reach equilibrium; however, fewer calculations are needed in the slow learning mode.

There are three basic types of unsupervised ART networks: ART1 (binary-valued input vectors), ART2 (continuous-valued input vectors), and fuzzy ART (both binary and continuous-valued input vectors). We will give details for only the ART1 network.

4.4.1 ART1

ART1 is designed to cluster binary vectors [20], and the basic architecture is shown in Figure 4.16. The explanation given here describing the operation of the ART1 network is based on discrete-time events for the architecture units [21], as opposed to differential equations [20]. The equilibrium weights of the network can be easily determined without resorting to iterative solutions of differential equations. The F_1 layer (feature representation field) and the F_2 layer (category representation field) are fully connected to each other in both directions, and the output nodes (neurons) are also connected to each other in both directions so as to realize the winner-take-all subsystem. However, for the sake of simplicity, these connections are not shown in Figure 4.16. The gain control unit G provides a control signal for the F_1 layer and has two states; that is, $G = 1$ if there is no unit at the output layer that is "on", otherwise $G = 0$. Therefore, the gain G can be realized by a threshold logic unit (TLU) (cf. Sect. 2.2) where the gain is computed as

$$G = T_{\text{binary}}\left(\sum_{i=1}^{n} x_i - n\sum_{j=1}^{m} z_j - 0.5\right) \tag{4.14}$$

where

$$T_{\text{binary}}(\lambda) = \begin{cases} 1 & \text{if } \lambda > 0 \\ 0 & \text{if } \lambda \leq 0 \end{cases} \tag{4.15}$$

Assuming $\boldsymbol{x} \neq \mathbf{0}$, if an output unit is in the on state, the quantity inside the parentheses in (4.14) will be negative ($G = 0$); and if there are no output units in the on state, the quantity inside the brackets will be positive ($G = 1$).

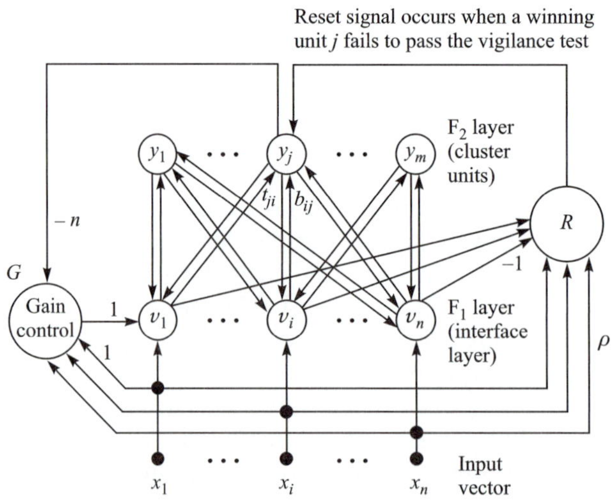

Figure 4.16 Architecture of the ART1 network, where t_{ji} are the *top-down weights* (initialized to 1s) and b_{ij} are the *bottom-up weights.*

In the F_1 layer, the state of the unit v_i is given by

$$v_i = T_{\text{binary}}\left(x_i + G + \sum_{j=1}^{m} t_{ji}z_j - 1.5\right) \tag{4.16}$$

If the gain $G = 1$, it is implied that no output unit is on and the summation in (4.16) will be zero. Therefore, if the ith element of the input vector x_i is a 1, then $v_i = 1$; however, if $x_i = 0$, then $v_i = 0$. On the other hand, when $G = 0$, since only one unit at the output can be in the on state, $\sum_{j=1}^{m} t_{ji}z_j = t_{ji}$ (i.e., unit j is the winning output unit and $z_j = 1$), and both x_i and t_{ji} must be a 1 for $v_i = 1$. Otherwise, if either x_i or t_{ji} (or both) is zero, then $v_i = 0$. In summary, from the above discussion, in order for $v_i = 1$ at least two of the three terms inside the parentheses in (4.16) must be present (i.e., a 1); this is the $^2\!/_3$ rule.

The F_2 layer (output layer) in Figure 4.15 is a winner-take-all competitive layer. All units in the output layer F_2 are off when an input vector is first presented to the network, and $G = 1$; therefore, from the previous discussion the input vector is replicated at the F_1 layer, that is, $v_i = x_i$ for $i = 1, 2, \ldots, n$. The bottom-up weights from the F_1 layer to the output unit j are computed according to

$$b_{ij}(k) = \frac{t_{ji}(k)}{\tau + \sum_{i=1}^{n} t_{ji}(k)} \tag{4.17}$$

where typically $\tau = 0.5$, k is the discrete-time index, and t_{ji}, the ith element of the top-down weight vector $\boldsymbol{t}_j$ (prototype vectors), is the connection weight from the jth unit at the F_2 layer to the ith unit at the F_1 layer.

We can next establish a measure for determining a "winning" cluster at the output, and then we decide how good the match is between the input $\boldsymbol{x}$ and the weight vector $\boldsymbol{t}_j$. This requires two similarity measures to be computed. Two measures are necessary because binary vectors are considered in the ART1 architecture and not bipolar [22]. The first similarity measure is computed as

$$\sigma_1 = \frac{\sum_{i=1}^{n} t_{ji}x_i}{\tau + \sum_{i=1}^{n} t_{ji}} \tag{4.18}$$

Among the enabled output units, the prototype vector $\boldsymbol{t}_j$ closest to the current input vector $\boldsymbol{x}$ can be found from the largest value computed according to (4.18). For a small value of τ, σ_1 in (4.18) is approximately the ratio of the number of overlapping 1s in the input vector $\boldsymbol{x}$ and the top-down weight vector $\boldsymbol{t}_j$ to the number of 1s in $\boldsymbol{t}_j$. For the case when two clusters at the output of the network have prototype vectors with the same number of overlapping 1s in the input $\boldsymbol{x}$, the one with the fewest 1s in its prototype vector will be selected as the winner. An input $\boldsymbol{x}$ to the network is compared to all existing clusters in order of decreasing similarity, and a new cluster is not formed unless all the existing clusters are sufficiently dissimilar. For the winning j cluster at the network output determined by the similarity measure in (4.18), $G = 0$, and a second similarity measure is used to determine whether the match between the input $\boldsymbol{x}$ and $\boldsymbol{t}_j$ is good enough, by computing

$$\sigma_2 = \frac{\sum_{i=1}^{n} t_{ji} x_i}{\sum_{i=1}^{n} x_i} \tag{4.19}$$

and then comparing this value to a threshold ρ (the *vigilance parameter*), where $0 \leq \rho \leq 1$. The comparison of the similarity measure σ_2 with the vigilance parameter ρ is carried out in the network by the reset unit in Figure 4.16 as

$$R = T_{\text{binary}}\left(\rho \sum_{i=1}^{n} x_i - \sum_{i=1}^{n} v_i\right) \tag{4.20}$$

However, the sum of the outputs v_i at the F_1 layer is $\sum_{i=1}^{n} v_i = \sum_{i=1}^{n} t_{ji} x_i$, and (4.20) can be written as

$$R = T_{\text{binary}}\left[\left(\frac{\rho}{\sigma_2} - 1\right) \sum_{i=1}^{n} v_i\right] \tag{4.21}$$

Therefore, from (4.21) we see that $R = 0$ if $\sigma_2 \geq \rho$ and $R = 1$ if $\sigma_2 < \rho$. When $R = 0$, the *resonance* condition exists in the network, and the top-down weights t_{ji} are updated according to

$$t_{ji}(k+1) = t_{ji}(k) + \mu(k) z_j(k)[v_i(k) - t_{ji}(k)] \tag{4.22}$$

where $0 \leq \mu(k) \leq 1$. When the reset value is $R = 1$, all output units in the F_2 layer are reset to the off state, and the current winning unit is disabled while the current input vector is being processed. When the winning unit is reset, the gain $G = 1$, and another winning unit is selected. This process is continued, and if there are no patterns in the learned cluster sufficiently similar to the input, they will be disabled one by one and the winning unit selected will be an unused unit.

EXAMPLE 4.5. This example illustrates how the vigilance parameter can affect the clustering of the ART1 network. The letters shown in Figure 4.17 are used as the nine inputs to the ART1 network. Specifically, each letter is transformed to a (column) vector representation by "stacking up" the consecutive *rows* of each letter image in the 7×5 array. This is assuming that the dark pixels are a 1, and the white pixels are a 0.[†] The maximum number of nodes in the category representation field, that is, the F_2 (competitive) layer, is assumed to be number of

---

[†]To display each binary letter map in MATLAB, the array bits would first be "toggled," then multiplied by 63, and **colormap gray** would be used along with the **image** function to display the letter images. In MATLAB: $0 \rightarrow$ black and $1 \rightarrow$ white. Scaling the images by 63 yields a 6-bit gray scale.

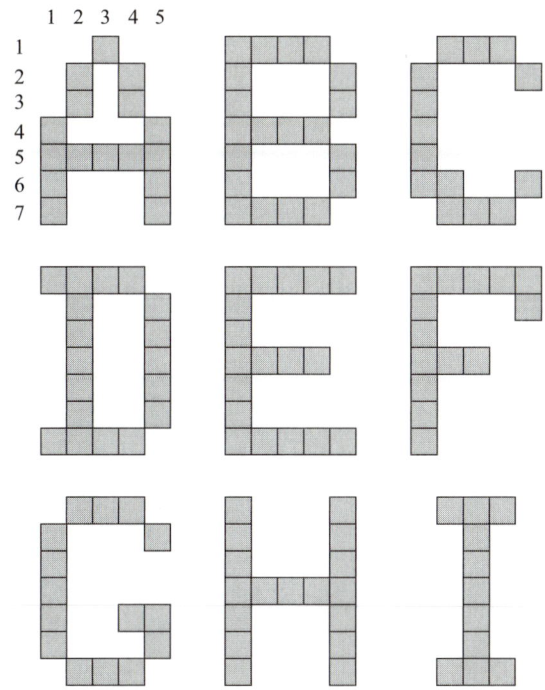

Figure 4.17 Letters used as inputs to an ART1 network in Example 4.5.

input patterns $m = 9$. The bottom-up and top-down weights are, respectively, initialized as

$$b_{ij} = \frac{L}{L - 1 + n} \text{ (where } L > 1\text{, typically } L = 2) \qquad t_{ij} = 1$$

Setting the vigilance parameter at $\rho = 0.3$ creates four categories or clusters. These are shown in Table 4.2. If the vigilance parameter is set at $\rho = 0.7$, two additional categories are formed. The six categories are shown in Table 4.3.

As seen from the results shown in Tables 4.2 and 4.3, when the vigilance parameter is relatively small, the clusters are coarser compared to a higher setting for the vigilance parameter, which impels the clustering to be finer. Raising the vigilance parameter to $\rho = 0.95$ places each pattern (letter) on its own cluster, allowing for pattern recall.

TABLE 4.2
ART1 clustering results for $\rho = 0.3$

Category	Pattern (letter)
1	A, B, C, G
2	D, E, F
3	H
4	I

TABLE 4.3
ART1 clustering results for $\rho = 0.7$

Category	Pattern (letter)
1	A
2	B, C, G
3	D
4	E, F
5	H
6	I

4.4.2 Fuzzy ART and Fuzzy ARTMAP

ART1 has several problems associated with it, for example, storage capacity inefficiency (the maximum number of clusters that are possible to learn is 2^n—for binary input vectors) and sensitivity to noise in the inputs, to name a few. Carpenter and Grossberg [23] developed ART2 for analog (or continuous) inputs to help overcome some of the ART1 problems. Fuzzy ART [24] is capable of learning categories in response to arbitrary sequences of *analog or binary* input patterns. The generalization of learning both analog and binary input patterns is accomplished by replacing the intersection operator ($\cap$) in the ART1 neural network architecture with the MIN operator ($\wedge$) of fuzzy set theory (cf. Sect. A.8). The MIN operator reduces to the intersection operator for the binary case, and thus fuzzy ART reduces to ART1 in response to binary input patterns. In fuzzy ART the input vectors are normalized according to a *complement coding* process that leads to a symmetric theory in which the MIN operator ($\wedge$) and the MAX operator ($\vee$) of fuzzy set theory [25] play complementary roles [24]. Complement coding achieves normalization of the input patterns while preserving amplitude information.

Fuzzy ARTMAP [26] is a generalization of the ARTMAP (adaptive resonance theory mapping) [27] neural network architecture. ARTMAP is also referred to as a predictive ART network, and it is capable of fast, yet stable, online recognition learning and hypothesis testing, given a stream of input patterns (vectors). The fuzzy ARTMAP architecture is shown in Figure 4.18. From the figure it can be seen that this architecture is made up of two fuzzy ART modules. During the supervised training of this network, ART_a receives a stream of input patterns $\{\boldsymbol{a}^{(p)}\}$ and ART_b receives a stream of output patterns $\{\boldsymbol{b}^{(p)}\}$ (where $\boldsymbol{b}^{(p)}$ is the correct response, given $\boldsymbol{a}^{(p)}$). The ART_a and ART_b modules are linked together by an associative learning network and an internal controller that ensures autonomous system operation in real time. The controller is designed to create the minimal number of ART_a recognition categories (or "hidden units") necessary to meet the accuracy criteria. This is accomplished by realizing a minimax learning rule that enables an ARTMAP system to learn quickly, efficiently, and accurately as the system conjointly minimizes predictive error and maximizes predictive generalization. This scenario will automatically link the predictive success to category size on

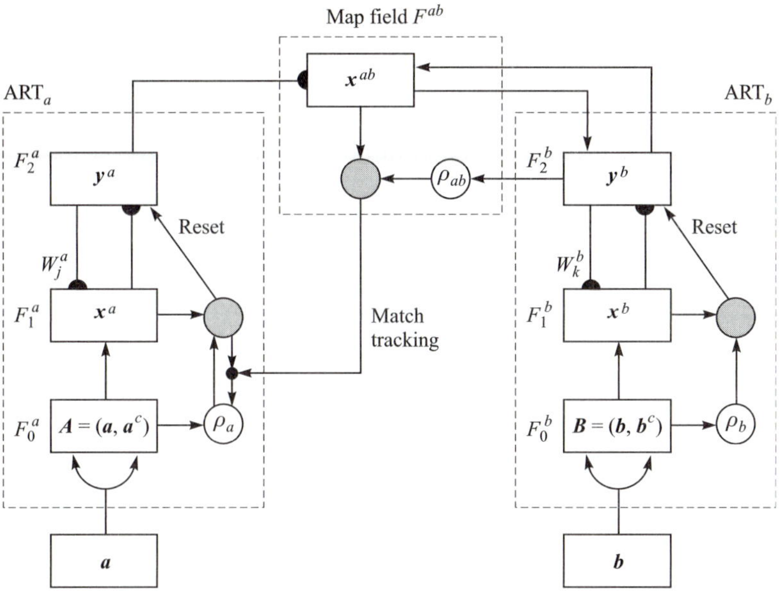

Figure 4.18 Fuzzy ARTMAP architecture. (*Source*: Reproduced with permission from Carpenter et al. [26], p. 699, © 1992 IEEE.)

a trial-by-trial basis using only operations at the local level. This is carried out by increasing the vigilance parameter ρ_a of ART$_a$ by the minimal amount necessary to correct a predictive error at ART$_b$. Lower values of the vigilance parameter ρ_a will result in larger categories to form (the lower ρ_a values will lead to broader generalization and higher code compression). When a predictive failure occurs at ART$_b$, ρ_a will be increased by the minimum amount necessary to initiate hypothesis testing at ART$_a$ using the *match-tracking* mechanism [27]. Match tracking sacrifices the minimum amount of generalization needed to correct a predictive error. Improved predictive performance can be achieved by training the network several times using different orderings of the input data set. This "voting strategy" is based on observing that ARTMAP fast learning typically leads to different adaptive weights and recognition categories for different orderings of the defined training set (even for different simulations yielding similar predictive performances). The performance of fuzzy ARTMAP has been compared to that of the nested generalized exemplar (NGE) [28, 29] and the fuzzy min-max classifier (FMMC) [30].

Laterally primed adaptive resonance theory (LAPART) [31] can perform incremental supervised learning of recognition categories and multidimensional maps in response to input vectors for static data as well as sequential data. In [31], Healy et al. emphasized its use in recognizing familiar sequences of patterns by verifying pattern pairs that are inferred from prior experience. The architecture consists of interconnected ART networks, where the interconnects enable LAPART to infer one pattern class from another to form a predictive sequence. LAPART predicts the next pattern class based on the recognition of a current pattern and then tests the prediction as new data become available. Fuzzy LAPART [32, 33] is a generalized LAPART architecture that can have both analog and binary inputs to the network. Slow learning with fast-commit and slow-recode options is incorporated into fuzzy LAPART for efficient coding of noisy input sets. Carpenter [34] has developed a class of ART models for learning, recognition, and prediction with arbitrary *distributed* code representations. These distributed networks combine fast stable learning capabilities of winner-take-all ART networks with the noise tolerance and code compression capabilities of multilayer perceptrons.

PROBLEMS

4.1. Generate 2,000 samples of a random vector $\boldsymbol{x} \in \Re^{2\times 1}$ having a probability density function given by

$$\wp_{x}(\boldsymbol{x}) = \frac{1}{2}\frac{1}{2\pi \det(\boldsymbol{Q}_1)}\exp\left[-\frac{1}{2}(\boldsymbol{x}-\boldsymbol{x}_1)^T\boldsymbol{Q}_1^{-1}(\boldsymbol{x}-\boldsymbol{x}_1)\right] + \frac{1}{2}\frac{1}{2\pi \det(\boldsymbol{Q}_2)}\exp\left[-\frac{1}{2}(\boldsymbol{x}-\boldsymbol{x}_2)^T\boldsymbol{Q}_2^{-1}(\boldsymbol{x}-\boldsymbol{x}_2)\right]$$

where

$$\boldsymbol{Q}_1 = \begin{bmatrix} 0.1^2 & 0 \\ 0 & 0.4^2 \end{bmatrix} \qquad \boldsymbol{x}_1 = [1 \quad 1]^T$$

$$\boldsymbol{Q}_2 = \begin{bmatrix} 0.5^2 & 0 \\ 0 & 0.3^2 \end{bmatrix} \qquad \boldsymbol{x}_2 = [-1 \quad -1]^T$$

Design a Kohonen SOM that has 100 neurons organized in both a square and a hexagonal grid to perform the classification of the input vector samples. Plot the final positions of the neurons, and compare them to the scatter plot of the samples of the input vector. For this problem write a computer program to implement the Kohonen SOM.

4.2. Generate 1,000 two-dimensional vectors $\boldsymbol{X} = [\boldsymbol{x}_1 \quad \boldsymbol{x}_2 \quad \cdots \quad \boldsymbol{x}_{1,000}] \in \Re^{2x1,000}$, with both components selected from a uniform distribution. The first component must be zero-mean with a variance of 3 ($\sigma^2 = 3$), and the second component must also be zero-mean but with unity variance ($\sigma^2 = 1$).†

(*a*) Plot the stochastic vectors in the *x*-*y* plane using

```
plot(X(1,:),X(2,:),'+'), axis([-4 4 -4 4])
```

(*b*) Design a Kohonen SOM that has 15 neurons organized in a *rectangular* 3×5 grid.

†Refer to Section A.7.4 for details of uniform distributions.

(*c*) How many training epochs does the network need to converge?
Hint: The three functions **initsm**, **nbman**, and **trainsm** from the MATLAB neural network toolbox can be used to solve this problem.

4.3. (*a*) Generate two sets of Gaussian random two-dimensional vectors. Both sets will have a variance of $\sigma^2 = 0.5$; however, the first set is to be centered at $(5, 0)$ and the second set centered at $(0, 5)$. The data can be generated in MATLAB using

```
X1N=sqrt(0.5)*randn(2,1000);
X1N(1,:)=X1N(1,:)+5*ones(1,1000);
X2N=sqrt(0.5)*randn(2,1000);
X2N(2,:)=X2N(2,:)+5*ones(1,1000);
```

Present these random vectors to a Kohonen SOM, and observe the topological ordering of the map, after determining that convergence of the network has been achieved. Assume the SOM lattice structure is square with 36 neurons. What can you conclude from the results?

(*b*) This part of the problem is similar to part (*a*); however, the random vectors to be generated are drawn from random numbers with a uniform distribution. Both sets of 1,000 two-dimensional vectors have a variance of $\sigma^2 = 0.5$ and again are centered at $(5, 0)$ and $(0, 5)$, respectively. Present these random vectors to a Kohonen SOM, and observe the topological ordering of the map, after determining that convergence of the network has been achieved. Again, assume the SOM lattice structure is square with 36 neurons. What can you conclude from the results? *Note*: Be careful how you generate the random vectors. Refer to Section A.7.4 for details of uniform distributions.

4.4. Similar to Example 4.2, this problem explores the topological ordering property of the Kohonen SOM; that is, a high-dimensional input space (three-dimensional data) is mapped to an abstract two-dimensional representation. Generate three sets of Gaussian random three-dimensional vectors. All three sets of random vectors are to have a variance of $\sigma^2 = 0.1$. The first set of vectors is to be centered at $(6, 0, 0)$, the second centered at $(0, 6, 0)$, and the third centered at $(0, 0, 6)$. Design a Kohonen SOM map with a two-dimensional square lattice structure with 64 neurons. What is a reasonable number of training epochs to ensure convergence of the network? What does the final two-dimensional map look like (plot the results)? What can you conclude?

4.5. This problem is similar to Problem 4.4 in that an additional set of Gaussian data is included. Specifically, generate four sets of random Gaussian vectors, each with a variance of $\sigma^2 = 0.1$. Center the consecutive data sets at $(7, 0, 7)$, $(7, 7, 7)$, $(0, 7, 7)$, and $(0, 0, 7)$, respectively. The following MATLAB commands will generate the data:

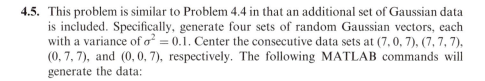

```
%variance = 0.1, centered at (7 0 7)
X1=sqrt(0.1)*randn(3,1000);
X1=detrend(X1')';
X1(1,:)=X1(1,:)+7*ones(1,1000);
X1(3,:)=X1(3,:)+7*ones(1,1000);
%variance = 0.1, centered at (7 7 7)
X2=sqrt(0.1)*randn(3,1000);
X2=detrend(X2')';
X2(1,:)=X2(1,:)+7*ones(1,1000);
X2(2,:)=X2(2,:)+7*ones(1,1000);
X2(3,:)=X2(3,:)+7*ones(1,1000);
%variance = 0.1, centered at (0 7 7)
X3=sqrt(0.1)*randn(3,1000);
X3=detrend(X3')';
X3(2,:)=X3(2,:)+7*ones(1,1000);
X3(3,:)=X3(3,:)+7*ones(1,1000);
%variance = 0.1, centered at (0 0 7)
X4=sqrt(0.1)*randn(3,1000);
X4=detrend(X4')';
X4(3,:)=X4(3,:)+7*ones(1,1000);
```

Design a Kohonen SOM with 64 neurons in a two-dimensional square lattice structure. How many training epochs are necessary to ensure convergence of the network. What does the final map look like (plot the results)? What can you conclude?

4.6. This problem addresses distinguishing between two overlapping two-dimensional Gaussian distributed patterns. That is, we have two *classes* of Gaussian distributed patterns, and these classes are labeled class 1 and class 2. The two sets of events are designated as C_{x_1} and C_{x_2}. The two sets of vectors can be generated in MATLAB as

```
X1=sqrt(0.1)*randn(2,1000);
X2(1,:)=2*randn(1,1000)+5*ones(1,1000);
X2(2,:)=2*randn(1,1000);
```

This will generate two sets of vectors. (1) In the set C_{x_1}, for class 1, will be two-dimensional vectors with a mean $\bar{\boldsymbol{x}}_1 = [0 \quad 0]^T$ and variance $\sigma_1^2 = 0.1$; therefore, the conditional probability density function (cf. Sect. A.7.4) for this class $\not{p}_{\boldsymbol{x}}(\boldsymbol{x}|C_{x_1})$ has mean $\bar{\boldsymbol{x}}_1$ and variance σ_1^2. (2) In the set C_{x_2}, for class 2, will be two-dimensional vectors with a mean $\bar{\boldsymbol{x}}_2 = [5 \quad 0]^T$ and variance $\sigma_2^2 = 4$; therefore, the conditional probability density function for this class $\not{p}_{\boldsymbol{x}}(\boldsymbol{x}|C_{x_2})$ has mean $\bar{\boldsymbol{x}}_2$ and variance σ_2^2.

(*a*) Plot the two sets of random vectors with different symbols, and qualitatively determine how much overlap exists.

(*b*) Using the Kohonen SOM, with 25 neurons in a square, sufficiently train the SOM, and plot the weight vectors. What can you determine from their formation in the *xy* plane?

4.7. In Example 4.3 seven four-dimensional binary vectors were used to determine the final weight vectors

$$w_1 = \begin{bmatrix} 1.2996 \\ 0.4952 \\ 0.4952 \\ 0.4848 \end{bmatrix} \quad \text{and} \quad w_2 = \begin{bmatrix} -0.1881 \\ 1 \\ 1 \\ 0.3603 \end{bmatrix}$$

(after 500 training epochs) using the LVQ1 program in Table 4.1. Given the following four vectors that were not used during the training process, determine the class to which each belongs.

$$x_{\text{test }1} = [1, 0, 1, 0]^T \qquad x_{\text{test }2} = [0, 0, 1, 0]^T$$
$$x_{\text{test }3} = [0, 1, 0, 0]^T \qquad x_{\text{test }4} = [1, 1, 0, 1]^T$$

4.8. Consider a classification problem depicted in Figure 4.19.

(*a*) Design an LVQ network to perform the classification.

(*b*) Test the classification boundaries placed by the network in the following way:

- Generate the test input points on a finer grid than the one shown in Figure 4.19.
- Present the test input points to the network.
- Let the network perform the classification.
- Generate a plot showing different classification regions.

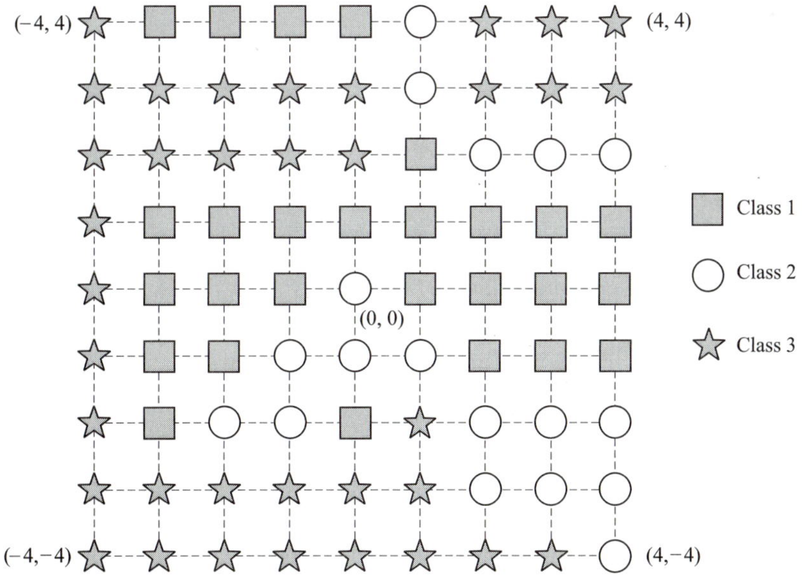

Figure 4.19 LVQ classification of nonlinearly separable classes.

4.9. You are given the following set of input vectors with the associated class:

Vector		**Class**
$x_1 = [0, 0, 1]^T$	(binary representation for 1)	1 (odd)
$x_2 = [0, 1, 0]^T$	(binary representation for 2)	2 (even)
$x_3 = [0, 1, 1]^T$	(binary representation for 3)	1 (odd)
$x_4 = [1, 0, 0]^T$	(binary representation for 4)	2 (even)
$x_5 = [1, 0, 1]^T$	(binary representation for 5)	1 (odd)
$x_6 = [1, 1, 0]^T$	(binary representation for 6)	2 (even)
$x_7 = [1, 1, 1]^T$	(binary representation for 7)	1 (odd)

Train an LVQ network with the data given above. What can you conclude?

4.10. After completing part (a) in Problem 4.3, use the hybrid adaptive pattern classification system shown in Figure 4.13 to fine tune the clusters, using the weight vectors from the SOM as inputs to an LVQ network. Develop a method for determining the proper classes for the SOM weight vectors to be used for training the LVQ network. What can you conclude from your results? Repeat this for part (*b*) of Problem 4.3.

Hint: You can use the LVQ1 program given in Table 4.1 directly, or modify it to randomly initialize the weight vectors.

4.11. Carry out the same procedure described in Problem 4.10 for using the hybrid adaptive pattern classification system shown in Figure 4.13, but now use the SOM output results from Problem 4.5.

4.12. Repeat Problem 4.11, but now use the SOM output results from Problem 4.6.

4.13. (*a*) In Section 4.4.1, details pertaining to learning in an ART1 network were presented. Based on the discussion in that section, write a step-by-step algorithm for training an ART1 network.

(*b*) From the algorithm you develop in part (*a*) of this problem, write a MATLAB function to implement the algorithm for ART1 learning.

4.14. This problem is an extension of Example 4.5.

(*a*) Generate all 26 letters of the alphabet, using the approach outlined in Example 4.5. For example, the (column) vectorized representation of the letter **L** can generated in MATLAB by

```
Lletter= [1 0 0 0 0 ...
          1 0 0 0 0 ...
          1 0 0 0 0 ...
          1 0 0 0 0 ...
          1 0 0 0 0 ...
          1 0 0 0 1 ...
          1 1 1 1 1]';
```

Note: Refer to the footnote on page 186 pertaining to plotting the letter images in MATLAB.

(*b*) Use the ART1 MATLAB function developed in Problem 4.13(*b*) to categorize the 26 letters of the alphabet with the vigilance parameter set to $\rho = 0.3$. How many clusters are formed? Which letters are categorized on which clusters? Now increase the vigilance parameter to $\rho = 0.7$. How does this affect the results? What is the minimum value of the vigilance parameter such that all 26 letters have their own cluster unit, that is, 26 total cluster units?

4.15. Figure 4.20 shows nine different images that can be vectorized by using the same procedure described in Example 4.5. As in Example 4.5, assume a dark pixel is a 1 and a white pixel is a 0.

Note: Refer to the footnote on page 186 pertaining to plotting the letter images in MATLAB.

(*a*) Using the ART1 MATLAB function you wrote in Problem 4.13(*b*), experiment with different values of the vigilance parameter and discuss the results. What can you conclude?

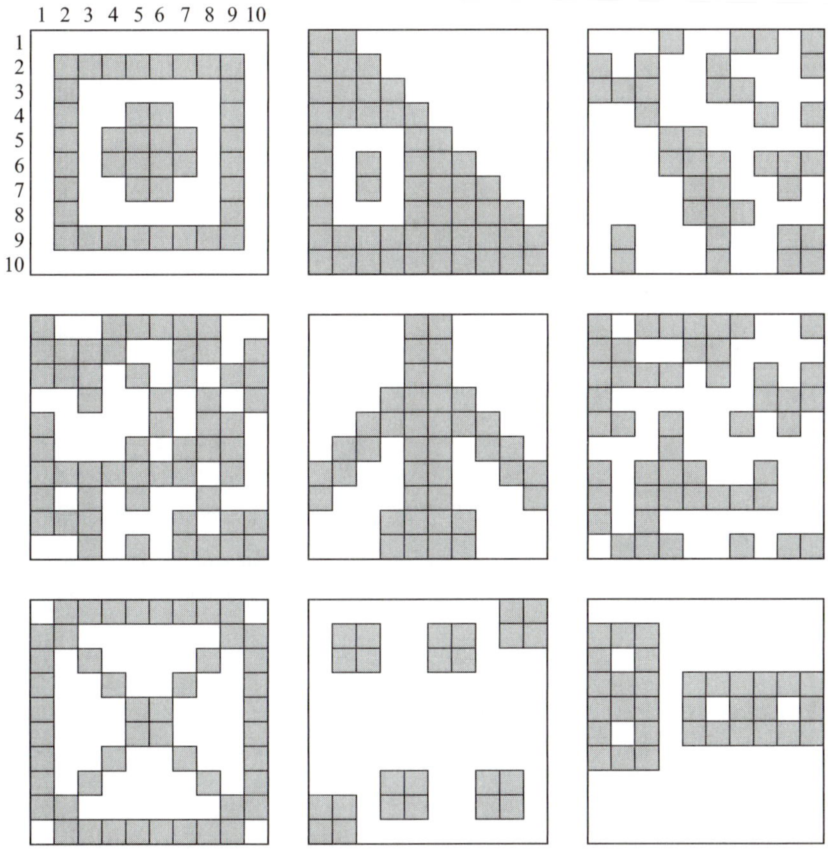

Figure 4.20 Images for clustering in Problem 4.15 using an ART1 neural network.

(*b*) Make the vigilance parameter large enough to create nine separate clusters on the output of the ART1 network. Selecting the first and fifth images, toggle 10 randomly selected bits in the images. Present these corrupted images to the trained network. Can the network still properly categorize the images?

REFERENCES

1. T. Kohonen, "Self-Organized Formation of Topologically Correct Feature Maps," *Biological Cybernetics*, vol. 43, 1982, pp. 59–69. Reprinted in 1988, Anderson and Rosenfeld [2], pp. 511–21.
2. J. A. Anderson and E. Rosenfeld, eds., *Neurocomputing: Foundations of Research*, Cambridge, MA: M.I.T. Press, 1988.
3. D. J. Willshaw and C. von der Malsburg, "How Patterned Neural Connections Can Be Set Up by Self-Organization," *Proceedings of the Royal Society of London, Series B*, vol. 194, 1976, pp. 431–45.
4. T. Kohonen, "The Self-Organizing Map," *IEEE Proceedings*, vol. 78, 1990, pp. 1464–80.
5. T. Kohonen, *Self-Organization and Associative Memory*, 3rd ed., Berlin: Springer-Verlag, 1989.
6. J. Hertz, A. Krogh, and R. G. Palmer, *Introduction to the Theory of Neural Computation*, Redwood City, CA: Addison-Wesley, 1991.
7. Y. Linde, A. Buzo, and R. M. Gray, "An Algorithm for Vector Quantizer Design," *IEEE Transactions on Communications*, vol. COM-28, 1980, pp. 84–95.
8. R. M. Gray, "Vector Quantization," *IEEE Acoustics, Speech, and Signal Processing Magazine*, vol. 1, 1984, pp. 4–29.
9. N. M. Nasrabadi and R. A. King, "Image Coding Using Vector Quantization: A Review," *IEEE Transactions on Communications*, vol. 36, 1988, pp. 957–71.
10. S. P. Luttrell, "Hierarchical Vector Quantization," *IEE Proceedings (London)*, vol. 136, Part I, 1989, pp. 405–13.
11. A. Buzo, A. H. Gray, R. M. Gray, and J. D. Markel, "Speech Coding Based upon Vector Quantization," *IEEE Transactions on Acoustics, Speech and Signal Processing*, vol. ASSP-28, 1980, pp. 562–74.
12. P. C. Mahalanobis, "On the Generalized Distance in Statistics," *Proceedings of Indian National Institute of Science (Calcutta)*, 1936, pp. 49–55.
13. F. Itakura and S. Saito, "Analysis Synthesis Telephony Based upon Maximum Likelihood Method," *Reports of the 6th International Congress on Acoustics*, ed. Y. Kohasi, Tokyo, Japan, 1968, C-5-5, pp. C17–20.
14. F. Itakura, "Maximum Prediction Residual Principle Applied to Speech Recognition," *IEEE Transactions on Acoustics, Speech, and Signal Processing*, vol. 23, 1975, pp. 67–72.
15. A. Gersho, "On the Structure of Vector Quantizers," *IEEE Transactions on Information Theory*, vol. IT-28, 1982, pp. 157–66.
16. T. Kohonen, "Learning Vector Quantization for Pattern Recognition," *Technical Report TKK-F-A601*, Helsinki University of Technology, Finland, 1986.
17. T. Kohonen, "Improved Versions of Learning Vector Quantization," *Proceedings of the International Joint Conference on Neural Networks*, San Diego, CA, vol. 1, 1990, pp. 545–50.
18. J. S. Baras and A. LaVigna, "Convergence of Kohonen's Learning Vector Quantization," *International Joint Conference on Neural Networks*, San Diego, CA, vol. 3, 1990, pp. 17–20.

19. R. O. Duda and P. E. Hart, *Pattern Classification and Scene Analysis*, New York: Wiley, 1973.
20. G. A. Carpenter and S. Grossberg, "A Massively Parallel Architecture for a Self-Organizing Neural Pattern Recognition Machine," *Computer Vision, Graphics, and Image Processing*, vol. 37, 1987, pp. 54–115.
21. B. Moore, "ART1 and Pattern Clustering," *Proceedings 1988 Connectionist Models Summer School, Pittsburgh*, eds., D. Touretsky, G. Hinton, and T. Sejnowski, San Mateo, CA: Morgan Kaufmann, 1988, pp. 174–85.
22. N. K. Bose and P. Liang, *Neural Networks Fundamentals with Graphs, Algorithms, and Applications*, New York: McGraw-Hill, 1996.
23. G. A. Carpenter and S. Grossberg, "ART2: Self-Organization of Stable Category Recognition Codes for Analog Input Patterns," *Applied Optics*, vol. 26, 1987, pp. 4919–30.
24. G. A. Carpenter, S. Grossberg, and D. B. Rosen, "Fuzzy ART: Fast Stable Learning and Categorization of Analog Patterns by an Adaptive Resonance System," *Neural Networks*, vol. 4, 1991, pp. 759–71.
25. L. A. Zadeh, "Fuzzy Sets," *Information and Control*, vol. 8, 1965, pp. 338–53.
26. G. A. Carpenter, S. Grossberg, N. Markuzon, J. H. Reynolds, and D. B. Rosen, "Fuzzy ARTMAP: A Neural Network Architecture for Incremental Supervised Learning of Analog Multidimensional Maps," *IEEE Transactions on Neural Networks*, vol. 3, 1992, pp. 698–713.
27. G. A. Carpenter, S. Grossberg, and J. H. Reynolds, "ARTMAP: Supervised Real-Time Learning and Classification of Non-Stationary Data by a Self-Organizing Neural Network," *Neural Networks*, vol. 4, 1991, pp. 565–88.
28. S. L. Salzberg, "Learning with Generalized Nested Exemplars," Ph.D. thesis, Technical Report TR-14-89, Cambridge, MA: Department of Computer Science, Harvard University, 1989.
29. S. L. Salzberg, "A Nearest Hyperrectangular Learning Method," *Machine Learning*, vol. 6, 1991, pp. 251–76.
30. P. Simpson, "Fuzzy Min-Max Classification with Neural Networks," *Heuristics: Journal of Knowledge Engineering*, vol. 4, 1991, pp. 1–9.
31. M. J. Healy, T. P. Caudell, and S. D. G. Smith, "A Neural Architecture for Pattern Sequence Verification through Inferencing," *IEEE Transactions on Neural Networks*, vol. 4, 1993, pp. 9–20.
32. G. Han, "A Fuzzy Neural Architecture with Binary and Analog Inputs for Supervised Learning through Inferencing," Ph.D. thesis, Melbourne, FL: Electrical Engineering Program, Florida Institute of Technology, 1993.
33. F. M. Ham, G. Han, and L. V. Fausett, "Fuzzy LAPART: A Neural Architecture for Supervised Learning through Inferencing for Stable Category Recognition," *Journal of Artificial Neural Networks*, vol. 2, 1995, pp. 241–64.
34. G. A. Carpenter, "Distributed Learning, Recognition, and Prediction by ART and ARTMAP Networks," *Neural Networks*, vol. 10, 1997, pp. 1473–94.

CHAPTER 5

Recurrent Networks and Temporal Feedforward Networks

5.1 INTRODUCTION

This chapter covers recurrent neural networks. In general, a network that has closed loops in its topological structure is considered a recurrent network. The temporal feedforward networks considered in this chapter are also recurrent networks; however, we make a distinction between these networks and those that are not multilayer feedforward networks. Therefore, this chapter consists of two basic parts. After an overview of recurrent neural networks in Section 5.2, the first part of the chapter addresses the discrete-time Hopfield network (Sect. 5.3) followed by simulated annealing (Sect. 5.4), and then the Boltzmann machine (Sect. 5.5). After an overview of temporal feedforward networks in Section 5.6, the second part of the chapter covers the simple recurrent network (SRN) (also referred to as the Elman network) (Sect. 5.7), time-delay networks (Sect. 5.8), and distributed time-lagged feedforward neural networks (Sect. 5.9).

5.2 OVERVIEW OF RECURRENT NEURAL NETWORKS

The feedforward networks previously studied implemented fixed-weight mappings from the input space to the output space. Because the networks have fixed weights, the *state* of any neuron is solely determined by the input to the unit and not the initial and past states of the neuron. This independence of initial and past states of the network neurons limits such networks because no *dynamics* are involved. To allow initial and past state involvement along with serial processing, *recurrent neural networks* utilize *feedback*. Recurrent neural networks are also characterized by use of nonlinear processing units; thus, such networks are nonlinear dynamic systems. Another important characteristic of recurrent networks is their relative insensitivity to the failure of individual devices (fault-tolerant). Hopfield refers to this as a *fail-soft* device [1]. Because recurrent networks have feedback paths, they are *sequential* rather than combinational; that is, they can demonstrate temporal behavior. These networks can be fully connected; in other words, all possible connections between the network neurons are allowed. Moreover, the connection weights in a recurrent neural network can be symmetric or asymmetric. In the case of symmetric connection weights $w_{ij}=w_{ji}\ \forall i,j$, and for asymmetric connection weights $w_{ij} \neq w_{ji}\ \forall i,j$.

In the symmetric case, the network always converges to stable point attractors (stable equilibrium points or states). However, these networks cannot accommodate temporal sequences of patterns. In the asymmetric case, the dynamics of the network can exhibit limit cycles and chaos [2], in addition to stable states, and with the proper selection of weights, temporal spatial patterns can be generated and stored in the network. These types of networks play an important role in high-level intelligent systems with application to, for example, symbolic reasoning. The development of recurrent neural networks is many times viewed to be inspired from notions in statistical physics [3, 4].

5.3 HOPFIELD ASSOCIATIVE MEMORY

In the classic paper by Hopfield [1], he states that physical systems consisting of a large number of simple elements (neurons) can exhibit *collective emergent properties*. Simply put, a collective property of a system cannot emerge from a single element, but it can emerge from local element interactions in the system. The model he describes has collective properties that produce a content-addressable memory that can correctly yield an entire memory from partial information. He describes additional emergent collective properties such as familiarity recognition, some capacity for generalization, categorization, error correction, and time sequence retention.

We consider the standard discrete-time version of the Hopfield neural network. It is considered a recurrent network because of feedback connections; however, a Hopfield network can also be viewed as a nonlinear *associative memory* (cf. Sect. 3.2) or a *content-addressable memory*. This network is intended to perform the function of data storage and retrieval. However, the network stores the information in a *dynamically* stable environment. A pattern that is stored in memory is to be retrieved in response to an input pattern that is a noisy (or incomplete) version of the stored pattern. Therefore, a content-addressable memory is error-correcting in the sense that it can perform reliable retrieval of patterns from memory even when incomplete or corrupted inputs are presented to the network.

The basic nature of the Hopfield network as a content-addressable memory (CAM) is to perform a mapping function; however, this is a dynamic mapping function. Some preliminary concepts need explanation before we go on to the details of the Hopfield network. An *attractor* is a state that the system will evolve toward in time, starting from a set of initial conditions. There is a set of initial conditions associated with each attractor. These initiate the evolution that terminates in that particular attractor. The set of initial conditions for a particular attractor is referred to as the *basin of attraction.* If an attractor is a unique point in the state space, it is called a *fixed point*. An attractor may have a more complicated structure, for example, *limit cycles.*

A *prototype state* (*prototype memory*) $\boldsymbol{\phi}_h$ is represented by a *fixed* (*stable*) *point* $\boldsymbol{\sigma}_h$ of the dynamic system. Therefore, $\boldsymbol{\phi}_h$, for $h = 1, 2, \ldots, r$, is mapped onto the stable points $\boldsymbol{\sigma}_h$ of the network. This mapping can be written as $\boldsymbol{\phi}_h \overset{\leftarrow}{\rightarrow} \boldsymbol{\sigma}_h$, where the forward direction (left to right) represents the encoding process and the backward direction (right to left) represents the decoding process. Therefore, the Hopfield network is an asynchronous nonlinear dynamical system, in which its phase space consists of stable points that are the prototype states (or prototype memories) of the network. This is the *emergent collective property* that Hopfield discusses in his paper [1]. Figure 5.1 illustrates the state space encoding/decoding performed by a Hopfield associative memory. During the recall process, a pattern is presented to the network. This pattern is assumed to contain information pertaining to one of the network's prototype memories. It could be that the input pattern only contains partial information in connection with a prototype memory. In response to this input, which is a starting point in phase space, if the starting point is "close" to the stable point that represents the memory to be retrieved, the dynamic system will evolve in time and converge to this memory state (i.e., the phase space flow of the state of the system will converge to this memory state). Thus, the dynamic system generates the appropriate memory. Therefore, a Hopfield neural network performs the function of storing information in a dynamically stable environment.

The discrete-time Hopfield network uses the McCulloch–Pitts model [5] for the neurons in the architecture, and we will consider only symmetric hard

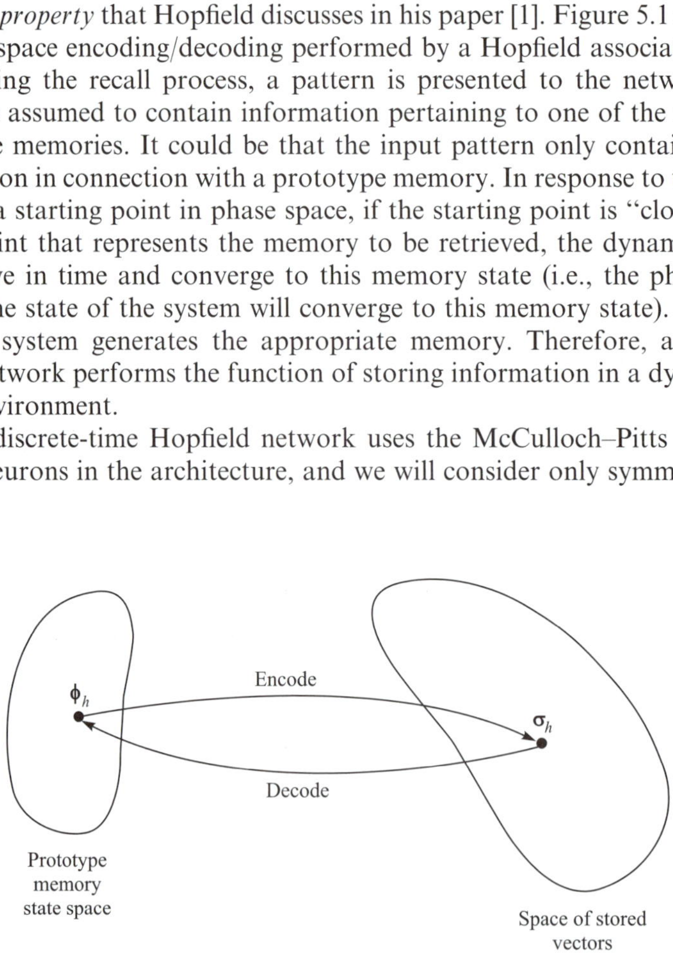

Figure 5.1 Encoding/decoding process performed by a Hopfield associative memory neural network.

limiter activation functions (cf. Sect. 2.3). Therefore, the state of the network at any time can only be +1 or −1. Using the discrete-time Hopfield model of the neuron shown in Figure 2.12, we can construct the Hopfield neural network shown in Figure 5.2.

For each neuron in Figure 5.2, the output of the linear combiner is written as

$$v_i = \sum_{j=1}^{n} w_{ij}x_j - \theta_i = \boldsymbol{w}_i^T \boldsymbol{x} - \theta_i, \quad \text{for} \quad i = 1, 2, \ldots, n \qquad (n \text{ neurons}) \tag{5.1}$$

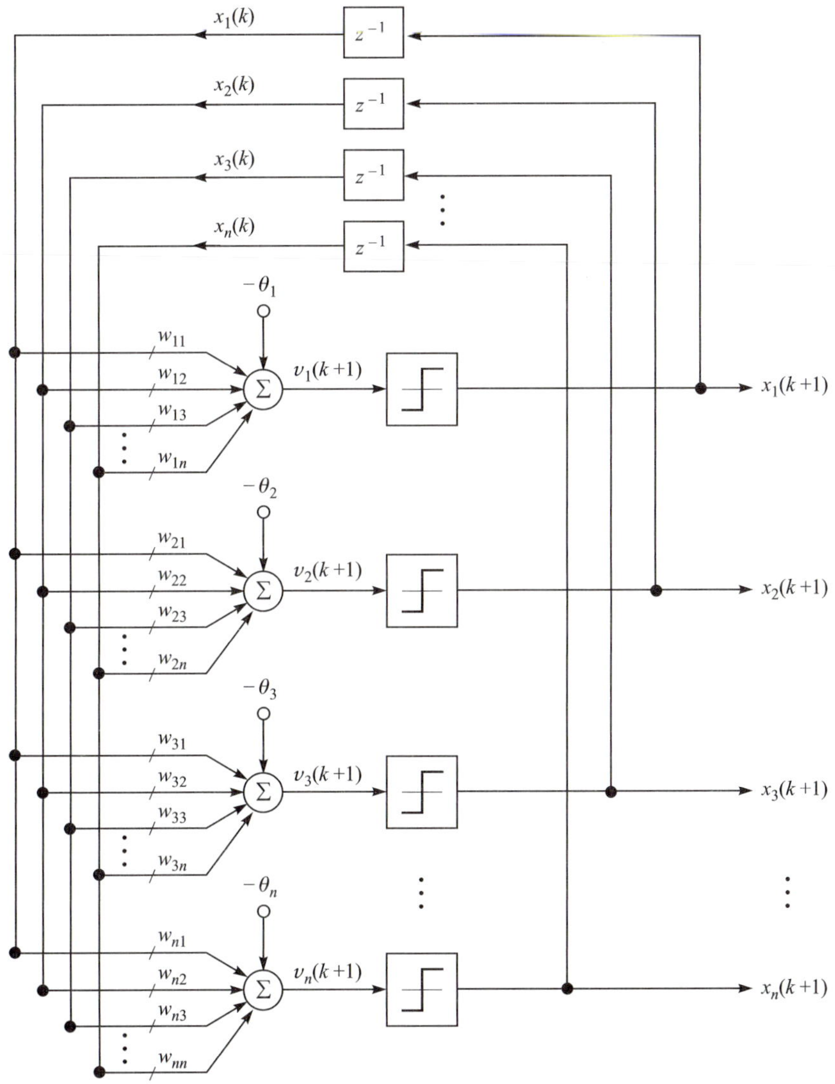

Figure 5.2 Discrete-time Hopfield neural network ($z^{-1} \equiv$ unit delay). It is assumed that $w_{ij} = w_{ji}$ (symmetric weight matrix), and $w_{ij} = 0$ for $i = j$ (i.e., no neuron output is fed back to itself, or no "self-loops").

where $\boldsymbol{x} = [x_1, x_2, \cdots, x_n]^T$ is the *state* (cf. Sect. A.2.12) of the network and θ_i is an externally applied threshold. For $i = 1, 2, \ldots, n$, each of the linear combiner outputs is presented to a symmetric hard limiter activation function, and the output of this is passed to a unit delay element. Then the unit delay outputs x_i, for $i = 1, 2, \ldots, n$, are fed back to the inputs of the neurons. The exception to this is that the output of a neuron is not fed back to itself; therefore, in Figure 5.2, $w_{ij} = 0$ for $i = j$. The state of each neuron is given by

$$x_j = \operatorname{sgn}(v_i) = \begin{cases} +1 & \text{for } v_i > 0 \\ -1 & \text{for } v_i < 0 \end{cases} \tag{5.2}$$

for $j = 1, 2, \ldots, n$, where sgn(•) is the *signum* function (cf. Sect. 2.3). If $v_i = 0$, the value of x_j will be defined as its previous state. We can write the threshold vector as $\boldsymbol{\theta} = [\theta_1, \theta_2, \cdots, \theta_n]^T$. Therefore, the vector-matrix form of (5.1) is given by

$$\boldsymbol{v} = \boldsymbol{W}\boldsymbol{x} - \boldsymbol{\theta} \tag{5.3}$$

where the network weight matrix is written as

$$\boldsymbol{W} = \begin{bmatrix} 0 & w_{12} & w_{13} & w_{14} & \cdots & w_{1n} \\ w_{21} & 0 & w_{23} & w_{24} & \cdots & w_{2n} \\ w_{31} & w_{32} & 0 & w_{34} & \cdots & w_{3n} \\ \cdots & \cdots & \cdots & \cdots & \cdots & \cdots \\ w_{n-11} & w_{n-12} & w_{n-13} & \cdots & 0 & w_{n-1n} \\ w_{n1} & w_{n2} & w_{n3} & \cdots & w_{nn-1} & 0 \end{bmatrix} \tag{5.4}$$

Each *row* in (5.4) is the associated *weight vector* for each neuron shown in Figure 5.2. The output of the network can now be written in vector-matrix form as

$$\boldsymbol{x}(k+1) = \operatorname{sgn}[\boldsymbol{W}\boldsymbol{x}(k) - \boldsymbol{\theta}] \tag{5.5}$$

or in scalar form

$$x_i(k+1) = \operatorname{sgn}\left(\sum_{j=1}^{n} w_{ij}x_j - \theta_i\right) \tag{5.6}$$

for $i = 1, 2, \ldots, n$. There are two basic operational phases associated with the Hopfield network: the *storage phase* and the *recall phase*.

During the storage phase, the associative memory (content-addressable memory) is built according to the outer-product rule given in Section 3.2.2 for correlation matrix memories. Given the set of r prototype memories $\{\boldsymbol{\phi}_1, \boldsymbol{\phi}_2, \ldots, \boldsymbol{\phi}_r\}$, the network weight matrix is computed as

$$\boldsymbol{W} = \frac{1}{n}\sum_{h=1}^{r} \boldsymbol{\phi}_h\boldsymbol{\phi}_h^T - \frac{r}{n}\boldsymbol{I} \tag{5.7}$$

The scaling factor $1/n$ in (5.7) is for convenience, and the second term $(r/n)/\boldsymbol{I}$, which is subtracted from the sum of the outer products of the prototype memory vectors, is to invoke the requirement $w_{ij} = 0$ for $i = j$. As stated in Section 3.2.2, (5.7) is a form of Hebbian learning (cf. Sect. 2.8.2). After the

CAM is built, the recall phase involves presenting the network with a test input vector $\boldsymbol{x}' \in \Re^{n\times 1}$, and the state of the network $\boldsymbol{x}(k)$ is *initialized* with the values of the unknown input, that is, $\boldsymbol{x}(k)|_{k=0} = \boldsymbol{x}(0) = \boldsymbol{x}'$. Using the expression in (5.6), the elements of the state vector $\boldsymbol{x}(k)$ are updated one at a time and randomly (i.e., *asynchronously*) until there is no significant change in the elements of the vector. When this condition is reached, the stable (equilibrium) state (or fixed point of the phase space of the system) $\boldsymbol{x}_e$ that results is the network output.

There are many versions of the discrete-time Hopfield neural network. However, based on the above discussion, the operational details for one version of the discrete-time Hopfield neural network can be summarized as follows:

Discrete-time Hopfield network training algorithm

Step 1. Given a set of prototype memories $\{\boldsymbol{\phi}_1, \boldsymbol{\phi}_2, \ldots, \boldsymbol{\phi}_r\}$, using the outer-product rule in (5.7), the synaptic weights of the network are calculated according to

$$w_{ij} = \begin{cases} \dfrac{1}{n}\displaystyle\sum_{n=1}^{r} \phi_{hi}\phi_{hj} & i \neq j \\ 0 & i = j \end{cases} \tag{5.8}$$

Step 2. Given an unknown input vector denoted by $\boldsymbol{x}'$, the Hopfield network is initialized by setting the state of the network $\boldsymbol{x}(k)$ at time $k = 0$ to $\boldsymbol{x}'$, that is,

$$\boldsymbol{x}(0) = \boldsymbol{x}' \tag{5.9}$$

Step 3. The elements of the state of the network $\boldsymbol{x}(k)$ are updated asynchronously according to (5.6), that is,

$$x_i(k+1) = \operatorname{sgn}\left(\sum_{j=1}^{n} w_{ij}x_j - \theta_i\right) \tag{5.10}$$

This iterative process is continued until it can be shown that the elements of the state vector do not change appreciably. When this condition is met, the network outputs the equilibrium state; that is, (5.10) will yield

$$\boldsymbol{x} = \boldsymbol{x}_e \tag{5.11}$$

One of the major problems associated with the Hopfield network is the possible generation of *spurious equilibrium states*† (or spurious attractors) [4, 6]. These are stable equilibrium states (or fixed points of the phase space of the system) that are not part of the design set of prototype memories. There are several reasons why spurious attractors can exist in Hopfield networks: (1) They can result from linear combinations of an odd number of patterns [6]. (2) For a large number of prototype memories to be stored in the content-addressable memory, there can exist local minima in the energy landscape. These local minima are not correlated to any of the prototype memories that are part of the network structure. (3) Spurious attractors can also result from the symmetric energy function $\mathscr{E}(\boldsymbol{x})$ that can be defined for Hopfield networks.

†Spurious equilibrium states are also referred to as *spin-glass states* [7, 8].

In the paper by Li et al. [9], a class of networks is analyzed that has the same basic structure as Hopfield neural networks. The design approach presented in this paper is based on a system of first-order linear ordinary differential equations that are defined on a closed hypercube of the state space. When solutions to these equations exist on the boundary of the hypercube, the system is said to be in a saturated mode. The approach presented is easier to analyze, synthesize, and implement than the classic Hopfield model [9]. One major advantage of the design methodology is that the number of spurious attractors is minimized. For example, the MATLAB neural network toolbox [10] has a function **newhop** that is based on this approach. The function **newhop** takes as input all the target vectors (patterns to be memorized by the network, only in bipolar form) to design the network.

Because the Hopfield network has symmetric weights and no neuron self-loops, an energy function, or Lyapunov function (cf. Sect. A.4), can be defined [11]. Also, because convergent flow to stable equilibrium states is the pertinent feature of a CAM, it is desirable to have a method to perform the convergence analysis. The mathematical condition that guarantees the state space flow is stable is $w_{ij} = w_{ji}$ for $i \neq j$ and $w_{ij} = 0$ for $i = j$ (symmetric weights and no neuron self-loops). The convergence proof follows from the choice of an appropriate energy function that is always (monotonically) decreasing with any state change. An energy function for the discrete-time Hopfield neural network can be written as [11, 12]

$$\mathscr{E} = -\frac{1}{2}\sum_{\substack{i=1 \\ i\neq j}}^{n}\sum_{j=1}^{n} w_{ij}x_i x_j - \sum_{i=1}^{n} x_i' x_i + \sum_{i=1}^{n} \theta_i x_i \tag{5.12}$$

where $\boldsymbol{x}$ is the state of the network, $\boldsymbol{x}'$ is an externally applied input presented to the network, $\boldsymbol{W} = [w_{ij}]$ is the synaptic weight matrix, and $\boldsymbol{\theta}$ is the threshold vector. The change in the energy function $\mathscr{E}$, denoted by $\Delta\mathscr{E}$, due to a change in the state of ith neuron x_i, denoted byΔx_i, is given by

$$\Delta\mathscr{E} = -\left(\sum_{\substack{j=1 \\ j\neq i}}^{n} w_{ij}x_j + x_i' - \theta_i\right)\Delta x_i \tag{5.13}$$

This relationship is valid for only one neuron changing its state at a time. Now according to the operation of the Hopfield network, a sign change in any state of the network is accompanied by the same sign change for the term inside the parentheses in (5.13). Therefore, the change in the energy function $\Delta\mathscr{E}$ in (5.13) will always be negative, and accordingly, the energy is always decreasing, which implies an asymptotically stable system (cf. Sect. A.4), leading to stable states of the system that do not further change with time. In other words, the operation of the Hopfield network leads to a monotonically decreasing energy function, and changes in the state of the network will continue until a local minimum of the energy landscape is reached. The attractors (basins of attraction) of the phase space are associated with the local minima of the energy landscape. These attractors correspond to the assigned memories of the network; however, as previously mentioned, spurious attractors can also exist. For no externally applied inputs, the energy function is given by

$$\mathscr{E} = -\frac{1}{2}\sum_{\substack{i=1\\i\neq j}}^{n}\sum_{j=1}^{n} w_{ij}x_i x_j \tag{5.14}$$

and the energy change is

$$\Delta\mathscr{E} = -\left(\sum_{\substack{j=1\\j\neq i}}^{n} w_{ij}x_j\right)\Delta x_i \tag{5.15}$$

The storage capacity of the Hopfield network was experimentally determined by Hopfield [1]. The maximum number of bipolar patterns he found that can be stored and recalled with reasonable accuracy is given approximately by

$$P_s \approx 0.15n \tag{5.16}$$

where n is the number of neurons in the network. However, requiring that most of the prototype memories be recalled perfectly leads to the maximum storage capacity of the network given by [4]

$$P_s = \frac{n}{2\ln n} \tag{5.17}$$

Moreover, if it is required that *all* r of the prototype memories are to be recalled perfectly, it is expected that the maximum storage capacity would be less than (5.17). In this case, it is required to get nr bits right with, say, 99 percent probability, and the maximum storage capacity is given by [4]

$$P_s = \frac{n}{4\ln n} \tag{5.18}$$

EXAMPLE 5.1. We consider a classic example that illustrates the operation of the Hopfield network. Figure 5.3 shows the three-neuron architecture with fixed weights. The threshold for each neuron is assumed to be zero. There are eight possible bipolar states that the network can assume. The weight matrix can be written directly from the state diagram in Figure 5.3 as

$$\boldsymbol{W} = \begin{bmatrix} 0 & -2/3 & 2/3 \\ -2/3 & 0 & -2/3 \\ 2/3 & -2/3 & 0 \end{bmatrix} \tag{5.19}$$

which satisfies the symmetry condition and the requirement of zero diagonal elements. The stability condition requires that

$$\boldsymbol{x} = \operatorname{sgn}(\boldsymbol{W}\boldsymbol{x} - \boldsymbol{\theta}) \tag{5.20}$$

However, of the eight possible bipolar vectors only two satisfy this stability condition, that is, $[-1, 1, -1]$ and $[1, -1, 1]$. The remaining states will transition to these stable states when presented to the network; this is illustrated in Figure 5.3. The weight matrix in (5.19) is made up of the two stable vectors (i.e., prototype memories). That is, using (5.7), we can write

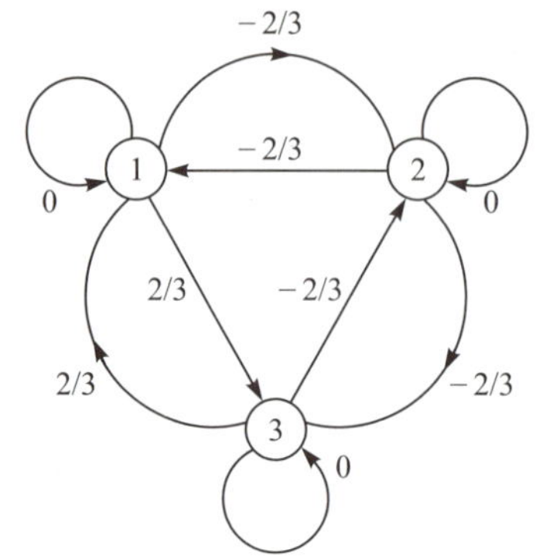

Figure 5.3 State diagram for a three-neuron Hopfield network with fixed weights.

$$\begin{aligned} W &= \frac{1}{3}\left(\begin{bmatrix} -1 \\ 1 \\ -1 \end{bmatrix}[-1, 1, -1] + \begin{bmatrix} 1 \\ -1 \\ 1 \end{bmatrix}[1, -1, 1]\right) - \frac{2}{3}\begin{bmatrix} 1 & 0 & 0 \\ 0 & 1 & 0 \\ 0 & 0 & 1 \end{bmatrix} \\ &= \begin{bmatrix} 0 & -2/3 & 2/3 \\ -2/3 & 0 & -2/3 \\ 2/3 & -2/3 & 0 \end{bmatrix} \end{aligned} \tag{5.21}$$

Another way to view this is in terms of the error-correcting capability of the Hopfield network. From Figure 5.4, we see that if the network is presented with $[-1, -1, 1]$, $[1, -1, -1]$, or $[1, 1, 1]$, the network converges to $[1, -1, 1]$. In all three cases there is a one bit error in the vector, and this is "corrected" by the network to yield the prototype vector $[1, -1, 1]$. There is a similar situation for the other prototype vector $[-1, 1, -1]$; see Figure 5.4.

Using the expression in (5.14) with the synaptic weights of the network shown in Figure 5.3, the energy function is given by

$$\mathscr{E} = \frac{2}{3}(x_1x_2 - x_1x_3 + x_2x_3) \tag{5.22}$$

For the eight possible states of the system, the two stable prototype memories yield the minimum values for the energy function in (5.22). Table 5.1 shows the results of calculating the eight values of the energy function along with the change in the energy function $\Delta\mathscr{E}$ due to the change in a single neuron in the network by using the expression in (5.15).

EXAMPLE 5.2. This example illustrates the capability of the Hopfield network to recognize characters. Figure 5.5 shows five characters generated in MATLAB, each character consisting of a 12×12 array of bipolar numbers. A $+1$ is considered to be black, and -1 is white. The Hopfield network requires $N = 144$ neurons with $N^2 = 20{,}736$ synaptic weights (although 144 of these weights are zero, associated with the diagonal elements of the weight matrix). The threshold value for each neuron in the network is assumed to be zero. Each of the characters is vectorized to form a single pattern. That is, if we consider a matrix

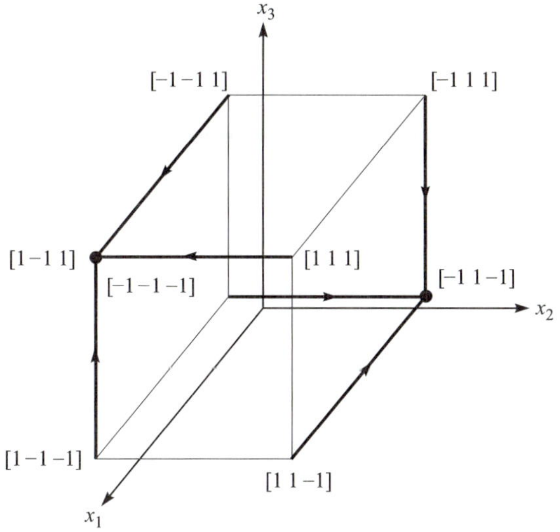

Figure 5.4 State flow for a three-neuron Hopfield network. The two stable states are indicated by the dots at the two corners of the cube.

$\boldsymbol{X}_h \in \Re^{12\times 12}$, for $h = 1, 2, \ldots, 5$, to be the matrix form of each character, then the vectorized form of the character is given by

$$\boldsymbol{\phi}_h = \text{vec}(\boldsymbol{X}_h) \in \Re^{144\times 1} \tag{5.23}$$

where the operation vec(•) forms a vector from a matrix by "stacking" the columns of the matrix (cf. Sect. A.2.17). These characters were created in a manner that would ensure good network performance [13]. This can be accomplished by observing the inner products of the vectors $\boldsymbol{\phi}_h$, for $h = 1, 2, \ldots, 5$. The closer the vectors are to being mutually orthogonal, the better the recognition results will be. The synaptic weight matrix is built by using the five prototype vectors from (5.23) and the expression in (5.7). The weight matrix $\boldsymbol{W} \in \Re^{144\times 144}$ is shown in Figure 5.6 as a gray-scale image.

TABLE 5.1
Values of energy function for Example 5.1

State of the network				
x_1	x_2	x_3	Energy function $\mathscr{E}$	Change in energy function $\Delta\mathscr{E}$ with respect to the initial state {1, 1, 1}
1	1	1	2/3	
−1	1	1	2/3	0
1	−1	1	−2	−8/3
−1	−1	1	2/3	8/3
1	1	−1	2/3	0
−1	1	−1	−2	−8/3
1	−1	−1	2/3	8/3
−1	−1	−1	2/3	0

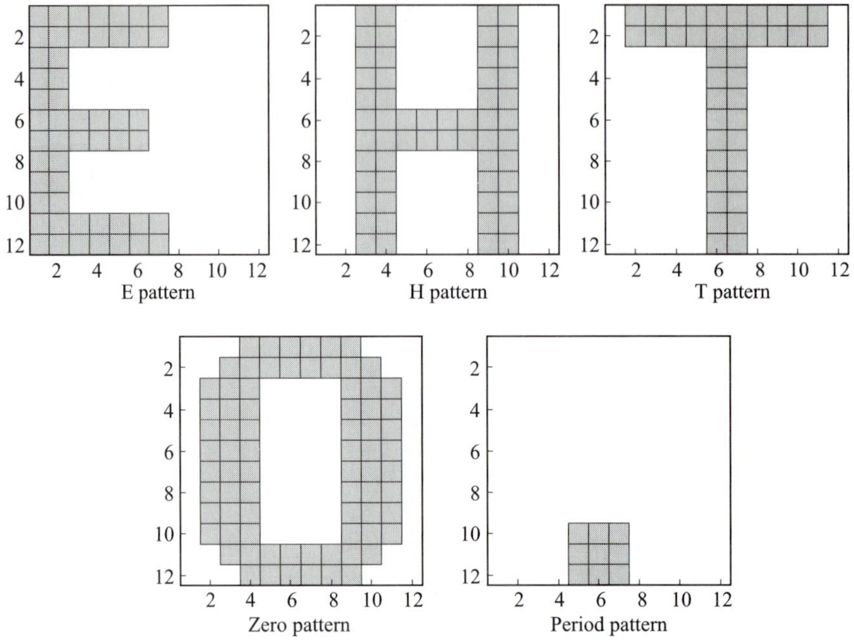

Figure 5.5 Five characters to be recognized by a Hopfield neural network.

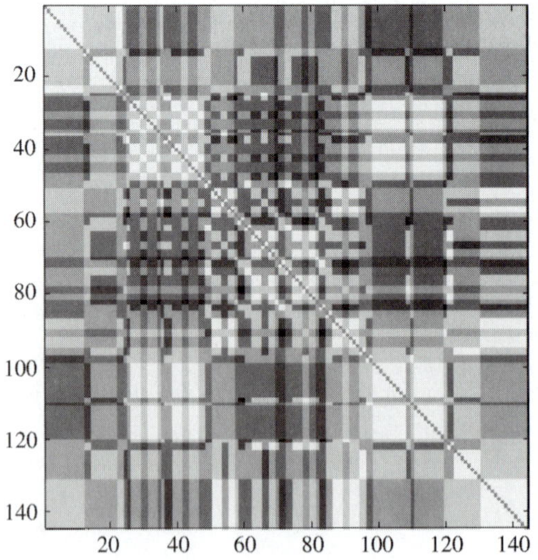

Figure 5.6 Synaptic weight matrix for Example 5.2 shown as a gray-scale image. The smaller weight elements show up as dark pixels, whereas the larger values show up lighter. The zero diagonal of the weight matrix is very evident in the image.

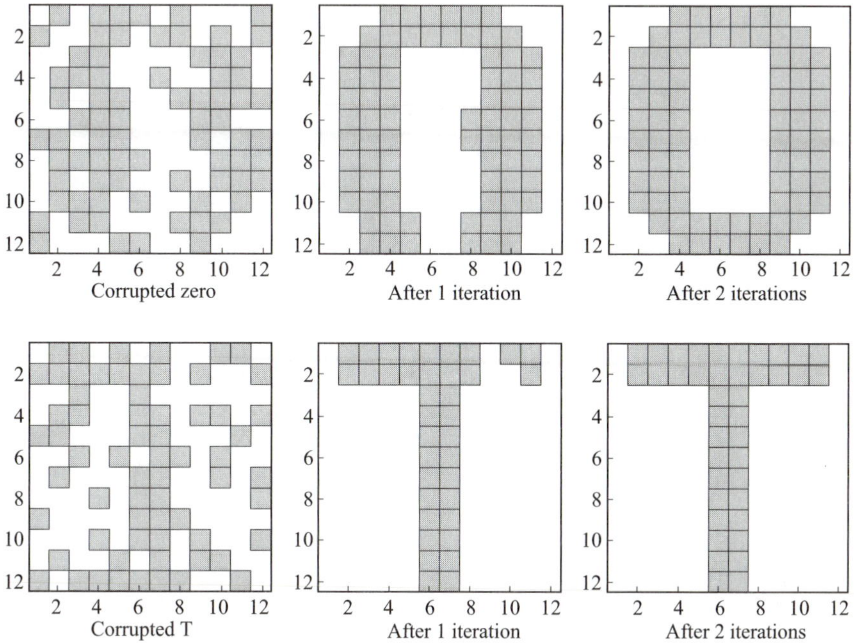

Figure 5.7 Simulation results for Example 5.2. In both cases the two characters were correctly identified after two iterations.

The first step in accessing the performance of the Hopfield network is to verify that the five prototype vectors (characters) can be recalled from the CAM. Applying the stability condition in (5.20) (with $\boldsymbol{\theta} = \mathbf{0}$), in each case the character could be recalled or recovered correctly from the information that is stored in the Hopfield network. To illustrate the error correction capability of the Hopfield network, each of the characters in Figure 5.5 was corrupted by noise. That is, random "toggling" of some of the "bits" was performed on each character (i.e., a random polarity change). The bit error rate (BER) was 30 percent (or the probability that a bit will be toggled is 0.3); therefore, on the average 30 percent of the bits in the pattern would have their sign changed (i.e., toggling from a black pixel to a white one and vice versa). Figure 5.7 shows the results of reconstructing two of the characters that were corrupted with noise. In both cases the characters could be completely reconstructed (or recalled) in two iterations by using (5.20) with the threshold set equal to zero. The remaining characters could be perfectly reconstructed after only one iteration. For a higher bit error rate than 30 percent, the results were sporadic. Most of the time the characters could not be completely reconstructed, or an incorrect character would be identified.

5.4 SIMULATED ANNEALING

In Section 5.3, we saw that the Hopfield neural network is capable of recalling stored patterns at the *local minima* of the network. However, many times it is desirable to have a network reach the global minimum, such as in optimization problems. In the case of the Hopfield network, the gradient descent rule used to retrieve stored patterns from the network allows the eventual convergence to a local minimum where it will remain (thus the term *stuck in a local minimum*). If it is desired to find the global minimum of an objective (cost) function associated with the global energy of the network, using only *local* information in the search routine, then some *randomness* must be added to the gradient descent approach to increase the probability of reaching the global minimum. This leads to our discussion of simulated annealing.

Simulated annealing was originally developed by Kirkpatrick et al. [14] and can be used to solve combinatorial optimization problems, or NP-complete (nondeterministic polynomial time complete) problems. Simulated annealing [14–19] differs from standard iterative improvement optimization methods (gradient descent) in that randomness is added during the search for the global minimum. This allows the system to "jump" out of a local minimum (see Figure 5.8) and continue the search for the global minimum. Without this "kick" the local minimum would be the final destination for an iterative improvement search method. Therefore, the algorithm does not have to get stuck in local minima, but to get out of a local minimum and into the global minimum, at least one ridge between them must be conquered (or climbed over). The "kick" provided by simulated annealing is a result of varying a *temperature* parameter in the algorithm, and thus transitions out of local minima are always feasible at nonzero temperatures. Another distinguishing feature of simulated annealing is that it exhibits an adaptive *divide-and-conquer* characteristic. Gross features of the state of the system will appear at higher temperatures, and finer details of the system state develop at lower temperatures.

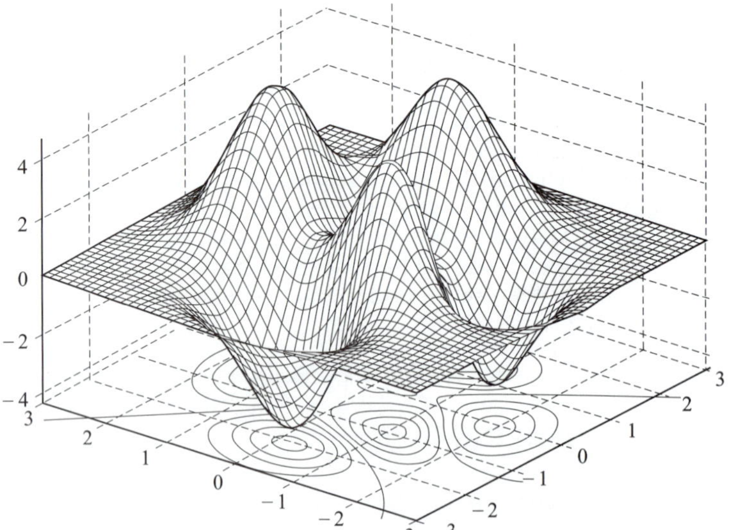

Figure 5.8 Plot of a function of two variables with multiple minima and maxima.

Simulated annealing draws on an analogy between the optimization of a cost function in a large number of system variables and a physical system with many degrees of freedom in thermal equilibrium at many different finite temperatures, as found in statistical mechanics [20]. Given a prescription for calculating the energy of a system and finding its low-temperature state is an optimization problem similar to those of combinatorial optimization. In condensed-matter physics [21], the term *annealing* refers to the physical process that involves heating a solid above its melting point to the liquid phase, and then lowering the temperature (slowly) so all particles are arranged in a lattice. The particles of the liquefied metal seek minimum energy configurations (or states) as it is allowed to cool. The cooling phase is carried out slowly (as opposed to quenching), so defects are minimized in the resulting crystalline lattice. The defect-free crystal state of the lattice corresponds to the *global* minimum energy of the structure. The simulated annealing process consists of first *melting* the system to be optimized at an effectively high temperature and then lowering the temperature in slow stages until the system *freezes* (i.e., no further changes take place).

In statistical mechanics an energy function $\mathscr{E}(\boldsymbol{x})$ is defined that measures the thermal energy of a collection of atoms (or a physical system) in a given state $\boldsymbol{x}$, where $\boldsymbol{x} \in \boldsymbol{\Phi}$ (set of all possible points). A basic result from physics is that the probability of occurrence of each state $\boldsymbol{x}$ at thermal equilibrium is given as

$$\Pr(\boldsymbol{x}) = \frac{1}{Z} e^{-\mathscr{E}(\boldsymbol{x})/k_B T} \tag{5.24}$$

where k_B is Boltzmann's constant [$k_B = 1.3806 \times 10^{-23}$ J/K (joule/kelvin)], T is the temperature, and Z is the *partition function*. The partition function is defined by

$$Z = \mathrm{Tr}(e^{-\mathscr{E}(\boldsymbol{x})/k_B T}) \tag{5.25}$$

where Tr denotes the sum over all possible configurations of the atoms in the sample system [14]. The partition function will bound $\Pr(\boldsymbol{x})$ in (5.24) between 0 and 1, and (5.24) can now be written as

$$\Pr(\boldsymbol{x}) = \frac{e^{-\mathscr{E}(\boldsymbol{x})/k_B T}}{\mathrm{Tr}(e^{-\mathscr{E}(\boldsymbol{x})/k_B T})} \tag{5.26}$$

and is known as the *Boltzmann–Gibbs distribution.* We can define a set of probabilities $\Pr(\boldsymbol{x} \to \boldsymbol{x}_p)$ associated with the stochastic state transitions, that is, transitions from any state $\boldsymbol{x}$ to some perturbed state $\boldsymbol{x}_p$. The set of transition probabilities $\Pr(\boldsymbol{x} \to \boldsymbol{x}_p)$ may not lead to thermal equilibrium, but since we are interested in thermal equilibrium and not limit cycles or chaotic behavior for the system, we need a sufficient condition on $\Pr(\boldsymbol{x} \to \boldsymbol{x}_p)$ that guarantees this. A sufficient condition on $\Pr(\boldsymbol{x} \to \boldsymbol{x}_p)$ that guarantees thermal equilibrium is that the probability of transitions from $\boldsymbol{x}$ to $\boldsymbol{x}_p$ equal the probability of transitions from $\boldsymbol{x}_p$ to $\boldsymbol{x}$ on average. This can be written as

$$\Pr(\boldsymbol{x})\Pr(\boldsymbol{x} \to \boldsymbol{x}_p) = \Pr(\boldsymbol{x}_p)\Pr(\boldsymbol{x}_p \to \boldsymbol{x}) \tag{5.27}$$

If (5.27) holds, the system will achieve equilibrium in accordance with the Boltzmann–Gibbs distribution. Rearranging (5.27) and using (5.26), we can write the following

$$\frac{\Pr(\boldsymbol{x} \to \boldsymbol{x}_p)}{\Pr(\boldsymbol{x}_p \to \boldsymbol{x})} = \frac{e^{-\mathscr{E}(\boldsymbol{x}_p)/k_B T}}{e^{-\mathscr{E}(\boldsymbol{x})/k_B T}} = e^{-[\mathscr{E}(\boldsymbol{x}_p)-\mathscr{E}(\boldsymbol{x})]/k_B T} = e^{-\Delta\mathscr{E}/k_B T} \tag{5.28}$$

where $\Delta\mathscr{E} = \mathscr{E}(\boldsymbol{x}_p) - \mathscr{E}(\boldsymbol{x})$, that is, the change in energy.

The Metropolis algorithm [22], which is a Monte Carlo technique, satisfies the condition in (5.27) and uses the following transition probability

$$\Pr(\boldsymbol{x} \to \boldsymbol{x}_p) = \begin{cases} 1 & \text{for } \Delta\mathscr{E} < 0 \\ e^{-\Delta\mathscr{E}/k_B T} & \text{for } \Delta\mathscr{E} \geq 0 \end{cases} \tag{5.29}$$

The Metropolis algorithm introduced a simple and efficient method for simulating the evolution of a collection of atoms in equilibrium at a given temperature. In each successive step of the algorithm the temperature is slowly lowered, an atom is given a small random perturbation (displacement), and the resulting change in the energy $\mathscr{E}$, denoted by $\Delta\mathscr{E}$, is then computed. From (5.29), if the energy change is negative, that is, $\Delta\mathscr{E} < 0$ (reduced energy), then the displacement is allowed and the configuration with the displaced atom is used as the initial condition to start the next step. If $\Delta\mathscr{E} \geq 0$ (transition from a lower to a higher energy state), the configuration may or may not be accepted, and will depend on the probability from (5.29), that is,

$$\Pr(\Delta\mathscr{E}) \triangleq e^{-\Delta\mathscr{E}/k_B T} \tag{5.30}$$

The randomization part of the algorithm is typically implemented by using uniformly distributed random numbers (cf. Sect. A.7.4) in the interval [0, 1]. A number is selected from this distribution and compared with $\Pr(\Delta\mathscr{E})$. If the number is less than $\Pr(\Delta\mathscr{E})$, the new (perturbed) configuration is retained; however, if the number is greater than or equal to $\Pr(\Delta\mathscr{E})$, the current configuration is unchanged and used for the next step. By repeating this basic step many times, the thermal motion of atoms in thermal contact with a heat bath at temperature T is simulated. Provided that the temperature is lowered sufficiently slowly, the collection of atoms can reach thermal equilibrium at each selected temperature. If the condition in (5.27) is satisfied, the system will evolve according to the Boltzmann–Gibbs distribution and reach equilibrium.

There are four basic components associated with a simulated annealing-based global search algorithm (which has its roots in the Metropolis algorithm used in statistical mechanics): (1) a concise description of the system configuration, (2) an objective or cost function (containing tradeoffs that must be made), (3) an exploration process, or random generator of "moves" or rearrangements of the system elements in a configuration, and (4) an annealing schedule of temperatures and defined time periods for which the system is to be evolved [14]. The basic idea is to go "downhill" *most* of the time instead of *always* going downhill [23].

A simulated annealing-based global search method, for locating a globally optimal solution of a given multivariate cost (objective) function $f(\boldsymbol{x})$ for $\boldsymbol{x} \in \boldsymbol{\Phi}$, is given in the following steps:

Step 1. Initialize the vector $\boldsymbol{x}$ to a random point in the set $\boldsymbol{\Phi}$.

Step 2. Select an annealing (cooling) schedule for the parameter T, and initialize T to a sufficiently large number.

Step 3. Compute $\boldsymbol{x}_p = \boldsymbol{x} + \Delta \boldsymbol{x}$ (where $\Delta \boldsymbol{x}$ is a proposed change in the system's state).

Step 4. Compute the change in the cost (energy) function $\Delta f = f(\boldsymbol{x}_p) - f(\boldsymbol{x})$.

Step 5. Use (5.29), associated with the Metropolis algorithm, to decide if $\boldsymbol{x}_p$ should be used as the new state of the system or keep the current state $\boldsymbol{x}$. Equation (5.29) is modified for the simulated annealing optimization algorithm as

$$\Pr(\boldsymbol{x} \rightarrow \boldsymbol{x}_p) = \begin{cases} 1 & \text{for } \Delta f < 0 \\ e^{-\Delta f / T} & \text{for } \Delta f \geq 0 \end{cases} \tag{5.31}$$

where T replaces $k_B T$. Functionally, in cases when $\Delta f \geq 0$, a random number η is selected from a uniform distribution in the range [0, 1]. If $\Pr(\boldsymbol{x} \rightarrow \boldsymbol{x}_p) > \eta$, then the perturbed state $\boldsymbol{x}_p$ is used as the new state (or search point); otherwise, the state (or search point) remains at $\boldsymbol{x}$.

Step 6. Steps 3 through 5 are repeated until the system reaches equilibrium, which is determined when the number of accepted transitions becomes insignificant, which will happen when the search point is at (or close to) a local minimum. Typically, steps 3 through 5 are carried out a predetermined number of times.

Step 7. The temperature T is updated according to the annealing schedule specified in step 2, and steps 3 through 6 are repeated. The process can be stopped when the temperature T reaches zero (the system *freezes*) or a predetermined small (positive) number.

The relative performance of this algorithm depends on the choice of the scheduling for the temperature parameter T. If the system is cooled down too quickly, convergence may be achieved too early and may result in a local minimum being found as the "solution." On the other hand, if the scheduling of T is set to be too slow, the algorithm will require a tremendous amount of computation time to converge. According to Geman and Geman [24], if the temperature parameter is varied as

$$T(k) = \frac{T(0)}{\log(1 + k)} \qquad k = 1, 2, \ldots \tag{5.32}$$

where k indicates the kth iterate of the search, $T(k)$ is the kth temperature, and $T(0)$ is the initial temperature (a sufficiently large positive constant), the simulated annealing algorithm is guaranteed to converge (with probability 1) to the global minimum of $f(\boldsymbol{x})$ as $k \rightarrow \infty$. The algorithm generates a Markov chain [25] that converges in accordance with a uniform distribution over the minimum energy configurations [24]. The temperature scheduling given in (5.32) is extremely slow, and this essentially renders it useless. There has been significant interest and work carried out to accelerate the simulated annealing search algorithm; for example, see the work by Szu [26]. Using a suboptimal approach may speed up convergence but one pays the price in performance; that is, the algorithm no longer guarantees convergence to the global minimum with probability 1. However, the suboptimal form of the algorithm may have many practical applications when near-optimal solutions are sufficient. One such suboptimal approach for the temperature scheduling is given by [14]

$$T(k) = \alpha T(k - 1) \qquad k = 1, 2, \ldots \tag{5.33}$$

where the decrementing factor α should be small and close to unity, typically $0.8 \leq \alpha \leq 0.99$. Equation (5.33) gives an exponential reduction in the temperature scheduling. To make this temperature scheduling algorithm even more practical, a finite number of transition attempts should be carried out at each finite temperature [14]. For example, at each temperature enough transitions are attempted so that 10 are accepted. If the desired number of acceptances is not achieved at three successive temperatures, then the system is defined to be *frozen* and the annealing stops.

EXAMPLE 5.3. To illustrate the use of simulated annealing for an optimization problem, we consider the famous "traveling salesman" problem. This problem may be posed as follows. On a sales trip the salesman needs to visit a given set of cities, N in fact. He wants to visit each of the cities only once, and in a sequence that would require the smallest amount of traveling (i.e., shortest *distance*). The optimization task is to determine the optimum sequence of cities that the salesman is to follow on the trip. The first step in the application of the simulated annealing approach to this problem is to identify the state space of possible solutions. In this particular case the solution is an ordered list of cities on the sales trip. In this approach we assume that any order of the cities is feasible, and therefore, if there are N cities to be visited, the possible number of different sequences is equal to $N!$. We see that the number of candidate solutions of the traveling salesman problem is proportional to the factorial of the number of cities. Even if some of the solutions can be easily discarded, the size of the problem can become too large for many of the more common optimization methods. Now suppose that we have an initial sequence of the cities. We need to specify the nature of the state perturbation, that is, the algorithm for "traveling" in the problem's configuration space. For this example, we assume that a new solution (i.e., a new sequence) is obtained by swapping the position of two cities in the current solution. Finally, we need to specify the cost function that facilitates fitness quantification of the proposed solution. Here the cost function is simply specified as the total distance traveled by the salesman.

Figure 5.9(*a*) shows the positions of 20 cities (randomly selected) that need to be visited. In our simulated annealing approach we may start from any initial sequence. One such sequence (generated randomly for this simulation) is shown in Figure 5.9(*b*). As shown, the sequence involves a substantial amount of traveling by the salesman and is obviously far from being optimal. At every step of the algorithm a new sequence is generated by swapping the positions of two randomly selected cities. By using the Metropolis criterion given in the simulated annealing-based global search algorithm, that is, (5.31), the newly proposed sequence is either accepted or rejected. The number of perturbations performed at a given temperature is set to $N(N-1)/2$. The temperature is decreased according to the exponential cooling schedule specified in (5.33) with the parameter $\alpha = 0.95$. The final sequence obtained after the optimization algorithm has terminated is shown in Figure 5.9(*c*). As shown, the sequence of cities involves a considerably smaller amount of traveling distance compared to the initial condition shown in Figure 5.9(*b*). Finally, in Figure 5.9(*d*) the cost function values (i.e., total distance traveled) that are evaluated along the optimization algorithm's state space trajectory are shown.

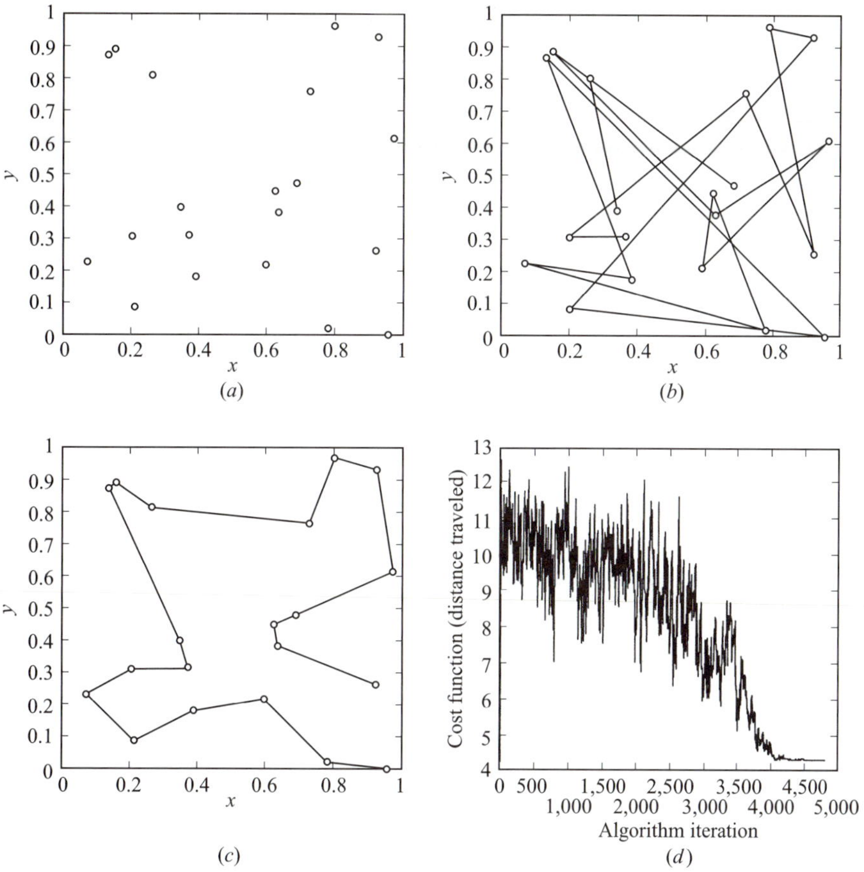

Figure 5.9 Traveling salesman problem solution using simulated annealing. (*a*) Position of 20 cities, (*b*) initial (condition) solution, (*c*) solution obtained by simulated annealing, and (*d*) cost function trajectory during the annealing process.

5.5 BOLTZMANN MACHINE

The Boltzmann machine is a parallel constraint satisfaction network based on simulated annealing (cf. Sect. 5.4) and uses *stochastic* neurons [27–29]. It is capable of learning the underlying constraints that are characteristic of a set of patterns by being shown examples from the set [28]. Because this network uses extensive internal feedback and stochastic neurons, it is considered a stochastic recurrent network. This neural network can be thought of as an extension of the Hopfield network (cf. Sect. 5.3). The main difference between the two networks is that the Boltzmann machine can have *hidden neurons*, whereas the Hopfield network does not permit this. There are two other major differences between the two networks: (1) As stated above, the Boltzmann machine uses stochastic neurons that "fire" in accordance with a probability distribution, whereas the Hopfield network has neurons based on the McCulloch–Pitts neuron model with a deterministic firing mechanism. (2) The Hopfield network operates unsupervised, whereas the Boltzmann machine may be trained in a supervised mode (it can also operate unsupervised). They have in common that (1) all synaptic weight connections are symmetric, (2) no self-feedback exists, (3) the processing units have bipolar states, and (4) the neurons are selected randomly and one at a time for updating.

The Boltzmann machine is well suited for constraint satisfaction tasks involving a large set of "weak" (or "soft") constraints [28, 30]. Constraint satisfaction searches [31] typically use "strong" constraints that must be satisfied by any solution. In problem domains involving games and puzzles the goal criteria very often have this strong-constraint characteristic; therefore, the strong constraints are the *rule* (do not confuse *legal* play with *good* play). However, for some practical problems many of the criteria are not all-or-nothing, and very often the *best* solution will violate some of the constraints [32]. Therefore, in this situation, an optimum solution is usually the "best" one that realistically fits the set of weak constraints as best as possible, but not exactly. Before we go on, an overview of stochastic neurons is given.

Stochastic neurons

We want to be able to account for the effects of synaptic noise in a neural network in some manner that is mathematically tractable. The most common way to do this is to use a probabilistic approach to dictate the firing of the neurons. We assume that neuron q will fire according to a probability rule; that is, the firing of the neuron is decided by the value of the activity level of the neuron $v_q = \sum_{j=1, j \neq q}^{n} w_{qj} x_j$ (cf. Sect. 2.2) according to the probability $\Pr(v_q)$ (cf. Sect. A.7.1). Thus, the neuron output y_q follows the probabilistic rule

$$y_q = \begin{cases} 1 & \text{with probability } \Pr(v_q) \\ -1 & \text{with probability } 1 - \Pr(v_q) \end{cases} \tag{5.34}$$

and if $v_q = 0$, $y_q = \pm 1$ each with probability $1/2$. A typical choice for the probability function $\Pr(v)$ is

$$\Pr(v) = \frac{1}{1 + e^{-2v/T}} \tag{5.35}$$

which is a sigmoid function (cf. Sect. 2.3), and T is a *temperaturelike* parameter (pseudotemperature) used to control the uncertainty associated with neuron firing. For the noiseless case, that is, as the temperature parameter T approaches zero, the probabilistic rule in (5.34) reverts to the deterministic rule given for the Hopfield network in (5.2). Figure 5.10 shows the probability distribution for the firing of a stochastic neuron and the noiseless limit ($T \to 0$) leading to the McCulloch–Pitts neuron binary activation function.

Like the Hopfield network, the Boltzmann machine has a symmetric synaptic weight structure, that is, $w_{ij} = w_{ji}$, and self-feedback is not allowed, therefore, $w_{ij} = 0 \; \forall i = j$. This type of neural network was called a Boltzmann machine because the probability of the states of the system is given by the Boltzmann–Gibbs distribution from statistical mechanics. The neurons in the Boltzmann machine are divided into two categories, visible and hidden. With

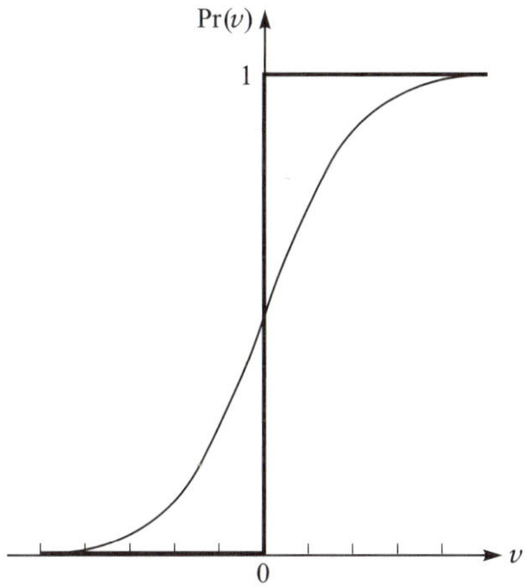

Figure 5.10 *Probability distribution function* (sigmoid) for a stochastic neuron firing, and the McCulloch–Pitts neuron activation function, denoted by the bold line.

hidden neurons in the Boltzmann machine, a similar situation exists as in feedforward networks, that is, the problem of determining the right connections to the hidden neurons without knowing from the training patterns what the hidden units should be representing. Figure 5.11 shows the distinction between the two types of neurons. If n_v represents the number of visible neurons and n_h the number of hidden neurons, the total number of connections in the network is $(n_v + n_h)(n_v + n_h - 1)$. In an unsupervised mode, there is no distinction made between input and output neurons; only visible neurons exist that directly interact with the "outside environment" and are *clamped* to environmental states. The hidden neurons are allowed to freely operate and serve the purpose of explicating the underlying constraints in the environmental inputs. This unsupervised learning process can be used to model a probability distribution that is specified by clamping patterns associated with the environment onto the visible neurons with the appropriate probabilities. If the network can learn the training probability distribution properly, it can perform what is called *pattern completion*. In a supervised mode of operation, the Boltzmann machine now has defined input and output neurons (see Figure 5.11), and it performs an association function. The supervised mode of training may involve providing a *probabilistic* correct response pattern for each of the input patterns.

The energy of the global network configuration is given by

$$\mathscr{E} = -\frac{1}{2}\sum_{\substack{i \\ i \neq j}}\sum_{j} w_{ij}x_i x_j + \sum_{i} \theta_i x_i \tag{5.36}$$

where x_i corresponds to the ith neuron output (state), θ_i is the ith neuron threshold, and w_{ij} is the synaptic weight connection between neurons i and j.

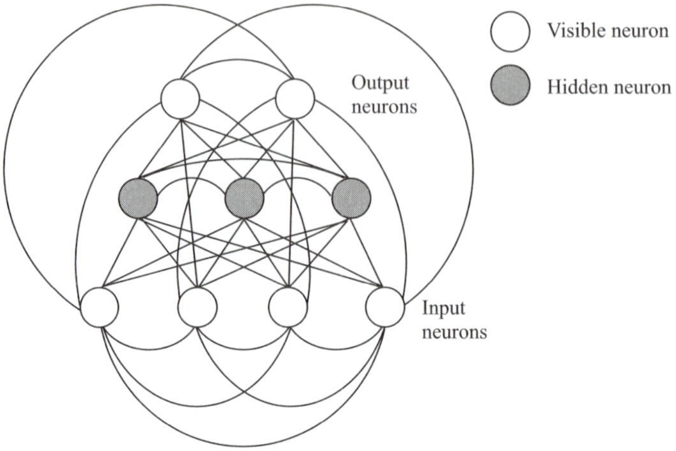

Figure 5.11 Example of a Boltzmann machine network showing the visible and hidden neurons.

The neuron state x_i is $+1$ if neuron i is on and -1 if it is off. The minima of (5.36) are the stable states of the network. The energy function in (5.36) can be written in vector-matrix form as

$$\mathscr{E} = -\frac{1}{2}\boldsymbol{x}^T \boldsymbol{W}\boldsymbol{x} + \boldsymbol{x}^T\boldsymbol{\theta} \tag{5.37}$$

where the symmetric weight matrix $\boldsymbol{W}$ has zero diagonal elements.

Updating the neurons in the Boltzmann machine is more involved than in the Hopfield network because of the hidden units; that is, we need a way to update the weights associated with the hidden units. However, in spite of this, the updating procedure is relatively straightforward. Generally speaking, there are two phases associated with the learning cycle in the Boltzmann machine, the positive phase and the negative phase that alternate; this is followed by synaptic weight adjustments. Specifically, a neuron i is chosen randomly, and the output is characterized by a state transition function for transitioning from a positive state x_i to the negative state $-x_i$. The state transition function is given by

$$\Pr(x_i \to -x_i) = \frac{1}{1 + e^{-\Delta_i \mathscr{E}/T}} \tag{5.38}$$

where $\Delta\mathscr{E}_i$ is the energy change due to the transitioning associated with neuron i and T is the pseudotemperature. Equation (5.38) will approach the "step function" associated with the deterministic McCulloch–Pitts neuron model used in the Hopfield network when $T \to 0$. Therefore, the Hopfield network is a special case of the Boltzmann machine when the temperature is zero and there are no hidden neurons. Comparing (5.38) to the probability function $\Pr(v)$ in (5.35) for the stochastic neuron, we see that they appear different but are equivalent in terms of their representation. To see this, we can view the change in the energy function given in (5.36) (assuming that the network contains no external biases, therefore, $\theta_i = 0\ \forall i$). Assuming again that neuron i changes its state from positive to negative, that is, $x_i \to -x_i$, the resulting change in the energy function will be given by

$$\Delta\mathscr{E} = \underbrace{-\Delta x_i}_{-[x_i-(-x_i)]} \underbrace{\sum_{j=1}^{n} w_{ij}x_j}_{v_i} = -2x_i v_i \tag{5.39}$$

where v_i is the activity level of neuron i. Substituting the result in (5.39) for the change in energy $\Delta\mathscr{E}$ in (5.38) gives

$$\Pr(x_i \to -x_i) = \frac{1}{1 + e^{2x_i v_i/T}} \tag{5.40}$$

Therefore, when the initial state of neuron i is $x_i = -1$, the probability that the neuron will transition or toggle to the opposite state, that is, $x_i \to 1$, is given by (5.40) as

$$\Pr(x_i \to 1) = \frac{1}{1 + e^{-2v_i/T}} \tag{5.41}$$

which is in agreement with (5.35). If the initial state of neuron i is $x_i = 1$, the probability that the neuron will toggle to the opposite state ($x_i \to -1$) is

$$\Pr(x_i \to -1) = \frac{1}{1 + e^{2v_i/T}} \tag{5.42}$$

However, it is easy to show that (5.42) can be written as

$$\Pr(x_i \to -1) = 1 - \frac{1}{1 + e^{-2v_i/T}} \tag{5.43}$$

Equations (5.41) and (5.43) together are exactly the probabilistic rule that is stated in (5.34) for a general stochastic neuron. In the Boltzmann machine there are a total of $n = n_v + n_h$ neurons (visible plus hidden neurons). If each neuron can take on one or the other of the bipolar states $x_i = \pm 1$, the total number of global states in the network is 2^n.

Simulated annealing is used in the Boltzmann machine to search the energy landscape to locate the global minimum of the energy function; thus, the network evolves according to Metropolis dynamics. In the case of the Hopfield network, a local minimum is usually desirable to locate because the local energy minima of this network are used to store information. For the Boltzmann machine, constraint satisfaction tasks are carried out, and it is thus desirable to "escape" from local minima to seek out the global minimum, given the current input. Therefore, the energy minimization process is carried out during Boltzmann learning by first operating the network at a high temperature and then slowly lowering the temperature until the network reaches thermal equilibrium. This is accomplished by following an annealing schedule, that is, a cooling schedule for the "temperature" parameter T (cf. Sect. 5.4). A coarse search of the global state space is carried out at high temperatures where small energy changes are essentially discounted, and the network tends to approach thermal equilibrium fairly rapidly and finds a good minimum. Lowering the temperature forces the network to perform a fine search of the energy landscape, and now it responds to small energy changes. During this fine-tuning process, the network seeks out a better minimum in the neighborhood of the minimum that was discovered by the coarse search of the energy landscape. Therefore, as the annealing process comes to an end, the network will settle into a minimum that is associated with the most plausible solution that fits a set of weak constraints (not exactly, but as best as possible).

The Boltzmann machine learning rule will not be derived here, but is presented in a step-by-step algorithmic form [4, 27]. The states of the visible neurons are labeled with α, and the hidden neurons are labeled with β. As before, we assume n_v visible neurons and n_h hidden neurons; therefore, α runs from 1 to 2^{n_v} and β runs from 1 to 2^{n_h}. A global state of the network is uniquely defined by α and β, with $2^{n_v+n_h}$ possible states. The probability of finding each of these global states $\Pr_{\alpha\beta}$ is given by the Boltzmann–Gibbs distribution (cf. Sect. 5.4). And $\Pr_\alpha$ is the probability of finding the visible neurons in state α independent of β. The probability $\Pr_{\alpha\beta}$ can be written in terms of the conditional probability (cf. Sect. A.7.1) $\Pr_{\beta|\alpha}$ of hidden state β given visible state α as

$$\Pr_{\alpha\beta} = \Pr_{\beta|\alpha} \Pr_\alpha \tag{5.44}$$

In the "freely running system," $\Pr_\alpha$ is the actual probability of finding the visible units in state α and is determined by the synaptic weights w_{ij} of the network. The *desired probabilities* for these states are given by the set R_α.

Operationally, the learning algorithm can be described by four nested loops [4]:

Learning algorithm for the Boltzmann machine

Loop 1. At the outermost loop, the synaptic weights of the network are updated many times to ensure convergence according to

$$\Delta w_{ij} = \mu\beta\left[\overline{\langle x_i x_j\rangle}_{\text{clamped}} - \langle x_i x_j\rangle_{\text{free}}\right] \tag{5.45}$$

where $\mu > 0$, and

$$\overline{\langle x_i x_j\rangle}_{\text{clamped}} = \sum_\alpha \sum_\beta R_\alpha P_{\beta|\alpha} x_{i|\alpha\beta} x_{j|\alpha\beta} \qquad i,j = 1, 2, \ldots, n;\ i \neq j \tag{5.46}$$

which is the correlation between the states of neurons i and j, conditioned on the visible neurons being *clamped* to the environment (this average is taken over all possible states). Therefore, $\overline{\langle x_i x_j\rangle}_{\text{clamped}}$ is the value of $\langle x_i x_j\rangle$ when the visible units are clamped in state α, averaged over the α's according to their desired probabilities R_α. In (5.46), $x_{i|\alpha\beta}$ is the state of neuron i given the visible neurons are in state α and the hidden neurons are jointly in state β. The values of α range from 1 to 2^{n_v} (for the visible neurons), the values of β range from 1 to 2^{n_h} (for the hidden neurons), and $n = n_v + n_h$ is the total number of neurons in the network. The first term in (5.45) is essentially a Hebbian term with the visible units clamped. The second term in (5.45) corresponds to Hebbian *unlearning* with the system free-running. This process converges when the free system unit/unit correlations $\langle x_i x_j\rangle$ are equal to the clamped ones.

Loop 2. For each iteration in loop 1 $\langle x_i x_j\rangle$ must be calculated in an *unclamped* state, and with the visible units clamped in each desired pattern. To operate the Boltzmann machine, the system must be in thermal equilibrium for some positive *temperature* $T > 0$. The state of the system $\boldsymbol{x}$ then fluctuates and the correlations $\langle x_i x_j\rangle$ are measured by taking the time average of $x_i x_j$. To obtain all the information that is necessary to compute the synaptic weight update rule in (5.45), this process must be carried out once with the visible neurons clamped in each of their states α for $R_\alpha > 0$, and once with the neurons unclamped. In each case, the system must repeatedly reach thermal equilibrium before an average can be taken.

Loop 3. For each of these averages in loop 2, thermal equilibrium must be reached using a *simulated annealing* temperature schedule $\{T(k)\}$, for a sufficiently large initial temperature $T(0)$, and then a gradual decrease in the temperature.

Loop 4. At each of these temperatures in loop 3, many neurons must be sampled and updated according to the rule from (5.34)

$$x_i = \begin{cases} 1 & \text{with probability } \Pr(v_i) \\ -1 & \text{with probability } 1 - \Pr(v_i) \end{cases} \tag{5.47}$$

where

$$\Pr(v_i) = \frac{1}{1 + e^{-2v_i/T}} \tag{5.48}$$

and v_i is the activity level of neuron i, that is,

$$v_i = \sum_{\substack{j=1 \\ j \neq i}}^{n} w_{ij} x_j \tag{5.49}$$

Using this learning strategy for updating the Boltzmann machine synaptic weights [see (5.45)] requires the difference between two averages that can both fluctuate. Therefore, dealing with a system that is poorly equilibrated, or using a short averaging time, can reduce the update cycle time, but poor weight updates $\{\Delta w_{ij}\}$ can occur and ultimately more updating cycles will be required. Although the Boltzmann machine is very slow, it has been very effective in solving complex problems. Since the Boltzmann machine is computationally intensive, and because of the simulated annealing process, several variations of the learning algorithm have been developed. For example, using a *mean-field approximation* from statistical mechanics [33], the excessive computation time can be reduced by replacing the stochastic bipolar-state neurons of the Boltzmann machine with deterministic, analog neurons [34].

5.6 OVERVIEW OF TEMPORAL FEEDFORWARD NETWORKS

The element of *time* in a neural network (continuous or discrete) can be a very important factor in the learning process. Many cognitive functions, for example, speech, vision, and motor control, rely on time for proper operation. When the element of time is included in a neural network, the network is then able to perform tasks that would otherwise not be feasible. For example, the network can track statistical variations in the input data that are associated with nonstationary processes. Following this section we discuss a selected number of neural networks for temporal processing. In each case, a *static* neural network architecture is considered, and the temporal nature of the network comes by way of certain *dynamic* properties that are incorporated into the architecture. Specifically, *short-term memory* is built into the network by way of *time delays*. The time delays allow the network to become a dynamic network. Instead of a time dependence associated with the network inputs there can be a spatial dependence. Therefore, our discussions actually relate to the *spatiotemporal* sensitivity of static networks. However, we will only speak in terms of a time dependence associated with the network of interest.

The following is a succinct overview of the most common types of temporal networks.

1. The *time-delay neural network* (TDNN) is a feedforward multilayer neural network with time delays included at each of the output layer neurons. It has been applied to speech recognition problems [35–37].
2. The *finite impulse response* (FIR) feedforward multilayer network is a generalization of the TDNN [38–40]. This network has an FIR digital filter replacing each of the weights in the TDNN architecture. This is the basis for the *distributed time-lagged feedforward neural network* (DTLFNN).
3. The simple recurrent network (SRN) or Elman network [41] is a single hidden-layer feedforward network that has feedback connections from the outputs of the hidden-layer neurons to the input of the network.
4. The real-time recurrent neural network (RTRNN) has two layers and is able to learn in real time [42]. It is similar to a Hopfield network, with the exception that Hopfield networks do not have hidden neurons.
5. The pipelined recurrent neural network (PRNN) [43] is a modular structure in which each module receives a different delayed rendition of the network input. Each of the modules is a fully connected recurrent network with one output neuron. This type of network has both feedforward connections as well as feedback, that is, recurrent, connections. This network was developed for the purpose of adaptively predicting nonstationary signals, for example, speech signals.
6. The nonlinear autoregressive moving average (NARMA) neural network is used for control and identification of nonlinear systems (cf. Sect. 10.6.2) [44–46].

5.7 SIMPLE RECURRENT NETWORK

The simple recurrent network, often referred to as the Elman network [41], is a single hidden-layer feedforward neural network. However, it has feedback connections from the outputs of the hidden-layer neurons to the input of the network. This network is similar to an architecture proposed by Jordan [47]. The SRN was originally developed to learn time-varying patterns or temporal sequences, specifically, character strings. The basic SRN architecture is shown in Figure 5.12. The upper portion of the network in Figure 5.12 contains the *context units*. The function of these units is to replicate the hidden-layer output signals at the *previous* time step.

The question is, What is the purpose of these context units? The answer is actually simple. The purpose of the context units is to deal with input pattern dissonance. In other words, pattern conflicts can possibly occur, resulting in

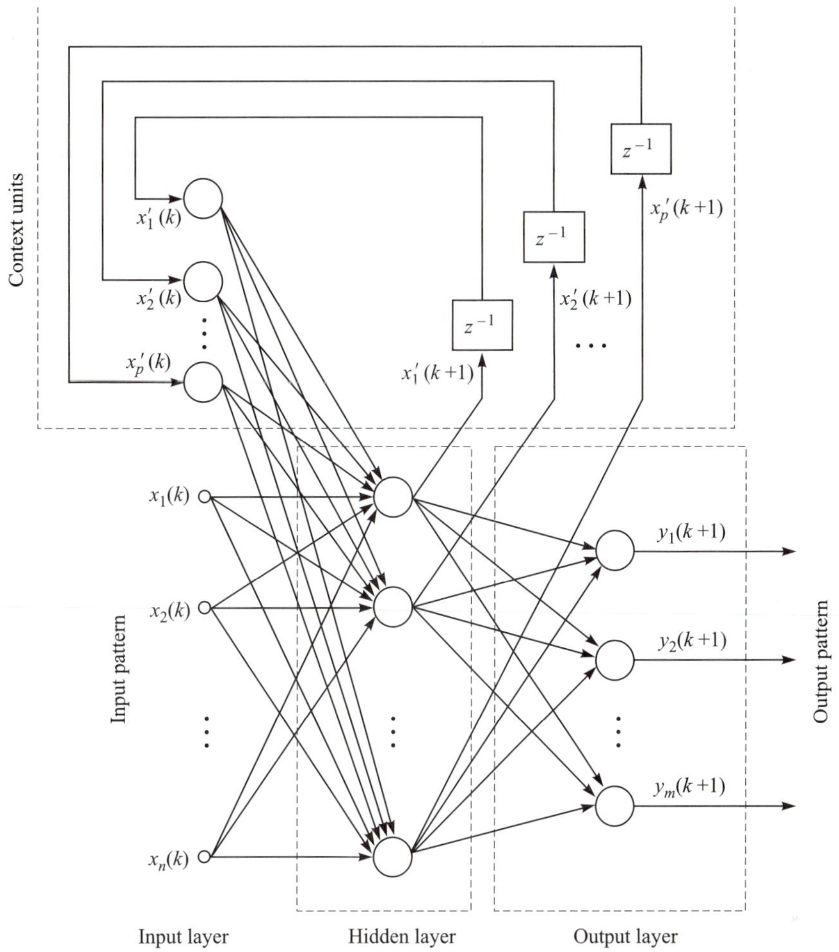

Figure 5.12 The SRN or Elman neural architecture (where z^{-1} is a unit time delay).

multiple outputs produced from a single input pattern. This could result in a perplexing situation for a standard backpropagation network. In the case of the SRN, the input to the network, that is, $\boldsymbol{x}(k)$ shown in Figure 5.12, is augmented with the hidden-layer outputs at the previous time step, that is, $\boldsymbol{x}'(k)$. Therefore, the feedback provided in the SRN, or $\boldsymbol{x}'(k)$, is basically establishing a context for the current input $\boldsymbol{x}(k)$. This can provide a mechanism within the network to discriminate between patterns occurring at different times that are essentially identical. The weights of the context units are fixed; however, the other network weights can be adjusted in a supervised training mode by using the error backpropagation algorithm with momentum (cf. Sect. 3.3.3).

EXAMPLE 5.4. We want to design an SRN to detect the peak amplitudes of infrasonic signals (cf. Example 10.8). Figure 5.13 shows two simulated infrasonic signals (sine waves) that could be the steady-state infrasound signatures of two volcano eruptions. Both of the signals have a fundamental frequency of 50 mHz; however, the first signal has a peak amplitude of 2.5 μbars, and the second signal has a peak amplitude of 1 μbar. The sampling frequency is assumed to be $f_s = 1$ Hz. The samples from both simulated infrasonic events are concatenated to form an input vector of length 204. Specifically, the input samples from each of the two signals are alternated (two replications each); that is, in the MATLAB vector designated to contain the input samples is **INPUTS**, defined as

```
INPUTS=[signal1 signal2 signal1 signal2];
```

where **INPUTS** $\in \Re^{1\times 204}$, **signal1** contains the samples from the first signal in Figure 5.13, and **signal2** contains the samples from the second signal in Figure 5.13. The target values associated with each of the signals reflect their respective *peak* amplitude. Therefore, for every sample of **signal1** (a total of 51), the associated target value is 2.5 (this is the vector **T1** used in MATLAB), and for **signal2**, the associated 51 values are all 1 (the vector **T2** in MATLAB). These are also concatenated to correspondingly match up with the values in the vector **INPUTS**; that is, in MATLAB the vector **TARGETS** contains the target values as

```
T1=2.5*ones(1,51);
T2=ones(1,51);
TARGETS=[T1 T2 T1 T2];
```

where **TARGETS** $\in \Re^{1\times 204}$.

The MATLAB neural network toolbox [10] is used to carry out the simulation. The main objective is to have the SRN (Elman network) perform as a peak-amplitude detector, that is, learn to distinguish between the two peak amplitudes using the training data {**INPUTS**, **TARGETS**}. In MATLAB the SRN (they refer to this network as the Elman network) has **tansig** activation functions (this is the hyperbolic tangent function f_{hts} discussed in Sect. 2.3) at the hidden-layer

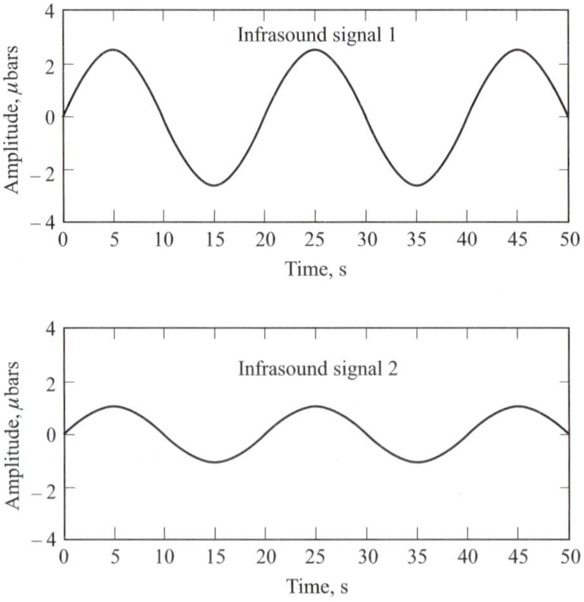

Figure 5.13 Two simulated infrasound signals, both with a frequency of 50 mHz.

neurons, and **purelin** (linear) activation functions at the output neurons. For this problem, the Elman neural network has a single input, a single output neuron, with 15 hidden (recurrent) neurons. The network processes the inputs sequentially; that is, the "time" samples in the vectors **INPUTS** and **TARGETS** are processed sequentially.

The initialization of the Elman network in MATLAB is carried out by using the function **initelm**. This function accepts as inputs (1) the input data (**INPUTS**), (2) the number of hidden (recurrent) **tansig** neurons, and (3) the target outputs (**TARGETS**). For the third argument, there is an option of giving the number of output **purelin** neurons instead of the target outputs to initialize the network. The outputs of the **initelm** function are the two initial weight matrices and biases for the two layers of the network. The MATLAB function **trainelm** is used to train the Elman network. This function uses the backpropagation learning rule with momentum (cf. Sect. 3.3.3) for multilayer feedforward networks. The function **trainelm** accepts as inputs the initial weight matrices and biases from the **initelm** function along with the inputs, target values, and an optional training parameter vector **TP**. After training, the outputs of the MATLAB function **trainelm** are the final network weight matrices and biases. The following are MATLAB commands used to initialize and train the network:

```
[W1,B1,W2,B2]=initelm(INPUTS,15,TARGETS);
TP=[10 5000 2 0.001 1.05 0.7 0.95 1.04];
[W1,B1,W2,B2]=trainelm(W1,B1,W2,B2,INPUTS,TARGETS,TP);
```

In the training parameter vector **TP**, all the values shown are default values except the first value of 10 (number of epochs between display), the second value of 5,000 (maximum number of epochs to train), and the third value of 2 (sum-squared error goal).

After 2,609 training epochs the error goal was satisfied, as shown in Figure 5.14. Figure 5.15 shows the results after training the Elman network, where the actual target values are plotted along with the estimates produced by the network. The

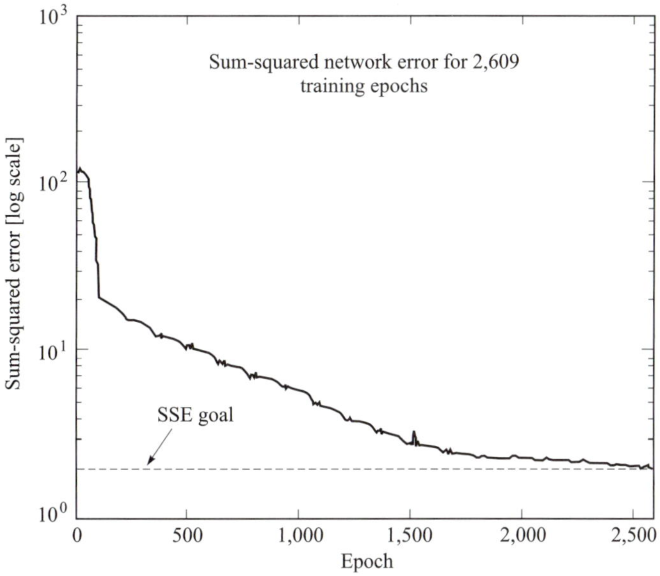

Figure 5.14 Sum-squared error of the network during training.

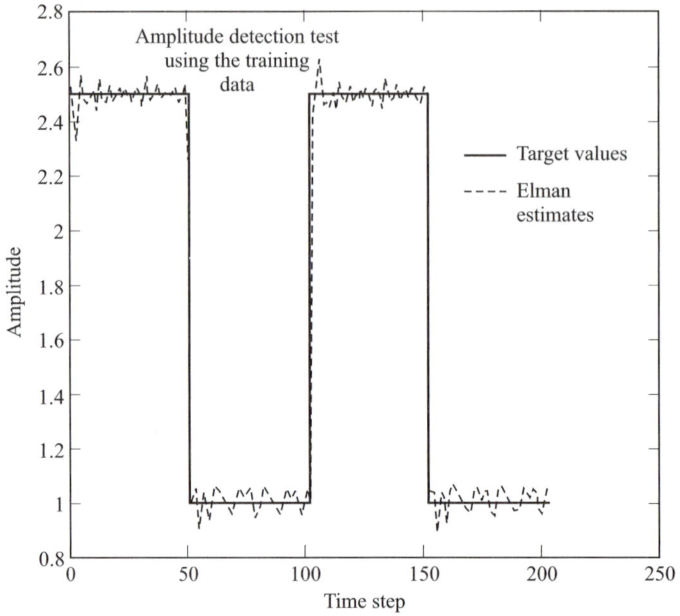

Figure 5.15 Peak-amplitude detection results for Example 5.4 using the training data.

MATLAB function **simuelm** is used to generate the outputs of the network after training using the input data in **INPUTS**, that is,

```
OUTPUTS=simuelm(INPUTS,W1,B1,W2,B2);
```

As can be seen from Figure 5.15, the Elman network performs rather well in detecting the proper peak amplitudes of the two signals.

However, an important property of any neural network is its ability to generalize. Suppose we present two additional infrasonic signals to the trained network, with the same frequency but with different peak amplitudes. These are two signals that the neural network has not seen before. The first signal has a peak amplitude of 0.75, and the second signal's peak amplitude is 1.75. Using again the MATLAB function **simuelm**, the outputs of the network can be generated using the "new" test input data. These results are shown in Figure 5.16. From the figure it is obvious that the network has not generalized. To improve the generalization capability of the network, more training inputs would be required to further train the network, that is, more training signals with different peak amplitudes.

5.8 TIME-DELAY NEURAL NETWORKS

The *time-delay neural network* uses time delays to perform temporal processing. It is actually a feedforward neural network, with the inputs to the network successively delayed in time. Figure 5.17 shows a single neuron with

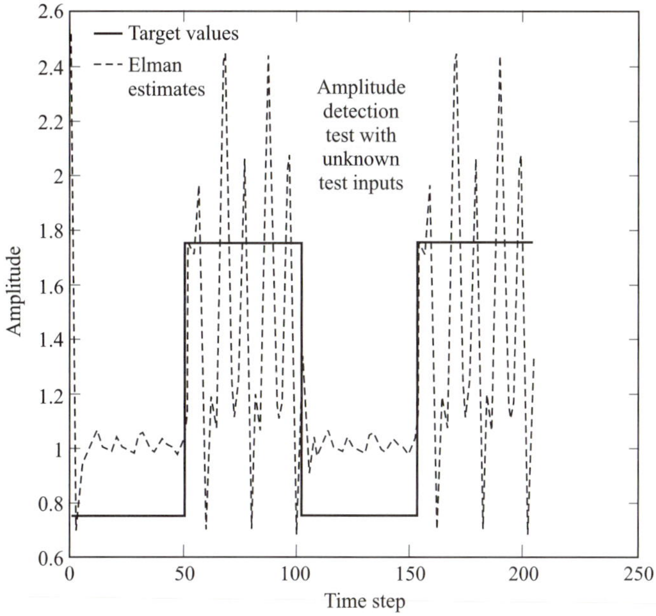

Figure 5.16 Peak-amplitude detection results for Example 5.4 using the test data.

multiple delays for each element of the input vector. This is a neuron "building block" for feedforward TDNNs (without loss of generality, a possible bias term has been omitted in the model). As the input vector $\boldsymbol{x}(k)$ evolves in time (k is the discrete-time index), the past p values are accounted for in the neuron. A temporal sequence for the input is established and can be expressed as

$$X = \{ \boldsymbol{x}(0), \boldsymbol{x}(1), \cdots, \boldsymbol{x}(m) \} \tag{5.50}$$

Thus the matrix $\boldsymbol{X}$ consists of the time-sensitive input (column) vector sequence to the network, that is, $\boldsymbol{x}(k)$, for $k = 0, 1, \ldots, m$. Within the structure of the neuron the past values of the input are established by way of the time delays shown in Figure 5.17 (for $p < m$). The total number of weights required for the single neuron is $(p+1)n$.

The single-neuron model can be extended to a multilayer structure. As previously mentioned, the structure of the TDNN incorporates a feedforward multilayer network with time delays. The typical structure of the TDNN is a layered architecture with only delays at the input of the network; however, it is possible to incorporate delays between the layers. The TDNN can be trained using a modified version of the standard backpropagation algorithm (cf. Sect. 3.3.1). Basically, if it is assumed that the training set consists of numerous shifted inputs corresponding to the target (output) data; then during training the network can learn the salient features of the input patterns. Each of the hidden units observes not only the inputs but also the p delayed versions of the inputs. Figure 5.18 shows an example of a three-layer TDNN used for speech recognition, specifically, for the recognition of phonemes [36]. This network is an example of a TDNN that uses time delays at the hidden layers as well as at

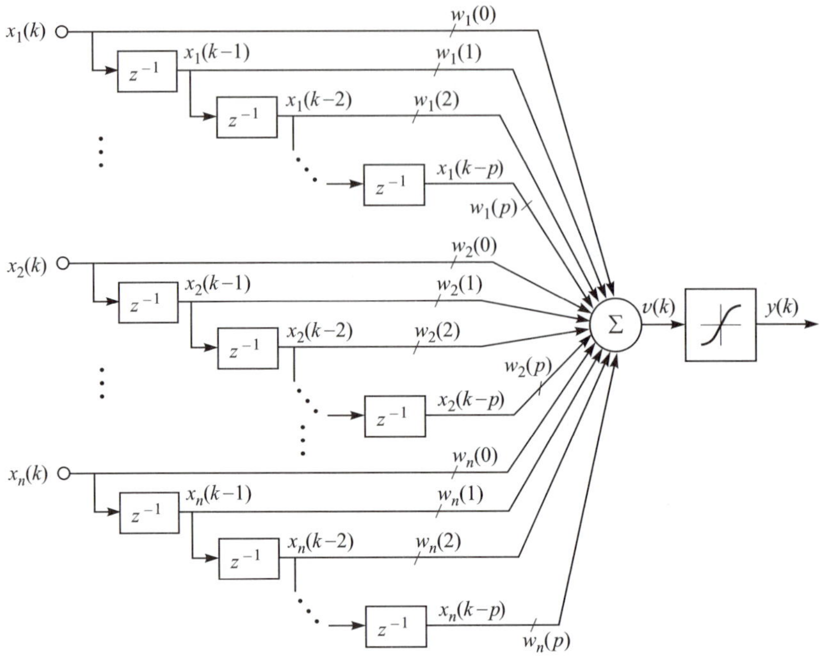

Figure 5.17 Basic TDNN neuron with n inputs and p delays for each input (k is the discrete-time index).

the input to the network. In Figure 5.18, the inputs are 16 melscale coefficients computed from the power spectrum of the signals (cf. Sect. 10.10.2).

5.9 DISTRIBUTED TIME-LAGGED FEEDFORWARD NEURAL NETWORKS

A *distributed time-lagged feedforward neural network* (DTLFNN) is distributed in the sense that the element of time is *distributed* throughout the entire network. The basic building block of the DTLFNN is the simple nonlinear neuron filter shown in Figure 5.19. Interestingly, this is actually the same structure for the TDNN neuron shown in Figure 5.17. For the multiple-input neuron in Figure 5.17, each input to the neuron, that is, $x_i(k)$, for $i = 1, 2, \ldots, n$, is *filtered* with a *finite impulse response* filter (since we are considering discrete-time processing) [48]. Specifically, referring to Figure 5.17, the output of the linear combiner is given by

$$v(k) = v_1(k) + v_2(k) + \cdots + v_n(k) = \sum_{i=1}^{n} v_i(k) \tag{5.51}$$

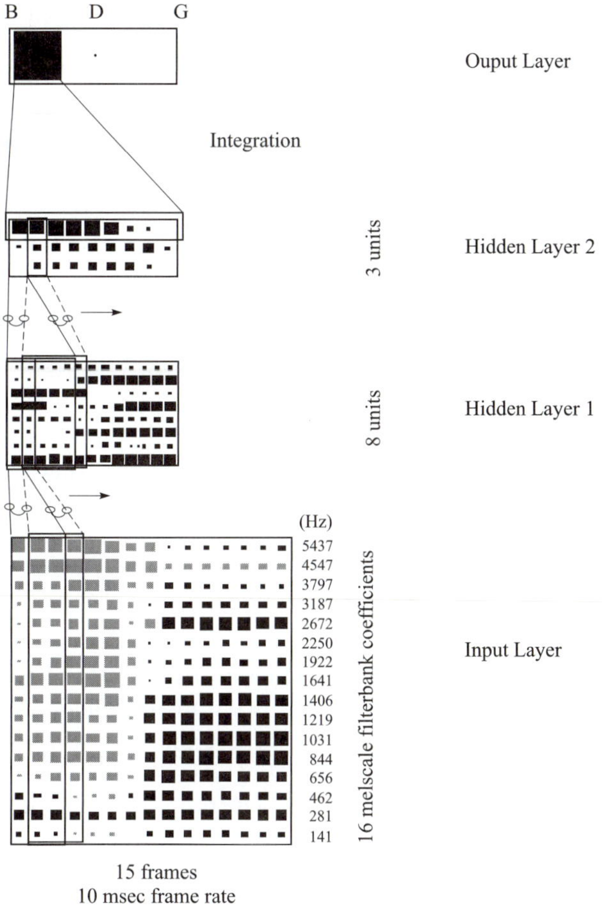

Figure 5.18 (Three) Layered TDNN architecture used by Waibel et al. [36] for the recognition of phonemes. (*Source*: Reproduced with permission from Waibel et al. [36] p. 330, © 1989 IEEE.)

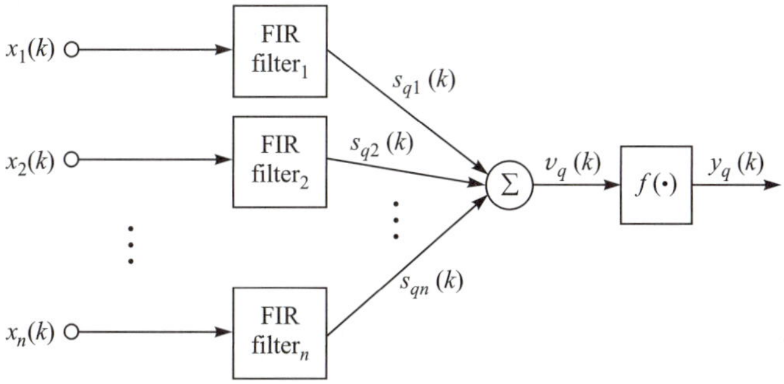

Figure 5.19 Nonlinear neuron filter.

where

$$v_i(k) = w_i(0)x_i(k) + w_i(1)x_i(k-1) + w_i(2)x_i(k-2) + \cdots + w_i(p)x_i(k-p)$$
$$= \sum_{r=0}^{p} w_i(r)x_i(k-r) \tag{5.52}$$

for $i = 1, 2, \ldots, n$. The sum in (5.52) is referred to as a convolution sum. In the z domain we can write from (5.52)

$$V_i(z) = w_i(0)X_i(z) + w_i(1)z^{-1}X_i(z) + w_i(2)z^{-2}X_i(z) + \cdots + w_i(p)z^{-p}X_i(z) \tag{5.53}$$

or as a transfer function we have

$$H_i(z) = \frac{V_i(z)}{X_i(z)} = w_i(0) + w_i(1)z^{-1} + w_i(2)z^{-2} + \cdots + w_i(p)z^{-p} \tag{5.54}$$

or

$$H_i(z) = \frac{V_i(z)}{X_i(z)} = \frac{w_i(0)z^p + w_i(1)z^{p-1} + w_i(2)z^{p-2} + \cdots + w_i(p)}{z^p} \tag{5.55}$$

In (5.54) or (5.55), $H_i(z)$ is the transfer function of an FIR filter. Therefore, each of the FIR filter blocks in Figure 5.19 is a transfer function $H_i(z)$ for the specific FIR filter given in (5.54) or (5.55), where the weights are actually the coefficients of the numerator polynomial that dictate where the *zeros* of the transfer function lie in the complex z plane. The output of the linear combiner in Figure 5.19 for the qth neuron of the network is given by

$$v_q(k) = s_{q1}(k) + s_{q2}(k) + \cdots + s_{qn}(k) = \sum_{i=1}^{n} s_{qi}(k) \tag{5.56}$$

Comparing (5.56) and (5.51), for the qth neuron, we see the individual filtered inputs are given by $s_{qi}(k) = v_{qi}(k)$. Specifically, each filtered input in Figure 5.19 expressed in the time domain is given by the convolution sum

$$s_{ji}(k) = \sum_{r=0}^{p} w_{ji}(r)x_i(k-r) \tag{5.57}$$

for $i = 1, 2, \ldots, n$; $j = 1, 2, \ldots, q$; and $r = 0, 1, \cdots, p$, where p is the total number of delays. Therefore, the output of the jth neuron in the network is given by

$$y_j(k) = f[v_j(k)] = f\left[\sum_{i=1}^{n} s_{ji}(k)\right] = f\left[\sum_{i=1}^{n}\sum_{r=0}^{p} w_{ji}(r)x_i(k-r)\right] \tag{5.58}$$

A DTLFNN consists of layers of neurons where the neurons have the form shown in Figure 5.19 and the neuron outputs have the form given in (5.58).

A DTLFNN is trained using a supervised learning algorithm, more specifically, a temporal backpropagation algorithm [38]. This training algorithm is a temporal generalization of the standard backpropagation training algorithm. From [38], the generalized temporal backpropagation algorithm can be summarized as follows:

Update the appropriate network weight (column) vector according to

$$\boldsymbol{w}_{ji}^{(s)}(k+1) = \boldsymbol{w}_{ji}^{(s)}(k) - \mu^{(s)}\delta_j^{(s+1)}(k)\boldsymbol{x}_i^{(s)}(k) \tag{5.59}$$

where

$$\delta_j^{(s)}(k) = \begin{cases} -e_j(k)f'[v_j(k)] & \text{for neuron } j \text{ in output layer} \\ f'[v_j(k)]\displaystyle\sum_{h=1}^{q_{s+1}} \boldsymbol{\Delta}_h^{s+1}(k)\boldsymbol{w}_{hj}^{s+1} & \text{for neuron } j \text{ in } s \text{ hidden layer} \end{cases} \tag{5.60}$$

In (5.60) $e_j(k)$ is the instantaneous error, and

$$\boldsymbol{\Delta}_h^s(k) = \begin{bmatrix} \delta_h^s(k) & \delta_h^s(k+1) & \cdots & \delta_h^s(k+p) \end{bmatrix}$$

PROBLEMS

5.1. Consider a simple Hopfield neural network consisting of two neurons each with a zero threshold. The synaptic weight matrix of the network is given by

$$\boldsymbol{W} = \begin{bmatrix} 0 & -1 \\ -1 & 0 \end{bmatrix}$$

There are four possible states that the network can be in, given by

$$\boldsymbol{x}_1 = \begin{bmatrix} 1 \\ 1 \end{bmatrix} \quad \boldsymbol{x}_2 = \begin{bmatrix} -1 \\ 1 \end{bmatrix} \quad \boldsymbol{x}_3 = \begin{bmatrix} -1 \\ -1 \end{bmatrix} \quad \boldsymbol{x}_4 = \begin{bmatrix} 1 \\ -1 \end{bmatrix}$$

These are shown in Figure 5.20, where $\boldsymbol{x}_2$ and $\boldsymbol{x}_4$ (the black "dots") are stable equilibrium states, and $\boldsymbol{x}_1$ and $\boldsymbol{x}_3$ are not (the white "dots").

(*a*) To show this, present the four inputs above to the Hopfield network and observe the outputs; that is, use the asynchronous update expression given in (5.10). What can you conclude?

(*b*) Using the energy function in (5.14), can the same conclusions be drawn as in part (a)?

5.2. Assume the following five bipolar prototype memories:

$$\boldsymbol{\phi}_1 = [-1, -1, 1, 1, 1, 1, 1, 1, 1, -1, -1, 1, 1, -1, 1, 1]^T$$

$$\boldsymbol{\phi}_2 = [-1, 1, 1, -1, -1, -1, 1, -1, 1, 1, 1, 1, -1, 1, 1, 1]^T$$

$$\boldsymbol{\phi}_3 = [1, 1, 1, -1, 1, 1, 1, 1, 1, 1, 1, -1, -1, -1, -1, -1]^T$$

$$\boldsymbol{\phi}_4 = [-1, 1, 1, -1, -1, 1, 1, 1, -1, -1, -1, 1, 1, -1, 1, -1]^T$$

$$\boldsymbol{\phi}_5 = [1, -1, -1, 1, 1, 1, 1, 1, 1, 1, 1, -1, -1, -1, -1, 1]^T$$

(*a*) With these five prototype memories, build a 16-neuron Hopfield content-addressable memory.

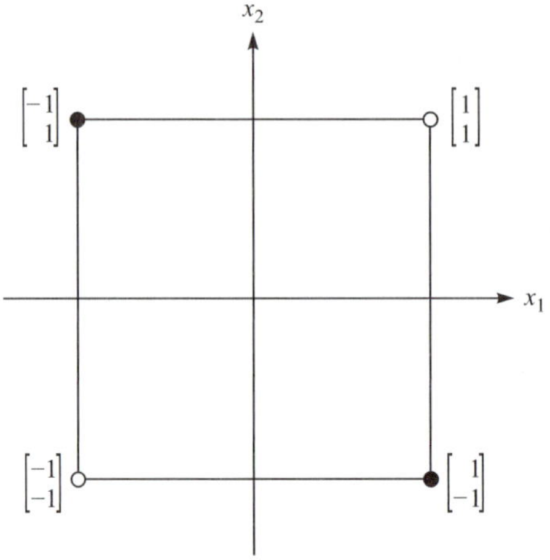

Figure 5.20 States for the two-neuron Hopfield network (Prob. 5.1).

(*b*) Using each of the prototype memories $\{\boldsymbol{\phi}_1, \boldsymbol{\phi}_2, \boldsymbol{\phi}_3, \boldsymbol{\phi}_4, \boldsymbol{\phi}_5\}$ as inputs to the Hopfield network designed in part (*a*), that is, $\boldsymbol{x}_1 = \boldsymbol{\phi}_1, \boldsymbol{x}_2 = \boldsymbol{\phi}_2, \ldots$, show that the prototype vectors can be recalled in one time step by using the asynchronous update expression in (5.10).

(*c*) Figure 5.21 shows a MATLAB function that can generate a noisy version of the input vector according to a user-supplied value for the bit error rate (BER). The MATLAB function **noise** shown in Figure 5.21 yields at the output not only the corrupted version of the input vector, but also the indices in the original input vector where the elements have been "toggled." Using each of the inputs, that is, $\boldsymbol{x}_1, \boldsymbol{x}_2, \ldots$, experimentally determine a reasonable value for the BER such that the corrupted prototypes (1) can always be recovered, (2) can be recovered on average 95 percent of the time, (3) can be recovered on average 90 percent of the time.

5.3. This problem is a variation of Example 5.2. The same five characters in Figure 5.5 are used as prototypes for developing a Hopfield network, except the "period" is replaced by the letter M shown in Figure 5.22.

(*a*) Repeat the steps carried out in Example 5.2, using the letter M in place of the period to build the 144-neuron Hopfield neural network. Display the synaptic weight matrix as a gray-scale image similar to the one shown in Figure 5.6.

(*b*) Show that each character can be recalled in one time step by using the asynchronous update expression in (5.10) with the synaptic weight matrix developed from part (*a*). Each input to the network will be the *vectorized* version of the character image; however, display the output of the network as the "reshaped" character image. This can be carried out by using the MATLAB function **reshape.**

(*c*) Introduce 30 percent errors in each of the character images. That is, on average 30 percent of the pixels in the character images will have a polarity change, that is, changing a black pixel to a white pixel, or vice versa. Use the MATLAB function **noise2** shown in Figure 5.23 to "corrupt" each image. *Note*: This MATLAB function is similar to the **errors** function in Problem 3.9 (Figure 3.18) and the **noise** function in Problem 5.2 (Figure

```
function [XC,I] = noise (X,BER)
% [XC,I] = NOISE(X,BER) generates a noisy
% version of the input vector X, i.e., XC.
% INPUTS:
% X:     Uncorrupted input vector, can only have
%        bipolar values, i.e., [-1,1].
% BER:   bit-error-rate given in percentage,
%        for example, BER=20 means that on the
%        average 20% of the elements in the
%        vector will be ''toggled'' (or the polarity
%        changed) .
% OUTPUTS:
% XC:    Corrupted version of the input vector X
% I:     Indices in the original input vector X
%        where the elements were ''toggled.''
%

N=rand(size(X));
XC=X;
T=1-BER/100;
  for i=1:length(X)
        if N(i)>=T
          if X(i)==1
             XC(i)=-1;
          else
             XC(i)=+1;
          end
        end
      end
I=find([X-XC]);
```

Figure 5.21 MATLAB function **noise** that generates a noisy version of the input vector with a defined bit error rate.

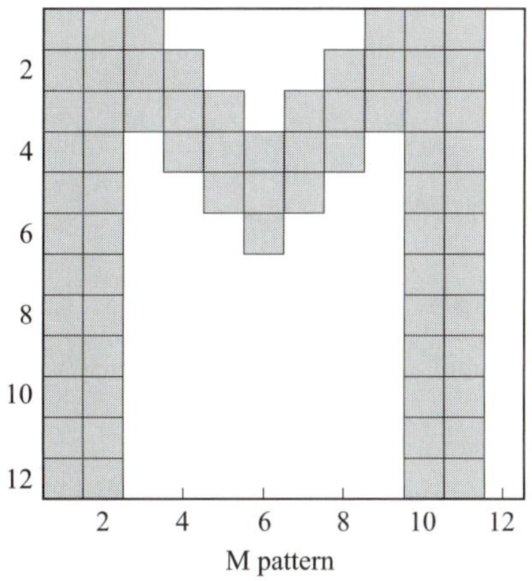

Figure 5.22 The letter M to replace the "period" character in Figure 5.5 in Example 5.2.

```
function [XC,I,J] = noise2(X,BER)
% [XC,I,J] = NOISE2(X,BER) generates a noisy
% version of the input matrix X, i.e., XC.
% INPUTS:
% X:     Uncorrupted input matrix, can only have
%        bipolar values, i.e., [-1,1].
% BER:   bit-error-rate given in percentage,
%        for example, BER=20 means that on the
%        average 20% of the elements in the
%        matrix will be ''toggled'' (or the polarity
%        changed).
% OUTPUTS:
% XC:    Corrupted version of the input matrix X
% I:     vector of row indices in the original input
%        matrix X where the elements were ''toggled.''
% J:     vector of column indices in the original input
%        matrix X where the elements were ''toggled.''

[nr,nc]=size(X);
N=rand(size(X));
XC=X;
T=1-BER/100;
       for i=1:nr
           for j=1:nc
               if N(i,j)>=T
                 if X(i,j)==1
                    XC(i,j)=-1;
                 else
                    XC(i,j)=+1;
                 end
              end
           end
       end
[I,J]=find([X-XC]);
```

Figure 5.23 MATLAB function **noise2** that generates a noisy version of the input matrix with a defined bit error rate.

5.21). In fact, the function **noise2** is a two-dimensional version of **noise** in Problem 5.2. Using the asynchronous update expression in (5.10) and the corrupted inputs generated using **noise2**, how many iterations does it take to recall the proper character stored in the Hopfield network memory?

5.4. Consider a Hopfield network consisting of five neurons designed to store two prototype memories given as:

$$\boldsymbol{\phi}_1 = [1, 1, -1, 1, 1]^T \qquad \text{and} \qquad \boldsymbol{\phi}_2 = [1, -1, 1, -1, 1]^T$$

(*a*) Build the Hopfield network using $\boldsymbol{\phi}_1$ and $\boldsymbol{\phi}_2$ and observe the synaptic weight matrix.

(*b*) Use the asynchronous update expression in (5.10), and show that the two prototype memories can be recovered from memory in one iteration each.

(*c*) Now consider the two vectors

$$x_3 = [1, 1, -1, -1, -1]^T \quad \text{and} \quad x_4 = [-1, 1, -1, 1, -1]^T$$

A vector pattern that was not used to build the Hopfield network, but nonetheless is still stored in the content-addressable memory, can be considered a fundamental memory of the network. Can either (or both) of these two vectors be implicitly stored in the Hopfield memory developed in part (*a*)? Give the details of your analysis to justify your conclusions.

5.5. Consider an optimization problem defined as follows: Find the minimum of the function

$$f(x_1, x_2) = (x_1 - 30)^2 + (x_2 - 20)^2 + 40 \sin^2(x_1 x_2) + \exp\left[-\frac{(x_1 - 10)^2 + (x_2 - 10)^2}{100}\right]$$

in a rectangle defined by $-40 \le x_1 \le 40$ and $-40 \le x_2 \le 40$ subject to a constraint that both x_1 and x_2 have to be integer numbers. Use the simulated annealing algorithm described in Section 5.4 to accomplish the optimization task. The function $f(x_1, x_2)$ can be thought of as an energy function, and since the optimization calls for integer values for the variables x_1 and x_2, step 3 of the algorithm needs to be modified as

$$\boldsymbol{x}_p = \boldsymbol{x} + \Delta \boldsymbol{x}$$

where $\Delta \boldsymbol{x}$ is a randomly selected vector from the set

$$S = \left\{ \begin{bmatrix} 1 \\ 0 \end{bmatrix}, \begin{bmatrix} 1 \\ 1 \end{bmatrix}, \begin{bmatrix} 0 \\ 1 \end{bmatrix}, \begin{bmatrix} -1 \\ 1 \end{bmatrix}, \begin{bmatrix} -1 \\ 0 \end{bmatrix}, \begin{bmatrix} -1 \\ -1 \end{bmatrix}, \begin{bmatrix} 0 \\ -1 \end{bmatrix}, \begin{bmatrix} 1 \\ -1 \end{bmatrix} \right\}$$

Use a suboptimal "cooling schedule" suggested in (5.33) with the decremental factor set to $\alpha = 0.99$. Perform the optimization for different starting points, and record the number of iterations required for the algorithm to find the solution.

5.6. Simulated annealing is frequently used to solve multidimensional combinatorial optimization problems. Typically, in these problems the goal of the optimization is to determine an n-dimensional vector which minimizes some energy function $\mathscr{E}(\boldsymbol{x})$ under the fundamental constraint

$$x_k = 1 \qquad \text{or} \qquad x_k = 0$$

that is,

$$x_k \in \{0, 1\} \qquad \text{where } k = 1, 2, \ldots, n$$

Very often some more constraints can be added to the optimization problem. Since all elements of the vector $\boldsymbol{x}$ are either 1 or 0, it is referred to as a *binary vector*. A set consisting of all binary vectors of size n is referred to as the *configuration space*, and the problem of combinatorial optimization can be regarded as a search through the configuration space for the *point* (i.e., vector $\boldsymbol{x}^*$) where the energy function reaches the global minimum.

As an example of a simple combinatorial optimization problem, consider minimization of the energy function

$$\mathscr{E}(\boldsymbol{x}) = 12x_1 + 14x_2 + 22x_3 + 38x_4 + 15x_5 + 13x_6 + 17x_7 + 28x_8 + 4x_9$$

subject to

$$x_i \in \{0, 1\} \qquad i = 1, 2, \cdots, 9 \tag{c.1}$$

$$x_1 + x_2 + 2x_3 + 6x_4 + 7x_5 + 8x_6 + 9x_7 + 3x_8 + 11x_9 \le 35 \tag{c.2}$$

Write a computer program that uses the method of simulated annealing to solve the above optimization problem. The algorithm should be as follows:

(*a*) Randomly initialize the starting point. The starting point must satisfy constraints (c.1) and (c.2).
(*b*) Set the initial temperature and select the cooling schedule.
(*c*) Perform the transition in configuration space by randomly selecting one of the vector components and changing its value from 0 to 1 or from 1 to 0.
(*d*) Accept or reject the proposed change on the basis of Metropolis criterion (cf. Sect. 5.4).
(*e*) Continue the configuration space search until
- The global minimum is reached or
- The temperature falls below some small predetermined value, and no further transitions in the configuration space are accepted or
- The maximum number of iterations is exceeded.

5.7. Most neural network training algorithms are based on *local search* techniques (steepest descent, conjugate gradient, Newton's method, etc.). All the algorithms based on a local search have a tendency to converge to a local minimum of the error performance surface and consequently provide a suboptimal problem solution. One of the techniques used to help network training algorithms to escape from a local minimum is based on adding artificial high-frequency noise to the energy cost function. For example, the additive noise term can be added to the energy function according to

$$\tilde{\mathscr{E}}(\boldsymbol{w}, \boldsymbol{N}) = \mathscr{E}(\boldsymbol{w}) + T(k)\boldsymbol{N}^T \boldsymbol{x}$$

where $\tilde{\mathscr{E}}(\boldsymbol{w}, \boldsymbol{N})$ is a perturbed energy function, $\boldsymbol{w}$ is the vector of network weights, $\boldsymbol{N}$ is a noise vector where each component is generated as a random white noise process, and $T(k)$ is a parameter that determines the magnitude of the energy function perturbations. Commonly the magnitude of $T(k)$ is decreased over the course of training.

(*a*) Show that a steepest descent learning rule for adjusting the weights that is based on the perturbed cost function has the form

$$\boldsymbol{w}(k+1) = \boldsymbol{w}(k) + \mu\left[\frac{\partial \mathscr{E}(\boldsymbol{w})}{\partial \boldsymbol{w}} + T(k)\boldsymbol{N}\right]$$

(b) Consider the problem of finding the global minimum of the energy function defined by

$$\mathscr{E}(w) = \cos^2(3w) + 0.4|w+1| \qquad |w| < 10$$

The graph of the function is presented in Figure 5.24. As we can see, the function has several local minima. Apply the *perturbed energy function* technique along with steepest descent to find the global minimum of the given function.

5.8. Design a simple recurrent network, that is, an Elman network, that can learn to detect the peak amplitudes of three signals. The signals are similar to those in Example 5.4. Specifically, these are three sinusoidal signals, all with a frequency of 30 mHz (simulated infrasound signals), sampled at 1 Hz over the time period from 0 to 100 s. The first signal has a peak amplitude of 5, the second signal has a peak amplitude of 2.5, and the third signal has a peak amplitude of 1. The

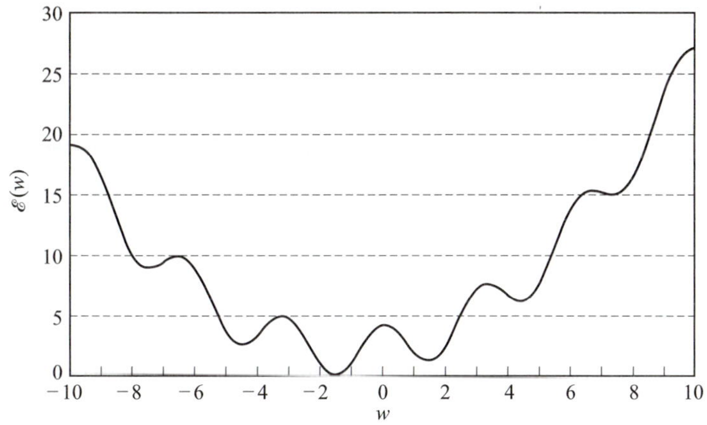

Figure 5.24 Graph of the energy function to be minimized in Problem 5.7.

following MATLAB commands will generate the appropriate signals and associated target values:

Training signals:

```
t=[0:100];
signal1=5*sin(2*pi*0.03*t);
signal2=2.5*sin(2*pi*0.03*t);
signal3=sin(2*pi*0.03*t);
```

Target values:

```
T1=5*ones(1,101);
T2=2.5*ones(1,101);
T3=ones(1,101);
```

As in Example 5.4, replicate the signals twice so the SRN is trained each epoch on sequences formed by repeating each waveform twice (obviously the associated target values are appropriately repeated).

(*a*) Use the MATLAB neural network toolbox [10] functions **initelm**, **trainelm**, and **simuelm** to initialize, train, and test the network. Experiment with the number of recurrent (hidden) neurons and the training parameters. How many training epochs are necessary to obtain reasonable results? That is, using the MATLAB function **simuelm** in the neural network toolbox, use the training data as the network inputs and obtain estimates of the signal's peak amplitudes. Plot both the target values and the estimates of the peak amplitudes of the signals on the same graph. What can you conclude?

(*b*) Generate three 30-mHz sinusoidal test signals, and associated target values, similar to the training data, except the peak amplitude of the first signal is 5.5, the peak amplitude of the second signal is 3, and the peak amplitude of the third signal is 1.75. These are three signals that the SRN trained in part (*a*) has not seen before. Use the MATLAB function **simuelm** and present the signals (in replicated form as before) to the trained SRN. Plot both the target values and the estimates of the peak amplitudes of the signals on the same graph. What can you conclude? Has the SRN generalized? If the results are not satisfactory, state what you would do to improve the results.

5.9. Design an SRN to perform frequency discrimination.

(*a*) In MATLAB generate two sinusoidal signals each with unity amplitude; however, the first signal has a frequency of 70 mHz, and the second signal has a frequency of 30 mHz. Both signals are assumed to be sampled at 1 Hz over the period from 0 to 50 s.

(*b*) Choose an appropriate representation for the frequency target values. Using the MATLAB neural network toolbox Elman network functions, train the SRN to a reasonable level of accuracy using 15 recurrent-layer (hidden) neurons. Experiment with the various training parameters. What can you conclude from the results you obtained?

(*c*) Experiment with (1) different numbers of recurrent-layer neurons, (2) different representations for the frequency target values, and (3) the training parameters. Did these results show any improvement over the results you obtained from part (*b*)?

(*d*) After you are satisfied with training of the SRN for frequency estimation, using this network, experiment with unity-amplitude sinusoidal signals with different frequencies, in the range from 20 to 60 mHz. What can you conclude about the network's ability to recognize other frequencies?

REFERENCES

1. J. J. Hopfield, "Neural Networks and Physical Systems with Emergent Collective Computational Abilities," *Proceedings of the National Academy of Sciences, USA*, vol. 79, 1982, pp. 2554–8. Reprinted in 1988, Anderson and Rosenfeld [23], pp. 460-4.
2. J. E. Slotine and W. Li, *Applied Nonlinear Control*, Englewood Cliffs, NJ: Prentice-Hall, 1991.
3. S. Haykin, *Adaptive Filter Theory*, 3rd ed., Upper Saddle River, NJ: Prentice-Hall, 1996.
4. J. Hertz, A. Krogh, and R. G. Palmer, *Introduction to the Theory of Neural Computation*, Redwood City, CA: Addison-Wesley, 1991.
5. W. S. McCulloch and W. Pitts, "A Logical Calculus of the Ideas Immanent in Nervous Activity," *Bulletin of Mathematical Biophysics*, vol. 5, 1943, pp. 115–33. Reprinted in 1988, Anderson and Rosenfeld [23], pp. 18–27.
6. D. J. Amit, *Model Brain Function: The World of Attractor Neural Networks*, New York: Cambridge University Press, 1989.
7. D. J. Amit, H. Gutfreund, and H. Sompolinsky, "Spin Glass Models of Neural Networks," *Physical Review A*, vol. 32, 1985, pp. 1007–18.
8. D. J. Amit, H. Gutfreund, and H. Sompolinsky, "Storing Infinite Numbers of Patterns in a Spin-Glass Model of Neural Networks," *Physical Review Letters*, vol. 55, 1985, pp. 1530–3.
9. J. H. Li, A. N. Michel, and W. Porod, "Analysis and Synthesis of a Class of Neural Networks: Linear Systems Operating on a Closed Hypercube," *IEEE Transactions on Circuits and Systems*, vol. 36, 1989, pp. 1405–22.
10. H. Demuth and M. Beale, *Neural Network Toolbox—For Use with MATLAB*, version 3, Natick, MA: The Mathworks, Inc., 1998.
11. J. J. Hopfield, "Neurons with Graded Response Have Collective Computational Properties like Those of Two-State Neurons," *Proceedings of the National Academy of Sciences*, vol. 81, 1984, pp. 3088–92. Reprinted in 1988, Anderson and Rosenfeld [23], pp. 579–83.

12. C. M. Marcus and R. M. Westervelt, "Dynamics of Iterated-Map Neural Networks," *Physical Review A*, vol. 40, 1989, pp. 501–4.
13. R. P. Lippmann, "An Introduction to Computing with Neural Nets," *IEEE Acoustics, Speech, and Signal Processing Magazine*, April 1987, pp. 4–22.
14. S. Kirkpatrick, C. D. Gelatt, Jr., and M. P. Vecchi, "Optimization by Simulated Annealing," *Science*, vol. 220, 1983, pp. 671–80. Reprinted in 1988, Anderson and Rosenfeld [23], pp. 554–67.
15. R. A. Rutenbar, "Simulated Annealing Algorithms: An Overview," *IEEE Circuits and Devices Magazine*, January 1989, pp. 19–26.
16. P. J. M. van Laarhoven and E. H. L. Aarts, *Simulated Annealing: Theory and Applications*, Boston: Kluwer Academic Publishers, 1988.
17. B. Hajek, "A Tutorial Survey of Theory and Applications of Simulated Annealing," *Procedings of the 24th IEEE Conference on Decision and Control*, vol. 2, 1985, pp. 755–60.
18. R. V. V. Vidal, ed., *Applied Simulated Annealing*, New York: Springer-Verlag, 1988.
19. B. Müller and J. Reinhardt, *Neural Networks: An Introduction*, Berlin: Springer-Verlag, 1990.
20. E. Shrödinger, *Statistical Thermodynamics*, London: Cambridge University Press, 1946.
21. K. Binder, ed., *The Monte Carlo Method in Condensed Matter Physics*, Topics in Applied Physics, vol. 71, New York: Springer-Verlag, 1992.
22. N. Metropolis, A. W. Rosenbluth, M. N. Rosenbluth, A. H. Teller, and E. Teller, "Equation of State Calculations by Fast Computing Machines," *Journal of Chemical Physics*, vol. 21, 1953, pp. 1087–92.
23. J. A. Anderson and E. Rosenfeld, eds., *Neurocomputing: Foundations of Research*, Cambridge, MA: M.I.T. Press, 1988.
24. S. Geman and D. Geman, "Stochastic Relaxation, Gibbs Distributions, and the Baysian Restoration of Images," *IEEE Transactions on Pattern Analysis and Machine Intelligence*, vol. PAMI-6, 1984, pp. 721–41.
25. D. L. Isaacson and R. W. Madsen, *Markov Chains: Theory and Applications*, New York: Wiley, 1976.
26. H. Szu, "Fast Simulated Annealing," in *Neural Networks for Computing*, Snowbird 1986, ed. J. S. Denker, New York: American Institute of Physics, 1986, pp. 420–5.
27. G. E. Hinton and T. J. Sejnowski, " Optimal Perceptual Inference," *Proceedings of the IEEE Computer Society Conference on Computer Vision and Pattern Recognition*, Washington, DC, New York: IEEE, 1983, pp. 448–53.
28. D. H. Ackley, G. E. Hinton, and T. J. Sejnowski, "A Learning Algorithm for Boltzmann Machines," *Cognitive Science*, vol. 9, 1985, pp. 147–69. Reprinted in 1988, Anderson and Rosenfeld [23], pp. 638–49.
29. G. E. Hinton and T. J. Sejnowski, "Learning and Relearning in Boltzmann Machines," in *Parallel Distributed Processing: Explorations in the Microstructure of Cognition*, eds. D. E. Rumelhart and J. L. McClelland, and the PDP Research Group, vol. 1: *Foundations*, Cambridge, MA: M.I.T. Press, 1986, pp. 282–317.
30. G. E. Hinton, T. J. Sejnowski, and D. H. Ackley, "Boltzmann Machines: Constraint Satisfaction Networks that Learn," Technical Report: CMU-CS-84-119, Pittsburgh, PA: Carnegie-Mellon University, May 1984.
31. P. H. Winston, *Artificial Intelligence*, 2nd ed., Reading, MA: Addison-Wesley, 1984.

32. G. E. Hinton, "Relaxation and Its Role in Vision," Ph.D thesis, University of Edinburgh, Scotland, 1977.
33. R. J. Glauber, "Time-Dependent Statistics of the Ising Model," *Journal of Mathematical Physics*, vol. 4, 1963, pp. 294–307.
34. C. Peterson and J. R. Anderson, "A Mean Field Theory Learning Algorithm for Neural Networks," *Complex Systems*, vol. 1, 1987, pp. 995–1019.
35. K. J. Lang and G. E. Hinton, "The Development of the Time-Delay Neural Network Architecture for Speech Recognition," Technical Report CMU-CS-88-152, Pittsburgh, PA: Carnegie-Mellon University, 1988.
36. A. Waibel, T. Hanazama, G. Hinton, K. Shikano, and K. L. Lang, "Phoneme Recognition Using Time Delay Neural Networks," *IEEE Transactions on Acoustics, Speech, and Signal Processing*, vol. ASSP-37, 1989, pp. 328–39.
37. A. Waibel, H. Sawai, and K. Shikano, "Modularity and Scaling in Large Phonemic Neural Networks," *IEEE Transactions on Acoustics, Speech, and Signal Processing*, vol. ASSP-37, 1989, pp. 1888–98.
38. E. A. Wan, "Temporal Backpropagation for FIR Neural Networks," *IEEE International Joint Conference on Neural Networks*, vol. 1, San Diego, CA, 1990, pp. 575–80.
39. E. A. Wan, "Temporal Backpropagation: An Efficient Algorithm for Finite Impulse Response Neural Networks," in *Proceedings of the 1990 Connectionist Models Summer Schools*, eds. D. S. Touretzky, J. L. Elman, T. J. Sejnowski, and G. E. Hinton, San Mateo, CA: Morgan Kaufmann, 1990, pp. 131–40.
40. E. A. Wan, "Time Series Prediction by Using a Connectionist Network with Internal Time Delays," in *Time Series Prediction: Forecasting the Future and Understanding the Past*, eds. A. S. Weigend and N. A. Gershenfeld, Reading, MA: Addison-Wesley, 1994, pp. 195–217.
41. J. L. Elman, "Finding Structure in Time," *Cognitive Science*, vol. 14, 1990, pp. 179–211.
42. R. J. Williams and D. Zipser, "A Learning Algorithm for Continually Running Fully Recurrent Neural Networks," *Neural Computation*, vol. 1, 1989, pp. 270–80.
43. S. Haykin and L. Li, "Nonlinear Adaptive Prediction of Nonstationary Signals," *IEEE Transactions on Signal Processing*, vol. 37, 1995, pp. 526–35.
44. K. S. Narendra and K. Parthasarathy, "Identification and Control of Dynamical Systems Using Neural Networks," *IEEE Transactions on Neural Networks*, vol. 1, 1990, pp. 4–27.
45. A. V. Levin and K. S. Narendra, "Control of Nonlinear Dynamical Systems Using Neural Networks, II: Observability and Identification," Technical Report 9116, New Haven, CT: Center for Systems Science, Yale Univesity, 1992.
46. K. S. Narendra, *Neural Networks for Identification and Control*, NIPS-95, Tutorial Program, Denver, CO, 1995, pp. 1–46.
47. M. I. Jordan, "Serial Order: A Parallel Distributed Processing Approach," Institute for Cognitive Science Report 8604, San Diego: University of California, 1986.
48. J. G. Proakis and D. G. Manolakis, *Digital Signal Processing: Principles, Algorithms, and Applications*, 3rd ed., Upper Saddle River, NJ: Prentice-Hall, 1996.

PART II

Applications of Neurocomputing

CHAPTER 6

Neural Networks for Optimization Problems

6.1 INTRODUCTION

The main goal of this chapter is to demonstrate the use of neural networks for solving several constrained optimization problems. In general, constrained optimization problems assume the minimization of some objective cost function subject to various constraints imposed on the independent variables. Mathematically, the constrained optimization problem can be formulated as follows:

Minimize

$$f(x_1, x_2, \ldots, x_n) \tag{6.1}$$

subject to

$$r_i(x_1, x_2, \ldots, x_n) = 0 \qquad i = 1, 2, \ldots, m \tag{6.2}$$

These types of optimization problems are encountered frequently in various areas of science and engineering including signal processing, regression analysis, statistics, operations research, and many others. Because of their practical importance, they have been studied extensively, and many numerical methods are found and are readily available in the existing literature [1–10]. However, most of these methods require an extensive amount of computations and are not very well suited for the applications that require the real-time or close to real-time optimization results. A neural network approach to the constrained optimization problems offers a somewhat different prospective and shows that even some relatively complex optimization problems can be resolved in real time by using highly parallel computational capabilities of relatively simple neural network architectures.

In this chapter we present several neural network algorithms for solving some important classes of constrained optimization problems:

1. Linear programming,
2. Quadratic programming.
3. Nonlinear continuous constrained optimization problems.

6.2 NEURAL NETWORKS FOR LINEAR PROGRAMMING PROBLEMS

Linear programming (LP) is the simplest form of the constrained optimization problem. It assumes that both the objective function and the constraint equations are expressed as linear combinations of independent variables. Referring to (6.1) and (6.2), we can rewrite the equations of the general constrained optimization problem for the LP case as follows:

Minimize

$$f(x_1, x_2, \ldots, x_n) = \sum_{i=1}^{n} c_i x_i \tag{6.3}$$

subject to

$$a_{11}x_1 + a_{12}x_2 + \cdots + a_{1n}x_n = b_1 \tag{6.4}$$

$$a_{21}x_1 + a_{22}x_2 + \cdots + a_{2n}x_n = b_2 \tag{6.5}$$

$$\vdots$$

$$a_{m1}x_1 + a_{m2}x_2 + \cdots + a_{mn}x_n = b_m \tag{6.6}$$

and

$$x_1 \geq 0,\ x_2 \geq 0,\ \ldots,\ x_n \geq 0 \tag{6.7}$$

where $m < n$; a_{ij}, b_i, and c_i represent real constants; and x_i are independent variables whose values are to be determined so that the objective function is minimized. Because there are a variety of common notations, the LP problem may be stated in many different forms. The form given in (6.3) through (6.7) is commonly referred to as the *standard form* of the LP problem. In a more compact vector-matrix notation, the standard form can be rewritten as follows:

Minimize

$$f(\boldsymbol{x}) = \boldsymbol{c}^T \boldsymbol{x} \tag{6.8}$$

subject to

$$\boldsymbol{A}\boldsymbol{x} = \boldsymbol{b} \tag{6.9}$$

and

$$\boldsymbol{x} > 0 \tag{6.10}$$

where $\boldsymbol{x}, \boldsymbol{c} \in \Re^{n\times 1}$, $\boldsymbol{A} \in \Re^{m\times n}$, and $\boldsymbol{b} \in \Re^{m\times 1}$.

When formulating the LP problem, we often find that the constraints can be represented as a mixture of linear equations and inequalities. In addition, the independent variables may not be required to satisfy (6.10). However, it can be shown that regardless of its initial formulation, every LP problem can be converted to the standard form. The conversion to the standard form can be carried out as follows:

1. Maximization of the objective function $f(\boldsymbol{x}) = \boldsymbol{c}^T\boldsymbol{x}$ can be substituted by the minimization of $f(\boldsymbol{x}) = -\boldsymbol{c}^T\boldsymbol{x}$.
2. Inequality constraints of the form

$$\sum_{j=1}^{n} a_{ij}x_j \leq b_i \tag{6.11}$$

   can be rewritten as

$$\sum_{j=1}^{n} a_{ij}x_j + x_{n+1} = b_i \tag{6.12}$$

   where $x_{n+1} > 0$ is a new variable, commonly referred to as the *surplus* variable.
3. Inequality constraints of the form

$$\sum_{j=1}^{n} a_{ij}x_j \geq b_i \tag{6.13}$$

   can be rewritten as

$$\sum_{j=1}^{n} a_{ij}x_j - x_{n+1} = b_i \tag{6.14}$$

   where $x_{n+1} > 0$ is a new variable commonly referred to as the *slack* variable.
4. If the constraint $x_i \geq 0$ does not apply, the variable x_i can be substituted by two new variables so that

$$x_i = x_i^{(1)} - x_i^{(2)} \tag{6.15}$$

   and

$$x_i^{(1)} \geq 0 \qquad x_i^{(2)} \geq 0$$

Using (6.11) through (6.15), we can convert every LP problem to the standard form. However, we can see that each of the conversion equations introduces an additional variable, and therefore increases the dimension of the problem. For that reason, in the neural network approach it may be beneficial to consider the LP problem in its original form rather than to perform the conversion to its standard counterpart.

Solution to the LP problem

Solving the LP problem can be viewed as a search through the vector space for the optimal vector of independent variables $\tilde{x}$. The optimal vector needs to satisfy all the constraints while minimizing the objective function at the same time. Constrained equations given in (6.9) determine m hyperplanes in the n-dimensional vector space of the independent-variable vector. Hyperplanes in (6.9) and nonnegativity constraints in (6.10) form a multidimensional polyhedron which is commonly referred to as the *feasible* region. Every independent variable vector which satisfies (6.9) and (6.10) is said to be a *feasible solution,* and the task of the LP problem is to find a feasible vector that minimizes the objective function. Further analysis of the geometric interpretation of the LP problem [1, 4] reveals that the optimum solution vector always resides in a vertex of the multidimensional polyhedron with at least $n - m$ components equal to zero, and the remaining variables taking nonzero positive values.

In general, the LP problem can have four possible solution types:

1. *Unique solution.* There is only one solution that satisfies all constraints, and the objective function reaches a minimum within the feasible region.
2. *Nonunique solution.* There are several feasible solutions where the objective function reaches a minimum.
3. *An unbounded solution.* The objective function is not bounded in the feasible region and it approaches $-\infty$.
4. *No feasible solution.* Constraints provided in (6.9) and (6.10) are too restrictive, and the set of feasible solutions is empty.

Although theoretically valid, cases 3 and 4 appear rarely in engineering and scientific applications. Furthermore, they can be easily detected, and in the further consideration of the LP problem we will assume that it is formulated in such a way that there exists at least one feasible solution.

Dual form of the LP problem

The LP problem formulated in (6.8) through (6.10) is commonly referred to as the *primal* LP problem. To every primal problem we can associate another LP problem called the *dual* LP problem. The dual LP problem has the following form:

Maximize

$$g(y) = \boldsymbol{b}^T \boldsymbol{y} \tag{6.16}$$

subject to

$$\boldsymbol{A}^T \boldsymbol{y} \leq \boldsymbol{c} \tag{6.17}$$

where $\boldsymbol{A} \in \Re^{m \times n}$, $\boldsymbol{b} \in \Re^{m \times 1}$, and $\boldsymbol{c} \in \Re^{n \times 1}$ have the same values as defined in (6.8) through (6.10), and $\boldsymbol{y} \in \Re^{m \times 1}$ is a vector of dual independent variables. Note that in the dual problem, independent variables $\boldsymbol{y}$ do not have to be nonnegative.

The relationship between the primal and dual LP problems can be stated in the following theorem. For the sake of brevity the proof of the theorem is omitted. The interested reader can find the proof in Gass [1].

THEOREM 6.1. Consider an LP problem defined in (6.8) through (6.10) and its dual formulation in (6.16) and (6.17). Exactly one of the following statements is true:

(*a*) Both the primal and dual problems have optimal solutions $\tilde{x}$ and $\tilde{y}$, and the following equation is satisfied

$$c^T\tilde{x} = b^T\tilde{y} \tag{6.18}$$

In other words, at the point of optimal solution the values of the objective functions for the LP and dual LP problems are the same.

(*b*) One of the problems is unbounded, in which case the other one is nonfeasible.

(*c*) Both problems are nonfeasible.

In addition, it can be proved that for any pair of feasible solutions of the primal and dual LP problems, the *duality gap* satisfies

$$c^T x - b^T y \geq 0 \tag{6.19}$$

Since the duality gap is a nonnegative quantity that decreases to zero at the point of the optimal solution, it can be taken as a measure of how well we have solved the LP optimization problem.

6.2.1 Neural Networks for the Standard Form of the LP Problem

The first step in a neural network implementation for solving the LP problem is to define an energy function that can be optimized in an unconstrained fashion. To accomplish this, the linear constraints in (6.9) and the nonnegativity constraints in (6.10) are appended to the objective function in some convenient way. Commonly, the constraints are incorporated as penalty terms that, whenever violated, increase the value of the energy function. Two energy functions that can be derived using the Lagrange multiplier method (cf. Sect. A.6.2) are defined as [2]

$$E_1(x) = L_1(x, \boldsymbol{\lambda}) = c^T x + \frac{K}{2}(Ax - b)^T(Ax - b) + \boldsymbol{\lambda}^T(Ax - b) \tag{6.20}$$

$$E_2(x) = L_2(x, \boldsymbol{\lambda}) = c^T x + \frac{K}{2}(Ax - b)^T(Ax - b) + \boldsymbol{\lambda}^T(Ax - b) - \alpha\boldsymbol{\lambda}^T\boldsymbol{\lambda} \tag{6.21}$$

with $K, \alpha \geq 0$, $\boldsymbol{\lambda} \in \Re^{m\times 1}$, and $x \geq 0$. The last term on the right-hand side of (6.21) is referred to as the regularization term. This term improves the stability of the Lagrange multiplier method in cases of an ill-conditioned system of constraints [2].

Applying the method of steepest descent in discrete time, we compute the gradient of the energy function in (6.21) with respect to x and obtain

$$\nabla_x E_2 = \frac{\partial E_2(\boldsymbol{x}, \boldsymbol{\lambda})}{\partial \boldsymbol{x}}$$
$$= \frac{\partial}{\partial \boldsymbol{x}}\left[\boldsymbol{c}^T\boldsymbol{x} + \frac{K}{2}(\boldsymbol{x}^T\boldsymbol{A}^T\boldsymbol{A}\boldsymbol{x} - \boldsymbol{x}^T\boldsymbol{A}^T\boldsymbol{b} - \boldsymbol{b}^T\boldsymbol{A}\boldsymbol{x} + \boldsymbol{b}^T\boldsymbol{b}) + \boldsymbol{\lambda}^T(\boldsymbol{A}\boldsymbol{x} - \boldsymbol{b}) - \alpha\boldsymbol{\lambda}^T\boldsymbol{\lambda}\right] \tag{6.22}$$

and

$$\frac{\partial E_2(\boldsymbol{x}, \boldsymbol{\lambda})}{\partial \boldsymbol{x}} = \boldsymbol{c} + K\boldsymbol{A}^T(\boldsymbol{A}\boldsymbol{x} - \boldsymbol{b}) + \boldsymbol{A}^T\boldsymbol{\lambda} = \boldsymbol{c} + \boldsymbol{A}^T(K\boldsymbol{r} + \boldsymbol{\lambda}) \tag{6.23}$$

where $\boldsymbol{r} \in \Re^{m\times 1}$ is defined as $\boldsymbol{r} = \boldsymbol{r}(\boldsymbol{x}) = \boldsymbol{A}\boldsymbol{x} - \boldsymbol{b}$. In a similar manner we have

$$\frac{\partial E_2}{\partial \boldsymbol{\lambda}} = \boldsymbol{A}\boldsymbol{x} - \boldsymbol{b} - \alpha\boldsymbol{\lambda} = \boldsymbol{r} - \alpha\boldsymbol{\lambda} \tag{6.24}$$

Based on (6.23) and (6.24), a set of update equations using steepest descent can be formulated as

$$x_i(k+1) = \begin{cases} x_i(k) - \mu(k)\left\{c_i + \sum_{j=1}^{m} a_{ji}[Kr_j(k) + \lambda_j(k)]\right\} & \text{if } x_i(k+1) > 0 \\ 0 & \text{if } x_i(k+1) \le 0 \end{cases} \tag{6.25}$$

and

$$\boldsymbol{\lambda}(k+1) = \boldsymbol{\lambda}(k) + v(k)[\boldsymbol{r}(k) - \alpha\boldsymbol{\lambda}(k)] \tag{6.26}$$

where $\boldsymbol{r}(k) = \boldsymbol{A}\boldsymbol{x}(k) - \boldsymbol{b}$, $K, \alpha \ge 0$, and $\mu(k)$, $v(k) > 0$ are learning rate parameters. Note that the update equations for the independent-variable vector in (6.25) guarantee that all the components remain nonnegative. A neural network architecture realization of this process is presented in Figure 6.1.

EXAMPLE 6.1. It is desired to solve the following LP problem:
Maximize

$$f(\boldsymbol{x}) = \boldsymbol{c}^T\boldsymbol{x} = x_1 + x_2 \tag{6.27}$$

subject to

$$-2x_1 + x_2 \le 3 \tag{6.28}$$

$$x_1 + 3x_2 \le 16 \tag{6.29}$$

$$4x_1 + x_2 \le 20 \tag{6.30}$$

$$x_1,\ x_2 \ge 0 \tag{6.31}$$

From the problem statement it can be seen that the LP problem is not in the standard form. By adding surplus variables x_3, x_4 and x_5, the LP problem can be transformed in the standard form as follows:

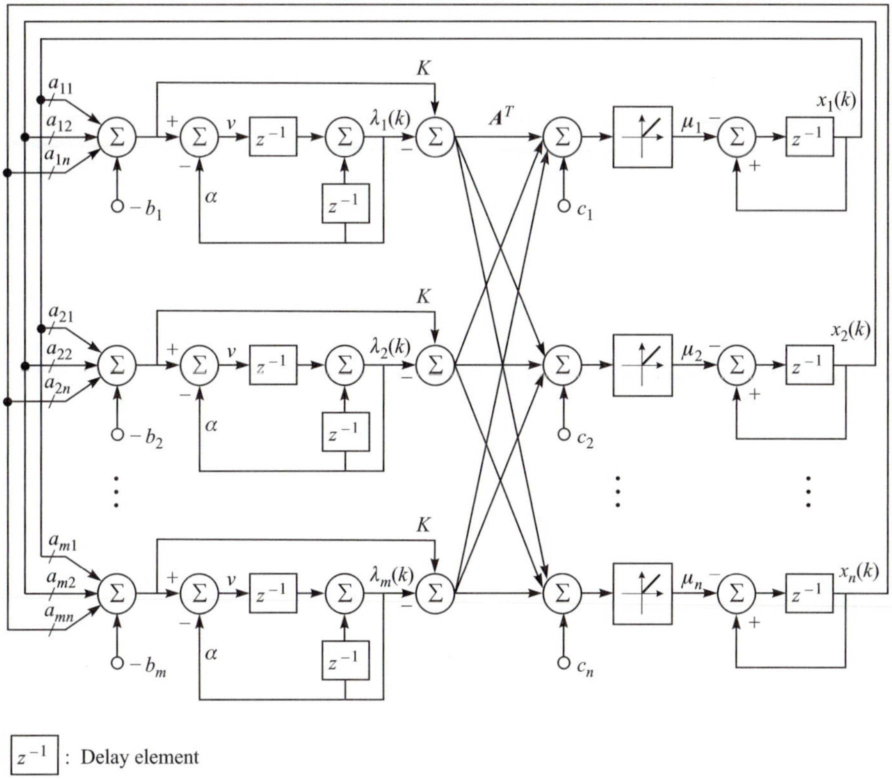

Figure 6.1 Discrete-time neural network for solving the LP problem in the standard form; implementation of Equations (6.25) and (6.26).

Maximize

$$f(\boldsymbol{x}) = x_1 + x_2 + 0x_3 + 0x_4 + 0x_5 \tag{6.32}$$

subject to

$$-2x_1 + x_2 + x_3 = 3 \tag{6.33}$$

$$x_1 + 3x_2 + x_4 = 16 \tag{6.34}$$

$$4x_1 + x_2 + x_5 = 20 \tag{6.35}$$

To solve this LP problem, the neural network in Figure 6.1 is simulated. The parameters of the network were chosen as $\mu = 0.01$, $\eta = 0.01$, $K = 0$, and $\alpha = 0$. Zero initial conditions were assumed both for $\boldsymbol{x}$ and $\boldsymbol{\lambda}$. Figure 6.2 shows the trajectories for each of the five independent variables. As can be seen, the network converges in approximately 3,000 iterations. The solution to the LP is given as $\tilde{\boldsymbol{x}} = [4.0042 \;\; 3.9953]^T$, which is within the learning rate parameter accuracy from the exact solution $\tilde{\boldsymbol{x}}^* = [4 \;\; 4]^T$. Note that the neural network approach provides the solution for the surplus variables as well.

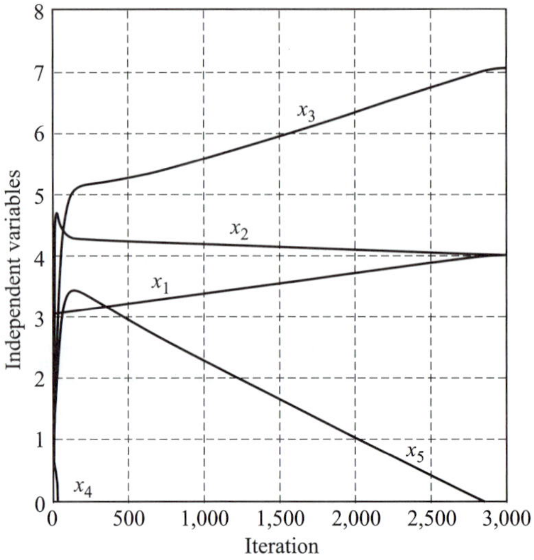

Figure 6.2 Trajectories of the independent variables for the neural network solution of the LP problem in Example 6.1.

6.2.2 Neural Networks for the Nonstandard Form of the LP Problem

In the previous section we demonstrated the use of neural networks for solving LP problems in the standard form. From Example 6.1 we can see the basic tradeoff in this approach. If the LP problem that we are trying to solve has a nonstandard form, the transformation to the standard form increases the problem dimension, and hence its complexity increases as well. In this section we will show that it can be computationally advantageous to consider the LP problem in its original form rather than to solve its standard equivalent form.

Neural network architectures for LP problems with nonequality constraints

Let us consider an LP problem in which all constraints are formulated in the form of inequalities:

Minimize

$$f(\boldsymbol{x}) = \boldsymbol{c}^T \boldsymbol{x} \tag{6.36}$$

subject to

$$a_{i1}x_1 + a_{i2}x_2 + \cdots + a_{in}x_n - b_i \le 0, \qquad \text{for} \quad i = 1, 2, \ldots, m \tag{6.37}$$

and

$$x_1 \ge 0, \ x_2 \ge 0, \ \ldots, \ x_n \ge 0 \tag{6.38}$$

It is common practice to abbreviate the constraints given in (6.37) using the notation

$$r_i(\boldsymbol{x}) = a_{i1}x_1 + a_{i2}x_2 + \cdots + a_{in}x_n - b_i, \qquad \text{for} \quad i = 1, 2, \ldots, m \tag{6.39}$$

Similar to the approach taken for the standard form LP problem in the previous section, we seek to formulate a suitable energy function that can be easily solved by a neural network. However, in this case the energy function needs to be constructed so that it penalizes every violation of the inequality constraints. After the energy function is formulated, the steepest descent, or any other unconstrained optimization technique, can be used to solve the LP problem.

Consider an energy function defined as

$$E(\boldsymbol{x}, K) = \boldsymbol{c}^T\boldsymbol{x} + K\sum_{i=1}^{m}\boldsymbol{\Phi}[r_i(\boldsymbol{x})] \tag{6.40}$$

where

$$\boldsymbol{\Phi}[r_i(\boldsymbol{x})] \begin{matrix} = 0 & \text{if } r_i(\boldsymbol{x}) \le 0 \\ > 0 & \text{if } r_i(\boldsymbol{x}) > 0 \end{matrix} \tag{6.41}$$

and

$$x_1 \ge 0,\ x_2 \ge 0,\ \ldots,\ x_n \ge 0 \tag{6.42}$$

This function consists of two terms. The first term on the right-hand side in (6.40) is the objective function of the LP problem that is to be minimized. The second term in (6.40) penalizes the violations of the constraints. The function $\boldsymbol{\Phi}(v)$ can be chosen as any piecewise differentiable function which has properties described in (6.41). Positive parameter K controls how well the unconstrained optimization problem in (6.40) to (6.42) approximates the original LP problem in (6.36) to (6.38). It can be easily shown that the two problems become equivalent as K approaches positive infinity. For that reason, the parameter K is commonly selected as a sufficiently large positive number. Applying the steepest descent technique, we have

$$\boldsymbol{x}(k+1) = \boldsymbol{x}(k) - \mu\frac{\partial E(\boldsymbol{x})}{\partial \boldsymbol{x}} \tag{6.43}$$

Taking the partial derivative of the energy function in (6.40) yields

$$\frac{\partial E(\boldsymbol{x})}{\partial \boldsymbol{x}} = \frac{\partial}{\partial \boldsymbol{x}}\left\{\boldsymbol{c}^T\boldsymbol{x} + K\sum_{i=1}^{m}\boldsymbol{\Phi}[r_i(\boldsymbol{x})]\right\} = \boldsymbol{c} + K\sum_{i=1}^{m}\boldsymbol{\Psi}[r_i(\boldsymbol{x})]\frac{\partial}{\partial \boldsymbol{x}}[r_i(\boldsymbol{x})] \tag{6.44}$$

where $\boldsymbol{\Psi}(v) = \dfrac{d\boldsymbol{\Phi}(v)}{dv} = \boldsymbol{\Phi}'(v)$. Combining (6.39) and (6.44), we have

$$\frac{\partial E(\boldsymbol{x})}{\partial \boldsymbol{x}} = \boldsymbol{c} + K\sum_{i=1}^{m}\boldsymbol{\Psi}[r_i(\boldsymbol{x})]\begin{bmatrix} a_{i1} \\ a_{i2} \\ \vdots \\ a_{in} \end{bmatrix} \tag{6.45}$$

Considering the nonnegativity constraints in (6.38), and substituting (6.45) into (6.43), we obtain a set of update equations as

$$x_j(k+1) = \begin{cases} x_j(k) - \mu_j\left\{c_j + K\sum_{i=1}^{m} \boldsymbol{\Psi}[r_i(\boldsymbol{x})]a_{ij}\right\} & \text{if } x_j(k+1) \geq 0 \\ 0 & \text{if } x_j(k+1) < 0 \end{cases} \tag{6.46}$$

Based on (6.46), we can construct a corresponding neural network architecture as shown in Figure 6.3. For the sake of simplicity, the function $\boldsymbol{\Phi}(v)$ is commonly selected as

$$\boldsymbol{\Phi}(v) = \begin{cases} \frac{1}{2}v^2 & \text{for } v > 0 \\ 0 & \text{for } v \leq 0 \end{cases} \tag{6.47}$$

An alternative energy function used to solve the LP problem with nonequality constraints can be defined as [2]

$$E(\boldsymbol{x}) = \boldsymbol{c}^T\boldsymbol{x} + K\sum_{i=1}^{m} \max\{0, r_i(\boldsymbol{x})\} \tag{6.48}$$

where $K > 0$. The gradient of (6.48) with respect to $\boldsymbol{x}$ is given by

$$\frac{\partial E(\boldsymbol{x})}{\partial \boldsymbol{x}} = c + K\sum_{i=1}^{m} S_i[a_{i1}, a_{i2}, \cdots, a_{in}]^T \tag{6.49}$$

Therefore, using (6.43) the discrete-time learning is

$$x_j(k+1) = x_j(k) - \mu_j\left(c_j + K\sum_{i=1}^{m} S_i a_{ij}\right) \tag{6.50}$$

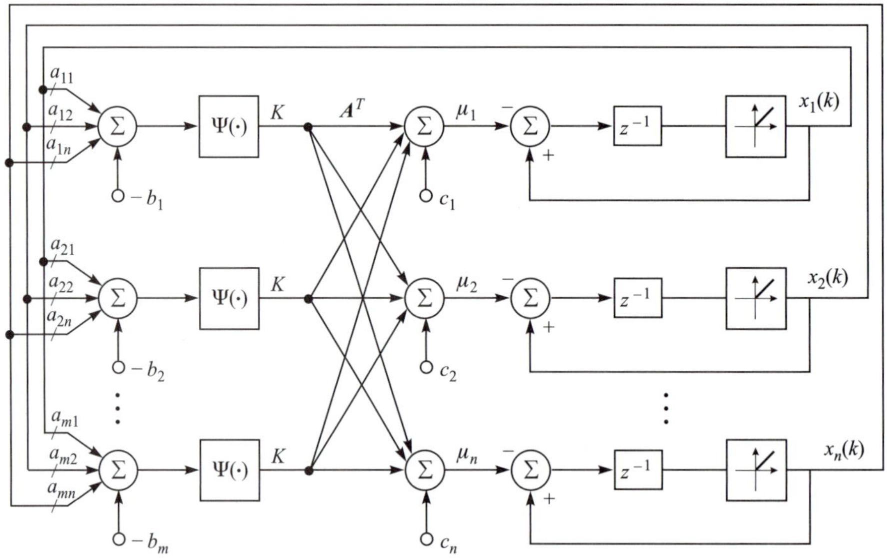

Figure 6.3 Discrete-time neural network solving the LP problem in the nonstandard form; implementation of Equation (6.46).

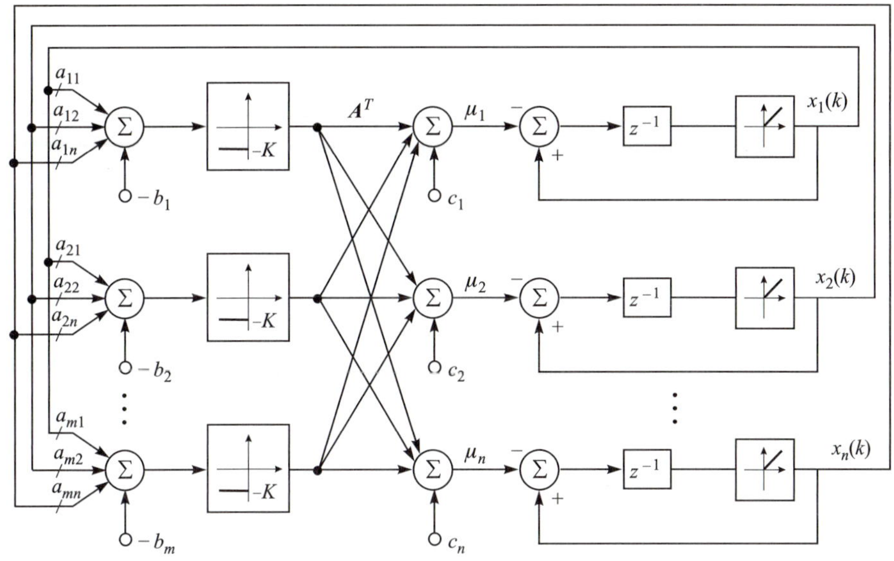

Figure 6.4 Discrete-time neural network architecture for solving the LP problem in the nonstandard form; implementation of Equations (6.50) and (6.51).

where

$$S_i = \begin{cases} 1 & \text{if } r_i(\boldsymbol{x}) > 0 \\ 0 & \text{if } r_i(\boldsymbol{x}) \le 0 \end{cases} \tag{6.51}$$

Of course the constraints in (6.42) have to be satisfied, and at every discrete time step k we have

$$x_j(k) = \max\{x_j(k), 0\} \tag{6.52}$$

The neural network architecture implementing Equations (6.50) and (6.51) is presented in Figure 6.4. Note that it has a very similar structure to the network shown in Figure 6.3.

EXAMPLE 6.2. Neural network architectures in Figures 6.3 and 6.4 were used to solve the LP problem given in Example 6.1. Parameters for the network in Figure 6.3 were chosen as $\mu_0 = 0.005$ and $K = 5$. For the second network the parameters were $\mu_0 = 0.005$ and $K = 5$ as well. The trajectories of the independent variables are shown in Figure 6.5. The initial conditions in both cases were set as $\boldsymbol{x}_0 = [1 \;\; 2]^T$. During the training, the learning rate was decreased according to the schedule

$$\mu = \frac{\mu_0}{\log(1+k)} \tag{6.53}$$

Note that both networks converge much faster than the one used in Example 6.1 where the LP problem was converted to its standard form. The solution obtained from the first network is $\tilde{\boldsymbol{x}}_1^T = [4.0127 \; 4.0218]$, and in the case of the second network the solution is $\tilde{\boldsymbol{x}}_2^T = [4.0060 \;\; 3.9976]$. Both solutions are of an accuracy comparable to the size of the learning rate parameter.

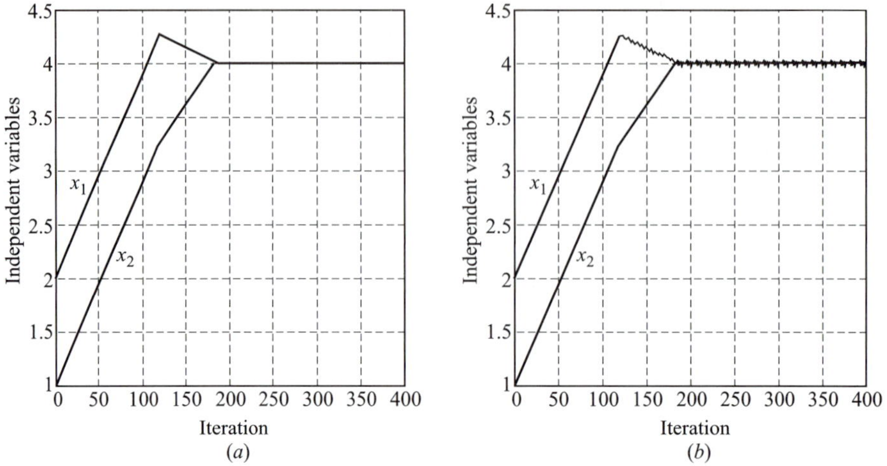

Figure 6.5 Trajectories of the independent variables for the neural network solution of the LP problem in Example 6.1: (*a*) using the network architecture shown in Figure 6.3; (*b*) using the network architecture shown in Figure 6.4.

Neural networks for LP problems with mixed constraints

The previous sections address the use of neural network architectures in solving LP problems in two important cases: when all constraints are equality constraints (standard form of the LP problem) and when all constraints are inequality constraints. In general, an LP problem can be formulated using both types of constraints. By adding surplus or slack variables, any LP problem can be transformed to its standard form. However, it is often beneficial not to do so. In other words, in the neural network approach to linear programming, a significant advantage can be gained from addressing the problem in its original form.

Consider an LP problem with mixed constraints given as follows:
Minimize

$$f(\boldsymbol{x}) = \boldsymbol{c}^T \boldsymbol{x} = \sum_{i=1}^{n} c_i x_i \tag{6.54}$$

subject to

$$a_{11}x_1 + a_{12}x_2 + \cdots + a_{1n}x_n = b_1 \tag{6.55}$$

$$\vdots$$

$$a_{p1}x_1 + a_{p2}x_2 + \cdots + a_{pn}x_n = b_p \tag{6.56}$$

$$a_{p+1,1}x_1 + a_{p+1,2}x_2 + \cdots + a_{p+1,n}x_n \le b_{p+1} \tag{6.57}$$

$$\vdots \tag{6.58}$$

$$a_{m1}x_1 + a_{m2}x_2 + \cdots + a_{mn}x_n \leq b_m \tag{6.59}$$

and

$$x_1, x_2, \ldots, x_n \geq 0 \tag{6.60}$$

The set of m constraints can be divided into two subsets. The first subset consists of p equality constraints, and the second subset has $m - p$ inequality constraints. Since the two sets are disjoint, the energy function can be formulated as a composite of energy functions given in (6.20), (6.21), (6.40), and (6.48). Depending on how the composite energy function is formed, several different learning algorithms and associated neural network architectures can be derived. For example, using (6.21) and (6.40), we can formulate the energy function according to

$$\begin{aligned} E(\boldsymbol{x}) = \boldsymbol{c}^T\boldsymbol{x} + \frac{K_1}{2}(\boldsymbol{A}_p\boldsymbol{x} - \boldsymbol{b}_p)^T(\boldsymbol{A}_p\boldsymbol{x} - \boldsymbol{b}_p) + \lambda_p^T(\boldsymbol{A}_p\boldsymbol{x} - \boldsymbol{b}_p) \\ - \alpha\boldsymbol{\lambda}_p^T\boldsymbol{\lambda}_p + \frac{K_2}{2}\sum_{i=p+1}^{m}\boldsymbol{\Phi}[r_i(\boldsymbol{x})] \end{aligned} \tag{6.61}$$

where

$$K_1, K_2, \alpha \geq 0 \tag{6.62}$$

$$\boldsymbol{A}_p = \begin{bmatrix} a_{11} & a_{12} & \cdots & a_{1n} \\ \cdots & \cdots & \cdots & \cdots \\ a_{p1} & a_{p2} & \cdots & a_{pn} \end{bmatrix} \qquad \boldsymbol{b}_p = [b_1, b_2, \cdots, b_p]^T \tag{6.63}$$

$$\boldsymbol{\lambda}_p = [\lambda_1, \lambda_2, \cdots, \lambda_p]^T \tag{6.64}$$

and the function $\boldsymbol{\Phi}(v)$ is any piecewise differentiable function satisfying (6.41).

The energy function in (6.61) consists of three different types of terms. The first term is

$$T_1 = \boldsymbol{c}^T\boldsymbol{x} \tag{6.65}$$

which is an LP objective function to be minimized. The second group of terms

$$T_2 = \frac{K_1}{2}(\boldsymbol{A}_p\boldsymbol{x} - \boldsymbol{b}_p)^T(\boldsymbol{A}_p\boldsymbol{x} - \boldsymbol{b}_p) + \boldsymbol{\lambda}_p^T(\boldsymbol{A}_p\boldsymbol{x} - \boldsymbol{b}_p) - \alpha\boldsymbol{\lambda}_p^T\boldsymbol{\lambda}_p \tag{6.66}$$

penalizes every violation of the equality constraints, and finally, the third term

$$T_3 = \frac{K_2}{2}\sum_{i=p+1}^{m}\boldsymbol{\Phi}[r_1(\boldsymbol{x})] \tag{6.67}$$

penalizes the violations of the inequality constraints. Applying the steepest descent approach, the update equations can be written as

$$\boldsymbol{\lambda}_p(k+1) = \boldsymbol{\lambda}_p(k) + v(k)[\boldsymbol{r}_p(k) - \alpha\boldsymbol{\lambda}_p(k)] \tag{6.68}$$

and

$$\boldsymbol{x}(k+1) = \boldsymbol{x}(k) - \mu(k)\left\{\boldsymbol{c} + \boldsymbol{A}_p^T\left[K_1\boldsymbol{r}_p(k) - \boldsymbol{\lambda}_p(k)\right] + K_2 \sum_{i=p+1}^{m} \boldsymbol{\Psi}[r_i(\boldsymbol{x})] \cdot \begin{bmatrix} a_{i1} \\ a_{i2} \\ \vdots \\ a_{in} \end{bmatrix}\right\} \tag{6.69}$$

where

$$\boldsymbol{r}_p = \boldsymbol{A}_p\boldsymbol{x} - \boldsymbol{b}_p \tag{6.70}$$

$$\boldsymbol{\Psi}(v) = \frac{\mathrm{d}\boldsymbol{\Phi}(v)}{\mathrm{d}v} \tag{6.71}$$

and $\mu(k)$, $v(k) > 0$ are learning rate parameters.

A neural network architecture realization of this process is presented in Figure 6.6. Note the enforcement of the positivity constraints on the independent variables.

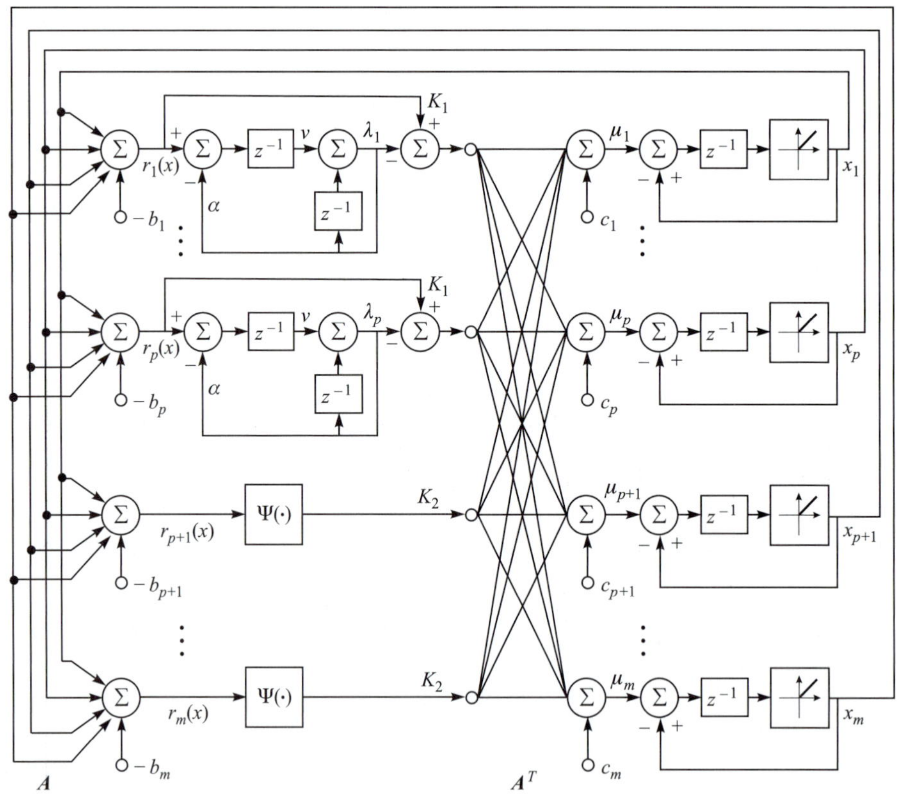

Figure 6.6 Discrete-time neural network for solving the LP problem in the nonstandard form; implementation of Equations (6.68) and (6.69).

6.3 NEURAL NETWORKS FOR QUADRATIC PROGRAMMING PROBLEMS

The quadratic programming (QP) problem is a special case of the nonlinear optimization problem in which a quadratic objective function is to be optimized subject to linear constraint conditions. However, owing to its importance and frequent occurrence in everyday practice, it is regarded as a separate problem. In this section we consider a neural network approach of solving three kinds of the QP problems defined relative to the form of the linear constraints. Before considering various forms of the QP problem, we take a closer look at the general quadratic form.

Quadratic form

A quadratic form is a part of the objective function which is optimized in all QP problems. In this section we define the quadratic form and briefly describe some of its properties, which are related to the QP problems.

Consider a vector $\boldsymbol{x} \in \Re^{n\times 1}$. A function defined as

$$f(\boldsymbol{x}) = \sum_{i=1}^{n}\sum_{j=1}^{n} q_{ij}x_i x_j \tag{6.72}$$

is called a quadratic form of vector $\boldsymbol{x}$. Frequently the coefficients q_{ij} are arranged in a matrix, and (6.72) can be rewritten in a more compact form as

$$f(\boldsymbol{x}) = \boldsymbol{x}^T\boldsymbol{Q}\boldsymbol{x} \tag{6.73}$$

where $\boldsymbol{Q} \in \Re^{n\times n}$. Recognizing that for every $\boldsymbol{x}$ and $\boldsymbol{Q}$, the product $\boldsymbol{x}^T\boldsymbol{Q}\boldsymbol{x}$ is a scalar quantity, we have

$$\boldsymbol{x}^T\boldsymbol{Q}\boldsymbol{x} = \left(\boldsymbol{x}^T\boldsymbol{Q}\boldsymbol{x}\right)^T = \boldsymbol{x}^T\boldsymbol{Q}^T\boldsymbol{x} \tag{6.74}$$

Therefore,

$$\boldsymbol{x}^T\boldsymbol{Q}\boldsymbol{x} = \frac{1}{2}\left(\boldsymbol{x}^T\boldsymbol{Q}\boldsymbol{x} + \boldsymbol{x}^T\boldsymbol{Q}^T\boldsymbol{x}\right) = \boldsymbol{x}^T\frac{\boldsymbol{Q}+\boldsymbol{Q}^T}{2}\boldsymbol{x} = \boldsymbol{x}^T\bar{\boldsymbol{Q}}\boldsymbol{x} \tag{6.75}$$

Therefore, matrix $\bar{\boldsymbol{Q}}$ is a symmetric matrix having elements given by

$$\bar{q}_{ij} = \frac{1}{2}\left(q_{ij} + q_{ji}\right) \tag{6.76}$$

where q_{ij} and q_{ji} are elements of $\boldsymbol{Q}$. From (6.75) and (6.76), it is obvious that for every matrix $\boldsymbol{Q}$, an equivalent quadratic form can be formulated with a symmetric coefficient matrix $\bar{\boldsymbol{Q}}$. For that reason, without any loss of generality, we can assume that $\boldsymbol{Q}$ is a real symmetric matrix.

A quadratic form is called positive definite (cf. Sect. A.2.6) if

$$\boldsymbol{x}^T\boldsymbol{Q}\boldsymbol{x} > 0 \tag{6.77}$$

for every nonzero $\boldsymbol{x} \in \Re^{n\times 1}$. Also, a quadratic form is called positive semidefinite if

$$x^T Q x \geq 0 \tag{6.78}$$

for all nonzero $x \in \Re^{n \times 1}$, and there exists at least one vector $x \neq \mathbf{0}$ for which $x^T Q x = 0$. Negative definite and negative semidefinite forms can also be defined by appropriate reversal of the inequality signs in (6.77) and (6.78).

A fundamental mathematical concept related to nonlinear programming is that of a convex function (cf. Definition A.11 in Sect. A.3.1). A function $f(x)$ defined over a convex set D in $\Re^{n \times 1}$ is said to be convex if for any two points x_1 and x_2 in D and for any $0 \leq \lambda \leq 1$ we have

$$f[\lambda x_1 + (1-\lambda) x_2] \leq \lambda f(x_1) + (1-\lambda) f(x_2) \tag{6.79}$$

The function is said to be concave if $-f(x)$ is a convex function. The concept of a convex function when $x \in \Re$ is illustrated in Figure 6.7. If the function is convex, its value in an arbitrary point on the segment (x_1, x_2) is smaller than the value on the straight line connecting points $f(x_1)$ and $f(x_2)$. The following theorem presents an important property of the convex functions. For the sake of brevity, the proof of the theorem is omitted and can be found in Wilde and Beightler [3].

THEOREM 6.2. If $f(x)$ is convex on a convex set D, then $f(x)$ has at most one local minimum. If there is such a minimum, it is a global minimum and is obtained on the convex set D.

Theorem 6.2 has far-reaching importance in optimization theory. It addresses a common problem of many iterative algorithms based on some sort of gradient technique—*escaping local minima of the error (energy) surface*. In essence, the theorem states that if the energy function is convex, the problem of local minima does not exist and the gradient-based minimization is guaranteed to terminate at a point of the global minimum.

The quadratic form in (6.73) is a common part of many energy functions used in nonlinear programming. Apart from the fact that many nonlinear programming problems involve the quadratic form by their very nature, the following theorem reveals the reasons why this form enjoys great popularity.

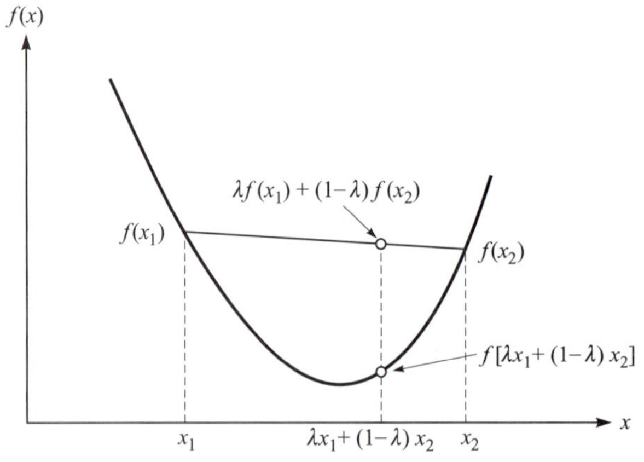

Figure 6.7 An example of the convex function.

THEOREM 6.3. Let $\boldsymbol{x} \in \Re^{n\times 1}$ and let $f(\boldsymbol{x}) = \boldsymbol{x}^T \boldsymbol{Q}\boldsymbol{x}$ be a quadratic form defined on a convex set $\boldsymbol{D} \subset \Re^{n\times 1}$. If matrix $\boldsymbol{Q}$ is semidefinite, the quadratic form $f(\boldsymbol{x})$ is convex on set $\boldsymbol{D}$.

Proof. Based on the definition of convex functions given in (6.79) we desire to demonstrate for $0 \leq \lambda \leq 1$ and all $\boldsymbol{x}_1, \boldsymbol{x}_2 \in \boldsymbol{D}$ that

$$f[\lambda \boldsymbol{x}_1 + (1-\lambda)\boldsymbol{x}_2] - \lambda f(\boldsymbol{x}_1) - (1-\lambda) f(\boldsymbol{x}_2) \leq 0 \tag{6.80}$$

Since $\boldsymbol{Q}$ is a symmetric matrix we have $\boldsymbol{x}_1^T\boldsymbol{Q}\boldsymbol{x}_2 = \boldsymbol{x}_2^T\boldsymbol{Q}\boldsymbol{x}_1$. By substituting the expression for the quadratic form, the left-hand side of (6.80) can be rewritten as

$$\begin{aligned}
&[\lambda \boldsymbol{x}_1 + (1-\lambda)\boldsymbol{x}_2]^T \boldsymbol{Q}[\lambda \boldsymbol{x}_1 + (1-\lambda)\boldsymbol{x}_2] - \lambda \boldsymbol{x}_1^T\boldsymbol{Q}\boldsymbol{x}_1 - (1-\lambda)\boldsymbol{x}_2^T\boldsymbol{Q}\boldsymbol{x}_2 \\
&\quad = \lambda^2 \boldsymbol{x}_1^T\boldsymbol{Q}\boldsymbol{x}_1 + (1-\lambda)^2\boldsymbol{x}_2^T\boldsymbol{Q}\boldsymbol{x}_2 + 2\lambda(1-\lambda)\boldsymbol{x}_1^T\boldsymbol{Q}\boldsymbol{x}_2 - \lambda \boldsymbol{x}_1^T\boldsymbol{Q}\boldsymbol{x}_1 - (1-\lambda)\boldsymbol{x}_2^T\boldsymbol{Q}\boldsymbol{x}_2 \\
&\quad = (\lambda^2 - \lambda)\boldsymbol{x}_1^T\boldsymbol{Q}\boldsymbol{x}_1 + (1-\lambda)[(1-\lambda)\boldsymbol{x}_2^T\boldsymbol{Q}\boldsymbol{x}_2 - \boldsymbol{x}_2^T\boldsymbol{Q}\boldsymbol{x}_2] + 2\lambda(1-\lambda)\boldsymbol{x}_1^T\boldsymbol{Q}\boldsymbol{x}_2 \\
&\quad = \lambda(\lambda-1)\boldsymbol{x}_1^T\boldsymbol{Q}\boldsymbol{x}_1 + \lambda(\lambda-1)\boldsymbol{x}_2^T\boldsymbol{Q}\boldsymbol{x}_2 - 2\lambda(\lambda-1)\boldsymbol{x}_1^T\boldsymbol{Q}\boldsymbol{x}_2 \\
&\quad = \lambda(\lambda-1)[\boldsymbol{x}_1^T\boldsymbol{Q}\boldsymbol{x}_1 + \boldsymbol{x}_2^T\boldsymbol{Q}\boldsymbol{x}_2 - 2\boldsymbol{x}_1^T\boldsymbol{Q}\boldsymbol{x}_2] \\
&\quad = \lambda(\lambda-1)[(\boldsymbol{x}_1 - \boldsymbol{x}_2)^T\boldsymbol{Q}(\boldsymbol{x}_1 - \boldsymbol{x}_2)]
\end{aligned} \tag{6.81}$$

Since $\boldsymbol{Q}$ is a semidefinite matrix, we have

$$(\boldsymbol{x}_1 - \boldsymbol{x}_2)^T\boldsymbol{Q}(\boldsymbol{x}_1 - \boldsymbol{x}_2) \geq 0 \tag{6.82}$$

On the other side, since $0 \leq \lambda \leq 1$

$$\lambda(\lambda - 1) \leq 0 \tag{6.83}$$

and therefore the inequality in (6.80) holds, which was to be proved.

Theorem 6.3 shows that the semidefinite quadratic form is a convex function. Therefore, according to Theorem 6.2, it has a unique minimum that can be easily found by using a gradient-based iterative technique. However, if the matrix $\boldsymbol{Q}$ is indefinite, there is a possibility that the quadratic form may have multiple local minima. From a practical standpoint, this is a rarely encountered limitation since the vast majority of quadratic programming problems can be formulated with $\boldsymbol{Q}$ being positive semidefinite. For this reason, in the sections that follow we are going to assume that $\boldsymbol{Q}$ is a symmetric and a positive semidefinite matrix. If this is not the case, some of the neural network algorithms described below may fail to converge to the global minimum.

Neural networks for the QP problem in standard form

Similar to the case of the LP problem, the QP problem can be formulated in standard form. The standard form of the QP problem is as follows:

Minimize

$$f(\boldsymbol{x}) = \boldsymbol{c}^T\boldsymbol{x} + \frac{1}{2}\boldsymbol{x}^T\boldsymbol{Q}\boldsymbol{x} \tag{6.84}$$

subject to

$$\boldsymbol{A}\boldsymbol{x} = \boldsymbol{b} \tag{6.85}$$

and

$$x_1, x_2, \ldots, x_n \geq 0 \tag{6.86}$$

where $\boldsymbol{x} \in \Re^{n\times 1}$, $\boldsymbol{c} \in \Re^{n\times 1}$, $\boldsymbol{Q} \in \Re^{n\times n}$, $\boldsymbol{A} \in \Re^{m\times n}$, $\boldsymbol{b} \in \Re^{m\times 1}$, $m < n$, and $\boldsymbol{Q}$ is assumed to be a symmetric positive definite matrix.

In order for a neural network approach to be used, a convenient energy function needs to be defined. Using the augmented Lagrange multiplier method (cf. Sect. A.6.2), we can define an energy function as [2]

$$E(\boldsymbol{x}, \boldsymbol{\lambda}) = \boldsymbol{c}^T\boldsymbol{x} + \frac{1}{2}\boldsymbol{x}^T\boldsymbol{Q}\boldsymbol{x} + \boldsymbol{\lambda}^T(\boldsymbol{A}\boldsymbol{x} - \boldsymbol{b}) + \frac{K}{2}(\boldsymbol{A}\boldsymbol{x} - \boldsymbol{b})^T(\boldsymbol{A}\boldsymbol{x} - \boldsymbol{b}) \tag{6.87}$$

where $\boldsymbol{\lambda} = \left[\lambda_1, \lambda_2, \cdots, \lambda_m\right]^T \in \Re^{m\times 1}$ and $K \geq 0$ is the penalty parameter.

Applying a gradient method, we can form the network update equations as

$$\boldsymbol{x}(k+1) = \boldsymbol{x}(k) - \mu\nabla_x E(\boldsymbol{x}, \boldsymbol{\lambda}) \tag{6.88}$$

and

$$\boldsymbol{\lambda}(k+1) = \boldsymbol{\lambda}(k) + \eta\nabla_\lambda E(\boldsymbol{x}, \boldsymbol{\lambda}) \tag{6.89}$$

where μ, $\eta > 0$ are the learning rate parameters. After we determine the gradient in (6.88) and (6.89), the resulting learning rules are

$$\boldsymbol{x}(k+1) = \boldsymbol{x}(k) - \mu\{\boldsymbol{c} + \boldsymbol{Q}\boldsymbol{x}(k) + \boldsymbol{A}^T\boldsymbol{\lambda}(k) + K\boldsymbol{A}^T[\boldsymbol{A}\boldsymbol{x}(k) - \boldsymbol{b}]\} \tag{6.90}$$

and

$$\boldsymbol{\lambda}(k+1) = \boldsymbol{\lambda}(k) + \eta(\boldsymbol{A}\boldsymbol{x} - \boldsymbol{b}) \tag{6.91}$$

A neural network architecture realization of this process is presented in Figure 6.8.

As a final note, consider the Hessian matrix (cf. Sect. A.3.5) associated with the energy function defined in (6.87)

$$\boldsymbol{H} = \frac{\partial^2 E(\boldsymbol{x}, \boldsymbol{\lambda})}{\partial \boldsymbol{x}^2} = \boldsymbol{Q} + K\boldsymbol{A}^T\boldsymbol{A} \tag{6.92}$$

From (6.92) we see that the Hessian matrix of the energy function is positive semidefinite, even for $K = 0$ if $\boldsymbol{Q}$ is positive semidefinite. Since that is the case in most QP problems, the network in Figure 6.8 is guaranteed to converge to a unique global minimum. However, even in the case when $\boldsymbol{Q}$ is not positive semidefinite, the Hessian matrix can be forced into positive definiteness, if the parameter K is set as a sufficiently large positive value. For that reason the penalty term $(K/2)(\boldsymbol{A}\boldsymbol{x} - \boldsymbol{b})^T(\boldsymbol{A}\boldsymbol{x} - \boldsymbol{b})$ tends to improve the convergence properties of the network in cases when some of the eigenvalues of $\boldsymbol{Q}$ are relatively small positive numbers or even in cases when $\boldsymbol{Q}$ is not positive definite.

EXAMPLE 6.3. Consider a QP problem given by the following:
Maximize

$$f(\boldsymbol{x}) = x_1 + x_2 - x_1^2 - 3x_2^2 \tag{6.93}$$

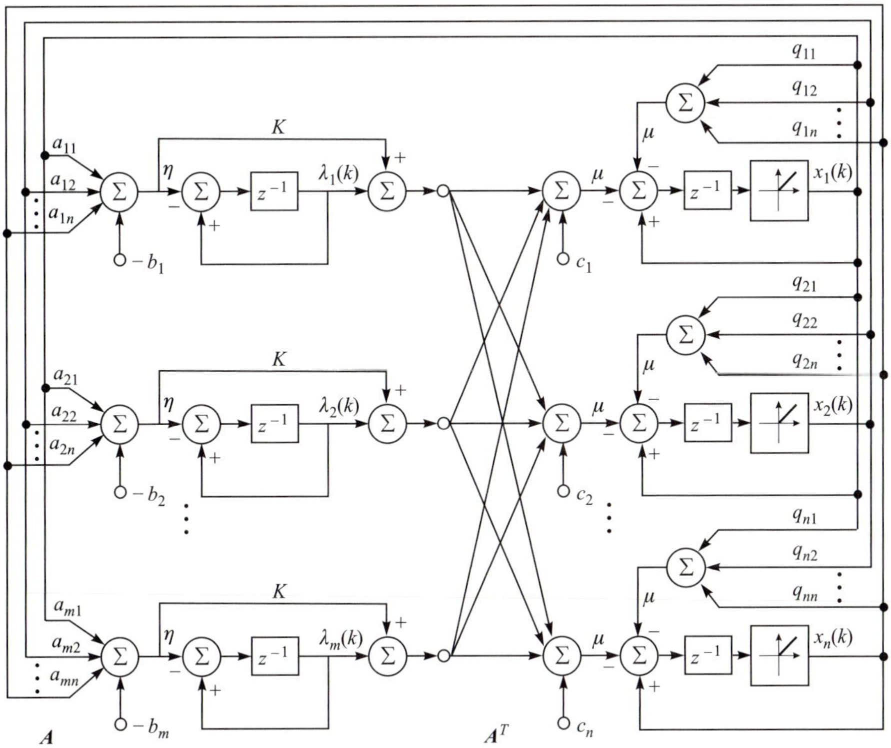

Figure 6.8 Discrete-time neural network for solving the QP problem in standard form: implementation of Equations (6.88) and (6.89).

subject to

$$-2x_1 + x_2 \le 3 \tag{6.94}$$

$$x_1 + 3x_2 \le 16 \tag{6.95}$$

$$4x_1 + x_2 \le 20 \tag{6.96}$$

$$x_1, x_2 \ge 0 \tag{6.97}$$

Since this problem is not in standard form, we can perform a transformation by adding appropriate surplus variables. An equivalent problem in standard form is given by the following:

Minimize

$$f(\boldsymbol{x}) = \boldsymbol{c}^T\boldsymbol{x} + \frac{1}{2}\boldsymbol{x}^T\boldsymbol{Q}\boldsymbol{x} = -x_1 - x_2 + x_1^2 + 3x_2^2 \tag{6.98}$$

subject to:

$$-2x_1 + x_2 + x_3 = 3 \tag{6.99}$$

$$x_1 + 3x_2 + x_4 = 16 \tag{6.100}$$

$$4x_1 + x_2 + x_5 = 20 \tag{6.101}$$

$$x_1, x_2, \ldots, x_5 \geq 0 \tag{6.102}$$

The neural network shown in Figure 6.8 was used to solve this QP problem. Learning rates were chosen as $\mu = 0.01$ and $\eta = 0.01$. Figure 6.9(*a*) and (*b*) shows the trajectories of the independent variables x_i, $i = 1, 2, \ldots, 5$, for two different values of the parameter K. We can see that in this case the nonzero value for K significantly improves the convergence properties of the network. An exact solution of the QP problem is given as $x_1 = 0.5$ and $x_2 = 0.1667$. In both cases, the network converged to values within the learning rate accuracy of the exact ones.

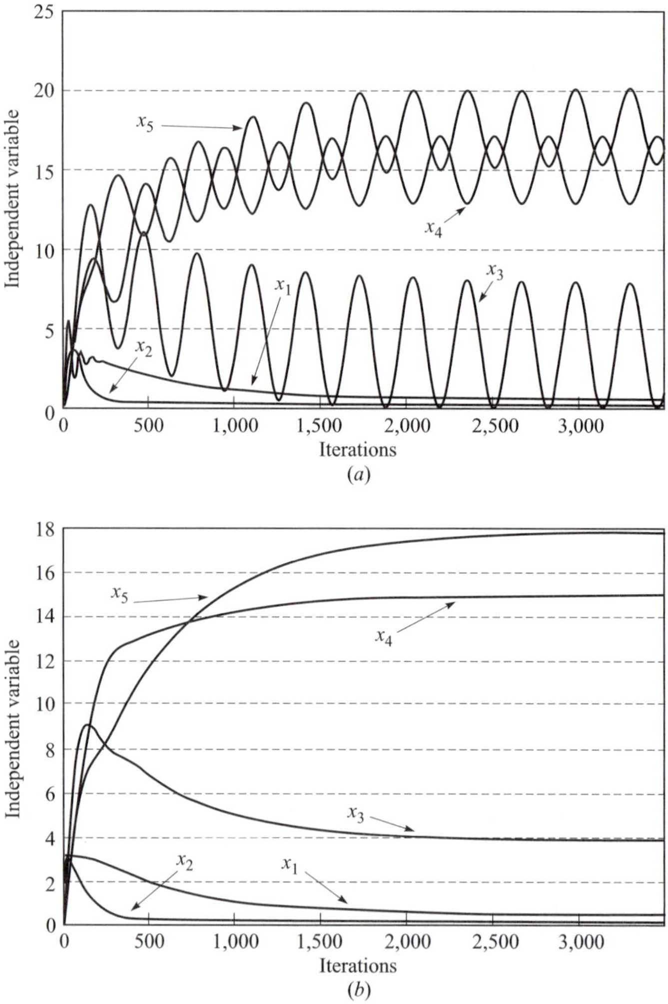

Figure 6.9 Trajectories of the independent variables for the QP problem in Example 6.3. (*a*) $K = 0$; network converged to $x_1 = 0.5176$ and $x_2 = 0.1642$. (*b*) $K = 1$; network converged to $x_1 = 0.5069$ and $x_2 = 0.1673$.

Neural networks for QP problems with inequality constraints

In the previous section we have examined the neural network approach to solving the QP problem in a standard form. As was demonstrated in Example 6.3, even if the problem is not in standard form, it can always be transformed by adding new independent variables. The price paid for the transformation is an increase of the problem's dimension and size of the corresponding neural network. In this section we address a neural network approach for solving the QP problem which has constraints given in terms of inequalities.

The form of the QP problem with inequality constraints is given as follows:

Minimize

$$f(\boldsymbol{x}) = \boldsymbol{c}^T\boldsymbol{x} + \frac{1}{2}\boldsymbol{x}^T\boldsymbol{Q}\boldsymbol{x} \tag{6.103}$$

subject to

$$\boldsymbol{A}\boldsymbol{x} \leq \boldsymbol{b} \tag{6.104}$$

and

$$\boldsymbol{x} \geq 0 \tag{6.105}$$

where $\boldsymbol{x}, \boldsymbol{c} \in \Re^{n\times 1}$, $\boldsymbol{A} \in \Re^{m\times n}$, $\boldsymbol{b} \in \Re^{m\times 1}$, and $\boldsymbol{Q} \in \Re^{n\times n}$ is a symmetric positive semidefinite matrix.

To solve the QP problem defined in (6.103) to (6.105), we can extend the method used in the LP problems with inequality constraints. We can define

$$r_i(\boldsymbol{x}) = \sum_{j=1}^{n} a_{ij}x_j - b_i \tag{6.106}$$

and formulate two different forms of the energy function. The first form can be written as

$$E_1(\boldsymbol{x}, K) = \boldsymbol{c}^T\boldsymbol{x} + \frac{1}{2}\boldsymbol{x}^T\boldsymbol{Q}\boldsymbol{x} + K\sum_{i=1}^{m} \boldsymbol{\Phi}[r_i(\boldsymbol{x})] \tag{6.107}$$

where $\boldsymbol{\Phi}(v)$ is any piecewise differentiable function satisfying following properties:

$$\boldsymbol{\Phi}[r_i(\boldsymbol{x})] \begin{array}{ll} = 0 & \text{if } r_i(\boldsymbol{x}) \leq 0 \\ > 0 & \text{if } r_i(\boldsymbol{x}) > 0 \end{array} \tag{6.108}$$

The summation term on the right-hand side of (6.107) is the penalty term that increases the value of the energy function whenever one of the inequality constraints is violated. The second form of the energy function for the QP problem can be written as

$$E_2(\boldsymbol{x}, K) = \boldsymbol{c}^T\boldsymbol{x} + \frac{1}{2}\boldsymbol{x}^T\boldsymbol{Q}\boldsymbol{x} + K\sum_{i=1}^{m} \max\{r_i(\boldsymbol{x}), 0\} \tag{6.109}$$

As can be seen, in both cases the energy function is generated by appending constraints to the QP problem's objective function. The parameter $K > 0$ is commonly referred to as the penalty parameter, and in both cases it needs to be chosen as a fairly large number to guarantee that the minimization is performed in a fashion that ensures that all the constraints are satisfied. By taking the steepest descent approach, the gradient of the energy function can be computed, and the learning established in discrete time steps.

The gradient of (6.107) with respect to $\boldsymbol{x}$ is given by

$$\frac{\partial E_1(\boldsymbol{x}, K)}{\partial \boldsymbol{x}} = \boldsymbol{c} + \boldsymbol{Q}\boldsymbol{x} + K \sum_{i=1}^{m} \boldsymbol{\Psi}[r_i(\boldsymbol{x})] \begin{bmatrix} a_{i1} \\ a_{i2} \\ \vdots \\ a_{in} \end{bmatrix} \tag{6.110}$$

Keeping in mind the nonnegativity constraints in (6.105), we can now write the update equations as

$$x_j(k+1) = \begin{cases} x_j(k) - \mu_j \left\{ c_j + \sum_{i=1}^{n} q_{ji} x_i + K \sum_{i=1}^{m} \boldsymbol{\Psi}[r_i(\boldsymbol{x})] a_{ij} \right\} & \text{if } x_j(k+1) \geq 0 \\ 0 & \text{if } x_j(k+1) < 0 \end{cases} \tag{6.111}$$

where $j = 1, 2, \ldots, n$, $\boldsymbol{\Psi}(v) = d\Phi(v)/dv$, and μ_j is the learning rate parameter.

For the energy function in (6.109) the gradient is given by

$$\frac{\partial E_2(\boldsymbol{x}, K)}{\partial \boldsymbol{x}} = \boldsymbol{c} + \boldsymbol{Q}\boldsymbol{x} + K \sum_{i=1}^{m} S_i \begin{bmatrix} a_{i1} \\ a_{i2} \\ \vdots \\ a_{in} \end{bmatrix} \tag{6.112}$$

where

$$S_i = \begin{cases} 0 & \text{if } r_i(\boldsymbol{x}) \leq 0 \\ 1 & \text{if } r_i(\boldsymbol{x}) > 0 \end{cases} \tag{6.113}$$

Again, keeping in mind the nonnegativity constraints in (6.105), we see the update equations become

$$x_j(k+1) = \begin{cases} x_j(k) - \mu_j \dfrac{\partial E_2(\boldsymbol{x}, K)}{\partial \boldsymbol{x}} & \text{if } x_j(k+1) \geq 0 \\ 0 & \text{if } x_j(k+1) < 0 \end{cases} \tag{6.114}$$

Neural network architectures based on the two different energy functions are shown in Figures 6.10 and 6.11. As can be seen, both networks have virtually identical structures, and the only significant difference is the nonlinear processing of the constraints in the first layer. Comparing these neural architectures with the network in Figure 6.8 reveals that the greatest benefit of using networks for QP problems with inequality constraints comes from the ability to address the original problem without the addition of slack or surplus variables.

(6.105), and the energy function can be formed as a composite of the energy functions defined in (6.87), (6.107), and (6.109). For example, if we define

$$\boldsymbol{A}_p = \begin{bmatrix} a_{11} & a_{12} & & a_{1n} \\ \cdots & \cdots & \cdots & \cdots \\ a_{p1} & a_{p2} & \cdots & a_{pn} \end{bmatrix} \tag{6.122}$$

$$\boldsymbol{b}_p = \begin{bmatrix} b_1 & b_2 & \cdots & b_p \end{bmatrix}^T \tag{6.123}$$

and

$$\boldsymbol{\lambda}_p = \begin{bmatrix} \lambda_1 & \lambda_2 & \cdots & \lambda_p \end{bmatrix}^T \tag{6.124}$$

the energy function can bc formulated using (6.87) and (6.107) as

$$\begin{aligned} E(\boldsymbol{x}, \boldsymbol{\lambda}_p, K_1, K_2) &= \boldsymbol{c}^T\boldsymbol{x} + \frac{1}{2}\boldsymbol{x}^T\boldsymbol{Q}\boldsymbol{x} + \boldsymbol{\lambda}_p^T(\boldsymbol{A}_p\boldsymbol{x} - \boldsymbol{b}_p) \\ &+ \frac{K_1}{2}(\boldsymbol{A}_p\boldsymbol{x} - \boldsymbol{b}_p)^T(\boldsymbol{A}_p\boldsymbol{x} - \boldsymbol{b}_p) + K_2\sum_{i=p+1}^{m}\boldsymbol{\Phi}[r_i(\boldsymbol{x})] \end{aligned} \tag{6.125}$$

Using the steepest descent gradient approach, the update equations may be written as

$$\boldsymbol{x}(k+1) = \boldsymbol{x}(k) - \mu\,\frac{\partial E(\boldsymbol{x}, K_1, K_2)}{\partial \boldsymbol{x}} \tag{6.126}$$

and

$$\boldsymbol{\lambda}_p(k+1) = \boldsymbol{\lambda}_p(k) + \eta\,\frac{\partial E(\boldsymbol{x}, \boldsymbol{\lambda}_p, K_1, K_2)}{\partial \boldsymbol{\lambda}_p} \tag{6.127}$$

The appropriate partial derivatives can be evaluated as

$$\begin{aligned} \frac{\partial E(\boldsymbol{x}, \boldsymbol{\lambda}_p, K_1, K_2)}{\partial \boldsymbol{x}} &= \boldsymbol{c} + \boldsymbol{Q}\boldsymbol{x} + \boldsymbol{A}_p^T\boldsymbol{\lambda}_p + K_1\boldsymbol{A}_p^T(\boldsymbol{A}_p\boldsymbol{x} - \boldsymbol{b}_P) \\ &+ K_2\sum_{i=p+1}^{m}\boldsymbol{\Psi}[r_i(\boldsymbol{x})]\begin{bmatrix} a_{i1} \\ a_{i2} \\ \vdots \\ a_{in} \end{bmatrix} \end{aligned} \tag{6.128}$$

and

$$\frac{\partial E(\boldsymbol{x}, \boldsymbol{\lambda}_p, K_1, K_2)}{\partial \boldsymbol{\lambda}_p} = \boldsymbol{A}_p\boldsymbol{x} - \boldsymbol{b}_p \tag{6.129}$$

By taking the nonnegativity requirements in (6.110) into consideration, the set of update equations in scalar form may be formulated as

$$x_j(k+1) = \begin{cases} x_j(k) - \mu\Big\{c_j + \sum_{i=1}^{n} q_{ji}x_i(k) \\ \qquad + \sum_{i=1}^{p} a_{ij}[K_1 r_i(\boldsymbol{x}) + \lambda_i(k)] \\ \qquad + K_2 \sum_{i=p+1}^{m} \boldsymbol{\Psi}[r_i(\boldsymbol{x})]a_{ij}\Big\} & \text{if } x_j(k+1) \geq 0 \\ 0 & \text{if } x_j(k+1) < 0 \end{cases} \tag{6.130}$$

and

$$\lambda_j(k+1) = \lambda_j(k) + \eta\left(\sum_{i=1}^{n} a_{ji}x_i - b_{pj}\right) \tag{6.131}$$

6.4 NEURAL NETWORKS FOR NONLINEAR CONTINUOUS CONSTRAINED OPTIMIZATION PROBLEMS

In previous sections of this chapter we have presented two important cases of constrained optimization—LP and QP problems. We now turn our attention back to the general settings of the constrained optimization problem as described in (6.1) and (6.2). Frequently, to emphasize the fact that both objective function and constraints may be nonlinear, this kind of problem is referred to as nonlinear programming (NP).

NP problems have enormous practical applications, and as such they have been the subject of extensive research from both the theoretical and the practical standpoint. For that reason, many excellent references are readily available, and numerous nonlinear programming algorithms have been and are still being developed. The main goal of this section is to provide several illustrations of how neural networks can be used as a computationally efficient and relatively simple tool for implementing some of the well-known nonlinear programming techniques, namely, penalty function methods, barrier function methods, ordinary Lagrange multiplier methods, and augmented Lagrange multiplier methods.

Before we address the neural implementation of the NP algorithms, let us revisit the definition of the problem. Based on the form of the constraints, we can define three different forms of the NP minimization problems [2]:

NP1 (NP problem with equality constraints):
Minimize

$$f(\boldsymbol{x}) = f(x_1, x_2, \ldots, x_n) \tag{6.132}$$

subject to

$$h_i(\boldsymbol{x}) = 0 \qquad \text{where } i = 1, 2, \ldots, m \tag{6.133}$$

NP2 (NP problem with inequality constraints:
Minimize

$$f(\boldsymbol{x}) = f(x_1, x_2, \ldots, x_n) \tag{6.134}$$

subject to

$$g_i(\boldsymbol{x}) \le 0 \qquad \text{where } i = 1, 2, \ldots, m \tag{6.135}$$

NP3 (NP problem with mixed constraints):
Minimize

$$f(\boldsymbol{x}) = f(x_1, x_2, \ldots, x_n) \tag{6.136}$$

subject to

$$h_i(\boldsymbol{x}) = 0 \qquad \text{where } i = 1, 2, \ldots, p \tag{6.137}$$

and

$$g_i(\boldsymbol{x}) \le 0 \qquad \text{where } i = p + 1, p + 2, \ldots, m \tag{6.138}$$

where $\boldsymbol{x} \in \Re^{n \times 1}$ is the vector of the independent variables, $f(\boldsymbol{x})$: $\Re^{n \times 1} \to \Re$ is the objective function, and functions $h_i(\boldsymbol{x})$, $g_i(\boldsymbol{x})$: $\Re^{n \times 1} \to \Re$ represent constraints. To simplify the derivations of the algorithms we will assume that both the objective function and the constraints are smooth differentiable functions of independent variables.

Using surplus variables, the inequality constraints in problems NP2 and NP3 can be converted to equality constraints. Similarly, each of the equality constraints can be converted to a pair of inequality constraints according to

$$h_i(\boldsymbol{x}) = 0 \quad \Leftrightarrow \quad h_i(\boldsymbol{x}) \le 0 \qquad \text{and} \qquad h_i(\boldsymbol{x}) \ge 0 \tag{6.139}$$

Therefore, every NP problem can be transformed into either the NP1 or NP2 form. However, it is preferred from the computational standpoint to consider the NP problem in its original form. Finally, note that the NP1 or NP2 form of the problem can be regarded as special cases of a more general NP3 form.

6.4.1 Neural Networks for Penalty Function NP Methods

Methods using penalty functions make an attempt to transform the NP problem to an equivalent unconstrained optimization problem, or to a sequence of constrained optimization problems. This transformation is accomplished through modification of the objective function so that it includes terms that penalize every violation of the constraints. In general, the modified objective function takes the following form:

$$f_A(\boldsymbol{x}) = f(\boldsymbol{x}) + \sum_{i=1}^{p} K_i^{(1)} \boldsymbol{\Phi}_i^{(1)}[h_i(\boldsymbol{x})] + \sum_{i=p+1}^{m} K_i^{(2)} \boldsymbol{\Phi}_i^{(2)}[g_i(\boldsymbol{x})] \tag{6.140}$$

Functions $\Phi_i^{(1)}$ and $\Phi_i^{(2)}$ are called penalty functions, and they are designed to increase the value of the modified objective function $f_A(\mathbf{x})$ whenever the vector of independent variables violates a constraint, or in other words whenever it is outside the *feasible* region. Penalty functions are commonly selected as at least one-time differentiable functions satisfying the following requirements:

1. For equality constraints

$$\Phi_i^{(1)} \begin{array}{ll} > 0 & \text{for } h_i(\mathbf{x}) \neq 0 \\ = 0 & \text{for } h_i(\mathbf{x}) = 0 \end{array} \tag{6.141}$$

2. For inequality constraints

$$\Phi_i^{(2)} \begin{array}{ll} > 0 & \text{for } g_i(\mathbf{x}) > 0 \\ = 0 & \text{for } g_i(\mathbf{x}) \leq 0 \end{array} \tag{6.142}$$

Some common choices for the penalty functions are given as follows: For the case of equality constraints

1. $\Phi_i^{(1)}(v) = \frac{1}{2}v^2$ (6.143)
2. $\Phi_i^{(1)}(v) = \frac{1}{\rho}|v|^\rho \qquad \rho > 0$ (6.144)
3. $\Phi_i^{(1)}(v) = \cosh v - 1$ (6.145)
4. $\Phi_i^{(1)}(v) = \ln\left[\frac{1}{2}(e^v + e^{-v})\right]$ (6.146)

and for the case of the inequality constraints

1. $\Phi_i^{(2)}(v) = \max\{0, v\}$ (6.147)
2. $\Phi_i^{(2)}(v) = (\max\{0, v\})^2$ (6.148)

For example, the typical modified objective function for the NP3 problem can be written as

$$f_A(\mathbf{x}) = f(\mathbf{x}) + \sum_{i=1}^{p} \frac{K_i^{(1)}}{\rho_1} |h_i(\mathbf{x})|^{\rho_1} + \sum_{i=p+1}^{m} K_i^{(2)} \max\{0, g_i(\mathbf{x})\}^{\rho_2} \tag{6.149}$$

where $\rho_1, \rho_2 \geq 0$. Parameters $K_i^{(1)}, K_i^{(2)} \geq 0$ are commonly referred to as *penalty parameters* or *penalty multipliers*, and in (6.149) we have assumed that a separate penalty parameter is associated with each of the penalty functions. In practice this is rarely the case, and commonly there is only one parameter multiplying the entire penalty term, that is,

$$\begin{aligned} f_A(\mathbf{x}) &= f(\mathbf{x}) + K\left[\sum_{i=1}^{p} \frac{1}{\rho_1} |h_i(\mathbf{x})|^{\rho_1} + \sum_{i=p+1}^{m} \max\{0, g_i(\mathbf{x})\}^{\rho_2}\right] \\ &= f(\mathbf{x}) + K\mathrm{P}(\mathbf{x}) \end{aligned} \tag{6.150}$$

where $P(\mathbf{x})$ represents the penalty term.

There are two fundamental issues that need to be addressed in the practical application of penalty functions. First, we need to be aware that (6.150) represents merely an approximation of the original problem in (6.136) through (6.138). The question is, How close is the approximation? The second issue involves a design of a computationally efficient neural network algorithm that can successfully solve the unconstrained problem in a timely manner.

From the form of the augmented objective function in (6.150), it should be obvious that the solution resides in the region where the value of the penalty function $P(\boldsymbol{x})$ is small. As a matter of fact, if K is increased toward infinity, the solution of the unconstrained problem will be forced into the feasible region of the original NP problem. Remember that if the point is in the feasible region, all the constraints are satisfied and the penalty function equals zero. In the limiting case, when $K \to \infty$, the two problems become equivalent. In short, the equivalence of the two problems can be summarized as in the following theorem.

THEOREM 6.4. Consider an NP problem given as follows:

Minimize

$$f(\boldsymbol{x}) = f(x_1, x_2, \ldots, x_n) \tag{6.151}$$

subject to

$$\boldsymbol{x} \in S, \qquad S \subset \Re^{n \times 1} \tag{6.152}$$

where S is a constrained set (i.e., the feasible region) defined by a number of either equality or inequality constraints. Define a sequence of unconstrained optimization problems as follows:

Minimize

$$q(K_j, \boldsymbol{x}) = f(\boldsymbol{x}) + K_j P(\boldsymbol{x}) \tag{6.153}$$

where P is a penalty function satisfying

$$P(\boldsymbol{x}) \geq 0 \qquad \text{for } \boldsymbol{x} \in \Re^{n \times 1} \tag{6.154}$$

$$P(\boldsymbol{x}) = 0 \qquad \text{if and only if } \boldsymbol{x} \in S \tag{6.155}$$

and K_j, $j = 1, 2, \ldots$, is a sequence of real numbers satisfying

$$K_j > 0 \qquad \forall j \tag{6.156}$$

$$K_{j+1} > K_j \qquad \forall j \tag{6.157}$$

$$K_j \to \infty \text{ as } j \to \infty \text{ in an arbitrary way} \tag{6.158}$$

Let $\{\tilde{\boldsymbol{x}}_j\}$, $j = 1, 2, \ldots$, be the sequence of solutions to the unconstrained optimization problems given in (6.153). Then the limit point of the sequence $\{\tilde{\boldsymbol{x}}_j\}$ is the solution of the NP problem given in (6.151) and (6.152).

For the sake of brevity, the proof of the Theorem 6.4 is omitted. The interested reader can find more details in Luenberger [4]. Theorem 6.4 essentially outlines the methodology of the penalty function methods. As can be seen, a sequence of unconstrained optimization problems is generated according to (6.153) to (6.158), and the solution to the generated sequence converges to the solution of the original NP problem. From the neural network standpoint, solving the sequence of unconstrained problems is clearly unacceptable. Because of that, practical implementation of the penalty methods is usually accomplished in either of the following two ways:

1. Penalty parameter K is made time-varying, and it is increased over the course of network training.
2. Penalty parameter K is selected as a sufficiently large positive number that ensures that the unconstrained problem represents a close approximation to its NP counterpart.

Once the modified objective function is specified, any one of many gradient techniques can be used to perform the minimization task. For the sake of simplicity, we will demonstrate the use of the steepest descent method. However, the conjugate gradient, Newton's, and quasi-Newton methods can offer significantly faster convergence rates [4], although at the price of increased computational complexity.

Applying the steepest descent approach, we can generate the update equations in accordance with

$$\boldsymbol{x}(k+1) = \boldsymbol{x}(k) - \mu \frac{\partial f_A(\boldsymbol{x})}{\partial \boldsymbol{x}} \tag{6.159}$$

where $\mu > 0$ is the learning rate parameter and the gradient term on the right-hand side of (6.159) depends on the penalty function selection. For example, when the form of the energy function is as given in (6.150), with $\rho_1 = 2$ and $\rho_2 = 1$, we have

$$\frac{\partial f_A(\boldsymbol{x})}{\partial \boldsymbol{x}} = \frac{\partial f(\boldsymbol{x})}{\partial \boldsymbol{x}} + K\left[\sum_{i=1}^{p} \frac{\partial h_i(\boldsymbol{x})}{\partial \boldsymbol{x}} h_i(\boldsymbol{x}) + \sum_{i=p+1}^{m} \frac{\partial}{\partial \boldsymbol{x}} \max\{0, g_i(\boldsymbol{x})\}\right] \tag{6.160}$$

After substituting (6.160) into (6.159), we have for the learning rule

$$\boldsymbol{x}(k+1) = \boldsymbol{x}(k) - \mu\left[\frac{\partial f(\boldsymbol{x})}{\partial \boldsymbol{x}} + K\sum_{i=1}^{p} \frac{\partial h_i(\boldsymbol{x})}{\partial \boldsymbol{x}} h_i(\boldsymbol{x}) + K\sum_{i=p+1}^{m} \frac{\partial}{\partial \boldsymbol{x}} \max\{0, g_i(\boldsymbol{x})\}\right] \tag{6.161}$$

The neural network architecture realization of this process is presented in Figure 6.11. Note that only a portion of the network for computing a single component of the independent-variable vector is presented. Also note that the network corresponds to the general case of the NP3 problem; that is, it accommodates both equality and inequality constraints. The appropriate networks for the NP1 and NP2 problems can be derived from the network in Figure 6.12 by eliminating portions responsible for either inequality constraints (in the case of NP1) or equality constraints (in the case of NP2).

EXAMPLE 6.4. Consider the following NP problem:

Minimize

$$f(\boldsymbol{x}) = \exp[(x_1 - 1.5)^2 + x_2^2] \tag{6.162}$$

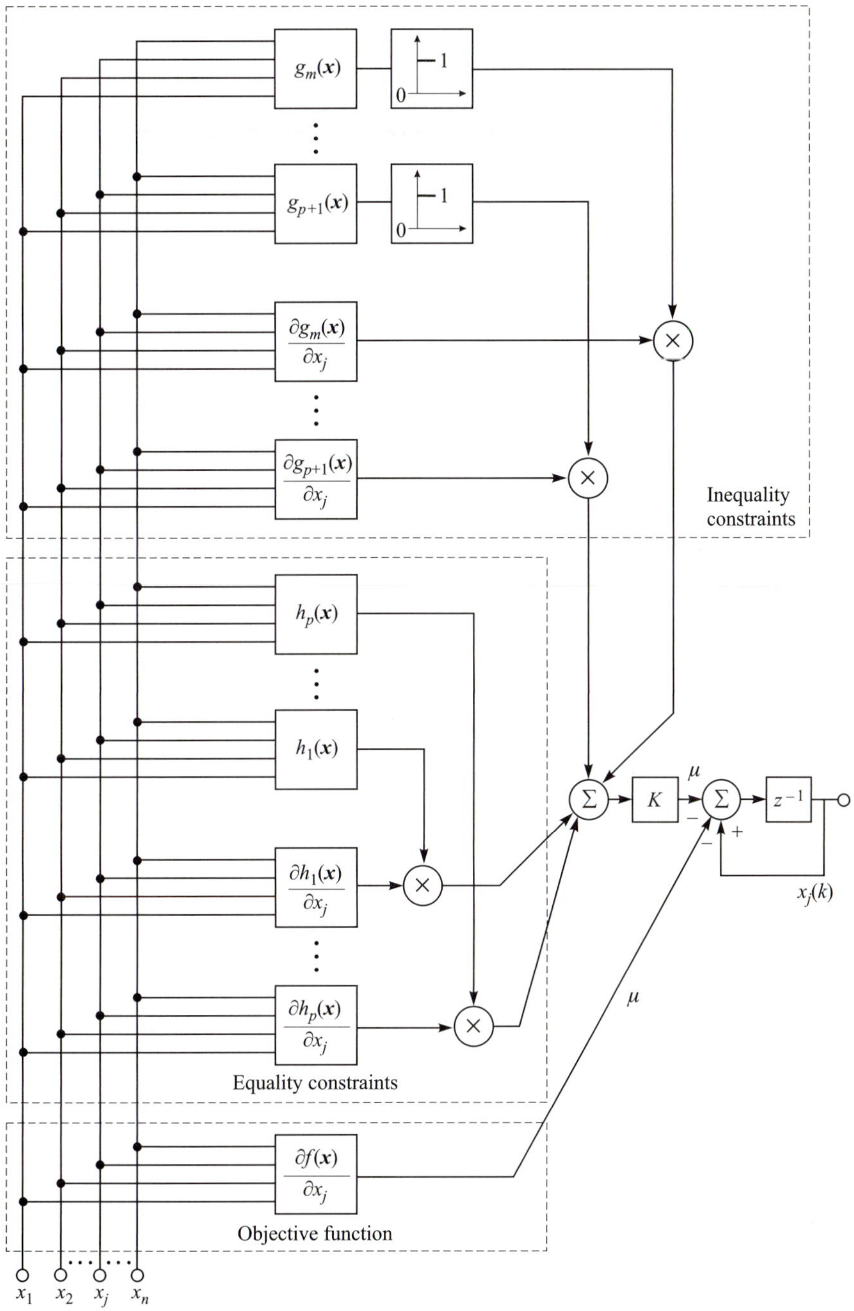

Figure 6.12 Discrete-time network for NP3 problems implementing penalty function method; implementation of Equation (6.161).

subject to

$$x_1^2 + x_2^2 - 1 \leq 0 \tag{6.163}$$

The above problem is obviously an NP problem with inequality constraints (i.e., NP2). The modified penalty function can be formed as:

$$f_A(\boldsymbol{x}) = \exp\left[(x_1 - 1.5)^2 + x_2^2\right] + \frac{K}{2}\max\left\{0, x_1^2 + x_2^2 - 1\right\} \tag{6.164}$$

By using the steepest descent method, the update equations can be computed as

$$\begin{aligned} x_1(k+1) = x_1(k) - \mu\Big\{&2(x_1 - 1.5)\exp\left[(x_1 - 1.5)^2 + x_2^2\right] \\ &+ K\left[\operatorname{sgn}(x_1^2 + x_2^2 - 1) + 1\right]x_1\Big\} \end{aligned} \tag{6.165}$$

and

$$x_2(k+1) = x_2(k) - \mu\left\{2x_2\exp\left[(x_1 - 1.5)^2 + x_2^2\right] + K\left[\operatorname{sgn}\left(x_1^2 + x_2^2 - 1\right) + 1\right]x_2\right\} \tag{6.166}$$

where $\operatorname{sgn}(v) = v/|v|$ is the sign function. The neural network architecture shown in Figure 6.12 was used to determine the solution of the NP problem. Parameters of the network were chosen as $K = 5$, and $\mu = 0.01$, and initial solution was set as $\boldsymbol{x} = [0 \;\; 1.5]^T$. The network converged in approximately 1,900 iterations, and the trajectory of the solution is shown in Figure 6.13.

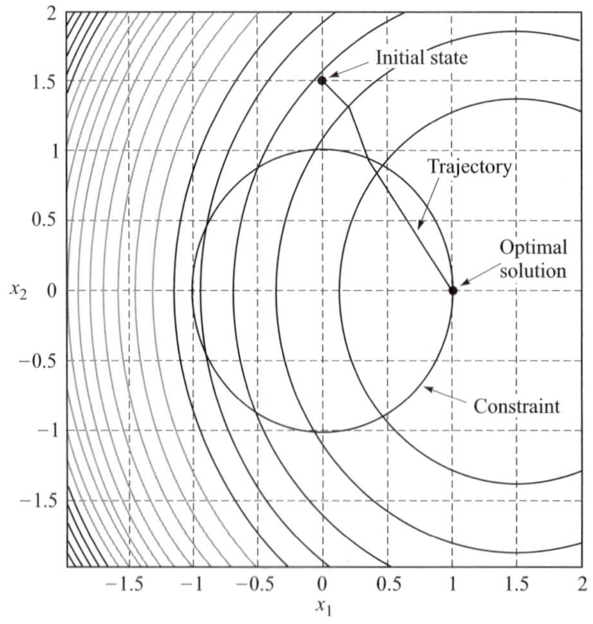

Figure 6.13 Trajectory of the solution for the NP problem in Example 6.4.

Computational difficulties in using penalty function

As we have discussed earlier, the solution to the unconstrained optimization problem (6.150) can be made arbitrarily close to the original NP problem provided that the value of the penalty parameter is sufficiently large. However, if a very large value for K is chosen, the Hessian of the penalty function may become ill-conditioned. In a situation when the solution of the NP problem resides on the hypersurface formed by the constraints (cf. Example 6.4), a large value of the penalty constant may cause oscillations of the algorithm about the boundary. If that type of behavior is detected, the algorithm must decrease the learning rate parameter μ to a very small value in order to achieve convergence.

The update equations derived from the penalty function method are based on some type of gradient learning method, for example, steepest descent, conjugate gradient, Newton's methods, etc. As such, they suffer from the common problem of all gradient methods—local minima. To prevent the algorithm's search from falling into a local minimum of the modified penalty function $f_A(\boldsymbol{x})$, we can use the method of stochastic annealing. The update equations based on this method can be derived according to

$$\boldsymbol{x}(k+1) = \boldsymbol{x}(k) - \mu \frac{\partial f_A(\boldsymbol{x})}{\partial \boldsymbol{x}} + \sigma^2 \boldsymbol{n} \tag{6.167}$$

where $\boldsymbol{n} \in \Re^{n \times 1}$ is a random generated white noise vector with zero mean and unity variance, that is,

$$E\{\boldsymbol{n}^T \boldsymbol{n}\} = 1 \tag{6.168}$$

and σ^2 is a parameter decreased toward zero over the course of network training. A larger value of σ^2 allows the algorithm to escape local minima, while the smaller value allows "fine-tuning" about the optimal solution.

6.4.2 Neural Networks for Barrier Function NP Methods

Similar to the penalty function methods, barrier methods are used to transform a constrained NP problem to an equivalent unconstrained problem or to a sequence of unconstrained problems. To accomplish the transformation, the original cost function of the NP problem is augmented with barrier functions which prevent the solution point from leaving the feasible region. To clarify this, consider the nonlinear programming problem of type NP2 as given in (6.134) and (6.135). The barrier function approach transforms the problem as follows:

Minimize

$$f_A(\boldsymbol{x}) = f(\boldsymbol{x}) + \frac{1}{K} \mathrm{B}(\boldsymbol{x}) \tag{6.169}$$

where $K > 0$ is a parameter used to control the barrier functions and $\mathrm{B}(\boldsymbol{x})$ is any function having the following properties:

1. $B(\boldsymbol{x}) = 0$ when $\boldsymbol{x} \in S$, where $S \subset \Re^{n\times 1}$ is the feasible region determined by the inequality conditions given in (6.135).
2. $B(\boldsymbol{x}) \to +\infty$ as the vector $\boldsymbol{x}$ approaches the border of the feasible region.

Ideally, the barrier function should have zero value everywhere in the feasible region except in the vicinity of the border, where it should approach infinity. In practice, the most commonly used barrier functions are

1. $$B(\boldsymbol{x}) = \sum_{i=1}^{m} \frac{1}{g_i(\boldsymbol{x})} \tag{6.170}$$

2. $$B(\boldsymbol{x}) = -\sum_{i=1}^{m} \ln\left[g_i(\boldsymbol{x})\right] \tag{6.171}$$

Since barrier functions in (6.170) and (6.171) have nonzero values over the feasible region, the parameter K needs to be a fairly large value to minimize the impact of the barrier terms on the NP cost function everywhere except at the boundary. However, large values of K may cause the problem to become ill-conditioned, and common practice is to increase its value over the course of network training. Note that both barrier functions given in (6.170) and (6.171) have discontinuities on the boundary of the feasible region. This imposes two serious restrictions on their use in NP problems. First, since the functions are discontinuous on the boundary, they cannot be used for the NP problems with equality constraints. Second, for many NP problems with inequality constraints the solution resides on the boundary itself (cf. Example 6.4). Some disadvantages of the barrier functions can be removed by applying the mixed penalty method as described in Cichocki and Unbehauen [2] and Nash and Sofer [9].

Unlike penalty function methods, barrier functions do not allow trial point $\boldsymbol{x}$ to escape the feasible region. This may be useful in the cases of NP problems where optimization may be terminated before the optimum point $\tilde{\boldsymbol{x}}$ is reached, for example, as soon as the cost function falls below some predetermined value. After the modified penalty function is constructed, any one of the unconstrained optimization methods can be used to determine the optimal solution of (6.169). Similar to the penalty function method, it can be shown that the solution of (6.169) converges to the solution of the original NP2 problem as $K \to \infty$ [7].

6.4.3 Neural Networks for Ordinary Lagrange Multiplier NP Methods

Similar to the penalty and barrier function approaches, the Lagrange multiplier method handles the constraints by incorporating them in a modified objective function. In developing the neural network implementation of the Lagrange multiplier method, we will first consider NP problems of the form NP1 (i.e., NP problems with equality constraints). The same approach will be extended to address NP problems with inequality constraints.

Lagrange multipliers for NP1 problems

The application of the Lagrange multiplier method (cf. Sect. A.6.2) for NP problems involves transformation of the NP problem to an unconstrained optimization problem. The unconstrained problem is formulated by appending the constraints to the objective function with Lagrange multipliers used as scaling factors. The new objective function is called the Lagrangian, and it is of the form

$$L(\boldsymbol{x}, \boldsymbol{\lambda}) = f(\boldsymbol{x}) + \sum_{i=1}^{m} \lambda_i h_i(\boldsymbol{x}) \tag{6.172}$$

where $\boldsymbol{x} \in \Re^{n\times 1}$ and $\boldsymbol{\lambda} = \left[\lambda_1, \lambda_2, \cdots, \lambda_m\right]^T \in \Re^{m\times 1}$.

The unconstrained problem in (6.172) has a total of $n+m$ unknowns and is higher in dimension than the original constrained problem; that is, the optimization is conducted in $\Re^{n+m}$ space. At the minimum, an objective function for the unconstrained optimization problem has to satisfy stationarity conditions. For the Lagrangian in (6.172), the stationarity conditions can be written as

$$\frac{\partial L(\boldsymbol{x}, \boldsymbol{\lambda})}{\partial \boldsymbol{x}} = \frac{\partial f(\boldsymbol{x})}{\partial \boldsymbol{x}} + \sum_{i=1}^{m} \lambda_i \frac{\partial h_i(\boldsymbol{x})}{\partial \boldsymbol{x}} = \boldsymbol{0} \tag{6.173}$$

and

$$\frac{\partial L(\boldsymbol{x}, \boldsymbol{\lambda})}{\partial \boldsymbol{\lambda}} = \left[h_1(\boldsymbol{x}), h_2(\boldsymbol{x}), \cdots, h_m(\boldsymbol{x})\right]^T = \boldsymbol{0} \tag{6.174}$$

where $\boldsymbol{0}$ denotes the zero (or null) vector of appropriate dimension. Equations (6.173) and (6.174) form a system of $n+m$ equations having $n+m$ unknowns that need to be solved for both the optimal values of independent variables and the Lagrange multipliers, that is, $(\tilde{\boldsymbol{x}}, \tilde{\boldsymbol{\lambda}})$. Note that (6.174) guarantees that all the constraints are satisfied at the optimal solution; in other words at the optimal point $\tilde{\boldsymbol{x}}$

$$L(\tilde{\boldsymbol{x}}, \boldsymbol{\lambda}) = f(\tilde{\boldsymbol{x}}) \tag{6.175}$$

For the neural network approach, the system of equations in (6.173) and (6.174) is solved in an iterative manner using some type of gradient-based technique. For the steepest descent approach we have for the two learning rules [2]

$$\boldsymbol{x}(k+1) = \boldsymbol{x}(k) - \mu_x \left[\frac{\partial f(\boldsymbol{x})}{\partial \boldsymbol{x}} + \sum_{i=1}^{m} \lambda_i \frac{\partial h_i(\boldsymbol{x})}{\partial \boldsymbol{x}}\right] \tag{6.176}$$

and

$$\boldsymbol{\lambda}(k+1) = \boldsymbol{\lambda}(k) + \mu_\lambda \left[h_1(\boldsymbol{x}), h_2(\boldsymbol{x}), \cdots, h_m(\boldsymbol{x})\right]^T \tag{6.177}$$

or in scalar form

$$x_j(k+1) = x_j(k) - \mu_x\left[\frac{\partial f(\boldsymbol{x}(k))}{\partial x_j} + \sum_{i=1}^{m} \lambda_i(k)\frac{\partial h_i(\boldsymbol{x}(k))}{\partial x_j}\right] \tag{6.178}$$

and

$$\lambda_j(k+1) = \lambda_j(k) + \mu_\lambda h_j(\boldsymbol{x}(k)) \tag{6.179}$$

where μ_x, $\mu_\lambda > 0$ are the learning rate parameters.

There are several problems associated with the practical implementation of the Lagrange multiplier method. In Section 6.3 we have seen that if the objective function is not convex, it can have multiple local minima and the minimization procedure may be easily trapped in one of them. In the case of Lagrange multipliers, even if the original cost function $f(\boldsymbol{x})$ is convex, there are no guarantees that the Lagrangian in (6.172) is convex as well. In addition, in some cases the iterative method in (6.178) and (6.179) may exhibit an oscillatory behavior about a local minimum. To prevent oscillations, and force the algorithm's convergence, an additional damping factor may be added to (6.177)

$$\boldsymbol{\lambda}(k+1) = \boldsymbol{\lambda}(k) + \mu_\lambda\left[\frac{\partial L(\boldsymbol{x}(k), \boldsymbol{\lambda}(k))}{\partial \boldsymbol{\lambda}} - \alpha\boldsymbol{\lambda}(k)\right] \tag{6.180}$$

where $0 \le \alpha \le 1$, called the *damping parameter*.

A neural network architecture for the process is presented in Figure 6.14.

Lagrange multipliers for NP2 problems

In NP2 problems, the constraints are given in the form of inequalities. By adding appropriate surplus variables the inequalities in (6.135) can be converted to equalities, and we can use the methods described in the previous section. The approach is illustrated by the following example.

EXAMPLE 6.5. Consider an NP2 problem given as follows:
Minimize

$$f(\boldsymbol{x}) = x_1^2 - 1.4x_1 + x_2^2 \tag{6.181}$$

subject to

$$x_1^2 + x_2^2 \le 1 \tag{6.182}$$

By adding a surplus variable θ^2, the original problem is transformed to this:
Minimize:

$$f(\boldsymbol{x}) = x_1^2 - 1.4x_1 + x_2^2 \tag{6.183}$$

subject to

$$x_1^2 + x_2^2 + \theta^2 = 1 \tag{6.184}$$

or

$$x_1^2 + x_2^2 + \theta^2 - 1 = 0 \tag{6.185}$$

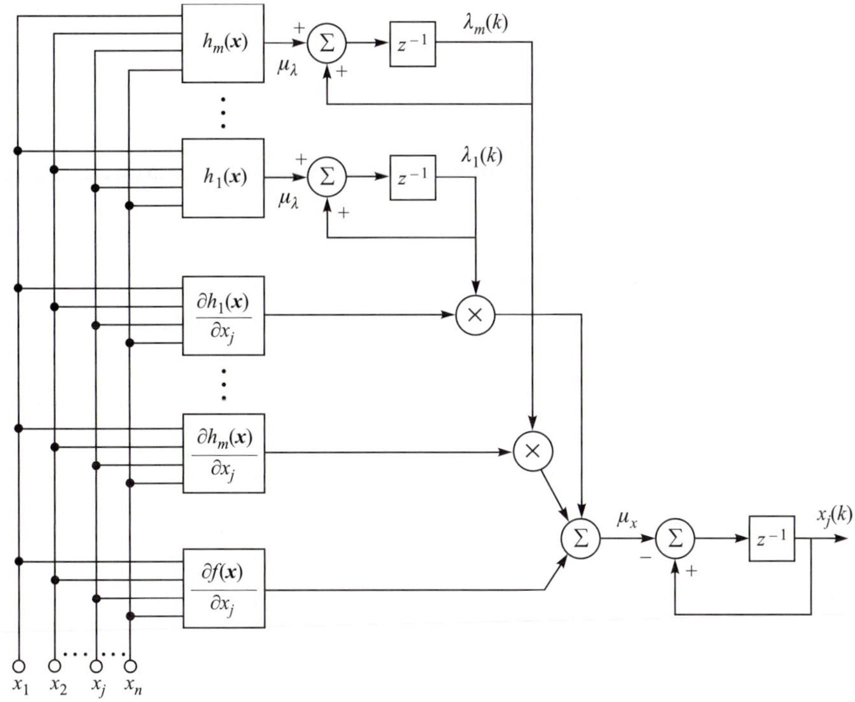

Figure 6.14 Discrete-time implementation of the Lagrange multiplier neural network. Equations (6.178) and (6.179) show the learning rule.

Now the Lagrangian formulation for the above problem can be formed as

$$L(\mathbf{x}, \lambda, \theta) = x_1^2 - 1.4x_1 + x_2^2 + \lambda(x_1^2 + x_2^2 + \theta^2 - 1) \tag{6.186}$$

Applying the stationarity conditions, we have

$$\frac{\partial L(\mathbf{x}, \lambda, \theta)}{\partial x_1} = 2(1 + \lambda)x_1 - 1.4 = 0 \tag{6.187}$$

$$\frac{\partial L(\mathbf{x}, \lambda, \theta)}{\partial x_2} = 2(1 + \lambda)x_2 = 0 \tag{6.188}$$

$$\frac{\partial L(\mathbf{x}, \lambda, \theta)}{\partial \lambda} = x_1^2 + x_2^2 + \theta^2 - 1 = 0 \tag{6.189}$$

$$\frac{\partial L(\mathbf{x}, \lambda, \theta)}{\partial \theta} = 2\lambda\theta = 0 \tag{6.190}$$

From (6.190) either $\lambda = 0$ or $\theta = 0$, or both $\lambda = 0$ and $\theta = 0$. Substituting $\lambda = 0$ into (6.187) through (6.189) yields a solution $x_1 = 0.7$, $x_2 = 0$, and $\theta^2 = 0.57$. The value of the cost function can be evaluated as $f(x_1, x_2) = -0.49$. Alternately, substituting $\theta = 0$ into (6.187) through (6.189), the solution of the system becomes $x_1 = 1$, $x_2 = 0$, and $\lambda = -0.3$. The value of the cost function at the point of the second solution is $f(x_1, x_2) = -0.4$. Finally, the system of equations obtained after setting both $\lambda = 0$ and $\theta = 0$ does not have a solution. Comparing the values of the cost function at the two stationary points, we see that the minimization is achieved at $x_1 = 0.7$ and $x_2 = 0$. The second solution gives the point of the local minimum for the Lagrangian.

In general, for NP2 problems, the Lagrangian function is formed as

$$L(\boldsymbol{x}, \lambda, \theta) = f(\boldsymbol{x}) + \sum_{j=1}^{m} \lambda_j [g_j(\boldsymbol{x}) + \theta_j^2] \tag{6.191}$$

Applying the stationary conditions gives

$$\frac{\partial L(\boldsymbol{x}, \lambda, \theta)}{\partial x_i} = \frac{\partial f(\boldsymbol{x})}{\partial x_i} + \sum_{j=1}^{m} \lambda_j \frac{\partial g_j(\boldsymbol{x})}{\partial x_i} = 0 \qquad i = 1, 2, \ldots, n \tag{6.192}$$

$$\frac{\partial L(\boldsymbol{x}, \lambda, \theta)}{\partial \lambda_i} = g_i(\boldsymbol{x}) - \theta_i^2 = 0 \qquad i = 1, 2, \ldots, m \tag{6.193}$$

$$\frac{\partial L(\boldsymbol{x}, \lambda, \theta)}{\partial \theta_i} = 2\lambda_i \theta_i = 0 \qquad i = 1, 2, \ldots, m \tag{6.194}$$

From (6.194) we can recognize three different cases: either $\lambda_i = 0$ or $\theta_i = 0$, or both; more specifically,

1. If $\lambda_i = 0$ and $\theta_i \neq 0$, then the constraint $g_i(\boldsymbol{x}) \leq 0$ is redundant and can be ignored. In other words, the presence of the constraint $g_i(\boldsymbol{x}) \leq 0$ does not change the results of the optimization.
2. If $\lambda_i \neq 0$ and $\theta_i = 0$, the optimal solution resides on the boundary of the feasible region, or on the curve determined by $g_i(\boldsymbol{x}) = 0$
3. If both $\lambda_i = 0$ and $\theta_i = 0$, then the corresponding boundary of the feasible region given by $g_i(\boldsymbol{x}) = 0$ passes through the global minimum of the cost function.

In the neural network approach, the system of equations in (6.192) through (6.194) is solved in an iterative manner using one of the gradient techniques. Several issues associated with this approach are outlined in the previous section.

Transformation of the inequality constraints to equivalent equality constraints increases the dimension of the nonlinear programming problem. For each transformed equation we introduce an additional variable. Although this does not present a problem for relatively small cases, in the case of large problems with many inequalities this may significantly increase the computational burden. For that reason, in Golub and Van Loan [2] and Press et al. [13], the regular Lagrange multiplier method is extended to inequality constraints according to the following equations

$$x_j(k+1) = x_j(k) - \mu \left[\frac{\partial f(\boldsymbol{x})}{\partial x_j} + \sum_{i=1}^{m} \lambda_i \frac{\partial g_i(\boldsymbol{x})}{\partial x_j} \right] \qquad j = 1, 2, \ldots, n \tag{6.195}$$

and

$$\lambda_i(k+1) = \max\{0, \lambda_i(k) + \eta g_i(\boldsymbol{x})\} \qquad i = 1, 2, \ldots, m \tag{6.196}$$

where $\eta > 0$ denotes the leaning rate parameter.

The neural network architecture realization of this process is shown in Figure 6.15. As can be seen from the figure, it is very similar to the architec-

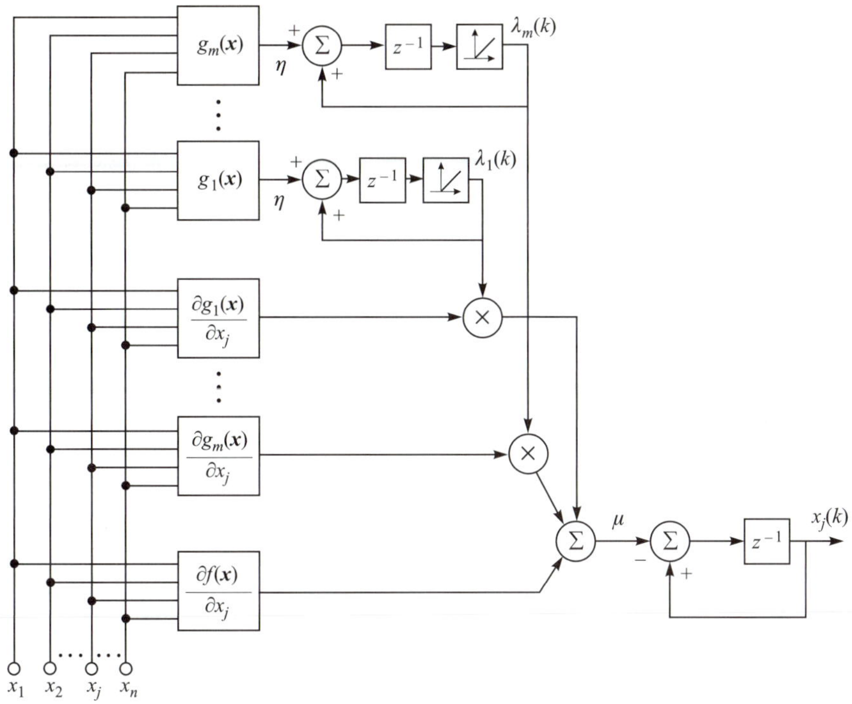

Figure 6.15 Discrete-time implementation of the Lagrange multiplier neural network for NP problems with inequality constraints. Equations (6.195) and (6.196) show the learning rules.

ture in Figure 6.14. The only difference is the nonlinear element evaluating the max function in (6.196).

6.4.4 Neural Networks for the Augmented Lagrange Multiplier Method

The augmented Lagrange multiplier method is one of the most effective general approaches to solving NP problems. It was derived independently by Hestens [14] and Powell [15]. Gill et. al. [17] provide additional references on quite extensive research conducted for these types of optimization problems. The augmented Lagrange multiplier method is applicable to all three forms of NP problems given in (6.132) through (6.138). For the sake of simplicity, we first describe its use in NP problems with equality constraints and then discuss its extension to NP problems having inequality or mixed constraints.

Augmented Lagrange methods for NP1 problems

According to the previous section, the Lagrangian function associated with the constrained NP defined in (6.132) and (6.133) is given by

$$L(\boldsymbol{x}, \lambda) = f(\boldsymbol{x}) + \sum_{i=1}^{m} \lambda_i h_i(\boldsymbol{x}) \tag{6.197}$$

The augmented Lagrangian is formed from (6.197) by addition of extra penalty terms. The most popular form of the augmented Lagrangian is given by [17, 18]

$$L_A(\boldsymbol{x}, \lambda, \boldsymbol{k}) = f(\boldsymbol{x}) + \sum_{i=1}^{m} \lambda_i h_i(\boldsymbol{x}) + \sum_{i=1}^{m} k_i h_i^2(\boldsymbol{x}) \tag{6.198}$$

where $\boldsymbol{\lambda} = [\lambda_1, \lambda_2, \cdots, \lambda_m]^T \in \Re^{m\times 1}$ denotes the vector of the Lagrange multipliers and $\boldsymbol{k} = [k_1, k_2, \cdots, k_m]^T \in \Re^{m\times 1}$ is the vector of positive penalty parameters.

It can be shown that the addition of the quadratic penalty term in (6.198) increases the positive definiteness of the Lagrangian's Hessian matrix [16, 17]. Furthermore, if the coefficients in $\boldsymbol{k}$ are sufficiently large, the Hessian matrix of the Lagrangian can be forced to have all eigenvalues greater then zero, which is of great importance from the neural network implementation standpoint, since the solution has to be sought in an iterative manner. We have already mentioned (cf. Theorem 6.2) that if a function is convex on a set, gradient-based search methods are guaranteed to converge to the local minimum. The fact that the Hessian matrix is positive definite ensures existence of a neighborhood of the solution in which the function is convex; and therefore, if the initial point in the optimization process is properly chosen, the algorithm will converge to the global minimum. After the augmented Lagrangian is formed, the optimal solution is found as its unconstrained minimum. This can be done by using any of the unconstrained optimization techniques. In the most simplistic approach, we can use steepest descent where the minimization of the Lagrangian is converted to a system of difference equations of the form

$$\boldsymbol{x}(k+1) = \boldsymbol{x}(k) - \mu_x \frac{\partial L(\boldsymbol{x}, \boldsymbol{\lambda}, \boldsymbol{k})}{\partial \boldsymbol{x}} \tag{6.199}$$

and

$$\boldsymbol{\lambda}(k+1) = \boldsymbol{\lambda}(k) + \mu_\lambda \frac{\partial L(\boldsymbol{x}, \boldsymbol{\lambda}, \boldsymbol{k})}{\partial \boldsymbol{\lambda}} \tag{6.200}$$

Evaluating the gradients specified in (6.199) and (6.200) yields the following update equations:

$$x_j(k+1) = x_j(k) - \mu_x \left\{ \frac{\partial f(\boldsymbol{x}(k))}{\partial x_j} + \sum_{i=1}^{m} [\lambda_i(k) + 2k_i h_i(\boldsymbol{x}(k))] \frac{\partial h_i(\boldsymbol{x}(k))}{\partial x_j} \right\} \tag{6.201}$$

and

$$\lambda_j(k+1) = \lambda_j(k) + \mu_\lambda h_j(\boldsymbol{x}(k)) \tag{6.202}$$

where $\mu_x, \mu_\lambda > 0$ represent the learning rate parameters.

A neural network architecture realization of (6.201) and (6.202) is shown in Figure 6.16. The network consists of two separate modules. The first mod-

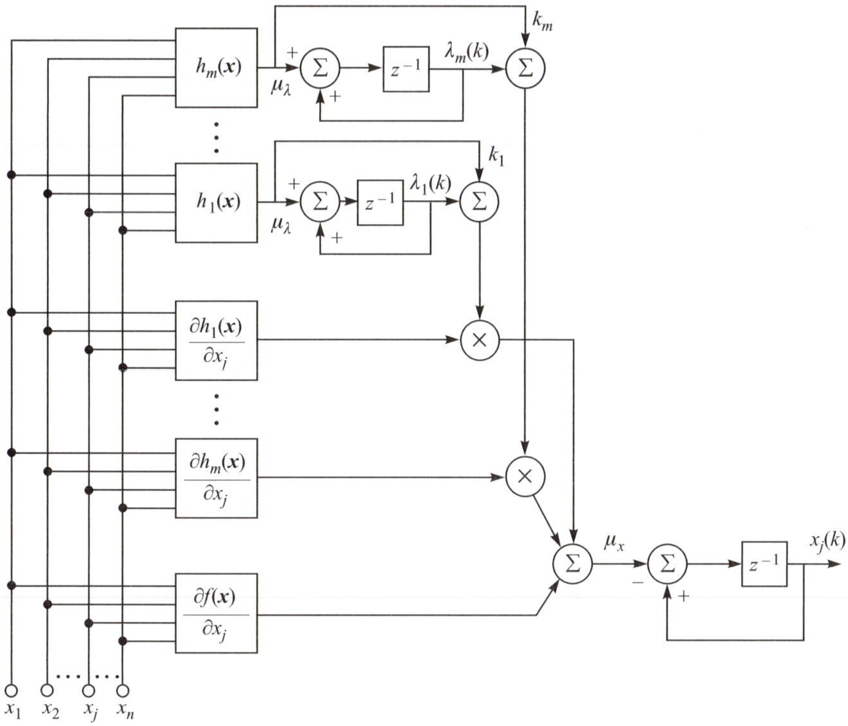

Figure 6.16 Discrete-time implementation of the neural network for augmented Lagrange multiplier method. Learning rules are given in (6.201) and (6.202).

ule performs the update of the solution, and the second module updates the Lagrange multipliers. Earlier versions of the augmented Lagrange method suggested different update rates for the solution $\boldsymbol{x}$ and Lagrange multipliers [15, 17]. Furthermore, in these early versions a complete unconstrained minimization was performed relative to $\boldsymbol{x}$ before each update of the Lagrange multipliers. This approach proved to be quite inefficient since the accuracy of the solution depends on the accuracy of the Lagrange multiplier estimates, and the algorithm will not converge to the optimal value of $\boldsymbol{x} = \tilde{\boldsymbol{x}}$ until the multipliers converge to the optimal value $\lambda = \tilde{\lambda}$. Based on these observations, different update strategies have been proposed [2, 15–18]. The algorithm presented in (6.201) and (6.202) is an extreme case in which both the solution estimate and the estimate of the Lagrange multipliers are computed in each iteration. In essence, the algorithm is posing a different optimization problem at every iterative step, and it is closely related to the so-called QP-based projected Lagrangian method [17].

There are several important properties of the augmented Lagrange multiplier method that need to be taken into consideration [17]:

1. *Local-minimum property*. Similar to the penalty function approach, the augmented Lagrange multiplier method guarantees convergence to the local minimum of the augmented Lagrangian. The local minimum of the augmented Lagrangian converges to the constrained minimum of the

objective function only in the limiting case when the penalty parameters in $\boldsymbol{k}$ are sufficiently large.

2. *Choice of the penalty parameter*. In general, the penalty parameters need to be chosen so that the Hessian matrix of the augmented Lagrangian is positive definite. If the values of the penalty parameters are too small, the algorithm may fail to converge or it may converge to a value that is a local minimum of the augmented Lagrangian but does not minimize the objective function itself. On the other hand, if the parameters are chosen too large, the algorithm may exhibit oscillatory behavior in the vicinity of the solution.
3. *Convergence of the Lagrange multipliers*. For the algorithm in (6.201) and (6.202), to find the optimal solution it is necessary that both $\boldsymbol{x}$ and $\boldsymbol{\lambda}$ converge to their optimal values $\tilde{\boldsymbol{x}}$ and $\tilde{\boldsymbol{\lambda}}$. In some cases the augmented Lagrangian can be very sensitive to the values for the multipliers, and it may take a considerable number of iterations before convergence is achieved [17].

Augmented Lagrange methods for NP2 problems

The method of the augmented Lagrange multipliers in the previous section can be extended to NP problems with inequality constraints. To accomplish this, the augmented Lagrangian is modified according to

$$L(\boldsymbol{x}, \lambda) = f(\boldsymbol{x}) + \sum_{i=1}^{m} \lambda_i \max\{0, g_i(\boldsymbol{x})\} + \sum_{i=1}^{m} \frac{K_i}{2} \max\{0, g_i(\boldsymbol{x})\}^2 \tag{6.203}$$

where λ_i, $i = 1, 2, \ldots, m$ are Lagrange multipliers and K_i, $i = 1, 2, \ldots, m$, are the penalty parameters. As can be seen from (6.203), any violation of the constraints increases the value of the Lagrangian; that is, only violated constraints are considered to be active. In a more compact form, (6.203) can be written as

$$L(\boldsymbol{x}, \lambda) = f(\boldsymbol{x}) + \sum_{i=1}^{m} S_i \left[\lambda_i g_i(\boldsymbol{x}) + \frac{K_i}{2} g_i(\boldsymbol{x})^2 \right] \tag{6.204}$$

where

$$S_i = \begin{cases} 0 & \text{if } g_i(\boldsymbol{x}) \le 0 \\ 1 & \text{if } g_i(\boldsymbol{x}) > 0 \end{cases} \tag{6.205}$$

To derive a corresponding neural network, the update equations can be obtained according to

$$\boldsymbol{x}(k+1) = \boldsymbol{x}(k) - \mu_x \frac{\partial L(\boldsymbol{x}, \lambda)}{\partial \boldsymbol{x}} \tag{6.206}$$

and

$$\boldsymbol{\lambda}(k+1) = \boldsymbol{\lambda}(k) + \mu_\lambda \frac{\partial L(\boldsymbol{x}, \boldsymbol{\lambda})}{\partial \boldsymbol{\lambda}} \tag{6.207}$$

After substitution of the appropriate gradients of (6.204) into (6.206) and (6.207), we have

$$x_j(k+1) = x_j(k) - \mu_x \left\{ \frac{\partial f(\boldsymbol{x}(k))}{\partial x_j} + \sum_{i=1}^{m} S_i[\lambda_i + K_i g_i(\boldsymbol{x}(k))] \frac{\partial g_i(\boldsymbol{x}(k))}{\partial x_j} \right\} \tag{6.208}$$

and

$$\lambda_j(k+1) = \lambda_j(k) + \mu_\lambda S_j g_j(\boldsymbol{x}(k)) \tag{6.209}$$

A neural network architecture realization of this process is shown in Figure 6.17.

EXAMPLE 6.6. Let us consider the problem in Example 6.4 using the method of augmented Lagrange multipliers. For the problem in (6.133) and (6.134) the augmented Lagrangian can be formed as

$$L(\boldsymbol{x}, \boldsymbol{\lambda}) = \exp\left[(x_1 - 1.5)^2 + x_2^2\right] + \lambda \max\{0,\ x_1^2 + x_2^2 - 1\} + \frac{K}{2} \max\{0,\ x_1^2 + x_2^2 - 1\}^2 \tag{6.210}$$

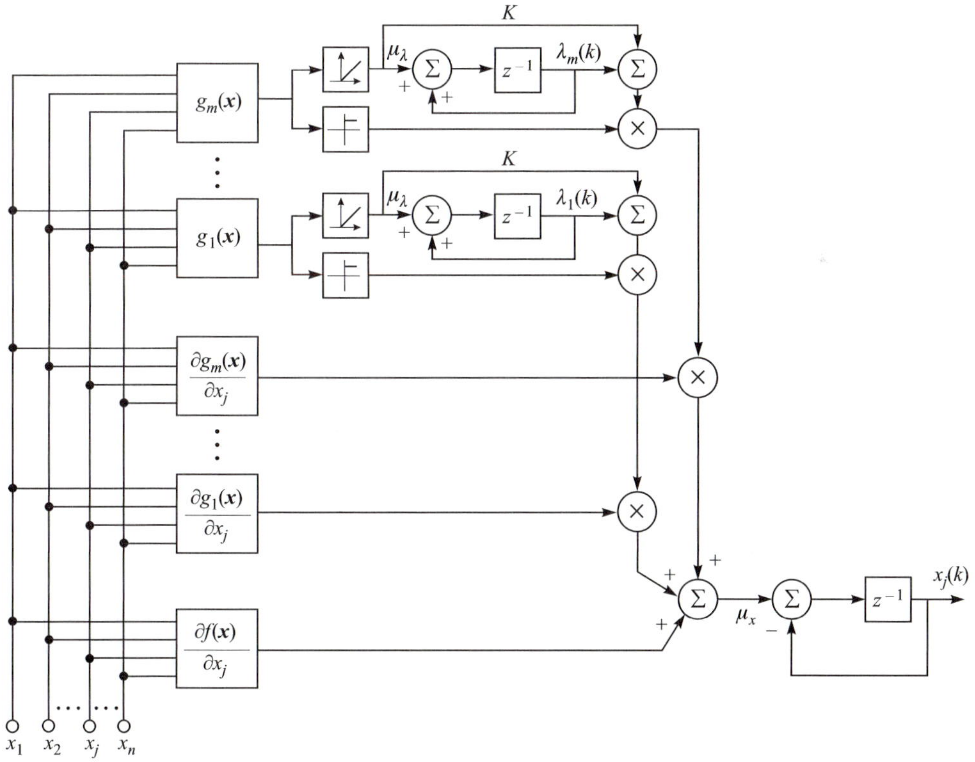

Figure 6.17 Discrete-time realization of the neural network for implementation of augmented Lagrange multiplier method for NP problems with inequality constraints. Equations (6.208) and (6.209) show the two learning rules.

The appropriate derivatives can be computed as

$$\frac{\partial L(\boldsymbol{x}, \boldsymbol{\lambda})}{\partial x_1} = 2(x_1 - 1.5)\exp\left[(x_1 - 1.5)^2 + x_2^2\right]$$
$$+ 2x_1\left[\operatorname{sgn}\left(x_1^2 + x_2^2 - 1\right) + 1\right]\left[\lambda + K\left(x_1^2 + x_2^2 - 1\right)\right] \quad (6.211)$$
$$\frac{\partial L(\boldsymbol{x}, \boldsymbol{\lambda})}{\partial x_2} = 2x_2\exp\left[(x_1 - 1.5)^2 + x_2^2\right]$$
$$+ 2x_2\left[\operatorname{sgn}\left(x_1^2 + x_2^2 - 1\right) + 1\right]\left[\lambda + K\left(x_1^2 + x_2^2 - 1\right)\right] \quad (6.212)$$

and

$$\frac{\partial L(\boldsymbol{x}, \boldsymbol{\lambda})}{\partial \lambda} = \left[\operatorname{sgn}\left(x_1^2 + x_2^2 - 1\right) + 1\right]\left(x_1^2 + x_2^2 - 1\right) \quad (6.213)$$

The network from Figure 6.17 was used to perform the optimization task. The parameters of the network were set to the values used in Example 6.4, that is, $K = 5$ and $\mu = 0.01$. Initial conditions were chosen as $x_1(0) = 0$, $x_2(0) = 0$, and $\lambda(0) = 1$. The network converged in approximately 700 iterations to the solution $\tilde{x} = [1.0088, 0.0002]^T$.

PROBLEMS

6.1. Write a MATLAB program to implement the neural network approach shown in Figure 6.1. Test the routine by solving the following LP problems:

(*a*) Maximize $f(\boldsymbol{x}) = 2x_1 + 4x_2$
subject to $x_1 - x_2 \le 1$
$3x_1 + 2x_2 \le 12$
$2x_1 + 3x_2 \le 3$
$-2x_1 + 3x_2 \le 9$
$x_1, x_2 \ge 0$

(*b*) Maximize $f(\boldsymbol{x}) = x_1 + 2x_2$
subject to $-x_1 + x_2 \le 1$
$-x_1 + x_2 \ge -1$
$x_1 + x_2 \le 4$
$x_1, x_2 \ge 0$

(*c*) Minimize $f(\boldsymbol{x}) = -5x_1 - x_2$
subject to $x_1 - x_2 \le 2$
$x_1 + 2x_2 \le 8$
$x_1, x_2 \ge 0$

(*d*) Maximize $f(\boldsymbol{x}) = 10x_1 + 19x_2 + 9x_3$
subject to $2x_1 + 3x_2 + 2x_3 \le 10$
$x_1, x_2, x_3 \ge 0$

For each of the problems:
- Convert the problem to the standard form.
- Use the neural network approach to solve the optimization problem.
- Solve the problem graphically and compare the results to the one obtained by using the neural network.

6.2. Write a MATLAB program to implement the neural network approach shown in Figure 6.3.

(*a*) Use the neural network approach to solve Problem 6.1.
(*b*) Solve the problems graphically, and verify the results obtained by the neural network.
(*c*) Compare the performance of the networks in Figures 6.1 and 6.3 from the viewpoint of complexity and convergence speed.

6.3. Repeat Problem 6.2, using the neural network given in Figure 6.4.

6.4. Consider an LP problem with inequality constraints defined as

$$\begin{aligned}&\text{Minimize} \quad f(\boldsymbol{x}) = \boldsymbol{c}^T\boldsymbol{x}\\ &\text{subject to} \quad \boldsymbol{A}\boldsymbol{x} \geq \boldsymbol{b}\end{aligned}$$

and its dual problem, formulated as

$$\begin{aligned}&\text{Maximize} \quad g(\boldsymbol{\lambda}) = \boldsymbol{b}^T\lambda\\ &\text{subject to} \quad \boldsymbol{A}^T\boldsymbol{\lambda} = \boldsymbol{c} \qquad \boldsymbol{\lambda} \geq 0\end{aligned}$$

where $\boldsymbol{A} \in \Re^{m\times n}$, $\boldsymbol{x}, \boldsymbol{c} \in \Re^{n\times 1}$, $\boldsymbol{b}, \boldsymbol{\lambda} \in \Re^{m\times 1}$. Consider the duality theorem (cf. Theorem 6.1 in Section 6.1) and the energy function defined as

$$E(\boldsymbol{x}, \boldsymbol{\lambda}) = \frac{1}{2}\left(\boldsymbol{c}^T\boldsymbol{x} - \boldsymbol{b}^T\boldsymbol{\lambda}\right)^2 + \frac{K_1}{2}\left(\boldsymbol{A}^T\boldsymbol{\lambda} - \boldsymbol{c}\right)^T\left(\boldsymbol{A}^T\boldsymbol{\lambda} - \boldsymbol{c}\right) + \frac{K_2}{2}\|\boldsymbol{\Phi}[\boldsymbol{A}\boldsymbol{x} - \boldsymbol{b}]\|_2^2$$

where for $\boldsymbol{v} \in \Re^{n\times 1}$

$$\boldsymbol{\Phi}(\boldsymbol{v}) = \begin{bmatrix}\max\{0, v_1\}\\ \max\{0, v_2\}\\ \vdots\\ \max\{0, v_n\}\end{bmatrix}$$

The duality theorem can be used to derive an iterative process for simultaneously solving both the primal and dual LP problems.

(*a*) Using the gradient approach and the above-defined energy function, derive the update equations for $\boldsymbol{x}$ and $\boldsymbol{\lambda}$. Note that $\boldsymbol{x}$ should be updated in accordance with steepest descent while the steepest ascent should be used for the update of $\boldsymbol{\lambda}$.
(*b*) Design a neural network implementing the equations in part (*a*).
(c) Write a MATLAB program to simulate the neural network in part (*b*). Test the code on the LP problem given by

$$\boldsymbol{b} = \begin{bmatrix}-3 & -16 & -20\end{bmatrix}^T \qquad \boldsymbol{A} = \begin{bmatrix}2 & 1\\ -1 & -3\\ -4 & -1\end{bmatrix} \quad \text{and} \quad \boldsymbol{c} = [-1 \quad -1]^T$$

6.5. Using Newton's gradient method (cf. Sect. A.5.3) requires computing the inverse of the Hessian matrix. Consider the energy function defined in (6.20) for the LP problem with equality constraints.

(*a*) Show that the Hessian matrix of the function in (6.20) can be computed as

$$\boldsymbol{H} = \nabla_x^2 E_1(\boldsymbol{x}) = K\boldsymbol{A}^T\boldsymbol{A}$$

(*b*) Prove that the Hessian matrix of the function in (6.20) has to be singular.

(*c*) The Levenberg–Marquardt algorithm (cf. Sect. 3.4.4) can be used to provide fast convergence, even if the Hessian matrix is singular. The update equations for this method are formed in accordance with

$$\boldsymbol{x}(k) = \boldsymbol{x}(k) - (\mu \boldsymbol{I} + \boldsymbol{H})^{-1}\boldsymbol{g}$$

where $E(\boldsymbol{x})$ is the energy function to be minimized, $\boldsymbol{H} = \nabla_x^2 E(\boldsymbol{x})$, $\boldsymbol{g} = \nabla_x E$, $\boldsymbol{I}$ is the identity matrix of appropriate dimension, and $\mu > 0$.

- Derive the Levenberg–Marquardt update equations for the energy function given in (6.20).
- Write a MATLAB program to implement the Levenberg–Marquardt method.
- Test the performance of your routine using LP problems given in Problem 6.1.

6.6. Equation (6.61) shows the energy function for the LP problem, with mixed constraints. It was formed using (6.21) and (6.40) as one of the possible combinations.

(*a*) Formulate an alternative energy function using (6.20) and (6.48).
(*b*) Based on the energy function in part (*a*) and the steepest descent approach, derive the update equations.
(*c*) Design a neural network implementing the equations in part (*b*).
(*d*) Write a MATLAB program to implement the equations in part (*c*), and test it using the LP problems given in Problem 6.1.

6.7. Update equations for the QP problem with equality constraints in (6.90) were derived using the steepest descent method. Show that an alternative set of equations can be formulated using

(*a*) Newton's method:

$$\boldsymbol{x}(k+1) = \boldsymbol{x}(k) - \left(\boldsymbol{Q} + k\boldsymbol{A}^T\boldsymbol{A}\right)^{-1}\left\{\boldsymbol{c} + \boldsymbol{Q}\boldsymbol{x}(k) + \boldsymbol{A}^T\boldsymbol{\lambda}(k) + K\boldsymbol{A}^T[\boldsymbol{A}\boldsymbol{x}(k) - \boldsymbol{b}]\right\}$$

(b) the Levenberg–Marquardt method:

$$\begin{aligned}\boldsymbol{x}(k+1) = \boldsymbol{x}(k) - (\mu\boldsymbol{I} + \boldsymbol{Q} + k\boldsymbol{A}^T\boldsymbol{A})^{-1}\{&c + \boldsymbol{Q}\boldsymbol{x}(k) + \boldsymbol{A}^T\boldsymbol{\lambda}(k) \\ &+ K\boldsymbol{A}^T[\boldsymbol{A}\boldsymbol{x}(k) - \boldsymbol{b}]\}\end{aligned}$$

where $\mu > 0$.

6.8. To illustrate the use of the penalty function method, consider the following NP problem:

$$\begin{aligned}\text{Minimize} \quad & f(\boldsymbol{x}) = -x_1x_2 \\ \text{subject to} \quad & h(\boldsymbol{x}) = 3x_1 + x_2 - 7 = 0\end{aligned}$$

with the exact solution given as $x_1 = {}^7\!/_6$ and $x_2 = {}^7\!/_2$.

(*a*) Form the energy function in accordance with

$$f_A(\boldsymbol{x}) = f(\boldsymbol{x}) + Kh(\boldsymbol{x})^2 \qquad K > 0$$

and prove that the unconstrained minimum is reached for

$$x_1 = \frac{14K}{12K-1} \qquad \text{and} \qquad x_2 = \frac{42K}{12K-1}$$

(*b*) Find the Hessian matrix of the energy function, and determine its condition number as a function of K . The condition number is defined as

$$\rho = \frac{\lambda_{\text{max}}}{\lambda_{\text{min}}}$$

where λ_{max} and λ_{min} are the largest and smallest eigenvalues of the matrix, respectively.

(*c*) Discuss the impact of parameter K from the standpoint of the solution accuracy as well as the viewpoint of the stability of the gradient search procedure.

(*d*) Write a MATLAB program to implement the neural network in Figure 6.12, and use it to solve the above NP problem. Experiment with different values for K.

6.9. Consider an NP problem defined as

$$\begin{aligned} \text{Minimize} \quad & f(\boldsymbol{x}) = e^{x_1} - x_1 x_2 + x_2^2 \\ \text{subject to} \quad & x_1^2 + x_2^2 = 4 \\ & 2x_1 + x_2 \leq 2 \end{aligned}$$

This is a problem of the NP3 form; that is, it has both equality and inequality constraints. One way to approach this problem is to form an augmented cost function using the mixed penalty-barrier method. For example, a mixed penalty-barrier function can be formed as

$$f_A(\boldsymbol{x}) = f(\boldsymbol{x}) + K \sum_{i=1}^{p} |h_i(\boldsymbol{x})|^{\rho_1} + \frac{1}{K} \sum_{i=p+1}^{m} \frac{1}{|g_i(\boldsymbol{x})|^{\rho_2}}$$

where $h_i(\boldsymbol{x})$ are equality constraints and $g_i(\boldsymbol{x})$ are inequality constraints and ρ_1, ρ_2 are positive numbers.

(*a*) Form the mixed penalty-barrier augmented cost function for the NP problem given above.

(*b*) Use the steepest descent gradient approach and the cost function in part (*a*) to derive the update equations.

(*c*) Write a MATLAB program to implement the equations given in part (*b*), and use your program to solve the above NP problem. Make sure that the initial point of the iterative procedure falls within the feasible region relative to the inequality constraints.

(*d*) Experiment with different values of parameter K, and discuss the convergence and the accuracy of the result as a function of K. Develop an update scheme to automatically increase K during the iterative process, and modify your MATLAB code in part (*c*) accordingly.

6.10. Consider the following NP problem:

$$\begin{aligned} \text{Minimize} \quad & f(\boldsymbol{x}) = x_1^2 + x_2^2 \\ \text{subject to} \quad & x_1 + x_2 = 1 \\ & x_1 + x_2 = 2 \end{aligned}$$

Clearly, the problem has a conflicting set of constraints and cannot be solved. However, if we apply the penalty method, an augmented cost function can be formed according to

$$f_A(\mathbf{x}) = x_1^2 + x_2^2 + K\left[(x_1 + x_2 - 1)^2 + (x_1 + x_2 - 2)^2\right]$$

(*a*) Ignoring the fact that the problem has no solution, derive the update equations for unconstrained optimization of the augmented cost function.
(*b*) Write MATLAB code to implement equations in part (*a*) and perform the unconstrained optimization.
(*c*) Explain the results.

6.11. Consider the following QP problem with inequality constraints:

$$\begin{aligned} &\text{Minimize} \quad f(x, y) = x^2 - xy + y^2 - 3x \\ &\text{subject to} \quad x + y \le 4 \qquad x, y \ge 0 \end{aligned}$$

(*a*) Restate the problem in vector-matrix form.
(*b*) Write a MATLAB program to implement the neural network shown in Figure 6.8. Transform the above QP problem to standard form, and use your MATLAB program to solve the problem.
(*c*) Write a MATLAB program to implement the neural network shown in Figure 6.10. Test your program by solving the QP problem given above.
(*d*) Repeat part (*c*) using the neural network approach shown in Figure 6.11.

6.12. Use the method of augmented Lagrange multipliers to solve the following NP problem

$$\begin{aligned} &\text{Minimize} \quad f(\mathbf{x}) = x_1^3 + x_2^2 \\ &\text{subject to} \quad x_1^2 + x_2^2 - 10 = 0 \\ &\qquad\qquad\quad 1 - x_1 \le 0 \\ &\qquad\qquad\quad 1 - x_2 \le 0 \end{aligned}$$

REFERENCES

1. S. I. Gass, *Linear Programming, Methods and Applications*, 5th ed., New York: McGraw-Hill, 1984.
2. A. Cichocki and R. Unbehauen, *Neural Networks for Optimization and Signal Processing*, Chichester, England: Wiley, 1993.
3. D. J. Wilde and C. S. Beightler, *Foundation of Optimization*, Englewood Cliffs, NJ: Prentice-Hall, 1967.
4. D. G. Luenberger, *Linear and Nonlinear Programming*, 2nd ed., Reading, MA: Addison-Wesley, 1989.
5. D. M. Himmelblau, *Applied Nonlinear Programming*, New York: McGraw-Hill, 1972.
6. W. I. Zangwill, *Nonlinear Programming, A Unified Approach*, Englewood Cliffs, NJ: Prentice-Hall, 1969.
7. M. S. Bazaraa and C. M. Shetty, *Nonlinear Programming, Theory and Applications*, New York: Wiley, 1979.
8. N. K. Kwak and M. C. Schniederjans, *Mathematical Programming*, Malabar, FL: Krieger, 1987.
9. S. G. Nash and A. Sofer, *Linear and Nonlinear Programming*, New York: McGraw-Hill, 1996.

10. A. Ravindran, D. T. Philips, and J. J. Solberg, *Operations Research, Principles and Practice*, New York: Wiley, 1987.
11. D. P. Gaver and G. L. Thompson, *Programming and Probability Models in Operations Research*, Belmont, CA: Wadsworth, 1973.
12. J. M. Ortega and W. C. Rheinboldt, *Iterative Solution of Nonlinear Equations in Several Variables*, New York: Academic Press, 1970.
13. K. J. Arrow, L. Hurwicz, and H. Uzawa, *Studies in Linear and Nonlinear Programming*, Stanford, CA: Stanford University Press, 1958.
14. M. R. Hestens, "Multiplier Gradient Methods," *Journal of Optimization Theory and Applications*, vol. 4, 1969, pp. 303–20.
15. M. J. D. Powell, "A Method for Nonlinear Constraints in Minimization Problems," In *Optimization*, ed. R. Fletcher, London: Academic, 1969, pp. 283–98.
16. R. Fletcher, *Practical Methods of Optimization*, Chichester, England: Wiley, 1987.
17. P. E. Gill, W. Murray, and M. H. Wright, *Practical Optimization*, London: Academic, 1981.

CHAPTER 7

Solving Matrix Algebra Problems with Neural Networks

7.1 INTRODUCTION

In this chapter several important concepts in matrix algebra are presented. The intention is to not only extend the reader's knowledge of matrix algebra concepts, but also formulate the problem setting in such a manner that a neurocomputing approach can be taken using *structured neural networks*. That is, by formulating the specific problem in a way that a structured neural network can be used to solve the problem, this neurocomputing approach can possibly lead to efficient and robust algorithms, specifically, those that are amenable to real-time (or online) applications. The parallel computing nature of neural networks allows the massively parallel learning algorithms for these types of problems to be realized in a relatively straightforward manner. The mathematical concepts presented are not new; however, in some cases the neurocomputing algorithms presented are unique and have practical applications. Moreover, a unified approach is taken to derive the learning rules in this chapter.

Structured neural networks, first introduced by Wang and Mendel [1], are special neural architectures that are *customized* to fit the specific matrix algebra application. For example, determining the inverse of a matrix can be carried out by any number of methods [2]. However, if the inverse of a matrix must be repeatedly computed, then more efficient methods should be employed that can take advantage of the *parallel* structure associated with the algorithms. In this chapter, several methods are given to solve a wide variety of matrix algebra problems, for example, matrix inversion, LU decomposition, QR factorization, Schur decomposition, symmetric eigenvalue problem, singular-value decomposition, solving the algebraic matrix Lyapunov equation, and solving the algebraic matrix Riccati equation.

The algorithms presented are associated with neuronlike adaptive signal processing systems. Also, all the learning algorithms presented are based on error backpropagation methods that were discussed in Chapter 3. Moreover, the majority of the neural networks considered in this chapter for solving matrix algebra problems consist of linear processing units. Linear neurons are typically considered as uninteresting from the viewpoint that only linear functions can be processed in the linear network, and a linear network with several layers can always be transformed to an architecture with one layer of linear processing units by multiplying the weights in the proper fashion [3]. However, we will see in this chapter linear multilayer neural networks are very useful for many computational tasks. Therefore, in this chapter linear neural networks (structured neural networks) are used to solve a relatively large class of matrix algebra problems.

The basic approach in this chapter for solving certain matrix algebra problems using neural networks involves four phases. The approach is similar to the methods found in [1, 4, 5, 6] and is as follows:

1. The first phase consists of constructing an appropriate error cost function for the particular type of problem to be solved. The error cost function is based on a defined error variable(s) which is typically formulated from a functional network for the particular problem. Thus, the problem in general is represented by a *structured* multilayer neural network.
2. The second phase is an optimization step which involves deriving the appropriate *learning rule* for the structured neural network using the defined error cost function from step (1) above. This typically involves deriving the learning rule in its *batch* (or vector-matrix) form. Once the vector-matrix form for the learning rule is derived, the scalar form can be formulated in a relatively straightforward manner.
3. The third phase involves using the learning rule developed in step (2) above to *train* the neural network to match some set of desired patterns, that is, input/output signal pairs. Therefore, the network is essentially *optimized* to minimize the associated error cost function. That is, the training phase involves adjusting the network's synaptic weights according to the derived learning rule in order to minimize the associated error cost function. For many of the learning rules presented in this chapter, the convergence speed can be improved by utilizing the search-then-converge scheduling strategy for the learning rate parameter given in (2.36).
4. The fourth and final phase is actually the *application phase* in which the appropriate output signals are collected from the structural neural network for a particular set of inputs to solve a specific problem.

7.2
INVERSE AND THE PSEUDOINVERSE OF A MATRIX

Computing the inverse of a matrix is probably one of the most important problems in linear algebra [2]. Many methods have been developed to compute the inverse of a matrix, or the pseudoinverse of a matrix. However, they are not usually conducive to *online* computations, which in certain situations is necessary for some real-time applications (e.g., are adaptive signal processing, robotics, and automatic control). Therefore, our goal is to develop methods that can be implemented in real time to compute the inverse (or pseudoinverse) of a matrix. One approach to accomplish this goal is to apply neurocomputing techniques for matrix inversion, using neuronlike processors.

Method 1: Matrix inversion

The first method is probably the most straightforward approach and assumes that the matrix whose inverse is to be found is given as $\boldsymbol{A} \in \Re^{n \times n}$ [6]. The matrix $\boldsymbol{A}$ is obviously assumed to be square, and we also assume that $\boldsymbol{A}$ is nonsingular. It will be shown shortly that the restrictions that $\boldsymbol{A}$ be square and nonsingular are actually not necessary. It is desired to compute the inverse of $\boldsymbol{A}$ written as $\boldsymbol{C} = \boldsymbol{A}^{-1}$, and thus we can write

$$\boldsymbol{AC} = \boldsymbol{CA} = \boldsymbol{I} \tag{7.1}$$

where $\boldsymbol{I} \in \Re^{n \times n}$ is the $n \times n$ identity matrix. As with many of the neurocomputing methods that can be developed to compute the inverse of $\boldsymbol{A}$, a defined error cost function (or energy function) will be integral in dictating the specific algorithm (or learning rule) to *train* the neural network. The first method involves a linear least-squares approach in which the error cost function is defined as

$$\mathscr{E}(\boldsymbol{C}) = \frac{1}{2} \operatorname{trace}(\boldsymbol{E}\boldsymbol{E}^T) \tag{7.2}$$

and the *error* matrix $\boldsymbol{E} \in \Re^{n \times n}$ is written from (7.1) as

$$\boldsymbol{E} = \boldsymbol{AC} - \boldsymbol{I} \tag{7.3}$$

Note that $\boldsymbol{E}$ can also be defined as $\boldsymbol{E} = \boldsymbol{CA} - \boldsymbol{I}$ from (7.1); however, even though the resulting learning rules will be different, the results (i.e., the inverse of $\boldsymbol{A}$) will be the same.

A simple continuous-time learning rule based on a steepest descent approach leads to the set of matrix differential equations given as

$$\frac{\mathrm{d}\boldsymbol{C}(t)}{\mathrm{d}t} = -\mu \nabla_{\boldsymbol{C}} \mathscr{E}(\boldsymbol{C}) = -\mu \frac{\partial \mathscr{E}(\boldsymbol{C})}{\partial \boldsymbol{C}} \tag{7.4}$$

where $\mu > 0$ is the learning rate parameter that must be selected small enough to ensure convergence to the inverse $\boldsymbol{C}$. The discrete-time form of (7.4) can be written in terms of a set of difference equations as

$$\boldsymbol{C}(k+1) = \boldsymbol{C}(k) - \mu \nabla_{\boldsymbol{C}} \mathscr{E}(\boldsymbol{C}) \tag{7.5}$$

where is the discrete-time index, that is, $k = 0, 1, 2, \ldots$. Therefore, the gradient of the error cost function in (7.2) must be computed. This is given as

$$\begin{aligned}\frac{\partial \mathscr{E}(\boldsymbol{C})}{\partial \boldsymbol{C}} &= \frac{\partial}{\partial \boldsymbol{C}}\left[\frac{1}{2}\text{ trace }(\boldsymbol{E}\boldsymbol{E}^T)\right] = \frac{\partial}{\partial \boldsymbol{C}}\left\{\frac{1}{2}\text{ trace }[(\boldsymbol{A}\boldsymbol{C}-\boldsymbol{I})(\boldsymbol{C}^T\boldsymbol{A}^T-\boldsymbol{I})]\right\}\\ &= \frac{\partial}{\partial \boldsymbol{C}}\left[\frac{1}{2}\text{ trace }(\boldsymbol{A}\boldsymbol{C}\boldsymbol{C}^T\boldsymbol{A}^T-\boldsymbol{A}\boldsymbol{C}-\boldsymbol{C}^T\boldsymbol{A}^T+\boldsymbol{I})\right]\\ &= \boldsymbol{A}^T\boldsymbol{A}\boldsymbol{C}-\boldsymbol{A}^T = \boldsymbol{A}^T\underbrace{(\boldsymbol{A}\boldsymbol{C}-\boldsymbol{I})}_{\boldsymbol{E}} = \boldsymbol{A}^T\boldsymbol{E}\end{aligned} \tag{7.6}$$

which is obtained by using two general results for differentiating a scalar with respect to a matrix, that is,

$$\frac{\partial}{\partial \boldsymbol{A}}\text{ trace }(\boldsymbol{B}\boldsymbol{A}\boldsymbol{C}) = \boldsymbol{B}^T\boldsymbol{C}^T \tag{7.7}$$

and

$$\frac{\partial}{\partial \boldsymbol{A}}\text{trace }\left(\boldsymbol{B}\boldsymbol{A}^T\boldsymbol{C}\right) = \boldsymbol{C}\boldsymbol{B} \tag{7.8}$$

and the proper chain rule (cf. Sect. A.3.4.2). Therefore, the discrete-time learning rule from (7.5) is given as

$$\boldsymbol{C}(k+1) = \boldsymbol{C}(k) + \mu\boldsymbol{A}^T[\boldsymbol{I}-\boldsymbol{A}\boldsymbol{C}(k)] \tag{7.9}$$

using the gradient result in (7.6). An alternate form for the learning rule, in terms of the error matrix $\boldsymbol{E}$, is given as

$$\boldsymbol{C}(k+1) = \boldsymbol{C}(k) - \mu\boldsymbol{A}^T\boldsymbol{E}(k) \tag{7.10}$$

Also, if the error matrix $\boldsymbol{E}$ is defined as $\boldsymbol{E} = \boldsymbol{C}\boldsymbol{A} - \boldsymbol{I}$, the associated learning rule is given as

$$\boldsymbol{C}(k+1) = \boldsymbol{C}(k) + \mu[I - \boldsymbol{C}(k)\boldsymbol{A}]\boldsymbol{A}^T \tag{7.11}$$

or written in terms of the error matrix $\boldsymbol{E}$, the learning rule is

$$\boldsymbol{C}(k+1) = \boldsymbol{C}(k) - \mu\boldsymbol{E}(k)\boldsymbol{A}^T \tag{7.12}$$

As previously stated, the learning rules in (7.9) and (7.11), or (7.10) and (7.12), will yield identical results.

Two immediate conclusions can be drawn based on the learning rule given in (7.9) or (7.11): (1) No divisions are required to compute the inverse of $\boldsymbol{A}$, only multiplications and additions (or subtractions). Therefore, the iterative process will yield a *solution* regardless of whether matrix $\boldsymbol{A}$ is singular or not. In fact, $\boldsymbol{A}$ can in general be $\boldsymbol{A} \in \Re^{m\times n}$, with $m = n$, an exactly determined system; that is, for a system of algebraic equations $\boldsymbol{A}\boldsymbol{x} = \boldsymbol{b}$, $\boldsymbol{x} \in \Re^{n\times 1}$, $b \in \Re^{m\times 1}$, or $m > n$ (overdetermined system), or $m < n$ (underdetermined system). Moreover, the learning rules in (7.9) and (7.11), or (7.10) and (7.12), can be used in general to compute the pseudoinverse of a matrix $\boldsymbol{A} \in \Re^{m\times n}$. This is explained in greater detail below. (2) From (7.10) or (7.12) it can be seen that the learning rule to train the neural network is based on error backpropagation. This is typical of most structured neural networks used to solve matrix algebra problems.

For the case where we seek to calculate the pseudoinverse of a matrix $\boldsymbol{A} \in \Re^{m \times n}$, the error matrix must be defined as

$$\boldsymbol{E} = \boldsymbol{ACA} - \boldsymbol{A} \tag{7.13}$$

where $\boldsymbol{E} \in \Re^{m \times n}$, $\boldsymbol{C} = \boldsymbol{A}^+ \in \Re^{n \times m}$, and $\boldsymbol{A}^+ \equiv$ pseudoinverse of $\boldsymbol{A}$. The pseudoinverse (or Moore-Penrose generalized inverse) of $\boldsymbol{A}$ (cf. Sect. A.2.7) can be written as

$$\boldsymbol{C} = \boldsymbol{A}^+ = (\boldsymbol{A}^T\boldsymbol{A})^{-1}\boldsymbol{A}^T, \quad \text{for} \quad m > n \tag{7.14}$$

or

$$\boldsymbol{C} = \boldsymbol{A}^+ = \boldsymbol{A}^T(\boldsymbol{A}\boldsymbol{A}^T)^{-1}, \quad \text{for} \quad m < n \tag{7.15}$$

Now if the error cost function is again written as

$$\mathscr{E}(\boldsymbol{C}) = \frac{1}{2} \text{ trace } (\boldsymbol{E}\boldsymbol{E}^T) \tag{7.16}$$

with the error matrix $\boldsymbol{E}$ taken as in (7.13), the calculated gradient of $\mathscr{E}(\boldsymbol{C})$ is given as

$$\nabla_{\boldsymbol{C}}\mathscr{E}(\boldsymbol{C}) = \frac{\partial \mathscr{E}(\boldsymbol{C})}{\partial \boldsymbol{C}} = \boldsymbol{A}^T\boldsymbol{ACAA}^T - \boldsymbol{A}^T\boldsymbol{AA}^T \tag{7.17}$$

Desiring the gradient in (7.17) to be minimal (i.e., in the vicinity of the global minimum the error cost function approaches zero) leads to

$$\boldsymbol{A}^T\boldsymbol{ACAA}^T - \boldsymbol{A}^T\boldsymbol{AA}^T = \boldsymbol{A}^T\boldsymbol{A}(\boldsymbol{CA} - \boldsymbol{I})\boldsymbol{A}^T = \boldsymbol{0} \tag{7.18}$$

or

$$\boldsymbol{A}^T\boldsymbol{ACAA}^T - \boldsymbol{A}^T\boldsymbol{AA}^T = \boldsymbol{A}^T(\boldsymbol{AC} - \boldsymbol{I})\boldsymbol{AA}^T = \boldsymbol{0} \tag{7.19}$$

If both sides of (7.18) are premultiplied by $(\boldsymbol{A}^T\boldsymbol{A})^{-1}$ (assuming this inverse exists), the resulting learning rule based on the method of steepest descent as in (7.5) is identical to (7.11). On the other hand, if (7.19) is postmultiplied by $(\boldsymbol{AA}^T)^{-1}$ (again assuming the inverse exists), the resulting learning rule is identical to (7.9). Therefore, the learning rules given in (7.9) and (7.11), which were initially developed for the special case when we assumed that $\boldsymbol{A}$ was square and nonsingular, are valid for the general case to compute the pseudoinverse of $\boldsymbol{A}$. However, to assess the *error* in the general case, the expression in (7.13) must be used. This recursive method for computing the pseudoinverse of a matrix is similar to the singular-value decomposition (SVD) (cf. Sect. A.2.14) approach.

The vector-matrix form of the learning rule for computing the generalized inverse of a matrix can be written from (7.9) as

$$\boldsymbol{C}(k+1) = \boldsymbol{C}(k) + \mu\boldsymbol{A}^T[\boldsymbol{I} - \boldsymbol{AC}(k)] = \boldsymbol{C}(k) + \mu\Delta\boldsymbol{C}(k) \tag{7.20}$$

The scalar form of this learning rule can be derived by first writing the update term $\Delta\boldsymbol{C}(k) = \boldsymbol{A}^T[\boldsymbol{I} - \boldsymbol{AC}(k)]$ in (7.20) as

$$\Delta\boldsymbol{C}(k) = \boldsymbol{A}^T - \boldsymbol{A}^T\boldsymbol{AC}(k) \tag{7.21}$$

and then writing the "fixed" term $\boldsymbol{W} \equiv \boldsymbol{A}^T\boldsymbol{A}$ in (7.21) in scalar form as

$$w_{ih} = \sum_{q=1}^{m} a_{qi}a_{qh} \tag{7.22}$$

for $i = 1, 2, \ldots, n$, and $h = 1, 2, \ldots, n$. Therefore, the scalar form of the learning rule given in (7.9) can be written using (7.22) as

$$c_{ij}(k+1) = c_{ij}(k) + \mu[a_{ij} - \sum_{h=1}^{n} w_{ih}c_{hj}(k)] \tag{7.23}$$

for $j = 1, 2, \ldots, m$. As a final note, it can be shown that the continuous-time learning rule given in (7.4) can be replaced by a nonlinear matrix differential equation without affecting the final results and accuracy of the solution [6, 7]. The nonlinearity can be an appropriate function $\xi(\bullet)$, whose argument is the gradient shown in (7.4), that is, $\xi[\nabla_C \mathscr{E}(\boldsymbol{C})]$.

Method 2: Matrix inversion

So far we have introduced a learning rule that can recursively compute the pseudoinverse of a matrix; however, the associated neural network architecture has not been presented. We now do this, but for another neural network method that can be used to compute the pseudoinverse of a square matrix. This method was first introduced by Wang and Mendel in 1992 [1], and also presented by Cichocki and Unbehauen [6]. Given the matrix $\boldsymbol{A} \in \Re^{n\times n}$, it is desired to find the inverse (or pseudoinverse if $\boldsymbol{A}$ is singular), that is, $\boldsymbol{C}=\boldsymbol{A}^{-1}$. The neural network approach is an *identity mapping* method that can be depicted in block diagram form shown in Figure 7.1. From Figure 7.1, we can see that the neural network performs the linear transformation $\boldsymbol{u} = \boldsymbol{A}\boldsymbol{x}$ and $\boldsymbol{y} = \boldsymbol{C}\boldsymbol{u}$; therefore, after training

$$\boldsymbol{y} = \boldsymbol{C}\boldsymbol{u} = \boldsymbol{C}\boldsymbol{A}\boldsymbol{x} = \boldsymbol{I}\boldsymbol{x} = \boldsymbol{x} \tag{7.24}$$

The first question is, What is the vector $\boldsymbol{x}$? This is the external excitatory input signal (or from Figure 7.1 also the desired signal $\boldsymbol{d} = \boldsymbol{x}$); and thus, after the neural network has been trained, the output signal $\boldsymbol{y}$ will equal the input signal $\boldsymbol{x}$. Therefore, the *mapping error* $\boldsymbol{e} = \boldsymbol{x} - \boldsymbol{y}$ in Figure 7.1 will be zero. The simplest inputs that can be used for the external excitatory signals are a set of linearly independent binary vectors given as $\boldsymbol{x}_1 = [1, 0, 0, \cdots, 0]^T$, $\boldsymbol{x}_2 = [0, 1, 0, \cdots, 0]^T$, ..., $\boldsymbol{x}_n = [0, 0, 0, \cdots, 1]^T$. However, to achieve faster conver-

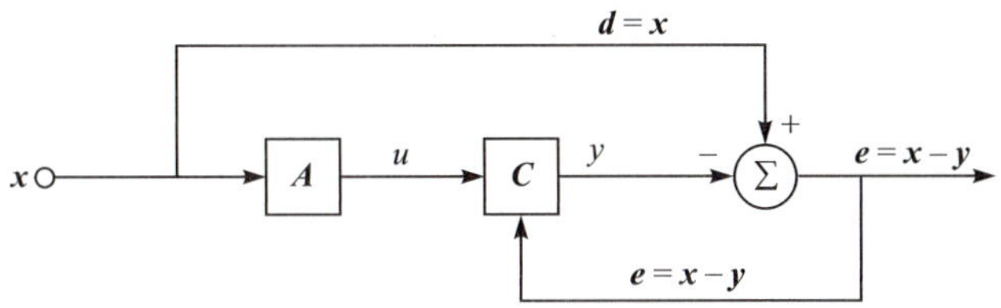

Figure 7.1 Block diagram of the structured neural network for matrix inversion using an identity mapping approach.

gence during training, it is best to use a set of linearly independent bipolar vectors given as $\boldsymbol{x}_1 = [1, -1, -1, \cdots, -1]^T$, $\boldsymbol{x}_2 = [-1, 1, -1, \cdots, -1]^T$, ..., $\boldsymbol{x}_n = [-1, -1, -1, \cdots, 1]^T$ [6]. This neural network method for matrix inversion is similar to the classic Gauss–Jordan elimination method [8] to compute a matrix inverse, which is based on the augmented matrix

$$\boldsymbol{A}_{\text{aug}} = \{\boldsymbol{A} \vdots \boldsymbol{I}_{n\times n}\} \tag{7.25}$$

Gauss–Jordan elimination simply involves performing elementary row (column) operations on the augmented matrix $\boldsymbol{A}_{\text{aug}}$, until $\boldsymbol{A}$ becomes the identity matrix and the original identity matrix becomes the inverse of $\boldsymbol{A}$, that is, $\{\boldsymbol{A} \vdots \boldsymbol{I}_{n\times n}\} \sim \cdots \sim \{\boldsymbol{I}_{n\times n} \vdots \boldsymbol{A}^{-1}\}$. The vector-matrix form of the learning rule can be derived from the error cost function

$$\mathscr{E}(\boldsymbol{C}) = \frac{1}{2}\|\boldsymbol{e}\|_2^2 = \frac{1}{2}\boldsymbol{e}^T\boldsymbol{e} \tag{7.26}$$

where $\boldsymbol{e} = \boldsymbol{x} - \boldsymbol{y}$ (the *mapping error*), as shown in Figure 7.1. The objective is to minimize $\mathscr{E}(\boldsymbol{C})$ in (7.26) with respect to $\boldsymbol{C}$. We can write (7.26) as

$$\begin{aligned}\mathscr{E}(\boldsymbol{C}) &= \frac{1}{2}\boldsymbol{e}^T\boldsymbol{e} = \frac{1}{2}(\boldsymbol{x}^T - \boldsymbol{y}^T)(\boldsymbol{x} - \boldsymbol{y}) = \frac{1}{2}(\boldsymbol{x}^T\boldsymbol{x} - \boldsymbol{x}^T\underbrace{\boldsymbol{y}}_{\boldsymbol{Cu}} - \boldsymbol{y}^T\boldsymbol{x} + \boldsymbol{y}^T\boldsymbol{y}) \\ &= \frac{1}{2}(\boldsymbol{x}^T\boldsymbol{x} - \boldsymbol{x}^T\boldsymbol{C}\boldsymbol{u} - \boldsymbol{u}^T\boldsymbol{C}^T\boldsymbol{x} + \boldsymbol{u}^T\boldsymbol{C}^T\boldsymbol{C}\boldsymbol{u})\end{aligned} \tag{7.27}$$

A continuous-time learning rule based on the method of steepest descent can be written as a set of matrix differential equations given by

$$\frac{d\boldsymbol{C}(t)}{dt} = -\mu\nabla_{\boldsymbol{C}}\mathscr{E}(\boldsymbol{C}) \tag{7.28}$$

Computing the necessary gradient $\nabla_{\boldsymbol{C}}\mathscr{E}(\boldsymbol{C})$ in (7.28) requires the use of the general results in (7.7) and (7.8). Therefore, computing the gradient of (7.27) gives

$$\nabla_{\boldsymbol{C}}\mathscr{E}(\boldsymbol{C}) = -\boldsymbol{x}\boldsymbol{u}^T + \boldsymbol{C}\boldsymbol{u}\boldsymbol{u}^T = -(\boldsymbol{x} - \underbrace{\boldsymbol{C}\boldsymbol{u}}_{\boldsymbol{y}})\boldsymbol{u}^T = -(\boldsymbol{x} - \boldsymbol{y})\boldsymbol{u}^T \tag{7.29}$$

and the continuous-time learning rule is given as

$$\frac{d\boldsymbol{C}(t)}{dt} = \mu[\boldsymbol{x}(t) - \boldsymbol{y}(t)]\boldsymbol{u}^T(t) \tag{7.30}$$

where $\mu > 0$ is the learning rate parameter. In discrete-time form the learning rule is given as

$$\boldsymbol{C}(k+1) = \boldsymbol{C}(k) + \mu[\boldsymbol{x}(k) - \boldsymbol{y}(k)]\boldsymbol{u}^T(k) \tag{7.31}$$

where

$$\boldsymbol{u}(k) = \boldsymbol{A}\boldsymbol{x}(k) \qquad \text{and} \qquad \boldsymbol{y}(k) = \boldsymbol{C}(k)\boldsymbol{u}(k) \tag{7.32}$$

from the block diagram in Figure 7.1. The scalar forms of the learning rule in (7.31) and (7.32) can, respectively, be written as

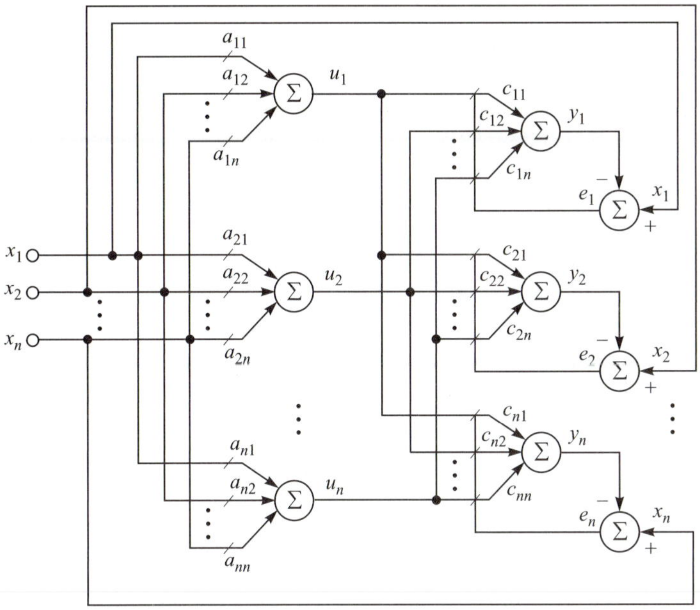

Figure 7.2 Neural network architecture for matrix inversion using the *mapping error* method. The first layer has linear neurons with *fixed* weights (a_{ij}, for $i, j = 1, 2, \ldots, n$) which are the *rows* of matrix $\boldsymbol{A}$. The second layer has linear neurons with *adaptive* weights (c_{ij}, for $i, j = 1, 2, \ldots, n$) which are the *rows* of $\boldsymbol{A}^{-1}$.

$$c_{ij}(k+1) = c_{ij}(k) + \mu[x_{ip} - y_{ip}(k)]u_{jp} \tag{7.33}$$

$$u_{ip} = \sum_{h=1}^{n} a_{ih}x_{hp} \quad \text{and} \quad y_{ip}(k) = \sum_{h=1}^{n} C_{ih}(k)u_{hp} \tag{7.34}$$

where $i, j = 1, 2, \ldots, n$, $k = 0, 1, 2, \ldots$, and $p = 1, 2, \ldots, n$ (i.e., n excitatory bipolar input signals as previously described). For each training epoch, the n excitatory input signals will be presented once. These n linearly independent bipolar vectors can be randomized according to their order of presentation to the neural network each training epoch. Figure 7.2 shows the neural network architecture for matrix inversion using the *mapping error* method.

EXAMPLE 7.1. The following example illustrates the ability of both neural network methods to compute the inverse of a nonsingular matrix. Given the matrix

$$\boldsymbol{A} = \begin{bmatrix} 1 & 1 & 2 \\ 3 & 2 & 3 \\ 1 & 1 & 1 \end{bmatrix} \tag{7.35}$$

the computed inverse $\boldsymbol{C}^{NN}$, using both neural network methods, yielded

$$\boldsymbol{C}^{NN} = \boldsymbol{A}^{-1} = \begin{bmatrix} -1.0000 & 1.0000 & -1.0000 \\ 0.0000 & -1.0000 & 3.0000 \\ 1.0000 & 0.0000 & -1.0000 \end{bmatrix} \tag{7.36}$$

TABLE 7.1
Neural network training details for Example 7.1

Methods used for matrix inversion	Learning rate parameter	Number of training epochs	$\|C^M - C^{NN}\|_2$
1	$\mu = 0.065$	2,500 $[\mathbf{C}(k=0) = \mathbf{0}]$	8.0048×10^{-6}
2	$\mu = 0.015$	2,500 $[\mathbf{C}(k=0) = \mathbf{0}]$	4.7986×10^{-5}
2 (with search-then-converge scheduling)	$\mu_0 = 0.06$, $\tau = 500$	1,200 $[\mathbf{C}(k=0) = \mathbf{0}]$	5.9149×10^{-5}

out to four decimal places. The same results shown in (7.36) were obtained using the MATLAB **inv** function, that is, $\boldsymbol{C}^M$. Table 7.1 summarizes the training results for both neural networks used to compute the matrix inverse. The $\|\bullet\|_2$ norm used, shown in Table 7.1, is the spectral norm of a matrix (cf. Sect. A.2.13). As shown in Section A.2.13, the spectral norm can be computed as the largest singular value of the matrix. For both methods, a fixed learning rate parameter was used, as indicated in Table 7.1. The search-then-converge scheduling strategy for the learning rate parameter given in (2.36) was also used with method 2. The same results were obtained as shown in (7.36) out to four decimal places (this scheduling strategy was not effective with method 1). As shown in Table 7.1, convergence of the method 2 algorithm was improved using this scheduling strategy for the learning rate parameter. The number of training epochs was reduced by more than one-half. Figure 7.3 shows the mean-square error for the two neural networks (including method 2 with the search-then-converge scheduling), as the networks were training to compute the inverse of $\boldsymbol{A}$ in (7.35).

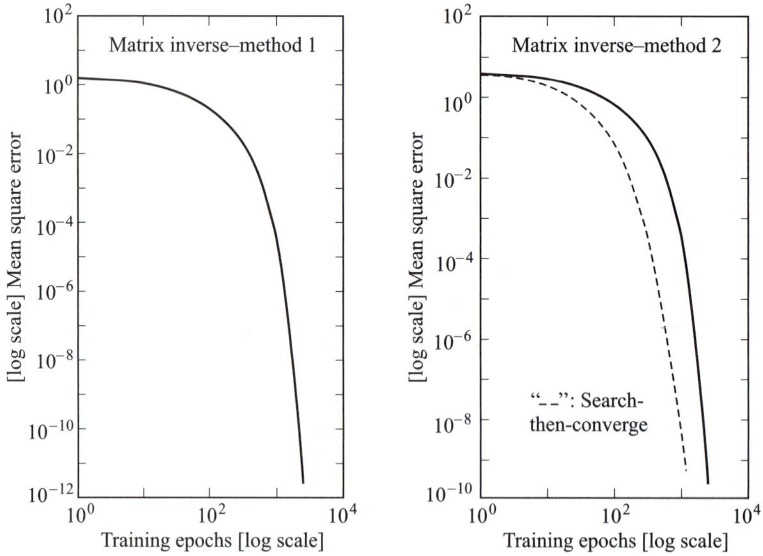

Figure 7.3 Mean-square error during training for both neural networks used to compute the inverse of a matrix.

7.3 LU DECOMPOSITION

The LU decomposition (or factorization) method is one of the most important matrix factorizations [8]. The reason is that a system of simultaneous linear algebraic equations, that is, $\boldsymbol{Ax} = \boldsymbol{b}$ (where we seek to compute the solution $\boldsymbol{x}$), can be readily solved if the matrix $\boldsymbol{A} \in \Re^{n \times n}$ can be decomposed into the product of a lower triangular matrix $\boldsymbol{L} \in \Re^{n \times n}$ and an upper triangular matrix $\boldsymbol{U} \in \Re^{n \times n}$, that is,

$$\boldsymbol{A} = \boldsymbol{LU} \tag{7.37}$$

The lower triangular matrix $\boldsymbol{L}$ is usually assumed to be a *unit* lower triangular matrix because the diagonal elements are unity. After matrix $\boldsymbol{A}$ has been decomposed into $\boldsymbol{L}$ and $\boldsymbol{U}$, then *two triangular systems* can be solved by using backsubstitution [8], that is,

$$\begin{aligned} &\text{First solve for } \boldsymbol{y}: \quad && \boldsymbol{Ly} = \boldsymbol{b} \\ &\text{Next solve for } \boldsymbol{x}: \quad && \boldsymbol{Ux} = \boldsymbol{y} \end{aligned} \tag{7.38}$$

From (7.37) it is obvious that we could write the LU decomposition of $\boldsymbol{A}$ as

$$\boldsymbol{A} = \boldsymbol{L}\underbrace{\boldsymbol{DD}^{-1}}_{\boldsymbol{I}}\boldsymbol{U} = (\underbrace{\boldsymbol{LD}}_{\hat{L}})(\underbrace{\boldsymbol{D}^{-1}\boldsymbol{U}}_{\hat{U}}) = \hat{\boldsymbol{L}}\hat{\boldsymbol{U}} \tag{7.39}$$

where $\boldsymbol{D} \in \Re^{n \times n}$ is a nonsingular diagonal matrix, assuming that $\boldsymbol{L}$ is only lower triangular and not *unit* lower triangular. Therefore, we can conclude that the LU decomposition of a matrix $\boldsymbol{A}$ is not necessarily unique. However, under certain conditions imposed on matrix $\boldsymbol{A}$, a unique LU decomposition does exist [2, 9]. This can be stated as follows (from Golub and Van Loan [2]):

> Given the matrix $\boldsymbol{A} \in \Re^{n \times n}$, an LU factorization of $\boldsymbol{A}$ exists if $\det(\boldsymbol{A}(1:k, 1:k)) \neq 0$ for $k = 1, 2, \ldots, n-1$, where $\boldsymbol{L} \in \Re^{n \times n}$ is unit lower triangular and $\boldsymbol{U} \in \Re^{n \times n}$ is upper triangular. If the LU factorization of $\boldsymbol{A}$ exists and if $\rho(\boldsymbol{A}) = n$, then the LU factorization is unique, and furthermore $\det(\boldsymbol{A})$ = product of the diagonal elements of $\boldsymbol{U}$.

The proof for this can be found in Golub and Van Loan [2].

If we assume a diagonal matrix $\boldsymbol{D} \in \Re^{n \times n}$, where $\rho(\boldsymbol{D}) = n$, then we can write a LDU factorization of matrix $\boldsymbol{A}$ as

$$\boldsymbol{A} = \boldsymbol{LDU} \tag{7.40}$$

Thus, from (7.40) we can write

$$\boldsymbol{L}(\boldsymbol{DU}) = \boldsymbol{A} \tag{7.41}$$

and

$$(\boldsymbol{LD})\boldsymbol{U} = \boldsymbol{A}$$

which are both possible factorizations of $\boldsymbol{A}$.

A special case exists when $\boldsymbol{A}$ is symmetric (i.e., $\boldsymbol{A} = \boldsymbol{A}^T$). Here the factorization can be written as [2, 9]

$$A = LDL^T \tag{7.42}$$

Furthermore, if $A > 0$ (i.e., A is positive definite), then the diagonal elements of D will be all nonzero positive values, and the factorization can be further simplified. That is, if we assume that $D = \Delta^2$, then Δ is called the *square root* of $D(\Delta = \sqrt{D})$, and the LDL^T factorization of A can be written as

$$A = LDL^T = L\Delta^2 L^T = (L\Delta)(\Delta L^T) = L_c L_c^T \tag{7.43}$$

where L_c is a lower triangular matrix. The decomposition of A in (7.43) is called the Cholesky factorization [2, 8, 9]. The Cholesky factorization of a matrix has many applications in signal processing; one of particular importance is the factorization of an autocorrelation matrix in *spectral factorization* [10]. In control theory the Cholesky factorization of a matrix is used in Lyapunov stability theory [11], and it is also used in square root Kalman filtering [12].

In general, matrix A can be rectangular, that is, $A \in \Re^{m\times n}$. This would result in an LU decomposition with $L \in \Re^{m\times m}$ and $U \in \Re^{m\times n}$; however, we restrict our discussions to the case when A is square. Many numerical methods have been developed to compute the LU decomposition of a matrix; the most basic method is based on Gaussian elimination and variations thereof, for example, Crout and Doolittle [2, 9]. Other numerical methods can be found in Press et al. [13] or [14]. Our concern here is to develop a neurocomputing approach for computing the LU decomposition of a matrix, but with the expressed desire to use this approach in a real-time setting. Figure 7.4 shows the block diagram of the structured neural network to perform the LU decomposition, and Figure 7.5 shows the two-layer neural network architecture.

From Figure 7.4 the desired signal (or target signal) is given as $d = Ax$, for a defined external excitatory input signal x. These input signals will be taken as the same linearly independent bipolar vectors previously described in Section 7.2. Also, from Figure 7.4 we can write the vector error signal e as

$$e = d - \underbrace{y}_{Lz} = Ax - L\underbrace{z}_{Ux} = Ax - LUx = (A - LU)x \tag{7.44}$$

Therefore, after the neural network has learned the elements of the two triangular matrices L and U, for an appropriate excitatory signal x, from (7.44) we

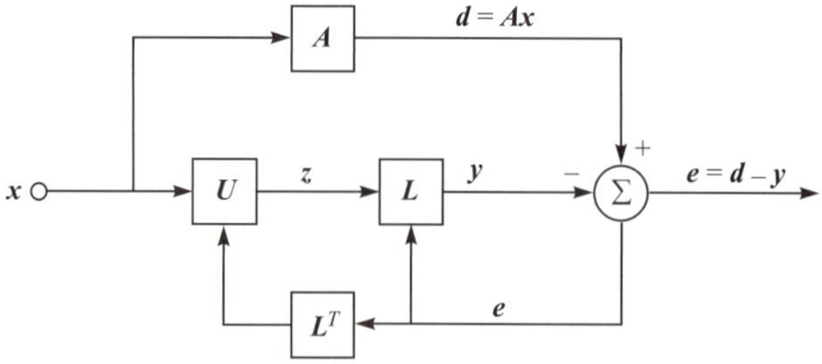

Figure 7.4 Block diagram of the structured neural network for LU decomposition. Note that the neural network is trained by error backpropagation.

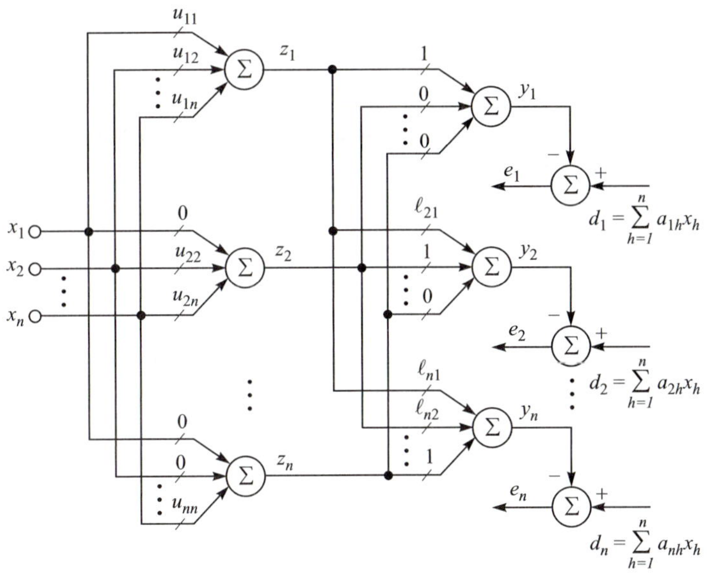

Figure 7.5 A two-layer neural network architecture for LU decomposition trained by error backpropagation.

can see that the error vector will be zero because $A = LU$. We can now define an error cost function, which is a function of both matrices L and U in the decomposition, given as

$$\begin{aligned} \mathscr{E}(L, U) &= \frac{1}{2}\|e\|_2^2 = \frac{1}{2}e^T e = \frac{1}{2}[(A - LU)x]^T[(A - LU)x] \\ &= \frac{1}{2}(x^T A^T Ax - x^T A^T LUx - x^T U^T L^T Ax + x^T U^T L^T LUx) \end{aligned} \tag{7.45}$$

using (7.44). From (7.45) we can derive two learning rules (for computing both L and U) based on the steepest descent gradient method. In discrete-time form the learning rules are given as

$$L(k+1) = L(k) - \mu \nabla_L \mathscr{E}(L, U) \tag{7.46}$$

and

$$U(k+1) = U(k) - \mu \nabla_U \mathscr{E}(L, U) \tag{7.47}$$

where $\mu > 0$ is the learning rate parameter. Therefore, we must compute the two gradient terms $\nabla_L \mathscr{E}(L, U)$ and $\nabla_U \mathscr{E}(L, U)$ in (7.46) and (7.47), respectively, from the error cost given in (7.45). The gradient of (7.45) with respect to matrix L can be determined by using the general results in (7.7) and (7.8) and the appropriate chain rule. The result is given as

$$\nabla_L \mathscr{E}(L, U) = -Axx^T U^T + LUxx^T U^T = \underbrace{(-Ax + LUx)}_{-e}\underbrace{x^T U^T}_{z^T} = -ez^T \tag{7.48}$$

Therefore, the learning rule for $\boldsymbol{L}$, from (7.46) and (7.48), is given by

$$\boldsymbol{L}(k+1) = \boldsymbol{L}(k) + \mu \boldsymbol{e}(k)\boldsymbol{z}^T(k) \tag{7.49}$$

where

$$\boldsymbol{e}(k) = [\boldsymbol{A} - \boldsymbol{L}(k)\boldsymbol{U}(k)]\boldsymbol{x}(k) \tag{7.50}$$

from (7.44), and

$$\boldsymbol{z}(k) = \boldsymbol{U}(k)\boldsymbol{x}(k) \tag{7.51}$$

from the block diagram in Figure 7.4. The gradient of the error cost function in (7.45) with respect to matrix $\boldsymbol{U}$ is given as

$$\nabla_U \mathscr{E}(\boldsymbol{L}, \boldsymbol{U}) = -\boldsymbol{L}^T \boldsymbol{A}\boldsymbol{x}\boldsymbol{x}^T + \boldsymbol{L}^T\boldsymbol{L}\boldsymbol{U}\boldsymbol{x}\boldsymbol{x}^T = \boldsymbol{L}^T \underbrace{(-\boldsymbol{A}\boldsymbol{x} + \boldsymbol{L}\boldsymbol{U}\boldsymbol{x})}_{-\boldsymbol{e}} \boldsymbol{x}^T = -\boldsymbol{L}^T \boldsymbol{e}\boldsymbol{x}^T \tag{7.52}$$

Therefore, the discrete-time learning rule for computing $\boldsymbol{U}$, using (7.47) and (7.52), is given as

$$\boldsymbol{U}(k+1) = \boldsymbol{U}(k) + \mu \boldsymbol{L}^T(k+1)\boldsymbol{e}(k)\boldsymbol{x}^T(k) \tag{7.53}$$

where $\boldsymbol{e}(k)$ is given as in (7.50). The vector-matrix form of the learning rules given in (7.49) to (7.51) and (7.53) will not give an LU decomposition of matrix $\boldsymbol{A}$ because no constraints have been imposed on $\boldsymbol{L}$ and $\boldsymbol{U}$. That is, $\boldsymbol{L}$ is not constrained to be lower triangular with unity elements on the diagonal, and $\boldsymbol{U}$ is not constrained to be upper triangular. Therefore, we now write the scalar form of the learning rules from (7.49) to (7.51) and (7.53), and we impose the appropriate constraints on the elements of the two matrices. The scalar form of the learning rule for the unit lower triangular matrix $\boldsymbol{L}$ is given as

$$\ell_{ij}(k+1) = \ell_{ij}(k) + \mu e_{ip}(k) z_{jp}(k) \tag{7.54}$$

for $k = 0, 1, 2, \ldots$, and $p = 1, 2, \ldots, n$ (i.e., n bipolar excitatory input signals as previously described). The constraints on the ℓ_{ij} elements of $\boldsymbol{L}$ are: (1) $\ell_{ij} = 1$ for $i = j$; (2) $\ell_{ij} = 0$, for $i < j$; and (3) for $i > j$ the remaining elements (connection weights) in $\boldsymbol{L}$ are adapted according to (7.54), $\forall i, j = 1, 2, \ldots, n$. In (7.54) the scalar forms for $\boldsymbol{e}_p$ and $\boldsymbol{z}_p$ can be computed as

$$e_{ip} = d_{ip} - y_{ip} \tag{7.55}$$

where

$$d_{ip} = \sum_{h=1}^{n} a_{ih} x_{hp} \tag{7.56}$$

$$y_{ip} = \sum_{h=1}^{n} \ell_{ih} z_{hp} \tag{7.57}$$

and

$$z_{ip} = \sum_{h=1}^{n} u_{ih} x_{hp} \tag{7.58}$$

where $i = 1, 2, \ldots, n$.

The scalar form of the learning rule for the upper triangular matrix $\boldsymbol{U}$ is given by

$$u_{ij}(k+1) = u_{ij}(k) + \mu \left[\sum_{h=i}^{n} \ell_{hi}(k+1) e_{hp}(k) \right] x_{jp} \tag{7.59}$$

for $k = 0, 1, 2, \ldots$, and $p = 1, 2, \ldots, n$. The constraints on the u_{ij} elements of $\boldsymbol{U}$ are: (1) $u_{ij} = 0$ for $i > j$, and (2) for $i \leq j$ the remaining elements (connection weights) in $\boldsymbol{U}$ are adapted according to (7.59), $\forall i, j = 1, 2, \ldots, n$. Imposing the constraints on $\boldsymbol{L}$ and $\boldsymbol{U}$ given above will ensure that the learning rules yield the correct LU decomposition. Observing the scalar forms of the learning rules in (7.54) and (7.59), we can see that there exists coupling between the two learning rules. In (7.59), the learning rule for $\boldsymbol{U}$, the elements of the lower triangular matrix $\boldsymbol{L}$ appears in the summation as the result of computing the *updated* values first for the elements of $\boldsymbol{L}$, and next the elements of $\boldsymbol{U}$ are updated.

7.4 QR FACTORIZATION

The QR factorization (or orthogonal triangularization) is another very important matrix decomposition that is used extensively in engineering and science. This method has extensive applications in computing the complete set of eigenvectors of a matrix [2, 9]. There are many other applications of QR factorization, especially in the area of signal processing [12, 15, 16]. For example, the basis of square root adaptive filtering methods [12, 16] and least-squares lattice filters [12] is QR factorization.

The QR factorization of a matrix $\boldsymbol{A} \in \Re^{m \times n}$ (assuming $m \geq n$) can be written as [2, 6, 12, 17, 18]

$$\boldsymbol{A} = \boldsymbol{Q} \begin{bmatrix} \boldsymbol{R} \\ \boldsymbol{0} \end{bmatrix} \tag{7.60}$$

where $\boldsymbol{Q} \in \Re^{m \times m}$ is orthogonal (i.e., $\boldsymbol{Q}^T \boldsymbol{Q} = \boldsymbol{I}$), $\boldsymbol{R} \in \Re^{n \times n}$ is an upper triangular matrix, and $\boldsymbol{0} \in \Re^{(m-n) \times n}$ is a null matrix. However, we will study the special case when $\boldsymbol{A}$ is a square matrix, that is, $\boldsymbol{A} \in \Re^{n \times n}$. In this case the QR factorization of $\boldsymbol{A}$ is simply written as

$$\boldsymbol{A} = \boldsymbol{Q}\boldsymbol{R} \tag{7.61}$$

where $\boldsymbol{Q} \in \Re^{n \times n}$ is an orthogonal matrix (i.e., $\boldsymbol{Q}^T \boldsymbol{Q} = \boldsymbol{Q}\boldsymbol{Q}^T = \boldsymbol{I}$), and $\boldsymbol{R} \in \Re^{n \times n}$ is an upper triangular matrix. Many algorithms have been developed to factor $\boldsymbol{A}$ into matrices $\boldsymbol{Q}$ and $\boldsymbol{R}$, for example, the modified Gram–Schmidt, fast Givens, block Householder, and Hessenberg methods [2]. However, the neural network approaches we take in this section to compute the QR factorization of a square matrix are motivated by online computational requirements for a particular application. The intention is to develop a neurocomputing approach that will perform online updates of $\boldsymbol{Q}$ and $\boldsymbol{R}$ for situations when some of the elements of matrix $\boldsymbol{A}$ change slowly with time. The initial weights of the neural network are considered to be set at their nominal values according to the nominal $\boldsymbol{A}$ matrix, using any one of the standard numerical methods for QR factorization [2, 9]. Two neural network architectures are presented that are structured to solve the QR factorization problem. The neurocomputing approaches are based on the original work by Wang and Mendel [4] and the material presented in Cichocki and Unbehauen [6].

Method 1: QR factorization

Figure 7.6 shows the block diagram of the first structured neural network to perform QR factorization. As shown in the figure, two error vectors $\bar{\boldsymbol{e}}$ and $\hat{\boldsymbol{e}}$ are defined. The error vector $\bar{\boldsymbol{e}}$ can be expressed as

$$\bar{\boldsymbol{e}} = \bar{\boldsymbol{d}} - \boldsymbol{y} = \boldsymbol{A}\boldsymbol{x} - \boldsymbol{Q}\boldsymbol{u} = \boldsymbol{A}\boldsymbol{x} - \boldsymbol{Q}\boldsymbol{R}\boldsymbol{x} = (\boldsymbol{A} - \boldsymbol{Q}\boldsymbol{R})\boldsymbol{x} \tag{7.62}$$

We can see from (7.62) that as the neural network learns the factorization ($\boldsymbol{Q}$ and $\boldsymbol{R}$), the error vector $\bar{\boldsymbol{e}}$ will approach zero for an appropriate external excitatory input signal $\boldsymbol{x}$. For the development of the learning rules to train the neural network, from (7.62) we will define the error vector $\bar{\boldsymbol{e}}$ as

$$\bar{\boldsymbol{e}} = \boldsymbol{A}\boldsymbol{x} - \boldsymbol{Q}\boldsymbol{u} \tag{7.63}$$

where we do not substitute $\boldsymbol{u} = \boldsymbol{R}\boldsymbol{x}$, but retain $\boldsymbol{u}$ as an auxiliary variable. Also from the block diagram in Figure 7.6, we can express the error vector $\hat{\boldsymbol{e}}$ as

$$\hat{\boldsymbol{e}} = \boldsymbol{v} - \boldsymbol{u} = \boldsymbol{Q}^T\bar{\boldsymbol{d}} - \boldsymbol{u} = \boldsymbol{Q}^T\boldsymbol{A}\boldsymbol{x} - \boldsymbol{R}\boldsymbol{x} = (\boldsymbol{Q}^T\boldsymbol{A} - \boldsymbol{R})\boldsymbol{x} \tag{7.64}$$

In (7.64), as the neural network learns the QR factorization, the error vector $\hat{\boldsymbol{e}}$ will approach zero for an appropriate external excitatory input signal $\boldsymbol{x}$. This is not obvious from (7.64); however, from the QR factorization of $\boldsymbol{A}$, that is, $\boldsymbol{A} = \boldsymbol{Q}\boldsymbol{R}$, if we premultiply by $\boldsymbol{Q}^T$, this gives $\boldsymbol{Q}^T\boldsymbol{A} = \underbrace{\boldsymbol{Q}^T\boldsymbol{Q}}_{\boldsymbol{I}}\boldsymbol{R} = \boldsymbol{R}$, which now allows us to see the result. From (7.64) we will define the error vector as

$$\hat{\boldsymbol{e}} = \boldsymbol{Q}^T\bar{\boldsymbol{d}} - \boldsymbol{u} \tag{7.65}$$

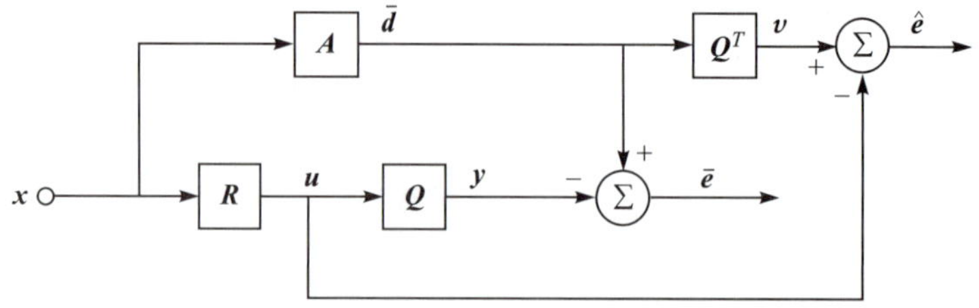

Figure 7.6 Block diagram of the structured neural network for QR factorization (method 1).

where again $\boldsymbol{u}$ is retained as an auxiliary variable, as in (7.63). The best excitatory input signals to use are the set of linearly independent bipolar vectors presented in Section 7.2. A total error cost function can be defined as

$$\mathscr{E}(\bar{e}, \hat{e}) = \mathscr{E}[\bar{e}(\boldsymbol{Q}, \boldsymbol{R}), \hat{e}(\boldsymbol{Q}, \boldsymbol{R})] = \frac{1}{2}\|\bar{e}\|_2^2 + \frac{\nu}{2}\|\hat{e}\|_2^2 \tag{7.66}$$

where $\nu > 0$ is a penalty parameter that can provide an appropriate proportion between the two terms in the cost function; however, typically $\nu = 1$. Minimizing the first term in (7.66) ensures that matrix $\boldsymbol{A}$ is decomposed into two matrices, that is, $\boldsymbol{Q}$ and $\boldsymbol{R}$. Minimizing the second term in (7.66) forces the $\boldsymbol{Q}$ matrix to be orthogonal (i.e., $\boldsymbol{Q}^T\boldsymbol{Q} = \boldsymbol{I}$). Our goal here is to derive two learning rules for Q and R based on the gradient steepest descent method. The two matrix differential equations that will result are given as

$$\frac{d\boldsymbol{Q}(t)}{dt} = -\mu\nabla_{\boldsymbol{Q}}\mathscr{E}(\bar{e},\hat{e}) = -\mu\frac{\partial\mathscr{E}(\bar{e},\hat{e})}{\partial\boldsymbol{Q}} \tag{7.67}$$

and

$$\frac{d\boldsymbol{R}(t)}{dt} = -\mu\nabla_{\boldsymbol{R}}\mathscr{E}(\bar{e},\hat{e}) = -\mu\frac{\partial\mathscr{E}(\bar{e},\hat{e})}{\partial\boldsymbol{R}} \tag{7.68}$$

where $\mu > 0$ is the learning rate parameter. By approximating the derivatives in (7.67) and (7.68), we can write the discrete-time forms of the learning rules as matrix difference equations, given, respectively as,

$$\boldsymbol{Q}(k+1) = \boldsymbol{Q}(k) - \mu\nabla_{\boldsymbol{Q}}\mathscr{E}(\bar{e},\hat{e}) \tag{7.69}$$

and

$$\boldsymbol{R}(k+1) = \boldsymbol{R}(k) - \mu\nabla_{\boldsymbol{R}}\mathscr{E}(\bar{e},\hat{e}) \tag{7.70}$$

Therefore, we must compute the two gradient terms in (7.69) and (7.70). To compute the gradient $\nabla_{\boldsymbol{Q}}\mathscr{E}(\bar{e},\hat{e})$, we can write the total error cost function in (7.66) as

$$\begin{aligned}\mathscr{E}(\bar{e},\hat{e}) &= \frac{1}{2}(\boldsymbol{Ax} - \boldsymbol{Qu})^T(\boldsymbol{Ax} - \boldsymbol{Qu}) + \frac{\nu}{2}(\boldsymbol{Q}^T\bar{\boldsymbol{d}} - \boldsymbol{u})^T(\boldsymbol{Q}^T\bar{\boldsymbol{d}} - \boldsymbol{u}) \\ &= \frac{1}{2}(\boldsymbol{x}^T\boldsymbol{A}^T\boldsymbol{Ax} - \boldsymbol{x}^T\boldsymbol{A}^T\boldsymbol{Qu} - \boldsymbol{u}^T\boldsymbol{Q}^T\boldsymbol{Ax} + \boldsymbol{u}^T\boldsymbol{Q}^T\boldsymbol{Qu}) \\ &\quad + \frac{\nu}{2}(\bar{\boldsymbol{d}}^T\boldsymbol{QQ}^T\bar{\boldsymbol{d}} - \bar{\boldsymbol{d}}^T\boldsymbol{Qu} - \boldsymbol{u}^T\boldsymbol{Q}^T\bar{\boldsymbol{d}} + \boldsymbol{u}^T\boldsymbol{u})\end{aligned} \tag{7.71}$$

using the expressions for the two error vectors given in (7.63) and (7.65). The gradient $\nabla_{\boldsymbol{Q}}\mathscr{E}(\bar{e},\hat{e})$ can be found from (7.71) by using the two general results in (7.7) and (7.8) and the appropriate chain rule (cf. Sect. A.3.4.2), and it is given as

$$\begin{aligned}\nabla_{\boldsymbol{Q}}\mathscr{E}(\bar{e},\hat{e}) &= -\boldsymbol{Axu}^T + \boldsymbol{Quu}^T + \nu\bar{\boldsymbol{d}}\bar{\boldsymbol{d}}^T\boldsymbol{Q} - \nu\bar{\boldsymbol{d}}\boldsymbol{u}^T \\ &= -(\underbrace{\boldsymbol{Ax} - \boldsymbol{Qu}}_{\bar{e}})\boldsymbol{u}^T + \nu\bar{\boldsymbol{d}}(\underbrace{\boldsymbol{Q}^T\bar{\boldsymbol{d}} - \boldsymbol{u}}_{\hat{e}})^T = -\bar{e}\boldsymbol{u}^T + \nu\bar{\boldsymbol{d}}\hat{e}^T\end{aligned} \tag{7.72}$$

Therefore, the batch (vector-matrix) form of the discrete-time learning rule for $\boldsymbol{Q}$ can be written, using (7.69) and (7.72), as

$$\boldsymbol{Q}(k+1) = \boldsymbol{Q}(k) + \mu[\bar{\boldsymbol{e}}(k)\boldsymbol{u}^T(k) - \nu\bar{\boldsymbol{d}}(k)\hat{\boldsymbol{e}}^T(k)] \tag{7.73}$$

where $\mu > 0$ and $\nu > 0$, $\bar{\boldsymbol{e}}(k)$ and $\hat{\boldsymbol{e}}(k)$ are given in (7.63) and (7.65), respectively, and $\boldsymbol{u}(k) = \boldsymbol{R}\boldsymbol{x}(k)$. The scalar form of the learning rule for $\boldsymbol{Q}$ in (7.73) can be written as

$$q_{ij}(k+1) = q_{ij}(k) + \mu[\bar{e}_i(k)u_j(k) - \nu\bar{d}_i(k)\hat{e}_j(k)] \tag{7.74}$$

where $i, j = 1, 2, \ldots, n$. The scalar forms for $\bar{\boldsymbol{e}}$, $\hat{\boldsymbol{e}}$, $\bar{\boldsymbol{d}}$, and $\boldsymbol{u}$ can be computed as

$$\bar{e}_i = \sum_{h=1}^{n} a_{ih}x_{hp} - \sum_{h=1}^{n} q_{ih}u_{hp} \tag{7.75}$$

where

$$u_{ip} = \sum_{h=1}^{n} r_{ih}x_{hp} \tag{7.76}$$

and

$$\hat{e}_i = \sum_{h=1}^{n} q_{hi}\bar{d}_{hp} - u_{ip} \tag{7.77}$$

where

$$\bar{d}_{ip} = \sum_{h=1}^{n} a_{ih}x_{hp} \tag{7.78}$$

for $i = 1, 2, \ldots, n$.

Now to derive the learning rule for $\boldsymbol{R}$, we substitute $\boldsymbol{u} = \boldsymbol{R}\boldsymbol{x}$ (from the block diagram in Figure 7.6) for the auxiliary variable $\boldsymbol{u}$ in the expressions for $\bar{\boldsymbol{e}}$ and $\hat{\boldsymbol{e}}$ in (7.63) and (7.65), that is, $\bar{\boldsymbol{e}} = \boldsymbol{A}\boldsymbol{x} - \boldsymbol{Q}\boldsymbol{R}\boldsymbol{x}$ and $\hat{\boldsymbol{e}} = \boldsymbol{Q}^T\bar{\boldsymbol{d}} - \boldsymbol{R}\boldsymbol{x}$. The total error cost function is now written as

$$\begin{aligned}\mathscr{E}(\bar{e},\hat{e}) &= \frac{1}{2}(\boldsymbol{A}\boldsymbol{x} - \boldsymbol{Q}\boldsymbol{R}\boldsymbol{x})^T(\boldsymbol{A}\boldsymbol{x} - \boldsymbol{Q}\boldsymbol{R}\boldsymbol{x}) + \frac{\nu}{2}(\boldsymbol{Q}^T\bar{\boldsymbol{d}} - \boldsymbol{R}\boldsymbol{x})^T(\boldsymbol{Q}^T\bar{\boldsymbol{d}} - \boldsymbol{R}\boldsymbol{x}) \\ &= \frac{1}{2}(\boldsymbol{x}^T\boldsymbol{A}^T\boldsymbol{A}\boldsymbol{x} - \boldsymbol{x}^T\boldsymbol{A}^T\boldsymbol{Q}\boldsymbol{R}\boldsymbol{x} - \boldsymbol{x}^T\boldsymbol{R}^T\boldsymbol{Q}^T\boldsymbol{A}\boldsymbol{x} + \boldsymbol{x}^T\boldsymbol{R}^T\boldsymbol{Q}^T\boldsymbol{Q}\boldsymbol{R}\boldsymbol{x}) \\ &\quad + \frac{\nu}{2}(\bar{\boldsymbol{d}}^T\boldsymbol{Q}\boldsymbol{Q}^T\bar{\boldsymbol{d}} - \bar{\boldsymbol{d}}^T\boldsymbol{Q}\boldsymbol{R}\boldsymbol{x} - \boldsymbol{x}^T\boldsymbol{R}^T\boldsymbol{Q}^T\bar{\boldsymbol{d}} + \boldsymbol{x}^T\boldsymbol{R}^T\boldsymbol{R}\boldsymbol{x})\end{aligned} \tag{7.79}$$

The gradient of the error cost function in (7.79) with respect to $\boldsymbol{R}$ is given as

$$\begin{aligned}\nabla_{\boldsymbol{R}}\mathscr{E}(\bar{e},\hat{e}) &= -\boldsymbol{Q}^T\boldsymbol{A}\boldsymbol{x}\boldsymbol{x}^T + \boldsymbol{Q}^T\boldsymbol{Q}\underbrace{\boldsymbol{R}\boldsymbol{x}}_{\boldsymbol{u}}\boldsymbol{x}^T - \nu\boldsymbol{Q}^T\bar{\boldsymbol{d}}\boldsymbol{x}^T + \nu\underbrace{\boldsymbol{R}\boldsymbol{x}}_{\boldsymbol{u}}\boldsymbol{x}^T \\ &= -\boldsymbol{Q}^T(\underbrace{\boldsymbol{A}\boldsymbol{x} - \boldsymbol{Q}\boldsymbol{u}}_{\bar{e}})\boldsymbol{x}^T - \nu(\underbrace{\boldsymbol{Q}^T\bar{\boldsymbol{d}} - \boldsymbol{u}}_{\hat{e}})\boldsymbol{x}^T \\ &= -\boldsymbol{Q}^T\bar{\boldsymbol{e}}\boldsymbol{x}^T - \nu\hat{\boldsymbol{e}}\boldsymbol{x}^T = -(\boldsymbol{Q}^T\bar{\boldsymbol{e}} + \nu\hat{\boldsymbol{e}})\boldsymbol{x}^T\end{aligned} \tag{7.80}$$

Therefore, the vector-matrix form of the discrete-time learning rule for $\boldsymbol{R}$ can be written, using (7.70) and (7.80), as

$$\boldsymbol{R}(k+1) = \boldsymbol{R}(k) + \mu[\boldsymbol{Q}^T(k+1)\bar{e}(k) + \nu\hat{e}(k)]\boldsymbol{x}^T(k) \tag{7.81}$$

where $\mu > 0$ and $\nu > 0$. However, there are additional constraints that must be imposed on $\boldsymbol{R}$ for it to be an upper triangular matrix. Therefore, for the r_{ij} elements $(i, j = 1, 2, \ldots, n)$ of the $\boldsymbol{R}$ matrix, the learning rule in (7.81) adaptively determines the r_{ij} elements for $i \leq j$, and $r_{ij} = 0$ for $i > j$. The scalar form of the learning rule in (7.81) can be written as

$$r_{ij}(k+1) = r_{ij}(k) + \mu\left[\sum_{h=1}^{n} q_{hi}(k+1)\bar{e}_h(k) + \nu\hat{e}_i(k)\right]x_j(k) \tag{7.82}$$

for $i \leq j$, $r_{ij} = 0$ for $i > j$ and $i, j = 1, 2, \ldots, n$. The scalar forms for the two error terms are given in (7.75) through (7.78).

Method 2: QR factorization

An alternate neurocomputing approach to adaptively perform QR factorization is summarized in block diagram form in Figure 7.7. The corresponding multilayer structured neural network architecture is shown in Figure 7.8. From the block diagram of the structured neural network in Figure 7.7, we see that the architecture is very similar to method 1 for QR factorization. In fact, the learning rules are very similar to those in method 1. However, this architecture is more easily realizable as a multilayer neural network that can be trained using error backpropagation.

The total error cost function is defined exactly as in method 1, see (7.66); however, from Figure 7.7 the two error vectors are written as

$$\bar{e} = \boldsymbol{Ax} - \boldsymbol{Qu} \tag{7.83}$$

and

$$\hat{e} = \boldsymbol{u} - \boldsymbol{Q}^T\boldsymbol{y} \tag{7.84}$$

To derive the discrete-time learning rule for the orthogonal matrix $\boldsymbol{Q}$, the same steepest descent gradient form is used in (7.69) for method 1. In (7.84), no substitution is made for $\boldsymbol{y}$ because this vector is considered to be constant. The justification is that the neural network learns the orthogonal matrix $\boldsymbol{Q}$ rela-

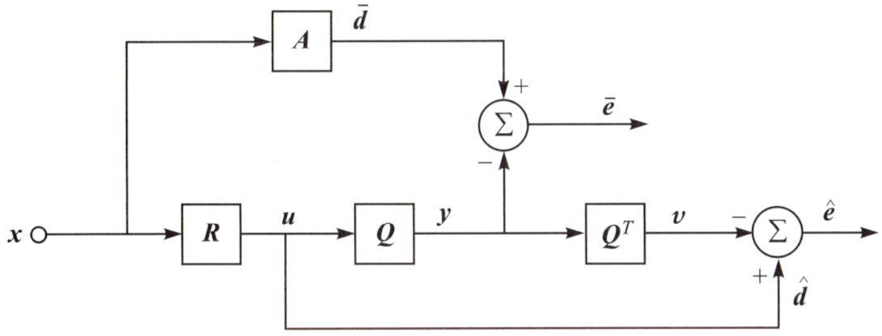

Figure 7.7 Alternate architecture for QR factorization (method 2).

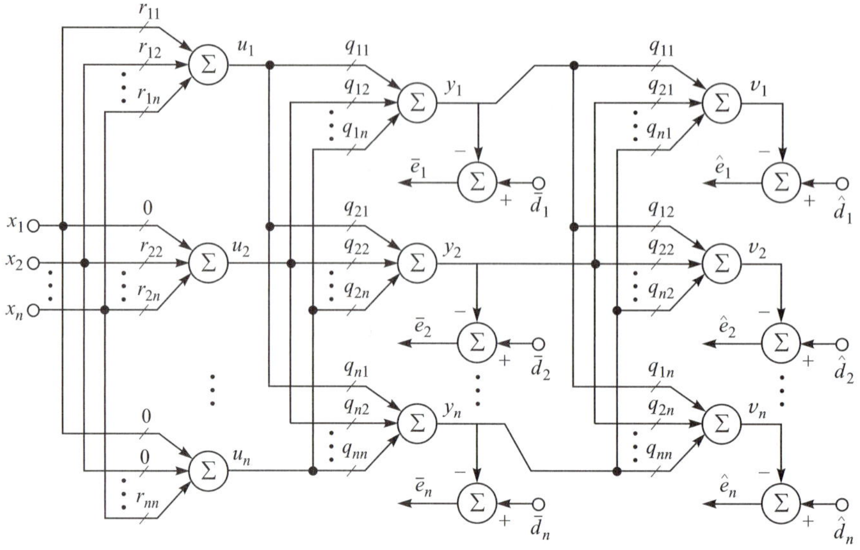

Figure 7.8 Multilayer structured neural network for QR factorization (method 2).

tively quickly compared to the rate of convergence for $\boldsymbol{R}$. Substituting (7.83) and (7.84) into (7.66) and computing the gradient of the total error cost function with respect to $\boldsymbol{Q}$ lead to the discrete-time learning rule for $\boldsymbol{Q}$ in vector-matrix form as

$$\boldsymbol{Q}(k+1) = \boldsymbol{Q}(k) + \mu[\bar{\boldsymbol{e}}(k)\boldsymbol{u}^T(k) + \nu \boldsymbol{y}(k)\hat{\boldsymbol{e}}^T(k)] \tag{7.85}$$

where $\mu > 0$ and $\nu > 0$, $\bar{\boldsymbol{e}}(k)$ and $\hat{\boldsymbol{e}}(k)$ are given in (7.83) and (7.84), respectively, and $\boldsymbol{u}(k) = \boldsymbol{R}\boldsymbol{x}(k)$, and $\boldsymbol{y}(k) = \boldsymbol{Q}\boldsymbol{u}(k)$.

The discrete-time learning rule for $\boldsymbol{R}$ can be derived from the total error cost function used for $\boldsymbol{Q}$, except $\boldsymbol{u}$ is replaced by $\boldsymbol{R}\boldsymbol{x}$ from the block diagram in Figure 7.7. The same steepest descent gradient form in (7.70) for method 1 is used for $\boldsymbol{R}$. The resulting discrete-time learning rule for $\boldsymbol{R}$ in vector-matrix form is given as

$$\boldsymbol{R}(k+1) = \boldsymbol{R}(k) + \mu[\boldsymbol{Q}^T(k+1)\bar{\boldsymbol{e}}(k) - \nu\hat{\boldsymbol{e}}(k)]\boldsymbol{x}^T(k) \tag{7.86}$$

where $\mu > 0$ and $\nu > 0$. The appropriate constraints on the elements of $\boldsymbol{R}$ are the same as those in method 1.

The scalar forms of the learning rules for $\boldsymbol{Q}$ and $\boldsymbol{R}$ can be written as

$$\begin{aligned} q_{ij}(k+1) &= q_{ij}(k) + \mu[\bar{e}_i(k)u_j(k) + \nu y_i(k)\hat{e}_j(k)] \\ r_{ij}(k+1) &= r_{ij}(k) + \mu\left[\sum_{h=1}^{n} q_{hi}(k+1)\bar{e}_h(k) - \nu\hat{e}_i(k)\right]x_j(k) \end{aligned} \tag{7.87}$$

The r_{ij} elements of $\boldsymbol{R}$ are adapted according to the learning rule in (7.87) for $i \le j$, and $r_{ij} = 0$ for $i > j$, and $i, j = 1, 2, \ldots, n$. The scalar forms for $\bar{\boldsymbol{e}}$, $\hat{\boldsymbol{e}}$, $\boldsymbol{y}$, and $\boldsymbol{u}$ can be computed in a similar manner to these quantities in method 1.

7.5 SCHUR DECOMPOSITION

The Schur decomposition [2] is another matrix factorization method that decomposes a matrix $\boldsymbol{A} \in \Re^{n \times n}$ into an orthogonal matrix $\boldsymbol{Q} \in \Re^{n \times n}$ and an upper triangular matrix $\boldsymbol{R} \in \Re^{n \times n}$ such that $\boldsymbol{A}$ can be written as

$$\boldsymbol{A} = \boldsymbol{Q}\boldsymbol{R}\boldsymbol{Q}^T \tag{7.88}$$

The elements down the diagonal of the upper triangular matrix $\boldsymbol{R}$ are the eigenvalues of the $\boldsymbol{A}$ matrix. From (7.88) matrix $\boldsymbol{R}$ can be written as

$$\boldsymbol{R} = \boldsymbol{D} + \boldsymbol{N} = \boldsymbol{Q}^T\boldsymbol{A}\boldsymbol{Q} \tag{7.89}$$

where $\boldsymbol{D} \in \Re^{n \times n}$ is a diagonal matrix with the eigenvalues of $\boldsymbol{A}$ on the diagonal and $\boldsymbol{N} \in \boldsymbol{R}^{n \times n}$ is strictly upper triangular [2].

Wang and Mendel [4] originally introduced a structured neural network for Schur decomposition of a square matrix. The approach given here is similar to their method. Figure 7.9 shows the network block diagram for Schur decomposition. From the figure the two error vectors can be written as

$$\bar{\boldsymbol{e}} = \boldsymbol{A}\boldsymbol{x} - \boldsymbol{Q}\boldsymbol{R}\boldsymbol{b} \tag{7.90}$$

where

$$\boldsymbol{b} = \boldsymbol{Q}^T\boldsymbol{x} \tag{7.91}$$

and

$$\hat{\boldsymbol{e}} = \boldsymbol{x} - \boldsymbol{Q}\boldsymbol{b} \tag{7.92}$$

From (7.90) and (7.91) it is obvious that the error vector $\bar{\boldsymbol{e}}$ will converge to zero, for an appropriate external excitatory input signal, when the neural network learns the Schur decomposition of $\boldsymbol{A}$ (i.e., $\boldsymbol{Q}$ and $\boldsymbol{R}$). In addition, from (7.91) and (7.92) it can be seen that as the neural network learns the orthogonal matrix $\boldsymbol{Q}$ in the Schur decomposition, the error vector $\hat{\boldsymbol{e}}$ will converge to zero for an appropriate external excitatory input signal.

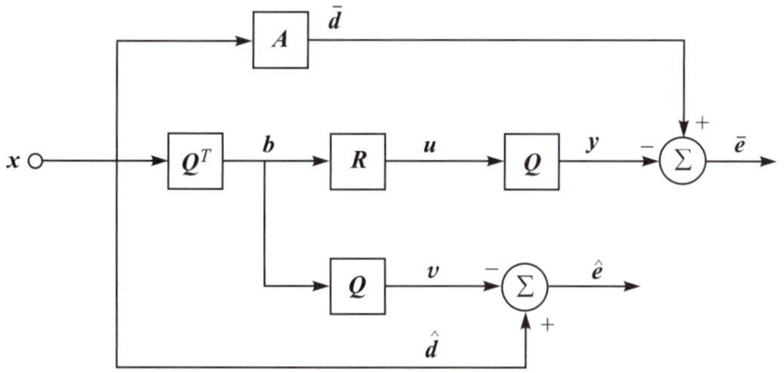

Figure 7.9 Block diagram of the structured neural network, trained by error backpropagation, for Schur decomposition.

Using the two error vectors in (7.90) and (7.92), a total error cost function can be defined as in (7.66) for the QR factorization. The discrete-time learning rules for training the neural network are identical in form to the QR factorization and are given in (7.69) and (7.70). The total error cost function is given as

$$\begin{aligned}\mathscr{E}(\bar{e},\hat{e}) &= \frac{1}{2}\|\bar{e}\|_2^2 + \frac{\nu}{2}\|\hat{e}\|_2^2 = \frac{1}{2}(\boldsymbol{Ax} - \boldsymbol{QRb})^T(\boldsymbol{Ax} - \boldsymbol{QRb}) + \frac{\nu}{2}(\boldsymbol{x} - \boldsymbol{Qb})^T(\boldsymbol{x} - \boldsymbol{Qb}) \\ &= \frac{1}{2}(\boldsymbol{x}^T\boldsymbol{A}^T\boldsymbol{Ax} - \boldsymbol{x}^T\boldsymbol{A}^T\boldsymbol{QRb} - \boldsymbol{b}^T\boldsymbol{R}^T\boldsymbol{Q}^T\boldsymbol{Ax} + \boldsymbol{b}^T\boldsymbol{R}^T\boldsymbol{Q}^T\boldsymbol{QRb}) \\ &\quad + \frac{\nu}{2}(\boldsymbol{x}^T\boldsymbol{x} - \boldsymbol{x}^T\boldsymbol{Qb} - \boldsymbol{b}^T\boldsymbol{Q}^T\boldsymbol{x} + \boldsymbol{b}^T\boldsymbol{Q}^T\boldsymbol{Qb})\end{aligned} \tag{7.93}$$

where $\nu > 0$ is the penalty parameter. From (7.93), and the general results given in (7.7) and (7.8) for differentiating scalar quantities with respect to matrices, the gradient terms in (7.69) and (7.70) can be computed. These are given as

$$\begin{aligned}\nabla_{\boldsymbol{Q}}\mathscr{E}(\bar{e},\hat{e}) &= -\boldsymbol{Axb}^T\boldsymbol{R}^T + \boldsymbol{QRbb}^T\boldsymbol{R}^T - \nu\boldsymbol{xb}^T + \nu\boldsymbol{Qbb}^T \\ &= -(\underbrace{\boldsymbol{Ax} - \boldsymbol{QRb}}_{\bar{e}})\boldsymbol{b}^T\boldsymbol{R}^T - \nu(\underbrace{\boldsymbol{x} - \boldsymbol{Qb}}_{\hat{e}})\boldsymbol{b}^T \\ &= -\bar{e}\underbrace{\boldsymbol{b}^T\boldsymbol{R}^T}_{\boldsymbol{u}^T} - \nu\hat{e}\boldsymbol{b}^T = -\bar{e}\boldsymbol{u}^T - \nu\hat{e}\boldsymbol{b}^T\end{aligned} \tag{7.94}$$

and

$$\nabla_{\boldsymbol{R}}\mathscr{E}(\bar{e},\hat{e}) = -\boldsymbol{Q}^T\boldsymbol{Axb}^T + \boldsymbol{Q}^T\boldsymbol{QRbb}^T = -\boldsymbol{Q}^T(\underbrace{\boldsymbol{Ax} - \boldsymbol{QRb}}_{\bar{e}})\boldsymbol{b}^T = -\boldsymbol{Q}^T\bar{e}\boldsymbol{b}^T \tag{7.95}$$

Therefore, the two learning rules for $\boldsymbol{Q}$ and $\boldsymbol{R}$ are given as

$$\boldsymbol{Q}(k+1) = \boldsymbol{Q}(k) + \mu[\bar{e}(k)\boldsymbol{u}^T(k) + \nu\hat{e}(k)\boldsymbol{b}^T(k)] \tag{7.96}$$

and

$$\boldsymbol{R}(k+1) = \boldsymbol{R}(k) + \mu\boldsymbol{Q}^T(k+1)\bar{e}(k)\boldsymbol{b}^T(k) \tag{7.97}$$

respectively. Where $\mu > 0$, $\nu > 0$ (typically, $\nu = 1$), $\boldsymbol{u} = \boldsymbol{Rb}$, $\boldsymbol{b} = \boldsymbol{Q}^T\boldsymbol{x}$, and $\bar{e}$ and $\hat{e}$ are given in (7.90) and (7.92), respectively. Finally, since $\boldsymbol{R}$ is restricted to be upper triangular, the learning rule in (7.97) applies to the adaptation of the r_{ij} elements (connection weights) of $\boldsymbol{R}$ for $i \le j$, and $r_{ij} = 0$ for $i > j$ $(i, j = 1, 2, \ldots, n)$. The scalar forms of the learning rules for $\boldsymbol{Q}$ and $\boldsymbol{R}$ are given, respectively, as

$$q_{ij}(k+1) = q_{ij}(k) + \mu[\bar{e}_{ip}(k)u_j(k) + \nu\hat{e}_{ip}(k)b_{jp}(k)] \tag{7.98}$$

for $i, j = 1, 2, \ldots, n$, and

$$r_{ij}(k+1) = r_{ij}(k) + \mu\left[\sum_{h=1}^{n} q_{hi}(k+1)\bar{e}_{hp}(k)\right]b_{jp}(k) \tag{7.99}$$

for $i \le j$, $r_{ij} = 0$ for $i > j$, and $p = 1, 2, \ldots, n$ for the n bipolar excitatory input vectors previously described. The scalar forms of the auxiliary variables are given as

$$u_i = \sum_{h=i}^{n} r_{ih} b_{hp} \tag{7.100}$$

and

$$b_{ip} = \sum_{h=1}^{n} q_{hi} x_{hp} \tag{7.101}$$

for $i = 1, 2, \ldots, n$. The scalar forms of the two error terms are given as

$$\bar{e}_{ip} = \sum_{h=1}^{n} a_{ih} x_{hp} - \sum_{h=1}^{n} q_{ih} u_h \tag{7.102}$$

and

$$\hat{e}_{ip} = x_{ip} - \sum_{h=1}^{n} q_{ih} b_{hp} \tag{7.103}$$

for $i = 1, 2, \ldots, n$. For the special case when $\boldsymbol{A}$ is symmetric (i.e., $\boldsymbol{A}^T = \boldsymbol{A}$), the best external excitatory input signal is to let $\boldsymbol{x}$ be a stochastic vector, that is, zero-mean Gaussian white noise typically with unity variance, instead of the bipolar vectors.

7.6 SPECTRAL FACTORIZATION—EIGENVALUE DECOMPOSITION (EVD) (SYMMETRIC EIGENVALUE PROBLEM)

If a square matrix $\boldsymbol{A} \in \Re^{n \times n}$ has distinct eigenvalues $\lambda_1, \lambda_2, \ldots, \lambda_n$ with the associated eigenvectors $\boldsymbol{v}_1, \boldsymbol{v}_2, \ldots, \boldsymbol{v}_n$ (column vectors), then a similarity (nonsingular) transformation can be formed as

$$\boldsymbol{V} = [\boldsymbol{v}_1, \boldsymbol{v}_2, \ldots, \boldsymbol{v}_n] \tag{7.104}$$

and matrix $\boldsymbol{A}$ can be written as

$$\boldsymbol{A} = \boldsymbol{V} \boldsymbol{\Lambda} \boldsymbol{V}^{-1} \tag{7.105}$$

where $\boldsymbol{\Lambda}$ is a diagonal matrix with the eigenvalues of $\boldsymbol{A}$ on the diagonal [2, 8, 17, 18], that is,

$$\boldsymbol{\Lambda} = \text{diag}[\lambda_1, \lambda_2, \ldots, \lambda_n] \tag{7.106}$$

Under certain conditions, if $\boldsymbol{A}$ has nondistinct (repeated) eigenvalues, the matrix can still be diagonalized [11]. However, we restrict our discussions to the case where $\boldsymbol{A}$ has distinct eigenvalues.

The result in (7.105) can be obtained from the standard eigenvalue problem (cf. Sect. A.2.9)

$$(\lambda_i \boldsymbol{I} - \boldsymbol{A})\boldsymbol{v}_i = \boldsymbol{0} \tag{7.107}$$

where λ_i for $i = 1, 2, \ldots, n$ are the eigenvalues of $\boldsymbol{A}$ and $\boldsymbol{v}_i$ for $i = 1, 2, \ldots, n$ are the eigenvectors of $\boldsymbol{A}$. Writing (7.107) as

$$\lambda_i \boldsymbol{v}_i = \boldsymbol{A}\boldsymbol{v}_i \tag{7.108}$$

for $i = 1, 2, \ldots, n$, we have

$$[\lambda_1 \boldsymbol{v}_1, \lambda_2 \boldsymbol{v}_2, \ldots, \lambda_n \boldsymbol{v}_n] = [\boldsymbol{A}\boldsymbol{v}_1, \boldsymbol{A}\boldsymbol{v}_2, \ldots, \boldsymbol{A}\boldsymbol{v}_n] \tag{7.109}$$

or

$$\underbrace{[\boldsymbol{v}_1, \boldsymbol{v}_2, \ldots, \boldsymbol{v}_n]}_{\boldsymbol{V}} \underbrace{\text{diag}[\lambda_1, \lambda_2, \ldots, \lambda_n]}_{\Lambda} = \boldsymbol{A} \underbrace{[\boldsymbol{v}_1, \boldsymbol{v}_2, \ldots, \boldsymbol{v}_n]}_{\boldsymbol{V}} \tag{7.110}$$

or

$$\boldsymbol{V}\boldsymbol{\Lambda} = \boldsymbol{A}\boldsymbol{V} \tag{7.111}$$

Now postmultiplying (7.111) by $\boldsymbol{V}^{-1}$, we can write

$$\boldsymbol{A} = \boldsymbol{V}\boldsymbol{\Lambda}\boldsymbol{V}^{-1} \tag{7.112}$$

which is precisely what was stated in (7.105). If we restrict $\boldsymbol{A}$ to be symmetric (i.e., $\boldsymbol{A}^T = \boldsymbol{A}$), this leads to the symmetric eigenvalue problem, and matrix $\boldsymbol{V}$ will be orthogonal, that is,

$$\boldsymbol{V}^T \boldsymbol{V} = \boldsymbol{V}\boldsymbol{V}^T = \boldsymbol{I} \tag{7.113}$$

moreover,

$$\boldsymbol{V}^{-1} = \boldsymbol{V}^T \tag{7.114}$$

Therefore, every real symmetric matrix with distinct eigenvalues can be diagonalized by an orthogonal similarity transformation. Another way to state the orthogonal nature of the eigenvectors of $\boldsymbol{A}$ for the symmetric case is

$$\boldsymbol{v}_i^T \boldsymbol{v}_j = \left\{ \begin{matrix} 1 & i = j \\ 0 & i \neq j \\ & \forall\, i, j \end{matrix} \right\} = \delta_{ij} \text{ (Kronecker delta)} \tag{7.115}$$

From (7.112) and (7.114) we can write

$$\begin{aligned} \boldsymbol{A} = \boldsymbol{V}\boldsymbol{\Lambda}\boldsymbol{V}^T &= [\boldsymbol{v}_1, \boldsymbol{v}_2, \ldots, \boldsymbol{v}_n] \text{diag}[\lambda_1, \lambda_2, \ldots, \lambda_n][\boldsymbol{v}_1, \boldsymbol{v}_2, \ldots, \boldsymbol{v}_n]^T \\ &= \lambda_1 \boldsymbol{v}_1 \boldsymbol{v}_1^T + \lambda_2 \boldsymbol{v}_2 \boldsymbol{v}_2^T + \cdots + \lambda_n \boldsymbol{v}_n \boldsymbol{v}_n^T = \sum_{i=1}^{n} \lambda_i \boldsymbol{v}_i \boldsymbol{v}_i^T \end{aligned} \tag{7.116}$$

which is the eigenvalue decomposition of $\boldsymbol{A}$. From (7.116) it is evident that the rank of $\boldsymbol{A}$ is the sum of the ranks of n matrices. However, these are all outer-product matrices (formed from the eigenvectors of $\boldsymbol{A}$) that have rank 1, and each outer-product matrix is multiplied by the associated eigenvalue. Therefore, if all eigenvalues are nonzero, $\boldsymbol{A}$ will be full rank; however, if any eigenvalue of $\boldsymbol{A}$ is zero, then $\boldsymbol{A}$ will be rank-deficient.

7.7 NEURAL NETWORK APPROACH FOR THE SYMMETRIC EIGENVALUE PROBLEM

The original neural network approach for the symmetric eigenvalue problem can be found in the paper by Wang and Mendel [4]. Other work in this area can be found [6, 19]. The neurocomputing approach taken here is very similar to the Schur decomposition structured neural network shown in the previous section. That is, the structured neural network for the symmetric eigenvalue problem, shown in Figure 7.10, has the same form as the one for the Schur decomposition, shown in Figure 7.9. In fact, for the Schur decomposition of $\boldsymbol{A} \in \Re^{n \times n}$, and for $\boldsymbol{A}$ symmetric (i.e., $\boldsymbol{A}^T = \boldsymbol{A}$), we can write from (7.88)

$$\boldsymbol{A} = \boldsymbol{Q}\boldsymbol{R}\boldsymbol{Q}^T = \boldsymbol{A}^T = \boldsymbol{Q}\boldsymbol{R}^T\boldsymbol{Q}^T \Rightarrow \boldsymbol{R} = \boldsymbol{R}^T \tag{7.117}$$

From (7.117) we can conclude that matrix $\boldsymbol{R}$ cannot be upper triangular, but must be a diagonal matrix with the eigenvalues of $\boldsymbol{A}$ on the diagonal, and the column vectors of $\boldsymbol{Q}$ are the associated eigenvectors of $\boldsymbol{A}$. Therefore, for $\boldsymbol{A}$ symmetric, $\boldsymbol{R} = \boldsymbol{\Lambda}$, $\boldsymbol{Q} = \boldsymbol{V}$, and the Schur decomposition of $\boldsymbol{A}$ becomes $\boldsymbol{A} = \boldsymbol{V}\boldsymbol{\Lambda}\boldsymbol{V}^T$, which is the eigenvalue decomposition given in (7.116). The detailed neural network architecture is shown in Figure 7.11.

Therefore, from the discussion above, we would expect the learning rules for $\boldsymbol{V}$ and $\boldsymbol{\Lambda}$, in the case of the symmetric eigenvalue problem, to have the same basic form as those for $\boldsymbol{Q}$ and $\boldsymbol{R}$ for the Schur decomposition. The only difference is the restriction on the elements of the $\boldsymbol{\Lambda}$ matrix for the symmetric eigenvalue problem, which must be diagonal. The form of the total error cost function is the same as in (7.93) for the Schur decomposition neurocomputing approach. Therefore, the vector-matrix forms of the discrete-time learning rules are given as

$$\boldsymbol{V}(k+1) = \boldsymbol{V}(k) + \mu[\bar{\boldsymbol{e}}(k)\boldsymbol{u}^T(k) + \nu\hat{\boldsymbol{e}}(k)\boldsymbol{b}^T(k)] \tag{7.118}$$

and

$$\boldsymbol{\Lambda}(k+1) = \boldsymbol{\Lambda}(k) + \mu\boldsymbol{V}^T(k+1)\bar{\boldsymbol{e}}(k)\boldsymbol{b}^T(k) \tag{7.119}$$

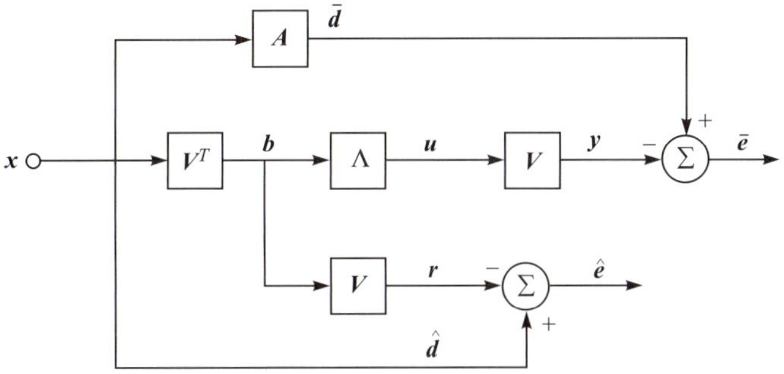

Figure 7.10 Block diagram of the structured neural network, trained by error backpropagation, for the symmetric eigenvalue problem.

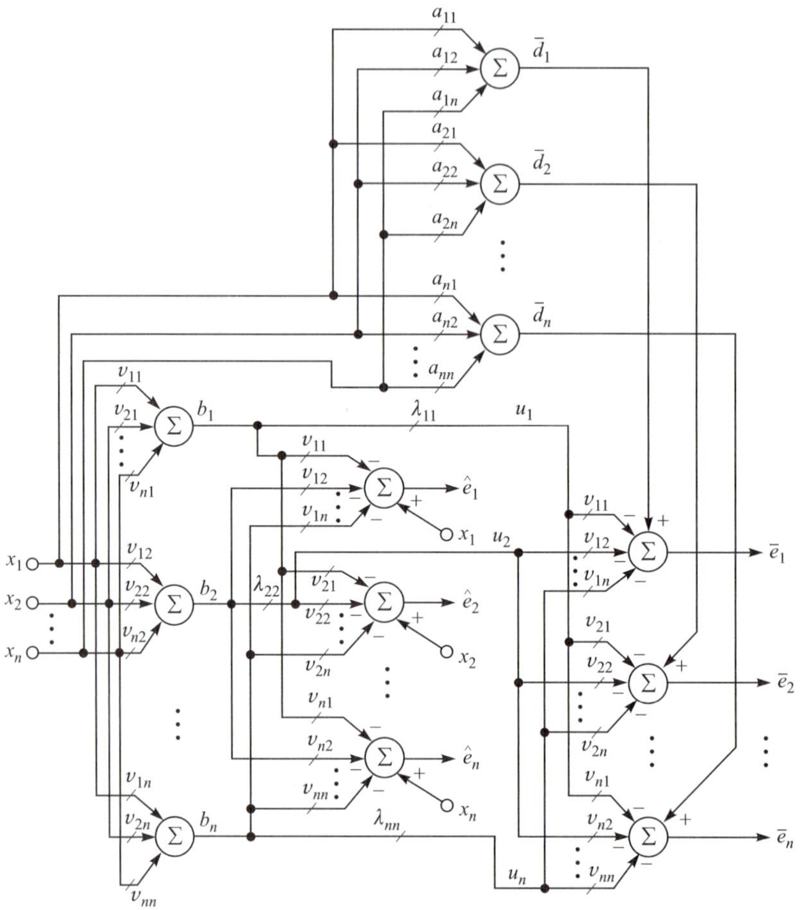

Figure 7.11 Detailed neural network architecture for the symmetric eigenvalue problem.

for $\mu > 0$, $\nu > 0$ (typically, $\nu = 1$), where, from the block diagram in Figure 7.10,

$$\bar{e} = Ax - Vu \tag{7.120}$$

with

$$u = \Lambda b \tag{7.121}$$

where

$$b = V^T x \tag{7.122}$$

and

$$\hat{e} = x - Vb \tag{7.123}$$

The best external excitatory input x is a stochastic vector, that is, zero-mean Gaussian white noise typically with unity variance. The scalar forms of the learning rules in (7.118) and (7.119) are given as

$$v_{ij}(k+1) = v_{ij}(k) + \mu[\bar{e}_i(k)u_j(k) + \nu\hat{e}_i(k)b_j(k)] \tag{7.124}$$

and

$$\lambda_{ij}(k+1) = \lambda_{ij}(k) + \mu\left[\sum_{h=1}^{n} v_{hi}(k+1)\bar{e}_h(k)\right]b_j(k) \tag{7.125}$$

respectively, for $i, j = 1, 2, \ldots, n$.

In (7.125) the learning rule only applies for the case when $i = j$, for $i \neq j$ $\lambda_{ij} = 0$, because $\boldsymbol{\Lambda}$ is a diagonal matrix. It is straightforward to determine the scalar forms of the auxiliary variables and the error terms based on the results for the Schur decomposition. In Cichocki and Unbehauen [6] an economical realization of the neural network block diagram in Figure 7.10 is given. Note in Figure 7.10 that $\boldsymbol{V}$ is *shared* between the computation of the two error terms. They exploited this fact which led to a simplified structured network that uses a timesharing (multiplexing) technique. The simplified neural network based on the timesharing method has computing units of the output layer described by the connection weight matrix $\boldsymbol{V}$ that are timeshared (or overlapped) by two computing channels to compute both $\bar{\boldsymbol{e}}$ and $\hat{\boldsymbol{e}}$.

Min/Max eigenvalue problem

Many times it is desirable to determine an extremal (i.e., minimal or maximal) real eigenvalue and the corresponding eigenvector of a real symmetric matrix $\boldsymbol{A} \in \Re^{n\times n}$, $\boldsymbol{A}^T = \boldsymbol{A}$. Finding the min/max eigenvalue of $\boldsymbol{A}$ corresponds to obtaining the minimal value of the function $\mathscr{R}(\boldsymbol{v})$ (the Rayleigh quotient [2, 20, 21]) defined as

$$\mathscr{R}(\boldsymbol{v}) = \pm\frac{\langle \boldsymbol{A}\boldsymbol{v}, \boldsymbol{v}\rangle}{\langle \boldsymbol{v}, \boldsymbol{v}\rangle} = \pm\frac{\boldsymbol{v}^T\boldsymbol{A}\boldsymbol{v}}{\boldsymbol{v}^T\boldsymbol{v}} \tag{7.126}$$

where $\boldsymbol{v} \in \Re^{n\times 1}$ is assumed to never be identically zero. The Rayleigh quotient in (7.126) is the unique number that minimizes $\|\boldsymbol{A}\boldsymbol{v} - \lambda\boldsymbol{v}\|_2^2$. If $\boldsymbol{v}$ is an eigenvector of $\boldsymbol{A}$, then the Rayleigh quotient $\mathscr{R}(\boldsymbol{v})$ is the associated eigenvalue of $\boldsymbol{A}$ with the $+$ sign for the minimum eigenvalue and the $-$ sign for the maximum eigenvalue.

A neurocomputing approach for determining the min/max eigenvalue of $\boldsymbol{A}$ can be formulated as a constrained optimization problem given by

$$\text{Minimize} \quad \pm\frac{1}{2}\boldsymbol{v}_i^T\boldsymbol{A}\boldsymbol{v}_i \qquad \text{for } i = 1, 2, \ldots, n \tag{7.127}$$

subject to the constraints

$$(\boldsymbol{A} - \lambda_i\boldsymbol{I})\boldsymbol{v}_i = \boldsymbol{0} \tag{7.128}$$

and

$$\boldsymbol{v}_i^T\boldsymbol{v}_i - 1 = 0 \tag{7.129}$$

A total error cost function can be formulated using a penalty method [22, 23] as

$$\mathscr{E}(\boldsymbol{v}_i, \lambda_i) = \frac{1}{2}\left[\pm\alpha\boldsymbol{v}_i^T\boldsymbol{A}\boldsymbol{v}_i + \boldsymbol{e}_i^T\boldsymbol{e}_i + \frac{\beta}{2}(\boldsymbol{v}_i^T\boldsymbol{v}_i - 1)^2\right] \tag{7.130}$$

where

$$e_i = Av_i - \lambda_i v_i \tag{7.131}$$

from the standard eigenvalue problem, $\alpha > 0$ and $\beta > 0$ are penalty parameters, and $i = 1, 2, \ldots, n$. Two learning rules can be derived by minimizing the total error cost function in (7.130) to determine the min/max eigenvalue λ_i and the associated eigenvector v_i. That is, by using a steepest descent gradient approach, the discrete-time learning rules (in vector-matrix form) are given by

$$\lambda_i(k+1) = \lambda_i(k) + \mu v_i^T(k) e_i(k) \tag{7.132}$$

and

$$v_i(k+1) = v_i(k) - \mu\{\pm\alpha A v_i(k) + A e_i(k) - \lambda_i(k) e_i(k) + \beta[v_i^T(k) v_i(k) - 1] v_i(k)\} \tag{7.133}$$

where $\mu > 0$, and the penalty parameters $\alpha > 0$ and $\beta > 0$ can be adjusted in an ad hoc manner; however, it is advisable to gradually decrease the penalty parameter α as the neural network trains (especially during the final stages of training) [6]. Again, the + sign corresponds to a minimum eigenvalue, and the − sign will yield a maximum eigenvalue. The scalar form of the discrete-time learning rule in (7.133) for the eigenvector can be written as

$$\begin{aligned} v_{ji}(k+1) = v_{ji}(k) - \mu\Bigg\{&\pm\alpha \sum_{h=1}^{n} a_{ih} v_{hi}(k) + \sum_{h=1}^{n} a_{ih} e_{hi}(k) \\ &- \lambda_i(k) e_{ji}(k) + \beta\left[\sum_{h=1}^{n} v_{hi}^2(k) - 1\right] v_{ji}(k)\Bigg\} \end{aligned} \tag{7.134}$$

for $j = 1, 2, \ldots, n$.

EXAMPLE 7.2. In this example it is desired to find the eigenvalues and eigenvectors of the matrix

$$A = \begin{bmatrix} 1 & 1 & \frac{1}{2} & -1 \\ 1 & -5 & \frac{1}{2} & -3 \\ \frac{1}{2} & \frac{1}{2} & -4 & \frac{1}{2} \\ -1 & -3 & \frac{1}{2} & -5 \end{bmatrix} \tag{7.135}$$

By using the **eig** function in MATLAB, the following eigenvalues and eigenvectors were obtained for A (out to four decimal places):

$$\Lambda^M = \text{diag}[-3.9408 \quad -2.5376 \quad 1.6005 \quad -8.1221] \tag{7.136}$$

$$V^M = \begin{bmatrix} 0.1245 & -0.3524 & -0.9275 & -0.0097 \\ -0.1840 & 0.6436 & -0.2620 & -0.6952 \\ -0.9734 & -0.1285 & -0.0836 & 0.1702 \\ -0.0558 & -0.6671 & 0.2532 & -0.6983 \end{bmatrix} \tag{7.137}$$

Using the neural network approach developed in this section, for a learning rate of $\mu = 0.00255$, $\nu = 1$, and $V(k=0) = \Lambda(k=0) = I$, the eigenvalues and eigenvectors were

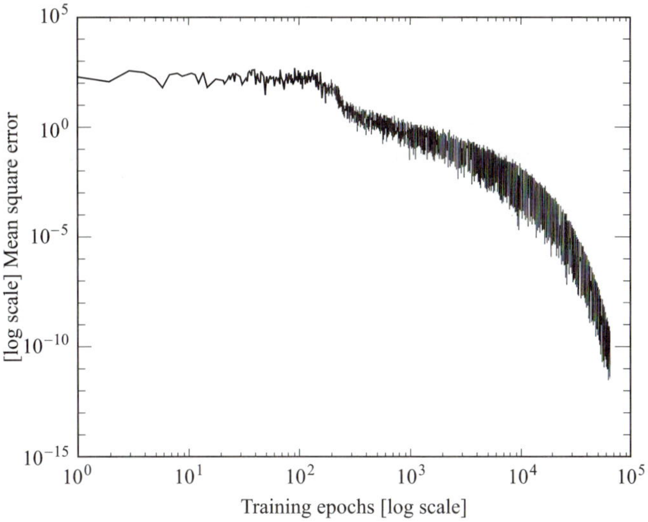

Figure 7.12 Mean-square error during training for computing the eigenvalues and eigenvectors of a symmetric matrix in Example 7.2

$$\boldsymbol{\Lambda}^{NN} = \text{diag}\left[1.6005, -3.9408, -8.1221, -2.5376\right] \tag{7.138}$$

$$\boldsymbol{V}^{NN} = \begin{bmatrix} -0.9275 & 0.1245 & -0.0097 & -0.3524 \\ -0.2620 & -0.1840 & -0.6952 & 0.6436 \\ -0.0836 & -0.9734 & 0.1702 & -0.1285 \\ 0.2532 & -0.0558 & -0.6983 & -0.6671 \end{bmatrix} \tag{7.139}$$

after 65,000 training steps. The amount of error involved in computing the eigenvalues and eigenvectors using the neural network approach can be quantified by calculating the norm of the difference between matrix $\boldsymbol{A}$ and its eigenvalue decomposition $\boldsymbol{V}\boldsymbol{\Lambda}\boldsymbol{V}^T$, given in (7.116). This was computed as

$$\text{error} = \|\boldsymbol{A} - \boldsymbol{V}\boldsymbol{\Lambda}\boldsymbol{V}^T\|_2 = 1.0969 \times 10^{-6} \tag{7.140}$$

where the $\|\bullet\|_2$ norm of the matrix is the largest singular value of $\boldsymbol{A} - \boldsymbol{V}\boldsymbol{\Lambda}\boldsymbol{V}^T$. These results are identical to the MATLAB results, with the exception of the eigenvalue/eigenvector ordering and some of the eigenvectors are within a minus sign of each other. Figure 7.12 shows the progress of the neural network learning, in terms of the mean-square error, to compute the eigenvalues and eigenvectors.

EXAMPLE 7.3. If it desired to calculate only the *largest* eigenvalue (and associated eigenvector) of $\boldsymbol{A}$, we can use the neurocomputing approach derived in this section. The matrix for which it is desired to find the largest eigenvalue and eigenvector is

$$\boldsymbol{A} = \begin{bmatrix} 0.45 & 0.55 \\ 0.55 & 0.45 \end{bmatrix} \tag{7.141}$$

Using MATLAB, the calculated eigenvalues and eigenvectors are given as

$$\boldsymbol{\Lambda}^M = \text{diag}\left[-0.1, 1\right] \qquad \boldsymbol{V}^M = \begin{bmatrix} 0.7071 & 0.7071 \\ -0.7071 & 0.7071 \end{bmatrix} \tag{7.142}$$

The training process for the neural network, to find the largest eigenvalue and eigenvector of the matrix in (7.141), was carried out in the following manner: (1) 5,000 training epochs for $\mu = 2 \times 10^{-4}$, and $\alpha = \beta = 0.5$, then using the synaptic weights from the first part of the training process; (2) 5,000 additional training epochs for $\mu = 2 \times 10^{-4}$, $\beta = 0.5$, and starting with an initial value of $\alpha = 0.5$, α was decreased by 0.1 percent every training step. The second part of the neural network training process was repeated eight more times, that is, using the final synaptic weights from the previous training process for the initial weights of the next stage. Therefore, after a total of 50,000 training steps, the neural network converged to the maximum eigenvalue of $\boldsymbol{A}$ as

$$\lambda_{\max}^{NN} = 1.0000 \tag{7.143}$$

and the associated eigenvector is

$$\boldsymbol{v}_{\max}^{NN} = [0.7071, 0.7071]^T \tag{7.144}$$

which corresponds to the results from MATLAB given in (7.141) and (7.142).

7.8 SINGULAR-VALUE DECOMPOSITION

The singular-value decomposition (SVD) [2, 8, 9, 17, 18, 20, 24] of a matrix (cf. Sect. A.2.14) is one of the most important matrix decomposition methods. It has many applications in signal processing, control theory, and parametric modeling of data, and specifically it is the method of choice for solving most linear least-squares problems, especially for ill-conditioned matrices [2]. As explained in Section A.2.14, the basic purpose of the SVD is to decompose a matrix $\boldsymbol{A} \in \Re^{m\times n}$ into two orthogonal matrices $\boldsymbol{U} \in \Re^{m\times m}(\boldsymbol{U}^T\boldsymbol{U} = \boldsymbol{U}\boldsymbol{U}^T = \boldsymbol{I})$ and $\boldsymbol{V} \in \Re^{n\times n}(\boldsymbol{V}^T\boldsymbol{V} = \boldsymbol{V}\boldsymbol{V}^T = \boldsymbol{I})$ and a pseudodiagonal matrix $\boldsymbol{S} \in \Re^{m\times n}$, that is, $\boldsymbol{S} = \text{pseudodiag}(\sigma_1, \sigma_2, \ldots, \sigma_p)$, where $p = \min(m, n)$, and the real nonnegative numbers $\sigma_1 \geq \sigma_2 \geq \ldots \geq \sigma_p \geq 0$ are called the singular values of $\boldsymbol{A}$. Therefore, $\boldsymbol{A}$ can be written as

$$\boldsymbol{A} = \boldsymbol{U}\boldsymbol{S}\boldsymbol{V}^T \tag{7.145}$$

and conversely $\boldsymbol{S}$ can be written as

$$\boldsymbol{S} = \boldsymbol{U}^T\boldsymbol{A}\boldsymbol{V} \tag{7.146}$$

If the rank of $\boldsymbol{A}$ is r, that is, $\rho(\boldsymbol{A}) = r$ (also referred to as the index of the smallest singular value), then the SVD of $\boldsymbol{A}$ can be written as

$$\boldsymbol{A} = \boldsymbol{U}\boldsymbol{S}\boldsymbol{V}^T = \boldsymbol{U}_r\boldsymbol{S}_r\boldsymbol{V}_r^T = \sum_{i=1}^{r} \sigma_i \boldsymbol{u}_i \boldsymbol{v}_i^T \tag{7.147}$$

where $\boldsymbol{u}_i$, for $i = 1, 2, \ldots, r$, are the first r columns of $\boldsymbol{U}$; $\boldsymbol{v}_i$, for $i = 1, 2, \ldots, r$, are the first r columns of $\boldsymbol{V}$; and σ_i, for $i = 1, 2, \ldots, r$, are the first r singular values from the pseudodiagonal matrix $\boldsymbol{S}$ (all the remaining singular values are zero). Therefore, $\boldsymbol{U}_r \in \Re^{m\times r}$, $\boldsymbol{V}_r \in \Re^{n\times r}$, and $\boldsymbol{S}_r \in \Re^{r\times r}$.

Our objective is to compute the two orthogonal matrices $\boldsymbol{U}$ and $\boldsymbol{V}$ and the pseudodiagonal matrix $\boldsymbol{S}$, using a neurocomputing approach [1, 4, 6, 25, 26].

Without loss of generality, we will assume $m \geq n$ and $\boldsymbol{U} \in \Re^{m \times n}(\boldsymbol{U}^T \boldsymbol{U} = \boldsymbol{I})$, $\boldsymbol{V} \in \Re^{n \times n}(\boldsymbol{V}^T \boldsymbol{V} = \boldsymbol{V}\boldsymbol{V}^T = \boldsymbol{I})$, and $\boldsymbol{S} = \mathrm{diag}[\sigma_1, \sigma_2, \ldots, \sigma_n] \in \Re^{n \times n}$, where the singular values are ordered such that $\sigma_1 \geq \sigma_2 \geq \cdots \geq \sigma_n \geq 0$. The approach taken can be viewed from the block diagram of the structured neural network shown in Figure 7.13. A total error cost function can be defined as

$$\mathscr{E}(\bar{\boldsymbol{e}}, \hat{\boldsymbol{e}}, \tilde{\boldsymbol{e}}) = \frac{1}{2}\|\bar{\boldsymbol{e}}\|_2^2 + \frac{\nu_1}{2}\|\hat{\boldsymbol{e}}\|_2^2 + \frac{\nu_2}{2}\|\tilde{\boldsymbol{e}}\|_2^2 \tag{7.148}$$

where $\nu_1 > 0$ and $\nu_2 > 0$ are the penalty parameters, typically, $0 < \nu_1, \nu_2 \leq 10$. Minimizing the first term in (7.148) provides a factorization of matrix $\boldsymbol{A}$ into the three matrices $\boldsymbol{U}$, $\boldsymbol{S}$, and $\boldsymbol{V}$. Minimization of the second term in (7.148) ensures that the matrix $\boldsymbol{V}$ (whose columns are *right singular vectors*) is orthogonal, and minimizing the third term provides an orthogonal matrix $\boldsymbol{U}$ (whose columns are the *left singular vectors*). The forms of the three vector-matrix discrete-time learning rules that result from a steepest descent gradient optimization approach to compute $\boldsymbol{V}$, $\boldsymbol{S}$, and $\boldsymbol{U}$ are given as

$$\boldsymbol{V}(k+1) = \boldsymbol{V}(k) - \mu_1 \nabla_{\boldsymbol{V}} \mathscr{E}(\bar{\boldsymbol{e}}, \hat{\boldsymbol{e}}, \tilde{\boldsymbol{e}}) \tag{7.149}$$

$$\boldsymbol{S}(k+1) = \boldsymbol{S}(k) - \mu_2 \nabla_{\boldsymbol{S}} \mathscr{E}(\bar{\boldsymbol{e}}, \hat{\boldsymbol{e}}, \tilde{\boldsymbol{e}}) \tag{7.150}$$

$$\boldsymbol{U}(k+1) = \boldsymbol{U}(k) - \mu_3 \nabla_{\boldsymbol{U}} \mathscr{E}(\bar{\boldsymbol{e}}, \hat{\boldsymbol{e}}, \tilde{\boldsymbol{e}}) \tag{7.151}$$

where $\mu_1 > 0$, $\mu_2 > 0$, and $\mu_3 > 0$ are the three independent learning rate parameters. From the block diagram in Figure 7.13, three expressions for the error vectors can be written as

$$\begin{aligned} \bar{\boldsymbol{e}} &= \bar{\boldsymbol{d}} - \boldsymbol{f} = \boldsymbol{A}\boldsymbol{x} - \boldsymbol{U}\underbrace{\boldsymbol{r}}_{\boldsymbol{S}\boldsymbol{b}} = \boldsymbol{A}\boldsymbol{x} - \boldsymbol{U}\boldsymbol{S}\underbrace{\boldsymbol{b}}_{\boldsymbol{V}^T\boldsymbol{x}} = \boldsymbol{A}\boldsymbol{x} - \boldsymbol{U}\boldsymbol{S}\boldsymbol{V}^T\boldsymbol{x} \\ &= (\boldsymbol{A} - \boldsymbol{U}\boldsymbol{S}\boldsymbol{V}^T)\boldsymbol{x}, \qquad (\bar{\boldsymbol{e}} \in \Re^{m \times 1}) \end{aligned} \tag{7.152}$$

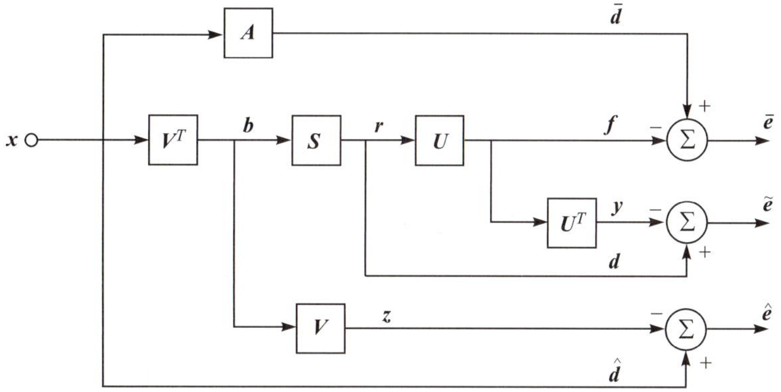

Figure 7.13 Structured neural network architecture for computing the SVD of a matrix.

$$\hat{e} = \underbrace{\hat{d}}_{x} - \underbrace{z}_{Vb} = x - V\underbrace{b}_{V^T x} = x - VV^T x = (I_{n\times n} - VV^T)x \qquad (\hat{e} \in \Re^{n\times 1}) \tag{7.153}$$

and

$$\tilde{e} = \underbrace{\tilde{d}}_{r} - \underbrace{y}_{U^T f} = r - U^T \underbrace{f}_{Ur} = r - U^T Ur = (I_{n\times n} - U^T U)\underbrace{r}_{SV^T x} = (I_{n\times n} - U^T U)SV^T x \qquad (\tilde{e} \in \Re^{n\times 1}) \tag{7.154}$$

Therefore, from (7.152), (7.153), and (7.154) we can see that for an appropriate external excitatory input $x \in \Re^{n\times 1}$, and as the columns of U converge to the left singular vectors of A, the diagonal elements of S converge to the singular values of A, and the columns of V converge to the right singular vectors of A, the three error vectors will converge to zero and the neural network will be properly trained.

To derive each of the three discrete-time learning rules in (7.149), (7.150), and (7.151), to compute the three matrices V, S, and U, respectively, the three gradients $\nabla_V \mathscr{E}(\bar{e}, \hat{e}, \tilde{e})$, $\nabla_S \mathscr{E}(\bar{e}, \hat{e}, \tilde{e})$, and $\nabla_U \mathscr{E}(\bar{e}, \hat{e}, \tilde{e})$ must be determined. To first derive the gradient $\nabla_V \mathscr{E}(\bar{e}, \hat{e}, \tilde{e})$, we can write the total error cost function in (7.148) as

$$\begin{aligned}
\mathscr{E}(\bar{e},\hat{e},\tilde{e}) &= \frac{1}{2}\|\bar{e}\|_2^2 + \frac{\nu_1}{2}\|\hat{e}\|_2^2 + \frac{\nu_2}{2}\|\tilde{e}\|_2^2 = \frac{1}{2}\bar{e}^T\bar{e} + \frac{\nu_1}{2}\hat{e}^T\hat{e} + \frac{\nu_2}{2}\tilde{e}^T\tilde{e} \\
&= \frac{1}{2}(\underbrace{x^T A^T - x^T VS^T U^T}_{\bar{e}^T})(\underbrace{Ax - USV^T x}_{\bar{e}}) \\
&\quad + \frac{\nu_1}{2}(\underbrace{x^T - x^T VV^T}_{\hat{e}^T})(\underbrace{x - VV^T x}_{\hat{e}}) + \frac{\nu_2}{2}(\underbrace{r^T - f^T U}_{\tilde{e}^T})(\underbrace{r - U^T f}_{\tilde{e}}) \\
&= \frac{1}{2}(x^T A^T Ax - x^T A^T USV^T x - x^T VS^T U^T Ax \\
&\quad + x^T VS^T U^T USV^T x) \\
&\quad + \frac{\nu_1}{2}(x^T x - 2x^T VV^T x + x^T VV^T VV^T x) \\
&\quad + \frac{\nu_2}{2}(r^T r - r^T U^T f - f^T Ur + f^T UU^T f)
\end{aligned} \tag{7.155}$$

The gradient of (7.155) with respect to matrix V can now be computed by using the general results for differentiation of a scalar with respect to a matrix given in (7.7) and (7.8) along with the proper chain rule. The result is

$$\nabla_V \mathscr{E}(\bar{e},\hat{e},\tilde{e}) = \frac{\partial \mathscr{E}(\bar{e},\hat{e},\tilde{e})}{\partial V} = -x\bar{e}^T US - \nu_1(x\hat{e}^T V + \hat{e}b^T) \tag{7.156}$$

where

$$b = V^T x \tag{7.157}$$

from the block diagram in Figure 7.13. Therefore, the discrete-time learning rule for $\boldsymbol{V}$ in vector-matrix form, using (7.149) and the gradient result in (7.156), is

$$\boldsymbol{V}(k+1) = \boldsymbol{V}(k) + \mu_1\{\boldsymbol{x}(k)\bar{\boldsymbol{e}}^T(k)\boldsymbol{U}(k)\boldsymbol{S}(k) + \nu_1[\boldsymbol{x}(k)\hat{\boldsymbol{e}}^T(k)\boldsymbol{V}(k) + \hat{\boldsymbol{e}}(k)\boldsymbol{b}^T(k)]\} \tag{7.158}$$

where

$$\bar{e} = \boldsymbol{A}\boldsymbol{x} - \boldsymbol{U}\boldsymbol{r} \tag{7.159}$$

$$\hat{e} = \boldsymbol{x} - \boldsymbol{V}\boldsymbol{b} \tag{7.160}$$

and

$$\boldsymbol{r} = \boldsymbol{S}\boldsymbol{b} \tag{7.161}$$

The external excitatory inputs to the network $\boldsymbol{x}$ should be taken as the set of linearly independent bipolar vectors, as previously described.

Next we can compute the gradient of the total error cost function with respect to matrix $\boldsymbol{S}$, which is required in the learning rule given in (7.150). The total error cost function in (7.148) can now be written as

$$\begin{aligned}
\mathscr{E}(\bar{e},\hat{e},\tilde{e}) &= \frac{1}{2}(\underbrace{\boldsymbol{x}^T\boldsymbol{A}^T - \boldsymbol{x}^T\boldsymbol{V}\boldsymbol{S}^T\boldsymbol{U}^T}_{\bar{e}^T})(\underbrace{\boldsymbol{A}\boldsymbol{x} - \boldsymbol{U}\boldsymbol{S}\boldsymbol{V}^T\boldsymbol{x}}_{\bar{e}}) \\
&\quad + \frac{\nu_1}{2}(\underbrace{\boldsymbol{x}^T - \boldsymbol{b}^T\boldsymbol{V}^T}_{\hat{e}^T})(\underbrace{\boldsymbol{x} - \boldsymbol{V}\boldsymbol{b}}_{\hat{e}}) + \frac{\nu_2}{2}(\underbrace{\boldsymbol{r}^T - \boldsymbol{f}^T\boldsymbol{U}}_{\tilde{e}^T})(\underbrace{\boldsymbol{r} - \boldsymbol{U}^T\boldsymbol{f}}_{\tilde{e}}) \\
&= \frac{1}{2}(\boldsymbol{x}^T\boldsymbol{A}^T\boldsymbol{A}\boldsymbol{x} - \boldsymbol{x}^T\boldsymbol{A}^T\boldsymbol{U}\boldsymbol{S}\boldsymbol{V}^T\boldsymbol{x} - \boldsymbol{x}^T\boldsymbol{V}\boldsymbol{S}^T\boldsymbol{U}^T\boldsymbol{A}\boldsymbol{x} + \boldsymbol{x}^T\boldsymbol{V}\boldsymbol{S}^T\boldsymbol{U}^T\boldsymbol{U}\boldsymbol{S}\boldsymbol{V}^T\boldsymbol{x}) \\
&\quad + \frac{\nu_1}{2}(\boldsymbol{x}^T\boldsymbol{x} - \boldsymbol{x}^T\boldsymbol{V}\boldsymbol{b} - \boldsymbol{b}^T\boldsymbol{V}^T\boldsymbol{x} + \boldsymbol{b}^T\boldsymbol{V}^T\boldsymbol{V}\boldsymbol{b}) \\
&\quad + \frac{\nu_2}{2}(\boldsymbol{r}^T\boldsymbol{r} - \boldsymbol{r}^T\boldsymbol{U}^T\boldsymbol{f} - \boldsymbol{f}^T\boldsymbol{U}\boldsymbol{r} + \boldsymbol{f}^T\boldsymbol{U}\boldsymbol{U}^T\boldsymbol{f})
\end{aligned} \tag{7.162}$$

The gradient of (7.162) respect to $\boldsymbol{S}$ is given as

$$\nabla_{\boldsymbol{S}}\mathscr{E}(\bar{e},\hat{e},\tilde{e}) = \frac{\partial\mathscr{E}(\bar{e},\hat{e},\tilde{e})}{\partial\boldsymbol{S}} = -\boldsymbol{U}^T\bar{\boldsymbol{e}}\boldsymbol{b}^T \tag{7.163}$$

Therefore, the discrete-time learning rule for $\boldsymbol{S}$ in vector-matrix form, using (7.150) and the gradient result in (7.163), is given as

$$\boldsymbol{S}(k+1) = \boldsymbol{S}(k) + \mu_2\boldsymbol{U}^T(k)\bar{\boldsymbol{e}}(k)\boldsymbol{b}^T(k) \tag{7.164}$$

which is only applicable for the adaptation of the diagonal elements of $\boldsymbol{S}$. All other elements in $\boldsymbol{S}$ are zero because $\boldsymbol{S}$ is considered to be a diagonal matrix.

Finally, we can compute the gradient of the total error cost function with respect to matrix $\boldsymbol{U}$. The total error cost function in (7.148) can now be written as

$$\mathscr{E}(\bar{e},\hat{e},\tilde{e}) = \frac{1}{2}(\underbrace{x^T A^T - x^T V S^T U^T}_{\bar{e}^T})(\underbrace{Ax - USV^T x}_{\bar{e}})$$
$$+ \frac{\nu_1}{2}(\underbrace{x^T - b^T V^T}_{\hat{e}^T})(\underbrace{x - Vb}_{\hat{e}})$$
$$+ \frac{\nu_2}{2}(\underbrace{x^T V S^T - x^T V S^T U^T U}_{\tilde{e}^T})(\underbrace{S V^T x - U^T U S V^T x}_{\tilde{e}})$$
$$= \frac{1}{2}(x^T A^T A x - x^T A^T U S V^T x - x^T V S^T U^T A x + x^T V S^T U^T U S V^T x)$$
$$+ \frac{\nu_1}{2}(x^T x - x^T V b - b^T V^T x + b^T V^T V b)$$
$$+ \frac{\nu_2}{2}(x^T V S^T S V^T x - 2x^T V S^T U^T U S V^T x$$
$$+ x^T V S^T U^T U U^T U S V^T x) \tag{7.165}$$

The gradient of (7.165) with respect to U is given as

$$\nabla_U \mathscr{E}(\bar{e},\hat{e},\tilde{e}) = \frac{\partial \mathscr{E}(\bar{e},\hat{e},\tilde{e})}{\partial U} = -\bar{e} r^T - \nu_2 (U \tilde{e} r^T + f \tilde{e}^T) \tag{7.166}$$

where

$$\tilde{e} = r - U^T f \tag{7.167}$$

and

$$f = Ur$$

Therefore, the discrete-time learning rule for U in vector-matrix form, using (7.151) and the gradient result in (7.166), is given as

$$U(k+1) = U(k) + \mu_3 \{\bar{e}(k) r^T(k) + \nu_2 [U(k)\tilde{e}(k) r^T(k) + f(k)\tilde{e}^T(k)]\} \tag{7.168}$$

The scalar forms of the learning rules are given as

$$v_{ij}(k+1) = v_{ij}(k) + \mu_1 \left(x_i(k) \left\{ \left[\sum_{h=1}^{m} \bar{e}_h(k) u_{hj}(k) \right] \sigma_j(k) \right\} + \nu_1 \left[x_i(k) \sum_{h=1}^{n} \hat{e}_h(k) v_{hj}(k) + \hat{e}_i(k) b_j(k) \right] \right) \tag{7.169}$$

for $i, j = 1, 2, \ldots, n$,

$$\sigma_i(k+1) = \sigma_i(k) + \mu_2 \left\{ \left[\sum_{h=1}^{m} \bar{e}_h(k) u_{hi}(k) \right] b_i(k) \right\} \tag{7.170}$$

for $i = 1, 2, \ldots, n$, and

$$u_{qj}(k+1) = u_{qj}(k) + \mu_3\Bigg(\bar{e}_q(k)r_j(k) + \nu_2\left\{\left[\sum_{h=1}^{n} u_{qh}(k)\tilde{e}_h(k)\right]r_j(k) + f_q(k)\tilde{e}_j(k)\right\}\Bigg) \tag{7.171}$$

for $q = 1, 2, \ldots, m$, and $j = 1, 2, \ldots, n$. The three error terms can also be written in scalar form as

$$\bar{e}_q = \sum_{h=1}^{n} a_{qh}x_h - \sum_{h=1}^{n} u_{qh}r_h \qquad q = 1, 2, \ldots, m \tag{7.172}$$

$$\hat{e}_i = x_i - \sum_{h=1}^{n} v_{ih}b_h \qquad i = 1, 2, \ldots, n \tag{7.173}$$

$$\tilde{e}_i = r_i - \sum_{h=1}^{m} u_{hi}f_h \qquad i = 1, 2, \ldots, n \tag{7.174}$$

where

$$r_i = \sigma_i b_i \qquad i = 1, 2, \ldots, n \tag{7.175}$$

$$b_i = \sum_{h=1}^{n} v_{hi}x_h \qquad i = 1, 2, \ldots, n \tag{7.176}$$

$$f_q = \sum_{h=1}^{n} u_{qh}r_h \qquad q = 1, 2, \ldots, m \tag{7.177}$$

As with all the previous neurocomputing approaches for solving certain matrix algebra problems, the SVD neural network is best suited for the case when $\boldsymbol{A}$ contains elements that can change slowly as a function of time. In this case, the initial weights of the network would be set at values which correspond to the SVD of $\boldsymbol{A}$ as computed by a highly robust numerical algorithm such as LAPACK [27]. Then when necessary, the neural network would update the appropriate synaptic connection weights to reflect the associated changes in the elements of matrix $\boldsymbol{A}$.

The previous structured neural network is a general architecture that is not unique. Depending on the specific properties of $\boldsymbol{A}$, for example, whether it is square, symmetric, etc., a more simplified neural network and learning rules can result. Some of these specialized structured architectures for SVD are given in block diagram form in the book by Cichocki and Unbehauen [6].

7.9 A NEUROCOMPUTING APPROACH FOR SOLVING THE ALGEBRAIC LYAPUNOV EQUATION

The algebraic Lyapunov equation plays a very important role in many applications (cf. Sect. A.7.8), especially in control theory [28, 29]. This equation is a special form of the linear matrix equation known as the Sylvester equation

$$\boldsymbol{AX} + \boldsymbol{XB} = -\boldsymbol{C} \tag{7.178}$$

where it is assumed that we know $\boldsymbol{A} \in \Re^{n\times n}$, $\boldsymbol{B} \in \Re^{m\times m}$, and $\boldsymbol{C} \in \Re^{n\times m}$ and seek the unique solution $\boldsymbol{X} \in \Re^{n\times m}$ [29]. If $\boldsymbol{B} = \boldsymbol{A}^T$, then $\boldsymbol{C} \in \Re^{n\times n}$, and the resulting equation is known as the *Lyapunov equation*, given as

$$\boldsymbol{AX} + \boldsymbol{XA}^T = -\boldsymbol{C} \tag{7.179}$$

where a unique solution exists for (7.179) if and only if $\lambda_i(\boldsymbol{A}) + \lambda_j(\boldsymbol{A}) \neq 0, \forall i, j = 1, 2, \ldots, n$ (cf. Sect. A 2.17). For the Lyapunov equation in (7.179), both the solution $\boldsymbol{X} \in \Re^{n\times n}$ and $\boldsymbol{C} \in \Re^{n\times n}$ are symmetric matrices and positive semidefinite, that is, $\boldsymbol{X}^T = \boldsymbol{X}, \boldsymbol{X} \geq 0$, and $\boldsymbol{C}^T = \boldsymbol{C}, \boldsymbol{C} \geq 0$.

A neurocomputing approach can be taken to solve the Lyapunov equation in (7.179) [6]. However, the real advantage of using a structured neural network to solve the Lyapunov equation is seen when some of (or all) the elements in the $\boldsymbol{A}$ matrix change slowly with time. In this case the initial synaptic weights of the network (i.e., the solution $\boldsymbol{X}$) could be set at values obtained from solving the Lyapunov equation using a highly robust numerical routine. Figure 7.14 shows the structured neural network architecture for solving the algebraic matrix Lyapunov equation. From the block diagram, the error term can be written as

$$\boldsymbol{e} = \boldsymbol{d} - \boldsymbol{y} = -\boldsymbol{Cu} - \boldsymbol{AXu} - \boldsymbol{Xz} \tag{7.180}$$

where

$$\boldsymbol{z} = \boldsymbol{A}^T\boldsymbol{u} \tag{7.181}$$

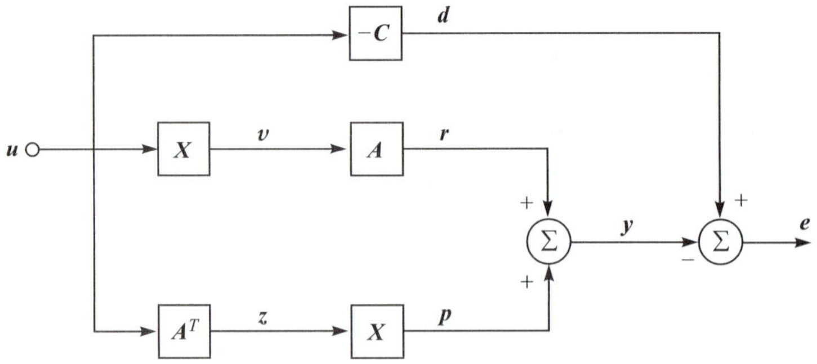

Figure 7.14 Structured neural network architecture for solving the algebraic matrix Lyapunov equation. The network is trained by error backpropagation.

Substituting (7.181) into (7.180) gives

$$e = d - y = -Cu - AXu - XA^T u = (-C - AX - XA^T)u \qquad (7.182)$$

Therefore, from (7.182) we see that given an appropriate external excitatory input signal $\boldsymbol{u}$, causing $\boldsymbol{X}$ to approach the solution to the Lyapunov equation in (7.179), the error vector $\boldsymbol{e}$ will approach zero. The basis for the continuous-time learning rule for the neural network is the steepest descent gradient method, which leads to a set of matrix differential equations

$$\frac{dX(t)}{dt} = -\mu \nabla_X \mathscr{E}(X) \qquad (7.183)$$

where the error cost function $\mathscr{E}(X)$ is defined through the error vector $\boldsymbol{e}$, given in (7.180). Thus, the mean-square error cost function can be written as

$$\begin{aligned}\mathscr{E}(X) &= \frac{1}{2}\|e\|_2^2 = \frac{1}{2}(Cu + AXu + Xz)^T(Cu + AXu + Xz)\\ &= \frac{1}{2}(u^T C^T Cu + u^T C^T AXu + u^T C^T Xz\\ &\quad + u^T X^T A^T Cu + u^T X^T A^T AXu + u^T X^T A^T Xz\\ &\quad + z^T X^T Cu + z^T X^T AXu + z^T X^T Xz) \end{aligned} \qquad (7.184)$$

To compute the gradient in (7.183), that is, $\nabla_X \mathscr{E}(X)$, we need the two general results used previously, (7.7) and (7.8), and the proper chain rule (cf. Sect. A.3.4.2). Carrying out the gradient calculation leads to the result

$$\begin{aligned}\nabla_X \mathscr{E}(X) &= A^T Cuu^T + A^T AXuu^T + A^T Xzu^T + Cuz^T + AXuz^T + Xzz^T\\ &= A^T(\underbrace{Cu + AXu + Xz}_{-e})u^T + (\underbrace{Cu + AXu + Xz}_{-e})z^T = -A^T eu^T - ez^T\end{aligned} \qquad (7.185)$$

Therefore, from (7.183) and the gradient in (7.185), we can write the continuous-time learning rule as

$$\frac{dX(t)}{dt} = \mu[A^T e(t)u^T(t) + e(t)z^T(t)] \qquad (7.186)$$

where $\mu > 0$ is the learning rate parameter. In discrete-time form, the learning rule is given as

$$X(k+1) = X(k) + \mu[A^T e(k)u^T(k) + e(k)z^T(k)] \qquad (7.187)$$

As previously stated, the solution to the Lyapunov equation, $\boldsymbol{X}$, is symmetric. However, in (7.187), the second term on the right-hand side is not symmetric. This is not a problem in terms of the learning rule computing the positive semidefinite, symmetric solution to (7.179). However, this form of the learning rule will converge more slowly than a modified version of (7.187) that forces symmetry every training step. There are two different ways of performing forced symmetry which require the same number of additional computations, compared to (7.187). The first modification involves using (7.187) as shown above and then performing an additional computational step every training step to force the symmetry, that is,

$$X(k+1) \leftarrow \frac{X(k+1) + X^T(k+1)}{2} \tag{7.188}$$

The second approach would first involve defining $\Delta X(k) = A^T e(k) u^T(k) + e(k) z^T(k)$, and then rewriting the learning rule in (7.187) as

$$X(k+1) = X(k) + \frac{\mu}{2}[\Delta X(k) + \Delta X^T(k)] \tag{7.189}$$

Taking the approach given in either (7.188) or (7.189) will result in faster convergence to the solution X, compared to the learning rule given in (7.187).

The scalar form of the learning rule in (7.187) can be written as

$$x_{ij}(k+1) = x_{ij}(k) + \mu\{e_i(k)z_j(k) + [\sum_{h=1}^{n} a_{hi}e_h(k)]u_j(k)\} \tag{7.190}$$

for $i, j = 1, 2, \ldots, n$, and

$$e_i = -\sum_{h=1}^{n} c_{ih}u_h - \sum_{h=1}^{n} a_{ih}v_h - \sum_{h=1}^{n} x_{ih}z_h \tag{7.191}$$

where

$$v_i = \sum_{h=1}^{n} x_{ih}u_h \tag{7.192}$$

and

$$z_i = \sum_{h=1}^{n} a_{hi}u_h \tag{7.193}$$

for $i = 1, 2, \ldots, n$. Equation (7.190) can be modified to invoke symmetry to facilitate faster convergence. The modified learning rule is given as

$$x_{ij}(k+1) = x_{ij}(k) + \frac{\mu}{2}\{e_i(k)z_j(k) + e_j(k)z_i(k) + [\sum_{h=1}^{n} a_{hi}e_h(k)]u_j(k) + [\sum_{h=1}^{n} a_{hj}e_h(k)]u_i(k)\} \tag{7.194}$$

for $i, j = 1, 2, \ldots, n$.

In Cichocki and Unbehauen [6], the external excitatory inputs are taken to be sinusoidal signals, that is, $u_l(t) = \sin(l\omega_0 t)$ for $l = 1, 2, \ldots, n$. However, the set of linearly independent bipolar vectors is a better choice for the external signal because an increase in convergence speed can generally be attained.

7.10 A NEUROCOMPUTING APPROACH FOR SOLVING THE ALGEBRAIC RICCATI EQUATION

The algebraic matrix Riccati equation plays a major role in optimal control and optimal estimation theory [28, 29]. The standard form of the equation is given as

$$A^T X + XA - XRX + Q = 0 \tag{7.195}$$

which is a nonlinear matrix equation, as evidenced by the third term on the left-hand side (which is nonlinear). Given $A \in \Re^{n \times n}$, $R \in \Re^{n \times n}$ ($R > 0$, $R = R^T$), and $Q \in \Re^{n \times n}(Q \geq 0, Q = Q^T)$, we seek the solution to (7.195), $X \in \Re^{n \times n}(X \geq 0, X = X^T)$.

A structured neural network approach for solving the algebraic matrix Riccati equation was first presented by Ham and Collins [30, 31]. Figure 7.15 shows the structured neural network architecture in block diagram form for solving the algebraic Riccati equation. The error vector e can be written from the block diagram as

$$e = b - y = XRv - A^T Xz - XAz - Qz \tag{7.196}$$

where

$$v = Xz \tag{7.197}$$

Substituting (7.197) into (7.196) yields

$$e = \left(XRX - A^T X - XA - Q\right)z \tag{7.198}$$

Therefore, from (7.198) we see that for an appropriate external excitatory signal z the error vector e will approach zero as X approaches the solution to the Riccati equation given in (7.195). We can develop a learning rule for training the neural network by taking a steepest descent optimization approach and defining the continuous-time learning rule as a set of matrix differential equations given as

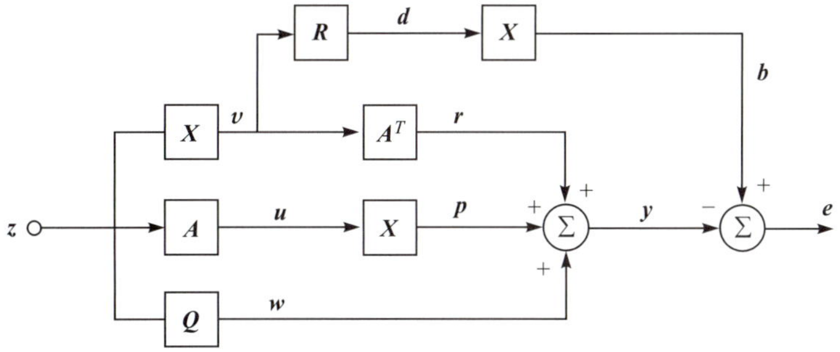

Figure 7.15 Structured network for solving the algebraic matrix Riccati equation. The network is trained by error backpropagation.

$$\frac{\mathrm{d}\boldsymbol{X}(t)}{\mathrm{d}t} = -\mu\nabla_{\boldsymbol{X}}\mathscr{E}(\boldsymbol{X}) \tag{7.199}$$

where $\mathscr{E}(\boldsymbol{X})$, the mean-square error cost function, is defined as

$$\mathscr{E}(\boldsymbol{X}) = \frac{1}{2}\|\boldsymbol{e}\|_2^2 \tag{7.200}$$

Substituting (7.196) into (7.200) gives

$$\begin{aligned}\mathscr{E}(\boldsymbol{X}) = \frac{1}{2}(&\boldsymbol{v}^T\boldsymbol{R}^T\boldsymbol{X}^T\boldsymbol{X}\boldsymbol{R}\boldsymbol{v} - \boldsymbol{v}^T\boldsymbol{R}^T\boldsymbol{X}^T\boldsymbol{A}^T\boldsymbol{X}\boldsymbol{z} - \boldsymbol{v}^T\boldsymbol{R}^T\boldsymbol{X}^T\boldsymbol{X}\boldsymbol{A}\boldsymbol{z} - \boldsymbol{v}^T\boldsymbol{R}^T\boldsymbol{X}^T\boldsymbol{Q}\boldsymbol{z}\\ &- \boldsymbol{z}^T\boldsymbol{X}^T\boldsymbol{A}\boldsymbol{X}\boldsymbol{R}\boldsymbol{v} + \boldsymbol{z}^T\boldsymbol{X}^T\boldsymbol{A}\boldsymbol{A}^T\boldsymbol{X}\boldsymbol{z} + \boldsymbol{z}^T\boldsymbol{X}^T\boldsymbol{A}\boldsymbol{X}\boldsymbol{A}\boldsymbol{z} + \boldsymbol{z}^T\boldsymbol{X}^T\boldsymbol{A}\boldsymbol{Q}\boldsymbol{z}\\ &- \boldsymbol{z}^T\boldsymbol{A}^T\boldsymbol{X}^T\boldsymbol{X}\boldsymbol{R}\boldsymbol{v} + \boldsymbol{z}^T\boldsymbol{A}^T\boldsymbol{X}^T\boldsymbol{A}^T\boldsymbol{X}\boldsymbol{z} + \boldsymbol{z}^T\boldsymbol{A}^T\boldsymbol{X}^T\boldsymbol{X}\boldsymbol{A}\boldsymbol{z} + \boldsymbol{z}^T\boldsymbol{A}^T\boldsymbol{X}^T\boldsymbol{Q}\boldsymbol{z}\\ &- \boldsymbol{z}^T\boldsymbol{Q}^T\boldsymbol{X}\boldsymbol{R}\boldsymbol{v} + \boldsymbol{z}^T\boldsymbol{Q}^T\boldsymbol{A}^T\boldsymbol{X}\boldsymbol{z} + \boldsymbol{z}^T\boldsymbol{Q}^T\boldsymbol{X}\boldsymbol{A}\boldsymbol{z} + \boldsymbol{z}^T\boldsymbol{Q}^T\boldsymbol{Q}\boldsymbol{z})\end{aligned} \tag{7.201}$$

To compute the gradient in (7.199), that is, $\nabla_{\boldsymbol{X}}\mathscr{E}(\boldsymbol{X})$, we must use the two general results given in (7.7) and (7.8) and the proper chain rule (cf. Sect. A.3.4.2). Computing the gradient of (7.201) with respect to $\boldsymbol{X}$ gives

$$\begin{aligned}\nabla_{\boldsymbol{X}}\mathscr{E}(\boldsymbol{X}) &= [\underbrace{\boldsymbol{X}\boldsymbol{R}\boldsymbol{v} - \boldsymbol{A}^T\boldsymbol{X}\boldsymbol{z} - \boldsymbol{X}\boldsymbol{A}\boldsymbol{z} - \boldsymbol{Q}\boldsymbol{z}}_{\boldsymbol{e}}]\boldsymbol{v}^T\boldsymbol{R}^T\\ &\quad - [\underbrace{\boldsymbol{X}\boldsymbol{R}\boldsymbol{v} - \boldsymbol{A}^T\boldsymbol{X}\boldsymbol{z} - \boldsymbol{X}\boldsymbol{A}\boldsymbol{z} - \boldsymbol{Q}\boldsymbol{z}}_{\boldsymbol{e}}]\boldsymbol{z}^T\boldsymbol{A}^T\\ &\quad - \boldsymbol{A}[\underbrace{\boldsymbol{X}\boldsymbol{R}\boldsymbol{v} - \boldsymbol{A}^T\boldsymbol{X}\boldsymbol{z} - \boldsymbol{X}\boldsymbol{A}\boldsymbol{z} - \boldsymbol{Q}\boldsymbol{z}}_{\boldsymbol{e}}]\boldsymbol{z}^T\\ &= \boldsymbol{e}\boldsymbol{v}^T\boldsymbol{R} - \boldsymbol{e}\boldsymbol{z}^T\boldsymbol{A}^T - \boldsymbol{A}\boldsymbol{e}\boldsymbol{z}^T\end{aligned} \tag{7.202}$$

Substituting the result in (7.202) into (7.199) gives the continuous-time learning rule as

$$\frac{\mathrm{d}\boldsymbol{X}(t)}{\mathrm{d}t} = \mu\big[\boldsymbol{A}\boldsymbol{e}(t)\boldsymbol{z}^T(t) + \boldsymbol{e}(t)\boldsymbol{z}^T(t)\boldsymbol{A}^T - \boldsymbol{e}(t)\boldsymbol{v}^T(t)\boldsymbol{R}\big] \tag{7.203}$$

where $\mu > 0$ is the learning rate parameter. In discrete-time form, the learning rule is given as

$$\boldsymbol{X}(k+1) = \boldsymbol{X}(k) + \mu[\boldsymbol{A}\boldsymbol{e}(k)\boldsymbol{z}^T(k) + \boldsymbol{e}(k)\boldsymbol{z}^T(k)\boldsymbol{A}^T - \boldsymbol{e}(k)\boldsymbol{v}^T(k)\boldsymbol{R}] \tag{7.204}$$

where k is the discrete-time index, that is, $k = 0, 1, 2, \ldots$. As was the case for the neurocomputing approach for solving the Lyapunov equation, the update term in (7.204), that is,

$$\Delta\boldsymbol{X}(k) = \boldsymbol{A}\boldsymbol{e}(k)\boldsymbol{z}^T(k) + \boldsymbol{e}(k)\boldsymbol{z}^T(k)\boldsymbol{A}^T - \boldsymbol{e}(k)\boldsymbol{v}^T(k)\boldsymbol{R} \tag{7.205}$$

is not symmetric, but the learning rule in (7.204) will still converge to the positive semidefinite, symmetric solution even though $\Delta\boldsymbol{X}(k)$ in (7.205) is not symmetric. However, this form of the learning rule will converge more slowly than a modified version of (7.204) that forces symmetry every training step. There are two different ways of performing forced symmetry which require the same number of additional computations, compared to (7.204).

The first modification involves using (7.204) as shown above and then performing an additional computational step every training step to force the symmetry, that is,

$$\boldsymbol{X}(k+1) \leftarrow \frac{\boldsymbol{X}(k+1)+\boldsymbol{X}^T(k+1)}{2} \tag{7.206}$$

The second approach involves using the update expression in (7.205) to rewrite the learning rule in (7.204) as

$$\boldsymbol{X}(k+1) = \boldsymbol{X}(k) + \frac{\mu}{2}[\Delta\boldsymbol{X}(k) + \Delta\boldsymbol{X}^T(k)] \tag{7.207}$$

Taking the approach given in either (7.206) or (7.207) will result in faster convergence to the solution $\boldsymbol{X}$, than the learning rule given in (7.204).

The scalar form of the learning rule (with the appropriate terms to force symmetry) is given as

$$\begin{aligned} x_{ij}(k+1) = x_{ij}(k) + \frac{\mu}{2}\Bigg\{&\left[\sum_{h=1}^{n} a_{ih}e_h(k)\right]z_j(k) + z_i(k)\left[\sum_{h=1}^{n} e_h(k)a_{hi}\right] \\ &+ e_i(k)\left[\sum_{h=1}^{n} z_h(k)a_{hi}\right] + \left[\sum_{h=1}^{n} a_{ih}z_h(k)\right]e_j(k) \\ &- e_i(k)\left[\sum_{h=1}^{n} v_h(k)r_{hj}\right] - \left[\sum_{h=1}^{n} r_{ih}v_h(k)\right]e_j(k)\Bigg\} \end{aligned} \tag{7.208}$$

where

$$\begin{aligned} e_i(k) = &\left[\sum_{h=1}^{n} x_{ih}(k)r_{hj}\right]v_i(k) - \left[\sum_{h=1}^{n} a_{hi}x_{hj}(k)\right]z_i(k) \\ &- \left[\sum_{h=1}^{n} x_{ih}(k)a_{hj}\right]z_i(k) - \left[\sum_{h=1}^{n} q_{ih}z_h(k)\right] \end{aligned} \tag{7.209}$$

and

$$v_i(k) = \sum_{h=1}^{n} x_{ih}(k)z_h(k) \tag{7.210}$$

for $i, j = 1, 2, \ldots, n$. The external excitatory vector input signals $\boldsymbol{z}$ should be taken as the set of n linearly independent bipolar vectors (previously described). Figure 7.16 shows the multilayer neural network architecture for solving the algebraic matrix Riccati equation. This network is trained by error backpropagation.

EXAMPLE 7.4. This example illustrates the capability of the neurocomputing approach for solving the algebraic matrix Riccati equation in (7.195) for the following conditions:

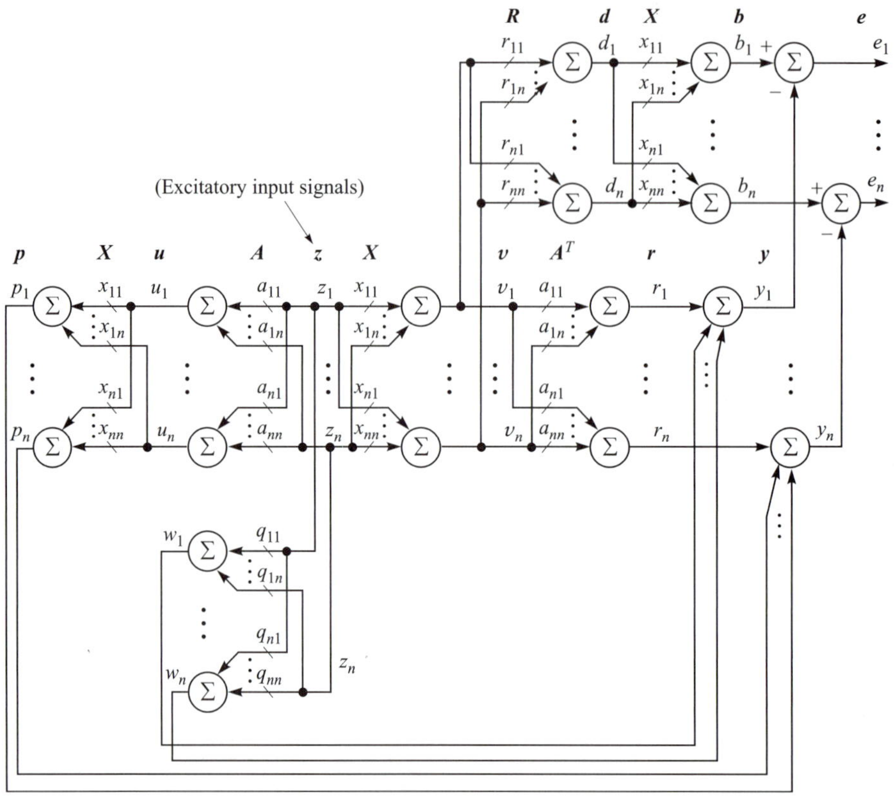

Figure 7.16 Multilayer neural network architecture for solving the algebraic matrix Riccati equation. The a_{ij}, q_{ij}, and r_{ij} for $i, j = 1, 2, \ldots, n$ elements are fixed and x_{ij} for $i, j = 1, 2, \ldots, n$ are adaptive.

$$\boldsymbol{A} = \begin{bmatrix} 0 & 1 & 0 \\ 0 & 0 & 1 \\ -6 & -11 & -6 \end{bmatrix} \qquad \boldsymbol{Q} = \begin{bmatrix} 1 & 0 & 0 \\ 0 & 0.25 & 0 \\ 0 & 0 & 0.5 \end{bmatrix}$$

$$\boldsymbol{R} = \begin{bmatrix} 10 & 0 & 0 \\ 0 & 3.3333 & 0 \\ 0 & 0 & 14.2857 \end{bmatrix} \tag{7.211}$$

The MATLAB control systems toolbox [32] function **`lqr2`** is first used to solve the steady-state algebraic matrix Riccati equation. The **`lqr2`** function in the MATLAB control systems toolbox solves the linear quadratic regulator problem using the Schur decomposition method. Therefore, this function requires as inputs the matrices $\boldsymbol{A}$, $\boldsymbol{Q}$, $\tilde{\boldsymbol{R}}$, and $\boldsymbol{B}$, where $\tilde{\boldsymbol{R}} = \boldsymbol{R}^{-1}$ (given above) and the $\boldsymbol{R}$ matrix is internally computed as $\boldsymbol{B}\tilde{\boldsymbol{R}}^{-1}\boldsymbol{B}^T$, where $\boldsymbol{B} = \boldsymbol{I}_{n \times n}$(identity matrix). The solution $\boldsymbol{X}^M$ yielded by the Schur method for solving the Riccati equation in MATLAB is given by

$$X^M = \begin{bmatrix} 0.3324 & 0.1094 & -0.0123 \\ 0.1094 & 0.3790 & -0.0059 \\ -0.0123 & -0.0059 & 0.0388 \end{bmatrix} \tag{7.212}$$

The neural network solution, X^{NN}, using a learning rate parameter $\mu = 0.00275$, $N = 500$ (or $N \times n = 1{,}500$ iterations, where $n = 3$), and with the initial conditions $X^{NN}(k = 0) = \mathbf{0}$ (the null matrix), yielded the same results as the MATLAB solution (out to four decimal places), as shown in (7.212). The external excitatory input signals were taken as the set of n linearly independent bipolar vectors. The three elements of the error vector e are shown plotted separately in Figure 7.17(*a*). The accuracy of the neural network computational method for solving the Riccati equation can be quantified by comparing it to the solution using the MATLAB **lqr2** function as

$$\|X^{NN} - X^M\|_2 = 9.5081 \times 10^{-6} \tag{7.213}$$

A comparison is now made to the case when sinusoidal signals are used as the external excitatory inputs to the neural network for the same learning rate parameter (i.e., $\mu = 0.00275$) and the same number of total training steps (i.e., $N' = N \times n = 1{,}500$) as used for the bipolar inputs. The general form of the sinusoidal inputs is given in discrete-time form as

$$z_l(kT) = \sin(kl\omega_0 T) \tag{7.214}$$

where $l = 1, 2, 3$, $\omega_0 = 500$ rad/s, $T = 5 \times 10^{-4}$ s (sampling period), and $k = 0, 1, \ldots, (N' - 1)$. The three elements of the error vector e are shown plotted separately in Figure 7.17(*b*) for the sinusoidal inputs. Figure 7.17(*b*) shows that

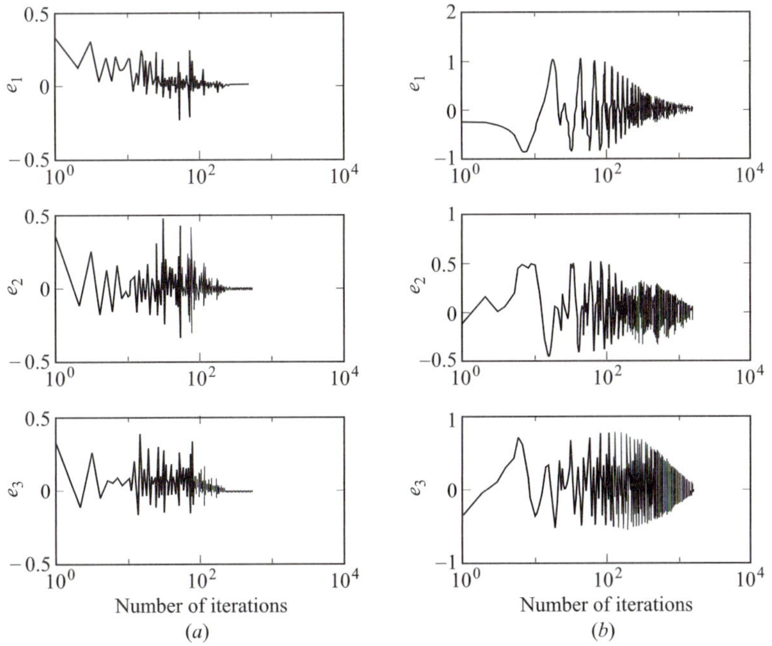

Figure 7.17 (*a*) Neural network learning mean-square error for the external excitatory inputs taken as the bipolar vectors. (*b*) Neural network learning mean-square error for the external excitatory inputs taken as sinusoidal signals.

the neural network has not converged after 1,500 training steps for the sinusoidal inputs, that is,

$$\| \boldsymbol{X}^{NN} - \boldsymbol{X}^{M} \|_2 = 0.0371 \tag{7.215}$$

An additional 3,500 training steps are required for the sinusoidal input case to achieve a comparable level of accuracy for the bipolar vector input neural network; see (7.213).

PROBLEMS

7.1. For matrices $\boldsymbol{A}, \boldsymbol{B}$, and $\boldsymbol{C}$ of appropriate dimensions, prove

(*a*) $\dfrac{\partial}{\partial \boldsymbol{A}} \text{trace}(\boldsymbol{BAC}) = \boldsymbol{B}^T \boldsymbol{C}^T$

(*b*) $\dfrac{\partial}{\partial \boldsymbol{A}} \text{trace}(\boldsymbol{BA}^T \boldsymbol{C}) = \boldsymbol{CB}$

7.2. Write a computer program to implement the neural network learning rule given in vector-matrix (batch) form in (7.9), method 1 for computing the pseudoinverse of a matrix, and compute the pseudoinverse of the following matrices:

(*a*) $\boldsymbol{A} = \begin{bmatrix} -1 & 2 & 3 \\ 4 & 5 & 6 \\ 4 & -1 & -1 \end{bmatrix}$ (*b*) $\boldsymbol{A} = \begin{bmatrix} 1 & 2 & 3 \\ 4 & 5 & 6 \\ 7 & 8 & 9 \end{bmatrix}$

(*c*) $\boldsymbol{A} = \begin{bmatrix} -1 & 2 & 3 & 5 \\ 2 & -4 & 6 & 3 \\ 2 & -3 & 1 & 7 \end{bmatrix}$ (*d*) $\boldsymbol{A} = \begin{bmatrix} -1 & 2 & 3 \\ 2 & -4 & 6 \\ 2 & -3 & 1 \\ 3 & 5 & -7 \end{bmatrix}$

In each case plot the error cost function according to the expression given in (7.16). Define a stopping criterion to terminate the training process. Your results can be compared to the pseudoinverse computed using the MATLAB function **`pinv`**.

7.3. Repeat Problem 7.2(*a*) and (*b*), using method 2 for computing the pseudoinverse of a matrix. Implement the vector-matrix (batch) form of the neural network learning rule given in (7.31). Use the bipolar vectors as the external excitatory inputs to the structured neural network. In both cases plot the error cost function according to the expression given in (7.26). Define a stopping criterion to terminate the training process. Repeat the problem, but use the search-then-converge scheduling for adjusting the learning rate parameter as in (2.36) instead of using a fixed learning rate parameter.

7.4. Using the structured neural network shown in Figure 7.4, compute the LU decomposition of the following matrices:

(*a*) $\boldsymbol{A} = \begin{bmatrix} 1 & 2 & -7 \\ -2 & 4 & 5 \\ -1 & -1 & 4 \end{bmatrix}$ (*b*) $\boldsymbol{A} = \begin{bmatrix} 0 & 0 & 4 \\ 3 & 2 & -6 \\ -1 & 2 & -1 \end{bmatrix}$

Use the vector-matrix (batch) form of the learning rules given in (7.49) and (7.53). Monitor the progress of the training by plotting the error cost function using the expression given in (7.45), and define a stopping criterion to terminate the training.

7.5. Use the method 1 structured neural network (Figure 7.6) for computing the QR factorization of a matrix, and find the factorization of the matrix

$$A = \begin{bmatrix} 1 & 4 & -3 \\ 4 & 5 & 2 \\ 5 & -3 & 2 \end{bmatrix}$$

Implement the vector-matrix form of the learning rules given in (7.73) and (7.81). Compare training times, in terms of the total number of training epochs, for two different external excitatory inputs. For the first input use the bipolar vectors and train the neural network. For the second input, use sinusoidal functions in discrete time and retrain the neural network. The discrete-time form of the sinusoidal inputs should be taken as $x_l(kT) = sin(kl\omega_0 T)$ where $l = 1, 2, 3$, $\omega_0 = 500$ rad/s, $T = 1 \times 10^{-4}$ s (sampling period), and $k = 0, 1, \ldots, (N-1)$, where N is the total number of training epochs. Define a stopping criterion to terminate the training.

7.6. The structured neural network for computing the Schur decomposition of a matrix $\boldsymbol{A} \in \Re^{n \times n}$ is depicted in block diagram form in Figure 7.18 (cf. Section 7.5). Define, as in Section 7.5, the two error vectors as $\bar{\boldsymbol{e}} = \boldsymbol{Ax} - \boldsymbol{QRb}$ and $\hat{\boldsymbol{e}} = \boldsymbol{x} - \boldsymbol{Qb}$, the total error cost function as $\mathscr{E}(\bar{\boldsymbol{e}}, \hat{\boldsymbol{e}}) = {}^1\!/_2\|\bar{\boldsymbol{e}}\|_2^2 + \nu/2\|\hat{\boldsymbol{e}}\|_2^2$, and the discrete-time vector-matrix form of the learning rules as $\boldsymbol{Q}(k+1) = \boldsymbol{Q}(k) - \mu\nabla_{\boldsymbol{Q}}\mathscr{E}(\bar{\boldsymbol{e}},\hat{\boldsymbol{e}})$ and $\boldsymbol{R}(k+1) = \boldsymbol{R}(k) - \mu\nabla_{\boldsymbol{R}}\mathscr{E}(\bar{\boldsymbol{e}},\hat{\boldsymbol{e}})$. Derive the two learning rules for $\boldsymbol{Q}$ and $\boldsymbol{R}$ by computing the necessary gradient terms $\nabla_{\boldsymbol{Q}}\mathscr{E}(\bar{\boldsymbol{e}},\hat{\boldsymbol{e}})$ and $\nabla_{\boldsymbol{R}}\mathscr{E}(\bar{\boldsymbol{e}},\hat{\boldsymbol{e}})$.

7.7. In method 2 for computing the QR factorization of a matrix $\boldsymbol{A} \in \Re^{n \times n}$, presented in Section 7.4, the two error vectors are defined as $\bar{\boldsymbol{e}} = \boldsymbol{Ax} - \boldsymbol{Qu}$ and $\hat{\boldsymbol{e}} = \boldsymbol{u} - \boldsymbol{Q}^T\boldsymbol{y}$. Derive the two discrete-time vector-matrix forms of the learning rules to compute $\boldsymbol{Q}$ and $\boldsymbol{R}$ based on the same total error cost function and steepest descent gradient forms of the learning rules previously used for method 1. *Hint*: When deriving the learning rule for $\boldsymbol{R}$, substitute $\boldsymbol{u} = \boldsymbol{Rx}$ in the total error cost function before computing the gradient.

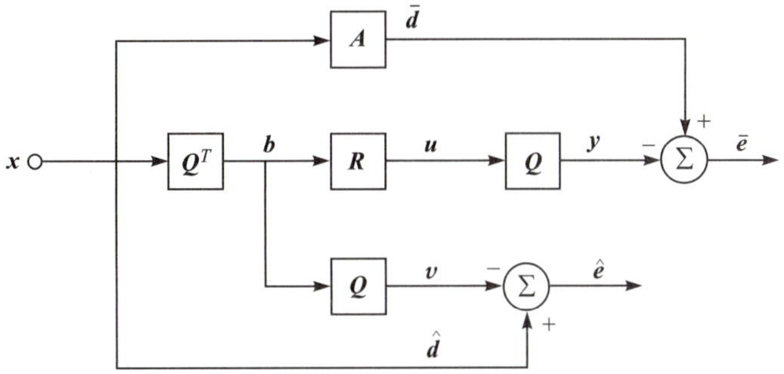

Figure 7.18 Block diagram of the structured neural network for the Schur decomposition in Section 7.5.

7.8. For the Schur decomposition structured neural network given in block diagram form in Figure 7.9 (and repeated in Figure 7.18), write a computer program to implement the vector-matrix form of the associated learning rules given in (7.96) and (7.97). Set $\nu = 1$, and be sure to impose the upper triangular constraint on the $\boldsymbol{R}$ matrix. Using the bipolar vectors for the external excitatory inputs to the network, compute the Schur decomposition (i.e., compute $\boldsymbol{Q}$ and $\boldsymbol{R}$) for the matrices

$$(a)\quad \boldsymbol{A} = \begin{bmatrix} 0.7562 & 0.3750 & -2.3775 \\ 0.4005 & 1.1252 & -0.2738 \\ -1.3414 & 0.7286 & -0.3229 \end{bmatrix} \qquad (b)\quad \boldsymbol{A} = \begin{bmatrix} -1 & -4 & 0 \\ -4 & 5 & 0 \\ 0 & 0 & 0 \end{bmatrix}$$

Monitor the mean-square error during training by plotting the error cost function using the expression given in (7.93), and define a stopping criterion to terminate the training.

7.9. Write a computer program to implement the vector-matrix form of the structured neural network (Figure 7.11) learning rules for computing the eigenvalues and eigenvectors of a symmetric matrix, given in (7.118) and (7.119). Using Gaussian random vectors as the external excitatory inputs to the network, compute the eigenvalues and eigenvectors of these matrices:

$$(a)\ \boldsymbol{A} = \begin{bmatrix} 1 & 0.5 & -1 \\ 0.5 & 2 & 1 \\ -1 & 1 & 3 \end{bmatrix} \quad (b)\ \boldsymbol{A} = \begin{bmatrix} 0 & 1 \\ 1 & 1 \end{bmatrix} \quad (c)\ \boldsymbol{A} = \begin{bmatrix} -1 & -1 \\ -1 & -3 \end{bmatrix}$$

$$(d)\ \boldsymbol{A} = \begin{bmatrix} -1.2384 & 1.1546 & -0.4880 & 0.4191 \\ 1.1546 & -0.6638 & -0.8822 & 0.0437 \\ -0.4880 & -0.8822 & 1.0290 & -0.3905 \\ 0.4191 & 0.0437 & -0.3905 & -0.1419 \end{bmatrix} \quad (e)\ \boldsymbol{A} = \begin{bmatrix} 1 & -1 & 1 \\ -1 & 1 & -1 \\ 1 & -1 & 1 \end{bmatrix}$$

Your results can be compared to the eigenvalue routine in MATLAB, that is, the `eig` function.

7.10. Every real square matrix $\boldsymbol{A} \in \Re^{n \times n}$, can be factored into $\boldsymbol{A} = \boldsymbol{QH}$, where $\boldsymbol{Q} \in \Re^{n \times n}$ is orthogonal ($\boldsymbol{QQ}^T = \boldsymbol{Q}^T\boldsymbol{Q} = \boldsymbol{I}$), and $\boldsymbol{H} \in \Re^{n \times n}$ is a symmetric ($\boldsymbol{H}^T = \boldsymbol{H}$) and positive semidefinite ($H \geq 0$) matrix. If $\boldsymbol{A}$ is invertible [$\rho(\boldsymbol{A}) = n$], then $\boldsymbol{H}$ is positive definite ($\boldsymbol{H} > 0$). This is called the *polar decomposition* of $\boldsymbol{A}$ [8].

Prove the above statement.

7.11. Write a computer program to implement the vector-matrix form of the structured neural network (Figure 7.13) learning rules for computing the singular-value decomposition of a matrix, given in (7.158), (7.164), and (7.168). Test your program by computing the SVD of the matrix

$$\boldsymbol{A} = \begin{bmatrix} 1 & 0 & 0 \\ -1 & 2 & -1 \\ 0 & 0 & 3 \end{bmatrix}$$

7.12. In Cichocki and Unbehauen [6], eight alternative structured neural network architectures are given for computing the singular-value decomposition of a square matrix $\boldsymbol{A} \in \Re^{n \times n}$. One of these architectures is depicted in Figure 7.19 in block diagram form. Derive the three discrete-time learning rules based on

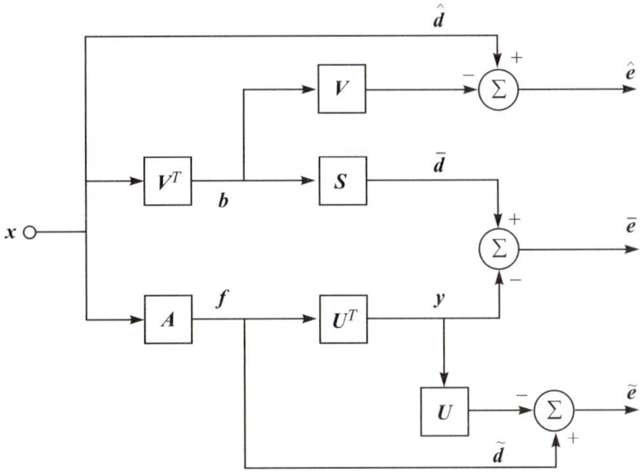

Figure 7.19 Alternate structured neural network architecture for computing the SVD of a square matrix.

the steepest descent gradient method to compute the SVD for $\boldsymbol{U}^T\boldsymbol{A} = \boldsymbol{S}\boldsymbol{V}^T$, where $\boldsymbol{V}^T\boldsymbol{V} = \boldsymbol{I}$ and $\boldsymbol{U}\boldsymbol{U}^T = \boldsymbol{I}$. Define the total error cost function similar to (7.148).

7.13. Using the same structured neural network to compute the SVD of a matrix $\boldsymbol{A} \in \Re^{m\times n}$, $m \geq n$, given in block diagram form in Figure 7.13 (Sect. 7.8), define the three error vectors as $\bar{\boldsymbol{e}} = (\boldsymbol{A} - \boldsymbol{U}\boldsymbol{S}\boldsymbol{V}^T)\boldsymbol{x}$, $\hat{\boldsymbol{e}} = (\boldsymbol{I} - \boldsymbol{V}\boldsymbol{V}^T)\boldsymbol{x}$, and $\tilde{\boldsymbol{e}} = (\boldsymbol{I} - \boldsymbol{U}^T\boldsymbol{U})\boldsymbol{r}$, where $\boldsymbol{U} \in \Re^{m\times n}(\boldsymbol{U}^T\boldsymbol{U} = \boldsymbol{I})$, $\boldsymbol{V} \in \Re^{n\times n}(\boldsymbol{V}\boldsymbol{V}^T = \boldsymbol{I})$, and $\boldsymbol{S} = \text{diag}[\sigma_1, \sigma_2, \ldots, \sigma_n] \in \Re^{n\times n}$. Using these expressions for the three error vectors, the total error cost function given in (7.148), and the discrete-time vector-matrix forms of the three learning rules for $\boldsymbol{V}$, $\boldsymbol{S}$, and $\boldsymbol{U}$ given in (7.149), (7.150), and (7.151), respectively, derive the three new learning rules. Compare these learning rules to the learning rules for $\boldsymbol{V}$, $\boldsymbol{S}$, and $\boldsymbol{U}$ in (7.158), (7.164), and (7.168), respectively, derived in Section 7.8.

7.14. For the polar decomposition of a matrix discussed in Problem 7.10, derive the discrete-time learning rules (in vector-matrix form) using a structured neural network approach to compute $\boldsymbol{Q}$ and $\boldsymbol{H}$. The learning rules to compute both $\boldsymbol{Q}$ and $\boldsymbol{H}$ should be based on the steepest descent gradient method and should have the general discrete-time forms $\boldsymbol{Q}(k+1) = \boldsymbol{Q}(k) - \mu\nabla_{\boldsymbol{Q}}\mathscr{E}(\bar{\boldsymbol{e}}, \hat{\boldsymbol{e}})$ and $\boldsymbol{H}(k+1) = \boldsymbol{H}(k) - \mu\nabla_{\boldsymbol{H}}\mathscr{E}(\bar{\boldsymbol{e}}, \hat{\boldsymbol{e}})$, where $\bar{\boldsymbol{e}}$ and $\hat{\boldsymbol{e}}$ are appropriately defined error vectors. Design the structured neural network by constructing an appropriate block diagram with an external excitatory input and the two error vectors $\bar{\boldsymbol{e}}$ and $\hat{\boldsymbol{e}}$. The total error cost function that should be used is $\mathscr{E}(\bar{\boldsymbol{e}}, \hat{\boldsymbol{e}}) = \mathscr{E}[\bar{\boldsymbol{e}}(\boldsymbol{Q}, \boldsymbol{H}), \hat{\boldsymbol{e}}(\boldsymbol{Q}, \boldsymbol{H})] = {}^1\!/_2\|\bar{\boldsymbol{e}}\|_2^2 + \nu/2\|\hat{\boldsymbol{e}}\|_2^2$. Determine a set of external excitatory signals that will yield the best convergence through a series of computer simulations. Try binary vectors, bipolar vectors, sinusoidal input signals, and Gaussian white noise. Test your computer program by finding the polar decomposition of these matrices: M

$$(a)\quad A = \begin{bmatrix} 1 & 2 & 3 \\ 5 & 6 & 7 \\ 7 & 8 & 9 \end{bmatrix} \qquad (b)\quad A = \begin{bmatrix} 0 & 1 & 0 \\ 0 & 0 & 1 \\ -6 & -11 & -6 \end{bmatrix}$$

Hint: The structured neural network should be a modified version of the method 1 QR factorization network given in Section 7.4.

7.15. Design a structured neural network which performs the following factorizations of a symmetric matrix $A \in \Re^{n\times n}(A^T = A)$

$$(a)\quad A = QTQ^T \qquad (b)\quad A = LDL^T$$

where $Q \in \Re^{n\times n}$ is an orthogonal matrix, $T \in \Re^{n\times n}$ is a symmetric tridiagonal matrix, $L \in \Re^{n\times n}$ is lower triangular matrix, and $D \in \Re^{n\times n}$ is a diagonal matrix. Test your computer programs (using the vector-matrix form of the learning rules) on the matrices given in Problem 7.9(*a*), (*c*), and (*e*).

7.16. Write a computer program to implement the discrete-time vector-matrix form of the learning rule to solve the Lyapunov equation discussed in Section 7.9. Test your computer program on the following matrices:

$$(a)\quad A = \begin{bmatrix} 0 & 1 & 0 \\ 0 & 0 & 1 \\ -1 & -2 & -2 \end{bmatrix}, \quad C = \begin{bmatrix} 2 & 1 & 0 \\ 1 & 1 & 1 \\ 0 & 1 & 1 \end{bmatrix}$$

$$(b)\quad A = \begin{bmatrix} 1 & 2 & 3 \\ 6 & 5 & 4 \\ 7 & 0 & 8 \end{bmatrix}, \quad C = \begin{bmatrix} 7 & 2 & 4 \\ 2 & 5 & 1 \\ 4 & 1 & 6 \end{bmatrix}$$

Use both bipolar vectors and the sinusoidal functions described in Example 7.4 as the external excitatory inputs to the network and compare the two results.

7.17. Write a computer program to implement the discrete-time vector-matrix form of the learning rule to solve the Riccati equation discussed in Section 7.10. Test your computer program on the following matrices:

$$A = \begin{bmatrix} 0 & 1 & 0 \\ 0 & 0 & 1 \\ -6 & -11 & -6 \end{bmatrix} \qquad Q = \begin{bmatrix} 2 & 0 & 0 \\ 0 & 0.25 & 0 \\ 0 & 0 & 0.5 \end{bmatrix}$$

$$R = \begin{bmatrix} 0.1 & 0 & 0 \\ 0 & 0.3 & 0 \\ 0 & 0 & 0.07 \end{bmatrix}^{-1}$$

Compare your final results to the MATLAB control system toolbox function **`lqr2`** that can be used to solve the linear quadratic regular problem. This function uses the Schur decomposition to solve the Riccati equation. As discussed in Example 7.4, matrices B and R must be presented to the **`lqr2`** function. We can define the weighting matrix R as $\tilde{R}$ in the MATLAB function **`lqr2`**, and then to properly compare the results using both methods, in the **`lqr2`** MATLAB function let $B = I_{n\times n}$ and $\tilde{R} = R^{-1}$. Use the bipolar vectors as the external excitatory inputs to the network. Also use the sinusoidal functions described in Example 7.4 as the external excitatory inputs, and compare the two results.

7.18. Repeat Problem 7.17 except write your computer program to compute the solution to the Riccati equation $\boldsymbol{X}$ one column at a time.

7.19. In digital control theory the discrete-time Lyapunov equation is given by $\boldsymbol{\Phi X \Phi}^T + \boldsymbol{C} = \boldsymbol{X}$, where the matrices $\boldsymbol{\Phi} \in \Re^{n \times n}$ and $\boldsymbol{C} \in \Re^{n \times n}$ are known and $\boldsymbol{X} \in \Re^{n \times n}$ is the desired solution to the equation. Design an appropriate structured neural network to compute the solution to the discrete-time Lyapunov equation. Determine the best external excitatory input signals to use in the network.

7.20. Design a structured neural network to solve the Sylvester equation given in Section 7.9, that is, $\boldsymbol{AX} + \boldsymbol{XB} = -\boldsymbol{C}$, where it is assumed we know $\boldsymbol{A} \in \Re^{n \times n}$, $\boldsymbol{B} \in \Re^{m \times m}$, and $\boldsymbol{C} \in \Re^{n \times m}$ and seek the solution $\boldsymbol{X} \in \Re^{n \times m}$.

7.21. For a matrix $\boldsymbol{A} \in \Re^{n \times n}$, with distinct eigenvalues (i.e., no repeated eigenvalues), a similarity transformation exists (which consists of the associated eigenvectors of $\boldsymbol{A}$) which will diagonalize $\boldsymbol{A}$ (cf. Sect. 7.6). However, if $\boldsymbol{A}$ has repeated eigenvalues, it is not similar to a diagonal matrix unless it has an independent full set of eigenvectors. If the eigenvectors are not independent, then $\boldsymbol{A}$ is said to be *defective* [2]. For such matrices, the *generalized* eigenvalue problem is addressed to determine the nontrivial solutions of the equation $\boldsymbol{A v}_i = \lambda_i \boldsymbol{B v}_i$, for $i = 1, 2, \ldots, n$. In addition to $\boldsymbol{A}$, the matrix $\boldsymbol{B} \in \Re^{n \times n}$ is given, and $\lambda_i \in \Re$ are the *generalized eigenvalues* and $\boldsymbol{v}_i \in \Re^{n \times 1}$ the corresponding *generalized right eigenvectors* for $i = 1, 2, \ldots, n$. It is easy to see that λ is a root of the characteristic equation $\det(\boldsymbol{A} - \lambda \boldsymbol{B}) = 0$. (*Note*: If $\boldsymbol{B} = \boldsymbol{I}_{n \times n}$, the resulting characteristic equation is for the standard eigenvalue problem.) The matrix $\boldsymbol{A} - \lambda \boldsymbol{B}$ is called a *matrix pencil* [20]. The QZ algorithm [2, 20] is a standard method for solving the generalized eigenvalue problem. Similar to the distinct eigenvalue case, the generalized eigenvalues and eigenvectors satisfy the factorization $\boldsymbol{AV} = \boldsymbol{BV\Lambda}$, where $\boldsymbol{\Lambda} = \mathrm{diag}[\lambda_1, \lambda_2, \ldots, \lambda_n]$, and $\boldsymbol{V} = [\boldsymbol{v}_1, \boldsymbol{v}_2, \ldots, \boldsymbol{v}_n]$. Design a neural network that solves the generalized eigenvalue problem, that is, computes λ_i and $\boldsymbol{v}_i$.

REFERENCES

1. L.-X. Wang and J. M. Mendel, "Structured Trainable Networks for Matrix Algebra," *Proceedings of the International Joint Conference on Neural Networks*, San Diego, CA, vol. 2, 1990, pp. 125–32.
2. G. H. Golub and C. F. Van Loan, *Matrix Computations*, 3rd ed., Baltimore, MD: Johns Hopkins, 1996.
3. P. Baldi and K. Hornik, "Neural Networks and Principal Component Analysis: Learning from Examples without Local Minima," *Neural Networks*, vol. 2, 1989, pp. 53–8.
4. L.-X. Wang and J. M. Mendel, "Parallel Structured Networks for Solving a Wide Variety of Matrix Algebra Problems," *Journal of Parallel and Distributed Computing*, vol. 14, 1992, pp. 236–47.
5. M. M. Polycarpou and P. A. Ioannou, "Learning and Convergence Analysis of Neural-Type Structured Networks," *IEEE Transactions on Neural Networks*, vol. 3, 1992, pp. 39–50.
6. A. Cichocki and R. Unbehauen, *Neural Networks for Optimization and Signal Processing*, New York: Wiley, 1993.

7. J.-S. Jong, S.-Y. Lee and S.-Y. Skin, "An Optimization Network for Matrix Inversion," *Neural Information Processing Systems*, ed. D. Z. Anderson, College Park, MD: American Institute of Physics, 1988, pp. 397–401.
8. G. Strang, *Introduction to Linear Algebra*, Wellesley, MA: Wellesley-Cambridge Press, 1993.
9. D. S. Watkins, *Fundamentals of Matrix Computations*, New York: Wiley, 1991.
10. S. J. Orfanidis, *Optimum Signal Processing: An Introduction*, 2nd ed., New York: McGraw-Hill, 1988.
11. C. T. Chen, *Linear System Theory and Design*, New York: Holt, Rinehart and Winston, 1984.
12. S. Haykin, *Adaptive Filter Theory*, 3rd ed., Upper Saddle River, NJ: Prentice-Hall, 1996.
13. W. H. Press, S. A. Teukolsky, W. T. Vetterling, and B. P Flannery, *Numerical Recipes in C: The Art of Scientific Computing*, 2nd ed., New York: Cambridge University Press, 1992.
14. W. H. Press, S. A. Teukolsky, W. T. Vetterling, and B. P Flannery, *Numerical Recipes in Fortran: The Art of Scientific Computing*, 2nd ed., New York: Cambridge University Press, 1992.
15. C. W. Therrien, *Discrete Random Signals and Statistical Signal Processing*, Englewood Cliffs, NJ: Prentice-Hall, 1992.
16. B. D. O. Anderson and J. B. Moore, *Optimal Filtering*, Englewood Cliffs, NJ: Prentice-Hall, 1979.
17. G. W. Stewart, *Introduction to Matrix Computations*, New York: Academic, 1973.
18. G. Strang, *Linear Algebra and Its Applications*, 2nd ed., New York: Academic, 1980.
19. A. Cichocki and R. Unbehauen, "Neural Networks for Computing Eigenvalues and Eigenvectors," *Biological Cybernetics*, vol. 68, 1992, pp. 155–64.
20. B. N. Datta, *Numerical Linear Algebra and Applications*, Pacific Grove, CA: Brooks/Cole, 1995.
21. R. S. Varga, *Matrix Iterative Analysis*, Englewood Cliffs, NJ: Prentice-Hall, 1962.
22. S. G. Nash and A. Sofer, *Linear and Nonlinear Programming*, New York: McGraw-Hill, 1996.
23. D. G. Luenberger, *Linear and Nonlinear Programming*, 2nd ed., Reading, MA: Addison-Wesley, 1984.
24. N. J. Higham, *Accuracy and Stability of Numerical Algorithms*, Philadelphia, PA: Society for Industrial and Applied Mathematics (SIAM), 1996.
25. H. Bourlard and Y. Camp, "Auto-Association by Multilayer Perceptrons and Singular Value Decomposition," *Biological Cybernetics*, vol. 59, 1988, pp. 291–4.
26. J. A. Sirat, "A Fast Algorithm for Principal Component Analysis and Singular Value Decomposition," *International Journal of Neural Systems*, vol. 2, nos. 1 and 2, 1991, pp. 147–55.
27. E. Anderson, Z. Bai, C. H. Bischof, J. W. Demmel, J. J. Dongarra, J. J. Du Croz, A. Greenbaum, S. J. Hammarling, A. McKenney, S. Ostrouchov, and D. C. Sorensen, *LAPACK Users' Guide, Release 2.0*, 2nd ed., Philadelphia, PA: Society for Industrial and Applied Mathematics (SIAM), 1995.
28. H. Kwakernaak and R. Sivan, *Linear Optimal Control Systems*, New York: Wiley-Interscience, 1972.
29. K. Zhou, J. C. Doyle, and K. Glover, *Robust and Optimal Control*, Upper Saddle River, NJ: Prentice-Hall, 1996.
30. F. M. Ham and E. Collins, "A Neural Network Architecture for Solving the Algebraic Matrix Riccati Equation," *Proceedings of the SPIE International Conference on Aerospace/Sensing and Controls (Applications and Science of*

Artificial Neural Networks II), eds. S. K. Rogers and D. W. Ruck, Orlando, FL, SPIE vol. 2760, 1996, pp. 294–301.
31. F. M. Ham and E. Collins, "A Neurocomputing Approach for Solving the Algebraic Matrix Riccati Equation," *Proceedings of the IEEE International Conference on Neural Networks*, Washington, vol. 1, 1996, pp. 617–22.
32. A. Grace, A. J. Laub, J. N. Little, and C. M. Thompson, *Control System Toolbox—For Use with MATLAB*, Natick, MA: The Mathworks, Inc., 1993.

CHAPTER 8

Solution of Linear Algebraic Equations Using Neural Networks

8.1 INTRODUCTION

Many problems encountered in science and engineering involve solving systems of linear algebraic equations, for example, in signal processing and robotics. In principle, solving systems of equations is equivalent to computing the inverse (or pseudoinverse) of a matrix (cf. Sect. A.2.7). In this chapter we want to extend some of the ideas from Chapter 7 to solve systems of equations, and we are interested in the development of *real-time* or *online* processing strategies. We do not intend for the methods presented in this chapter to compete with standard numerical methods for solving systems of equations *off-line* [1]. Many excellent numerical methods already exist [1]. Typically, time constraints for solving systems of equations off-line are not important, but instead a good solution is paramount. However, if it is necessary to solve systems of equations repeatedly online, or in real time, and if the time constraints for solving these equations are more demanding than a typical digital computer can provide, then it is necessary to explore alternative methods. One possibility is to use systolic (or wavefront) arrays [2–5]. Another approach is to use *analog* artificial neural networks because of their inherent parallel architectures [6–8].

Our motivation is to develop various learning strategies for online solution of systems of linear equations. The basic philosophy here is similar to that of Chapter 7 for solving certain matrix algebra problems with neural networks. That is, using a neurocomputing approach to solve a system of linear equations can be computationally advantageous for systems that must be repeatedly solved when certain system parameters change very slowly with time (which is the reason why the system of equations must be repeatedly solved). Because of the small system parameter variations, the "new" solution is not "far" from the "old"solution, and as a result, few "learning" steps will be required to reach the new solution. We will not address in-depth issues of hardware implementation of the resulting neural architectures for such systems. This is obviously an important topic that must be addressed in order to realize a particular system, and it is left to other excellent sources on the subject [e.g., 6–8].

8.2 SYSTEMS OF SIMULTANEOUS LINEAR ALGEBRAIC EQUATIONS

Given the set of linear algebraic equations with constant coefficients

$$\begin{aligned} a_{11}x_1 + a_{12}x_2 + \cdots + a_{1n}x_n &= b_1 \\ a_{21}x_1 + a_{22}x_2 + \cdots + a_{2n}x_n &= b_2 \\ &\vdots \\ a_{m1}x_1 + a_{m2}x_2 + \cdots + a_{mn}x_n &= b_m \end{aligned} \tag{8.1}$$

it is desired to solve for the unknown quantities $x_1, x_2, \ldots, x_n$ [9–24], given the coefficients a_{ij} for $i = 1, 2, \ldots, m$ and $j = 1, 2, \ldots, n$ and b_i for $i = 1, 2, \ldots, m$. Equations (8.1) can be written in a more convenient (compact) form, that is, a vector-matrix form as

$$\boldsymbol{A}\boldsymbol{x} = \boldsymbol{b} \tag{8.2}$$

where it is assumed that $\boldsymbol{A} \in \Re^{m\times n}$, $\boldsymbol{x} \in \Re^{n\times 1}$, $\boldsymbol{b} \in \Re^{m\times 1}$, and

$$\boldsymbol{A} = \begin{bmatrix} a_{11} & a_{12} & \cdots & a_{1n} \\ a_{21} & a_{22} & \cdots & a_{2n} \\ \cdots & \cdots & \cdots & \cdots \\ a_{m1} & a_{m2} & \cdots & a_{mn} \end{bmatrix} \tag{8.3}$$

$$\boldsymbol{x} = [x_1, x_2, \cdots, x_n]^T \tag{8.4}$$

$$\boldsymbol{b} = [b_1, b_2, \cdots, b_m]^T \tag{8.5}$$

There are three cases that can exist: (1) If $m > n$, this is a common situation (more equations than unknowns), and the system of equations is referred to as being *overdetermined* [25]. (2) If $n > m$ (more unknowns than equations), the system of equations is called *underdetermined* [25]. (3) When $m = n$, there are as many equations as unknowns (exactly determined system). Matrix $\boldsymbol{A}$ is sometimes referred to as the *data matrix*, and the $\boldsymbol{b}$ vector is occasionally called the *observation vector* [25].

The simplest example to illustrate the formulation of a set of linear equations is to consider two straight lines in the xy plane, for example,

$$\begin{aligned} y &= -2x + 2 \\ y &= x \end{aligned} \tag{8.6}$$

The question would be: Is there a common point in the xy plane that the two lines share; or equivalently, is there a point of intersection in the xy plane? Our approach in answering this question will be to formulate a set of algebraic equations and then solve them. We can rewrite the equations in (8.6) as

$$\begin{aligned} 2x + y &= 2 \\ -x + y &= 0 \end{aligned} \tag{8.7}$$

or in vector-matrix form

$$\begin{bmatrix} 2 & 1 \\ -1 & 1 \end{bmatrix} \begin{bmatrix} x \\ y \end{bmatrix} = \begin{bmatrix} 2 \\ 0 \end{bmatrix} \tag{8.8}$$

This is an *exactly determined* system; that is, $\boldsymbol{A}$ is a square matrix. Solving this system of algebraic equations (which requires finding the inverse of $\boldsymbol{A}$) will give us the intersection of the two straight lines in the xy plane

$$\begin{bmatrix} x \\ y \end{bmatrix} = \begin{bmatrix} 2 & 1 \\ -1 & 1 \end{bmatrix}^{-1} \begin{bmatrix} 2 \\ 0 \end{bmatrix} = \begin{bmatrix} \frac{1}{3} & \frac{-1}{3} \\ \frac{1}{3} & \frac{2}{3} \end{bmatrix} \begin{bmatrix} 2 \\ 0 \end{bmatrix} = \begin{bmatrix} \frac{2}{3} \\ \frac{2}{3} \end{bmatrix} \tag{8.9}$$

which is precisely what is discovered when the two straight lines are plotted in the xy plane and the intersection is observed; see Figure 8.1. Because the two lines had an intersection, the inverse of the square matrix $\boldsymbol{A}$ existed. Had there not been an intersection of the two straight lines (i.e., if they had been parallel lines), $\boldsymbol{A}$ would be singular.

The types of systems that we are interested in are much more complex than illustrated in this example. We want to study large-scale overdetermined and underdetermined systems, ill-conditioned systems, and systems that have uncertainty associated with them.

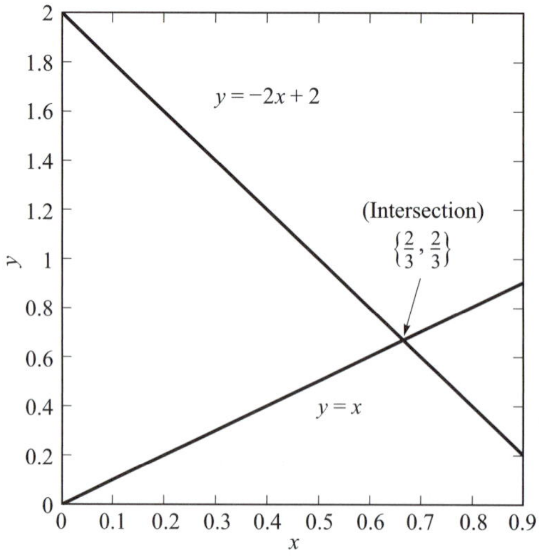

Figure 8.1 Two intersecting lines.

8.3 LEAST-SQUARES SOLUTION OF SYSTEMS OF LINEAR EQUATIONS

Assuming the general set of linear equations given in (8.2) with $m \geq n$ (overdetermined system), we will first consider the full-rank least-squares problem [25]. In other words, we assume that matrix $\boldsymbol{A}$ has full rank† $\rho(\boldsymbol{A}) = n$ and seek a solution $\boldsymbol{x}$. To accomplish this, we strive to *minimize* the scalar cost function

$$\mathscr{E}(\boldsymbol{x}) = \|\boldsymbol{A}\boldsymbol{x} - \boldsymbol{b}\|_p \tag{8.10}$$

for some suitably chosen p. Depending on the choice of p (i.e., different norms) different optimal solutions will result. If the 1-norm (L_1-norm) and ∞-norm (L_∞-norm) (or Chebyshev norm) (cf. Sect. A.2.13) is used, the minimization of (8.10) is difficult because the function $\mathscr{E}(\boldsymbol{x}) = \|\boldsymbol{A}\boldsymbol{x} - \boldsymbol{b}\|_p$ is not differentiable for these values of p [26]. When $p = 1$, this is referred to as the *least-absolute-deviations problem.* For $p = 2$, this is called the *linear least-squares problem.* And when $p = \infty$, this is referred to as the *minimax problem.* The type of norm that is used in (8.10) greatly depends on the distribution of the errors in the data and the type of application that is being addressed [18–20]. For the case when the errors have a double exponential distribution (or Laplace distribution) [27, 28], the L_1-norm can be used. When the error distribution has markedly sharp transitions, for example, a uniform distribution, then the L_∞-norm (or Chebyshev norm) should be used. However, if the error distribution is Gaussian (normal), then it is best to use the Euclidean norm (or L_2-norm). Using the Euclidean norm, the resulting *least-squares* problem

$$\min_{\forall \boldsymbol{x}} \mathscr{E}(\boldsymbol{x}) = \min_{\forall \boldsymbol{x}} \|\boldsymbol{A}\boldsymbol{x} - \boldsymbol{b}\|_2 \tag{8.11}$$

is more tractable than the other cases. If $\boldsymbol{A}$ is full-rank, (8.2) has a unique solution [25]. The scalar cost function that is typically used can be written as

$$\mathscr{E}(\boldsymbol{x}) = \frac{1}{2}\|\boldsymbol{A}\boldsymbol{x} - \boldsymbol{b}\|_2^2 = \frac{1}{2}\|\boldsymbol{e}\|_2^2 \tag{8.12}$$

where

$$\boldsymbol{e} = \boldsymbol{A}\boldsymbol{x} - \boldsymbol{b} \tag{8.13}$$

is the solution error vector (cf. Sect. A.2.7). Computing the gradient of (8.12) with respect to the vector $\boldsymbol{x}$, and setting the result equal to zero, that is, $\nabla_{\boldsymbol{x}}\mathscr{E}(\boldsymbol{x}) = 0$, lead to the normal equations

$$\boldsymbol{A}^T\boldsymbol{A}\boldsymbol{x} - \boldsymbol{A}^T\boldsymbol{b} = \boldsymbol{0} \tag{8.14}$$

Solving for $\boldsymbol{x}$ directly from (8.14) gives

$$\boldsymbol{x} = (\boldsymbol{A}^T\boldsymbol{A})^{-1}\boldsymbol{A}^T\boldsymbol{b} \tag{8.15}$$

---

†This restriction is actually not necessary when we take a neurocomputing approach to solve a system of linear equations. The details of this will be pointed out at the appropriate time.

where

$$A^{+} = (A^{T}A)^{-1}A^{T} \tag{8.16}$$

is defined as the *pseudoinverse* of A (cf. Sect. A.2.7). The solution in (8.15) is the "batch" solution to the problem in (8.2). If $m < n$ (underdetermined problem) and $\rho(A) = m$ (i.e., A has full rank), the solution to (8.2) is given by

$$x = A^{T}(AA^{T})^{-1}b \tag{8.17}$$

where

$$A^{+} = A^{T}(AA^{T})^{-1} \tag{8.18}$$

is now the pseudoinverse of A.

8.4 A LEAST-SQUARES NEUROCOMPUTING APPROACH FOR SOLVING SYSTEMS OF LINEAR EQUATIONS

Let us assume that we require a solution to be found for (8.2) that is based not on a batch method but instead on an approach that can "learn" the solution. One straightforward method is based on using the method of steepest gradient descent. Our approach follows the methods used in Chapter 7. The continuous-time (analog) learning rule is given by

$$\frac{dx}{dt} = -\mu\nabla_{x}\mathscr{E}(x) \tag{8.19}$$

where $\mu \in \Re^{n\times n}(\mu = [\mu_{ij}]$, for $i, j = 1, 2, \ldots, n)$ is a positive definite matrix which is typically chosen to be diagonal, that is, $\mu_{ij} = \mu_j\delta_{ij}$, where $\mu_j > 0$ and δ_{ij} is the Kronecker delta (cf. Sect. A.2.8) and $x(0) = x_0$. The gradient shown in (8.19) has already been calculated and is the left side of (8.14). Substituting this gradient into (8.19) gives the learning rule for solving a system of equations as

$$\frac{dx}{dt} = -\mu(A^{T}Ax - A^{T}b) = -\mu A^{T}(Ax - b) \tag{8.20}$$

In Section A.2.7, we define the solution error to be

$$e = Ax - b \tag{8.21}$$

Substituting (8.21) into (8.20) gives the continuous-time learning rule

$$\frac{dx}{dt} = -\mu A^{T}e \tag{8.22}$$

for solving the system of equations in (8.2), with $x(0) = x_0$ as the initial conditions. Equations (8.21) and (8.22) together are adequate for solving any system of equations regardless of whether the system is exactly determined ($m = n$), overdetermined ($m > n$), underdetermined ($m < n$), or if A is rank-deficient, or even if A is ill-conditioned. Many times it is convenient to use the normalized matrix equation $\hat{A}x = \hat{b}$, where

$$\hat{a}_{ij} = \frac{a_{ij}}{\|\boldsymbol{a}_i^T\|_2} \quad \text{and} \quad \hat{b}_i = \frac{b_i}{\|\boldsymbol{a}_i^T\|_2} \quad \text{for } i = 1, 2, \ldots, m \text{ and } j = 1, 2, \ldots, n$$

Therefore, $\boldsymbol{a}_i$ is the ith row of matrix $\boldsymbol{A}$, and $\|\bullet\|_2$ is the L_2-norm (or Euclidean norm).

By following similar procedures presented in Chapter 7 for the development of structured neural networks, a neural architecture could be developed to realize the least-squares neurocomputing approach given above for solving systems of linear equations. A block diagram of a differential system is shown in Figure 8.2 that would be the basis for an analog system realization of the neurocomputing approach, and Figure 8.3 shows the detailed neural network architecture. This is a three-layer network with a bank of integrators at the output. The integrator initial conditions are the initial conditions associated with the continuous-time learning rule given in (8.22). The feedback connections provide the "inputs" to the network along with the elements of the observation vector. If the integrators (operational amplifiers) operate in their saturated region, the feedback signals are essentially passed through a sigmoidal nonlinearity. This provides a natural saturation characteristic for the output feedback signals.

Given the error vector as defined in (8.21), this error vector formulation is only applicable for the *exactly determined* case (i.e., $m = n$). We must recast the error expression as

$$\boldsymbol{e} = \boldsymbol{A}^T\boldsymbol{A}\boldsymbol{x} - \boldsymbol{A}^T\boldsymbol{b} \tag{8.23}$$

in order to evaluate the erorrs during the learning process for any case. Note that the right side of (8.23) being set equal to zero results in the normal equations in (8.14). Using (8.23) in the error cost function

$$\mathscr{E}(\boldsymbol{x}) = \frac{1}{2}\|\boldsymbol{e}\|_2^2 = \frac{1}{2}\boldsymbol{e}^T\boldsymbol{e} \tag{8.24}$$

and computing the gradient of (8.24)[†] with respect to $\boldsymbol{x}$ give the new continuous-time learning rule from (8.19) as

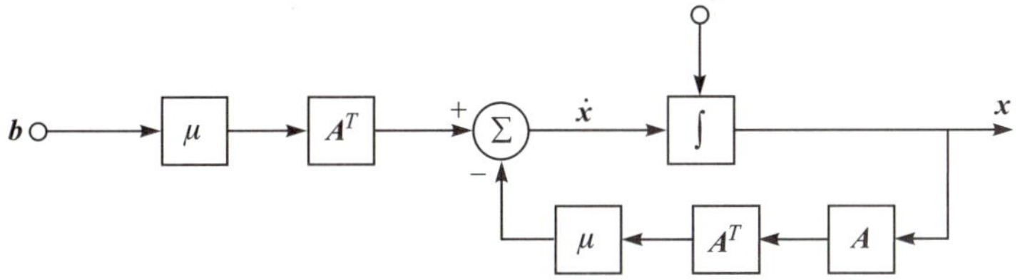

Figure 8.2 Block diagram of a differential system to realize the least-squares neurocomputing approach for solving systems of linear equations.

[†]Taking the gradient of (8.24) with respect to $\boldsymbol{x}$ results in $\nabla_{\boldsymbol{x}}\mathscr{E}(\boldsymbol{x}) = \boldsymbol{A}^T\boldsymbol{A}(\boldsymbol{A}^T\boldsymbol{A}\boldsymbol{x} - \boldsymbol{A}^T\boldsymbol{b}) = \boldsymbol{A}^T\boldsymbol{A}\boldsymbol{e}$; however, by applying the same rationale used in Section 7.2, the gradient utilized is $\nabla_{\boldsymbol{x}}\mathscr{E}(\boldsymbol{x}) = \boldsymbol{e}$. See the explanation pertaining to (7.18) and (7.19).

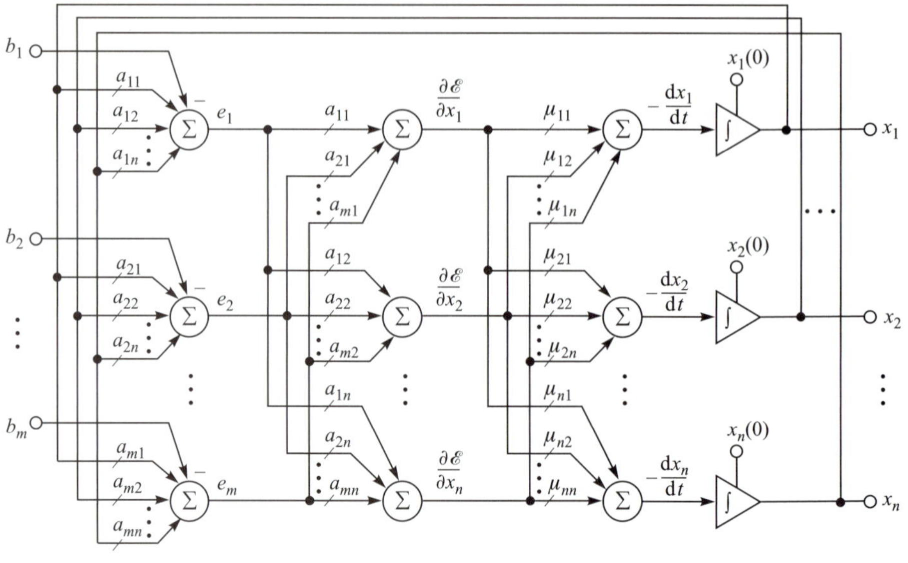

Figure 8.3 Neural network architecture for solving a system of linear equations (least-squares approach).

$$\frac{d\boldsymbol{x}}{dt} = -\boldsymbol{\mu}\boldsymbol{e} \tag{8.25}$$

with $\boldsymbol{x}(0) = \boldsymbol{x}_0$ as the initial conditions. Therefore, (8.23) and (8.25) together give a general learning rule for solving a system of simultaneous linear algebraic equations. Moreover, it is not necessary to make the "full-rank" assumption, that is, $\rho(\boldsymbol{A}) = \min(m, n)$, as was the case in Section 8.3. This is obvious from the learning rule itself. As we can see, there are no divisions necessary, only multiplications and additions. The results obtained by using the learning rule in (8.23) and (8.25) are the same as those obtained by using the singular-value decomposition (SVD) (cf. Sect. A.2.14) to determine the solution of a system of linear equations.

The choice of elements in the $\boldsymbol{\mu}$ matrix is critical for stability reasons. It is obviously desirable to ensure that the vector-differential equation in (8.25) is stable. The choice of elements in $\boldsymbol{\mu}$ also dictates the speed of convergence to the equilibrium state (i.e., the solution of $\boldsymbol{A}\boldsymbol{x} = \boldsymbol{b}$). It is actually straightforward to derive the conditions for stability of the learning rule, that is, the vector-differential equation in (8.25). From Section A.4, if it can be shown that the total derivative of a suitably chosen *Lyapunov (energy) function* is negative definite, then this shows *asymptotic stability* of the dynamical system. In this case the Lyapunov or energy function is the error cost function in (8.24), and the dynamical system is the continuous-time learning rule in (8.25). Therefore, the total derivative of the energy function can be calculated according to (A.197) and is given by

$$\dot{\mathscr{E}}(\boldsymbol{x}) = \nabla_x^T \mathscr{E}(\boldsymbol{x}) \frac{d\boldsymbol{x}}{dt} \tag{8.26}$$

However, if we substitute the general form of the continuous-time learning rule given in (8.19) into (8.26), we have

$$\dot{\mathscr{E}}(\boldsymbol{x}) = -\nabla_{\boldsymbol{x}}^T \mathscr{E}(\boldsymbol{x})\boldsymbol{\mu}\nabla_{\boldsymbol{x}}\mathscr{E}(\boldsymbol{x}) \tag{8.27}$$

Using $\nabla_{\boldsymbol{x}}\mathscr{E}(\boldsymbol{x}) = \boldsymbol{e}$, as previously explained in (8.27), we obtain

$$\dot{\mathscr{E}}(\boldsymbol{x}) = -\boldsymbol{e}^T\boldsymbol{\mu}\boldsymbol{e} \tag{8.28}$$

Recall that we assumed that matrix $\boldsymbol{\mu}$ is positive definite, that is, $\boldsymbol{\mu} > 0$; therefore,

$$\dot{\mathscr{E}}(\boldsymbol{x}) = -\boldsymbol{e}^T\boldsymbol{\mu}\boldsymbol{e} < 0 \tag{8.29}$$

(providing $\boldsymbol{e} \neq 0$) which says that the total derivative of the energy function is always *negative definite*. Therefore, from Section A.4, (8.25) is *asymptotically stable*.

The scalar form of the learning rule can be written directly from (8.25) and (8.23) as

$$\frac{dx_i}{dt} = -\sum_{h=1}^{n} \mu_{ih} e_h \qquad x_i(0) = x_{i0} \tag{8.30}$$

for $i = 1, 2, \ldots, n$. Again the choice of the μ_{ih} parameters must be carefully made to ensure stability of the differential equations and reasonable convergence speed to the equilibrium state (i.e., the solution of $\boldsymbol{Ax} = \boldsymbol{b}$). The e_h terms in (8.30) are the individual elements of the error vector given in (8.23). These scalar elements of the error vector can be written by first defining the matrix

$$\boldsymbol{W} \triangleq \boldsymbol{A}^T\boldsymbol{A} \qquad \boldsymbol{W} \in \Re^{n\times n} \tag{8.31}$$

In scalar form, (8.31) becomes

$$w_{hr} = \sum_{q=1}^{m} a_{qh} a_{qr} \tag{8.32}$$

for $h = 1, 2, \ldots, n$ and $r = 1, 2, \ldots, n$. Using (8.32), the elements of the error vector in (8.23) can be written as

$$e_h = \sum_{r=1}^{n} w_{hr} x_r - \sum_{j=1}^{m} a_{jh} b_j \tag{8.33}$$

EXAMPLE 8.1. It is desired to solve the following system of equations (underdetermined system):

$$\begin{bmatrix} 1 & 4 & 7 & 10 \\ 2 & 5 & 8 & 11 \\ 3 & 6 & 9 & 12 \end{bmatrix} \begin{bmatrix} x_1 \\ x_2 \\ x_3 \\ x_4 \end{bmatrix} = \begin{bmatrix} -1 \\ 2 \\ 5 \end{bmatrix} \tag{8.34}$$

The discrete-time form of the learning rule in (8.25) and (8.23) is used to solve (8.34), that is,

$$\boldsymbol{x}(k+1) = \boldsymbol{x}(k) - \boldsymbol{\mu}\boldsymbol{e}(k) \tag{8.35}$$

where

$$e(k) = A^T A x(k) - A^T b \tag{8.36}$$

with a *scalar* $\mu = \mu_j = 0.003$ for $j = 1, 2, 3, 4$, and k is the discrete-time index. The initial conditions were set to $x(0) = [0, 0, 0, 0]^T$, and the training was carried out for 2,500 iterations. The solution produced by the learning rule in (8.35) and (8.36) is

$$x = [2.5000, 1.3333, 0.1667, -1.0000]^T \tag{8.37}$$

This is the same solution that is obtained by using the SVD (cf. Sect. A.2.14), out to four decimal places. The value of the cost function versus the iteration number is shown in Figure 8.4.

Relation between solving a system of linear equations and the pseudoinverse of a matrix

As stated in the introduction to this chapter, solving systems of linear equations is equivalent to computing the inverse (or pseudoinverse) of a matrix. This can be readily seen from the learning rule given in (8.35) and (8.36). Given a matrix $A \in \Re^{m \times n}$ whose pseudoinverse is to be found, we assume a succession of (m total) m-dimension b vectors as

$$\begin{aligned} b_1 &= [1, 0, 0, 0, \cdots, 0]^T = e_1 \qquad b_2 = [0, 1, 0, 0, \cdots, 0]^T = e_2 \\ b_3 &= [0, 0, 1, 0, \cdots, 0]^T = e_3 \qquad b_m = [0, 0, 0, 0, \cdots, 1]^T = e_m \end{aligned} \tag{8.38}$$

In other words, *collectively*, the set of *observation vectors* forms an $m \times m$ identity matrix

$$[b_1, b_2, \cdots, b_m] = [e_1, e_2, \cdots, e_m] = I \in \Re^{m \times m} \tag{8.39}$$

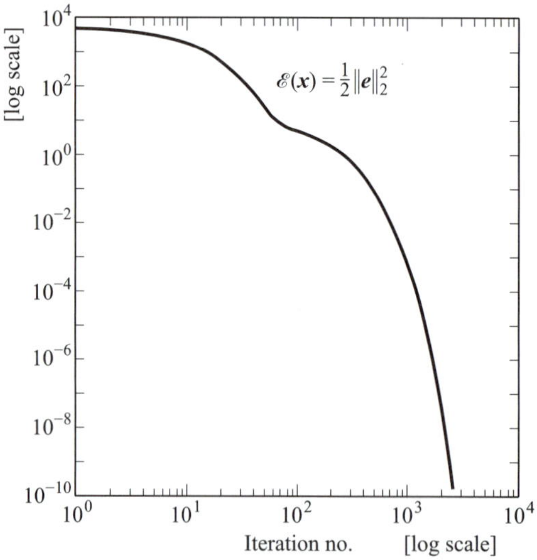

Figure 8.4 Convergence profile for the discrete-time learning rule given in (8.35) and (8.36) to solve the system of linear equations in (8.34).

Therefore, in (8.36) the $\boldsymbol{b}$ vector is replaced with the identity *matrix* given in (8.39), the vector $\boldsymbol{x}$ is replaced with the *matrix* $\boldsymbol{C} \in \Re^{n\times m}$, and the error vector $\boldsymbol{e}$ is replaced with the error *matrix* $\boldsymbol{E} \in \Re^{n\times m}$. This results in an expression for the error *matrix* as

$$\boldsymbol{E}(k) = \boldsymbol{A}^T\boldsymbol{A}\boldsymbol{C}(k) - \boldsymbol{A}^T \tag{8.40}$$

and the expression in (8.35) becomes (for a scalar learning rate parameter μ)

$$\boldsymbol{C}(k+1) = \boldsymbol{C}(k) - \mu\boldsymbol{E}(k) \tag{8.41}$$

Substituting (8.40) into (8.41), and again assuming a scalar learning rate parameter $\mu > 0$, gives

$$\boldsymbol{C}(k+1) = \boldsymbol{C}(k) + \mu\boldsymbol{A}^T[\boldsymbol{I} - \boldsymbol{A}\boldsymbol{C}(k)] \tag{8.42}$$

which is the learning rule given in (7.20) to determine the pseudoinverse of a matrix. Even though the error matrix in (8.40) is not the same as the one in (7.13), it can be shown that they are equivalent with respect to the form of the error cost function that is used.

8.5 CONJUGATE GRADIENT LEARNING RULE FOR SOLVING SYSTEMS OF LINEAR EQUATIONS

Convergence issues relating to any numerical method for solving systems of equations are very important [15, 29]. We now further develop the neurocomputing approach presented in the previous section. We use the conjugate gradient method [30–32] (cf. Sect. A.5.5) instead of the method of steepest descent to increase the speed of convergence. We follow the account given in Section A.5.5 for the Fletcher-Reeves conjugate gradient algorithm with a restart capability (i.e., a steepest descent step after every n iterations). Therefore, we want to solve $\boldsymbol{A}\boldsymbol{x} = \boldsymbol{b}$, where we assume $\boldsymbol{A} \in \Re^{m\times n}$, with $m \geq n$, that is, an overdetermined or exactly determined system.

The first order of business to incorporate the conjugate gradient method into our discrete-time learning rule in the previous section is to derive an expression for α_k in the update expression for the iterate $\boldsymbol{x}_k$. In step 4 of the Fletcher-Reeves conjugate gradient algorithm, the update of the solution is given by

$$\boldsymbol{x}_{k+1} = \boldsymbol{x}_k + \alpha_k\boldsymbol{d}_k \tag{8.43}$$

where

$$\alpha_k = \min_{\alpha\geq 0} \mathscr{E}(\boldsymbol{x}_k + \alpha\boldsymbol{d}_k) \tag{8.44}$$

The vector $\boldsymbol{d}_k$ is the current direction vector, and $\mathscr{E}(\bullet)$ is the objective function to be minimized. In our case the objective function is given by

$$\mathscr{E}(\boldsymbol{x}) = \frac{1}{2}\|\boldsymbol{e}\|_2^2 = \frac{1}{2}\boldsymbol{e}^T\boldsymbol{e} \tag{8.45}$$

where the solution error for solving $\boldsymbol{A}\boldsymbol{x} = \boldsymbol{b}$ is given by

$$e = Ax - b \tag{8.46}$$

Therefore,

$$\begin{aligned}\mathscr{E}(x_k + \alpha d_k) &= \frac{1}{2}[A(x_k + \alpha d_k) - b]^T[A(x_k + \alpha d_k) - b] \\ &= \frac{1}{2}(x_k^T A^T A x_k + \alpha x_k^T A^T A d_k - x_k^T A^T b + \alpha d_k^T A^T A x_k \\ &\quad + \alpha^2 d_k^T A^T A d_k - \alpha d_k^T A^T b - b^T A x_k - \alpha b^T A d_k + b^T b\end{aligned} \tag{8.47}$$

We must compute the gradient of (8.47) with respect to α and set the result equal to zero.

$$\nabla_\alpha \mathscr{E}(x_k + \alpha d_k) = \frac{\partial \mathscr{E}(x_k + \alpha d_k)}{\partial \alpha} = d_k^T A^T A x_k + \alpha d_k^T A^T A d_k - d_k^T A^T b = 0 \tag{8.48}$$

Solving for α from (8.48) and letting $\alpha \to \alpha_k$ gives

$$\alpha_k = \frac{d_k^T A^T b - d_k^T A^T A x_k}{d_k^T A^T A d_k} = \frac{d_k^T \overbrace{[A^T b - A^T A x_k]}^{-g_k}}{d_k^T A^T A d_k} = \frac{-d_k^T g_k}{d_k^T A^T A d_k} \tag{8.49}$$

or

$$\alpha_k = \frac{-g_k^T d_k}{d_k^T A^T A d_k} \tag{8.50}$$

where g_k is the gradient of (8.45),

$$g_k = \nabla_x \mathscr{E}(x_k) = A^T A x_k - A^T b \tag{8.51}$$

Therefore, the *Fletcher-Reeves conjugte gradient algorithm (with restart)* for solving $Ax = b$ is summarized in the following steps:

Fletcher-Reeves conjugate gradient algorithm (with restart) for solving $Ax = b$

Step 1. Set x_0.
Step 2. Compute $g_k\big|_{k=0} = g_0 = A^T A x_0 - A^T b$.
Step 3. Set $d_0 = -g_0$.
Step 4. Compute $x_{k+1} = x_k + \alpha_k d_k$, where $\alpha_k = -g_k^T d_k / (d_k^T A^T A d_k)$.
Step 5. Compute $g_{k+1} = A^T A x_{k+1} - A^T b$.
Step 6. Compute $d_{k+1} = -g_{k+1} + \beta_k d_k$, where $\beta_k = g_{k+1}^T g_{k+1} / (g_k^T g_k)$.
Steps 4 to 6 are carried out for $k = 0, 1, \ldots, n-1$.
Step 7. Replace x_0 by x_n and go to step 1.
Step 8. Continue until convergence is achieved; termination criterion could be $\|d_k\| < \varepsilon$ (where ε is an appropriate predetermined small number).

Figure 8.5 shows a discrete-time neural network architecture for the conjugate gradient algorithm for solving systems of linear equations.

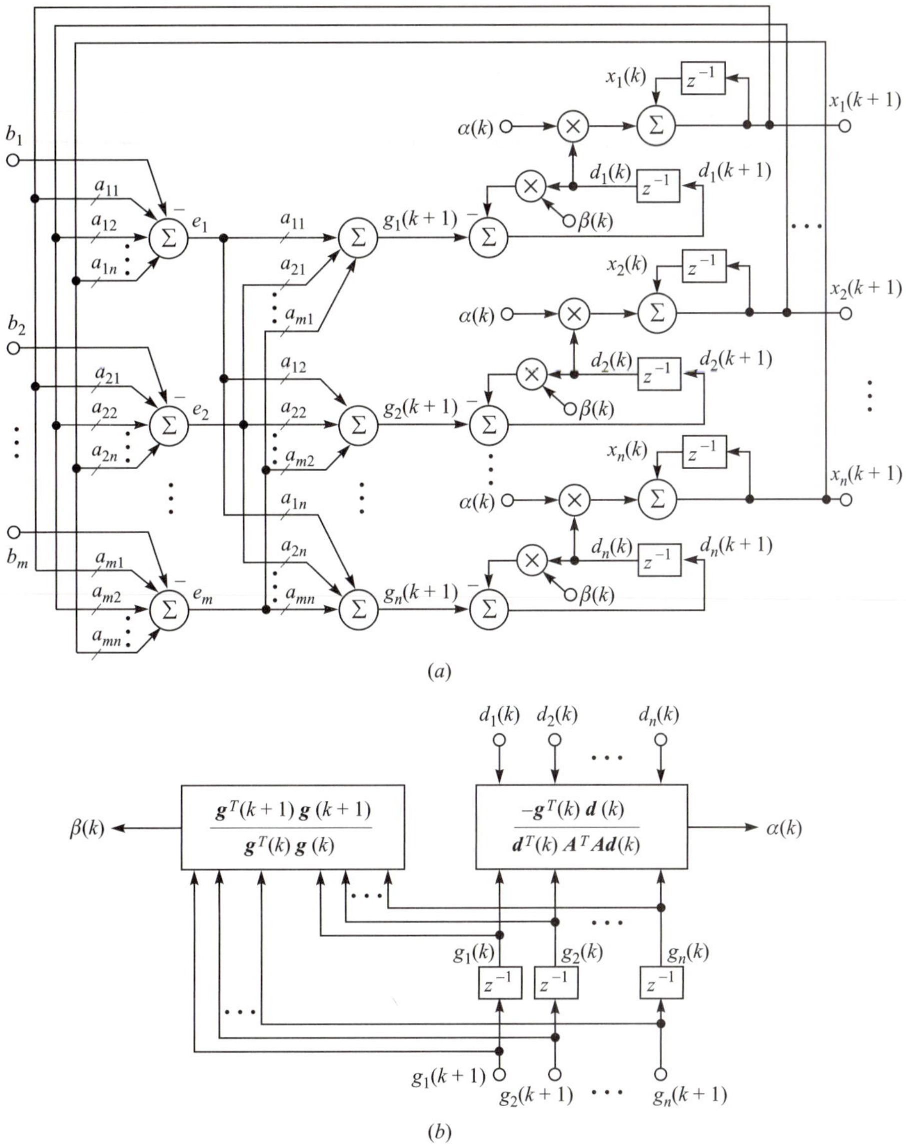

Figure 8.5 (*a*) Discrete-time neural network architecture for the conjugate gradient algorithm for solving systems of linear algebraic equations. (*b*) Portion of the neural architecture for computing α_k and β_k.

EXAMPLE 8.2. We seek the solution to the system of equations

$$\begin{bmatrix} 3 & 5 & 1 & 4 \\ 3 & -8 & -2 & 11 \\ -1 & 0 & 0 & -7 \\ -10 & -3 & -2 & -5 \\ 1 & -5 & -2 & 5 \\ 8 & -6 & 2 & -2 \end{bmatrix} \begin{bmatrix} x_1 \\ x_2 \\ x_3 \\ x_4 \end{bmatrix} = \begin{bmatrix} -7 \\ 2 \\ 8 \\ 4 \\ 10 \\ 3 \end{bmatrix} \tag{8.52}$$

For this overdetermined system of equations, both the discrete-time steepest descent method given in (8.35) and (8.36) and the discrete-time conjugate gradient method given above are used to solve this system for $\boldsymbol{x}$. Using the steepest descent gradient method, with a single learning rate parameter set at $\mu = 0.0055$ and a set of random numbers to initialize $\boldsymbol{x}$, the learning rule converged to the solution in 1,900 iterations. The solution is given as

$$\boldsymbol{x} = \left[1.8725, 0.0935, -7.4699, -1.5274\right]^T \tag{8.53}$$

This is the same solution obtained using the SVD (out to four decimal places). A plot of the convergence profile for each component in the solution vector is shown in Figure 8.6(*a*). Using the conjugate gradient learning rule presented above (with the same initial conditions used for the steepest descent learning rule), after *four* iterations the solution in (8.53) was reached! Therefore, the conjugate gradient-based learning rule required 475 times fewer training steps than the steepest descent-based method! For the conjugate gradient-based learning rule, a plot of the convergence profile for each component of the solution vector is shown in Figure 8.6(*b*).

8.6 A GENERALIZED ROBUST APPROACH FOR SOLVING SYSTEMS OF LINEAR EQUATIONS CORRUPTED WITH NOISE

Many times the distribution of the errors in the data is not Gaussian (or normal). In these cases, for example, if impulse ("spiky") noise exists, the standard least-squares performance criterion used in the previous section based on the L_2-norm will provide very poor estimates of the solution to $\boldsymbol{Ax} = \boldsymbol{b}$. Therefore, in the presence of very large errors (referred to as *outliers*), impulse noise, or colored noise, a different approach must be taken when the

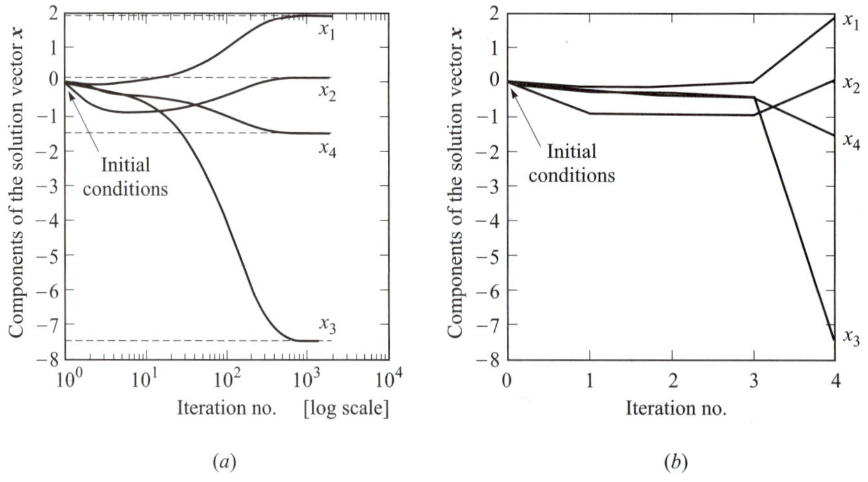

Figure 8.6 (*a*) Convergence profile for solving (8.52) using a steepest descent-based learning rule. (*b*) Convergence profile for solving (8.52) using a conjugate gradient-based learning rule.

system errors do not follow a Gaussian distribution. To mitigate the effects of non-Gaussian noise and provide *robust* estimates of the solution vector $\boldsymbol{x}$, a more general formulation of the error cost function is utilized. The ordinary least-squares problem is optimal only if all errors reside in the observation vector $\boldsymbol{b}$ and have a Gaussian distribution (i.e., uncorrelated errors with zero mean and equal variance). We now consider the case in which errors can exist in the data matrix $\boldsymbol{A}$ and they can have a non-Gaussian distribution.

The approach taken here is more general than the iteratively reweighted least-squares method (or robust least-squares criterion) [8, 13, 19, 20]. In general, we want to weight the outliers (i.e., the *heavy errors*) less than the smaller errors that can exist in the data. Therefore, this would require a *weighting* function that "grows" less than quadratically. The quadratic function that we refer to is the one used for the least-squares approach to the problem of solving $\boldsymbol{Ax} = \boldsymbol{b}$ presented in Section 8.4. Out at the "tails" of the quadratic function, the weighting imposed on large (or heavy) system errors is too large. Unless these errors are diminished, their influence on the solution to $\boldsymbol{Ax} = \boldsymbol{b}$ will be overwhelming, and the estimates of the solution vector $\boldsymbol{x}$ will be very poor. Therefore, as stated previously, we need a weighting function that *grows less than* quadratically.

The problem of solving $\boldsymbol{Ax} = \boldsymbol{b}$, $\boldsymbol{A} \in \Re^{m\times n}$, $\boldsymbol{b} \in \Re^{m\times 1}$, can now be stated in the following generalized minimization problem. Determine the vector

$$\boldsymbol{x} \in \Re^{n\times 1} \tag{8.54}$$

that minimizes the *weighted statistical error cost function* [33–35]

$$\mathscr{E}(\boldsymbol{x}) = \boldsymbol{1}^T \boldsymbol{S}\mathrm{E}\{f(\boldsymbol{e})\} \tag{8.55}$$

where $\boldsymbol{1} = [1 \quad 1 \quad \cdots \quad 1]^T$, $\boldsymbol{S} \in \Re^{m\times m}(\boldsymbol{S}^T = \boldsymbol{S}$ and $\boldsymbol{S} > 0)$, $\mathrm{E}\{\bullet\}$ is the expectation operator, $f\{\bullet\}$ is a convex weighting function,† and

$$\boldsymbol{e} = \boldsymbol{Ax} - \boldsymbol{b} \tag{8.56}$$

Because only *instantaneous* errors can be processed, the expectation operator can be dropped [33]. Therefore, the practical error cost function is written as

$$\mathscr{E}(\boldsymbol{x}) = \boldsymbol{1}^T \boldsymbol{S} f(\boldsymbol{e}) \tag{8.57}$$

where the positive definite, symmetric weight matrix $\boldsymbol{S}$ is usually taken as the identity matrix. To take a steepest descent gradient approach for the development of a learning strategy for solving $\boldsymbol{Ax} = \boldsymbol{b}$, we start with the same general form used in Section 8.4

$$\frac{\mathrm{d}\boldsymbol{x}}{\mathrm{d}t} = -\boldsymbol{\mu}\nabla_x \mathscr{E}(\boldsymbol{x}) \tag{8.58}$$

where $\boldsymbol{\mu} \in \Re^{n\times n}(\boldsymbol{\mu} = [\mu_{ij}]$, for $i, j = 1, 2, \ldots, n)$ is a positive definite matrix which is typically chosen to be diagonal, that is, $\mu_{ij} = \mu_j \delta_{ij}$, where $\mu_j > 0$. Therefore, we must determine the gradient in (8.58)

†Actually $f\{\bullet\}$ can be a function that is *nearly* convex.

$$\nabla_x \mathscr{E}(x) = \frac{\partial \mathscr{E}(x)}{\partial x} = \frac{\partial}{\partial x} \mathbf{1}^T S f(e) = \frac{\partial e}{\partial x} S g(e) \tag{8.59}$$

where $g(t) = \mathrm{d}f(t)/\mathrm{d}t$ and t is a dummy variable. The partial derivative $\partial e/\partial x$ in (8.59) requires an explanation. We can write the error vector from (8.56) as

$$e = Ax - b = \begin{bmatrix} a_1 \\ a_2 \\ \vdots \\ a_m \end{bmatrix} x - b \tag{8.60}$$

where $a_1, a_2, \ldots, a_m$ are the *rows* of A, that is, $a_j \in \Re^{1\times n}$, for $j = 1, 2, \ldots, m$. Therefore, (8.60) can be written as

$$e = \begin{bmatrix} a_1 x \\ a_2 x \\ \vdots \\ a_m x \end{bmatrix}_{m\times 1} - b \tag{8.61}$$

Taking the partial derivative of (8.61) with respect to the vector x gives (using the results in Sect. A.3.4.1)

$$\frac{\partial e}{\partial x} = \frac{\partial}{\partial x} \begin{bmatrix} a_1 x \\ a_2 x \\ \vdots \\ a_m x \end{bmatrix} = \begin{bmatrix} \frac{\partial}{\partial x} a_1 x \\ \frac{\partial}{\partial x} a_2 x \\ \vdots \\ \frac{\partial}{\partial x} a_m x \end{bmatrix} = \left[a_1^T, a_2^T, \cdots, a_m^T \right]_{n\times m} = A^T \tag{8.62}$$

Therefore, by using our result from (8.62), the gradient in (8.59) $\nabla_x \mathscr{E}(x)$ can be written as

$$\nabla_x \mathscr{E}(x) = \frac{\partial \mathscr{E}(x)}{\partial x} = \frac{\partial}{\partial x} \mathbf{1}^T S f(e) = A^T S g(e) \tag{8.63}$$

Using (8.63), the *robust* learning rule from (8.58) is given by

$$\frac{\mathrm{d}x}{\mathrm{d}t} = -\mu A^T S g(e) \tag{8.64}$$

To show this is a general result, if we let the weight matrix $S = I$ and let the weighting function be *quadratic*, that is, $f(t) = \frac{1}{2} t^2$ [$g(t) = \mathrm{d}f(t)/\mathrm{d}t = t$, that is, a linear function], then (8.64) becomes

$$\frac{\mathrm{d}x}{\mathrm{d}t} = -\mu A^T e \tag{8.65}$$

which is the result we obtained in (8.22) for our first least-squares solution to the problem. Figure 8.7 shows the detailed neural network architecture for the generalized robust approach for solving systems of linear equations. In the figure the weight matrix is assumed to be the identity matrix, that is, $S = I_{m\times m}$.

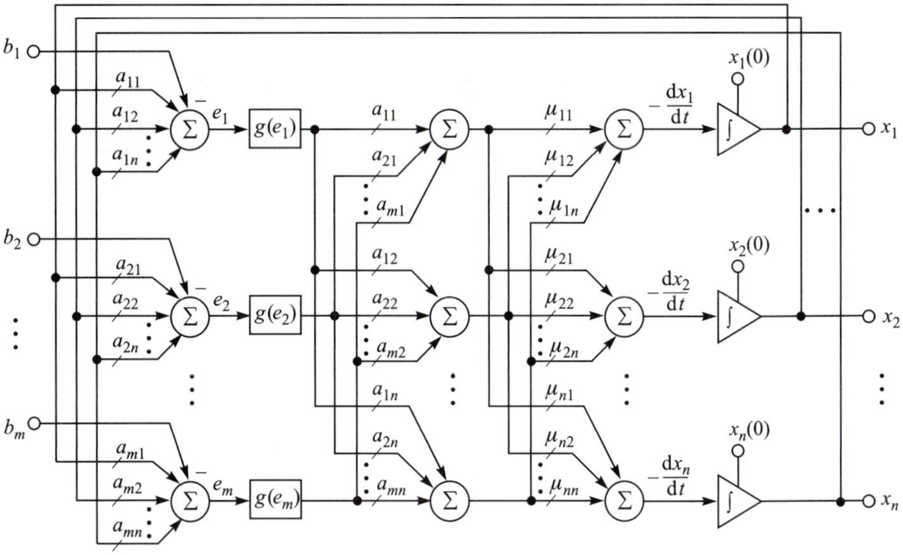

Figure 8.7 Neural network architecture for the generalized robust approach for solving systems of linear equations. It is assumed that $\boldsymbol{S} = \boldsymbol{I}_{m \times m}$, and $g(t) = \mathrm{d}f(t)/\mathrm{d}t$ (t is a dummy variable), where $f(\bullet)$ is the selected nonlinear weighting function.

We need to discuss the selection of the weighting function $f(\bullet)$ that appears in the error cost function in (8.57). As stated previously, this function must be convex or nearly convex (cf. Definition A.11, Sect. A.3.1). There are many choices for the selection of the weighting function, but four of the most popular ones are summarized in Table 9.2 and graphed in Figure 9.21. The associated derivatives are shown in Figure 9.22. The type of function that is used and the value of the cutoff parameter β both depend on the data. Therefore, a *trial-and-error* procedure is typically carried out to determine the "best" function to use and the "best" value for β. An interesting point can be illustrated by using one of the most common weighting functions, the logistic function, given by

$$f_L(t) = \beta^2 \ln[\cosh(t/\beta)] \tag{8.66}$$

Figure 8.8 shows a plot of this weighting function for three values of the cutoff parameter β and compared to the quadratic function. As seen from Figure 8.8, as the value of the cutoff parameter β increases, the shape of the curve approaches the quadratic function. Therefore, for large values of β the robustness capabilities of the learning algorithm diminish. Observing the figure, we see that between approximately $-2/3$ and $2/3$ for the error, all the curves provide about the same (quadratic) weighting. Outside this range for the error, depending on the cutoff parameter value, the weighting will be *less than* quadratic; however, care must be taken in selecting the value of β. For values of β that are too small, valuable information can be discarded; conversely, for values too large, outliers and noise can dominate the data and degrade the estimate of the solution vector (see Figure 8.8). The region shown in the figure

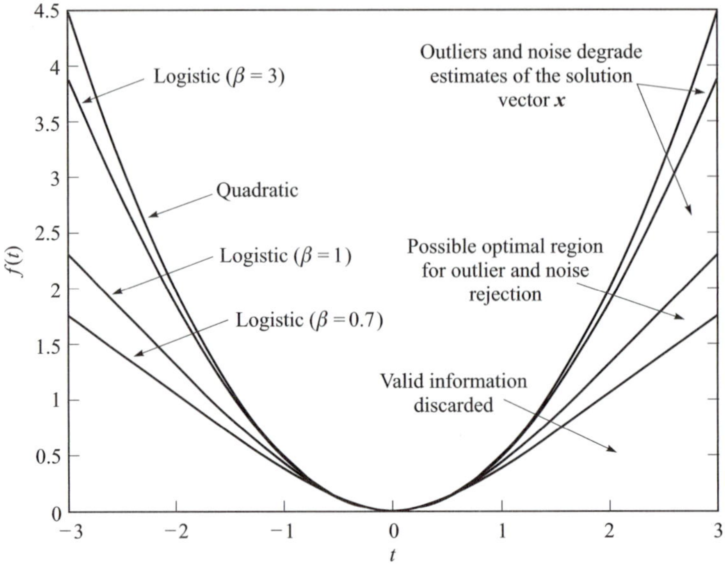

Figure 8.8 Logistic weighting function for three different values of the cutoff parameter β compared to the quadratic function $f_Q(t) = {}^1\!/_2 t^2$, where t is a dummy variable.

between the values of 0.7 and 1 for β illustrates a possible optimal region where good outlier and noise rejection occurs. Figure 8.9 shows the respective derivative functions.

The robust neurocomputing learning rule can be summarized in the following steps. The discrete-time, vector-matrix form of the learning rule is presented.

Robust learning algorithm for solving systems of equations with noise

Step 1. Select appropriate $\boldsymbol{\mu}$ and $\boldsymbol{S}$ matrices.
Step 2. Select an appropriate weighting function $f(\bullet)$ and compute the derivative $g(\bullet)$. Also select a reasonable value for the cutoff parameter β.
Step 3. Set $k = 0$ (where k is the discrete-time index), and choose an initial starting point $\boldsymbol{x}(0)$.
Step 4. Compute the solution error according to

$$\boldsymbol{e}(k) = \boldsymbol{A}\boldsymbol{x}(k) - \boldsymbol{b} \tag{8.67}$$

Step 5. Update the estimate of the solution vector according to

$$\boldsymbol{x}(k+1) = \boldsymbol{x}(k) - \boldsymbol{\mu}\boldsymbol{A}^T\boldsymbol{S}g[\boldsymbol{e}(k)] \tag{8.68}$$

Step 6. Stop if convergence is achieved; else $k \leftarrow k + 1$ and go to step 4.

EXAMPLE 8.3. This example illustrates how the discrete-time robust learning rule in (8.67) and (8.68) can mitigate the effects of impulse noise. It is desired to find the solution $\boldsymbol{x} \in \Re^{10\times 1}$ to a system of 10 equations and 10 unknowns, that is, $\boldsymbol{Ax} = \boldsymbol{b}$ where

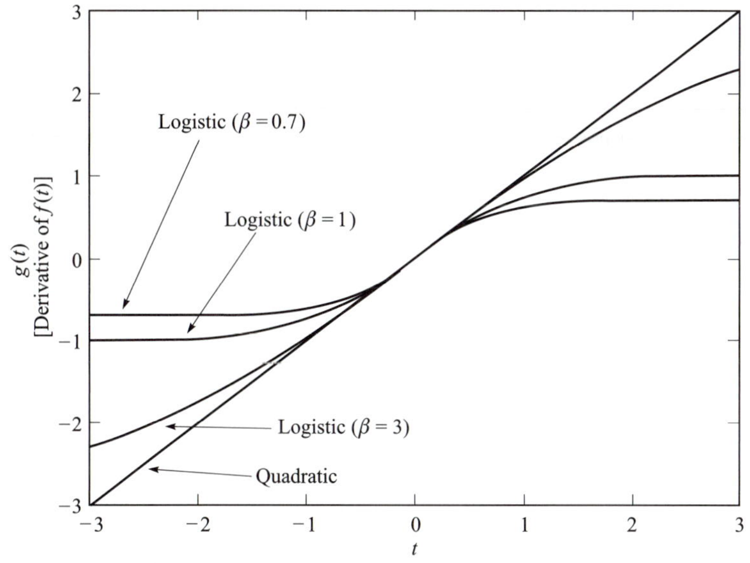

Figure 8.9 Derivatives of the weighting functions in Figure 8.8.

$$A = \begin{bmatrix} 0.1417 & -1.0075 & -0.7732 & -0.9113 & 0.4871 & -1.6601 & -1.3186 & 0.4035 & -0.6255 & -1.0275 \\ -1.2650 & 1.3090 & -0.1451 & -0.2353 & -1.4658 & 0.4176 & 1.1449 & 2.0468 & 0.8100 & -0.2398 \\ -0.2704 & -0.5756 & 0.2369 & 0.3593 & -0.6807 & 0.4390 & 0.4493 & -1.0552 & -1.0041 & -0.3516 \\ -0.5650 & -2.1317 & -2.3882 & -1.5344 & 0.1802 & 1.3482 & 1.1635 & -1.4374 & -1.1220 & -1.0924 \\ 1.8503 & -0.5428 & 2.4866 & 0.9782 & 0.2633 & 0.4738 & -1.0853 & 1.1041 & 0.2045 & -1.1709 \\ 0.8016 & 0.0237 & 0.3418 & -0.4727 & 1.6016 & -0.3046 & 1.9539 & -1.3561 & 0.8812 & -0.5963 \\ 1.0093 & -0.0254 & 0.0430 & -0.9378 & -1.4734 & 0.2955 & 0.2372 & 0.3510 & 0.5705 & 0.5723 \\ -0.2946 & -1.6822 & 0.1412 & -0.8454 & 1.1526 & -0.7930 & 0.1700 & -1.9511 & -1.3950 & -0.8927 \\ 0.5661 & -0.0328 & 0.6679 & 1.4221 & -0.3615 & 1.5595 & 0.1963 & 0.5752 & 0.4470 & -0.2336 \\ 0.1112 & 0.9667 & -0.7838 & -0.4777 & -0.8919 & 2.6388 & -0.2533 & -1.5569 & -0.1884 & -0.4363 \end{bmatrix} \tag{8.69}$$

and

$$\boldsymbol{b} = [-0.7800, -1.1331, 0.9693, -0.0026, 1.8051, 0.3402, 0.2821, 1.4632, 0.1786, 0.5643]^T \tag{8.70}$$

The $\boldsymbol{A}$ matrix in (8.69) is considered to be the "truth" array because the matrix that is actually used $\boldsymbol{A}_c = \boldsymbol{A} + \Delta\boldsymbol{A}$ is *corrupted* with noise. Therefore, we will actually solve

$$\boldsymbol{A}_c \boldsymbol{x}_c = \boldsymbol{b} \tag{8.71}$$

where $\boldsymbol{A}_c$ is the matrix in (8.69) with some of its elements corrupted with impulse noise. Specifically, random numbers are taken from a uniform distribution in the interval [0, 0.25], and these values are added to randomly selected entries in matrix $\boldsymbol{A}$ in (8.69) to form the corrupted array $\boldsymbol{A}_c$. Therefore, this uniform distribution is used as a model for impulse noise (or outliers). The array elements that are corrupted with the impulse noise are randomly selected with probability 0.15; that is, on average 15 percent of the elements in the $\boldsymbol{A}$ matrix in (8.69) will be corrupted with impulse noise with a variance $\sigma^2 = (0.15)(0.25)^2/12 =$

7.8125×10^{-4} (cf. Sect. A.7.4). The actual Δa_{ij} values and array entries of $\boldsymbol{A}$ to be corrupted are both shown in Table 8.1.

Therefore, the objective is to solve (8.71) for $\boldsymbol{x}_c$, and to use a method that forces the solution to be as close as possible to the true solution, that is, $\boldsymbol{x}_c \cong \boldsymbol{x}$. Using MATLAB to solve (8.71) for $\boldsymbol{x}_c$ gives

$$\boldsymbol{x}_c^M = [0.4295, 0.2243, 0.9805, -0.7051, 0.4362, 0.5609, 0.3072, 0.2087, -1.4279, 0.5178]^T \tag{8.72}$$

Because $\boldsymbol{A}_c$ is a full-rank matrix with a good condition number (cf. Sect. A.2.15), the solution in (8.72) was computed in a straightforward manner using the MATLAB built-in function **inv**, that is, $\boldsymbol{x}_c^M = inv(\boldsymbol{A}_c)^*\boldsymbol{b}$. The true solution (i.e., the solution of the "truth" system $\boldsymbol{A}\boldsymbol{x}^M = \boldsymbol{b}$) $\boldsymbol{x}^M$ was computed in the same way, that is, $\boldsymbol{x}^M = inv(\boldsymbol{A})^*\boldsymbol{b}$, to give

$$\boldsymbol{x}^M = [0.3177, 0.0486, 0.9002, -0.4987, 0.2073, 0.4242, 0.2029, -0.0008, -1.0085, 0.2863]^T \tag{8.73}$$

The robust neurocomputing learning rule given above in discrete-time form is used next to determine the *robust* solution. A total of 420 iterations were necessary for convergence, and the logistic weighting function was used with the cutoff parameter set to $\beta = 0.7$. This was determined by a trial-and-error procedure to be the best value to use. A scalar learning rate parameter was used, and $\mu = 0.01$ was determined by a trial-and-error procedure to be the best value. The weight matrix was set to $\boldsymbol{S} = \boldsymbol{I}$, and the initial conditions in $\boldsymbol{x}_c^N(0)$ were set to Gaussian random numbers with zero mean and unity variance. The solution yielded by the robust neurocomputing approach $\boldsymbol{x}_c^N$ is

$$\boldsymbol{x}_c^N = [0.2663, 0.0306, 0.9194, -0.5053, 0.1882, 0.4092, 0.1678, -0.0553, -0.8985, 0.2828]^T \tag{8.74}$$

TABLE 8.1

Randomly generated perturbations from a uniform distribution used as impulse noise to corrupt randomly selected elements in *A* matrix for Example 8.3

Array index	Δa_{ij}
(1, 3)	0.0564
(1, 10)	0.2393
(3, 6)	0.2084
(4, 1)	0.0837
(4, 3)	0.0137
(5, 1)	0.0104
(6, 8)	0.2120
(9, 3)	0.0817
(9, 7)	0.2353

The results are compared by computing the standard error of estimation (SEE) between (1) the solution of $A_c x_c = b$ using MATLAB and the solution to the truth system $Ax^M = b$ and (2) the solution of $A_c x_c = b$ using the robust neurocomputing approach and the solution to the truth system $Ax^M = b$. Using the solutions given in (8.72) and (8.73), the SEE for the MATLAB solution is

$$\mathrm{SEE}_M = \left[\frac{\sum_{i=1}^{10} (x_{ci}^M - x_i^M)^2}{10} \right]^{1/2} = 0.2116 \tag{8.75}$$

and using (8.73) and (8.74), the SEE for the robust neurocomputing solution is

$$\mathrm{SEE}_N = \left[\frac{\sum_{i=1}^{10} (x_{ci}^N - x_i^M)^2}{10} \right]^{1/2} = 0.0450 \tag{8.76}$$

Therefore, the SEE_M (using MATLAB) is approximately 4.7 times greater than the SEE_N (using the robust neurocomputing approach). The improvement to the solution of the system of equations provided by the robust neurocomputing approach over the MATLAB approach is substantial and can be seen by observing Figure 8.10.

We can observe the sensitivity of the solutions due to the perturbations in matrix A, that is, the perturbations given in Table 8.1, using the relationship given in Section A.2.15, Equation (A.115),

$$\frac{\|\Delta x\|}{\|x + \Delta x\|} \le \mathrm{cond}(A) \frac{\|\Delta A\|}{\|A\|} \tag{8.77}$$

where the condition number of A, denoted by cond(A), is given as in (A.107), that is, the ratio of the largest singular value of A to the smallest one. Given the results from this problem, the right side of (8.77) works out to be 1.8494. The left side of (8.77) for the MATLAB solution works out to be 0.3105, and for the neurocomputing approach it is 0.0940. Therefore, the inequality is satisfied, and again these results show that the robust neurocomputing approach yields the best results.

8.7 REGULARIZATION METHODS FOR ILL-POSED PROBLEMS WITH ILL-DETERMINED NUMERICAL RANK

When one is seeking a solution to

$$Ax = b \tag{8.78}$$

where $A \in \Re^{m\times n}$, $x \in \Re^{n\times 1}$, $b \in \Re^{m\times 1}$, and $m \ge n$, matrix A may be ill-conditioned and have ill-determined numerical rank [25]; that is, the singular values of A decay toward zero without a defined gap in the spectrum. Therefore, the focus of this discussion is on ill-conditioned least-squares problems. These types of problems can arise in many different situations, for example, the numerical solution of Fredholm integral equations of the first kind, which

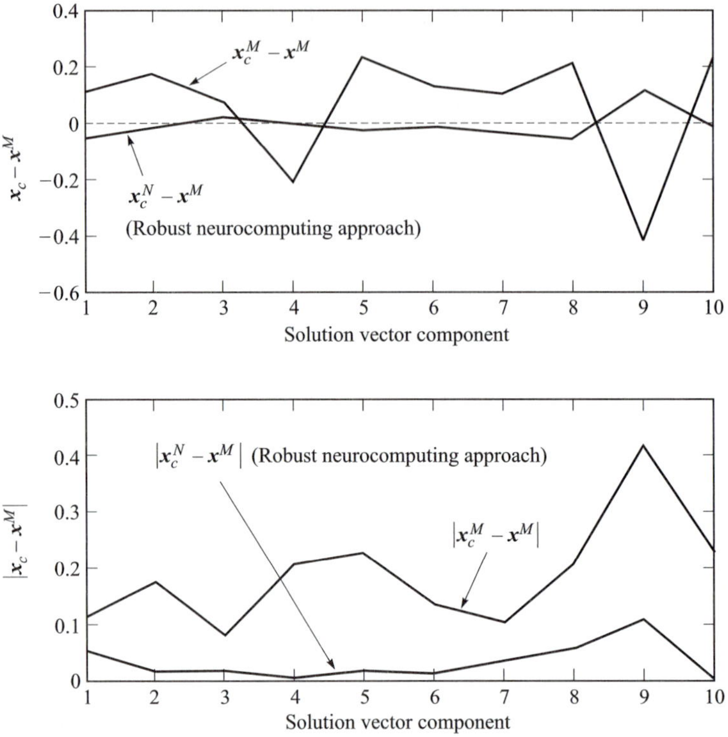

Figure 8.10 Comparison of the MATLAB solution residuals for the corrupted system with respect to the solution of the truth system, and the robust neurocomputing solution residuals for the corrupted system with respect to the solution of the truth system.

are classical examples of ill-posed problems. Several methods can be used to solve ill-conditioned problems, for example, truncated QR (factorization) regularization [38], truncated singular-value decomposition (SVD) regularization [39–41], and damped least squares, or *Tikhonov regularization* [42–53]. We will concentrate on the Tikhonov regularization method.

Our original scalar cost (energy) function defined in (8.12), which sets the stage for a least-squares solution to (8.78), was given by

$$\mathscr{E}(\boldsymbol{x}) = \frac{1}{2}\|\boldsymbol{A}\boldsymbol{x} - \boldsymbol{b}\|_2^2 \tag{8.79}$$

Computing the gradient of (8.79) with respect to $\boldsymbol{x}$ and setting the result equal to zero lead to the normal equations given in (8.14), repeated here:

$$\boldsymbol{A}^T\boldsymbol{A}\boldsymbol{x} - \boldsymbol{A}^T\boldsymbol{b} = \boldsymbol{0} \tag{8.80}$$

We will approach the solution to (8.80) in a slightly different manner than that shown in (8.15) where the Moore-Penrose inverse was used. Here we will use the SVD of $\boldsymbol{A}$ (cf. Sect. A.2.14), that is,

$$\boldsymbol{A} = \boldsymbol{U}\boldsymbol{S}\boldsymbol{V}^T \tag{8.81}$$

From Section A.2.14, the least-squares solution to (8.78) by way of the normal equations in (8.80) using the SVD of $\boldsymbol{A}$ is given by

$$\boldsymbol{x} = \boldsymbol{V}\boldsymbol{S}^{+}\boldsymbol{U}^{T}\boldsymbol{b} = \sum_{i=1}^{n} \frac{\boldsymbol{v}_i \boldsymbol{u}_i^T}{\sigma_i} \boldsymbol{b} \tag{8.82}$$

If $\boldsymbol{A}$ is rank-deficient, that is, $\rho(\boldsymbol{A}) < n$, then the summation limit in (8.82) must reflect the actual rank of $\boldsymbol{A}$. If we let $\rho(\boldsymbol{A}) = k \leq n$, then (8.82) is written as

$$\boldsymbol{x}_k = \boldsymbol{V}\boldsymbol{S}^{+}\boldsymbol{U}^{T}\boldsymbol{b} = \sum_{i=1}^{k} \frac{\boldsymbol{v}_i \boldsymbol{u}_i^T}{\sigma_i} \boldsymbol{b} \tag{8.83}$$

Equation (8.83) is considered the truncated SVD (TSVD) method, and k is considered the *truncation parameter* (it plays a role similar to λ in the Tikhonov regularization method [39]). A problem with this method occurs when $\boldsymbol{A}$ has ill-determined numerical rank; that is, the singular values of $\boldsymbol{A}$ decay toward zero without a defined gap in the spectrum.

We now define a *regularized* energy function, written as

$$\mathscr{E}(\boldsymbol{x}) = \frac{1}{2}\|\boldsymbol{A}\boldsymbol{x} - \boldsymbol{b}\|_2^2 + \frac{\lambda^2}{2}\|\boldsymbol{x}\|_2^2 \tag{8.84}$$

where $\lambda > 0$ is a free parameter referred to as the regularization parameter. This parameter controls the "smoothness" of the regularized solution. We seek a solution $\boldsymbol{x} = \boldsymbol{x}_\lambda$ of the problem

$$\min_{\boldsymbol{x} \in \Re^n} \mathscr{E}(\boldsymbol{x})$$

The additional term in the energy function is minimized simultaneously with the standard *error* term, that is, $\frac{1}{2}\|\boldsymbol{A}\boldsymbol{x} - \boldsymbol{b}\|_2^2$. The second term, $(\lambda^2/2)\|\boldsymbol{x}\|_2^2$, is considered the smoothness constraint; it is also referred to as the stabilizer energy. Computing the gradient of (8.84) with respect to $\boldsymbol{x}$ and setting the result equal to zero lead to the "modified" normal equations given by

$$(\boldsymbol{A}^T\boldsymbol{A} + \lambda^2\boldsymbol{I})\boldsymbol{x} = \boldsymbol{A}^T\boldsymbol{b} \tag{8.85}$$

The fundamental issue in (8.85) is the *condition* of the *regularized* matrix $\tilde{\boldsymbol{A}} = \boldsymbol{A}^T\boldsymbol{A} - \lambda^2\boldsymbol{I}$, that is, the *condition number* of $\tilde{\boldsymbol{A}}$ compared to the condition number of $\boldsymbol{A}^T\boldsymbol{A}$. It is straightforward to show that the condition number of $\tilde{\boldsymbol{A}}$ is given by

$$\text{cond}(\tilde{\boldsymbol{A}}) = \frac{\sigma_{\max}^2 + \lambda^2}{\sigma_{\min}^2 + \lambda^2} \tag{8.86}$$

where $\sigma_{\max}$ and $\sigma_{\min}$ are, respectively, the maximum and minimum singular values of $\boldsymbol{A}$. For example, if the maximum and minimum singular values of $\boldsymbol{A}$ are $\sigma_{\max} = 5$ and $\sigma_{\min} = 0.1$, respectively, then the condition number of $\boldsymbol{A}^T\boldsymbol{A}$ is 2,500. If the regularization parameter is set at $\lambda = 1/\sqrt{2}$, the condition number of $\tilde{\boldsymbol{A}}$ from (8.86) is 50. Thus the regularization parameter improves (reduces) the condition number by a factor of 50. We want to again use the SVD to determine the solution to (8.85). Substituting (8.81) into (8.85) gives

$$(VS^TSV^T + \lambda^2 I)x = VS^TU^Tb \tag{8.87}$$

Premultiplying both sides of (8.87) first by V^T, then by $(S^TS)^{-1}$ (assuming for the moment that A is full rank, so this inverse exists), and finally by V gives

$$[I + \lambda^2 V(S^TS)^{-1}V^T]x = VS^+U^Tb \tag{8.88}$$

where S^+ is the transpose of S with the reciprocal of the nonzero singular values on the pseudodiagonal (cf. Sect. A.2.14). The left side of (8.88) can be written as

$$\begin{aligned}[I + \lambda^2 V(S^TS)^{-1}V^T]x &= \left(\sum_{i=1}^{n} v_i v_i^T + \lambda^2 \sum_{i=1}^{n} \frac{v_i v_i^T}{\sigma_i^2}\right)x \\ &= \sum_{i=1}^{n} v_i \left(1 + \frac{\lambda^2}{\sigma_i^2}\right) v_i^T x = V\Sigma V^T x\end{aligned} \tag{8.89}$$

where

$$\Sigma = \begin{bmatrix} 1 + \dfrac{\lambda^2}{\sigma_1^2} & 0 & 0 & \cdots & 0 \\ 0 & 1 + \dfrac{\lambda^2}{\sigma_2^2} & 0 & \cdots & 0 \\ \cdots & \cdots & \cdots & \cdots & \cdots \\ 0 & 0 & \cdots & 0 & 1 + \dfrac{\lambda^2}{\sigma_n^2} \end{bmatrix}_{n \times n} \tag{8.90}$$

Therefore, using (8.89), Equation (8.88) can be written as

$$V\Sigma V^T x = VS^+U^Tb \tag{8.91}$$

Premultiplying (8.91) on both sides first by V^T, then by Σ^{-1}, and finally by V gives

$$x_\lambda = V\Sigma^{-1}S^+U^Tb = V \begin{bmatrix} \dfrac{1}{\sigma_1 + \dfrac{\lambda^2}{\sigma_1}} & 0 & 0 & \cdots \\ 0 & \dfrac{1}{\sigma_2 + \dfrac{\lambda^2}{\sigma_2}} & 0 & \cdots \\ \vdots & & \ddots & \end{bmatrix}_{n \times n} U^Tb \tag{8.92}$$

or

$$x_\lambda = \sum_{i=1}^{n} \frac{v_i u_i^T b}{\sigma_i + \lambda^2/\sigma_i} \tag{8.93}$$

or

$$\boldsymbol{x}_\lambda = \sum_{i=1}^{n} \left(\frac{\sigma_i^2}{\sigma_i^2 + \lambda^2} \right) \frac{\boldsymbol{v}_i \boldsymbol{u}_i^T}{\sigma_i} \boldsymbol{b} \tag{8.94}$$

If we define

$$\beta_i \triangleq \boldsymbol{u}_i^T \boldsymbol{b} \tag{8.95}$$

then (8.94) can also be written as

$$\boldsymbol{x}_\lambda = \sum_{i=1}^{n} \left(\frac{\beta_i}{\sigma_i + \lambda^2/\sigma_i} \right) \boldsymbol{v}_i \tag{8.96}$$

We can make several observations concerning the result in (8.94), or (8.96):

1. If the regularization parameter is set to zero, that is, $\lambda = 0$, then (8.94) reduces to the result in (8.82), that is, $\boldsymbol{x}_0 = \sum_{i=1}^{n} [(\boldsymbol{v}_i \boldsymbol{u}_i^T / \sigma_i) \boldsymbol{b}]$.
2. A set of solutions is generated from (8.94) for a defined range of values for the regularization parameter $\lambda > 0$. The ideal situation, then, is to define a criterion for the selection of an *appropriate* solution from the set of admissible solutions. One method that can be used for the selection of an appropriate solution, that is, selection of the Tikhonov parameter λ, is the *L*-curve method [44, 46, 50 52]. Other methods can be found [53–55].
3. If $\boldsymbol{b}$ in $\boldsymbol{Ax} = \boldsymbol{b}$ is unperturbed, with $\boldsymbol{A}$ ill-conditioned and having ill-determined numerical rank, it satisfies the *discrete Picard condition* (DPC) [45, 48, 53] if and only if $|\boldsymbol{u}_i^T \boldsymbol{b}|$, for $i = 1, 2, \ldots, n$, decays (on the average) to zero faster than the singular values σ_i, for $i = 1, 2, \ldots, n$, that is, the sequence

$$\frac{|\boldsymbol{u}_1^T \boldsymbol{b}|}{\sigma_1}, \frac{|\boldsymbol{u}_2^T \boldsymbol{b}|}{\sigma_2}, \ldots, \frac{|\boldsymbol{u}_n^T \boldsymbol{b}|}{\sigma_n} \tag{8.97}$$

   is (for the most part) monotone decreasing. The $\boldsymbol{u}_i$'s are the left singular vectors of $\boldsymbol{A}$.
4. The regularization method imposes a weak smoothness constraint on the possible solutions over the range of λ. Comparing (8.94) to (8.82), we see that the role of the regularization parameter λ is to dampen or filter the terms in the sum corresponding to singular values that are smaller than approximately λ. In any practical application, the regularization parameter will satisfy the inequality

$$\sigma_n \leq \lambda \leq \sigma_1 \tag{8.98}$$

   We can define the *filtering* factors as

$$f_i(\lambda) = \frac{\sigma_i^2}{\sigma_i^2 + \lambda^2} \tag{8.99}$$

   for $i = 1, 2, \ldots, n$. Therefore, by using (8.99) Equation (8.94) can be written as

$$\boldsymbol{x}_\lambda = \sum_{i=1}^{n} f_i(\lambda) \frac{\boldsymbol{v}_i \boldsymbol{u}_i^T}{\sigma_i} \boldsymbol{b} \tag{8.100}$$

When the Tikhonov parameter is properly chosen, the filtering factors in (8.99) are supposed to tend to zero in such a way that the contributions $(\beta_i/\sigma_i)\boldsymbol{v}_i$ (where $\beta_i = \boldsymbol{u}_i^T\boldsymbol{b}$) to the solution $\boldsymbol{x}_\lambda$ from the smaller σ_i's are *filtered* out. This will have the effect of eliminating the error contamination in the solution when the DPC fails to hold. The filtering out of the small singular values takes effect when $\sigma_i < \lambda$. Another way to view this is to write

$$\left(\frac{\sigma_i^2}{\sigma_i^2 + \lambda_2}\right)\frac{\boldsymbol{v}_i\boldsymbol{u}_i^T}{\sigma_i}\boldsymbol{b}$$

in (8.94) as

$$\frac{1/\sigma_i}{1 + (\lambda/\sigma_i)^2}\beta_i\boldsymbol{v}_i$$

and consider the case when one of the singular values σ_i is much smaller than λ. Then as $\sigma_i \to 0$, we can write (using l'Hôpital's rule)

$$\lim_{\sigma_i \to 0}\frac{1/\sigma_i}{1 + (\lambda/\sigma_i)^2}\beta_i\boldsymbol{v}_i = \lim_{\sigma_i \to 0}\frac{\sigma_i}{2\lambda^2}\beta_i\boldsymbol{v}_i = 0 \qquad \text{for } \sigma_i \ll \lambda \tag{8.101}$$

Therefore, the associated terms in the sum in (8.94) will approach $\mathbf{0}$, providing the appropriate filtering or *natural truncation* of the terms in the SVD sum, which shows the required continuity of the solution for a real physical system.

A variation of the Tikhonov regularization method involves an energy function of the form

$$\mathscr{E}(\boldsymbol{x}) = \frac{1}{2}\|\boldsymbol{A}\boldsymbol{x} - \boldsymbol{b}\|_2^2 + \frac{\lambda^2}{2}\|\boldsymbol{L}\boldsymbol{x}\|_2^2 \tag{8.102}$$

where $\boldsymbol{L}$ is the regularization matrix. Again we seek a solution $\boldsymbol{x} = \boldsymbol{x}_\lambda$ of the problem

$$\min_{\boldsymbol{x}\in\Re^n} \mathscr{E}(\boldsymbol{x})$$

The regularization matrix can be taken as a discrete approximation to a derivative operator, and in general $\boldsymbol{L}_p \in \Re^{(n-p)\times n}$, where p is the order of the derivative, and $\rho(\boldsymbol{L}_p) = n - p$. For example, a first-order derivative matrix would be

$$\boldsymbol{L}_1 = \begin{bmatrix} -1 & 1 & 0 & 0 & & \cdots & 0 \\ 0 & -1 & 1 & 0 & 0 & \cdots & 0 \\ 0 & 0 & -1 & 1 & 0 & \cdots & 0 \\ \cdots & \cdots & \cdots & \cdots & \cdots & \cdots & \cdots \\ 0 & 0 & 0 & \cdots & 0 & -1 & 1 \end{bmatrix}_{(n-1)\times n} \tag{8.103}$$

and the second-order derivative matrix is given by

$$L_2 = \begin{bmatrix} 1 & -2 & 1 & 0 & & & \cdots & 0 \\ 0 & 1 & -2 & 1 & 0 & & \cdots & 0 \\ 0 & 0 & 1 & -2 & 1 & 0 & \cdots & 0 \\ \cdots & \cdots & \cdots & \cdots & \cdots & \cdots & \cdots & \cdots \\ 0 & 0 & 0 & \cdots & 0 & 1 & -2 & 1 \end{bmatrix}_{(n-2)\times n} \tag{8.104}$$

Minimizing (8.102) with respect to x (i.e., if the gradient of (8.102) is computed with respect to x and the result set equal to zero) leads to another set of modified normal equations given by

$$(A^T A + \lambda^2 L^T L)x = A^T b \tag{8.105}$$

We would like to solve (8.105) for different values of λ using merely the SVD of A as we did previously for $L = I_{n\times n}$; however, this cannot be carried out. There are two methods that have been devised for dealing with this problem: *conversion to a standard regularization problem* and *using the generalized singular-value decomposition* (GSVD) [25]. An overview of these two approaches can be found in the papers by Varah [38] and Hansen [46].

EXAMPLE 8.4. The purpose of this example is to compare results obtained from the Tikhonov regularization method with the truncated SVD (TSVD) method for solving $Ax = b$ when A is ill-conditioned. We will let A be a 20×20 Hilbert matrix (cf. Sect. A.2.19) where the elements are given by $a_{ij} = 1/(i + j - 1)$, for $i, j = 1, 2, \ldots, n$, (for $n = 20$); therefore,

$$A = \begin{bmatrix} 1/1 & 1/2 & \cdots & 1/n \\ 1/2 & 1/3 & \cdots & 1/(n+1) \\ \cdots & \cdots & \cdots & \cdots \\ 1/n & 1/(n+1) & \cdots & 1/(2n+1) \end{bmatrix} \tag{8.106}$$

To assess the quality of the results from the two methods and make a meaningful comparison, we generate a known (exact) solution

$$x_e(j) = \sqrt{0.5j} \qquad j = 1, 2, \ldots, 20 \tag{8.107}$$

Therefore, from A in (8.106) and x_e in (8.107) vector b can be generated as $b = Ax_e$. The condition number of A (using MATLAB) is $\text{cond}(A) = \sigma_{\max}(A)/\sigma_{\min}(A) = 1.0675 \times 10^{19}$, and the computed rank is $\rho(A) = 13$.

We will first use the TSVD method to find a solution. Recall in regularization methods we desire a solution x such that $\|x\|_2$ is made relatively small along with $\|Ax - b\|_2$. Figure 8.11(a) shows a plot of the L_2-norm of x_k, that is, $\|x_k\|_2$, versus the SVD truncation parameter k, see (8.83), for $k = 1, 2, \ldots, 20$. From the figure we see that the curve is relatively flat up to (and including) $k = 14$; after $k = 14$ the curve begins to rise very rapidly. The corner at $k = 14$ divides the curve into two pieces. The left piece of the curve in Figure 8.11(a) (typically corresponding to *signal*) is relatively flat, and the right piece (corresponding to *noise*) is very steep. The flat portion of the curve corresponds to a value for $\|x_k\|_2$ that is very close to the L_2-norm of x_e, $\|x_e\|_2$. Therefore, it might be tempting to use a value of $k = 14$ for the truncation parameter; however, the smallest relative error actually occurs at $k = 11$. The relative error is plotted versus the truncation parameter in Figure 8.11(b). Therefore, $k = 11$ is the best value to use for the truncation parameter for TSVD, and the relative error is given by

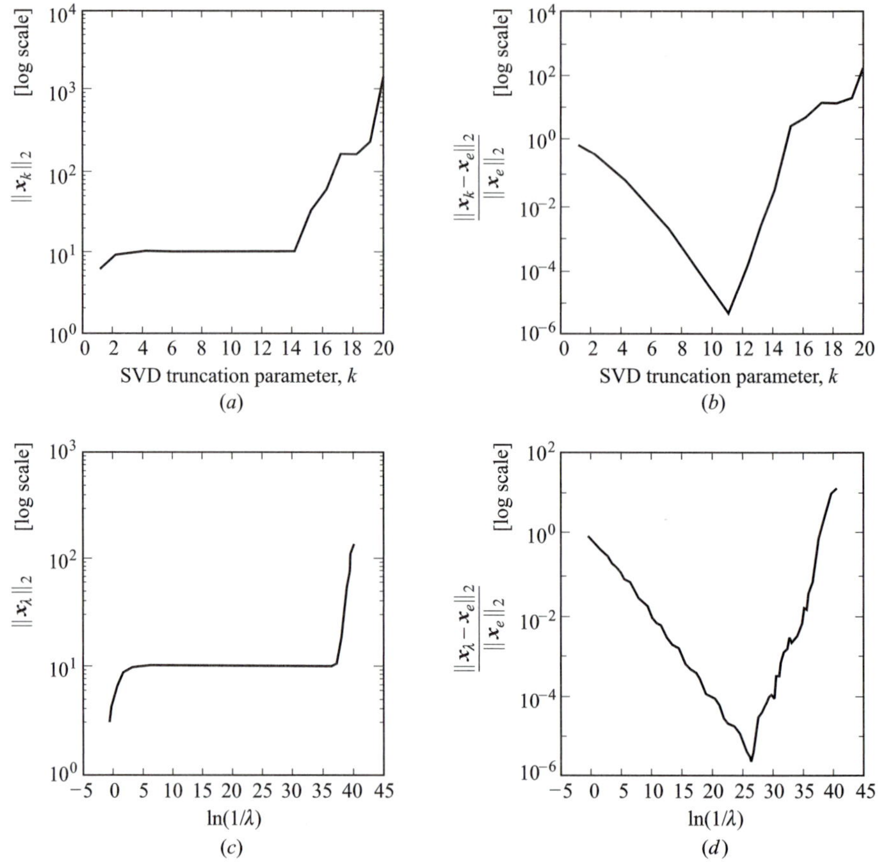

Figure 8.11 (*a*) Plot of the L_2-norm of $\boldsymbol{x}_k$, $\|\boldsymbol{x}_k\|_2$, versus the SVD truncation parameter $k\,(k = 1, 2, \ldots, 20)$ for the TSVD method. (*b*) Relative error versus the TSVD truncation parameter k. (*c*) Plot of the L_2-norm of $\boldsymbol{x}_\lambda$, $\|\boldsymbol{x}_\lambda\|_2$, versus $\ln(1/\lambda)$ (where λ is the regularization parameter) for the Tikhonov regularization method. (*d*) Relative error versus $\ln(1/\lambda)$ for the Tikhonov regularization method.

$$\left.\frac{\|\boldsymbol{x}_k - \boldsymbol{x}_e\|_2}{\|\boldsymbol{x}_e\|_2}\right|_{k=11} = 4.735258615542855 \times 10^{-6} \tag{8.108}$$

For the Tikhonov regularization method, a range of values must be generated for the regularization parameter λ. We want to fill in values between λ_1 and λ_ℓ. We let $\lambda_1 = \sigma_{\max}(\boldsymbol{A})$ and $\lambda_\ell < \max\{\sigma_{\min}(\boldsymbol{A}), \sigma_{\max}(\boldsymbol{A}) \cdot \varepsilon\}$, where ε is the machine roundoff unit. In MATLAB this is taken as $eps = 2^{-52} = 2.220446049250313 \times 10^{-16}$. Table 8.2 shows the singular values of $\boldsymbol{A}$ as computed by the MATLAB **svd** built-in function. We desire a mesh grid of $\ell = 100$ points for λ with the endpoints $\lambda_1 = \sigma_1 = 1.907134720407253$ and $\lambda_\ell = \sigma_1 \cdot eps/100 = 4.234689755316386 \times 10^{-18}$. The remaining 98 $(\ell - 2)$ points, that is, λ_2 through $\lambda_{\ell-1}$, can be generated from [50, 53]

$$\lambda_q = \lambda_1 \left(\frac{\lambda_\ell}{\lambda_1}\right)^{(q-1)/(\ell-1)} \qquad q = 1, 2, \ldots, \ell \tag{8.109}$$

TABLE 8.2
Singular values of A computed using the MATLAB (version 5.1) built-in function svd

i	σ_i
1	1.907134720407253
2	$4.870384065720490 \times 10^{-1}$
3	$7.559582130544090 \times 10^{-2}$
4	$8.961128614856439 \times 10^{-3}$
5	$8.676711091714799 \times 10^{-4}$
6	$7.033431473193232 \times 10^{-5}$
7	$4.830510048804696 \times 10^{-6}$
8	$2.827652055224344 \times 10^{-7}$
9	$1.413954758555360 \times 10^{-8}$
10	$6.036095327608074 \times 10^{-10}$
11†	$2.192890048019410 \times 10^{-11}$
12	$6.740801127335105 \times 10^{-13}$
13	$1.738400044807469 \times 10^{-14}$
14	$3.740758559585443 \times 10^{-16}$
15	$1.704893499875155 \times 10^{-17}$
16	$1.464577158596144 \times 10^{-17}$
17	$9.694815980744144 \times 10^{-18}$
18	$8.115826272543485 \times 10^{-18}$
19	$1.760134198986154 \times 10^{-18}$
20	$1.786569618132934 \times 10^{-19}$

†Optimal number of singular values to retain for TSVD method.

For these 100 values of λ, 100 "solutions" were generated using (8.94). Figure 8.11(c) shows a plot of the L_2-norm of $\boldsymbol{x}_\lambda$, that is, $\|\boldsymbol{x}_\lambda\|_2$, versus $\ln(1/\lambda)$. Notice that this curve looks very similar to the curve in Figure 8.11(a) for the TSVD method. Again the flat portion of the curve corresponds to a value for $\|\boldsymbol{x}_\lambda\|_2$ that is very close to the L_2-norm of $\boldsymbol{x}_e$, or $\|\boldsymbol{x}_e\|_2$. Figure 8.11(d) shows the plot of the relative error versus $\ln(1/\lambda)$ for the Tikhonov regularization method. The well-defined minimum shown in the figure corresponds to the *optimal* regularization parameter of $\lambda_o = 3.245947737964136 \times 10^{-12}$ and a relative error

$$\left.\frac{\|\boldsymbol{x}_\lambda - \boldsymbol{x}_e\|_2}{\|\boldsymbol{x}_e\|_2}\right|_{\lambda=\lambda_o} = 2.673028585143722 \times 10^{-6} \tag{8.110}$$

This is much better than the relative error given for the TSVD method shown in (8.108). Notice that the optimal regularization parameter lies between singular values 11 and 12 shown in Table 8.2. Therefore, the filtering of the terms in the sum in (8.94) will occur after this value because singular values 12 through 20 are smaller than λ_o.

If the MATLAB built-in function **pinv** is used to compute the pseudoinverse of $\boldsymbol{A}$ to compute the solution to $\boldsymbol{Ax} = \boldsymbol{b}$, that is, $\boldsymbol{x}^M = pinv(\boldsymbol{A})\boldsymbol{b}$, then the resulting relative error is

$$\frac{\|\boldsymbol{x}^M - \boldsymbol{x}_e\|_2}{\|\boldsymbol{x}_e\|_2} = 2.330150590018773 \times 10^{-3} \tag{8.111}$$

This result is 3 orders of magnitude worse than the relative errors for the TSVD method and the Tikhonov regularization method. The actual method used by

pinv to compute the pseudoinverse of A is the SVD. The default tolerance value used to truncate the singular values (less than this tolerance value) is computed as **max(size(A)) *norm(A) *eps**. Therefore, for this problem the default tolerance value is $20 * \sigma_1 * eps = 8.469379510632772 \times 10^{-15}$. In Table 8.2, this default tolerance values lies between singular values 13 and 14; therefore, 13 terms are used in the sum in (8.83) to compute the solution $\boldsymbol{x}^M$. The same default tolerance value is used to determine the rank of A in MATLAB using the **rank** function, which is based on the SVD of A.

8.8 MATRIX SPLITTINGS FOR ITERATIVE DISCRETE-TIME METHODS FOR SOLVING LINEAR EQUATIONS

There are four basic iterative discrete-time methods used to solve linear equations of the form

$$\boldsymbol{A}\boldsymbol{x} = \boldsymbol{b} \tag{8.112}$$

for $\boldsymbol{A} \in \Re^{n \times n}$ that are based on *matrix splitting* [25, 29, 56–58]. The solution vector $\boldsymbol{x}$ exists and is unique if and only if $\boldsymbol{A}$ is nonsingular. All the methods have the basic form

$$\boldsymbol{M}\boldsymbol{x}(k+1) = \boldsymbol{N}\boldsymbol{x}(k) + \boldsymbol{b} \tag{8.113}$$

where k is the discrete-time index. We can express matrix $\boldsymbol{A}$ as the matrix sum

$$\boldsymbol{A} = \boldsymbol{D} - \boldsymbol{E} - \boldsymbol{F} \tag{8.114}$$

where $\boldsymbol{D} \in \Re^{n \times n}$ is a diagonal matrix, that is, $\boldsymbol{d} = \text{diag}[a_{11}, a_{22}, \cdots, a_{nn}]$, and $\boldsymbol{E} \in \Re^{n \times n}$ and $\boldsymbol{F} \in \Re^{n \times n}$ are, respectively, strictly lower and upper triangular matrices. The entries of $\boldsymbol{E}$ and $\boldsymbol{F}$ are the negatives of the entries of $\boldsymbol{A}$, respectively, below and above the main diagonal of $\boldsymbol{A}$. The diagonal elements of $\boldsymbol{D}$ are all assumed to be nonzero.

Jacobi iterative method

Substituting (8.114) into (8.112) gives

$$(\boldsymbol{D} - \boldsymbol{E} - \boldsymbol{F})\boldsymbol{x} = \boldsymbol{b} \tag{8.115}$$

This is then *split* as

$$\boldsymbol{D}\boldsymbol{x} = (\boldsymbol{E} + \boldsymbol{F})\boldsymbol{x} + \boldsymbol{b} \tag{8.116}$$

and an iterative scheme can be written from (8.116) as

$$\boldsymbol{D}\boldsymbol{x}(k+1) = (\boldsymbol{E} + \boldsymbol{F})\boldsymbol{x}(k) + \boldsymbol{b} \tag{8.117}$$

Therefore, according to (8.113), $\boldsymbol{M} = \boldsymbol{D}$ and $\boldsymbol{N} = \boldsymbol{E} + \boldsymbol{F}$. Since the diagonal elements of $\boldsymbol{D}$ are nonzero, we can write (8.117) as

$$\boldsymbol{x}(k+1) = \boldsymbol{D}^{-1}(\boldsymbol{E} + \boldsymbol{F})\boldsymbol{x}(k) + \boldsymbol{D}^{-1}\boldsymbol{b} \tag{8.118}$$

where $k \geq 0$, and $\boldsymbol{x}(0)$ is the given initial-condition vector. This is the vector-matrix form of the *Jacobi iterative method* [56], and we call the matrix

$$B = D^{-1}(E + F) \tag{8.119}$$

the *Jacobi matrix*. The scalar form of (8.118) is given by

$$x_i(k+1) = -\sum_{\substack{j=1 \\ j \neq i}}^{n} \left(\frac{a_{ij}}{a_{ii}}\right) x_j(k) + \frac{b_i}{a_{ii}} \tag{8.120}$$

where $1 \leq i \leq n$, $k \geq 0$, and $x_i(0)$ is the given initial condition. We see from (8.120) that in general all the components of vector $\boldsymbol{x}(k)$ must be saved while computing the components of vector $\boldsymbol{x}(k+1)$. However, it seems plausible to use the latest estimates $x_i(k+1)$ of the components x_i of the solution vector $\boldsymbol{x}$ in all subsequent computations. This leads to the second method in this class.

Gauss-Seidel iterative method

Starting with (8.115), $(\boldsymbol{D} - \boldsymbol{E} - \boldsymbol{F})\boldsymbol{x} = \boldsymbol{b}$, we rearrange (i.e., split) this expression as

$$(\boldsymbol{D} - \boldsymbol{E})\boldsymbol{x} = \boldsymbol{F}\boldsymbol{x} + \boldsymbol{b} \tag{8.121}$$

where $\boldsymbol{D} - \boldsymbol{E}$ is a nonsingular lower triangular matrix. An iterative scheme can be written from (8.121) as

$$(\boldsymbol{D} - \boldsymbol{E})\boldsymbol{x}(k+1) = \boldsymbol{F}\boldsymbol{x}(k) + \boldsymbol{b} \tag{8.122}$$

or

$$\boldsymbol{x}(k+1) = (\boldsymbol{D} - \boldsymbol{E})^{-1}\boldsymbol{F}\boldsymbol{x}(k) + (\boldsymbol{D} - \boldsymbol{E})^{-1}\boldsymbol{b} \tag{8.123}$$

for $k \geq 0$, and $\boldsymbol{x}(0)$ is the given initial-condition vector. Comparing (8.122) with (8.113) reveals $\boldsymbol{M} = \boldsymbol{D} - \boldsymbol{E}$ and $\boldsymbol{N} = \boldsymbol{F}$. This is the vector-matrix form of the *Gauss-Seidel iterative method* [56], and the matrix $\boldsymbol{C} = (\boldsymbol{D} - \boldsymbol{E})^{-1}\boldsymbol{F}$ is called the *Gauss-Seidel matrix*. The scalar form of the Gauss-Seidel iterative method can be written from $\boldsymbol{D}\boldsymbol{x}(k+1) = \boldsymbol{E}\boldsymbol{x}(k+1) + \boldsymbol{F}\boldsymbol{x}(k) + \boldsymbol{b}$ and is given by

$$x_i(k+1) = -\frac{1}{a_{ii}}\sum_{j=1}^{i-1} a_{ij}x_j(k+1) - \frac{1}{a_{ii}}\sum_{j=i+1}^{n} a_{ij}x_j(k) + \frac{b_i}{a_{ii}} \tag{8.124}$$

where $1 \leq i \leq n$, $k \geq 0$, and $x_i(0)$ is the given initial condition.

Successive overrelaxation iterative method

In the case of the successive overrelaxation (SOR) iterative method, we write the split of the $\boldsymbol{A}$ matrix as

$$\boldsymbol{A} = \boldsymbol{M}_\omega - \boldsymbol{N}_\omega = \boldsymbol{D} - \boldsymbol{E} - \boldsymbol{F} \tag{8.125}$$

where

$$\boldsymbol{M}_\omega = \frac{1}{\omega}(\boldsymbol{D} - \omega\boldsymbol{E}) \tag{8.126}$$

and

$$N_\omega = \frac{1}{\omega}[(1-\omega)\boldsymbol{D} + \omega\boldsymbol{F}] \tag{8.127}$$

The parameter ω is called the *relaxation factor*. Therefore, substituting (8.126) and (8.127) into (8.113) gives

$$\frac{1}{\omega}(\boldsymbol{D} - \omega\boldsymbol{E})\boldsymbol{x}(k+1) = \frac{1}{\omega}[(1-\omega)\boldsymbol{D} + \omega\boldsymbol{F}]\boldsymbol{x}(k) + \boldsymbol{b} \tag{8.128}$$

Multiplying both sides of (8.128) by ω and then premultiplying both sides by $\boldsymbol{D}^{-1}$ give

$$(\boldsymbol{I} - \omega\boldsymbol{D}^{-1}\boldsymbol{E})\boldsymbol{x}(k+1) = [(1-\omega)\boldsymbol{I} + \omega\boldsymbol{D}^{-1}\boldsymbol{F}]\boldsymbol{x}(k) + \omega\boldsymbol{D}^{-1}\boldsymbol{b} \tag{8.129}$$

We now define $\boldsymbol{L} \triangleq \boldsymbol{D}^{-1}\boldsymbol{E}$ (strictly lower triangular) and $\boldsymbol{U} \triangleq \boldsymbol{D}^{-1}\boldsymbol{F}$ (strictly upper triangular) and substitute these into (8.129) to give

$$(\boldsymbol{I} - \omega\boldsymbol{L})\boldsymbol{x}(k+1) = [(1-\omega)\boldsymbol{I} + \omega\boldsymbol{U}]\boldsymbol{x}(k) + \omega\boldsymbol{D}^{-1}\boldsymbol{b}$$

and premultiplying both sides by $(\boldsymbol{I} - \omega\boldsymbol{L})^{-1}$ yields

$$\boldsymbol{x}(k+1) = (\boldsymbol{I} - \omega\boldsymbol{L})^{-1}[(1-\omega)\boldsymbol{I} + \omega\boldsymbol{U}]\boldsymbol{x}(k) + \omega(\boldsymbol{I} - \omega\boldsymbol{L})^{-1}\boldsymbol{D}^{-1}\boldsymbol{b} \tag{8.130}$$

for $k \geq 0$, and $\boldsymbol{x}(0)$ is the given initial-condition vector. Equation (8.130) is the vector-matrix form of the *SOR iterative method* [56]. The matrix

$$\mathscr{L}(\omega) = (\boldsymbol{I} - \omega\boldsymbol{L})^{-1}[(1-\omega)\boldsymbol{I} + \omega\boldsymbol{U}] \tag{8.131}$$

is called the *successive relaxation matrix*. If the relaxation factor lies in the range $0 \leq \omega \leq 1$, then this is *underrelaxation*. However, if $\omega > 1$, this is *overrelaxation*. Note that if the relaxation factor is set to $\omega = 1$, then (8.130) reverts to the vector-matrix form of the Gauss-Seidel method in (8.123). The scalar form of the SOR iterative method can be derived from (8.128) rewritten as $\boldsymbol{D}\boldsymbol{x}(k+1) = \omega\boldsymbol{E}\boldsymbol{x}(k+1) + (1-\omega)\boldsymbol{D}\boldsymbol{x}(k) + \omega\boldsymbol{F}\boldsymbol{x}(k) + \omega\boldsymbol{b}$. The scalar form of the SOR iterative method is given by

$$x_i(k+1) = \frac{\omega}{a_{ii}}\left[b_i - \sum_{j=1}^{i-1} a_{ij}x_j(k+1) - \sum_{j=i+1}^{n} a_{ij}x_j(k)\right] + (1-\omega)x_i(k) \tag{8.132}$$

where $1 \leq i \leq n$, $k \geq 0$, and $x_i(0)$ is the given initial condition. Figure 8.12 shows a neural network architecture realization of the SOR iterative method.

Using the definition of the successive relaxation matrix in (8.131), Equation (8.130) can be written as

$$\boldsymbol{x}(k+1) = \mathscr{L}(\omega)\boldsymbol{x}(k) + \boldsymbol{R}\boldsymbol{b} \tag{8.133}$$

where $\boldsymbol{R} \triangleq \omega(\boldsymbol{I} - \omega\boldsymbol{L})^{-1}\boldsymbol{D}^{-1}$. We now define an error vector as

$$\boldsymbol{e}(k) = \boldsymbol{x}(k) - \boldsymbol{x} \qquad k \geq 0 \tag{8.134}$$

where $\boldsymbol{x}$ is the unique vector solution of (8.112). For this error, from (8.133) we can write a homogeneous error difference equation as

$$\boldsymbol{e}(k+1) = \mathscr{L}(\omega)\boldsymbol{e}(k) \tag{8.135}$$

The relaxation factor ω can be chosen to minimize $\sigma_r[\mathscr{L}(\omega)]$ in order to make $\boldsymbol{x}(k)$ converge to $\boldsymbol{x}$ as rapidly as possible [58], where $\sigma_r(\bullet)$ is the *spectral radius*

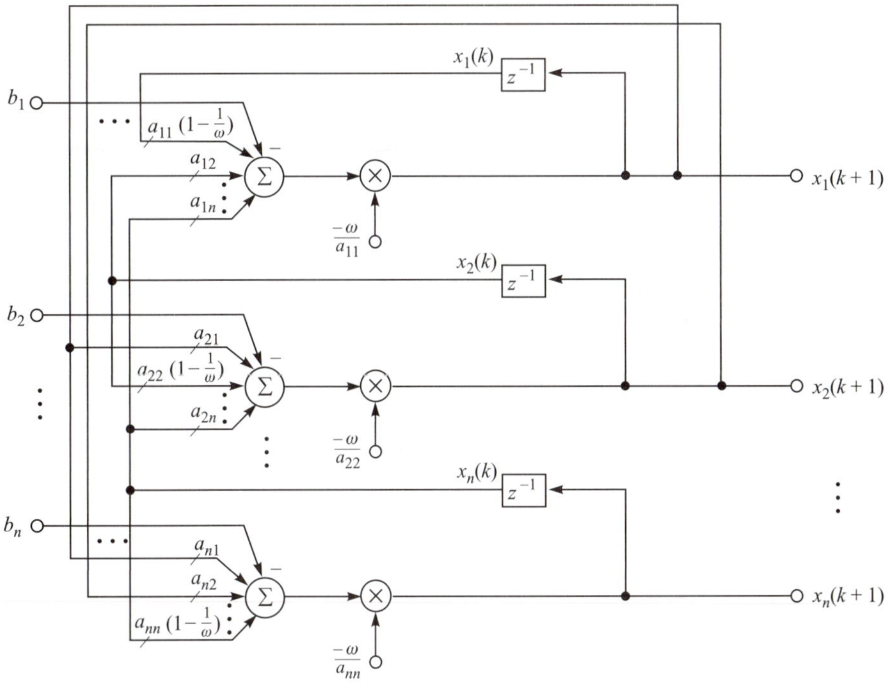

Figure 8.12 Linear neural network architecture for the SOR iterative method.

(cf. Sect. A.2.13) of $\mathscr{L}(\omega)$. We will call the optimal value of the relaxation factor ω^o. The calculation of ω^o can be difficult except in simple cases. Typically it is approximated by trying several values of ω and observing the effect on the speed of convergence. Even with the problem of calculating ω^o, the effort is worth it because of the resulting dramatic increase in the speed of convergence of $\boldsymbol{x}(k)$ to $\boldsymbol{x}$.

Richardson's iterative method

Another method that can be considered in this class of iterative techniques for solving systems of equations is *Richardson's iterative method* [57]. The basic idea is to iterate until the negative of the discrete-time approximation to the first derivative of the solution reaches zero, that is,

$$-\frac{\boldsymbol{x}(k+1)-\boldsymbol{x}(k)}{\beta} = \boldsymbol{A}\boldsymbol{x}(k)-\boldsymbol{b} \tag{8.136}$$

where $\boldsymbol{e}(k) = \boldsymbol{A}\boldsymbol{x}(k) - \boldsymbol{b}$ and $k \geq 0$. This expression can be rearranged as

$$\boldsymbol{x}(k+1) = \boldsymbol{x}(k) - \beta(k)[\boldsymbol{A}\boldsymbol{x}(k)-\boldsymbol{b}] \tag{8.137}$$

where $\beta = \beta(k)$, $\boldsymbol{x}(0)$ is the given initial-condition vector, and the *optimal iteration parameters* can be determined from [8]

$$\beta(k) = \frac{\boldsymbol{e}^T(k)\boldsymbol{e}(k)}{\boldsymbol{e}^T(k)\boldsymbol{A}\boldsymbol{e}(k)} \tag{8.138}$$

We can derive the scalar form of Richardson's iterative method from the vector-matrix form given in (8.137) as

$$x_i(k+1) = x_i(k) - \beta(k)\left[\sum_{j=1}^{n} a_{ij}x_j(k) - b_i\right] \tag{8.139}$$

where $1 \leq i \leq n$, $k \geq 0$, and $x_i(0)$ is the given initial condition. If we choose $\beta(k) = 1/a_{ii}$ in (8.139), we can write the iterative expression as

$$x_i(k+1) = x_i(k) - \frac{1}{a_{ii}}\left[\sum_{j=1}^{n} a_{ij}x_j(k) - b_i\right] \tag{8.140}$$

which in turn can be written as

$$x_i(k+1) = -\sum_{\substack{j=1 \\ j\neq i}}^{n}\left(\frac{a_{ij}}{a_{ii}}\right)x_j(k) + \frac{b_i}{a_{ii}} \tag{8.141}$$

where $1 \leq i \leq n$, $k \geq 0$, and $x_i(0)$ is the given initial condition. This is precisely the Jacobi iterative method given in (8.120).

EXAMPLE 8.5. This example compares the performances of three of the four methods presented in this section for solving linear equations of the form $\boldsymbol{Ax} = \boldsymbol{b}$. In this example the system is given by

$$\begin{bmatrix} 6 & 5 & 5 & 6 \\ 5 & 9 & 4 & 3 \\ 5 & 4 & 10 & 5 \\ 6 & 3 & 5 & 7 \end{bmatrix}\boldsymbol{x} = \begin{bmatrix} 55 \\ 47 \\ 63 \\ 55 \end{bmatrix} \tag{8.142}$$

Because the condition number for $\boldsymbol{A}$ is relatively small [cond($\boldsymbol{A}$) = 394.8742], the **inv** function in MATLAB can be used to solve the system of equations in (8.142). The result is given as

$$\boldsymbol{x}^M = inv(\boldsymbol{A})^*\boldsymbol{b} = \begin{bmatrix} 1.0000 \\ 2.0000 \\ 3.0000 \\ 4.0000 \end{bmatrix} \tag{8.143}$$

The successive overrelaxation method requires determining the *optimal* value for the relaxation parameter ω. Following the procedure given in this section on the SOR method, the minimum of $\sigma_r[\mathscr{L}(\omega)]$ as a function of ω must be found. Figure 8.13(a) shows a plot of $\sigma_r[\mathscr{L}(\omega)]$ as a function of ω, for the range of values for ω given by $0 \leq \omega \leq 2.5$ and $\Delta\omega = 1/10{,}000$. The well-defined minimum establishes the optimal value to use for the relaxation parameter given by $\omega^o = 1.7215$. Each of the three methods uses the same set of initial conditions given by $\boldsymbol{x}(0) = [0.0118, 0.0315, 0.1444, -0.0351]^T$. These initial conditions were generated by selecting four random numbers from a Gaussian distribution with zero mean and variance of 0.01. Using the optimal value for the relaxation parameter determined from Figure 8.13(a), the SOR algorithm was run, and the convergence profile is shown in Figure 8.13(b). It took 115 iterations for the algorithm to converge. The termination criterion was defined to be when the absolute error is less than 10^{-7} (that is, $\|\boldsymbol{x}^{SOR} - \boldsymbol{x}^M\|_2 < 10^{-7}$) then convergence is achieved. Figure 8.13(c) and (d) shows the convergence profiles for solving (8.142) using

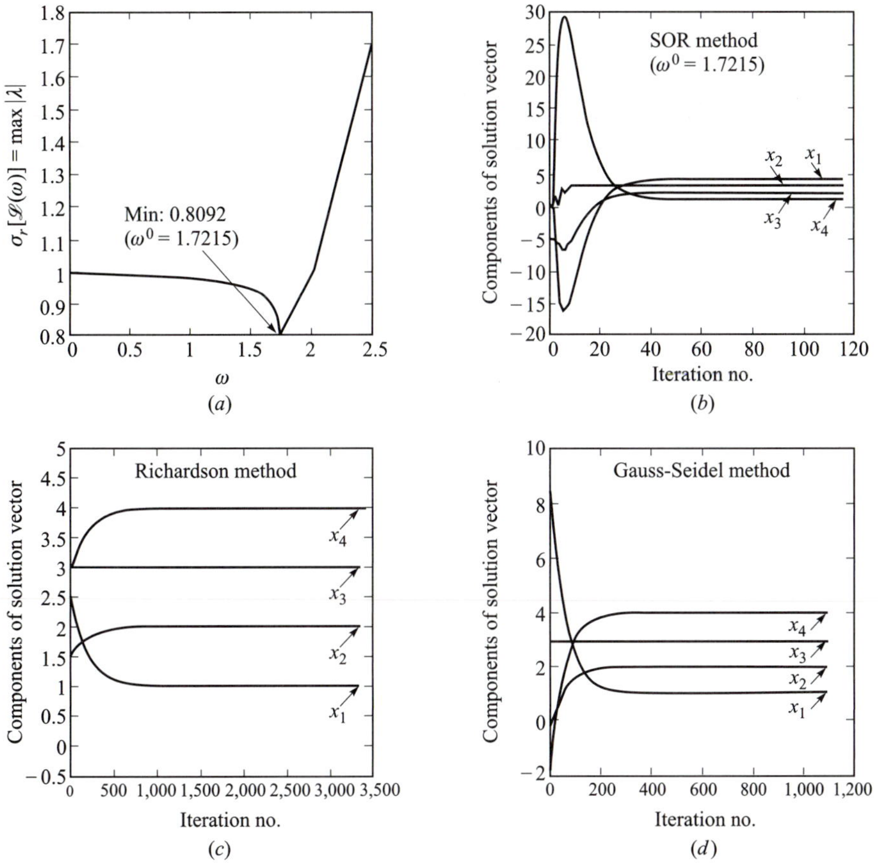

Figure 8.13 (*a*) Plot of the spectral radius of the successive relaxation matrix $\sigma_r[\mathscr{L}(\omega))]$ versus the relaxation factor ω to determine the optimal value of ω. From the plot, the minimum occurs at $\omega^o = 1.7215$. (*b*) Convergence profile for solving (8.142) by using the SOR method. (*c*) Convergence profile for solving (8.142) by using the Richardson method. (*d*) Convergence profile for solving (8.142) by using the Gauss-Seidel method.

the Richardson and Gauss-Seidel methods, respectively. The same termination criterion was used for both of these methods in order to properly compare all the results. Table 8.3 summarizes the simulation results. From Table 8.3 we see that the SOR iterative method gives the best results, that is, the fastest convergence. Comparing the SOR results with the next-best results (Gauss-Seidel, $\omega = 1$), we see that the SOR method is about 10 times faster.

8.9 TOTAL LEAST-SQUARES PROBLEM

If the errors are confined to the $\boldsymbol{b}$ vector of the system $\boldsymbol{A}\boldsymbol{x} = \boldsymbol{b}$, then the ordinary least-squares approach for solving the system is appropriate.

TABLE 8.3
Comparison of simulation (MATLAB version 5.1) results using three methods for solving (8.142)

Method	Absolute error $\|x - x^M\|_2$	Relative error $\frac{\|x - x^M\|_2}{\|x^M\|_2}$	Number of iterations required for convergence
Successive overrelaxation ($\omega^o = 1.7215$)	2.1999×10^{-8}	4.0165×10^{-9}	115
Gauss-Seidel	5.3028×10^{-8}	9.6816×10^{-9}	1,100
Richardson	2.7619×10^{-8}	5.0426×10^{-9}	3,400
Jacobi	Diverged	Diverged	Diverged

However, if errors occur in data matrix $\boldsymbol{A}$ as well as observation vector $\boldsymbol{b}$, then *total least squares* (TLS) [25, 59–62] can be utilized to yield a solution $\boldsymbol{x}$. It is typically unrealistic to assume the data matrix $\boldsymbol{A}$ is not corrupted by errors. It is more realistic to assume $\boldsymbol{A}$ does contain errors because many times the data in $\boldsymbol{A}$ consist of measurements that have been corrupted by noise. This measurement noise can be generated by the actual instrument itself that is responsible for collecting the data (i.e., the measurements). Modeling errors and quantization errors can also corrupt the data matrix. The main principle of the TLS approach can be introduced by first reformulating the standard least-squares problem [25].

A different view of the standard least-squares problem

Given an overdetermined system with m linear equations in the form

$$\boldsymbol{A}\boldsymbol{x} \approx \boldsymbol{b} \tag{8.144}$$

with n unknowns, that is, $\boldsymbol{A} \in \Re^{m\times n}$, $\boldsymbol{b} \in \Re^{m\times 1}$, $\boldsymbol{x} \in \Re^{n\times 1}$, and $m > n$, the least-squares problem seeks to

$$\underset{\tilde{\boldsymbol{b}} \in \Re^{m\times 1}}{\text{Minimize}} \; \|\boldsymbol{b} - \tilde{\boldsymbol{b}}\|_2 \tag{8.145}$$

$$\text{subject to} \quad \tilde{\boldsymbol{b}} \in \mathscr{R}(\boldsymbol{A}) \tag{8.146}$$

where $\mathscr{R}(\boldsymbol{A})$ is the *range*† of matrix $\boldsymbol{A}$. Vector $\tilde{\boldsymbol{b}}$ is given by

$$\tilde{\boldsymbol{b}} = \boldsymbol{b} - \tilde{\boldsymbol{e}} \quad (\tilde{\boldsymbol{e}} \perp \boldsymbol{A}\tilde{\boldsymbol{x}}) \tag{8.147}$$

where

$$\tilde{\boldsymbol{e}} = \boldsymbol{b} - \boldsymbol{A}\tilde{\boldsymbol{x}} \tag{8.148}$$

and

$$\tilde{\boldsymbol{x}} \text{ solves least-squares problem } \Leftrightarrow \boldsymbol{A}^T\boldsymbol{b} - \boldsymbol{A}^T\boldsymbol{A}\tilde{\boldsymbol{x}} = \boldsymbol{0} \tag{8.149}$$

†The *range* of a matrix $\boldsymbol{A}$ is defined as $\mathscr{R}(\boldsymbol{A}) = \{\boldsymbol{y} \in \Re^{m\times 1} : \boldsymbol{y} = \boldsymbol{A}\boldsymbol{x} \text{ for some } \boldsymbol{x} \in \Re^{n\times 1}\}$.

where $\tilde{\boldsymbol{b}}$ is the orthogonal projection of $\boldsymbol{b}$ onto $\mathscr{R}(\boldsymbol{A})$. Therefore, from (8.147) we see that

$$\tilde{\boldsymbol{e}} = \boldsymbol{b} - \tilde{\boldsymbol{b}} \tag{8.150}$$

Also, from (8.149), assuming $\rho(\boldsymbol{A}) = n$, we see that $\tilde{\boldsymbol{x}} = (\boldsymbol{A}^T\boldsymbol{A})^{-1}\boldsymbol{A}^T\boldsymbol{b}$ (a unique solution to $\underset{\boldsymbol{x}\in\Re^{n\times 1}}{\text{minimize}}\ \|\boldsymbol{Ax} - \boldsymbol{b}\|_2$), and premultiplying this by $\boldsymbol{A}$ gives

$$\boldsymbol{A}\tilde{\boldsymbol{x}} = \boldsymbol{A}(\boldsymbol{A}^T\boldsymbol{A})^{-1}\boldsymbol{A}^T\boldsymbol{b} \tag{8.151}$$

From (8.148) and (8.150) $\boldsymbol{A}\tilde{\boldsymbol{x}} = \tilde{\boldsymbol{b}}$, and (8.151) can be written as $\tilde{\boldsymbol{b}} = \boldsymbol{A}(\boldsymbol{A}^T\boldsymbol{A})^{-1}\boldsymbol{A}^T\boldsymbol{b}$; therefore,

$$\tilde{\boldsymbol{b}} = \boldsymbol{P}_A\boldsymbol{b} \tag{8.152}$$

and

$$\boldsymbol{P}_A = \boldsymbol{A}(\boldsymbol{A}^T\boldsymbol{A})^{-1}\boldsymbol{A}^T \tag{8.153}$$

is the *orthogonal projector* onto $\mathscr{R}(\boldsymbol{A})$. Therefore, $\tilde{\boldsymbol{b}}$ is the orthogonal projection of $\boldsymbol{b}$ onto $\mathscr{R}(\boldsymbol{A})$. If $\rho(\boldsymbol{A}) < n$, the least-squares problem $\underset{\boldsymbol{x}\in\Re^{n\times 1}}{\text{minimize}}\ \|\boldsymbol{Ax} - \boldsymbol{b}\|_2$ has an infinite number of solutions. However, for stability and minimal sensitivity, a unique solution having a minimal L_2-norm is selected from the set of all minimizers [59]

$$\mathscr{X} = \{\boldsymbol{x} \in \Re^{n\times 1} : \|\boldsymbol{Ax} - \boldsymbol{b}\|_2 = \min\} \tag{8.154}$$

This solution is denoted by $\tilde{\boldsymbol{x}}$ (in the full-rank case there is only one least-squares solution, and it must have a minimal L_2-norm), and as pointed out in Section A.2.14, $\tilde{\boldsymbol{x}}$ can be found via the SVD.

From (8.145) and (8.146), once a minimizing $\tilde{\boldsymbol{b}}$ is found, then any $\boldsymbol{x}$ that satisfies

$$\boldsymbol{Ax} = \tilde{\boldsymbol{b}} \tag{8.155}$$

is called a least-squares solution and

$$\Delta\tilde{\boldsymbol{b}} = \boldsymbol{b} - \tilde{\boldsymbol{b}} \tag{8.156}$$

is the corresponding *least-squares correction*. Equations (8.145) and (8.146) are satisfied if $\tilde{\boldsymbol{b}}$ is the orthogonal projection of $\boldsymbol{b}$ onto $\mathscr{R}(\boldsymbol{A})$. Therefore, the least-squares problem involves perturbing the observation vector $\boldsymbol{b}$ by the minimal amount $\Delta\tilde{\boldsymbol{b}}$ so that

$$\tilde{\boldsymbol{b}} = \boldsymbol{b} - \Delta\tilde{\boldsymbol{b}} \tag{8.157}$$

can be *predicted* by the columns of data matrix $\boldsymbol{A}$. As previously stated, the basic assumption in the standard least-squares problem is that errors only occur in the observation vector $\boldsymbol{b}$, and $\boldsymbol{A}$ is assumed to be known perfectly. However, this is typically not a very realistic assumption. Therefore, we want to now account for errors in both $\boldsymbol{b}$ and $\boldsymbol{A}$ and consider the total least-squares problem.

Basic total least-squares problem

Given an overdetermined system as in (8.144), the total least-squares problem seeks to

$$\underset{[\hat{A}|\hat{b}] \in \Re^{m\times(n+1)}}{\text{Minimize}} \quad \|[A|b] - [\hat{A}|\hat{b}]\|_F \tag{8.158}$$

$$\text{subject to} \qquad \hat{b} \in \mathscr{R}(\hat{A}) \tag{8.159}$$

where $[A|b]_{m\times(n+1)}$ is an augmented matrix; that is, the b column vector is "tacked" onto the "end" of the A matrix, and $\|\bullet\|_F$ is the Frobenius norm (cf. Sect. A.2.13). Once a minimizing $[\hat{A}|\hat{b}]$ is found, then any vector x that satisfies

$$\hat{A}x = \hat{b} \tag{8.160}$$

is referred to as a TLS "solution" and $[\Delta\hat{A}|\Delta\hat{b}] = [A|b] - [\hat{A}|\hat{b}]$ is the associated TLS correction. We will denote the TLS solution by $\hat{x}$. Another way to view this problem is find the vector $\hat{x}$ that seeks to

$$\text{Minimize} \qquad [\|\Delta\hat{A}\|_F^2 + \|\Delta\hat{b}\|_F^2] \tag{8.161}$$

$$\text{subject to} \qquad (A - \Delta\hat{A})\hat{x} = (b - \Delta\hat{b}) \tag{8.162}$$

The singular-value decomposition (cf. Sect. A.2.14) is the standard method used to solve the TLS problem. Equation (8.144) can be written in the form

$$[A|b][x^T|-1]^T = \mathbf{0} \tag{8.163}$$

The best rank-n TLS approximation $[\hat{A}|\hat{b}]$ of $[A|b]$ which minimizes the deviations in variance is given by

$$[\hat{A}|\hat{b}] = U\hat{S}V^T \tag{8.164}$$

with $\hat{S} = \text{diag}[\sigma_1, \sigma_2, \ldots, \sigma_n, 0]$. The minimal TLS correction is given by

$$\sigma_{n+1} = \min_{\rho([\hat{A}|\hat{b}])=n} \|[A|b] - [\hat{A}|\hat{b}]\|_F \tag{8.165}$$

and the corresponding TLS correction matrix

$$[\Delta\hat{A}|\Delta\hat{b}] = [A|b] - [\hat{A}|\hat{b}] = \sigma_{n+1}u_{n+1}v_{n+1}^T \tag{8.166}$$

solves the TLS problem in (8.158) and (8.159), and

$$\hat{x} = -\frac{1}{v_{n+1,n+1}}[v_{1,n+1}, v_{2,n+1}, \ldots, v_{n,n+1}]^T \tag{8.167}$$

is the unique solution to (8.162).

8.10 AN L_∞-NORM (MINIMAX) NEURAL NETWORK FOR SOLVING LINEAR EQUATIONS

We want to develop a neural network architecture in this section to solve $\boldsymbol{Ax} = \boldsymbol{b}(\boldsymbol{A} \in \Re^{m\times n}, \boldsymbol{x} \in \Re^{n\times 1}, \boldsymbol{b} \in \Re^{m\times 1})$ based on the L_∞-norm (or Chebyshev norm) of the errors

$$e_i(\boldsymbol{x}) = \sum_{j=1}^{n} a_{ij}x_j - b_i \qquad i = 1, 2, \ldots, m \tag{8.168}$$

This is also referred to as the *minimax problem* and is formulated as follows:

Find the solution vector $\boldsymbol{x}$ that minimizes the energy function given by

$$\mathscr{E}(\boldsymbol{x}) = \max_{1<i<m} \{|e_i(\boldsymbol{x})|\} \tag{8.169}$$

where $e_i(\boldsymbol{x})$ is given in (8.168). This can be written more compactly as

$$\min_{\boldsymbol{x}\in\Re^n} \quad \max_{1<i<m} \{|e_i(\boldsymbol{x})|\} \tag{8.170}$$

This minimax optimization problem can be recast as a linear programming problem with inequality constraints

$$\begin{aligned} &\text{Minimize} && x_0 \\ &\text{subject to} && |e_i(\boldsymbol{x})| \le x_0 \qquad \text{for } i = 1, 2, \ldots, m \\ &\text{and } x_0 \ge 0 \end{aligned} \tag{8.171}$$

The optimal value of x_0 (x_0^o) must satisfy (8.170)

$$x_0^o = \mathscr{E}(\boldsymbol{x}^o) = \min_{\boldsymbol{x}\in\Re^n} \max_{1<i<m} \{|e_i(\boldsymbol{x})|\} \tag{8.172}$$

The linear programming problem in (8.171) can be rewritten in a form that is more amenable for a solution, that is,

$$\begin{aligned} &\text{Minimize} && x_0 \\ &\text{subject to} && \begin{cases} f_{i1}(\hat{\boldsymbol{x}}) = x_0 + e_i(\boldsymbol{x}) \ge 0 \\ f_{i2}(\hat{\boldsymbol{x}}) = x_0 - e_i(\boldsymbol{x}) \ge 0 \end{cases} \\ &\text{and} && x_0 \ge 0 \end{aligned} \tag{8.173}$$

where $\hat{\boldsymbol{x}} = [x_0, x_1, x_2, \cdots, x_n]^T$. One approach for solving the linear programming problem in (8.173) is to first formulate an energy function based on the quadratic penalty function approach [30], that is,

$$\mathscr{E}(\hat{\boldsymbol{x}}) = \alpha_1 x_0 + \frac{\alpha_2}{2}\sum_{i=1}^{m}\left(\left\{[f_{i1}(\hat{\boldsymbol{x}})]_{\min}\right\}^2 + \left\{[f_{i2}(\hat{\boldsymbol{x}})]_{\min}\right\}^2\right) \tag{8.174}$$

where $[\xi]_{\min} = \min[0, \xi]$ and $\alpha_1, \alpha_2 > 0$ [8]. Using this energy function, a steepest descent gradient system can be formed consisting of two differential equations that constitute the neural network continuous-time learning rule. The two differential equations have the basic form

$$\frac{\mathrm{d}x_0}{\mathrm{d}t} = -\mu_0 \nabla_{x_0} \mathscr{E}(\hat{\boldsymbol{x}}) \tag{8.175}$$

$$\frac{\mathrm{d}x_j}{\mathrm{d}t} = -\mu_j \nabla_{x_j} \mathscr{E}(\hat{\boldsymbol{x}}) \tag{8.176}$$

where $j = 1, 2, \ldots, n$. The gradient in (8.175) can be computed as

$$\begin{aligned} \nabla_{x_0} \mathscr{E}(\hat{\boldsymbol{x}}) &= \frac{\partial \mathscr{E}(\hat{\boldsymbol{x}})}{\partial x_0} \\ &= \alpha_1 + \frac{\alpha_2}{2} \sum_{i=1}^{m} \left(\frac{\partial}{\partial x_0} \{[f_{i1}(\hat{\boldsymbol{x}})]_{\min}\}^2 + \frac{\partial}{\partial x_0} \{[f_{i2}(\hat{\boldsymbol{x}})]_{\min}\}^2 \right) \\ &= \alpha_1 + \alpha_2 \sum_{i=1}^{m} \{[x_0 + e_i(\boldsymbol{x})] T^1_{\text{binary},i} + [x_0 - e_i(\boldsymbol{x})] T^2_{\text{binary},i}\} \end{aligned} \tag{8.177}$$

where

$$T^1_{\text{binary},i} \triangleq T^1_{\text{binary},i}[f_{i1}(\hat{\boldsymbol{x}})] = \begin{cases} 0 & \text{if } x_0 + e_i(\boldsymbol{x}) \geq 0 \\ 1 & \text{otherwise} \end{cases} \tag{8.178}$$

and

$$T^2_{\text{binary},i} \triangleq T^2_{\text{binary},i}[f_{i2}(\hat{\boldsymbol{x}})] = \begin{cases} 0 & \text{if } x_0 - e_i(\boldsymbol{x}) \geq 0 \\ 1 & \text{otherwise} \end{cases} \tag{8.179}$$

The gradient in (8.176) can be computed as

$$\begin{aligned} \nabla_{x_j} \mathscr{E}(\hat{\boldsymbol{x}}) &= \frac{\partial \mathscr{E}(\hat{\boldsymbol{x}})}{\partial x_j} \\ &= \frac{\alpha_2}{2} \sum_{i=1}^{m} \left(\frac{\partial}{\partial x_j} \{[f_{i1}(\hat{\boldsymbol{x}})]_{\min}\}^2 + \frac{\partial}{\partial x_j} \{[f_{i2}(\hat{\boldsymbol{x}})]_{\min}\}^2 \right) \\ &= \alpha_2 \sum_{i=1}^{m} (a_{ij} \{[x_0 + e_i(\boldsymbol{x})] T^1_{\text{binary},i} + [x_0 - e_i(\boldsymbol{x})] T^2_{\text{binary},i}\} \end{aligned} \tag{8.180}$$

where $T^1_{\text{binary},i}$ and $T^2_{\text{binary},i}$ are defined in (8.178) and (8.179), respectively. By setting the two gradients equal to zero, (8.177) and (8.180) can be written as

$$\nabla_{x_0} \mathscr{E}(\hat{\boldsymbol{x}}) = \frac{\alpha_1}{\alpha_2} + \sum_{i=1}^{m} \{[x_0 + e_i(\boldsymbol{x})] T^1_{\text{binary},i} + [x_0 - e_i(\boldsymbol{x})] T^2_{\text{binary},i}\} \tag{8.181}$$

$$\nabla_{x_j} \mathscr{E}(\hat{\boldsymbol{x}}) = \sum_{i=1}^{m} a_{ij} \{[x_0 + e_i(\boldsymbol{x})] T^1_{\text{binary},i} + [x_0 - e_i(\boldsymbol{x})] T^2_{\text{binary},i}\} \tag{8.182}$$

Using (8.181) and (8.182), Equations (8.175) and (8.176) can be written as

$$\frac{\mathrm{d}x_0}{\mathrm{d}t} = -\mu_0 \left(\frac{\alpha_1}{\alpha_2} + \sum_{i=1}^{m} \{[x_0 + e_i(\boldsymbol{x})] T^1_{\text{binary},i} + [x_0 - e_i(\boldsymbol{x})] T^2_{\text{binary},i}\} \right) \tag{8.183}$$

$$\frac{\mathrm{d}x_j}{\mathrm{d}t} = -\mu_j \sum_{i=1}^{m} (a_{ij} \{[x_0 + e_i(\boldsymbol{x})] T^1_{\text{binary},i} + [x_0 - e_i(\boldsymbol{x})] T^2_{\text{binary},i}\}) \tag{8.184}$$

where $j = 1, 2, \ldots, n$, $\mu_0 > 0$, and $\mu_j > 0$. Figure 8.14 shows a neural network architecture for solving systems of linear equations using the minimax quadratic penalty function approach.

8.11 AN L_1-NORM (LEAST-ABSOLUTE-DEVIATIONS) NEURAL NETWORK FOR SOLVING LINEAR EQUATIONS

The L_1-norm is sometimes referred to as the *least-absolute-deviations* norm. This norm can provide a very useful alternative to the L_∞-norm (Chebyshev

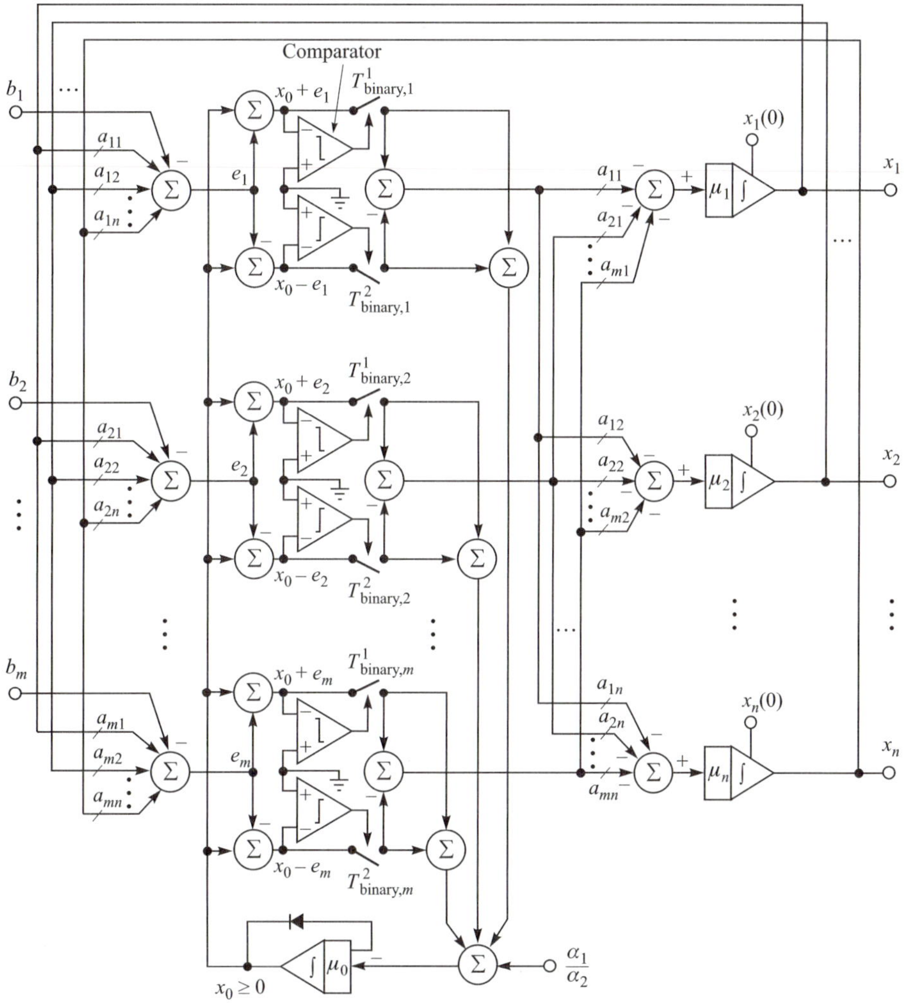

Figure 8.14 Neural network architecture using the minimax quadratic penalty function approach for solving systems of linear equations. (*Source*: Adapted with permission from Cichocki and Unbehauen [8], p. 265.)

norm) and the L_2-norm (least-squares norm) for solving systems of linear equations (and nonlinear equations). The L_1-norm and the L_∞-norm have found many applications in the field of signal processing [10, 12, 17, 21, 23, 24, 63]. The L_1-norm solution to systems of linear equations possesses properties that are not common to the L_2-norm least-squares solution to the same systems. For example, L_1-norm solutions tend to be robust, meaning these solutions are typically insensitive to relatively large errors in the data; L_1-norm solutions of overdetermined systems of linear equations always exist (however, they are not necessarily unique) compared to least-squares (L_2-norm) solutions of the same systems that are unique for full-rank systems; and L_1-norm problems are equivalent to linear programming problems, and conversely, linear programming problems can be formulated as L_1-norm problems.

The basic form of the L_1-norm (least-absolute-deviations) problem can be stated as follows. Given the system of linear algebraic equations

$$\boldsymbol{Ax} = \boldsymbol{b} \tag{8.185}$$

($\boldsymbol{A} \in \Re^{m \times n}$, $\boldsymbol{x} \in \Re^{n \times 1}$, $\boldsymbol{b} \in \Re^{m \times 1}$) along with the following constraints

$$\begin{aligned} \delta_i &= \boldsymbol{c}_i^T \boldsymbol{x} \qquad \text{for } i = 1, 2, \ldots, p \\ \delta_i &\leq \boldsymbol{c}_i^T \boldsymbol{x} \qquad \text{for } i = p+1, p+2, \ldots, q \end{aligned} \tag{8.186}$$

we want to determine the solution vector $\boldsymbol{x}$ that minimizes the energy function given by

$$\mathscr{E}(\boldsymbol{x}) = \|\boldsymbol{e}(\boldsymbol{x})\|_1 = \sum_{i=1}^{m} |e_i(\boldsymbol{x})| \tag{8.187}$$

where the *error vector* $\boldsymbol{e}(\boldsymbol{x})$ is given by $\boldsymbol{e}(\boldsymbol{x}) = \boldsymbol{Ax} - \boldsymbol{b}$, or in scalar form

$$e_i(\boldsymbol{x}) = \sum_{j=1}^{n} a_{ij} x_j - b_i \tag{8.188}$$

Therefore, using (8.188), the energy function given in (8.187) can be written as

$$\mathscr{E}(\boldsymbol{x}) = \sum_{i=1}^{m} \left| \sum_{j=1}^{n} a_{ij} x_j - b_i \right| \tag{8.189}$$

subject to the constraints given in (8.186). The constraint values are contained in the vector $\boldsymbol{\delta} = [\delta_1, \delta_2, \cdots, \delta_q]^T$, and $\boldsymbol{c}_i = [c_{i1}, c_{i2}, \cdots, c_{in}]^T$ are the constraint vectors. As stated in the introduction to this section, L_1-norm problems are equivalent to linear programming problems.

We now want to develop a neural network architecture for one realization of the L_1-norm solution to systems of linear equations [8]. We first consider the following unconstrained problem. Determine the solution vector $\boldsymbol{x}$ that minimizes the energy function

$$\mathscr{E}(\boldsymbol{x}) = \|\boldsymbol{e}(\boldsymbol{x})\|_i = \sum_{i=1}^{m} |e_i(\boldsymbol{x})| \tag{8.190}$$

where

$$e_i(\boldsymbol{x}) = \sum_{j=1}^{n} a_{ij}x_j - b_i \tag{8.191}$$

We can now modify this unconstrained problem to include auxiliary inequality and equality constraints as a linear programming problem [8], that is,

$$\text{Minimize} \quad \sum_{i=1}^{m} \sigma_i \tag{8.192}$$

$$\text{subject to} \quad -\sigma_i \leq e_i(\boldsymbol{x}) \leq \sigma_i \qquad i = 1, 2, \ldots, m \tag{8.193}$$

One method that can be used to solve this problem is the Lagrange multiplier technique (cf. Sect. A.6.2). The approach taken is actually an augmentation of the standard Lagrange function to include regularization terms. Therefore, the augmented Lagrange (energy) function is given by

$$\begin{aligned}\mathscr{L}(\boldsymbol{x}, \boldsymbol{\lambda}, \boldsymbol{\sigma}) = \mathscr{E}(\boldsymbol{x}) = \sum_{i=1}^{m}\Big[\sigma_i + \frac{\alpha_i}{2}\Big(\big\{[e_i(\boldsymbol{x}) + \sigma_i]_{\min}\big\}^2 + \big\{[e_i(\boldsymbol{x}) - \sigma_i]_{\min}\big\}^2\Big)\\ + \lambda_i\Big\{[e_i(\boldsymbol{x}) + \sigma_i]_{\min} + [e_i(\boldsymbol{x}) - \sigma_i]_{\min} - \frac{\sigma}{2}\lambda_i^2\Big\}\Big]\end{aligned} \tag{8.194}$$

where

$[\xi]_{\min}$ = $\min[0, \xi]$
$[\xi]_{\max}$ = $\max[0, \xi]$
$\alpha_i > 0$ = penalty parameters
λ_i = Lagrange multipliers
$\sigma \geq 0$ = regularization (and stabilization) parameter

Using the energy function in (8.194), a steepest descent gradient system can be formed consisting of three differential equations that constitute the neural network continuous-time learning rule. The three differential equations are

$$\frac{dx}{dt} = -\mu_j \sum_{i=1}^{m} a_{ij}(\lambda_i + \alpha_i \mathfrak{z}_i) \tag{8.195}$$

$$\frac{d\sigma_i}{dt} = -\mu_0(1 - |\lambda_i + \alpha_i \mathfrak{z}_i|) \tag{8.196}$$

$$\frac{d\lambda_i}{dt} = \beta_i(\mathfrak{z}_i - \sigma\lambda_i) \tag{8.197}$$

for $j = 1, 2, \ldots, n$ and $i = 1, 2, \ldots, m$, and

$$\mu_j > 0 \qquad \beta_i > 0 \tag{8.198}$$

$$\mathfrak{z}_i = \begin{cases} e_i(\boldsymbol{x}) + \sigma_i & \text{for } e_i(\boldsymbol{x}) < -\sigma_i \\ 0 & \text{for } -\sigma_i \leq e_i(\boldsymbol{x}) \leq \sigma_i \\ e_i(\boldsymbol{x}) - \sigma_i & \text{for } e_i(\boldsymbol{x}) > -\sigma_i \end{cases} \tag{8.199}$$

A neural network architecture is shown in Figure 8.15 that realizes the continuous-time (analog) process to solve systems of linear equations based on the L_1-norm (least-absolute-deviations norm).

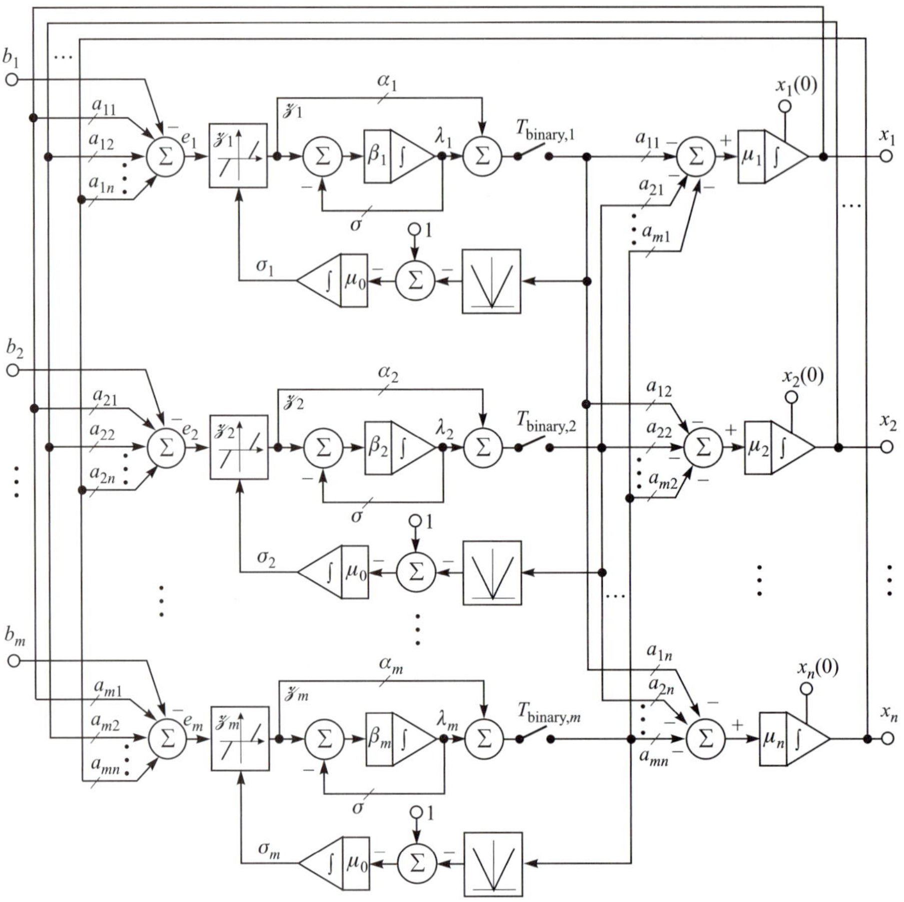

Figure 8.15 Artificial neural network architecture for solving systems of linear equations using the L_1-norm (least-absolute-deviations norm). (*Source*: Adapted with permission from Cichocki and Unbehauen [8], p. 276.)

PROBLEMS

8.1. Given the following overdetermined systems of linear equations ($\boldsymbol{Ax} = \boldsymbol{b}$)

$$(a)\ \begin{bmatrix} 0 & 1 & 0 \\ 2 & 3 & 0 \\ 2 & 0 & 0 \\ 0 & 2 & 1 \\ 1 & -1 & 1 \\ 1 & 5 & 6 \\ 5 & 4 & 4 \end{bmatrix} \begin{bmatrix} x_1 \\ x_2 \\ x_3 \end{bmatrix} = \begin{bmatrix} -2 \\ -8 \\ 1 \\ 1 \\ -6 \\ 6 \\ 6 \end{bmatrix} \qquad (b)\ \begin{bmatrix} 1 & 0 \\ 0 & 1 \\ 1 & 1 \\ -1 & 1 \\ 2 & -1 \end{bmatrix} \begin{bmatrix} x_1 \\ x_2 \end{bmatrix} = \begin{bmatrix} 1 \\ 0 \\ 0 \\ 1 \\ -1 \end{bmatrix}$$

solve each for $\boldsymbol{x}$, using the standard least-squares neurocomputing approach presented in Section 8.4. The termination criterion should be when $\|\boldsymbol{x}^N - \boldsymbol{x}^M\|_2 < 10^{-8}$, where $\boldsymbol{x}^N$ is the neural network solution and $\boldsymbol{x}^M$ is the

MATLAB solution (using the **pinv** built-in function). Initialize your network, using an appropriate set of random numbers. Optimize the learning rate parameter and number of training steps in each case, so the solution is achieved in a minimal number of iterations.

8.2. Given the following underdetermined systems of linear equations

$$(a)\quad \begin{bmatrix} 22 & 5 & 4 & 0 & 4 & 0 & 4 \\ 1 & 17 & 10 & 3 & 9 & 7 & 3 \\ 10 & 6 & 0 & 11 & 7 & 6 & 15 \\ 6 & 6 & 0 & 19 & 6 & 3 & 2 \end{bmatrix} \begin{bmatrix} x_1 \\ x_2 \\ x_3 \\ x_4 \\ x_5 \\ x_6 \\ x_7 \end{bmatrix} = \begin{bmatrix} 1 \\ 3 \\ 14 \\ 4 \end{bmatrix}$$

$$(b)\quad \begin{bmatrix} 6 & 2 & 4 & -9 & -12 & 2 & -12 & 0 & 1 \\ 8 & -10 & 1 & 8 & -22 & 0 & -11 & -11 & 7 \\ 9 & -7 & -6 & 6 & 10 & -10 & 15 & -13 & -12 \\ -10 & 11 & -6 & -8 & -5 & -9 & 1 & -3 & -5 \\ 2 & -1 & 4 & -3 & 3 & -4 & -12 & 10 & -3 \end{bmatrix} \begin{bmatrix} x_1 \\ x_2 \\ x_3 \\ x_4 \\ x_5 \\ x_6 \\ x_7 \\ x_8 \\ x_9 \end{bmatrix} = \begin{bmatrix} -12 \\ -13 \\ 9 \\ 0 \\ -6 \end{bmatrix}$$

solve each for $\boldsymbol{x}$, using the standard least-squares neurocomputing approach presented in Section 8.4. Use the same termination criterion as in Problem 8.1.

8.3. From the properties of the trace of a matrix in Section A.2.9, we can write $\text{trace}(\boldsymbol{AB}) = \text{trace}(\boldsymbol{BA})$, for $\boldsymbol{A} \in \Re^{n \times m}$ and $\boldsymbol{B} \in \Re^{m \times n}$. Using this property and other properties of the trace of a matrix, show that the following error cost functions are the same. The first is given by

$$\mathscr{E}_1(\boldsymbol{C}) = \frac{1}{2}\text{trace}(\boldsymbol{E}_1 \boldsymbol{E}_1^T)$$

where the error matrix $\boldsymbol{E}_1$ is associated with computing the pseudoinverse of a matrix presented in Equation (7.13), that is,

$$\boldsymbol{E}_1 = \boldsymbol{ACA} - \boldsymbol{A}$$

where $\boldsymbol{A} \in \Re^{m \times n}$, and $\boldsymbol{C} \in \Re^{n \times m}$ is the pseudoinverse of $\boldsymbol{A}$. The second error cost function is

$$\mathscr{E}_2(\boldsymbol{C}) = \frac{1}{2}\text{trace}(\boldsymbol{E}_2^T \boldsymbol{E}_2)$$

where the error matrix $\boldsymbol{E}_2$ is shown in (8.40), that is,

$$\boldsymbol{E}_2 = \boldsymbol{A}^T\boldsymbol{AC} - \boldsymbol{A}^T$$

This error matrix is also associated with computing the pseudoinverse inverse of a matrix; however, it was developed from the learning rule for solving a system of linear equations. Show that $\mathscr{E}_1(\boldsymbol{C}) = \mathscr{E}_2(\boldsymbol{C})$.

8.4. Write a computer program to implement the conjugate gradient neurocomputing approach for solving systems of equations presented in Section 8.5, that is, the Fletcher-Reeves conjugate gradient algorithm with the restart option. Using your program, find the solution $\boldsymbol{x}$ to the matrix equation

$$\begin{bmatrix} 10 & 7 & 8 & 7 \\ 7 & 5 & 6 & 5 \\ 8 & 6 & 11 & 8 \\ 7 & 5 & 8 & 11 \end{bmatrix} \begin{bmatrix} x_1 \\ x_2 \\ x_3 \\ x_4 \end{bmatrix} = \begin{bmatrix} 32 \\ 23 \\ 33 \\ 31 \end{bmatrix}$$

Initialize your network using an appropriate set of random numbers. What is the minimum number of training steps necessary to converge to the solution? Decide on a reasonable termination criterion for your program. Try using the computer program that you wrote for Problem 8.1 to implement the least-squares neurocomputing approach, and find the solution to the above matrix equation. Explain your results.

8.5. Consider the matrix equation

$$\begin{bmatrix} 10 & 7 & 8 & 7 \\ 7 & 5 & 6 & 5 \\ 8 & 6 & 11 & 8 \\ 7 & 5 & 8 & 11 \end{bmatrix} \begin{bmatrix} x_1 \\ x_2 \\ x_3 \\ x_4 \end{bmatrix} = \begin{bmatrix} 32.1 \\ 22.9 \\ 32.9 \\ 31.1 \end{bmatrix}$$

This is the same system considered in Problem 8.4 except the elements in the $\boldsymbol{b}$ vector have been perturbed by an amount ± 0.1. First check the condition number of the $\boldsymbol{A}$ matrix. Based on this condition number, would you expect to see a dramatic change in the elements of the solution vector $\boldsymbol{x}$ compared to what you found for the solution in Problem 8.4? Using your conjugate gradient program, determine the solution to the matrix equation. Was there a drastic change in the solution compared to the solution you found in Problem 8.4?

8.6. A complex linear matrix equation can be written as

$$(\boldsymbol{A} + j\boldsymbol{C})\boldsymbol{x} = \boldsymbol{b} + j\boldsymbol{d}$$

where the vector $\boldsymbol{x}$ is written as $\boldsymbol{x} = \boldsymbol{x}_R + j\boldsymbol{x}_I$ ($j = \sqrt{-1}$). Substituting this complex vector into the complex linear matrix equation above yields two equations

$$\begin{aligned} \boldsymbol{A}\boldsymbol{x}_R - \boldsymbol{C}\boldsymbol{x}_I &= \boldsymbol{b} \\ j(\boldsymbol{C}\boldsymbol{x}_R - \boldsymbol{A}\boldsymbol{x}_I) &= j\boldsymbol{d} \end{aligned}$$

which can be written as a real vector-matrix equation given by

$$\begin{bmatrix} \boldsymbol{A} & -\boldsymbol{C} \\ \boldsymbol{C} & \boldsymbol{A} \end{bmatrix} \begin{bmatrix} \boldsymbol{x}_R \\ \boldsymbol{x}_I \end{bmatrix} = \begin{bmatrix} \boldsymbol{b} \\ \boldsymbol{d} \end{bmatrix}$$

Using this transformation and the standard least-squares neurocomputing approach presented in Section 8.4, find the solution to

$$\begin{bmatrix} 1+j2 & 5-j7 & 3-j3 \\ 3-j5 & 1+j5 & j2 \\ 3 & 3-j3 & 3+j3 \end{bmatrix} \begin{bmatrix} x_1 \\ x_2 \\ x_3 \end{bmatrix} = \begin{bmatrix} 1+j2 \\ 1-j2 \\ 1+j2 \end{bmatrix}$$

Check your results by computing the solution in MATLAB, using standard methods.

8.7. Write a computer program to implement the generalized robust approach for solving systems of linear equations corrupted by noise presented in Section 8.6. Assume that the nonlinear weighting function is the logistic function, that is, $f_L(t) = \beta^2 \ln[\cosh(t/\beta)]$. A *nominal* linear matrix equation is given by

M

$$\begin{bmatrix} -1 & -4 & 5 & 6 & 10 \\ -1 & -1 & -9 & -2 & -16 \\ 5 & 15 & 0 & -21 & -1 \\ -6 & -6 & -6 & 1 & -7 \\ -1 & -13 & 5 & 16 & -10 \end{bmatrix} \begin{bmatrix} x_1 \\ x_2 \\ x_3 \\ x_4 \\ x_5 \end{bmatrix} = \begin{bmatrix} -12 \\ 3 \\ -4 \\ 1 \\ -4 \end{bmatrix} \quad \text{nominal system}$$

Use MATLAB to find the solution for this nominal system, that is, $\boldsymbol{x}^M$. Assume that data matrix $\boldsymbol{A}$ has been corrupted by impulse noise, and the resulting system is now given by

$$\begin{bmatrix} -1 & -4 & 5.2361 & 6 & 10 \\ -1 & -0.9178 & -9 & -1.9770 & -16 \\ 5 & 15 & 0.1781 & -21 & -1 \\ -6 & -6 & -6 & 1 & -7 \\ -1 & -12.8241 & 5 & 16 & -10 \end{bmatrix} \begin{bmatrix} x_1 \\ x_2 \\ x_3 \\ x_4 \\ x_5 \end{bmatrix}$$

$$= \begin{bmatrix} -12 \\ 3 \\ -4 \\ 1 \\ -4 \end{bmatrix} \quad \text{corrupted system}$$

The corruption of some of the $\boldsymbol{A}$ matrix elements resulted from randomly selecting 15 percent of the elements to be perturbed by random numbers from a uniform distribution with a variance of (0.15)*(0.005). Again using MATLAB, determine the solution to this system of equations. Use your computer program to find the robust solution, and compare this to the MATLAB solution. Compute the standard error of estimation (SEE) for the MATLAB solution and the robust neurocomputing solution, using (8.75) and (8.76), respectively. Plot your results in the same manner as in Figure 8.10 for Example 8.3.
Hint: Try different values of the β parameter as well as the learning rate parameter. Also, use as the initial weights for the robust neural network random numbers from a zero-mean Gaussian distribution with a variance of 25.

8.8. A Kahan matrix is an upper triangular matrix (cf. Sect. A.2.19). It can be generated as a very ill-conditioned matrix with ill-determined numerical rank. A MATLAB function to generate Kahan matrices is shown below. (*Note*: the angle α is entered in degrees.)

M

```
function A=kahan (n,alpha)
alpha=alpha*pi/180;
TMP1=zeros (n,n);
TMP2=eye (n,n);
TMP1 (1,1)=1;
for k=2:n
TMP1 (k,k)=(sin(alpha))^(k-1);
end
for k=1:n
  for j=1:n
     if j>k
        TMP2 (k, j)=cos(alpha);
      end
   end
end
A=TMP1*TMP2;
```

Using this function, generate a 10×10 Kahan matrix with an angle $\alpha = 1.0°$; this will be data matrix $\boldsymbol{A}$. The elements of the "exact solution" vector $\boldsymbol{x}_e$ are given by

$$x_{ei} = e^{2i} \qquad \text{for } i = 1, 2, \ldots, 10$$

From the Kahan $\boldsymbol{A}$ matrix that you generate and the exact solution vector $\boldsymbol{x}_e$, generate the observation vector $\boldsymbol{b} = \boldsymbol{A}\boldsymbol{x}_e$. What is the condition number of data matrix $\boldsymbol{A}$? Use the neurocomputing approach to solve $\boldsymbol{A}\boldsymbol{x} = \boldsymbol{b}$ that yields the best results, that is, the best results with respect to the relative error (cf. Example 8.4).

8.9. Generate a Hilbert matrix of dimension 15 in MATLAB, using the **`hilb`** function. The elements of the exact solution vector $\boldsymbol{x}_e$ are given by

$$x_{ei} = e^{i/2} \qquad \text{for } i = 1, 2, \ldots, 15$$

From the Hilbert matrix $\boldsymbol{A}$ and the exact solution vector $\boldsymbol{x}_e$, generate the observation vector $\boldsymbol{b}$ as $\boldsymbol{b} = \boldsymbol{A}\boldsymbol{x}_e$. What is the condition number of data matrix $\boldsymbol{A}$? Using both the truncated singular-value decomposition (TSVD) and Tikhonov regularization methods presented in Section 8.7, find the best solution to $\boldsymbol{A}\boldsymbol{x} = \boldsymbol{b}$ in the sense of the relative error (cf. Example 8.4). Determine the optimal regularization parameter.

8.10. Using the (*a*) Gauss-Seidel, (*b*) Jacobi, (*c*) Richardson, and (*d*) successive overrelaxation (SOR) iterative algorithms based on matrix splitting, find the solution to

$$\begin{bmatrix} 4 & 3 & 2 & 5 & 2 \\ 3 & 7 & 4 & 4 & 3 \\ 2 & 4 & 11 & 7 & 4 \\ 5 & 4 & 7 & 16 & 7 \\ 2 & 3 & 4 & 7 & 11 \end{bmatrix} \begin{bmatrix} x_1 \\ x_2 \\ x_3 \\ x_4 \\ x_5 \end{bmatrix} = \begin{bmatrix} -34 \\ -2 \\ -3 \\ -123 \\ -113 \end{bmatrix}$$

For each of the iterative methods, minimize the number of iterations to achieve convergence based on the relative-error criterion $\|\boldsymbol{x}^N - \boldsymbol{x}^M\|_2 / \|\boldsymbol{x}^M\|_2 < 10^{-8}$, where $\boldsymbol{x}^N$ is any of the iterative solutions listed in (*a*) through (*d*) and $\boldsymbol{x}^M$ is the MATLAB solution. Use the same initial random weights for each method.

8.11. Determine the parameters (a, b) (slope and ordinate intersect) of the straight line

$$y(x) = ax + b$$

that fit the data points $\{x_i, y_i\} = \{(0, 1), (2, 2), (3, 3), (4, 7), (6, 9)\}$. This problem can be formulated as an overdetermined system of linear equations given by

$$\begin{bmatrix} 0 & 1 \\ 2 & 1 \\ 3 & 1 \\ 4 & 1 \\ 6 & 1 \end{bmatrix} \begin{bmatrix} a \\ b \end{bmatrix} = \begin{bmatrix} 1 \\ 2 \\ 3 \\ 7 \\ 9 \end{bmatrix}$$

Apply the least-squares neurocomputing approach to solve this system of equations. Plot the data points and the least-squares straight line, using the values of (a, b) that you find, on the same plot.

8.12. Apply the L_1-norm (cf. Sect. 8.11) and L_∞-norm (cf. Sect. 8.10) criteria to Problem 8.11.

8.13. Find the parameters (a, b, c, d) of the polynomial

$$y(x) = ax^3 + bx^2 + cx + d$$

that fit the data points $\{x_i, y_i\} = \{(0, 1), (1, 3), (2, 4), (4, 5), (5, 8), (7, 9), (10, 11)\}$. Apply the least-squares neurocomputing approach, and plot the resulting least-squares polynomial using the values of (a, b, c, d) that you find and the data points on the same plot to observe the fit.
Hint: Use the MATLAB **`polyval`** function to evaluate the polynomial $y = ax^3 + bx^2 + cx + d$ for the coefficients (a, b, c, d) that you determine.

8.14. It is desired to approximate the function $f(x) = e^{-0.1x} \cos x$ for $\{x_i\} = \{0, 0.1, 0.2, \ldots, 10\}$ with a sixth-degree polynomial

$$y(x) = ax^6 + bx^5 + cx^4 + dx^3 + ex^2 + fx + g$$

Determine the coefficients of the polynomial $y(x)$, that is, (a, b, c, d, e, f, g), such that the polynomial fits the function $f(x)$ as best it can in a least-squares sense. Experiment with different-order polynomials to determine which order gives the best fit.

Hint: Use the conjugate gradient learning rule presented in Section 8.5. The MATLAB function shown below can be used to generate data matrix $\boldsymbol{A}$ and observation vector $\boldsymbol{b}$ from all the data points.

```
function [A,b]=points (x,y,n)
% [A,b]=points(x,y,n)
% Sets up: Ax=b
% n: degree of the polynomial
% x: vector of "x" points
% y: vector of "y" points
    k=length(x);
    for i=1:k
        for j=1:n+1
            A(i,j)=x(i)^(j-1);
        end
    end
    A=rot90(A',-1);
    b=y;
```

8.15. Assume a *nominal* linear matrix equation $\boldsymbol{Ax} = \boldsymbol{b}$, where the data matrix can be generated in MATLAB as

```
A = round(10*randn(10,10));
```

and the observation vector can be generated as

```
b=round(5*randn(10,1))
```

A corrupted data matrix A_c can be generated using the MATLAB function **uniferr** provided below. This function can be used to corrupt some of the elements of data matrix $\boldsymbol{A}$ in MATLAB randomly with impulse noise. The two elements in the argument of the function that are mandatory to provide are the nominal data matrix $\boldsymbol{A}$ and the *percentage* (on average) of elements in $\boldsymbol{A}$ to corrupt. The third element in the argument of the function is optional. For this problem use the default value. For the percentage of elements to be corrupted, use 15 percent.

Find the least-squares solution to the nominal system and the corrupted system, using the MATLAB function **pinv** (for the pseudoinverse of a matrix). Use the generalized robust approach for solving systems of equations presented in Section 8.6, and determine the robust solution to the corrupted system. Assume that the nonlinear weighting function is the logistic function $f_L(t) = \beta^2 \ln[\cosh(t/\beta)]$. Use as the initial weights for the robust neural network, random numbers from a zero-mean Gaussian distribution with a variance of 25. Try different values of β as well as the learning rate parameter. Compute the standard error of estimation for the MATLAB solution and the robust neurocomputing solution, using (8.75) and (8.76), respectively. Plot your results in the same manner as in Figure 8.10 for Example 8.3.

```
function [AC,N]=uniferr(A,per,s)
% [AC,N]=uniferr(A,per,s)
% A:     input matrix
% AC:    corrupted "output matrix"
% N:     noise matrix, random elements from
%        uniform distribution on the interval
%        (0.0,1.0), that is, if s=1
% per:   percentage of elements to be corrupted
% s:     interval scale factor, i.e.,
%        (0.0,s), the default value is
%        s=0.2449 resulting in a
%        variance of (per/100)*0.005
  if nargin>2
     scale=s;
  else
     scale=0.2449;
     end
  [nr,nc]=size(A);
  N=zeros(size(A));
  for i=1:nr
  for j=1:nc
  if (rand <= per/100)
  N(i,j)=scale*rand;
     end
   end
  end
  AC=A+N;
```

REFERENCES

1. E. Anderson, Z. Bai, C. H. Bischof, J. W. Demmel, J. J. Dongarra, J. J. Du Croz, A. Greenbaum, S. J. Hammarling, A. McKenney, S. Ostrouchov, and D. C. Sorenson, *LAPACK Users' Guide, Release 2.0*, 2nd ed., Philadelpha: Society for Industrial and Applied Mathematics, 1995.
2. G.-J. Li and B. W. Wah, "The Design of Optimal Systolic Arrays," *IEEE Transactions on Computers*, vol. C-34, 1985, pp. 66–77.
3. S. Y. Kung, *VLSI Array Processors*, Englewood Cliffs, NJ: Prentice-Hall, 1988.
4. C. Lehmann, M. Viredaz, and F. Blayo, "A Generic Systolic Array Building Block for Neural Networks with On-Chip Learning," *IEEE Transactions on Neural Networks*, vol. 4, 1993, pp. 400–7.
5. V. Kumar, A. Grama, A. Gupta, and G. Karypis, *Introduction to Parallel Computing: Design and Analysis of Algorithms*, Redwood City, CA: Benjamin/Cummings, 1994.
6. A. Cichocki and R. Unbehauen, "Neural Networks for Solving Systems of Linear Equations and Related Problems," *IEEE Transactions on Circuits and Systems*, vol. CAS-39, 1992, pp. 124–38.
7. A. Cichocki and R. Unbehauen, "Neural Networks for Solving Systems of Linear Equations—Part II: Minimax and Least Absolute Value Problems," *IEEE Transactions on Circuits and Systems*, vol. CAS-39, 1992, pp. 619–33.
8. A. Cichocki and R. Unbehauen, *Neural Networks for Optimization and Signal Processing*, New York: Wiley, 1993.

9. S. A. Ruzinsky and E. T. Olsen, "L_1 and L_∞ Minimization via a Variant of Karmarkar's Algorithm," *IEEE Transactions on Acoustics, Speech, and Signal Processing*, vol. 37, 1989, pp. 245–53.
10. J. W. Bandler, W. Kellerman, and K. Madsen, "A Nonlinear L_1 Optimization Algorithm for Design, Modeling, and Diagnosis of Networks," *IEEE Transactions on Circuits and Systems*, vol. CAS-34, 1987, pp. 174–81.
11. I. Barrodale and C. A. Zala, "L_1 and L_∞ Curve Fitting and Linear Programming Algorithms and Applications," in *Numerical Algorithms*, eds. J. L. Mohamed and J. E. Walsh, Oxford: Oxford University Press, 1986, pp. 220–38.
12. N. N. Abdelmalek, "Chebychev and L_1 Solution of Overdetermined Systems of Linear Equations with Bounded Variables," *Numerical Functional Analysis and Optimization*, vol. 8, 1986, pp. 399–418.
13. D. P. O'Leary, "Robust Regression Computation Using Iteratively Reweighted Least-Squares," *SIAM Journal on Matrix Analysis and Applications*, vol. 11, 1990, pp. 466–80.
14. R. H. Bartels, A. R. Conn, and Y. Li, "Primal Methods Are Better than Dual Methods for Solving Overdetermined Systems in the ℓ_∞ Sense?" *SIAM Journal on Numerical Analysis*, vol. 26, 1989, pp. 693–726.
15. A. Dax, "The Convergence of Linear Stationary Iterative Processes for Solving Singular Unstructured Systems of Linear Equations," *SIAM Review*, vol. 32, 1990, pp. 611–25.
16. A. Dax, "A Row Relaxation Method for Large L_1 Problems," *Linear Algebra and Its Applications*, vols. 154–156, 1991, pp. 793–818.
17. J. Schroeder, R. Yarlagadda, and J. Hershey, "L_p Normed Minimization with Applications to Linear Predictive Modeling for Sinusoidal Frequency Estimation," *Signal Processing*, vol. 24, 1991, pp. 193–216.
18. P. J. Huber, *Robust Statistical Procedures*, 2nd ed., Philadelphia: Society for Industrial and Applied Mathematics, 1996.
19. P. J. Huber, *Robust Statistics*, New York: Wiley, 1981.
20. F. R. Hampel, E. N. Ronchetti, P. Rousser, and W. A. Stachel, *Robust Statistics—The Approach Based on Influence Functions*, Philadelphia: Wiley, 1987.
21. C. A. Zala, I. Barrodale, and J. S. Kennedy, "High-Resolution Signal and Noise Field Estimation Using the L_1 Norm (Least Absolute Values)," *IEEE Transactions on Oceanic Engineering*, vol. OE-12, 1987, pp. 253–64.
22. P. Bloomfield and W. L. Steiger, *Least Absolute Deviations: Theory, Applications, and Algorithms*, Boston: Birkhäuser, 1983.
23. R. K. Ward, "An On-line Adaptation for Discrete ℓ_1 Linear Estimation," *IEEE Transactions on Automatic Control*, vol. AC-29, 1984, pp. 67–71.
24. S. Levy, C. Walker, T. J. Ulrych, and P. K. Fullagar, "A Linear Programming Approach to the Estimation of the Power Spectra of Harmonic Processes," *IEEE Transactions on Acoustics, Speech, and Signal Processing*, vol. ASSP-30, 1982, pp. 675–9.
25. G. H. Golub and C. F. Van Loan, *Matrix Computations*, 3rd ed., Baltimore, MD: Johns Hopkins University Press, 1996.
26. Y. Zhang, "A Primal-Dual Interior Point Approach for Computing the L_1 and L_∞ Solutions of Overdetermined Linear Systems," *Journal of Optimization Theory and Applications*, vol. 77, 1993, pp. 323–41.
27. A. M. Mood, F. A. Graybill, and D. C. Boes, *Introduction to the Theory of Statistics*, 3rd ed., New York: McGraw-Hill, 1974.
28. M. Abramowitz and C. A. Stegun (eds.), *Handbook of Mathematical Functions with Formulas, Graphs, and Mathematical Tables*, 9th printing, New York: Dover, 1972.

29. N. Higham, *Accuracy and Stability of Numerical Algorithms*, Philadelphia: Society for Industrial and Applied Mathematics, 1996.
30. S. G. Nash and A. Sofer, *Linear and Nonlinear Programming*, New York: McGraw-Hill, 1996.
31. D. G. Luenberger, *Linear and Nonlinear Programming*, 2nd ed., Reading, MA: Addison-Wesley, 1984.
32. M. Avriel, *Nonlinear Programming: Analysis and Methods*, Englewood Cliffs, NJ: Prentice-Hall, 1976.
33. F. M. Ham and T. M. McDowall, "Robust Learning in a Partial Least-Squares Neural Network," *Journal of Nonlinear Analysis, Theory, Methods & Applications*, Proceedings of the *Second World Congress of Nonlinear Analysts 1996*, July 10–17, Athens, Greece, vol. 30, no. 5, 1997, pp. 2903–14.
34. T. M. McDowall and F. M. Ham, "Robust Partial Least-Squares: A Modular Neural Network Approach," *Applications and Science of Artificial Neural Networks III*, ed. S. Rogers, Proceedings of SPIE, vol. 3077, 1997, pp. 344–55.
35. T. M. McDowall and F. M. Ham, "Robust Learning in a Generalized Partial Least Squares Regression Modular Neural Network," *Journal of Neural, Parallel and Scientific Computations*, vol. 6, 1998, pp. 391–415.
36. T. M. McDowall, "A Robust Partial Least Squares Neural Network," Ph.D. dissertation, Melbourne: Florida Institute of Technology, May 1997.
37. F. M. Ham and T. M. McDowall, "Inverse Partial Least Squares: A Robust Neural Network Approach," *Applications and Science of Computational Intelligence*, eds. S. K. Rogers, D. B. Fogel, J. C. Bezdek, and B. Bosacchi, Proceedings of SPIE, vol. 3390, 1998, pp. 36–47.
38. J. M. Varah, "A Practical Examination of Some Numerical Methods for Linear Discrete Ill-Posed Problems," *SIAM Review*, vol. 21, 1979, pp. 100–11.
39. P. C. Hansen, "Truncated Singular Value Decomposition Solutions to Discrete Ill-Posed Problems with Ill-Determined Numerical Rank," *SIAM Journal of Scientific Statistical Computing*, vol. 11, 1990, pp. 503–18.
40. P. C. Hansen, "The Truncated SVD as a Method for Regularization," *BIT*, vol. 27, 1987, pp. 534–53.
41. P. C. Hansen, T. Sekii, and H. Shibahashi, "The Modified Truncated SVD Method for Regularization in General Form," *SIAM Journal of Scientific Statistical Computing*, vol. 13, 1992, pp. 1142–50.
42. A. N. Tikhonov, "Solution of Incorrectly Formulated Problems and the Regularization Method," *Dokl. Akad. Nauk. SSSR*, vol. 151, 1963, pp. 501–4 (in Russian). [*Soviet Math. Dokl.*, vol. 4, 1963, pp. 1035–8 (in English).]
43. A. N. Tikhonov and V. Y. Arsenin, *Solutions of Ill-Posed Problems*, New York: Wiley, 1977.
44. A. N. Tikhonov and A. V. Goncharsky, *Ill-Posed Problems in the Natural Sciences*, Moscow: MIR Publishers, 1987.
45. J. M. Varah, "Pitfalls in the Numerical Solution of Linear Ill-Posed Problems," *SIAM Journal of Scientific Statistical Computing*, vol. 4, 1983, pp. 164–76.
46. P. C. Hansen, "Analysis of Discrete Ill-Posed Problems by Means of the *L*-Curve," *SIAM Review*, vol. 34, 1992, pp. 561–80.
47. P. C. Hansen, "Test Matrices for Regularization Methods," *SIAM Journal of Scientific Computing*, vol. 16, 1995, pp. 506–12.
48. P. C. Hansen and D. O'Leary, "The Use of the *L*-Curve in the Regularization of Discrete Ill-Posed Problems," *SIAM Journal of Scientific Computing*, vol. 14, 1993, pp. 1487–1503.

49. P. C. Hansen, "Regularization Tools: A MATLAB Package for Analysis and Solution of Discrete Ill-Posed Problems," *Numerical Algorithms*, vol. 6, 1994, pp. 1–35.
50. P. C. Hansen, "Regularization Tools: A MATLAB Package for Analysis and Solution of Discrete Ill-Posed Problems," version 2.0 for MATLAB 4.0, Technical Report UNI•C-92-03, Danish Computing Center for Research and Education, Technical University of Denmark, June 1992 (revised June 1993). [Available in PostScript via Netlib (netlib@research.att.com) from the director NUMERALGO.)
51. M. Hanke and P. C. Hansen, "Regularization Methods for Large-Scale Problems," *Surveys on Mathematics for Industry*, vol. 3, 1993, pp. 253–315.
52. M. Hanke, "Limitations of the *L*-Curve Method in Ill-Posed Problems," *BIT*, vol. 36, 1996, pp. 287–301.
53. L. Wu, "Regularization Methods and Algorithms for Least Squares and Kronecker Product Least Squares Problems," Ph.D. dissertation, Melbourne: Florida Institute of Technology, May 1997.
54. G. H. Golub, M. T. Heath, and G. Wahba, "Generalized Cross-Validation as a Method for Choosing a Good Ridge Parameter," *Technometrics*, vol. 21, 1979, pp. 215–23.
55. V. A. Morozov, *Methods for Solving Incorrectly Posed Problems*, New York: Springer-Verlag, 1984.
56. R. S. Varga, *Matrix Iterative Analysis*, Englewood Cliffs, NJ: Prentice-Hall, 1962.
57. D. M. Young, *Iterative Solution of Large Linear Systems*, New York: Academic, 1971.
58. K. E. Atkinson, *An Introduction to Numerical Analysis*, 2nd ed., New York: Wiley, 1989.
59. S. Van Huffel and J. Vandewalle, *The Total Least Squares Problem: Computational Aspects and Analysis*, Philadelphia: Society for Industrial and Applied Mathematics (*Frontiers in Applied Mathematics*, vol. 9), 1991.
60. G. H. Golub and C. F. Van Loan, "An Analysis of the Total Least Squares Problem," *SIAM Journal of Numerical Analysis*, vol. 17, 1980, pp. 883–93.
61. Y. Nievergelt, "Total Least Squares: State-of-the-Art Regression in Numerical Analysis," *SIAM Review*, vol. 36, 1994, pp. 258–64.
62. B. De Moor and J. Vandewalle, "A Unifying Theorem for Linear and Total Linear Least Squares," *IEEE Transactions on Automatic Control*, vol. 35, 1990, pp. 563–6.
63. S. A. Ruzinsky and E. T. Olsen, "L_1 and L_∞ Minimization via a Variant of Karmarkar's Algorithm," *IEEE Transactions on Acoustics, Speech, and Signal Processing*, vol. ASSP-37, 1989, pp. 245–53.

CHAPTER 9

Statistical Methods Using Neural Networks

9.1 INTRODUCTION

The purpose of this chapter is to present two very important statistical analysis and modeling methods that have utility in scientific and engineering problem solving, and to show how these methods can be realized as neurocomputing algorithms. Specifically, principal-component analysis (PCA) [1–5] is presented first. PCA has been used extensively in many engineering and scientific applications, and it is the basis for principal-component regression (PCR) [4, 6–8]. PCR is a factor analysis-based statistical modeling method. That is, given a set of empirical data (i.e., *training* data), a statistical calibration model can be developed based on an optimal number of PCR factors retained for building the model. This statistical calibration model can be used to predict (or estimate) the output, given an unknown input to the calibration model.

The second method is partial least-squares regression (PLSR) [7–14]. This is another factor analysis-based method for statistical model building. The main difference between PCR and PLSR lies in the manner in which the empirical data are used to develop the statistical calibration model. In the case of PCR, only the *measurement* data (i.e., the independent-variable block) is used in the process to develop the calibration model, whereas for PLSR both the *measurement* and *target* data (or dependent-variable block) are used to develop the calibration model. Typically, PLSR will give better predictions than PCR simply because more information is being used in the development of the statistical calibration model.

PCR and PLSR are two of many methods that constitute the field of *chemometrics* [8, 14]. Chemometrics is the discipline concerned with the application of statistical and mathematical methods, and methods based on mathematical logic, to chemistry. Only recently, some of the chemometric methods (i.e., PCR and PLSR) have been used in engineering disciplines and other scientific areas [9]. In both PCR and PLSR, the ability to select the essential features (factor analysis) of the training data during the development of the calibration model leads to much better predictive performance than, for example, a classical least-squares (CLS) approach [7]. That is, when the factor analysis is properly carried out, only the pertinent characteristics of the data are retained for model development. The remaining data, associated with noise, for example, that would result in reduced predictive performance, are discarded [9]. PCA, and PCR along with PLSR, can all be realized in various neural network architectures. Therefore, some of the neural network methods for PCA, PCR, and PLSR are presented in this chapter.

9.2 PRINCIPAL-COMPONENT ANALYSIS

Principal-component analysis, also known as the Hotelling transform [5] in digital image processing [15] and the Karhunen-Loève transformation in communication theory [16], is used in many engineering and scientific fields. Applications of PCA include, for example, data compression and coding (decoding), pattern recognition, image processing, adaptive beamforming, reduced-order controller design, and high-resolution spectral analysis (for frequency estimation), to name a few. As we will soon see, PCA is directly related to eigenvalue decomposition (EVD) (cf. Sec. 7.6).

In general, PCA is a statistical method which can determine an *optimal* linear transformation matrix $\boldsymbol{W} \in \Re^{m\times n} (m < n)$ such that given an input vector $\boldsymbol{x} \in \Re^{n\times 1}$, which is considered to be from a zero-mean wide-sense-stationary stochastic process, the data in $\boldsymbol{x}$ can be *compressed* according to

$$\boldsymbol{y} = \boldsymbol{W}\boldsymbol{x} \tag{9.1}$$

where $\boldsymbol{y} \in \Re^{m\times 1}$. Therefore, PCA projects the input data from the original n-dimensional vector space onto an m-dimensional output space, through the transformation matrix $\boldsymbol{W}$, where typically $m \ll n$. Dimensionality reduction is thus performed by PCA, where $\boldsymbol{y}$ contains (retains) most of the essential information that is resident in the input vector $\boldsymbol{x}$. That is, PCA can transform a very large amount of correlated input data to a set of statistically decorrelated components (or features). The *components* are usually ordered according to decreasing variance.

To illustrate this, let $\boldsymbol{x}$ be a zero-mean stochastic input vector, that is, $\boldsymbol{x} = [x_1, x_2, \ldots, x_n]^T$, with a covariance matrix (or correlation matrix since $\boldsymbol{x}$ is considered to be zero-mean) $\boldsymbol{C}_x = \mathrm{E}[\boldsymbol{x}\boldsymbol{x}^T]$, where $\boldsymbol{C}_x \in \Re^{n\times n}$ and $\mathrm{E}[\bullet]$ is the expectation operator (cf. Sec. A.7.4). Also, the vectors $[\boldsymbol{w}_1, \boldsymbol{w}_2, \ldots, \boldsymbol{w}_m]$ are an orthonormal set, that is, orthogonal with unity length, that is, $\boldsymbol{w}_i^T \boldsymbol{w}_j = \delta_{ij} (\forall i \text{ and } j)$, where δ_{ij} is the *Kronecker delta*, and the unit length of each vector can be expressed through the L_2 norm as $\|\boldsymbol{w}_i\|_2^2 = \boldsymbol{w}_i^T \boldsymbol{w}_i = 1$. The vectors $[\boldsymbol{w}_1, \boldsymbol{w}_2, \ldots, \boldsymbol{w}_m]$ are considered as the first m eigenvectors of the covariance matrix $\boldsymbol{C}_x$, such that $\boldsymbol{w}_1 = [w_{11}, w_{12}, \ldots, w_{1n}]^T$ corresponds to the largest eigenvalue (λ_1) of $\boldsymbol{C}_x$, $\boldsymbol{w}_2 = [w_{21}, w_{22}, \ldots, w_{2n}]^T$ corresponds to next-largest eigenvalue (λ_2), etc. Therefore, from the standard eigenvalue problem, the matrix equation can be written as

$$\boldsymbol{C}_x \boldsymbol{w}_j = \lambda_j \boldsymbol{w}_j \qquad \text{for} \quad j = 1, 2, \ldots, n \tag{9.2}$$

and is satisfied with $\lambda_1 \geq \lambda_2 \geq \cdots \geq \lambda_n \geq 0$ and $\|\boldsymbol{w}_i\|_2^2 = 1$. The first m eigenvectors $[\boldsymbol{w}_1, \boldsymbol{w}_2, \cdots, \boldsymbol{w}_m]$ of $\boldsymbol{C}_x$ are considered the *principal eigenvectors*. These are the *directions* in the n-dimensional vector space for which the input data have the largest variance (or predominant information content). Therefore, for a given input vector $\boldsymbol{x}$, the *m principal components* in $\boldsymbol{y}$ are defined by the transformation given in (9.1), where the transformation matrix $\boldsymbol{W}$ is

$$\boldsymbol{W} = [\boldsymbol{w}_1, \boldsymbol{w}_2, \ldots, \boldsymbol{w}_m]^T \tag{9.3}$$

In other words, the m-dimensional principal-component subspace of the input data space is defined as the subspace spanned by the *m principal eigenvectors* of the input covariance matrix $\boldsymbol{C}_x$.

If (9.2) is written for each eigenvalue and eigenvector, that is,

$$\begin{aligned} \boldsymbol{C}_x[\boldsymbol{w}_1, \boldsymbol{w}_2, \cdots, \boldsymbol{w}_n] &= [\lambda_1 \boldsymbol{w}_1, \lambda_2 \boldsymbol{w}_2, \ldots, \lambda_n \boldsymbol{w}_n] \\ \boldsymbol{C}_x \boldsymbol{W}^T &= \boldsymbol{W}^T \text{diag}[\lambda_1, \lambda_2, \ldots, \lambda_n] = \boldsymbol{W}^T \boldsymbol{\Lambda} \end{aligned} \tag{9.4}$$

Therefore,

$$\boldsymbol{C}_x \boldsymbol{W}^T = \boldsymbol{W}^T \boldsymbol{\Lambda} \tag{9.5}$$

and premultiplying both sides of (9.5) by $\boldsymbol{W}$ (where $\boldsymbol{W}\boldsymbol{W}^T = \boldsymbol{I}$, for $m \leq n$, because the rows of $\boldsymbol{W}$ are orthonormal), we can write

$$\boldsymbol{\Lambda} = \boldsymbol{W} \boldsymbol{C}_x \boldsymbol{W}^T \tag{9.6}$$

Also, if (9.5) is postmultiplied by $\boldsymbol{W}(\boldsymbol{W}^T \boldsymbol{W} = \boldsymbol{I}$, for $m = n$), we can write

$$\boldsymbol{C}_x = \boldsymbol{W}^T \boldsymbol{\Lambda} \boldsymbol{W} = \sum_{i=1}^{n} \lambda_i \boldsymbol{w}_i \boldsymbol{w}_i^T \tag{9.7}$$

which is the eigenvalue decomposition (EVD), or spectral factorization, of $\boldsymbol{C}_x$ (cf. Sect. 7.6). Therefore, in (9.6) $\boldsymbol{\Lambda} \in \Re^{m \times m}$ represents the covariance matrix of the output vector $\boldsymbol{y}$ in (9.1). Because $\boldsymbol{\Lambda}$ is diagonal with nonnegative diagonal elements, the elements of the output vector $\boldsymbol{y}$ are uncorrelated and have variances equal to the eigenvalues of the covariance matrix $\boldsymbol{C}_x$.

This can best be viewed by considering again the linear transformation given in (9.1). We will assume we are interested in the jth principal component y_j, and the ordering of all the principal components of interest is such that the first principal component is associated with the largest variance in the input data, the second principal component is associated with the second-largest variance in the input data, and so on (as stated above). Therefore, we will let y_j be a linear combination of the elements of the data input vector $\boldsymbol{x}$, that is,

$$y_j = w_{1j} x_1 + w_{2j} x_2 + \cdots + w_{nj} x_n = \boldsymbol{w}_j^T \boldsymbol{x} \tag{9.8}$$

where $w_j \in \Re^{n\times 1}$. Because x is considered to be a zero-mean stochastic vector, y_j is a zero-mean stochastic variable with an associated variance given by

$$\sigma_{y_j}^2 = \mathrm{E}[y_j^2] = \mathrm{E}[w_j^T x x^T w_j] = w_j^T \mathrm{E}[x x^T] w_j = w_j^T C_x w_j = \sum_{i=1}^{n} \sum_{h=1}^{n} w_{ij} w_{hj} c_{xxih} \tag{9.9}$$

and we require that $w_j^T w_j = 1$. Therefore, our intention is to maximize the variance associated with y_j, as given in (9.9), subject to the constraint $w_j^T w_j = 1$. This problem can be formulated using the Lagrange multiplier method (cf. Sect. A.6.2), that is,

$$\mathscr{L}(w_j) = \sigma_{y_j}^2 - \lambda_j (w_j^T w_j - 1) \tag{9.10}$$

and the *extrema* can be found by taking the partial derivative of (9.10) with respect to w_j and setting the result equal to zero (cf. Sect. A.3.4.1). That is,

$$\begin{aligned} \frac{\partial \mathscr{L}(w_j)}{\partial w_j} &= \frac{\partial}{\partial w_j}[\sigma_{y_j}^2 - \lambda_j(w_j^T w_j - 1)] \\ &= \frac{\partial}{\partial w_j}[w_j^T C_x w_j - \lambda_j(w_j^T w_j - 1)] \\ &= 2C_x w_j - 2\lambda_j w_j = \mathbf{0} \end{aligned} \tag{9.11}$$

or

$$(C_x - \lambda_j I) w_j = \mathbf{0} \tag{9.12}$$

which is the original eigenvalue problem given in (9.2). Furthermore, (9.12) has a non-trivial solution if and only if

$$|C_x - \lambda_j I| = 0 \tag{9.13}$$

for which the λ_j (variances), for $j = 1, 2, \ldots, n$, in (9.13) are the eigenvalues of C_x, and the w_j, for $j = 1, 2, \ldots, n$, in (9.13) are the associated (principal) eigenvectors. Moreover, if (9.12) is premultiplied on both sides by w_j^T, we obtain

$$w_j^T (C_x - \lambda_j I) w_j = \underbrace{w_j^T C_x w_j}_{\sigma_{y_j}^2} - \lambda_j \underbrace{w_j^T w_j}_{1} = \sigma_{y_j}^2 - \lambda_j = 0 \tag{9.14}$$

or

$$\lambda_j = \sigma_{y_j}^2 \tag{9.15}$$

Therefore, from (9.15) for $j = 1$, $\lambda_1 = \sigma_{y_1}^2$, which is the largest eigenvalue of the covariance matrix C_x, or the largest variance of the linear combination of the input data vector elements in x, and the associated eigenvector w_1 indicates the vector space *direction* of the largest variance.

Now if only the first m eigenvectors of C_x are retained to form the transformation matrix W given in (9.3), and then we write from (9.1) $y = Wx$, the elements of y are the first m principal components. If m is carefully selected, the majority of the information content in the input $\boldsymbol{x}$ is contained in $\boldsymbol{y}$. Furthermore, the linear least-squares estimate of $\boldsymbol{x}$ can be written as

$$\hat{\boldsymbol{x}} = \boldsymbol{C}_x \boldsymbol{W}^T (\boldsymbol{W}\boldsymbol{C}_x \boldsymbol{W}^T)^{-1} \boldsymbol{y} \tag{9.16}$$

which can be derived from the mean-square error function

$$J(\boldsymbol{x}) = \mathrm{E}\{(\boldsymbol{x} - \hat{\boldsymbol{x}})^T (\boldsymbol{x} - \hat{\boldsymbol{x}})\} \tag{9.17}$$

Equation (9.17) is minimized with respect to $\boldsymbol{x}$ when the rows of $\boldsymbol{W}$ span the first m principal eigenvectors of $\boldsymbol{C}_x$. We can also write an approximation to $\boldsymbol{x}$ in terms of the m principal eigenvectors $[\boldsymbol{w}_1, \boldsymbol{w}_2, \ldots, \boldsymbol{w}_m]$, as

$$\begin{aligned}
\hat{\boldsymbol{x}} &= \sum_{h=1}^{m} (\boldsymbol{x}^T \boldsymbol{w}_h)\boldsymbol{w}_h = \sum_{h=1}^{m} \boldsymbol{w}_h(\boldsymbol{x}^T \boldsymbol{w}_h) = \sum_{h=1}^{m} \boldsymbol{w}_h(\boldsymbol{w}_h^T \boldsymbol{x}) \\
&= (\boldsymbol{w}_1\boldsymbol{w}_1^T + \boldsymbol{w}_2\boldsymbol{w}_2^T + \cdots + \boldsymbol{w}_m\boldsymbol{w}_m^T)\boldsymbol{x} \\
&= (\underbrace{[\boldsymbol{w}_1, \boldsymbol{w}_2, \ldots, \boldsymbol{w}_m]}_{\boldsymbol{W}^T} \underbrace{[\boldsymbol{w}_1, \boldsymbol{w}_2, \ldots, \boldsymbol{w}_m]^T}_{\boldsymbol{W}})\boldsymbol{x} = \boldsymbol{W}^T \underbrace{\boldsymbol{W}\boldsymbol{x}}_{\boldsymbol{y}} = \boldsymbol{W}^T \boldsymbol{y}
\end{aligned} \tag{9.18}$$

Also, from (9.7) an approximation to the covariance matrix $\boldsymbol{C}_x$ can be written as $\boldsymbol{C}_x = \boldsymbol{W}^T \boldsymbol{\Lambda} \boldsymbol{W}$, where $\boldsymbol{W} \in \Re^{m \times n}$, and $\boldsymbol{\Lambda} \in \Re^{m \times m}$, where the first m eigenvalues of $\boldsymbol{C}_x$ are down the diagonal of $\boldsymbol{\Lambda}$ in descending order. Moreover, from (9.6) we can write $\boldsymbol{\Lambda} = \boldsymbol{W}\boldsymbol{C}_x\boldsymbol{W}^T$, and as previously stated, $\boldsymbol{\Lambda}$ is the covariance matrix of the output vector $\boldsymbol{y}$; and because $\boldsymbol{\Lambda}$ is diagonal with nonnegative entries, the elements of $\boldsymbol{y}$ are uncorrelated and have variance equal to the m eigenvalues of $\boldsymbol{C}_x$.

In practice, the covariance matrix $\boldsymbol{C}_x$ is usually not known, nor is the probability distribution of $\boldsymbol{x}$; however, typically a large number of sample data vectors $\boldsymbol{x}(k)$, for $k = 1, 2, \ldots, N$, can be *collected*. From these N sample vectors, that is, $\boldsymbol{X} = [\boldsymbol{x}(1), \boldsymbol{x}(2), \ldots, \boldsymbol{x}(N)]$, an estimate of the covariance matrix can be obtained by taking the time average over the N available sample vectors [17] as

$$\boldsymbol{C}_x \approx \hat{\boldsymbol{C}}_x = \frac{1}{N}\sum_{k=1}^{N} \boldsymbol{x}(k)\boldsymbol{x}^T(k) = \frac{1}{N}\boldsymbol{X}\boldsymbol{X}^T \tag{9.19}$$

In summary, the objective of PCA is to determine the set of m orthogonal principal eigenvectors $\boldsymbol{W} = [\boldsymbol{w}_1, \boldsymbol{w}_2, \ldots, \boldsymbol{w}_m]^T \in \Re^{m \times n}$ (for $m < n$) associated with the input data. Furthermore, these orthogonal vectors should span the input data space such that they account for the variance of the input data as much as possible. Thus, the information in $\boldsymbol{x}$ is *compressed* into outputs (principal components) contained in the vector $\boldsymbol{y}$. However, often the information $\boldsymbol{x}(k)$, for $k = 1, 2, \ldots, N$, necessary to estimate $\boldsymbol{C}_x$, as shown in (9.19), is not collectively available, but can be assimilated one vector measurement at a time. Therefore, for this common situation, adaptive methods must be used to *extract* the principal components associated with the unknown covariance matrix $\boldsymbol{C}_x$. Thus, adaptive extraction algorithms will be discussed next, and the methods presented are based on several neural network architectures that can be trained using Hebbian learning (cf. Sect. 2.8.2).

9.3
LEARNING ALGORITHMS FOR NEURAL NETWORK ADAPTIVE ESTIMATION OF PRINCIPAL COMPONENTS

Over the past several years many neural network approaches have been developed for adaptive extraction of principal components of the covariance matrix [18–95]. Presented in the following sections are four of the most well-known methods for adaptive extraction of principal components. These methods are not unrelated; in fact they can be systematically derived from the original formulation by Oja [18] of a single neuron with Hebbian learning as a principal-component analyzer. Therefore, Oja's single-neuron Hebbian learning principal-component analyzer will be presented first. The single-neuron case will then be extended to estimation of several principal components. The single-layer neural network architecture for multiple principal eigenvector extraction was proposed by Oja and Karhunen [20]. The typical heuristic argument that is given to extend the single principal-component case to estimation of several principal components is replaced with a mathematical derivation for the learning rule. From this symmetric subspace learning rule, two other learning rules can be directly derived, that is, the generalized hebbian algorithm (GHA) [21] and the stochastic gradient ascent (SGA) algorithm [26]. Finally, the APEX (adaptive principal-component extraction) algorithm of Kung et al. [30] is presented. There are basically two classes of PCA learning rules: *reestimation algorithms* and *decorrelating algorithms* [40]. The first four that we will investigate are considered to be reestimation algorithms, and the last one (APEX) is a decorrelating algorithm.

9.3.1 Estimation of the First Principal Component—Normalized Hebbian Learning Rule of Oja

Our objective here is to develop a simple (single-neuron) neural network that is capable of extracting the first principal component. In 1982, Oja [18] proposed a single linear processing element for this purpose, shown in Figure 9.1, and from the figure the expression

$$y_1 = \sum_{j=1}^{n} w_{1j} x_j = \boldsymbol{x}^T \boldsymbol{w}_1 = \boldsymbol{w}_1^T \boldsymbol{x} \tag{9.20}$$

can be written. An error expression can be written as $\boldsymbol{e} = \boldsymbol{x} - \hat{\boldsymbol{x}}$, where the estimate of $\boldsymbol{x}$, for the single-component case, is given as $\hat{\boldsymbol{x}} = \boldsymbol{w}_1 y_1$ from (9.18). Defining an error cost (Lyapunov) function as

$$\begin{aligned} L(\boldsymbol{w}_1) &= \frac{1}{2}\|\boldsymbol{e}\|_2^2 = \frac{1}{2}\|\boldsymbol{x} - \hat{\boldsymbol{x}}\|_2^2 = \frac{1}{2}(\boldsymbol{x} - \boldsymbol{w}_1 y_1)^T(\boldsymbol{x} - \boldsymbol{w}_1 y_1) \\ &= \frac{1}{2}(\boldsymbol{x}^T \boldsymbol{x} - 2 y_1 \boldsymbol{w}_1^T \boldsymbol{x} + y_1^2 \boldsymbol{w}_1^T \boldsymbol{w}_1) \end{aligned} \tag{9.21}$$

where $\|\bullet\|_2$ is the L_2 (Euclidean) norm, and computing the gradient with respect to the weight vector $\boldsymbol{w}_1$ give

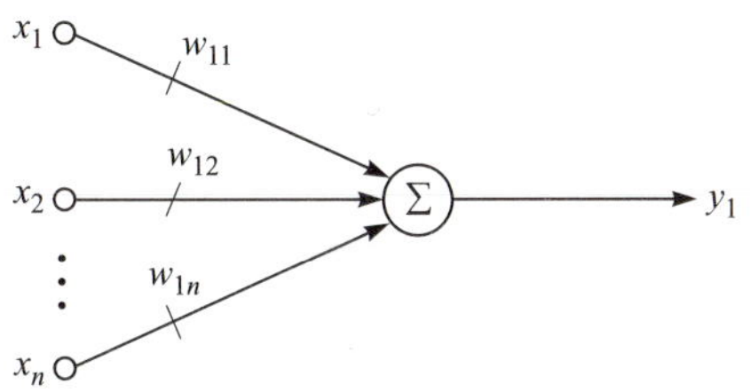

Figure 9.1 Oja's neuron model for estimating the first principal component.

$$\nabla_{w_1} L(w_1) = \frac{\partial L(w_1)}{\partial w_1} = \frac{1}{2}\frac{\partial}{\partial w_1}(x^T x - 2y_1 w_1^T x + y_1^2 w_1^T w_1) = -y_1 x + y_1^2 w_1 \quad (9.22)$$

using the results from Section A.3.4.1.

The continuous-time learning rule for extracting the first principal component can be expressed in terms of a steepest descent gradient approach [64, 65] as the vector differential equation

$$\frac{dw_1(t)}{dt} = -\mu \nabla_{w_1} L(w_1) \quad (9.23)$$

where $\mu > 0$ is the learning rate parameter. Using the gradient result in (9.22), the continuous-time learning rule can be written as

$$\frac{dw_1(t)}{dt} = \mu[y_1(t)x(t) - y_1^2(t)w_1(t)] \quad (9.24)$$

and the discrete-time learning rule is

$$w_1(k+1) = w_1(k) + \mu[y_1(k)x(k) - y_1^2(k)w_1(k)] \quad (9.25)$$

where k is the time index. This adaptive learning rule is referred to as the *normalized Hebbian* or *Oja's rule* [18], where the change in the weight vector can be written as

$$\Delta w_1(k) = w_1(k+1) - w_1(k) = \mu[y_1(k)x(k) - y_1^2(k)w_1(k)] \quad (9.26)$$

From (9.25) the scalar learning rule can be written as

$$w_{1j}(k+1) = w_{1j}(k) + \mu[y_1(k)x_j(k) - y_1^2(k)w_{1j}(k)] \qquad \text{for} \qquad j = 1, 2, \ldots, n \quad (9.27)$$

It has been proved that Oja's learning rule converges to the weight vector w_1 (i.e., the first principal eigenvector of the covariance matrix C_x[18, 20, 39]) with the following properties: For details of convergence properties of PCA analysis algorithms, see [94, 95].

1. The weight vector w_1 has unit length, that is, $\|w_1\|_2^2 = \sum_{h=1}^{n} w_{1h}^2 = w_1^T w_1 = 1$.
2. The weight vector w_1 is an eigenvector of the covariance matrix C_x.
3. The weight vector w_1 maximizes the variance of the output $y_1 = w_1^T x$; therefore, for zero-mean inputs, w_1 is the first principal eigenvector and y_1 is the first principal component.

The first term on the right-hand side of (9.26) is the *standard Hebbian co-occurrence* term (cf. Sect. 2.8.2), and the second term on the right-hand side is an active decay or forgetting term, which prevents the synaptic weight vector $\boldsymbol{w}_1$ from becoming unbounded during the training process.

It is important to discuss the learning rate parameter μ because it has a direct effect on the convergence of the algorithm given in (9.25). If the value of μ is too large, the learning rule will not converge; that is, the learning algorithm can become numerically unstable. On the other hand, if μ is set too small, convergence will be painfully slow. Typically, the learning rate parameter is considered to be time-varying, that is, $\mu = \mu(k)$, and should be initially set at a relatively large value and then gradually reduced until the desired accuracy is achieved. In fact, to ensure numerical stability of the algorithm, the learning rate must satisfy the inequality $0 < \mu(k) < 1/(1.2\lambda_1)$, where λ_1 is the largest eigenvalue of the covariance matrix $\boldsymbol{C}_x$ [20]. A good initial value to use can be based on the inputs $\boldsymbol{x}(k)$, that is, $\mu(k) = 1/[2\boldsymbol{x}^T(k)\boldsymbol{x}(k)]$. This will ensure good initial convergence, because the step sizes will be relatively large. Then the learning rate should be gradually decreased in order to *fine-tune* the weight updates to obtain an accurate estimate of the synaptic weight vector $\boldsymbol{w}_1$ [20, 38, 62]. Many times in numerical gradient search algorithms a *forgetting factor* [2] or leakage factor is introduced into an expression for the learning rate. In work by Cichocki and Unbehauen [38] an expression is given for a variable learning rate as

$$\mu_i(k) = \frac{1}{\dfrac{\gamma}{\mu_i(k-1)} + y_i^2(k)} \quad \text{and} \quad \mu_i(0) = \frac{1}{y_i^2(0)} \quad \text{for} \quad i = 1, 2, \ldots, m \tag{9.28}$$

where γ is the forgetting factor in the range $0 \leq \gamma \leq 1$. Typically the forgetting factor is in the range $0.9 \leq \gamma \leq 1$. For estimating the first principal component, in (9.28) $i = 1$. Therefore, (9.28) can be used to express individual learning rates for extraction of m principal components; or a single value for μ can be used for extracting the m principal components, see (9.48).

EXAMPLE 9.1. To illustrate Oja's learning rule for extraction of the first principal component, 1,000 zero-mean Gaussian random three-dimensional vectors are generated where the x component has unity variance and the y component and z component have variance of 0.002. Implementing the discrete-time learning rule given in (9.25), with the initial weight vector randomly selected, the network converged in 726 iterations (i.e., it was not necessary to present all 1,000 training vectors). Convergence was determined by computing the L_2-norm of the difference between the weight vector before and after an update. The learning rate parameter was adaptively adjusted according to (9.28) with $\gamma = 1$.

In Figure 9.2 the convergence path of the weight vector can be seen in three dimensions, where the circle indicates the initial random weight vector, and the final weight vector is shown as the star. The final weight vector as computed by the neural network is $\boldsymbol{w}_1^{NN} = [-1.0000, -0.0002, -0.0016]^T$. The eigenvector of $\boldsymbol{C}_x$, where the covariance matrix is computed according to (9.19), associated with the largest eigenvalue ($\lambda_1^M = 0.9843$) is $\boldsymbol{w}_1^M = [1.0000, -0.0002, 0.0011]^T$ using MATLAB. Notice that $\boldsymbol{w}_1^{NN}$ and $\boldsymbol{w}_1^M$ are basically the same to within a minus sign, which can typically occur because of the sign ambiguity associated with eigenvectors. As shown in Figure 9.2, after convergence to the *first principal*

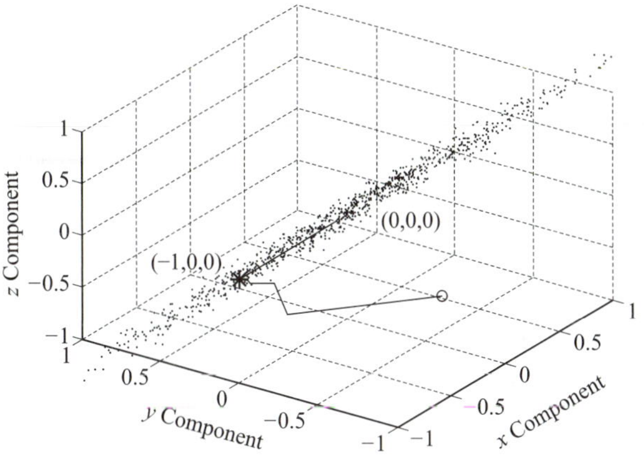

Figure 9.2 Sample vectors for Example 9.1, and the convergence path for the first principal eigenvector (weight vector) of the covariance matrix $\boldsymbol{C}_x$, as adaptively computed by the neural network (with the forgetting factor $\gamma = 1$).

eigenvector, the direction indicated (solid line in the figure) is the direction of maximum variance. Finally, the variance (i.e., eigenvalue) can be computed from the data in $\boldsymbol{X}$, which contains 1,000 sample vectors, and the weight vector computed from the neural network $\boldsymbol{w}_1^{NN}$, that is,

$$\lambda_1^{NN} = \text{var}(\boldsymbol{w}_1^{NN^T} \boldsymbol{X}) = 0.9852 \tag{9.29}$$

which agrees with the eigenvalue computed using MATLAB.

Before we go on to the estimation of several principal components, we need to discuss the issue of normalization of the learning rule [18, 40]. To show that Oja's learning rule given above is normalized Hebbian learning, we can start with a typical version of Hebbian learning (in discrete-time scalar form)

$$w_j(k+1) = w_j(k) + \mu(k)y(k)x_j(k) \tag{9.30}$$

for $j = 1, 2, \ldots, n$. This learning rule in its basic form leads to unlimited growth of the synaptic weights $w_j(k)$, for $j = 1, 2, \ldots, n$, which would not allow the weight vector $\boldsymbol{w}$ to converge. This can be overcome by incorporating some form of *normalization* (or *saturation*) in the learning rule for adaptively changing the synaptic weights. The normalization invokes competition among the synapses of the neuron which will lead to stabilization. A reasonable form of normalization, using (9.30), can be written as

$$w_j(k+1) = \frac{w_j(k) + \gamma\mu(k)y(k)x_j(k)}{\left\{\sum_{j=1}^{n} [w_j(k) + \gamma\mu(k)y(k)x_j(k)]^2\right\}^{1/2}} \tag{9.31}$$

where the summation in the denominator of (9.31) extends over the entire set of synapses associated with the neuron, and γ is the *plasticity coefficient*, which follows the approach of Oja [18]. Therefore, (9.31) represents a *normalized* learning rule that can be simplified by truncating a power series expansion of the expression. First, recall that the synaptic weight vector is to be of unit length, that is, $\|\boldsymbol{w}(k)\|_2 = [\sum_{j=1}^{n} w_j^2(k)]^{1/2} = 1$. Second, we can expand (9.31) in a power series about $\gamma = 0$ as

$$w_j(k+1) = \frac{w_j(k+1)\big|_{\gamma=0}}{0!} + \frac{\left[\dfrac{dw_j(k+1)}{d\gamma}\right]\Bigg|_{\gamma=0}}{1!}\gamma + O(\gamma^2) \tag{9.32}$$

and assuming $\gamma \ll 1$ only the first two terms in (9.32) will be retained. The term $O(\gamma^2)$ represents higher-order effects in γ and is ignored. The first term in (9.32) can be written as

$$\frac{w_j(k+1)\big|_{\gamma=0}}{0!} = \frac{w_j(k)}{\left[\sum_{j=1}^{n} w_j^2(k)\right]^{1/2}} = \frac{w_j(k)}{\underbrace{\|\boldsymbol{w}(k)\|_2}_{1}} = w_j(k) \tag{9.33}$$

using (9.31). For the second term in (9.32), we will define in (9.31)

$$\alpha = w_j(k) + \gamma\mu(k)y(k)x_j(k) \tag{9.34}$$

and

$$\beta = \left\{\sum_{j=1}^{n}[w_j(k) + \gamma\mu(k)y(k)x_j(k)]^2\right\}^{1/2} \tag{9.35}$$

therefore,

$$\begin{aligned}\frac{\left[\dfrac{dw_j(k+1)}{d\gamma}\right]\Bigg|_{\gamma=0}}{1!}\gamma &= \frac{\left(\beta\dfrac{d\alpha}{d\gamma}\right)\Bigg|_{\gamma=0} - \left(\alpha\dfrac{d\beta}{d\gamma}\right)\Bigg|_{\gamma=0}}{\beta^2\big|_{\gamma=0}} \\ &= \gamma\mu(k)[y(k)x_j(k) - w_j(k)y^2(k)]\end{aligned} \tag{9.36}$$

Substituting the results from (9.33) and (9.36) into (9.32), with $\gamma = 1$ [and $\mu(k)$ sufficiently small] and ignoring the term $O(\gamma^2)$, we obtain

$$w_j(k+1) = w_j(k) + \mu(k)[y(k)x_j(k) - y^2(k)w_j(k)] \qquad \text{for } j = 1, 2, \ldots, n \tag{9.37}$$

which is Oja's discrete-time scalar form of the learning rule for estimating the first principal eigenvector given in (9.27). The normalization in (9.31) could also have been carried out by first assuming that the *plasticity coefficient* $\gamma = 1$, and then assuming $\mu(k) \ll 1$, the approximation result in (9.37) would once again be obtained.

9.3.2 Estimation of Several Principal Components—Symmetric Subspace Learning Rule

A typical approach that is taken to present the case for adaptive estimation of several (m) principal components is to heuristically *extend* Oja's result for estimating the first principal component. This has been presented by replacing the weight vector in (9.25) by a weight matrix $\boldsymbol{W} \in \Re^{m \times n}$, and replacement of the scalar y_1 by a vector $\boldsymbol{y} \in \Re^{m \times 1}$. However, this will not be our approach. We will be more rigorous and derive the learning rule from a similar setting that was used to develop the learning algorithm for the single-principal-component case.

Shown in Figure 9.3 is a single-layer linear neural network first proposed by Karhunen and Oja [20, 26, 62], which is the basis for our derivation. The original PCA mapping $\boldsymbol{y} = \boldsymbol{W}\boldsymbol{x}$ given in (9.1), and shown in Figure 9.3, allows for estimation of multiple (m) principal components. We will define an error vector as $\boldsymbol{e} = \boldsymbol{x} - \hat{\boldsymbol{x}} \in \Re^{n \times 1}$, where $\hat{\boldsymbol{x}} = \boldsymbol{W}^T\boldsymbol{y} = \boldsymbol{W}^T\boldsymbol{W}\boldsymbol{x}$ as given in (9.18). We can also define an error cost (Lyapunov) function as

$$\begin{aligned} L(\boldsymbol{W}) &= \frac{1}{2}\|\boldsymbol{e}\|_2^2 = \frac{1}{2}\boldsymbol{e}^T\boldsymbol{e} = \frac{1}{2}(\boldsymbol{x} - \hat{\boldsymbol{x}})^T(\boldsymbol{x} - \hat{\boldsymbol{x}}) = \frac{1}{2}(\boldsymbol{x} - \boldsymbol{W}^T\boldsymbol{W}\boldsymbol{x})^T(\boldsymbol{x} - \boldsymbol{W}^T\boldsymbol{W}\boldsymbol{x}) \\ &= \frac{1}{2}(\boldsymbol{x}^T\boldsymbol{x} - 2\boldsymbol{x}^T\boldsymbol{W}^T\boldsymbol{W}\boldsymbol{x} + \boldsymbol{x}^T\boldsymbol{W}^T\boldsymbol{W}\boldsymbol{W}^T\boldsymbol{W}\boldsymbol{x}) \end{aligned} \tag{9.38}$$

The continuous-time learning rule for adaptively estimating m principal components (or principal eigenvectors) can again be expressed in terms of a steepest descent gradient approach [64, 65] as the matrix differential equation

$$\frac{d\boldsymbol{W}(t)}{dt} = -\mu\nabla_{\boldsymbol{W}}L(\boldsymbol{W}) \tag{9.39}$$

where μ is the learning rate parameter. Therefore, the gradient of $L(\boldsymbol{W})$ given in (9.38) must be computed. To accomplish this, we need two general results given in Section A.3.4.2 for differentiating a scalar with respect to a matrix, that is,

$$\frac{\partial}{\partial \boldsymbol{A}}\text{trace}(\boldsymbol{BAC}) = \boldsymbol{B}^T\boldsymbol{C}^T \tag{9.40}$$

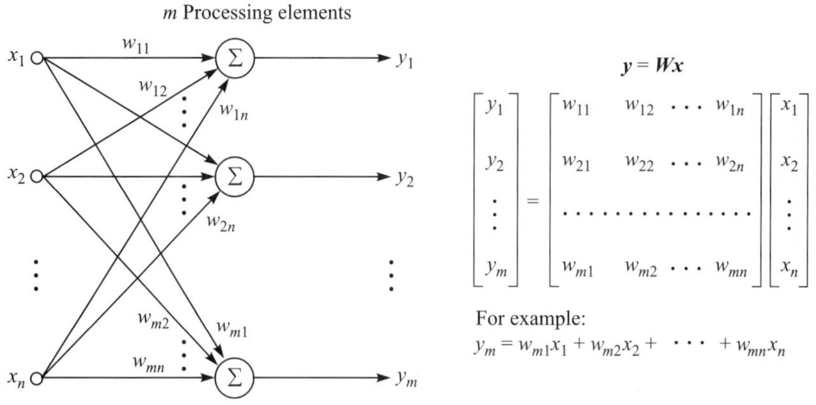

Figure 9.3 Karhunen and Oja's single-layer linear neural network for extraction of multiple (m) principal components.

and

$$\frac{\partial}{\partial A} \operatorname{trace}(\boldsymbol{B}\boldsymbol{A}^T\boldsymbol{C}) = \boldsymbol{C}\boldsymbol{B} \tag{9.41}$$

By using the general results in (9.40) and (9.41), and the appropriate chain rule, the gradient of $L(\boldsymbol{W})$ can be written as

$$\nabla_W L(\boldsymbol{W}) = \frac{\partial L(\boldsymbol{W})}{\partial \boldsymbol{W}} = -\boldsymbol{W}\boldsymbol{x}\boldsymbol{x}^T + \boldsymbol{W}\boldsymbol{x}\boldsymbol{x}^T\boldsymbol{W}^T\boldsymbol{W} - \boldsymbol{W}\boldsymbol{x}\boldsymbol{x}^T + \boldsymbol{W}\boldsymbol{W}^T\boldsymbol{W}\boldsymbol{x}\boldsymbol{x}^T \tag{9.42}$$

However, the two rightmost terms in (9.42) will approach zero very quickly because in the rightmost term $\boldsymbol{W}\boldsymbol{W}^T \to \boldsymbol{I} \in \Re^{m \times m}$, that is,

$$-\boldsymbol{W}\boldsymbol{x}\boldsymbol{x}^T + \underbrace{\boldsymbol{W}\boldsymbol{W}^T}_{\boldsymbol{I}}\boldsymbol{W}\boldsymbol{x}\boldsymbol{x}^T = \boldsymbol{0} \tag{9.43}$$

Therefore, using this approximation, the gradient from (9.42) can be substituted into (9.39) to give the continuous-time learning rule as

$$\frac{d\boldsymbol{W}(t)}{dt} = \mu \boldsymbol{W}(t)\boldsymbol{x}(t)\boldsymbol{x}^T(t)[\boldsymbol{I} - \boldsymbol{W}^T(t)\boldsymbol{W}(t)] \tag{9.44}$$

The discrete-time version of this learning rule is given as

$$\boldsymbol{W}(k+1) = \boldsymbol{W}(k) + \mu(k)\boldsymbol{W}(k)\boldsymbol{x}(k)\boldsymbol{x}^T(k)[\boldsymbol{I} - \boldsymbol{W}^T(k)\boldsymbol{W}(k)] \tag{9.45}$$

where we now consider the learning parameter to be adaptive, that is, $\mu(k)$. The discrete-time *batch* learning rule given in (9.45) is known as the Karhunen-Oja *symmetric subspace learning rule* [20, 41, 62]. The learning rule in (9.45) can also be written using the defined error vector $\boldsymbol{e} = \boldsymbol{x} - \hat{\boldsymbol{x}} = \boldsymbol{x} - \boldsymbol{W}^T\boldsymbol{W}\boldsymbol{x}$, and $\boldsymbol{y} = \boldsymbol{W}\boldsymbol{x}$, as

$$\boldsymbol{W}(k+1) = \boldsymbol{W}(k) + \mu(k)\boldsymbol{y}(k)\boldsymbol{e}^T(k) \tag{9.46}$$

which shows that the learning algorithm is based on error backpropagation. In (9.45) the first term in the weight update expression, that is, $\boldsymbol{W}\boldsymbol{x}\boldsymbol{x}^T$, is the standard Hebbian term which follows directly from the input data. The second term, $\boldsymbol{W}\boldsymbol{x}\boldsymbol{x}^T\boldsymbol{W}^T\boldsymbol{W}$, is a nonlinear decay term which ensures that the synaptic weight matrix $\boldsymbol{W}$ is close to being orthogonal [20, 41, 62].

The scalar discrete-time learning rule follows directly from (9.45) and can be written as

$$w_{ij}(k+1) = w_{ij}(k) + \mu(k)y_i(k)[x_j(k) - \sum_{h=1}^{m} w_{hj}(k)y_h(k)] \tag{9.47}$$

where $i = 1, 2, \ldots, m$ and $j = 1, 2, \ldots, n$, and $y_i(k) = \boldsymbol{w}_i^T(k)\boldsymbol{x}(k)$. In (9.47), if $i, m = 1$, the learning rule reduces to Oja's single processing unit learning rule, given in (9.27).

An interesting aspect of this symmetric subspace learning rule is that the rows of the weight matrix $\boldsymbol{W}$ do not converge to the actual principal eigenvectors of the covariance matrix, but to some linear combination of the first m principal eigenvectors of $\boldsymbol{C}_x$. Therefore, the neural network learns the space spanned by the m principal components; however, the final eigenvectors $\boldsymbol{w}_i$, for $i = 1, 2, \ldots, m$, are not in the actual principal eigenvector directions [37, 39]. In addition, the learning rule given in (9.45) or (9.47) will yield weight vectors w_i, for $i = 1, 2, \ldots, m$, that span the same subspace as the first m principal eigenvectors of $\boldsymbol{C}_x$; however, they will differ from trial to trial. That is, the initial conditions set before training the neural network and the actual data samples used for training will dictate the final set of orthogonal vectors extracted by the network. Also, the eigenvalues (variances) associated with these extracted eigenvectors using the Karhunen-Oja symmetric subspace learning rule tend to be distributed in a uniform fashion. Therefore, the computed variances are not the actual eigenvalues of the covariance matrix $\boldsymbol{C}_x$, but their sum is equal to the sum of the actual eigenvalues of $\boldsymbol{C}_x$.

In the previous section, we presented a method to adaptively change the learning rate $\mu(k)$ with a user-defined forgetting factor (γ) for extracting m principal components, shown in (9.28). This adaptive method can be used for the learning parameter in the symmetric subspace learning rule in (9.45) or (9.47). An alternate approach that requires only computing a single learning parameter and that can be used for extraction of all m principal components using (9.45) is given as

$$\mu(k) = \frac{1}{\dfrac{\gamma}{\mu(k-1)} + \|\boldsymbol{y}(k)\|_2^2}, \qquad \text{and} \qquad \mu(0) = \frac{1}{\|\boldsymbol{y}(0)\|_2^2} \tag{9.48}$$

$$\text{for } k = 1, 2, 3, \ldots, \text{ and } 0 \le \gamma \le 1$$

Therefore, in the learning rules, the learning rate parameter can be taken as either $\mu(k)$ or $\mu_i(k)$, depending on which type of update is used. In (9.48), $\|\boldsymbol{y}\|_2^2$ can be replaced with $\|\boldsymbol{y}\|_\infty$ (that is, $\max|y_i|$, for $i = 1, 2, \ldots, m$).

The next two PCA learning algorithms can be derived directly from the *scalar* symmetric subspace learning rule given in (9.47). In the next section, the scalar learning rule in (9.47) is first derived, and from this result the generalized hebbian algorithm follows directly. Then, from the GHA learning rule, Oja's stochastic gradient ascent algorithm can be derived. We should point out at this time that the scalar forms of all the learning rules are the forms that would actually be implemented in a parallel neural architecture. The *batch* forms of these algorithms are typically used for the purposes of analysis and performing simulations during development of the particular neural network.

9.3.3 Estimation of Several Principal Components—Generalized Hebbian Algorithm

In 1989, Sanger [21, 36] proposed a learning rule that would adaptively extract the first m principal eigenvectors of covariance matrix $\boldsymbol{C}_x$. This learning rule is called the generalized Hebbian algorithm, and it can be derived from the Karhunen-Oja discrete-time learning rule given in (9.45). We will first derive (9.47) by writing the *batch* vector-matrix form of the learning rule in (9.45) as

$$\begin{bmatrix} \boldsymbol{w}_1^T(k+1) \\ \boldsymbol{w}_2^T(k+1) \\ \vdots \\ \boldsymbol{w}_m^T(k+1) \end{bmatrix} = \begin{bmatrix} \boldsymbol{w}_1^T(k) \\ \boldsymbol{w}_2^T(k) \\ \vdots \\ \boldsymbol{w}_m^T(k) \end{bmatrix} + \mu(k) \underbrace{\begin{bmatrix} \boldsymbol{w}_1^T(k)\boldsymbol{x}(k) \\ \boldsymbol{w}_2^T(k)\boldsymbol{x}(k) \\ \vdots \\ \boldsymbol{w}_m^T(k)\boldsymbol{x}(k) \end{bmatrix}}_{\boldsymbol{y}(k)}$$

$$\cdot \left\{ \boldsymbol{x}^T(k) - \underbrace{[\boldsymbol{x}^T(k)\boldsymbol{w}_1(k), \boldsymbol{x}^T(k)\boldsymbol{w}_2(k), \cdots, \boldsymbol{x}^T(k)\boldsymbol{w}_m(k)]}_{\boldsymbol{y}^T(k) = [y_1(k), y_2(k), \cdots, y_m(k)]} \begin{bmatrix} \boldsymbol{w}_1^T(k) \\ \boldsymbol{w}_2^T(k) \\ \vdots \\ \boldsymbol{w}_m^T(k) \end{bmatrix} \right\}$$

$$= \begin{bmatrix} \boldsymbol{w}_1^T(k) \\ \boldsymbol{w}_2^T(k) \\ \vdots \\ \boldsymbol{w}_m^T(k) \end{bmatrix} + \mu(k) \begin{bmatrix} y_1(k) \\ y_2(k) \\ \vdots \\ y_m(k) \end{bmatrix}$$

$$\cdot \left\{ \boldsymbol{x}^T(k) - \underbrace{[y_1(k), y_2(k), \cdots, y_m(k)] \begin{bmatrix} w_{11}(k) & w_{12}(k) & \cdots & w_{1n}(k) \\ w_{21}(k) & w_{22}(k) & \cdots & w_{2n}(k) \\ \cdots & \cdots & \cdots & \cdots \\ w_{m1}(k) & w_{m2}(k) & \cdots & w_{mn}(k) \end{bmatrix}}_{\left[\sum_{h=1}^{m} w_{h1}(k)y_h(k), \sum_{h=1}^{m} w_{h2}k)y_h(k), \cdots, \sum_{h=1}^{m} w_{hn}(k)y_h(k)\right]} \right\} \tag{9.49}$$

Therefore, from (9.49) the scalar discrete-time form of Karhunen-Oja's symmetric subspace learning rule can be written as

$$w_{ij}(k+1) = w_{ij}(k) + \mu(k)y_i(k)[x_j(k) - \sum_{h=1}^{m} w_{hj}(k)y_h(k)] \tag{9.50}$$

for $i = 1,2,\ldots,m$ and $j = 1,2,\ldots,n$, which is the expression presented in (9.47). From the expression in (9.49) we can write this learning rule as

$$\begin{bmatrix} \boldsymbol{w}_1^T(k+1) \\ \boldsymbol{w}_2^T(k+1) \\ \vdots \\ \boldsymbol{w}_m^T(k+1) \end{bmatrix} = \begin{bmatrix} \boldsymbol{w}_1^T(k) \\ \boldsymbol{w}_2^T(k) \\ \vdots \\ \boldsymbol{w}_m^T(k) \end{bmatrix} + \mu(k) \left\{ \boldsymbol{y}(k)\boldsymbol{x}^T(k) - \boldsymbol{y}(k)\boldsymbol{y}^T(k) \begin{bmatrix} \boldsymbol{w}_1^T(k) \\ \boldsymbol{w}_2^T(k) \\ \vdots \\ \boldsymbol{w}_m^T(k) \end{bmatrix} \right\} \tag{9.51}$$

The learning algorithm of Sanger can be derived from (9.51) by "breaking" the symmetry associated with the matrix $\boldsymbol{y}(k)\boldsymbol{y}^T(k)$ in the weight update term. This can be accomplished by only retaining the lower triangular portion of the symmetric matrix $\boldsymbol{y}(k)\boldsymbol{y}^T(k)$, that is, applying the operator $\text{LT}\{\boldsymbol{y}(k)\boldsymbol{y}^T(k)\}$, where $\text{LT}\{\bullet\}$ selects the lower triangular portion of the matrix which includes the diagonal elements of the matrix [21]. Applying this operator to the matrix $\mathbf{y}(k)\mathbf{y}^T(k)$ in (9.51), we obtain

$$\begin{bmatrix} \mathbf{w}_1^T(k+1) \\ \mathbf{w}_2^T(k+1) \\ \vdots \\ \mathbf{w}_m^T(k+1) \end{bmatrix} = \begin{bmatrix} \mathbf{w}_1^T(k) \\ \mathbf{w}_2^T(k) \\ \vdots \\ \mathbf{w}_m^T(k) \end{bmatrix} + \mu(k)\left\{\mathbf{y}(k)\mathbf{x}^T(k) - \mathrm{LT}\{\mathbf{y}(k)\mathbf{y}^T(k)\}\begin{bmatrix} \mathbf{w}_1^T(k) \\ \mathbf{w}_2^T(k) \\ \vdots \\ \mathbf{w}_m^T(k) \end{bmatrix}\right\}$$

$$= \begin{bmatrix} \mathbf{w}_1^T(k) \\ \mathbf{w}_2^T(k) \\ \vdots \\ \mathbf{w}_m^T(k) \end{bmatrix} + \mu(k)\left\{\mathbf{y}(k)\mathbf{x}^T(k) - \mathrm{LT}\begin{bmatrix} y_1(k)y_1(k) & y_1(k)y_2(k) & \cdots & y_1(k)y_m(k) \\ y_2(k)y_1(k) & y_2(k)y_2(k) & \cdots & y_2(k)y_m(k) \\ \cdots & \cdots & \cdots & \cdots \\ y_m(k)y_1(k) & y_m(k)y_2(k) & \cdots & y_m(k)y_m(k) \end{bmatrix}\begin{bmatrix} \mathbf{w}_1^T(k) \\ \mathbf{w}_2^T(k) \\ \vdots \\ \mathbf{w}_m^T(k) \end{bmatrix}\right\}$$

$$= \begin{bmatrix} \mathbf{w}_1^T(k) \\ \mathbf{w}_2^T(k) \\ \vdots \\ \mathbf{w}_m^T(k) \end{bmatrix} + \mu(k)\left\{\mathbf{y}(k)\mathbf{x}^T(k) - \underbrace{\begin{bmatrix} y_1(k)y_1(k) & 0 & \cdots & 0 \\ y_2(k)y_1(k) & y_2(k)y_2(k) & \cdots & 0 \\ \cdots & \cdots & \cdots & \cdots \\ y_m(k)y_1(k) & y_m(k)y_2(k) & \cdots & y_m(k)y_m(k) \end{bmatrix}\begin{bmatrix} \mathbf{w}_1^T(k) \\ \mathbf{w}_2^T(k) \\ \vdots \\ \mathbf{w}_m^T(k) \end{bmatrix}}_{\begin{bmatrix} y_1(k)[y_1(k)\mathbf{w}_1^T(k)] \\ y_2(k)[y_1(k)\mathbf{w}_1^T(k) + y_2(k)\mathbf{w}_2^T(k)] \\ y_3(k)[y_1(k)\mathbf{w}_1^T(k) + y_2(k)\mathbf{w}_2^T(k) + y_3(k)\mathbf{w}_3^T(k)] \\ \vdots \\ y_m(k)[y_1(k)\mathbf{w}_1^T(k) + y_2(k)\mathbf{w}_2^T(k) + y_3(k)\mathbf{w}_3^T(k) + \cdots + y_m(k)\mathbf{w}_m^T(k)] \end{bmatrix}}\right\} \tag{9.52}$$

Therefore, from (9.52) Sanger's discrete-time scalar PCA learning rule can be written as

$$w_{ij}(k+1) = w_{ij}(k) + \mu(k)y_i(k)[x_j(k) - \sum_{h=1}^{i} w_{hj}(k)y_h(k)] \tag{9.53}$$

where $i = 1, 2, \ldots, m$ and $j = 1, 2, \ldots, n$.

Comparing Sanger's GHA learning rule in (9.53) to the symmetric subspace learning rule of Karhunen and Oja in (9.50), we see that the only difference is in the summation limit. By breaking the symmetry of $\mathbf{y}(k)\mathbf{y}^T(k)$, that is, only retaining the lower triangular portion of the matrix as shown in (9.52), as i increases from $1 \to m$, the number of terms in the summation in (9.53) increases accordingly, as shown in (9.52). For the symmetric subspace learning rule in (9.50), as i increases from $1 \to m$, the number of terms in the summation is constant, that is, m (the first m principal components). This difference in the summation limit in the two learning rules accounts for the Karhunen-Oja symmetric subspace learning rule's not extracting the actual principal eigenvectors, but instead a linear combination of them that spans the same subspace as the m actual principal eigenvectors, whereas Sanger's PCA learning rule (GHA) in (9.53) extracts the m actual principal eigenvectors of the covariance matrix $\boldsymbol{C}_x$ [21, 36]. Furthermore, it has been proved that the neural network trained with Sanger's GHA learning rule will converge from any initial set of random weights to the eigenvalues and eigenvectors of the covariance matrix $\boldsymbol{C}_x$ in decreasing eigenvalue order [39]. As with the Karhunen-Oja symmetric subspace learning rule, Sanger's GHA can have an adaptive learning parameter adjusted in time according to either (9.28) or (9.48). Therefore, in the learning rule (9.53), the learning rate parameter can be taken as either $\mu(k)$ or $\mu_i(k)$, depending on which type of update is used.

The GHA learning rule in (9.53) can also be written in a recursive vector form as

$$\boldsymbol{w}_i(k+1) = \boldsymbol{w}_i(k) + \mu_i(k) y_i(k) \tilde{\boldsymbol{x}}_i(k) \tag{9.54}$$

where

$$\tilde{\boldsymbol{x}}_i(k) = \tilde{\boldsymbol{x}}_{i-1}(k) - \boldsymbol{w}_i(k) y_i(k) \tag{9.55}$$

with

$$\tilde{\boldsymbol{x}}_0(k) = \boldsymbol{x}(k) \qquad \text{for } i = 1, 2, \ldots, m \tag{9.56}$$

9.3.4 Estimation of Several Principal Components—Stochastic Gradient Ascent Algorithm

Another PCA neural network learning algorithm, which is closely related to the GHA of Sanger, is the stochastic gradient ascent (SGA) algorithm proposed by Oja [26]. The SGA learning algorithm can be derived in a straightforward manner from the GHA. The GHA given in (9.53) has the summation limit as i; in the SGA algorithm the summation limit is taken to be $i-1$ by *breaking up* the summation in (9.53) as

$$\sum_{h=1}^{i} w_{hj}(k) y_h(k) \Rightarrow \sum_{h=1}^{i-1} w_{hj}(k) y_h(k) + w_{ij}(k) y_i(k) \tag{9.57}$$

Then if the summation on the right-hand side of (9.57) is weighted by a scalar α, the discrete-time scalar SGA algorithm of Oja can be written as

$$w_{ij}(k+1) = w_{ij}(k) + \mu(k) y_i(k) [x_j(k) - w_{ij}(k) y_i(k) - \alpha \sum_{h=1}^{i-1} w_{hj}(k) y_h(k)] \tag{9.58}$$

where $i = 1,2, \ldots, m$ and $j = 1,2, \ldots, n$ and $\alpha > 1$ (and is typically taken to be $\alpha = 2$). A major advantage of the SGA algorithm over the GHA is its behavior when extracting the less dominant components. That is, the speed of convergence is better for the SGA algorithm over the GHA when one is extracting the principal eigenvectors that are less dominant than the first few. Another advantage of the SGA algorithm is that it can be easily extended for extraction of minor components [26].

9.3.5 Estimation of Several Principal Components—Adaptive Principal-Component Extraction Algorithm

The adaptive principal-component extraction (APEX) algorithm for PCA is based on anti-Hebbian learning and is considered a *decorrelating*-type algorithm [40]. APEX, which has application to a new type of PCA called constrained PCA (CPCA) [66], was first proposed by Kung et al. [25, 30, 81], and it can concurrently estimate the principal eigenvectors of the covariance matrix $\boldsymbol{C}_x$ by successively extracting them from the input patterns in a parallel manner [67]. A related method developed by Bannour and Azimi-Sadjadi [68] is based on a deflation transformation [96, 97], as opposed to lateral connections between the output nodes, which renders their method essentially sequential. The linear APEX model is shown in Figure 9.4. The main difference between this architecture and the previous PCA networks lies in the additional lateral connections at the outputs of the network, as shown in Figure 9.4. These lateral connections provide feedback in the topology and account for orthogonalization of the synaptic weights. Another interesting feature of the APEX architecture is that it allows the size of the model to *grow* or *shrink* without need for retraining the *old* units. That is, if it is desired to extract one or more additional components, the appropriate number of additional neurons can be simply *attached* to the previous model.

For the model given in Figure 9.4, it is assumed that $\boldsymbol{x} \in \Re^{n\times 1}$, $\boldsymbol{y} \in \Re^{m\times 1}$, $\boldsymbol{C} \in \Re^{m\times(m-1)}$, and $\boldsymbol{W} \in \Re^{m\times n}$. Therefore, two different types of synaptic weights are utilized in the APEX model, feedforward connection weights $\boldsymbol{W}$ and lateral connection weights $\boldsymbol{C}$.

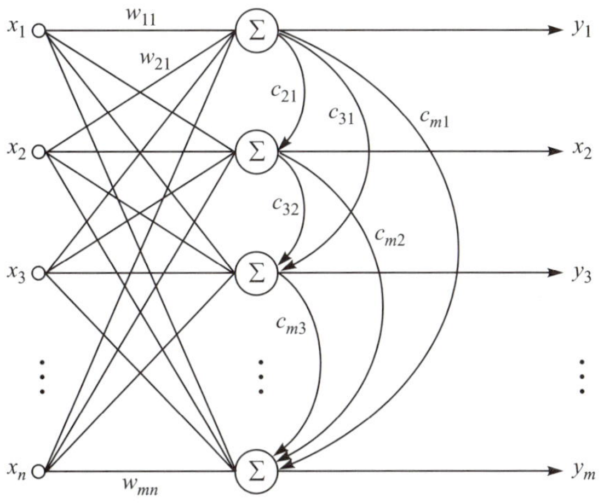

Figure 9.4 The parallel APEX model with linear processing units.

1. The *feedforward connection weights* $\boldsymbol{w}_i(k) = [w_{i1}(k) \;\; w_{i2}(k) \;\; \cdots \;\; w_{in}(k)]^T$ for $i = 1, 2, \cdots, m$ (where m is the desired number of principal components to be extracted) are the synapses associated with the branches from the inputs to the linear neurons. The adaptation of the feedforward synaptic weights is in accordance with standard Hebbian learning, where the connections constitute *excitatory* signals, thereby providing *self-amplification*.
2. The *lateral connection weights* are associated with the connection branches made from the individual neuron outputs to all successive neurons for $i = 1, 2, \ldots, m-1$, thereby providing feedback to the network (see Figure 9.4). The adaptation of the lateral connection weights is performed in accordance with anti-Hebbian learning.

The feedback (or lateral) connection weight matrix can be written as

$$\boldsymbol{C}(k) = \begin{bmatrix} 0 & 0 & 0 & 0 & 0 & \cdots & 0 \\ c_{21}(k) & 0 & 0 & 0 & 0 & \cdots & 0 \\ c_{31}(k) & c_{32}(k) & 0 & 0 & 0 & \cdots & 0 \\ c_{41}(k) & c_{42}(k) & c_{43}(k) & 0 & 0 & \cdots & 0 \\ c_{51}(k) & c_{52}(k) & c_{53}(k) & c_{54}(k) & 0 & \cdots & 0 \\ \cdots & \cdots & \cdots & \cdots & \cdots & \cdots & \cdots \\ c_{m1}(k) & c_{m2}(k) & c_{m3}(k) & c_{m4}(k) & c_{m5}(k) & \cdots & c_{mm-1}(k) \end{bmatrix} = \begin{bmatrix} \boldsymbol{c}_1(k) \\ \boldsymbol{c}_2(k) \\ \boldsymbol{c}_3(k) \\ \boldsymbol{c}_4(k) \\ \boldsymbol{c}_5(k) \\ \vdots \\ \boldsymbol{c}_m(k) \end{bmatrix} \tag{9.59}$$

where the $\boldsymbol{c}_i \in \Re^{1\times(m-1)}$, for $i = 1, 2, \ldots, m$, are the rows of the $\boldsymbol{C}$ matrix. The first row of $\boldsymbol{C}$ shown in (9.59) has all zero elements because there are no lateral connections associated with the first neuron. Moreover, the extraction of the first principal eigenvector ($\boldsymbol{w}_1$) is carried out in accordance with the single-neuron learning rule of Oja, as given in (9.25) (cf. Sect. 9.3.1). The lateral weights in (9.59) are used to successively subtract the principal-component information as each successive neuron is trained (see Figure 9.4) according to

$$y_i(k) = \boldsymbol{w}_i^T(k)\boldsymbol{x}(k) - \boldsymbol{c}_i(k)\boldsymbol{y}(k) \tag{9.60}$$

where $i = 1, 2, \ldots, m$, and $\boldsymbol{y}(k) = \boldsymbol{W}(k)\boldsymbol{x}(k)$. The *lateral connection weights* are updated according to the anti-Hebbian learning rule given as

$$\boldsymbol{c}_i(k+1) = \boldsymbol{c}_i(k) + \mu[y_i(k)\boldsymbol{y}^T(k) - y_i^2(k)\boldsymbol{c}_i(k)] \tag{9.61}$$

for $i = 2, 3, \ldots, m$ (i starts at 2 because the first row of the lateral connection weight matrix $\boldsymbol{C}$, that is, $\boldsymbol{c}_1$, is assumed to have all zero elements). The *feed-*

forward connection weights are updated in accordance with the standard Hebbian learning rule given as

$$w_i(k+1) = w_i(k) + \mu[y_i(k)x(k) - y_i^2(k)w_i(k)] \tag{9.62}$$

where $i = 1, 2, \ldots, m$. The learning rate parameter is assumed to be the same in both update equations, that is, (9.61) and (9.62). In Kung et al. [30], two different architectures are presented for adaptive extraction of principal components, that is, parallel APEX and sequential APEX. Given below is an algorithm for the parallel APEX model. This model is more consistent with *neural*-type learning, that is, parallel processing.

Summary of the parallel APEX algorithm

Step 1. Select the learning rate parameter $\mu > 0$. This can be a fixed value; that is, a reasonably small value can be selected for μ, or it can be precomputed (estimated) using methods given by Kung and Diamantaras [69], Kung et al. [30], and Haykin [40]. Alternatively, μ can be calculated iteratively using the method given in (9.28), which is also presented in Kung et al. [30].

Step 2. Initialize W and C to some random values, and set $k = 1$.

Step 3. For $i = 1$ compute

$$y_1(k) = w_1^T(k)x(k) \tag{9.63}$$

where w_1^T is the first row of the feedforward connection weight matrix W, and update the first synaptic weight vector according to the learning rule of Oja for the single-neuron model, that is,

$$w_1(k+1) = w_1(k) + \mu[y_1(k)x(k) - y_1^2(k)w_1(k)] \qquad (w_1 \in \Re^{n\times 1}) \tag{9.64}$$

Step 4. For $i = 2, 3, \ldots, m$ compute

$$y(k) = W(k)x(k) \tag{9.65}$$

where it is important to note that in this step of the algorithm $W(k) = [\,w_1(k) \quad w_2(k) \quad \cdots \quad w_{m-1}(k)\,]^T$, therefore, $y \in \Re^{(m-1)\times 1}$. Next from (9.60) compute

$$y_i(k) = w_i^T(k)x(k) - c_i(k)y(k) \tag{9.66}$$

and then compute the update for the feedforward connection weights according to (9.62), that is,

$$w_i(k+1) = w_i(k) + \mu[y_i(k)x(k) - y_i^2(k)w_i(k)] \tag{9.67}$$

and finally compute the update for the lateral connection weights according to (9.61), that is,

$$c_i(k+1) = c_i(k) + \mu[y_i(k)y^T(k) - y_i^2(k)c_i(k)] \tag{9.68}$$

Step 5. Increase k by 1, that is, $k \leftarrow k+1$ [and the next input $x(k)$ is used].

Step 6. If the weights in W have converged, then stop; else, go to step 3.

With the parallel APEX learning algorithm given above, the principal eigenvectors of the covariance matrix C_x will be extracted in decreasing eigenvalue order. Also, if a total of N zero-mean input vectors have been measured and stored, that is, $X = \{x(k)\} \in \Re^{n\times N}$ for $k = 1, 2, \ldots, N$, the learning algorithm above can be used *off-line*. That is, each training epoch would consist of presenting all N exemplars [$x(k)$, for $k = 1, 2, \ldots, N$], and this process can be repeated for step 2 through step 6 for a total of M training epochs until the weights in W have converged (or a total of NM repeated presentations of the

$\boldsymbol{x}$ input vectors). For the *off-line* training process, after the first training epoch, in step 2, $\boldsymbol{W}$ and $\boldsymbol{C}$ are set to the values from the previous training epoch.

The various PCA neural networks that have been presented can be considered as detectors of orthogonal features, which is a fundamental problem in statistical pattern recognition. This process is similar in the brain of mammals that can encode mutually independent aspects of large amounts of data contained in the input patterns presented to it. For example, when humans view a complex scene, the data space is transformed to a feature space. However, the transformation is designed such that the data set may be represented by a *reduced* number of *effective* features, which retains most of the intrinsic information content of the data, that is, a dimensionality reduction, which is precisely what PCA is accomplishing [40]. PCA is probably the oldest and the most well-known technique in multivariate analysis. It was first introduced by Pearson [98] who used PCA in a biological setting to reformulate linear regression analysis [4].

EXAMPLE 9.2. In this example, PCA is used for an image coding problem [21, 40, 99–104]. The technique used for coding an image is relatively simple and similar to the method used by Sanger [21], and also presented in Haykin [40]. Figure 9.5 is a diagram of the layout of a generic image that is decomposed into 8×8 nonoverlapping blocks. The digitized images used in this example consisted of 240×352 eight-bit pixels. Each pixel in the image is an integer value in the range [0, 255] constituting 256 gray levels, where 255 is white and 0 is black. For each of the 1,320 8×8 blocks in the image, depicted in Figure 9.5, that is, Ω_{ij}, for $i = 1, 2, \ldots, 30$ and $j = 1, 2, \ldots, 44$, a *training* vector is generated by

$$\boldsymbol{x}_q = \text{vec}(\Omega_{ij}^T) = [(\omega_{11}, \omega_{12}, \cdots, \omega_{18})(\omega_{21}, \omega_{22}, \cdots, \omega_{28},) \cdots (\omega_{81}, \omega_{82}, \cdots, \omega_{88})]^T \tag{9.69}$$

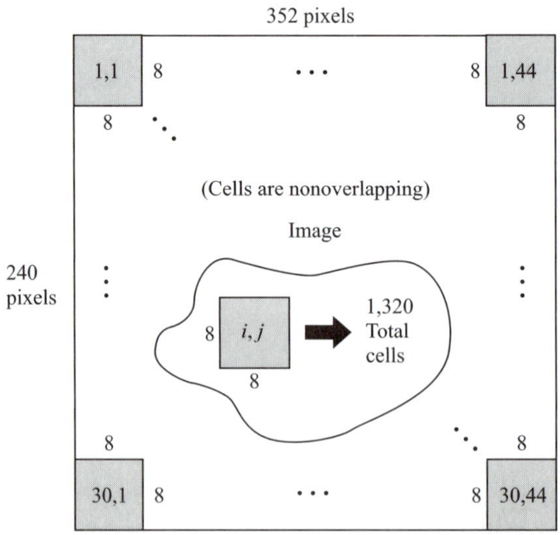

Figure 9.5 Nonoverlapping image coding process, where each 8-bit pixel has a 256 gray-level range.

for $q = 1, 2, \ldots, 1{,}320$, therefore, each of the 1,320 $\boldsymbol{x}_q$ vectors is 64×1 in dimension (see Sect. A.2.17). An approximation to the covariance matrix $\boldsymbol{C}_x$ can be computed according to (9.19), that is,

$$\boldsymbol{C}_x = \mathrm{E}[\boldsymbol{x}\boldsymbol{x}^T] \cong \frac{1}{N}\sum_{q=1}^{N} \boldsymbol{x}_q \boldsymbol{x}_q^T \qquad \text{for } N = 1{,}320 \tag{9.70}$$

Two PCA methods are used for extracting the principal eigenvectors of $\boldsymbol{C}_x$: a direct eigenvalue decomposition (EVD) approach (cf. Sects. 7.6 and 9.2) and a neural network approach, that is, using Oja's symmetric subspace learning algorithm. Regardless of which method is used, once the principal eigenvectors of $\boldsymbol{C}_x$ have been extracted, a set of these vectors is used to code the image cells. That is, a coding ratio of 16 : 1 is used to observe the effect of the coding on the image. For a 16:1 coding ratio only the first four principal eigenvectors, of the 64 for each cell, are used to *estimate* the image features within each of the cells. Therefore, the transformation matrix is given as $\boldsymbol{W} = [\boldsymbol{w}_1, \boldsymbol{w}_2, \ldots, \boldsymbol{w}_m]^T$ as in (9.3), for $m = 4$. Estimation of the image features from cell to cell is determined according to (9.18), that is,

$$\hat{\boldsymbol{x}}_q = \boldsymbol{W}^T \boldsymbol{W} \boldsymbol{x}_q \tag{9.71}$$

for $q = 1, 2, \ldots, 1{,}320$, that is, each of the image cells. After the image coding step is completed, the image must be reconstructed. This is carried out by simply performing an "inverse vec" operation, that is, the inverse operation of the process that is given in (9.69). When the inverse operation has been completed for each of the successive cells, the image will be reconstructed using an approximation for each cell, that is, $\hat{\boldsymbol{x}}_q \Rightarrow \hat{\Omega}_{ij}$.

Figure 9.6(*a*) shows the original tiger image before image coding, and in Figure 9.6(*b*) the tiger image is shown after applying the 16:1 (i.e., using the first four principal eigenvectors from the EVD method) coding process of each image cell and then reconstructing the image as explained above. Comparison of the two images in Figure 9.6 reveals a very discernible image for the coded image of the tiger in Figure 9.6(*b*), compared to the original image in Figure 9.6(*a*).

Figure 9.7 shows the first four 8×8 masks (principal eigenvectors) used to code the tiger image in Figure 9.6(*b*). The same results were obtained using Sanger's GHA, where a 64-input, four-neuron single-layer neural network (as shown in Figure 9.3) was used and trained with the learning rule in (9.53). As a comparison to the results shown above, the same single-layer neural network architecture was used but trained by the Karhunen-Oja symmetric subspace learning algorithm. The results for the coded tiger image are shown in Figure 9.8. Comparing Figure 9.6(*b*) using the EVD method (or Sanger's GHA) to the image in Figure 9.8 using the Karhunen-Oja symmetric subspace learning algorithm to encode the same image, we can see that the two coded images have similar quality. Figure 9.9 shows the first four 8×8 masks used to code the tiger image in Figure 9.6(*b*) using the Karhunen-Oja symmetric subspace learning rule. The similarities in the coded images shown in Figures 9.6(*b*) and 9.8 are evident in spite of the differences in the masks shown in Figures 9.7 and 9.9. Even though the Karhunen-Oja symmetric subspace learning rule does not yield the actual principal eigenvectors of covariance matrix $\boldsymbol{C}_x$, the extracted synaptic network weights are, nonetheless, linear combinations of the principal eigenvectors that span the same subspace that the principal eigenvectors span.

To illustrate the *generalization* capability of the neural network trained using Oja's symmetric subspace learning rule, the tiger masks shown in Figure 9.9 were

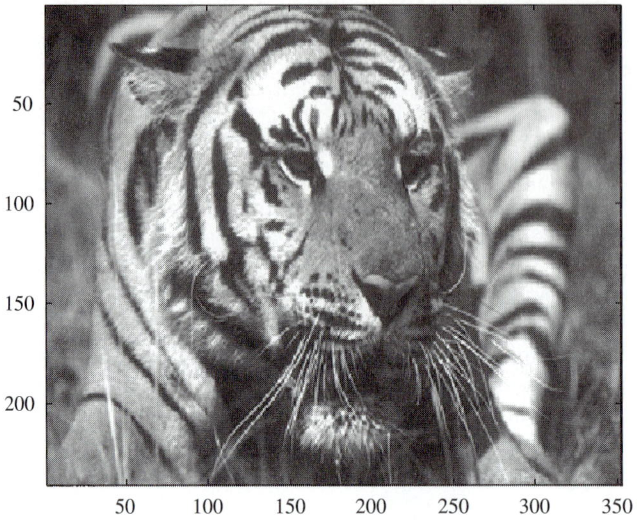

(a)

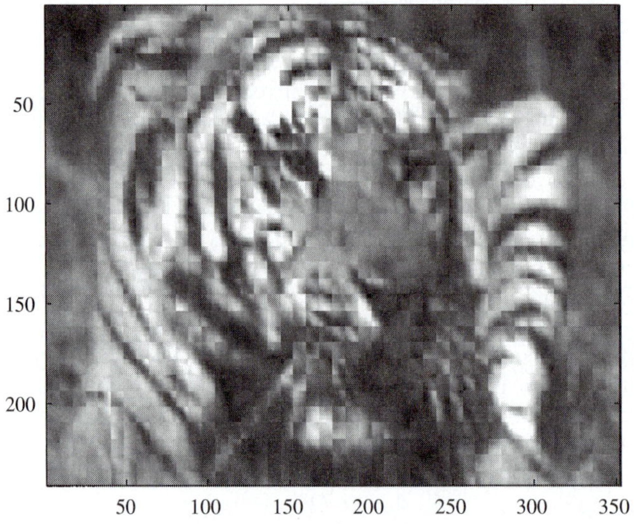

(b)

Figure 9.6 (*a*) Original tiger image has 240 × 352 eight-bit pixels used for the coding process. (*b*) Image of part (*a*) coded with a 16:1 coding ratio; that is, only four principal eigenvectors were retained for coding using the EVD method.

used to code a different image that the neural network had not previously seen. The results are shown in Figure 9.10, where Figure 9.10(*a*) shows the original fruit image and Figure 9.10(*b*) is the fruit image in Figure 9.10(*a*) coded using the tiger masks from the Karhunen-Oja symmetric subspace learning rule for a 16:1 coding ratio (see Figure 9.9). Apparently, the tiger and fruit images are statistically similar because of the relatively good coding performance that is evident in

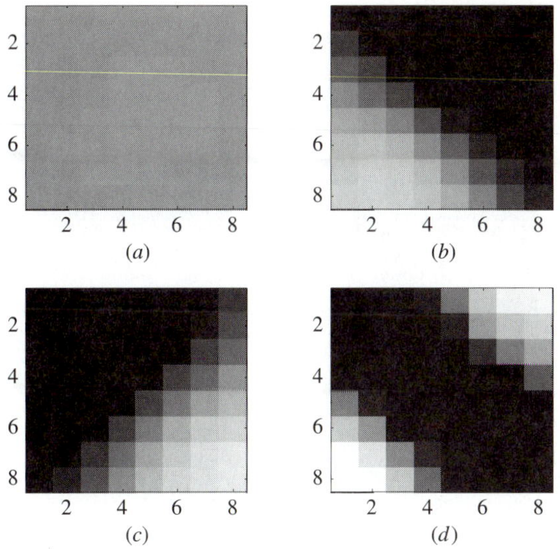

Figure 9.7 (*a*)–(*d*) Four 8×8 masks extracted by the EVD method for the tiger image shown in Figure 9.6(*a*). These are the first four principal eigenvectors of the covariance matrix given in (9.70).

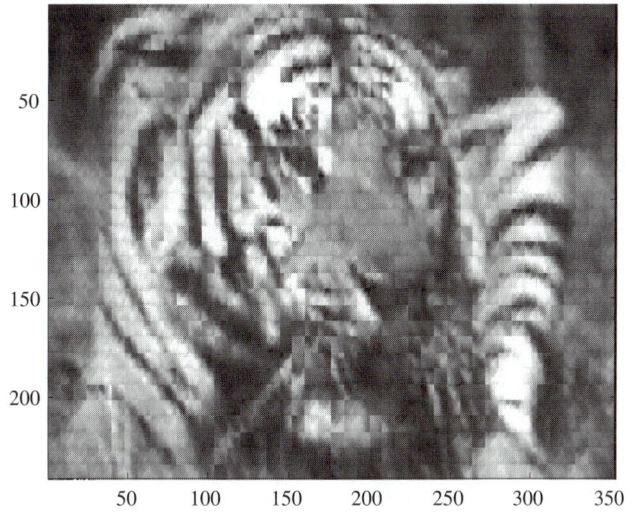

Figure 9.8 Image of Figure 9.6(*a*) (240×352 eight-bit pixels) which is coded with a 16:1 coding ratio; that is, only four principal eigenvectors are retained for coding using the Karhunen-Oja symmetric subspace learning algorithm. (*Copyright 2000 McGraw-Hill and its licensors. All rights reserved.*)

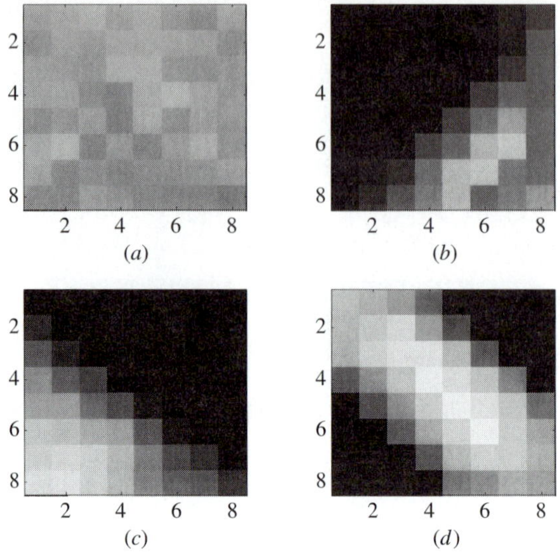

Figure 9.9 (*a*)–(*d*) Four 8 × 8 masks learned by the neural network trained using the Karhunen-Oja symmetric subspace learning rule for the tiger image shown in Figure 9.6(*a*).

Figure 9.10(*b*), when the tiger masks in Figure 9.9 were applied to the coding of the fruit image [shown in Figure 9.10(*a*)] resulting in the coded fruit image shown in Figure 9.10(*b*).

9.3.6 Nonlinear Principal-Component Analysis (NLPCA) and Robust PCA

Standard linear PCA presented in the previous sections can arise as an optimal solution to several different signal representation problems, which can be summarized as follows:

1. Maximization of linearly transformed variances $E\{[\boldsymbol{w}_i^T \boldsymbol{x}]^2\}$ or the outputs of a linear network under the orthogonality constraints, $\boldsymbol{W}\boldsymbol{W}^T = \boldsymbol{I} \in \Re^{m \times m}$.
2. Minimization of the mean-square representation error $E\{\|\boldsymbol{x} - \hat{\boldsymbol{x}}\|^2\}$, where the approximation to the input data $\boldsymbol{x}$ is given by the lower-dimensional linear subspace $\hat{\boldsymbol{x}} = \boldsymbol{W}^T \boldsymbol{W} \boldsymbol{x}$.
3. Uncorrelatedness of the linear network outputs $y_i = \boldsymbol{w}_i^T \boldsymbol{x}$, for $i = 1, 2, \ldots, m$, for different neurons after an orthogonal transform $\boldsymbol{W}\boldsymbol{W}^T = \boldsymbol{I}$.
4. Minimization of representation entropy.

Derivations of the optimal PCA solutions with associated assumptions and conditional constraints can be found in [41]. Karhunen and Joutsensalo [41] justify that linear PCA arises as an optimal solution to these different information representation problems due to the fact that the solution is based

Figure 9.10 (*a*) Original fruit image has 240 × 352 eight-bit pixels used for the coding process. (*b*) Image of part (*a*) coded with a 16:1 coding ratio using the tiger masks in Figure 9.9 learned by the neural network using Karhunen-Oja's learning rule. (*Copyright 2000 McGraw-Hill and its licensors. All rights reserved.*)

on only the second-order statistics of the inputs. Linear PCA neural networks, and the associated learning algorithms, have some limitations that diminish their attractiveness in certain situations [41]:

1. Standard PCA networks are able to realize only linear input/output mappings.

2. The eigenvectors needed in standard PCA can be computed efficiently using well-known numerical methods. Gradient-type neural PCA learning algorithms typically converge relatively slowly, and to achieve relatively good accuracy requires an excessive number of iterations for large problems.
3. Principal components are defined solely by the data covariances, or correlations. These second-order statistics of the inputs characterize completely only Gaussian data and stationary, linear processing operations.
4. Linear PCA networks cannot usually separate independent subsignals from their linear mixture.

If a PCA-type network contains nonlinearities, the setting becomes more suitable for a neural realization, and (1) the input/output mappings become in general nonlinear, which is usually a major justification for using neural networks, thereby processing of data is often more efficient; (2) neural algorithms become more competitive relative to classical PCA methods; (3) using nonlinearities implicitly introduces higher-order statistics into the computations; and (4) the outputs of standard PCA networks are usually at most uncorrelated but not independent, which would be more desirable in certain cases; adding nonlinearities to a PCA network increases the independence of the outputs so that the input signals can sometimes be essentially separated from their constituent mixture (Karhunen and Joutsensalo [41]). Generalizations of linear PCA can result in robust and nonlinear extensions of the standard neural PCA algorithms. The resulting learning algorithms derived by considering generalizations of the optimization problems leading to standard PCA can be divided into two classes [41]: *robust PCA algorithms* [70–73, 105] and *nonlinear PCA (NLPCA) algorithms* [42–47, 52, 60, 61, 63, 106, 107].

Robust PCA can be defined so that the criterion to be optimized grows less than quadratically (i.e., a nonquadratic optimization criterion), and the constraint conditions are the same as for the standard PCA solution. The nonquadratic optimization criterion introduces a nonlinearity into the gradient algorithms; however, this makes the results more robust against impulsive and colored noise, and data outliers [7, 12]. Therefore, the gradient algorithms resulting from the nonquadratic optimization criteria tend to be more robust than their linear (i.e., quadratic performance criteria) counterparts. More specifically, a quadratic performance criterion weighs heavily the large error values that are often due to strong noise components or outliers. Conversely, a nonquadratic criterion would be a better choice for the case of strong noise or outliers because it would grow more slowly than a quadratic function and would weigh less any large errors. Also, the nonquadratic optimization criterion allows for higher-order statistics [43, 107–110] associated with the inputs to be taken into account. There are several important problems that cannot be adequately solved using only second-order statistics. Typically, the weight vectors of the network neurons, that is, the basis vectors of the expansion, are required to be mutually orthogonal. Robust PCA problem settings usually lead to learning algorithms which are *mildly* nonlinear, where the nonlinearities appear in the network architecture at only selected locations [41]. That is, at least some of the neuron outputs of the robust PCA network will still have linear responses $y_i = \boldsymbol{w}_i^T \boldsymbol{x}$, for $i = 1, 2, \ldots, m$, where $\boldsymbol{w}_i$ is the weight vector of the ith neuron.

Conversely, in nonlinear PCA architectures, all the outputs of the network neurons have nonlinearities [that is, $g(y_i) = g(\boldsymbol{w}_i^T \boldsymbol{x})$, where $g(\bullet)$ is a selected nonlinearity], and the optimization criterion is still formulated as a quadratic function, as in the linear PCA case. A third type of nonlinear PCA neural network, more general than the first two types described above, is based on a nonquadratic optimization criterion as in *robust PCA algorithms*, and nonlinearities appear at all the output neurons, as in the case of *nonlinear PCA algorithms*. In Karhunen and Joutsensalo [41] it is stated that probably the most important single result is the derivation of the well-known generalized Hebbian algorithm (cf. Sect. 9.3.3) of Sanger [21], as well as its robust and nonlinear counterparts, from the variance maximization and mean-square error minimization problems. Related approaches for extraction of statistically independent features of data can be found in Parra et al. [111], which is a nonlinear version of independent-component analysis [112–114] (cf. Sect. 10.8), and symplectic nonlinear independent-component analysis [115].

One structure of the NLPCA network is shown in Figure 9.11 for the symmetric case, where the feedback connections (shown as dotted lines) are removed after the network has been trained; thus, the network is then strictly a feedforward architecture. Another NLPCA structure is the hierarchic architecture shown in Figure 9.12. As with the symmetric architecture, the feedback connections are removed after the network has been trained. There are drawbacks associated with PCA networks that contain nonlinearities: (1) The mathematical analysis of many learning rules is often inherently more difficult than that with linear PCA networks; therefore, properties of the network with

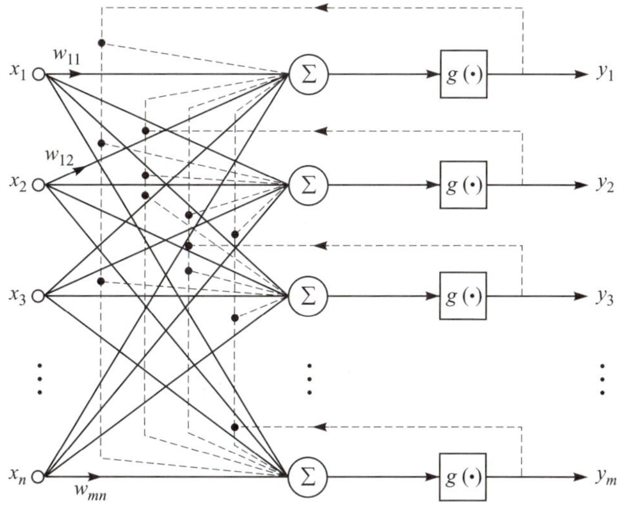

Figure 9.11 Architecture for the symmetric nonlinear PCA neural network. The feedback connections (dashed lines) are only needed for the training phase.

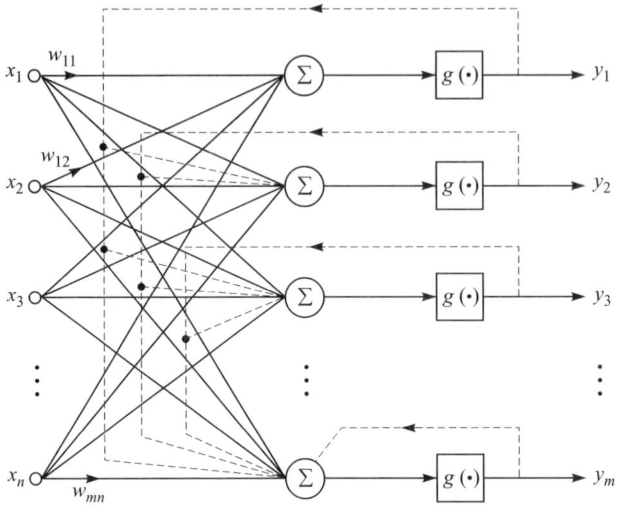

Figure 9.12 Architecture for the hierarchic nonlinear PCA neural network. The feedback connections (dashed lines) are only needed for the training phase.

nonlinearities are often less understood. (2) Nonlinear learning algorithms are more complicated and may sometimes result in convergence to a local minimum. (3) Adding nonlinearities to the network may not necessarily avoid certain problems. Therefore, introducing nonlinearities to a PCA network should not be done arbitrarily. Other structures have been proposed for NLPCA, for example, recurrent generalizations and nonlinear PCA architectures with lateral connections [45].

As previously stated, robust PCA neural networks have a nonlinearity introduced into the associated gradient algorithms which can have the effect of "robustifying" the neural networks with respect to impulsive and colored noise, and data outliers. That is, formulating a nonquadratic optimization criterion tends to reject the influence of outliers by placing less emphasis on the input data vectors, which can have relatively large errors. Nonlinear extensions of linear PCA neural networks were heuristically discussed in 1992 [58], but not until 1993 was the mathematical foundation actually set for robust PCA-type neural networks [42, 70-73]. After 1993 there were many research results reported for robust PCA [41, 43–48, 74]. The three types of robust PCA neural subspace learning algorithms that follow are attributed to Karhunen and Joutsensalo [41, 43, 44, 48, 70, 71, 74, 75].

In order for the presentation of each case to directly follow that of Karhunen and Joutsensalo, the synaptic weight matrix is now defined as $\boldsymbol{W} \in \Re^{n\times m}$, with the inputs $\boldsymbol{x} \in \Re^{n\times 1}$, which are considered to be zero-mean stochastic vectors, and the outputs are $\boldsymbol{y} \in \Re^{m\times 1}$. The three robust PCA subspace discrete-time learning algorithms can be derived from either the variance maximization problem or the minimum representation error criterion, and are written as

$$W(k+1) = W(k) + \mu(k)[I - W(k)W^T(k)]x(k)g[x^T(k)W(k)] \tag{9.72}$$

$$W(k+1) = W(k) + \mu(k)\{x(k)g[e^T(k)]W(k) + g[e(k)]x^T(k)W(k)\} \tag{9.73}$$

where

$$e(k) = x(k) - \hat{x}(k) = x(k) - W(k)W^T(k)x(k) = [I - W(k)W^T(k)]x(k) \tag{9.74}$$

and

$$W(k+1) = W(k) + \mu(k)g[e(k)]x^T(k)W(k) \tag{9.75}$$

In (9.72), (9.73), and (9.75), $\mu(k) > 0$ is the *gain* parameter controlling the learning rate. The learning parameter can be adaptively changed each step according to (9.28) or (9.48). These robust PCA learning algorithms can be derived from two separate performance criteria; for example, first the learning rule in (9.72) can be derived from the performance criterion associated with the variance maximization problem which is modified as (for each w_i, $i = 1, 2, \ldots, m$)

$$J_1(w_i) = \mathrm{E}\{f[x^T w_i]|w_i\} + \frac{1}{2}\sum_{j=1}^{m} \lambda_{ij}(w_i^T w_j - \delta_{ij}) \tag{9.76}$$

which is to be maximized. Only instantaneous inputs are considered; therefore, the expectation operator is dropped in (9.76) in computing the gradient. In (9.76), δ_{ij} is the Kronecker delta, the first term on the right-hand side (i.e., the conditional expectation) is to be maximized (on the condition that the weight vectors of the neurons be orthonormal), and the summation imposes the necessary orthonormality constraints (that is, $w_i^T w_j = \delta_{ij}$) via the Lagrange multipliers $\lambda_{ij} = \lambda_{ji}$. It is assumed that the function $f(t)$ is even, nonnegative, continuously differentiable (almost everywhere), and $f(t) \leq t^2/2$ with strict inequality for large $|t|$. Moreover, its only minimum occurs at $t = 0$, and $f(t_2) \geq f(t_1)$ if $|t_2| > |t_1|$. The criterion in (9.76) will *grow* less than quadratically, at least for large values, and can be written more compactly in vector-matrix form as

$$J_1(W) = \mathbf{1}^T \mathrm{E}\{f[W^T x]|W\} + \frac{1}{2}\mathrm{trace}[\Lambda(W^T W - I)] \tag{9.77}$$

where the 1s vector $\mathbf{1}^T = [1, \ldots, 1]$ has the appropriate dimension, and the elements of matrix Λ are the λ_{ij}. From (9.77), the robust PCA learning rule in (9.72) can be derived [70], and $g(t)$ is the derivative of $f(t)$, that is, $df(t)/dt$. Note that if $g(t) = t$, this is a special case which results in Karhunen-Oja's linear symmetric subspace learning rule given in (9.45) (cf. Sect. 9.3.2). However, the appropriate interpretation must be made according to the manner in which the synaptic weight matrix in (9.45) was defined, that is, $W \in \Re^{m \times n}$. Therefore, the criterion in (9.77) is a generalization of the variance maximization problem, and details of the derivation leading to (9.72) can be found in [41, 43, 60].

The robust PCA learning rules in (9.73) and (9.75) can be derived from a generalization of the representation error [41, 70, 71]; that is, the associated performance criterion to be minimized is given in vector-matrix form as

$$J_2(\boldsymbol{W}) = \mathbf{1}^T \mathrm{E}\{f[\boldsymbol{e}]\} = \mathbf{1}^T \mathrm{E}\{f[\boldsymbol{x} - \hat{\boldsymbol{x}}]\} = \mathbf{1}^T \mathrm{E}\{f[\boldsymbol{x} - \boldsymbol{W}^T \boldsymbol{W}\boldsymbol{x}] | \boldsymbol{W}\} \qquad (9.78)$$

The function $f(t)$ must satisfy similar conditions previously given. Details of the derivations leading to (9.73) and (9.75) from the generalized representation error criterion in (9.78) can be found in [43, 75]. Note that in (9.73) the first term in the update expression on the right-hand side, that is, $\boldsymbol{x}(k)g[\boldsymbol{e}^T(k)]\boldsymbol{W}(k)$, is for each weight vector proportional to only the data input vector; therefore, it can be eliminated, which leads directly to the simplified learning rule in (9.75) [70]. Also, in the robust PCA learning rule in (9.73) if $g(t) = t$, this learning rule then minimizes the mean-square error, which is a special case yielding the standard PCA subspace learning rule developed earlier by Xu [76] and Russo [77], independently.

The choice of the nonlinearity is arbitrary (providing it is even, nonnegative, and continuously differentiable); however, we will present three of the more common ones that have been used by Karhunen and Joutsensalo [43]. These are the quadratic function $f_1(t) = t^2/2$, the linear criterion $f_2(t) = |t|$, and the criterion $f_3(t) = \ln[\frac{1}{2}(e^t + e^{-t})]$. Figure 9.13 shows a plot of each of these three functions. The corresponding derivatives of each function, with respect to t, can be computed respectively, in closed form as $g_1(t) = t$, $g_2(t) = \mathrm{sgn}(t)$ [where $\mathrm{sgn}(t)$ denotes the signum function, or the sign of t], and $g_3(t) = \tanh t$. The derivative of each nonlinearity is shown in Figure 9.14.

EXAMPLE 9.3. In this example, the performances of the robust PCA subspace learning rule given in (9.72) and the Karhunen-Oja symmetric subspace learning rule given in (9.45) are compared. Five hundred three-dimensional zero-mean random Gaussian vectors are generated with variances $\sigma_x^2 = 5$, $\sigma_y^2 = 3$, and $\sigma_z^2 = 0.2$ in the x, y, and z directions, respectively. Theoretically, the associated covariance matrix is diagonal with its eigenvalues equal to the variances of the three

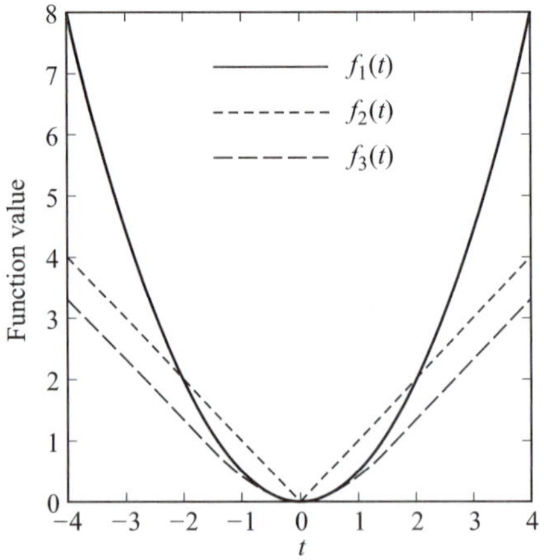

Figure 9.13 Typical nonlinearities ($f_\ell(t)$) that can be used in the performance criteria in (9.76) and (9.77), where $f_1(t) = t^2/2$, $f_2(t) = |t|$, and $f_3(t) = \ln[\frac{1}{2}(e^t + e^{-t})]$.

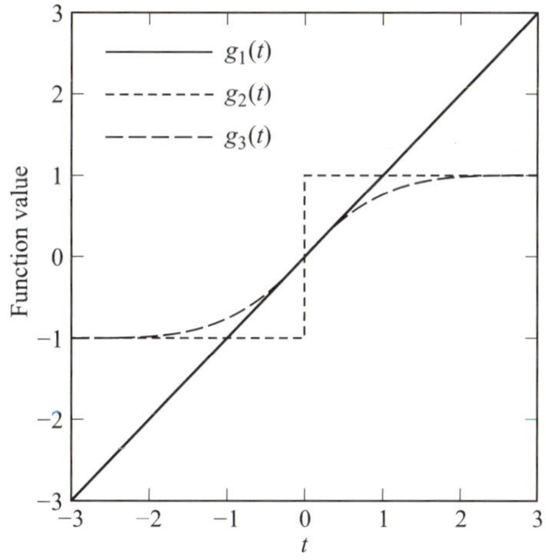

Figure 9.14 Derivatives ($g_\ell(t)$) of the three nonlinear functions shown in Figure 9.13, where $g_1(t) = t$, $g_2(t) = \text{sgn}(t)$, and $g_3(t) = \tanh t$.

components. The problem first involves computing the first two principal eigenvectors that span the subspace in the xy plane, using both Karhunen-Oja's symmetric subspace learning rule and the robust PCA subspace learning rule. Then the cosine of the angle subtended by the calculated normal component to the coplanar x, y vectors (learned by the two different neural networks) with the z axis in the reference coordinate system is computed. The angles are computed as an average of 300 independent trials of the 500 three-dimensional vectors; that is, the 500 three-dimensional vectors are randomly generated every trial. Also, for each trial the 500 random vectors are presented to the two neural networks twice, that is, two training epochs per trial. The cosine of the angle using Karhunen-Oja's symmetric subspace learning rule is $\cos(\angle_{KO}) = 0.9993$, and the cosine of the angle using the robust PCA subspace learning rule is $\cos(\angle_{ROB}) = 0.9743$. For both neural networks, the adaptive learning parameter approach given in (9.48) was used, with the forgetting factor set to 0.95.

However, this does not illustrate the *robustness* of the robust PCA subspace learning rule. Therefore, another similar experiment was performed in which the random vectors generated come from the same Gaussian distribution with probability of 0.9, but with probability 0.1 the Gaussian random vectors are corrupted with additive impulse noise from a uniform distribution in the interval [−10, 10]. Although the data variances are changed according to $\sigma_x^2 = 5 + (20^2)(0.1)/12$, $\sigma_y^2 = 3 + (20^2)(0.1)/12$, and $\sigma_z^2 = 0.2 + (20^2)(0.1)/12$, theoretically, the principal subspace remains unchanged, that is, the xy plane. After training the two neural networks (using the same approach in the first part of this problem, without impulse noise), the cosine of the angle between the computed normal to the extracted subspace and the z axis in the reference coordinate frame is computed for both results. Karhunen-Oja's symmetric subspace neural network results yielded $\cos(\angle_{KO}) = 0.9317$, and the cosine of the angle using the robust PCA subspace learning rule is $\cos(\angle_{ROB}) = 0.9865$. Therefore, for the case when impulse noise is corrupting the data, the robust PCA subspace neural network yielded better results than the Karhunen-Oja symmetric subspace neural network. Figure 9.15 shows the results for the robust PCA subspace neural network when the data are corrupted by impulse noise.

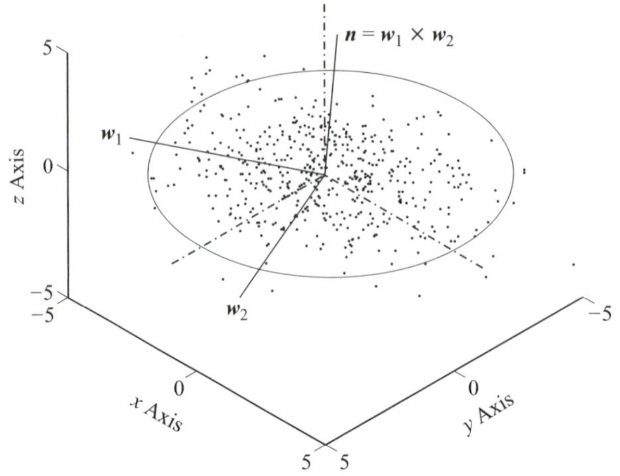

Figure 9.15 Robust PCA subspace neural network results when 10 percent of the Gaussian random data are corrupted by additive impulse noise from a uniform distribution in the interval [−10, 10]. The vectors $\boldsymbol{w}_1$ and $\boldsymbol{w}_2$ were learned by the neural network, and the computed cosine of the angle is given as $\cos(\angle_{ROB}) = 0.9865$, as compared to the results from the Karhunen-Oja symmetric subspace neural network, $\cos(\angle_{KO}) = 0.9317$. The "dash-dot" lines represent the reference coordinate system.

9.4 PRINCIPAL-COMPONENT REGRESSION

There is a large class of data regression methods that constitute the field of *chemometrics* [7, 8, 11, 14]. Chemometric methods are used extensively in analytical chemistry for quantitative analysis of spectroscopic data [116, 117]. Two of these multivariate regression methods will be presented here, principal-component regression (PCR) [4, 6–8] based on PCA, and partial least-squares regression (PLSR) [7–14], which is presented in the next section. The basic problem is to fit a *calibration model* to empirical data that are collected from a particular system. The intent is to develop the bilinear model [7], using either PCR or PLSR, by compressing the feature space of the data into a reduced data space that retains only the essential information in the data such that the calibration model is optimized with respect to its predictive performance. That is, both PCR and PLSR are considered as *factor analysis* (or rank reduction) based methods [7, 8, 10–13], which can yield a calibration model through a *training* process, which can then be used to predict certain quantities given test inputs to the model. Recently, Ham and Kostanic [9] have applied PLSR to engineering signal processing problems with much success. PCR has also been used in super-resolution algorithms for frequency estimation, that is, the multiple signal classification (MUSIC) and minimum-norm methods [2].

Classical least-squares (CLS)

Classical least-squares (CLS) [4, 6, 7, 12] is briefly discussed before PCR because comparisons will be made between these two methods, and in the next section CLS will also be compared to PLSR. Assume that there exists a linear relationship, that is, a vector $\boldsymbol{h} \in \Re^{n\times 1}$, between a set of independent variables (independent-variable block) given in matrix form as

$$\boldsymbol{X}(k) = \begin{bmatrix} x_{11}(k) & x_{12}(k) & \cdots & x_{1m}(k) \\ x_{21}(k) & x_{22}(k) & \cdots & x_{2m}(k) \\ \cdots & \cdots & \cdots & \cdots \\ x_{n1}(k) & x_{n2}(k) & \cdots & x_{nm}(k) \end{bmatrix} \tag{9.79}$$

where $\boldsymbol{X}(k) \in \Re^{n\times m}$, k is the time index, and the dependent variables (dependent-variable block) given in vector form as $\boldsymbol{y}(k) \in \Re^{m\times 1}$, such that we can write

$$\boldsymbol{y}(k) = \boldsymbol{X}^T(k)\boldsymbol{h} \tag{9.80}$$

Each column vector in matrix $\boldsymbol{X}(k)$ consists of a time sequence of a wide-sense-stationary process, and the columns of $\boldsymbol{X}(k)$ are not necessarily linearly independent. However, the columns of $\boldsymbol{X}(k)$ in (9.79) are assumed to be corrupted by Gaussian white noise $\boldsymbol{N}(k) \in \Re^{n\times m}$; that is, each column of $\boldsymbol{N}(k)$ is a white sequence with zero mean, and the covariance matrix of $\boldsymbol{N}(k)$ can be approximated by $\boldsymbol{C}_N \approx (1/m)\boldsymbol{N}(k)\boldsymbol{N}^T(k) \in \Re^{n\times n}$, using the expression in (9.19). Therefore, a *measurement* of $\boldsymbol{X}(k)$ in (9.79) is given as

$$\boldsymbol{Z}(k) = \boldsymbol{X}(k) + \boldsymbol{N}(k) \tag{9.81}$$

where $\boldsymbol{Z}(k) \in \Re^{n\times m}$. The information in the columns of $\boldsymbol{X}(k)$ is not known; however, $\boldsymbol{Z}(k)$ contains noisy measurements of $\boldsymbol{X}(k)$. An estimate of the dependent variable $\boldsymbol{y}(k)$ can be written using $\boldsymbol{Z}(k)$ as

$$\hat{\boldsymbol{y}}(k) = \boldsymbol{Z}^T(k)\boldsymbol{g} \tag{9.82}$$

where $\boldsymbol{g} \in \Re^{n\times 1}$ and $\hat{\boldsymbol{y}}(k) \in \Re^{m\times 1}$. Therefore, the objective is to find the linear relationship $\boldsymbol{g}$ in (9.82) which relates the measured variables $\boldsymbol{Z}(k)$ to an estimate of the dependent variable $\hat{\boldsymbol{y}}(k)$. To accomplish this, a classical least-squares approach can be taken by first defining an error vector given as

$$\boldsymbol{e}(k) = \boldsymbol{y}(k) - \hat{\boldsymbol{y}}(k) \tag{9.83}$$

where $\boldsymbol{e}(k) \in \Re^{m\times 1}$. Using (9.83), a quadratic performance measure is defined as

$$\boldsymbol{J}_g = \frac{1}{m}\boldsymbol{e}^T(k)\boldsymbol{W}\boldsymbol{e}(k) \tag{9.84}$$

which is to be minimized with respect to $\boldsymbol{g}$. However, the positive definite symmetric weight matrix in (9.84), $\boldsymbol{W} \in \Re^{m\times m}$, can be taken as the $m \times m$ identity matrix, that is, $\boldsymbol{W} = \boldsymbol{I}_m$, because all errors are considered to be equally weighted. Therefore, (9.84) can be written as

$$J_g = \frac{1}{m}\boldsymbol{e}^T(k)\boldsymbol{e}(k) = \frac{1}{m}[\boldsymbol{y}(k) - \hat{\boldsymbol{y}}(k)]^T[\boldsymbol{y}(k) - \hat{\boldsymbol{y}}(k)] \tag{9.85}$$

Substituting (9.80), (9.81), and (9.82) into (9.85) gives

$$\begin{aligned} J_g &= \frac{1}{m}[y(k) - \hat{y}(k)]^T [y(k) - \hat{y}(k)] \\ &= \boldsymbol{h}^T \left[\frac{1}{m}\boldsymbol{X}(k)\boldsymbol{X}^T(k)\right]\boldsymbol{h} - 2\boldsymbol{g}^T\left[\frac{1}{m}\boldsymbol{X}(k)\boldsymbol{X}^T(k)\right]\boldsymbol{h} \\ &\quad + \boldsymbol{g}^T\left[\frac{1}{m}\boldsymbol{X}(k)\boldsymbol{X}^T(k)\right]\boldsymbol{g} + \boldsymbol{g}^T\left[\frac{1}{m}\boldsymbol{N}(k)\boldsymbol{N}(k)^T\right]\boldsymbol{g} \\ &= \boldsymbol{h}^T\boldsymbol{C}_X\boldsymbol{h} - 2\boldsymbol{g}^T\boldsymbol{C}_X\boldsymbol{h} + \boldsymbol{g}^T\boldsymbol{C}_X\boldsymbol{g} + \boldsymbol{g}^T\boldsymbol{C}_N\boldsymbol{g} \end{aligned} \tag{9.86}$$

where $\boldsymbol{C}_X \approx (1/m)\boldsymbol{X}(k)\boldsymbol{X}^T(k) \in \Re^{n\times n}$ is an approximation to the covariance matrix of $\boldsymbol{X}(k)$, and it is assumed that $\boldsymbol{N}(k)$ and $\boldsymbol{X}(k)$ are uncorrelated, that is,

$$\mathrm{E}[\boldsymbol{n}(k)\boldsymbol{x}^T(k)] = \mathrm{E}[\boldsymbol{x}(k)\boldsymbol{n}^T(k)] = \boldsymbol{0} \tag{9.87}$$

Taking the partial derivative of (9.86) with respect to $\boldsymbol{g}$ and setting the result equal to zero give

$$\begin{aligned} \frac{\partial \boldsymbol{J}_g}{\partial \boldsymbol{g}} &= \frac{\partial}{\partial \boldsymbol{g}}(\boldsymbol{h}^T\boldsymbol{C}_X\boldsymbol{h} - 2\boldsymbol{g}^T\boldsymbol{C}_X\boldsymbol{h} + \boldsymbol{g}^T\boldsymbol{C}_X\boldsymbol{g} + \boldsymbol{g}^T\boldsymbol{C}_N\boldsymbol{g}) \\ &= -2\boldsymbol{C}_X\boldsymbol{h} + 2\boldsymbol{C}_X\boldsymbol{g} + 2\boldsymbol{C}_N\boldsymbol{g} = 0 \end{aligned} \tag{9.88}$$

Solving for $\boldsymbol{g}$ from (9.88) gives

$$\boldsymbol{g} = (\boldsymbol{C}_X + \boldsymbol{C}_N)^{-1}\boldsymbol{C}_X\boldsymbol{h} \tag{9.89}$$

To determine $\boldsymbol{g}$ from (9.89), we need knowledge of the covariance matrices for the uncorrupted measurements $\boldsymbol{C}_X$ and the measurement noise $\boldsymbol{C}_N$ and the linear relationship $\boldsymbol{h}$ between the uncorrupted measurements $\boldsymbol{X}(k)$ and the dependent variable $\boldsymbol{y}(k)$. However, these quantities are usually not known, but the CLS approach can yield $\boldsymbol{g}$, using only the information contained in corrupted measurements $\boldsymbol{Z}(k)$ and the dependent variable $\boldsymbol{y}(k)$ as

$$\boldsymbol{g} = [\boldsymbol{Z}(k)\boldsymbol{Z}^T(k)]^{-1}\boldsymbol{Z}(k)\boldsymbol{y}(k) \tag{9.90}$$

[4, 118] which can be directly related to (9.89). This can be shown by first writing an approximation to the covariance matrix of $\boldsymbol{Z}(k)$ as

$$\boldsymbol{C}_Z \approx \frac{1}{m}\boldsymbol{Z}(k)\boldsymbol{Z}^T(k) = \frac{1}{m}\boldsymbol{X}(k)\boldsymbol{X}^T(k) + \frac{1}{m}\boldsymbol{N}(k)\boldsymbol{N}^T(k) = \boldsymbol{C}_X + \boldsymbol{C}_N \tag{9.91}$$

using (9.81) and the assumption that $\boldsymbol{N}(k)$ and $\boldsymbol{X}(k)$ are uncorrelated. The inverse of (9.91) gives the first part of the expression in (9.89). Next, using the information in the dependent variable $\boldsymbol{y}(k)$ along with the measurement data $\boldsymbol{Z}(k)$, we can write

$$\frac{1}{m}\boldsymbol{Z}(k)\boldsymbol{y}(k) = \left[\frac{1}{m}\boldsymbol{X}(k)\boldsymbol{X}^T(k)\right]\boldsymbol{h} = \boldsymbol{C}_X\boldsymbol{h} \tag{9.92}$$

using (9.80) and (9.81) and again the assumption that $\boldsymbol{N}(k)$ and $\boldsymbol{X}(k)$ are uncorrelated. The expression in (9.92) gives the second part of (9.89). Therefore, from the results in (9.91) and (9.92), the relationship between (9.89) and (9.90) can be seen.

It is important to realize that the CLS result in (9.90) uses *all* the available information from the corrupted measurements $\boldsymbol{Z}(k)$ and the dependent variable $\boldsymbol{y}(k)$. However, there are many situations in which this is highly undesirable, and can result in *overfitting* of the data [7, 118]. That is, in the process of developing the model $\boldsymbol{g}$ in (9.90), the model parameters could be based on not only the essential cause-and-effect features associated with the empirical data $\boldsymbol{Z}(k)$ and $\boldsymbol{y}(k)$, but also unwanted effects in the data. This can result in a model with poor predictive capability. These unwanted effects could be associated with either measurement noise or any other features of the data associated with additional cause-and-effect phenomena that are not of any interest to the analyst. In a broad sense, these effects could also be considered as noise because the nature of the phenomenon may not be known a priori. Therefore, it is desirable to use only those features of the empirical data that will yield a calibration model with improved predictive performance, using a regression method that is capable of making this critical selection without prior knowledge of the obscuring effects in the data that can degrade the model's performance. Principal-component regression is one method that is capable of making this crucial selection through the factor analysis capability associated with this type of regression technique. Typically, the resulting calibration model developed via PCR will have much better predictive capabilities than a CLS model. The details of PCR are presented next, and in the following section details of partial least-squares regression will be given. Typically, PLSR will yield better predictive results than PCR [7, 10]. The underlying reasons for this performance enhancement will be explained in the next section (cf. Sect. 9.5).

Principal-component regression

Before PCR is presented, we define the data matrix, referred to as $\boldsymbol{Z}(k)$ before, to be $\boldsymbol{A} \in \Re^{m \times n}$, where each *row* of $\boldsymbol{A}$ is a particular measurement (i.e., a total of m measurements) and each measurement has the same number of components (i.e., a total of n components in each measurement). The use of $\boldsymbol{A}$ for the independent-variable block is more consistent with what is found in the literature, because $\boldsymbol{A}$ typically refers to spectroscopic absorption data [119], although this does *not* have to be the case. In fact, matrix $\boldsymbol{A}$ can actually contain any measurement data; therefore, the time index k has been dropped. However, if the data contained in $\boldsymbol{A}$ are spectroscopic data, then it is implied that each measurement consists of n spectral frequencies or wave numbers [119]. On the other hand, if each row in $\boldsymbol{A}$ is a time-domain measurement, then each row consists of n time samples, where k would be the time index. Also, the dependent-variable block, previously referred to as $\boldsymbol{y}(k)$, is now designated as $\boldsymbol{c} \in \Re^{m \times 1}$. This, again, is done to be more consistent with the notation found in the majority of the literature, because each element in the $\boldsymbol{c}$ vector usually corresponds to a *concentration* value, typically measured for each of the sample spectra collected spectrophotometrically [7]. However, the elements of $\boldsymbol{c}$ can be any *target values* that are independently determined which directly relate to the corresponding *row* measurements in the independent-variable block, that is, the $\boldsymbol{A}$ matrix. We will limit our discussions to only the *single-component case* [7], that is, the case when there is only one component of interest in the measured data in $\boldsymbol{A}$. This is the reason that the target values in $\boldsymbol{c}$ only form a column vector and not a matrix. This same notation will apply to the discussions relating to PLSR which follow in the next section.

As previously shown for CLS, the main problem with this method is associated with retaining all the information which resides in the empirical data and is used to develop the calibration model; see (9.90). There are at least two other problems associated with this method for developing a calibration model: (1) If collinearities [7] exist in data matrix $\boldsymbol{A}$ (i.e., linear dependence, where one of the independent variables can be written as an exact, or approximate, linear combination of the others), then $\boldsymbol{A}^T\boldsymbol{A}$ will be ill-conditioned [96]. In addition, for the case when there is no measurement noise, and their exist exact repeated measurements, $(\boldsymbol{A}^T\boldsymbol{A})^{-1}$ may not exist. (2) Even if $\boldsymbol{A}$ is not collinear, but if n (the number of sample components or frequencies) is very large, calculating the inverse of $\boldsymbol{A}^T\boldsymbol{A}$ can be computationally intensive. Therefore, it is highly desirable to have a method (or methods) which is (are) not sensitive to collinearities, not computationally intensive, and capable of only using the pertinent information in the empirical data, such that the calibration model does not include the unwanted effects, thereby resulting in enhanced prediction performance when compared to CLS. Both PCR and PLSR are such methods; that is, they are based on *factor analysis* (or rank reduction) [7].

In the case of PCR, we want to develop a calibration model that is capable of predicting concentrations $\hat{\boldsymbol{c}}$, given test inputs to the model that were not used in its development. Moreover, we want the PCR calibration model to possess better predictive capabilities than a CLS calibration model using the same data for model development. Therefore, we assume that the independent-variable block $\boldsymbol{A}$, which is available for developing a calibration model, has the form

$$\boldsymbol{A} = \begin{bmatrix} a_{11} & a_{12} & \cdots & a_{1n} \\ a_{21} & a_{22} & \cdots & a_{2n} \\ \cdots & \cdots & \cdots & \cdots \\ a_{m1} & a_{m2} & \cdots & a_{mn} \end{bmatrix} \tag{9.93}$$

where each row in $\boldsymbol{A}$ corresponds to a measurement that can be corrupted by noise. The target values (dependent-variable block) which are associated with each of the respective measurements are given in the column vector

$$\boldsymbol{c} = [c_1, c_2, \cdots, c_m]^T \tag{9.94}$$

If we assume that there exists a linear relationship between the independent- and dependent-variable blocks $\{\boldsymbol{A}, \boldsymbol{c}\}$, that is,

$$\boldsymbol{b}_f \in \Re^{n\times 1} \tag{9.95}$$

we can write

$$\boldsymbol{c} = \boldsymbol{A}\boldsymbol{b}_f + \boldsymbol{e}_c \tag{9.96}$$

where $\boldsymbol{e}_c \in \Re^{m\times 1}$ is the *error vector* which accounts for all errors, including the measurement noise. The CLS solution of (9.96) can be found by minimizing an error cost function (Lyapunov function) with respect to $\boldsymbol{b}_f$, that is,

$$L(\boldsymbol{b}_f) = \frac{1}{2}\|\boldsymbol{e}_c\|_2^2 = \frac{1}{2}\|\boldsymbol{c} - \boldsymbol{A}\boldsymbol{b}_f\|_2^2 \tag{9.97}$$

which leads to the solution

$$\hat{\boldsymbol{b}}_{f-\text{CLS}} = (\boldsymbol{A}^T\boldsymbol{A})^{-1}\boldsymbol{A}^T\boldsymbol{c} \tag{9.98}$$

where $\hat{\boldsymbol{b}}_{f\text{CLS}} \in \Re^{n\times 1}$ is the final CLS calibration model, and assuming $m \geq n$ (i.e., more, or an equal number of, measurements exist than components in each measurement). This is considered the *overdetermined* case [2, 96]. However, if $m < n$ (the *underdetermined* case [2, 96]), then the CLS calibration model is computed as $\hat{\boldsymbol{b}}_{f\text{CLS}} = \boldsymbol{A}^T(\boldsymbol{A}\boldsymbol{A}^T)^{-1}\boldsymbol{c}$. This model can now be used to predict the dependent variable given a set of *independent* data that were not used to develop the calibration model, given as $\{\boldsymbol{A}_{\text{test}}, \boldsymbol{c}_{\text{test}}\}$, where $\boldsymbol{A}_{\text{test}} \in \Re^{p\times n}$ and $\boldsymbol{c}_{\text{test}} \in \Re^{p\times 1}$, that is,

$$\hat{\boldsymbol{c}}_{\text{testCLS}} = \boldsymbol{A}_{\text{test}}\hat{\boldsymbol{b}}_{f\text{CLS}} \tag{9.99}$$

However, as stated previously, there are some basic problems associated with the CLS approach. The main problem is that all the features of the empirical data in $\boldsymbol{A}$ were used to develop the calibration model in (9.98), including possibly noise and other unwanted obscuring effects. Note that for convenience matrix $\boldsymbol{A}$, which is used for developing the calibration model (i.e., the *training* phase), has the "train" subscript omitted, which is implied. This also applies to the dependent-variable block $\boldsymbol{c}$ in (9.98).

PCR (and also PLSR), however, is based on the following representations of the empirical data; that is, the independent-variable block is written as a decomposition, given as

$$\boldsymbol{A} = \boldsymbol{T}\boldsymbol{B} + \boldsymbol{E}_A \tag{9.100}$$

In (9.100), $\boldsymbol{B} \in \Re^{h\times n}$ contains the *loading* vectors (or *loading spectra*), and the rows represent the *new* PCA basis set of h vectors (where h is the number of *factors* to be retained, which is explained below). In the case of PCA, the rows of $\boldsymbol{B}$ are the (principal) eigenvectors of $\boldsymbol{A}^T\boldsymbol{A}$ (the covariance matrix) and are orthonormal, assuming $m \geq n$ (the *overdetermined* case). Therefore, the rows of $\boldsymbol{B}$ can be obtained by using any of the methods presented in Sections 9.3.2 through 9.3.6 or the EVD method. And $\boldsymbol{T} \in \Re^{m\times h}$ is a matrix of *intensities* (or *scores* or *latent variables*) in the new coordinate system of the h PCA loading vectors for each of the m measurements. Finally, $\boldsymbol{E}_A \in \Re^{m\times n}$ is a matrix of measurement (or spectral) residuals not fit by the *optimal* model. Therefore, in (9.100) the $\boldsymbol{A}$ matrix can be written as a linear combination of the outer products of the columns of $\boldsymbol{T}$ with the rows of $\boldsymbol{B}$ plus the matrix of measurement residuals $\boldsymbol{E}_A$. That is, $\boldsymbol{A} = \sum_{q=1}^{h} \boldsymbol{t}_q\boldsymbol{b}_q + \boldsymbol{E}_A$, where $\boldsymbol{t}_q$ for $q = 1, 2, \ldots, h$ are the *intensity* column vectors of $\boldsymbol{T}$ and $\boldsymbol{b}_q$ for $q = 1, 2, \ldots, h$ are the *loading* row vectors of $\boldsymbol{B}$.

PCR actually consists of two basic steps: PCA (data compression), that is, selection of the optimal number of principal eigenvectors, and calibration model building using the PCA results, that is, the regression step. In general, $h < m, n$ in which case there will be a reduced number of *intensities*. However, to *optimize* the calibration model, the optimal number of PCR (or PLSR) factors must be determined. This is known as the *factor analysis* capability that both PCR and PLSR possess. When the optimal number of factors is selected, this will lead to compression of the data, causing the noise, and any other obscuring effects, to be minimized. Noise in the data is typically distributed throughout all loading vectors (mostly the higher-order ones), whereas the uncorrupted measurement content is generally concentrated in the first few.

The second general relationship for PCR (and PLSR) involves using the intensity (or score) matrix $\boldsymbol{T}$ in the new coordinate system, and writing a similar relationship to (9.96). That is, we now write

$$c = \boldsymbol{T}\boldsymbol{v} + \boldsymbol{e}_c \tag{9.101}$$

where $\boldsymbol{v} \in \Re^{h \times 1}$ are the regression coefficients (or inner relationships) [7, 10]. From (9.101), it is obvious that the problems that were associated with CLS do not exist. For the PCR approach and $m \geq n$, and assuming the residuals ($\boldsymbol{E}_A$) in (9.100) to be sufficiently small, we can write the following from (9.100) by postmultiplying both sides by $\boldsymbol{B}^T$:

$$\boldsymbol{T}\boldsymbol{B}\boldsymbol{B}^T = \boldsymbol{A}\boldsymbol{B}^T \tag{9.102}$$

However, because the rows of matrix $\boldsymbol{B}$ (which contains the loading vectors) are orthonormal, that is, $\boldsymbol{B}\boldsymbol{B}^T = \boldsymbol{I}_h$, Equation (9.102) can be written as

$$\boldsymbol{T} = \boldsymbol{A}\boldsymbol{B}^T \tag{9.103}$$

Substituting (9.103) into (9.101) yields

$$c = \boldsymbol{A}\boldsymbol{B}^T\boldsymbol{v} + \boldsymbol{e}_c \tag{9.104}$$

If we define the error cost function as

$$L(\boldsymbol{v}) = \frac{1}{2}\|\boldsymbol{e}_c\|_2^2 \tag{9.105}$$

and minimize this with respect to the regression coefficient vector $\boldsymbol{v}$, this will lead to the least-squares solution

$$\boldsymbol{v} = (\boldsymbol{B}\boldsymbol{A}^T\boldsymbol{A}\boldsymbol{B}^T)^{-1}\boldsymbol{B}\boldsymbol{A}^T\boldsymbol{c} \tag{9.106}$$

where $\boldsymbol{v}$ can be thought of as the calibration model in *factor space*. However, if we compare the model in (9.104) to the original model in (9.96) for the CLS problem, we see that

$$\hat{\boldsymbol{b}}_{f\text{PCR}} = \boldsymbol{B}^T\boldsymbol{v} = \boldsymbol{B}^T(\boldsymbol{B}\boldsymbol{A}^T\boldsymbol{A}\boldsymbol{B}^T)^{-1}\boldsymbol{B}\boldsymbol{A}^T\boldsymbol{c} \tag{9.107}$$

Comparing the term $(\boldsymbol{B}\boldsymbol{A}^T\boldsymbol{A}\boldsymbol{B}^T)^{-1} \in \Re^{h\times h}$ in (9.107) to the associated term in (9.98) for the CLS model $\hat{\boldsymbol{b}}_{f\text{CLS}}$, that is, $(\boldsymbol{A}^T\boldsymbol{A})^{-1} \in \Re^{n\times n}$, for $h < n$ computing the calibration model $\hat{\boldsymbol{b}}_{f\text{PCR}} \in \Re^{n\times 1}$ in (9.107) is less computationally intensive for the PCR approach than for CLS.

Therefore, if the optimal number of PCR factors (h) is selected and retained to compute the PCR calibration model according to (9.107), the resulting model will typically have better predictive performance than a calibration model developed using CLS, as in (9.98). As was the case with CLS, the PCR calibration model $\hat{\boldsymbol{b}}_{f\text{PCR}}$ can be used to predict the independent variable given a set of *independent* test data not used to develop the calibration model, given as $\{A_{\text{test}}, c_{\text{test}}\}$, where $A_{\text{test}} \in \Re^{p \times n}$ and $c_{\text{test}} \in \Re^{p \times 1}$, that is,

$$\hat{c}_{\text{testPCR}} = A_{\text{test}} \hat{b}_{f\text{PCR}} \tag{9.108}$$

When $m < n$ (the underdetermined case), the rows of $\boldsymbol{B}$ (which contains the loading vectors) can also be the (principal) eigenvectors of $A^T A$, and the expression in (9.107) can again be used to compute the PCR calibration model $\hat{b}_{f\text{PCR}}$. However, care must be taken in selecting the principal eigenvectors because for the underdetermined case the eigenvectors of $A^T A$ do not span the entire row space of A. Therefore, it is probably best to use the singular-value decomposition (cf. Sect. A. 2.14) to compute the eigenvectors of $A^T A$.

Two questions still remain unanswered. How is the optimal number of factors selected, and how can the predictive performance of the calibration model be quantitatively assessed? The second question will be answered first because the answer to the first question hinges on the second, that is, the method that will be shown to quantitatively assess the calibration model performance. We have been discussing predictive performances of CLS versus PCR; however, we have not presented a method to quantitatively determine performance. A common method that is used to determine how well the calibration model, for example, for PCR, can predict unknown concentrations, or the dependent variable $\boldsymbol{c}$, given a set of test data that were not used to compute the calibration model, is the *standard error of prediction* (SEP) [7]. Using an independent set of test data $\{A_{\text{test}}, c_{\text{test}}\}$, the SEP is defined as

$$\text{SEP} = \left[\sum_{i=1}^{m_{\text{test}}} (c_{i\,\text{test}} - \hat{c}_{i\,\text{test}})^2 / m_{\text{test}} \right]^{1/2} \tag{9.109}$$

where $c_{i\,\text{test}}$ is the reference (actual) pure component of interest for the test data, $\hat{c}_{i\,\text{test}}$ is the PCR (or PLS) prediction (estimate) of $c_{i\,\text{test}}$, and m_{test} is the total number of test measurements. For the sake of completeness, we also define a similar measure of performance that is computed for the training data, $\{A_{\text{train}}, c_{\text{train}}\}$ which is used to develop the calibration model. This is referred to as the *standard error of calibration* (SEC) [7], and it is defined for the training data $\{A_{\text{train}}, c_{\text{train}}\}$ as

$$\text{SEC} = \left[\sum_{i=1}^{m_{\text{train}}} \frac{(c_{i\,\text{train}} - \hat{c}_{i\,\text{train}})^2}{m_{\text{train}} - h - 1} \right]^{1/2} \tag{9.110}$$

where $c_{i\,\text{train}}$ is the reference (actual) pure component of interest for the training data, $\hat{c}_{i\,\text{train}}$ is the PLC prediction (estimate) of $c_{i\,\text{train}}$, m_{train} is the total number of training *measurements*, and h is the number of PCR (or PLSR) factors. The number of PCR (or PLSR) factors h is included in the denominator of (9.110) because a penalty must be placed on the prediction performance of the calibration model as additional factors are included for the training data. Therefore, (9.110) actually represents a weighted performance measure for the training data. In other words, by including h in the denominator of (9.110), as more PCR (or PLSR) factors are added to the model, the predictive performance of the model must account for this increase when regressing the training data, therefore, minimizing *overfitting* of the data [7,118].

Selection of optimal number of factors

Now the issue of optimal PCR factor selection can be addressed. The methods presented here can also be applied in a similar manner to PLSR. The SEP performance measure in (9.109) can be used to determine the *optimal* number of PCR factors h^o to retain for the development of the PCR calibration, that is, $\hat{\boldsymbol{b}}_{fPCR}$ given in (9.107). In the first method, which is probably the most common, a set of calibration models $\hat{\boldsymbol{b}}_{f\mathrm{PCR}}$ for $h = 1, 2, \ldots, q$ is generated for a range of PCR factors using the training data set $\{\boldsymbol{A}_{\mathrm{train}}, \boldsymbol{c}_{\mathrm{train}}\}$. Having the set of q calibration models, $\hat{\boldsymbol{b}}_{f\mathrm{PCR}}$ for $h = 1, 2, \ldots, q$ the test data set $\{\boldsymbol{A}_{\mathrm{test}}, \boldsymbol{c}_{\mathrm{test}}\}$ is used to assess the predictive performance of each calibration model. Using the relationship in (9.109), the SEP can be calculated for each prediction of $\boldsymbol{c}_{\mathrm{test}}$ that the calibration models yield, that is, $\hat{\boldsymbol{c}}_{\mathrm{test}}$ for $h = 1, 2, \ldots, q$ using $\boldsymbol{A}_{\mathrm{test}}$. The selection of h^o is determined by observing the SEP values as a function of the number of PCR factors q. Typically, the number of factors h associated with the *minimum* SEP that is observed will indicate the optimal number of PCR factors, that is, h^o, given as

$$h^o = \{h : \mathrm{SEP}_{\min} = \min\{\mathrm{SEP}(h)\} \; \forall h = 1, 2, \ldots, h^o, \ldots, q\} \tag{9.111}$$

This method for selecting the optimal number of PCR factors h^o will be referred to as *independent validation*. However, care must be taken in using (9.111) because the absolute minimum may give a result that would allow additional factors to be retained that are potentially associated with noise [7]. Therefore, each individual case must be assessed carefully to reconcile what is actually the *optimal* number of PCR factors to retain. In some cases, the absolute minimum from (9.111) is not the best choice, and many times one factor less than this absolute minimum will give the best overall performance [7].

Another method that can be used for selecting the number of PCR (or PLSR) factors to retain is called *cross-validation* [7, 8]. Cross-validation is sometimes referred to as a *leave-one-out-at-a-time* analysis approach because one measurement is left out of the data set $\{\boldsymbol{A}_{\mathrm{train}}, \boldsymbol{c}_{\mathrm{train}}\}$ and the PCR model is developed on the remaining measurements, then the measurement that was left out is tested to yield a prediction of c_{train}, that is, $\hat{c}_{\mathrm{train}}$. This process is repeated until all measurements are left out and used for prediction, and then the SEP is computed for the entire m_{train} measurements according to (9.109). Cross-validation is often used when the number of measurements is sparse, that is, not many measurements are available. In this case there are not enough data to form a training set and a test set. The results obtained using this method are not typically as good as those of the independent validation method. Ideally, it is desirable to have a statistically representative set of measurements such that separate training and test sets can be formed [4–7, 120, 121].

9.5 PARTIAL LEAST-SQUARES REGRESSION

Partial least-squares regression, is another *factor analysis*-based method like PCR. However, PLSR has one major advantage over PCR. During the "compression" step in PLSR, the target values (dependent-variable block) are used in addition to the independent-variable block, whereas in PCR only the independent-variable block is used in the PCA-based compression step. Using both the independent-variable block $\boldsymbol{A} \in \Re^{m \times n}$ and the dependent-variable block $\boldsymbol{c} \in \Re^{m \times 1}$ in the compression step can have a profound effect on the predictive performance results using PLSR, typically resulting in better performance than both CLS and PCR.

PLSR was originally introduced by Wold [122, 123], and it arose as a practical solution to data analytic problems in econometrics and social sciences. The basic problem is the same as that for PCR, that is, to *fit* a calibration model to empirical data, and use this model to predict certain quantities given a set of test data as input to the calibration model after the training phase. PLSR is sometimes referred to in the context of *abstract factor analysis* (rank reduction), because physically significant quantities within the system cannot always be identified directly with the mathematical modeling process [124]. However, if properly interpreted, the mathematical basis which constitutes the factor analysis method of PLSR can yield very powerful information related to certain key features within the physical setting from which the data were extracted and used in the PLSR modeling process [12]. Moreover, PLSR is considered a *full-spectrum* technique, in contrast to the inverse least-squares (ILS) method [12]; therefore, efficient outlier techniques are available using residues from the PLS process. Outliers are considered to be samples that are not representative of the calibration samples; therefore, subsequent estimated outputs must be carefully scrutinized [7, 12].

When PCR was presented, it was assumed that we would limit our discussions to the single-component case [7], which will be the same for PLSR. Here we will present the PLSR1 calibration algorithm (which is for the single-component case) and two different prediction algorithms; and in Section 9.6 a neural network realization of PLSR will be presented. As stated in the previous section, the underlying models given for PCR are also the basis for PLSR, that is, (9.100) and (9.101). However, as we will see in the presentation of the PLSR1 calibration algorithm, which follows the approach given by Haaland and Thomas [12], both $\boldsymbol{A}$ and $\boldsymbol{c}$ are used for the compression phase. There are many different approaches that have been developed for the PLSR algorithm and various presentations of the underlying principles of PLSR [7–13]; however, the explanation given below is probably the most straightforward, and thus the easiest to understand. Also, the two prediction methods presented follow the methods given by Haaland and Thomas [12].

Each of the main seven steps of the PLSR1 calibration algorithm given below is shown to be derived from a CLS approach. For the general explanation of the calibration algorithm, the data $\{\boldsymbol{A}, \boldsymbol{c}\}$ are used as *training* data, that is, $\{\boldsymbol{A}_{\text{train}}, \boldsymbol{c}_{\text{train}}\}$ for ultimately developing the calibration model; however, the "train" subscript is omitted from the PLS1 calibration algorithm explanation to avoid confusion.

PLSR1 calibration algorithm

Step 1: Mean-centering and variance-scaling of the data. The first step typically taken is pretreatment of the data, that is, mean-centering and variance-scaling [7, 10]. An exhaustive explanation of the underlying reasons for this type of pretreatment of the data will not be given; however, typical situations that require pretreatment of the data will be explained. For example, if the collected data (the measurements), that is, the rows of $\boldsymbol{A}$, have a *bias* associated with the data, then mean-centering of the data would be advisable. This process would also be carried out for the dependent variables, that is, $\boldsymbol{c}$. What this basically accomplishes is the elimination of the need to fit a nonzero intercept to the data, which often results in a decrease of the complexity of the calibration model. That is, a reduction in the number of PLSR factors by one required to model the data [12]. If the collected data are measured with different units, variance-scaling is then advisable. Mean-centering involves calculating the mean value of each column of $\boldsymbol{A}$ and subtracting the mean value for the particular column from each of the elements in the respective columns. If $\boldsymbol{A}$ is mean-centered, $\boldsymbol{c}$ should also be mean-centered. Variance-scaling involves calculating the standard deviation of each column in $\boldsymbol{A}$ and then dividing each element in the respective column by the associated standard deviation value. This same process should be carried out for the dependent variables in $\boldsymbol{c}$. However, this is usually only necessary for the multicomponent case. There is a continuing debate among statisticians relating to mean-centering and variance-scaling. Many individuals insist that the data should always be preconditioned, and others maintain that the data should never be mean-centered and variance-scaled [121]. We feel that pretreatment of the data is necessary for the reasons given above; however, if there is no compelling reason to mean-center or variance-scale the data, these processes should not be performed arbitrarily.

First, an index h (number count for the PLS factors) is initially set to 1.

Step 2: Forming the weight loading vector $\hat{\boldsymbol{w}}_h \in \Re^{n \times 1}$. This step is actually a CLS calibration, and the model used is given as

Model: $$\boldsymbol{A} = \boldsymbol{c}\boldsymbol{w}_h^T + \boldsymbol{E}_A \tag{9.112}$$

where the least-squares solution is given as

Least-squares solution: $$\hat{\boldsymbol{w}}_h = \boldsymbol{A}^T\boldsymbol{c}/\boldsymbol{c}^T\boldsymbol{c} \tag{9.113}$$

then normalize $\hat{\boldsymbol{w}}_h$, that is,

$$\hat{\boldsymbol{w}}_h \leftarrow \frac{\hat{\boldsymbol{w}}_h}{\|\hat{\boldsymbol{w}}_h\|_2}$$

In (9.113) each vector $\hat{\boldsymbol{w}}_h$, for each h increment, is the weight vector that is proportional to a weighted average of the row elements in matrix $\boldsymbol{A}$, where the weights in the average are proportional to the elements in $\boldsymbol{c}$. Each of the weight vectors $\hat{w}_h$ is normalized and constructed to be mutually orthogonal; therefore, the $\hat{w}_h$ vectors are orthonormal. This step is quite different from that with the PCR method [4, 6–8] because the information in the dependent variable $\boldsymbol{c}$ is used in addition to $\boldsymbol{A}$ to form the weight vectors in PLSR; in PCR only the information in matrix $\boldsymbol{A}$ is used in this step. The matrix $\boldsymbol{E}_A \in \Re^{m\times n}$ in (9.112) contains the residuals associated with $\boldsymbol{A}$.

Step 3: Generation of the score (latent variable) vector $\hat{t}_h \in \Re^{m\times 1}$. In this step, $\boldsymbol{A}$ is now written with respect to the latent variables or the scores as

Model: $$\boldsymbol{A} = \boldsymbol{t}_h \hat{\boldsymbol{w}}_h^T + \boldsymbol{E}_A \qquad (9.114)$$

where the least squares solution is given as

Least-squares solution: $$\hat{\boldsymbol{t}}_h = \boldsymbol{A}\hat{\boldsymbol{w}}_h / \hat{\boldsymbol{w}}_h^T \hat{\boldsymbol{w}}_h = \boldsymbol{A}\hat{\boldsymbol{w}}_h \qquad (9.115)$$

This step is also CLS, where the least-squares estimate of $\boldsymbol{t}_h$, $\hat{\boldsymbol{t}}_h$, is obtained by regressing $\boldsymbol{A}$ on $\hat{\boldsymbol{w}}_h$ as shown in (9.115). The individual elements of $\hat{\boldsymbol{t}}_h$ indicate how much of $\hat{\boldsymbol{w}}_h$ is contained in each row of data matrix $\boldsymbol{A}$. The vector $\hat{\boldsymbol{t}}_1$ represents the intensities (or amounts) of the first weight loading vector in the row data of $\boldsymbol{A}$ for the new PLS coordinate system. Since $\hat{\boldsymbol{w}}_1$ is a first-order attempt to represent the uncorrupted data from the corrupted (noisy) data contained in the rows of $\boldsymbol{A}$, $\hat{\boldsymbol{t}}_1$ represents a first-order attempt to determine the amount of the pure component of interest (i.e., the information contained in $\boldsymbol{c}$) in each of the associated rows of $\boldsymbol{A}$. Thus, in the PLSR method, each $\hat{\boldsymbol{t}}_h$ vector is related to both $\boldsymbol{A}$ and $\boldsymbol{c}$ rather than solely to $\boldsymbol{A}$, as in the case of PCR.

Step 4: Relating the score vector $\hat{t}_h$ to the elements of c. In this step the score vector $\hat{\boldsymbol{t}}_h$ (or the latent variable associated with key features relating to the pure component of interest contained in each row of $\boldsymbol{A}$) representing the intensities in the new PLSR coordinate system is related to the elements of vector $\boldsymbol{c}$ by using a linear least-squares regression. In PLSR, as opposed to the inverse least-squares method [12, 125] and PCR, a separate relation between the scores $\hat{\boldsymbol{t}}_h$ and the elements of the $\boldsymbol{c}$ vector (or the $\boldsymbol{c}$ residuals) is found after each weight vector is estimated. The relation between $\hat{\boldsymbol{t}}_h$ and $\boldsymbol{c}$ is modeled as

Model: $$\boldsymbol{c} = v_h \hat{\boldsymbol{t}}_h + \boldsymbol{e}_c \qquad (9.116)$$

and the least-squares solution is given as

Least-squares solution: $$\hat{v}_h = \hat{\boldsymbol{t}}_h^T \boldsymbol{c} / \hat{\boldsymbol{t}}_h^T \hat{\boldsymbol{t}}_h \qquad (9.117)$$

where for each h increment (9.117) gives an estimate of $v_h \in \Re$ which is the scalar *regression coefficient* (inner relationship) relating $\hat{\boldsymbol{t}}_h$ to the elements in $\boldsymbol{c}$. The vector $\boldsymbol{e}_c \in \Re^{m\times 1}$ in (9.116) contains the PLSR residuals associated with $\boldsymbol{c}$. The relation in (9.117) is similar to an ILS solution, in that the sum of the squared $\boldsymbol{c}$ errors is minimized. However, in this step, the solution which is similar to ILS is constructed element by element.

Step 5: Generation of $\hat{b}_h \in \Re^{n\times 1}$, the PLS loading vector for A. Orthogonal $\hat{\boldsymbol{t}}_h$ vectors are desirable in order to remove collinearities (i.e., linear dependence). Orthogonal $\hat{t}_h$ vectors can be obtained by forming a new model for A based on the latent variable $\hat{t}_h$. The new model is given as

Model: $$A = \hat{t}_h b_h^T + E_A \tag{9.118}$$

where the least-squares solution is given as

Least-squares solution: $$\hat{b}_h = A^T \hat{t}_h / \hat{t}_h^T \hat{t}_h \tag{9.119}$$

The vectors $\hat{b}_h$, for $h = 1, 2, \ldots$, are the PLSR loading vectors. This step along with the next one in the algorithm ensures that the $\hat{t}_h$ vectors will be mutually orthogonal. The least-squares regression is simultaneously performed for all samples in each row of A, as indicated in (9.119). Unlike the first PCA loading vector in PCR, the first PLSR loading vector $\hat{b}_1$ determined from (9.119) does not account for the maximum variance in the rows of A. However, it does represent an attempt to account for as much variation in A while simultaneously correlating with $\hat{t}_h$ which approximates c. Also unlike PCA, the $\hat{b}_h$ vectors are not mutually orthogonal. Moreover, since $\hat{t}_1$ is the first-order approximation to the c vector, column elements in A associated with the largest positive elements in $\hat{b}_1$ tend to indicate those column elements in A which exhibit the greatest dependence on the elements in c for that particular loading vector. However, the $\hat{w}_1$ vector, which is directly related to c, will exhibit this tendency better than $\hat{b}_1$, and therefore $\hat{w}_1$ will be more useful than $\hat{b}_1$ for extracting information from the PLSR1 analysis.

Step 6: Calculation of the residuals in A and c. The product of the scores $\hat{t}_h$ and the loading vectors $\hat{b}_h$ is the PLSR approximation to A. The residuals E_A in the matrix A are computed by subtracting the PLSR approximation to the rows of matrix A from the measurements in the rows of A as given in

A residuals: $$E_A = A - \hat{t}_h \hat{b}_h^T \tag{9.120}$$

c residuals: $$e_c = c - \hat{v}_h \hat{t}_h \tag{9.121}$$

Similarly, the portion of the information in vector c that has been modeled by PLSR can be removed to obtain the residuals in c, that is, e_c, as given in (9.121). The product $\hat{v}_h \hat{t}_h$ in (9.121) represents the PLSR estimate of c, $\hat{c}$, based on the information in matrix A.

Step 7: Increment h, substitute E_A for A and e_c for c in step 2, and continue for the desired number of loading vectors (or the optimal number of PLSR factors h^o).

Summary of the PLSR1 calibration algorithm

Step 1. Pretreatment of the data: mean-center $A \in \Re^{m\times n}$ and $c \in \Re^{m\times 1}$, and variance-scale A if necessary. Set the index h to 1 (where h is the number of PLS factors).

Step 2. Forming the weight loading vector $\hat{w}_h \in \Re^{n\times 1}$

Model: $$A = c w_h^T + E_A \qquad (E_A \in \Re^{m\times n} \text{ contains } A \text{ residuals})$$

Least-squares solution: $\hat{w}_h = \dfrac{A^T c}{c^T c}$

Normalize $\hat{w}_h$, that is, $\hat{w}_h \leftarrow \dfrac{\hat{w}_h}{\|\hat{w}_h\|_2}$

Step 3. Generation of the score (latent variable) vector $\hat{t}_h \in \Re^{m\times 1}$

Model: $A = \hat{t}_h w_h^T + E_A$

Least-squares solution: $\hat{t}_h = \dfrac{A\hat{w}_h}{\hat{w}_h^T \hat{w}_h} = A\hat{w}_h$

Step 4. Relating the score vector $\hat{t}_h$ to the elements of c.

Model: $c = v_h \hat{t}_h + e_c$ ($e_c \in \Re^{m\times 1}$ contains c residuals)

Least-squares solution: $\hat{v}_h = \dfrac{\hat{t}_h^T c}{\hat{t}_h^T \hat{t}_h}$

where $v_h \in \Re$ is the *scalar regression coefficient* (inner relationship) relating $\hat{t}_h$ to the elements in c.

Step 5. Generation of $\hat{b}_h \in \Re^{n\times 1}$, the loading vector for A.

Model: $A = \hat{t}_h b_h^T + E_A$

Least-squares solution: $\hat{b}_h = \dfrac{A^T \hat{t}_h}{\hat{t}_h^T \hat{t}_h}$

Step 6. Calculation of the residuals in A and c

A residuals: $E_A = A - \hat{t}_h \hat{b}_h^T$

c residuals: $e_c = c - \hat{v}_h \hat{t}_h$

Step 7. Increment h, substitute E_A for A and e_c for c in step 2, and continue for the desired number of loading vectors, or the optimal number of PLS factors h^o, where $h^o = m$ (for $m < n$) or $h^o = n$ (for $m \geq n$).

PLSR1 prediction algorithms

Prediction method 1. There are two different methods that can be applied to perform prediction; that is, given an unknown measurement, predict the desired dependent-variable component using the extracted information from the PLSR1 calibration phase. The first prediction algorithm presented here follows the method 2 explanation given in Haaland and Thomas [12]. The following procedure is easily understood and is given in two parts: calibration model development and prediction using the calibration model. One drawback to this method is that it does not allow determination of the residuals of A, and therefore, no diagnostic information about the quality of the calibration model fit to the data is available when predictions are obtained.

From the PLSR1 calibration algorithm given above, a calibration model $\hat{\boldsymbol{b}}_{f\text{PLSR}} \in \Re^{n\times 1}$, can be formed from the weight loading vectors $\hat{\boldsymbol{w}}_h$ for $h = 1, 2, \ldots, q$, the loading vectors for $\boldsymbol{A}$, $\hat{\boldsymbol{b}}_h$ for $h = 1, 2, \ldots, q$, and the inner relationships $\hat{v}_h$ for $h = 1, 2, \ldots, q$. If all the PLSR factors are generated, then $q = m$, for $m < n$, and $q = n$, for $m \geq n$. Ideally, we want to only use the optimal number of PLSR factors, that is, $q = h^o$. The optimal number of PLSR factors typically results in a calibration model $\hat{\boldsymbol{b}}_{f\text{PLSR}}$, which yields minimal predictive performance error based on independent *test* data and thus does not allow the model to overfit the data. The procedure to determine h^o is the same as the method previously given for PCR. To generate the calibration model, we first form the matrices

$$\hat{\boldsymbol{W}}^T = [\hat{\boldsymbol{w}}_1, \hat{\boldsymbol{w}}_2, \cdots, \hat{\boldsymbol{w}}_q] \tag{9.122}$$

and

$$\hat{\boldsymbol{B}}^T = [\hat{\boldsymbol{b}}_1, \hat{\boldsymbol{b}}_2, \cdots, \hat{\boldsymbol{b}}_q] \tag{9.123}$$

and

$$\hat{\boldsymbol{v}}^T = [\hat{v}_1, \hat{v}_2, \cdots, \hat{v}_q] \tag{9.124}$$

where $\hat{\boldsymbol{W}} \in \Re^{q\times n}$, $\hat{\boldsymbol{B}} \in \Re^{q\times n}$, and $\hat{\boldsymbol{v}} \in \Re^{q\times 1}$. From (9.122), (9.123), (9.124), and $q = h^o$, the final optimal calibration model (or final calibration coefficients) $\hat{\boldsymbol{b}}_{f\text{PLSR}}$ can be formed as

$$\hat{\boldsymbol{b}}_{f\text{PLSR}} = \hat{\boldsymbol{W}}^T(\hat{\boldsymbol{B}}\hat{\boldsymbol{W}}^T)^{-1}\hat{\boldsymbol{v}} \tag{9.125}$$

Summary of the PLSR1 prediction method 1

Step 1. From the PLSR1 calibration phase, form the following matrices (for the optimal or desired number of PLSR factors $h = h^o$);

$$\hat{\boldsymbol{W}}^T = [\hat{\boldsymbol{w}}_1, \hat{\boldsymbol{w}}_2, \cdots, \hat{\boldsymbol{w}}_q] \quad \text{where } \hat{\boldsymbol{W}} \in \Re^{h^o\times n}$$
$$\hat{\boldsymbol{B}}^T = [\hat{\boldsymbol{b}}_1, \hat{\boldsymbol{b}}_2, \cdots, \hat{\boldsymbol{b}}_q] \quad \text{where } \hat{\boldsymbol{B}} \in \Re^{h^0\times n}$$
$$\hat{\boldsymbol{v}}^T = [\hat{v}_1, \hat{v}_2, \cdots, \hat{v}_1] \quad \text{where } \hat{\boldsymbol{v}} \in \Re^{h^0\times 1}$$

Step 2. Compute the final regression coefficients, or the optimal calibration model $\hat{\boldsymbol{b}}_{f\text{PLSR}}$:

$$\hat{\boldsymbol{b}}_{fPLSR} = \hat{\boldsymbol{W}}^T(\hat{\boldsymbol{B}}\hat{\boldsymbol{W}}^T)^{-1}\hat{\boldsymbol{v}}$$

[Note: $\rho(\hat{\boldsymbol{B}}\hat{\boldsymbol{W}}^T) = h^o$, where $h^o \ll n, m$.]

Step 3. Given the set of measurements $\boldsymbol{A}_{\text{test}}$ (not used to develop the calibration model $\hat{\boldsymbol{b}}_{f\text{PLSR}}$), estimate the outputs (or dependent variables)

$$\hat{\boldsymbol{c}}_{\text{test}} = \boldsymbol{A}_{\text{test}}\hat{\boldsymbol{b}}_{f\text{PLSR}}$$

or if the training data were mean-centered (where the mean of the dependent reference data is given by $\bar{\boldsymbol{c}}_{\text{train}}$)

$$\hat{\boldsymbol{c}}_{\text{test}} = A_{\text{test}}\hat{\boldsymbol{b}}_{f\text{PLSR}} + \bar{\boldsymbol{c}}_{\text{train}}$$

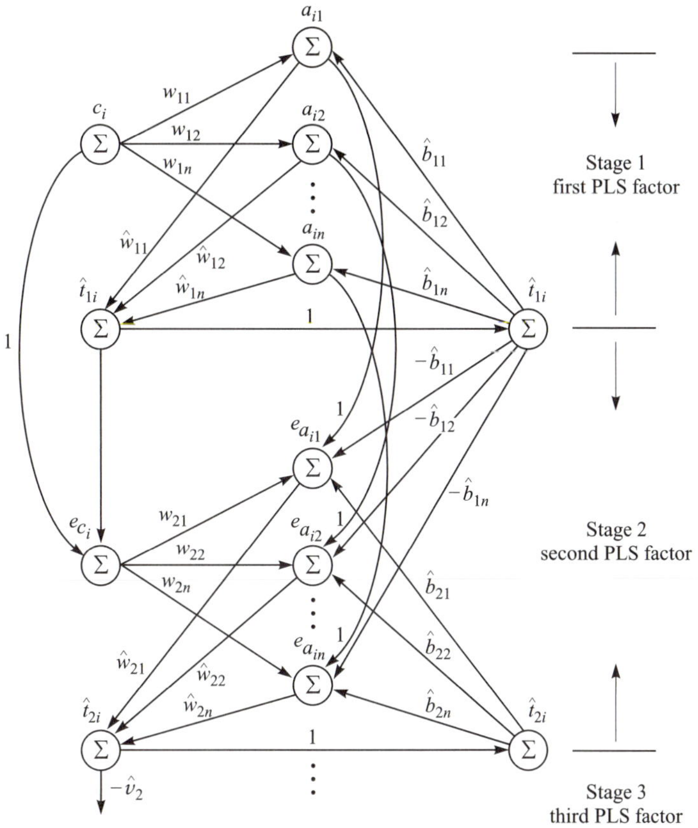

Figure 9.16 PLSNET-C architecture for adaptive extraction of the PLSR weight loading vectors $\hat{\boldsymbol{w}}_h$, regression coefficients $\hat{\upsilon}_h$, and loading vectors $\hat{\boldsymbol{b}}_h$, for $h = 1, 2, \ldots, h^o$ (where h^o is the optimal number of PLSR factors).

validation method explained above. Each additional factor that is required of PLSNET-C is simply carried out by adding a stage to the network. For example, after PLSNET-C is initially trained for a specified number of factors, and it is desired to add another factor, this can be accomplished by simply adding another stage to the network. The resulting network does not have to be completely retrained, only trained for the additional stage that was added, with the synaptic weights already set for the upper stages from the previous training phase. This staging for each PLSR factor is the *adaptive modular* nature of the PLSNET-C architecture. The architecture given in Figure 9.16 is similar to the APEX network for PCA, shown in Figure 9.4 (cf. Sect. 9.3.5), with the obvious difference that during the training process PLSNET-C includes the target, or dependent-variable, information, whereas APEX does not. Therefore, PLSNET-C is trained in a supervised mode, and APEX is an unsupervised trained neural network.

As previously stated, PLSNET-C is trained according to three standard Hebbian learning rules for the extraction of the PLSR weight loading vectors $\hat{\boldsymbol{w}}_h$, regression coefficients $\hat{\upsilon}_h$, and loading vectors $\hat{\boldsymbol{b}}_h$, for $h = 1, 2, \ldots, h^o$, where h^o is the optimal number of PLSR factors. Each of the three Hebbian learning rules can be derived from the PLSNET-C architecture shown in Figure 9.16; that is, Figure 9.17 shows each portion of the network architecture in Figure 9.16 that is associated with the extraction of the PLSR weight loading vectors, the regression coefficients, and the loading vectors.

We will first concentrate on deriving the learning rule for extracting the PLSR weight loading vectors, and the other two learning rules for the regression coefficients and the loading vectors can be derived in similar manner. Figure 9.17(a) shows the portion of the PLSNET-C network (for a single stage) that is responsible for extracting the PLSR weight loading vectors, and the underlying model is given as

$$\boldsymbol{a}_i = c_i \boldsymbol{w}_h + \boldsymbol{e}^{w}_{a_i} \tag{9.131}$$

for a single *measurement* or *observation* where $\boldsymbol{a}_i \in \Re^{n \times 1}$ (for convenience $\boldsymbol{a}_i$ is considered a column vector) and the error vector $\boldsymbol{e}^{w}_{a_i} \in \Re^{n \times 1}$, for $i = 1, 2, \ldots, m$. This is the same model structure that was given in (9.112).

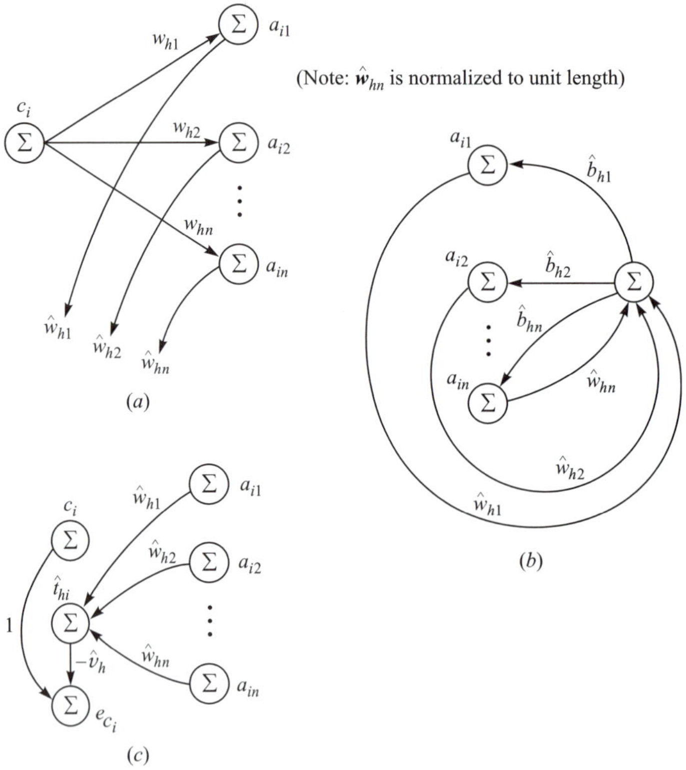

Figure 9.17 (a) Subnetwork of PLSNET-C architecture in Figure 9.16 for extraction of the PLS weight loading vectors $\hat{\boldsymbol{w}}_h$; (b) subnetwork of PLSNET-C architecture in Figure 9.16 for extraction of the PLS loading vectors $\hat{\boldsymbol{b}}_h$; and (c) subnetwork of PLSNET-C architecture in Figure 9.16 for extraction of the PLS scalar regression coefficients $\hat{v}_h$.

PLSNET-C is structured such that only one measurement is processed at a time. To derive the first learning rule, an error cost function (Lyapunov function) is written as

$$L_w(\boldsymbol{w}_h) = \frac{1}{2}\|\boldsymbol{e}^w_{a_i}\|_2^2 \tag{9.132}$$

where $\boldsymbol{e}^w_{a_i}$ is the error associated with the measurement as given in (9.131). The gradient of (9.132), with respect to the weight vector $\boldsymbol{w}_h$, can be computed and is given as

$$\nabla L_w(\boldsymbol{w}_h) = \frac{\partial L_w(\boldsymbol{w}_h)}{\partial \boldsymbol{w}_h} = -c_i\boldsymbol{a}_i + c_i^2\hat{\boldsymbol{w}}_h \tag{9.133}$$

Therefore, using the result by Amari [64], the discrete-time learning rule for the PLSR weight loading vectors is given as

$$\hat{\boldsymbol{w}}_h(k+1) - \hat{\boldsymbol{w}}_h(k) = -\mu_w \frac{\partial L_w(\boldsymbol{w}_h)}{\partial \boldsymbol{w}_h} = \mu_w[c_i\boldsymbol{a}_i - c_i^2\hat{\boldsymbol{w}}_h(k)] \tag{9.134}$$

or

$$\hat{\boldsymbol{w}}_h(k+1) = \hat{\boldsymbol{w}}_h(k) + \mu_w c_i[\boldsymbol{a}_i - c_i\hat{\boldsymbol{w}}_h(k)] \tag{9.135}$$

where k is the discrete time index and $\mu_w > 0$ is the learning parameter. The normalization of the PLS weight loading vectors shown in Figure 9.17(*a*) is not necessary. However, it has been included so PLSNET-C can be directly related to the classic PLSR1 calibration algorithm.

The other two learning rules for extracting the PLSR loading vectors ($\hat{\boldsymbol{b}}_h$) and the regression coefficients ($\hat{v}_h$) can be derived in a similar manner as for the weight loading vectors by referring to Figure 9.17(*b*) and (*c*), respectively. For each of the three learning rules the results are summarized in Table 9.1.

In each of the three learning rules given in Table 9.1 for computing the three sets of weights, that is, $\{\hat{\boldsymbol{w}}_h, \hat{\boldsymbol{b}}_h, \hat{v}_h\}$ for $h = 1, 2, \ldots, h^o$, it can be seen that the standard Hebbian co-occurence term is present (this is the first term on the right-hand side of each of the three expressions). The second term on the right side of each expression represents an active decay which essentially prevents the respective weights $\{\hat{\boldsymbol{w}}_h, \hat{\boldsymbol{b}}_h, \hat{v}_h\}$ from becoming unbounded during the training process. The three learning rules have the same form as Oja's normalized Hebbian learning rule given in (9.25) for learning the first principal eigenvector (cf. Sect. 9.3.1). It can be shown [126] that if the learning rate μ_w satisfies the inequality

$$0 < \mu_w < \left(\sum_{i=1}^{m} c_i^2\right)^{-1} \tag{9.136}$$

where c_i, for $i = 1, 2, \ldots, m$, are the training target values, and also

$$\mu_w \geq \mu_b \geq \mu_v \tag{9.137}$$

then PLSNET-C is guaranteed to converge. There have been other neural network realizations of PLSR, for example, Holcomb and Morari [128], where a linear feedforward neural network architecture is presented. Also,

TABLE 9.1
Summary of the PLSNET-C learning rules

PLS information	Error function	Error cost function	PLSNET-C learning rule
$\hat{w}_h \in \Re^{n\times 1}$ (Weight loading vector)	$e^{w}_{a_i} = a_i - c_i w_h$	$L_w(w_h) = \frac{1}{2}\|e^{w}_{a_i}\|_2^2$	$\hat{w}_h(k+1) = \hat{w}_h(k) + \mu_w c_i[a_i - c_i\hat{w}_h(k)]$
$\hat{b}_h \in \Re^{n\times 1}$ (Loading vector)	$e^{b}_{a_i} = a_i - \hat{t}_{hi} w_h$	$L_b(b_h) = \frac{1}{2}\|e^{b}_{a_i}\|_2^2$	$\hat{b}_h(k+1) = \hat{b}_h(k) + \mu_b \hat{t}_{hi}[a_i - \hat{t}_{hi}\hat{b}_h(k)]$
$\hat{v}_h \in \Re$ (Scalar regression coefficient or inner relationship)	$e^{v}_{c_i} = c_i - v_h \hat{t}_{hi}$	$L_v(v_h) = \frac{1}{2}\|e^{v}_{c_i}\|_2^2$	$\hat{v}_h(k+1) = \hat{v}_h(k) + \mu_v \hat{t}_{hi}[c_i - \hat{t}_{hi}\hat{v}_h(k)]$

I. Where the learning parameters must satisfy $\mu_w \geq \mu_b \geq \mu_v$
II. $h = 1, 2, \ldots, h^o$ (optimal number of PLS factors)
III. c_i are the training target values
IV. a_i are the training input patterns
V. $\hat{t}_{hi} = \hat{w}_h^T a_i$ (measurement intensity)
VI. $i = 1, 2, \ldots, m$ (number of measurements)

nonlinear PLSR neural network architectures have been proposed by Qin and McAvoy [129] and Malthouse [130]. However, the PLSR neural network presented here differs considerably from these architectures, mainly in the adaptive modular nature of PLSNET-C.

PLSNET-Prediction (PLSNET-P)

The PLSNET-P architecture is shown in Figure 9.18. The architecture is based on the PLSR1 prediction algorithm (method 2) given previously. This architecture is actually a subnetwork of the PLSNET-C architecture. The major difference in their operation is that PLSNET-P does not use target data, but predicts (or estimates) a value for the independent variable $\hat{c}$, given a test measurement input $a_{\text{test}\,i}$, for $i = 1, 2, \ldots,$ to the network.

Thus, the PLSNET-P neural network is not trained, but has its weights set according to the extracted information from the train PLSNET-C neural architecture, that is, $\{\hat{w}_h, \hat{b}_h, \hat{v}_h\}$, for $h = 1, 2, \ldots, h^o$. This can be seen in Figure 9.19, where the coupling is shown through the three sets of synaptic weights $\{\hat{w}_h, \hat{b}_h, \hat{v}_h\}$. Similarities between PLSNET-C and PLSNET-P mainly involve the calculation of residuals associated with the input measurements, and both must calculate the PLSR score (or latent variable). As shown in Figure 9.18, the test input measurement vector a_{test} is presented to stage 1 (note that the "test" subscript has been eliminated from the test input measurement samples in Figure 9.18 to avoid confusion). The *output* of PLSNET-P shown in Figure 9.19 is a prediction (or estimate) of the associated dependent variable (or response variable) $\hat{c}$, given a particular test input.

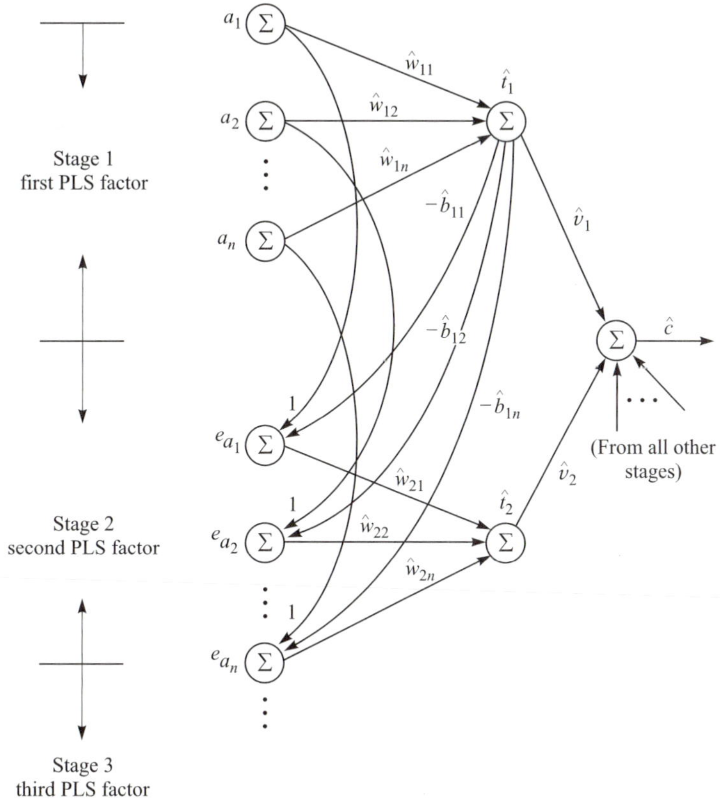

Figure 9.18 PLSNET = P architecture for prediction (or estimation) of the dependent variable ($\hat{c}$) given an independent test measurement or observation as the input.

EXAMPLE 9.4. In this example, the predictive performances of PLSNET and a CLS approach are compared. The data in this example consist of a set of 200 simulated near-infrared (NIR) spectra with associated concentration target values ranging from 2.7 milligrams per deciliter (mg/dL)) to 500 mg/dL for the components of interest in the spectra. The 200 spectra were generated from a cardinal component of interest [shown in Figure 9.20(*a*) as the lowest-amplitude spectrum] and a cardinal obscuring component [shown in Figure 9.20(*a*) as the higher-amplitude spectrum]. Both components were generated as Gaussian functions with two *absorption bands* each, as shown in Figure 9.20(*a*). The obscuring component is considered to be the simulated NIR absorption of water which dominates the NIR absorption of the component of interest [131]. Therefore, the 200 simulated NIR spectra were generated by adding the obscuring component to the component of interest with the obscuring component 3 orders of magnitude larger, and zero-mean Gaussian noise with a relative variance of $\sigma^2 = 9$ was added to the component of interest before adding the obscuring component with a random amplitude to simulate the nonideal nature of a spectrophotometer, that is, baseline variations [7]. Five of the 200 simulated NIR spectra are shown in Figure 9.20(*b*). Because of the dominant effect of the water absorption, all the spectra appear to be identical.

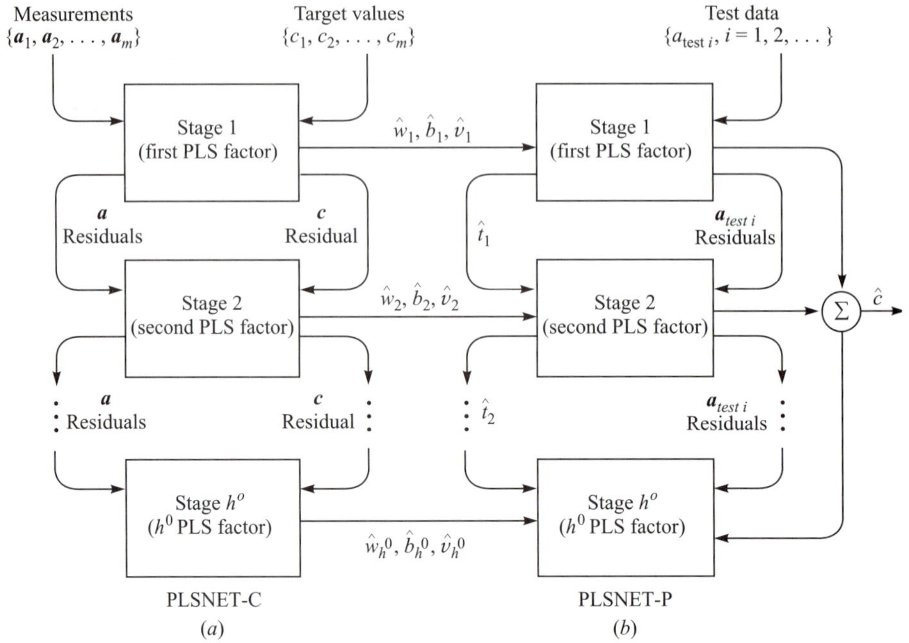

Figure 9.19 Coupling of PLSNET-C and PLSNET-P through the synaptic weights $\{\hat{w}_h, \hat{b}_h, \hat{v}_h\}$, for $h = 1, 2, \ldots, h^o$, extracted by PLSNET-C during training.

Under ideal conditions there exists a linear relationship between the amount of absorption and the concentration of the analyte (component) of interest in an aqueous solution according to the well-known Lambert-Beer law [119]. However, in many situations there can exist a near-linear (slightly nonlinear) relationship; therefore, the varying 200 reference concentration values were generated from

$$c_i = 10^3 \tan^{-1}\left(10^{-4} \sum_{j=1}^{n=100} a_{ij}^p\right) \tag{9.138}$$

where $i = 1, 2, \ldots, m = 200$, and the simulated pure-component NIR absorption spectra are contained in $A^p \in \Re^{m \times n}$ (m spectra and n spectral components or frequencies).

Therefore, the 200 composite simulated spectra contain 200 spectra of interest, that are to be identified and correlated to their associated reference values (i.e., *reference concentrations*) [132], where there exists a nonlinear relationship between the pure-component absorption spectra and the associated reference concentration, a highly dominant obscuring component, that is, water absorption, random noise, and baseline variations. The 200 composite spectra, and associated reference concentration values in ascending order, were divided into two equal data sets, $\{A_{\text{train}}, c_{\text{train}}\}$ and $\{A_{\text{test}}, c_{\text{test}}\}$, each with 100 spectra and concentrations. The training data set was taken as the odd samples, and the test set, the even samples. A CLS model ($\hat{b}_{f\text{CLS}}$) was developed from $\hat{b}_{f\text{CLS}} = (A_{\text{train}}^T A_{\text{train}})^{-1} A_{\text{train}}^T c_{\text{train}}$. Figure 9.20($c$) shows the CLS calibration model vector components as a function of the sample components (or simulated frequencies). As seen in Figure 9.20(c), the calibration model is very erratic; therefore, it is expected that predictions

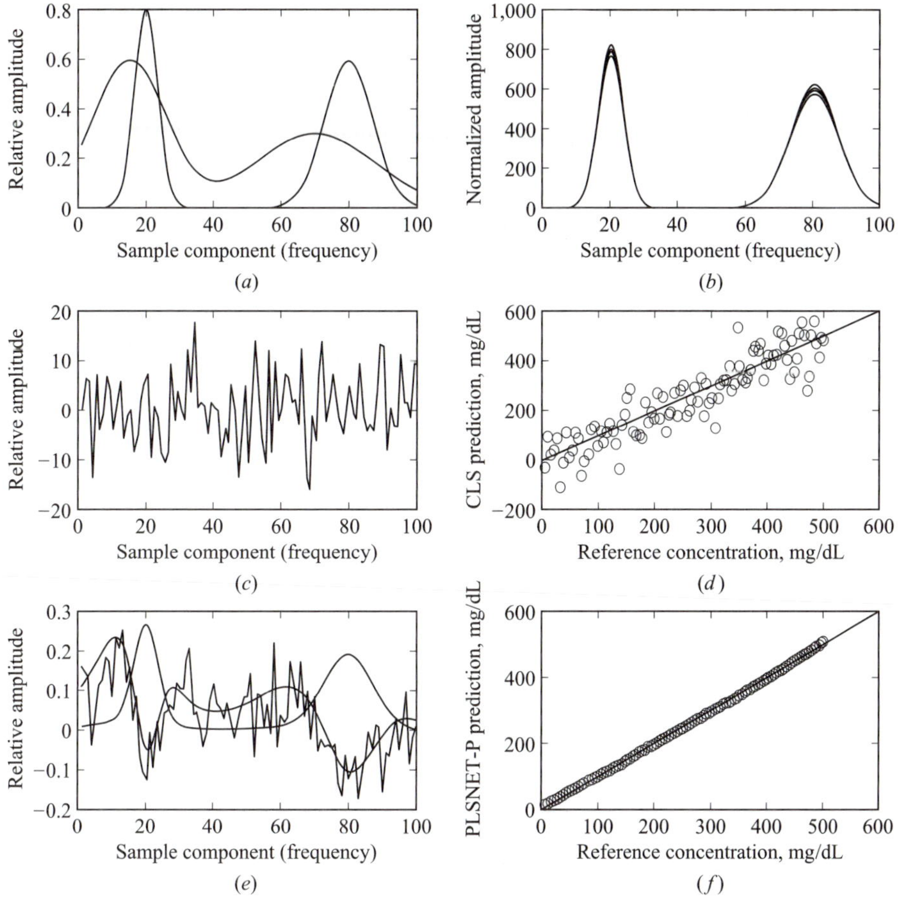

Figure 9.20 (*a*) Cardinal spectral components used to generate the 200 simulated NIR spectra, (*b*) five representative simulated NIR spectra, (*c*) CLS calibration model, (*d*) CLS concentration predictions, (*e*) first three PLS weight loading vectors extracted by PLSNET-C, and (*f*) PLSNET-P concentration predictions.

based on this model would be relatively poor. Projecting the test data absorption matrix $\boldsymbol{A}_{\text{test}}$ onto the CLS calibration model $\hat{\boldsymbol{b}}_{f\text{CLS}}$, the associated test concentrations ($\boldsymbol{c}_{\text{test}}$) can be predicted, that is, $\hat{\boldsymbol{c}}_{\text{testCLS}} = \boldsymbol{A}_{\text{test}}\hat{\boldsymbol{b}}_{f\text{CLS}}$. Using the expression in (9.109), the SEP is computed as 67.08 mg/dL, and the CLS test predictions are shown in Figure 9.20(*d*).

PLSNET-C was trained with $\{\boldsymbol{A}_{\text{train}}, \boldsymbol{c}_{\text{train}}\}$, with three stages (for three PLSR factors), the computed value for the learning rate parameter μ_w from (9.136) is 1.1205×10^{-8}, and $\mu_b = \mu_v = 0.05\mu_w$, which satisfied the inequality in (9.137). After 6,000 training epochs PLSNET-C converged, and the three weight loading vectors are shown in Figure 9.20(*e*). From observing Figure 9.20(*e*), it is obvious that only the first two PLSR factors should be retained for prediction using PLSNET-P. The third weight loading vector shown in Figure 9.20(*e*) (the relatively erratic signal compared to the first two) is associated with noise and should not be retained. Retention of only two PLSR factors was confirmed by using both the independent validation and cross-validation methods. Using the test set $\{A_{\text{test}}, c_{\text{test}}\}$, concentration predictions were generated using PLSNET-P, with the set of synpatic weights $\{\hat{w}_h, \hat{b}_h, \hat{v}_h\}$, for $h = 1, 2, (h^o)$ (i.e., the first two PLSR factors), extracted by PLSNET-C. Using the expression in (9.109), the SEP is calculated as 5.16 mg/dL, which is approximately 13 times less than the SEP computed for the CLS approach; Figure 9.20(*f*) shows the PLSNET-P predictions. It is interesting to compare the obscuring component shown in Figure 9.20(*a*) with the first PLSR weight loading vector shown in Figure 9.20(*e*) (i.e., the one that is similar to the obscuring component). This is not coincidental; in fact, it is typical of PLSR (or PLSNET-C) to extract this information from the data because $\hat{w}_1$ is the first-order approximation to the obscuring component. From stage 1 to stage 2 in PLSNET-C, the score (or latent variable) vector $\hat{t}_1$ and the loading vector $\hat{b}_1$ that are associated with the first weight loading vector $\hat{w}_1$ are used to generate the first *spectral residue*; see Figure 9.16. Thus, an approximation to the obscuring component is removed from the spectral data, and the process continues to the next stage in PLSNET-C. PLSNET is able to predict the concentrations for the test data much better than CLS because of the factor analysis capabilities of PLSR. That is, in PLSNET only 2 factors out of 100 were retained for prediction, whereas CLS used all 100 factors including those associated with noise, which increased the prediction error compared to PLSNET.

9.7 ROBUST PLSR: A NEURAL NETWORK APPROACH

As stated in Section 9.6, nonlinear PLSR neural networks have been developed [129, 130]. These networks include typical nonlinear activation functions, and thus, the training process is carried out to perform a nonlinear mapping. However, the robust PLSR approach presented here is based on the same *linear* PLSNET architecture previously described, but the learning rule is based on a nonquadratic cost function developed by Ham and McDowall [133–137]. That is, a statistical error cost function is utilized that grows less than quadratically which allows for higher-order statistics associated with the inputs to be taken into account. This generalization of PLSR serves to robustify the results when the empirical data contain impulsive and colored noise, and outliers. The robustification techniques developed by Karhunen and Joutsensalo [41] (for robust PCA) and Cichocki and Unbehauen [38] (for robustly solving systems of linear equations) can be applied directly to PLSNET. The resulting learning rules for the robust PLSR neural network are general in the sense that when the *weighting function* is assumed to be quadratic (i.e., a mean-square-error criterion) and the *weighting matrix* is assumed to be the identity matrix, the learning rules for PLSNET given in Section 9.6 can be recovered. Several nonlinear weighting functions are presented that can be used to perform robust PLSR. Some of these are the same as those presented in Section 9.3.6 for robust PCA.

We first derive the robust PLSNET learning rule for the weight loading vectors $\hat{w}_h$, for $h = 1, 2, \ldots, h^o$ (where h^o is the optimal number of PLSR factors). The statistical error cost function for the weight loading vectors can be written as

$$\bar{L}_w(\boldsymbol{w}_h) = \mathbf{1}^T \boldsymbol{S}_w \mathrm{E}\{f(\boldsymbol{e}_{a_i}^w)\} \tag{9.139}$$

where

$$\boldsymbol{e}_{a_i}^w = \boldsymbol{a}_i - c_i \boldsymbol{w}_h \tag{9.140}$$

from (9.131), $\boldsymbol{e}_{a_i}^w \in \Re^{n\times 1}$, where $\boldsymbol{a}_i \in \Re^{n\times 1}$ is a single measurement or observation, $i = 1, 2, \ldots, m$ (where m is the total number of measurements or observations), $c_i \in \Re$ is the target value associated with the measurement or observation, $\boldsymbol{S}_w \in \Re^{n\times n}$ is a positive definite ($\boldsymbol{S}_w > 0$) symmetric ($\boldsymbol{S}_w^T = \boldsymbol{S}_w$) weighting matrix, and in (9.139) $f(\bullet)$ is an appropriate nonlinear weighting function (convex or nearly convex), and $\mathbf{1}^T = [1, \ldots, 1] \in \Re^{1\times n}$ is the **1**s vector. The weight matrix $\boldsymbol{S}_w$ is introduced to allow for thc option to weight the elements of the error vector $\boldsymbol{e}_{a_i}^w$ differently. Taking a steepest descent approach to derive a discrete-time robust learning rule for the PLSR weight loading vectors, that is,

$$\hat{\boldsymbol{w}}_h(k+1) = \hat{\boldsymbol{w}}_h(k) - \mu_w \nabla_{\boldsymbol{w}} \bar{L}_w(\boldsymbol{w}_h) \tag{9.141}$$

we see the gradient $\nabla_{\boldsymbol{w}} \bar{L}_w(\boldsymbol{w}_h)$ must be computed. When the gradient of (9.139) is computed with respect to $\boldsymbol{w}_h$, only instantaneous values are considered; therefore, the expectation operator $\mathrm{E}\{\bullet\}$ is dropped, and the result is

$$\nabla_{\boldsymbol{w}} \bar{L}_w(\boldsymbol{w}_h) = \frac{\partial \bar{L}_w(\boldsymbol{w}_h)}{\partial \boldsymbol{w}_h} = -\, c_i \boldsymbol{S}_w g(\boldsymbol{e}_{a_i}^w) = -\, c_i \boldsymbol{S}_w g(\boldsymbol{a}_i - c_i \hat{\boldsymbol{w}}_h) \tag{9.142}$$

where $g(t) = \mathrm{d}f(t)/\mathrm{d}t$. Using (9.141) and (9.142), the instantaneous robust PLSNET learning rule for the weight loading vectors is given by

$$\hat{\boldsymbol{w}}_h(k+1) = \hat{\boldsymbol{w}}_h(k) + \mu_w \boldsymbol{S}_w c_i g[\boldsymbol{a}_i - c_i \hat{\boldsymbol{w}}_h(k)] \tag{9.143}$$

where $\mu_w > 0$ is the learning rate parameter and can be set according to (9.136) for the linear PLSNET case.

Before we discuss the two instantaneous robust learning rules for the loading vectors and the regression coefficients, there are several observations we can make regarding (9.143): (1) If $\boldsymbol{S}_w = \boldsymbol{I}_{n\times n}$ and $f(t) = \frac{1}{2}t^2$ (quadratic weighting function used for *linear* PLSNET, i.e., mean-square error criterion), then (9.143) is identical to (9.135) for the linear PLSNET case. (2) With the weight matrix $\boldsymbol{S}_w$ included in (9.143), the learning rule can be considered as weighted least-squares [38]. (3) If $\boldsymbol{S}_w = \mathrm{diag}[1/\sigma_i^2]$, where σ_i^2; for $i = 1, 2, \ldots, n$, are the variances of the columns of the measurement matrix, that is, $\boldsymbol{A} = [\boldsymbol{a}_1, \boldsymbol{a}_2, \ldots, \boldsymbol{a}_m]^T \in \Re^{m\times n}$, as described in Section 9.4 [see (9.93)], this would be the variance-scaling process discussed in Section 9.5 for the PLSR1 calibration algorithm. In this case matrix $\boldsymbol{S}_w$ would be applied only once to perform the variance-scaling. However, if not all the measurement data, that is, $\boldsymbol{A}$, are known a priori, this procedure cannot be carried out. (4) In (9.143) the term $c_i g(\boldsymbol{a}_i - c_i \hat{\boldsymbol{w}}_h)$ would be the negative of the gradient of the statistical error cost function in (9.139) if $\boldsymbol{S}_w = \boldsymbol{I}$. However, if $\boldsymbol{S}_w$ is the inverse of the Hessian matrix (cf. Sect. A.3.5) associated with $\bar{L}_w(\boldsymbol{w}_h) = \mathbf{1}^T \mathrm{E}\{f(\boldsymbol{e}_{a_i}^w)\}$, that is, $\boldsymbol{H}_w = \nabla_w^2 \bar{L}_w(\boldsymbol{w}_h)$, and the gradient, as previously stated, would be $\nabla_w \bar{L}_w(\boldsymbol{w}_h) = -\, c_i g(\boldsymbol{a}_i - c_i \hat{\boldsymbol{w}}_h)$, then (9.143) can be considered as Newton's method (cf. Sect. A.5.3). In this case, (9.143) would be written as

$$\hat{\boldsymbol{w}}_h(k+1) = \hat{\boldsymbol{w}}_h(k) + \mu_w \boldsymbol{H}_w^{-1} \nabla_w \bar{L}_w(\boldsymbol{w}_h) \tag{9.144}$$

The instantaneous robust PLSNET learning rules for the loading vectors $\hat{\boldsymbol{b}}_h \in \Re^{n \times 1}$ and the scalar regression coefficients $\hat{v}_h \in \Re$, for $h = 1, 2, \ldots, h^o$, can be derived using a procedure similar to that used to derive the learning rule for the weight loading vectors. The statistical error cost function for the loading vectors can be written as

$$\bar{L}_b(\boldsymbol{b}_h) = \mathbf{1}^T \boldsymbol{S}_b \mathrm{E}\{f(\boldsymbol{e}_{a_i}^b)\} \tag{9.145}$$

where $\boldsymbol{S}_b \in \Re^{n \times n}$, $\boldsymbol{S}_b > 0$, $\boldsymbol{S}_b^T = \boldsymbol{S}_b$, $\boldsymbol{e}_{a_i}^b \in \Re^{n \times 1}$ is written as

$$\boldsymbol{e}_{a_i}^b = \boldsymbol{a}_i - \hat{t}_{hi} \boldsymbol{b}_h \tag{9.146}$$

(see Table 9.1), and

$$\hat{t}_{hi} = \hat{\boldsymbol{w}}_h^T \boldsymbol{a}_i \tag{9.147}$$

is the measurement intensity as explained in Section 9.5. Using (9.145), the instantaneous robust PLSNET learning rule for the loading vectors is given as

$$\hat{\boldsymbol{b}}_h(k+1) = \hat{\boldsymbol{b}}_h(k) + \mu_b \boldsymbol{S}_b \hat{t}_{hi} g[\boldsymbol{a}_i - \hat{t}_{hi} \hat{\boldsymbol{b}}_h(k)] \tag{9.148}$$

where $\mu_b > 0$ and $\mu_b \leq \mu_w$. The statistical error cost function for the scalar regression coefficients can be written as

$$\bar{L}_v(v_h) = s_v \mathrm{E}\{f(e_{c_i}^v)\} \tag{9.149}$$

where $s_v > 0$, and $e_{c_i}^v \in \Re$ is written as

$$e_{c_i}^v = c_i - v_h \hat{t}_{hi} \tag{9.150}$$

(see Table 9.1). Using (9.149), the instantaneous robust PLSNET learning rule for the regression coefficients is given by

$$\hat{v}_h(k+1) = \hat{v}_h(k) + \mu_v s_v \hat{t}_{hi} g[c_i - \hat{t}_{hi} \hat{v}_h(k)] \tag{9.151}$$

where $\mu_v > 0$ and $\mu_v \leq \mu_b$. Therefore, the inequality in (9.137) for linear PLSNET is applicable for robust PLSNET.

Before we present the robust PLSNET algorithm, some discussion about selecting the weighting function $f(\bullet)$ is necessary. There exist several weighting functions that are appropriate for robust PLSNET calibration. The set of weighting functions presented here is not inclusive; however, the particular weighting functions discussed are nonlinear functions commonly used for robust processing [38, 41, 138–140]. In particular, the *M*-estimators [139, 140] weighting function has been applied to robust principal-component estimation using a neural network [138], and it has also been used by Chen and Jain [141] in a backpropagation neural network for robust function approximation. Table 9.2 shows the weighting functions $f(t)$ and their derivatives $g(t) = \mathrm{d}f(t)/\mathrm{d}t$. Figure 9.21 shows each of the weighting functions in Table 9.2 plotted for $\beta = 1$, and Figure 9.22 shows the derivative of each of the weighting functions. From Figure 9.21 it can be seen that for each of the weighting functions used for robust PLSNET ($\beta = 1$) the particular *weighting* for $t \in [-1, 1]$ that the functions provide is essentially *quadratic*. However, outside this range for t, the weighting functions grow less than quadratically for t

TABLE 9.2
Weighting functions for robust PLSNET

Weighting function $f(t)$ ($\beta > 0$)	Derivative $g(t) = \frac{df(t)}{dt}$
Logistic function $f_L(t) = \beta^2 \ln \cosh \frac{t}{\beta}$	$g_L(t) = \beta \tanh \frac{t}{\beta}$
Huber's function: $f_H(t) = \begin{cases} \frac{t^2}{2} & \text{for } \lvert t\rvert \leq \beta \\ \beta\lvert t\rvert - \frac{\beta^2}{2} & \text{for } \lvert t\rvert > \beta \end{cases}$	$g_H(t) = \begin{cases} -\beta & \text{for } t < -\beta \\ t & \text{for } \lvert t\rvert \leq \beta \\ \beta & \text{for } t > \beta \end{cases}$
Talwar's function: $f_T(t) = \begin{cases} \frac{t^2}{2} & \text{for } \lvert t\rvert \leq \beta \\ \frac{\beta^2}{2} & \text{for } \lvert t\rvert > \beta \end{cases}$	$g_T(t) = \begin{cases} 0 & \text{for } t < -\beta \\ t & \text{for } \lvert t\rvert \leq \beta \\ 0 & \text{for } t > \beta \end{cases}$
M estimator $f_M(t) = \frac{1 - e^{-\beta t^2}}{1 + e^{-\beta t^2}}$	$g_M(t) = \frac{4\beta t e^{-\beta t^2}}{(1 + e^{-\beta t^2})^2}$

increasing in magnitude. Therefore, erroneous training data (outliers) can be dramatically deemphasized or basically rejected altogether (see Figure 9.22), thus providing *robustness* for training the PLSNET calibration network and improving performance compared to the quadratic weighting case when impulsive or colored noise is present. The choice of the weighting function

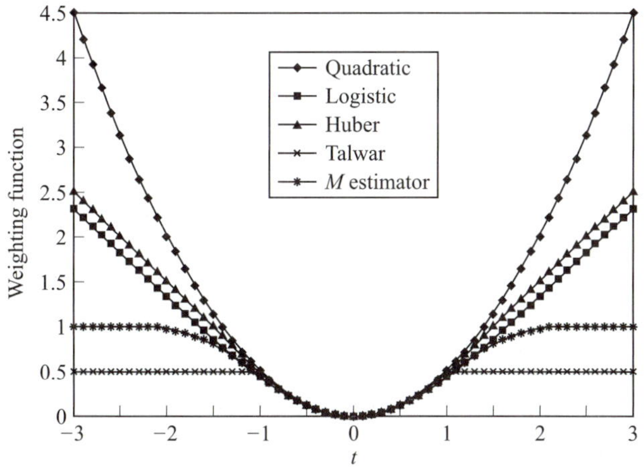

Figure 9.21 Weighting functions used for robust PLSNET ($\beta = 1$).

used in robust PLSNET depends on the characteristics of the training data. Therefore, the particular weighting function is typically selected experimentally. However, caution should be used when adjusting the value of β. For example, if the value of β is set too low for the logistic function (see Table 9.2), the weighting provided by the function could completely (or nearly) reject all the training inputs. Thus, the neural network would never be trained. On the other hand, if β is set too high, the weighting provided will approach the quadratic case and *robustness* will be lost. The value of β is typically selected in the range $0.7 \leq \beta \leq 8$.

Another formulation of the robust PLSR based on the inverse modeling has been developed [137]. The main advantage of this method compared to the forward modeling approach is fewer computational requirements.

The robust PLSNET forward model calibration process can be summarized in the following algorithm.

Robust PLSNET calibration algorithm

Training data set consists of the measurement/target value pairs:

$$\{\boldsymbol{a}_i, c_i\} \qquad \text{for } i = 1, 2, \ldots, m$$

Step 1. Select h^o (optimal number of factors), and set $h = 1$.
Step 2. Compute the learning rate parameters:

$$\mu_w = \eta\left(\sum_{i=1}^{m} c_i^2\right)^{-1} \qquad \eta < 1 \qquad \mu_b = \frac{\mu_w}{\alpha} \qquad \mu_v = \frac{\mu_b}{\alpha} \qquad (\text{typically } \alpha = 2)$$

Step 3. Set the values for the weight matrices: $\boldsymbol{S}_w$, $\boldsymbol{S}_b$, and s_v.
Step 4. Select an appropriate weighting function $f(t)$, and set the value of β.
Step 5. Initialize the weights $\hat{\boldsymbol{w}}_h$, $\hat{\boldsymbol{b}}_h$, and $\hat{v}_h$.
Step 6. Set $i = 1$.
Step 7. (*a*) $\hat{\boldsymbol{w}}_h \leftarrow \hat{\boldsymbol{w}}_h + \mu_w \boldsymbol{S}_w c_i g(\boldsymbol{a}_i - c_i \hat{\boldsymbol{w}}_h)$

(*b*) $\hat{\boldsymbol{w}}_h \leftarrow \dfrac{\hat{\boldsymbol{w}}_h}{\|\hat{\boldsymbol{w}}_h\|_2}$

(*c*) $\hat{t}_{hi} = \hat{\boldsymbol{w}}_h^T \boldsymbol{a}_i$
(*d*) $\hat{\boldsymbol{b}}_h \leftarrow \hat{\boldsymbol{b}}_h + \mu_b \boldsymbol{S}_b \hat{t}_{hi} g(\boldsymbol{a}_i - \hat{t}_{hi}\hat{\boldsymbol{b}}_h)$
(*e*) $\hat{v}_h \leftarrow \hat{v}_h + \mu_v s_v \hat{t}_{hi} g(c_i - \hat{t}_{hi}\hat{v}_h)$
If $i = m$, go to step 8; else $i \leftarrow i + 1$, then go to (*a*).
Step 8. If $h = h^o$, go to step 9; else $h \leftarrow h + 1$, then go to step 6.
Step 9. If convergence is achieved, stop; else, set $h = 1$, then go to step 6.

Convergence can be determined by computing the SEP, see (9.109), using an independent test data set.

In the robust PLSNET calibration algorithm shown above, if the weight matrix is chosen to be the inverse of the Hessian matrix, the best weighting function to use is the logistic function. Huber and Talwar weighting functions result in a second derivative of zero for values greater in magnitude than the cutoff parameter β. The *M* estimator weighting function results in a complicated expression for the second derivative. Even though Newton's method provides its own adaptive learning rate, the standard learning rate parameters

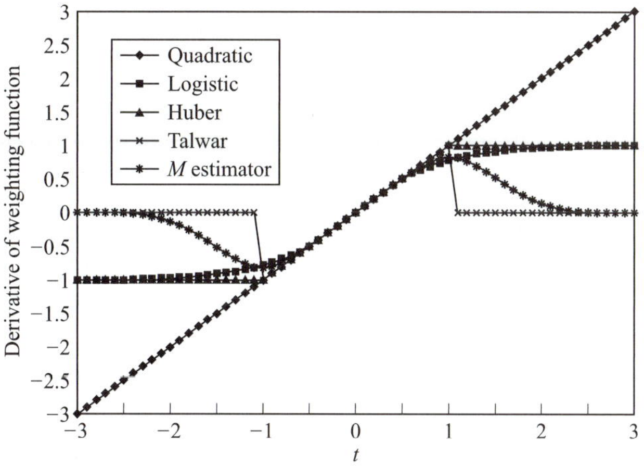

Figure 9.22 Derivatives of the weighting functions in Figure 9.21 ($\beta = 1$).

$\{\mu_w, \mu_b, \mu_v\}$ should be included in the learning rules because only one measurement is processed at a time. That is, Newton's method is typically used in a "batch-mode" operation. However, since PLSNET processes one measurement at a time, a complete solution appropriate for one measurement is not desired. This is the same philosophy that is used in the conjugate gradient approach for backpropagation for the feedforward multilayer perceptron in Section 3.4.1.

PROBLEMS

9.1. Design a single-unit neural network to implement Oja's normalized Hebbian learning rule to estimate the first principal eigenvector, given in discrete-time vector form in (9.25). Generate two independent sets of 5,000 random numbers from a zero-mean, unity-variance, normal (Gaussian) distribution. Scale the values of the first sequence such that the variance is 5 (i.e., multiply the numbers in the sequence by $\sqrt{5}$). From the two sequences of random numbers estimate the covariance matrix, using the expression in (9.19). Compute the eigenvalues and eigenvectors of the covariance matrix, using a standard eigenvalue routine, for example, the MATLAB **`eig`** function. Using Oja's normalized Hebbian learning rule in discrete-time form, compute the first principal eigenvector. Set the initial synaptic weight vector elements to random numbers. Use both an appropriate fixed learning rate parameter and the adaptive algorithm given in (9.28). Compare the two results with respect to the total number of training epochs necessary for convergence. Compare the neural network results to the eigenvector (associated with the largest eigenvalue) computed using the standard eigenvalue routine. Estimate the eigenvalue associated with the first principal eigenvector estimated using Oja's normalized Hebbian learning rule. *Hint*: Use the **`var`** (variance) function in MATLAB as demonstrated in Example 9.1; see (9.29).

9.2. Consider a wide-sense-stationary first-order discrete-time Markov process given by the stochastic difference equation

$$\boldsymbol{x}(k) = \alpha \boldsymbol{x}(k-1) + \boldsymbol{w}(k)$$

where $\alpha = 0.9$; $x, \boldsymbol{w} \in \Re^{3\times 1}$, and $\boldsymbol{w}$ is zero-mean Gaussian white noise with a covariance matrix given as

$$\boldsymbol{C}_v = \begin{bmatrix} 5 & 0 & 0 \\ 0 & 3 & 0 \\ 0 & 0 & 0.1 \end{bmatrix}$$

(*a*) Calculate the covariance matrix $\boldsymbol{C}_x = \mathrm{E}\{\boldsymbol{x}\boldsymbol{x}^T\}$.
(*b*) Calculate the theoretical eigenvalues and eigenvectors of $\boldsymbol{C}_x$.
(*c*) Construct a neural network to implement the discrete-time form of Karhunen-Oja's symmetric subspace learning rule in vector-matrix form given in (9.45). Use both an appropriate fixed learning rate parameter and the adaptive algorithm given in (9.48). Use the network that you develop to extract the two-dimensional principal subspace. Find the angle between "Oja's subspace" and the theoretical subspace, that is, the subspace spanned by the first two eigenvectors calculated in part (*b*).

9.3. Construct a neural network to implement Sanger's generalized Hebbian algorithm. The discrete-time form of the scalar learning rule is given in (9.53). Use the network that you have developed to extract the two-dimensional principal subspace [i.e., the first (and only) two principal eigenvectors] for the data generated in Problem 9.1. Use both an appropriate fixed learning rate parameter and the adaptive algorithm given in (9.48). Find the angle between "Sanger's subspace" and the theoretical subspace.

9.4. In Problem 9.2 (*c*) and Problem 9.3 use the adaptive learning rate parameter and experiment with the forgetting factor (γ). Observe its influence on the speed of convergence.

9.5. Consider the signal

$$\boldsymbol{x}(k) = \boldsymbol{s}(k) + \boldsymbol{w}(k), \qquad k = 0, 1, 2, \ldots$$

where

$$\boldsymbol{s} = \left[\sin\left(\frac{2\pi}{20}1\right), \sin\left(\frac{2\pi}{20}2\right), \ldots, \sin\left(\frac{2\pi}{20}20\right)\right]^T \in \Re^{20\times 1}$$

and

$$\boldsymbol{w} \in \Re^{20\times 1}$$

is zero-mean Gaussian white noise with covariance

$$\boldsymbol{C}_w = \mathrm{diag}(\sigma_w^2, \sigma_w^2, \cdots, \sigma_w^2) \in \Re^{20\times 20} \qquad \text{where } \sigma_w^2 = 0.1$$

(*a*) Use Oja's single-unit learning rule and extract the first principal eigenvector of $\boldsymbol{C}_x$. Use the adaptive algorithm for the learning rate parameter given in (9.28).
(*b*) Compare the estimated first principal eigenvector with the signal $\boldsymbol{s}$, and explain what happened.

9.6. Nonlinear iterative partial least-squares (NIPALS) method. The NIPALS method is an iterative method for the extraction of eigenvectors of a positive definite matrix $\boldsymbol{C}_x$ with distinct eigenvalues.

(*a*) Let $\boldsymbol{e}_1$ be the eigenvector associated with the largest eigenvalue of $\boldsymbol{C}_x$. Consider the iterative process

$$\boldsymbol{w}(k+1) = \boldsymbol{C}_x \boldsymbol{w}(k)$$

$$\boldsymbol{w}(k+1) = \frac{\boldsymbol{w}(k+1)}{\|\boldsymbol{w}(k+1)\|_2}$$

for $k = 0, 1, 2, \ldots$, where $\boldsymbol{w}$ is a random vector of appropriate dimension. Prove that

$$\lim_{k \to \infty} \boldsymbol{w}(k) = \boldsymbol{e}_1$$

(*b*) Show that the following deflation technique can be used for the extraction of the second principal eigenvector:

$$\boldsymbol{C}_x^{(D_1)} = (\boldsymbol{I} - \boldsymbol{e}_1\boldsymbol{e}_1^T)\boldsymbol{C}_x(\boldsymbol{I} - \boldsymbol{e}_1\boldsymbol{e}_1^T)$$

where D_1 is the designator of the first-order deflation of $\boldsymbol{C}_x$. This is called the *deflation transformation* of $\boldsymbol{C}_x$.

Note: Parts (*a*) and (*b*) together constitute the NIPALS method, with the continuation of the deflation process in (*b*) to extract higher-order principal eigenvectors.

9.7. Construct a neural network to implement the discrete-time form of Oja's stochastic gradient ascent (SGA) learning rule given in scalar form in (9.58). Use the adaptive algorithm for the learning rate parameter given in (9.48).

(*a*) Generate three independent sets of 1,000 random numbers from a zero-mean unity-variance normal distribution. Scale the values of the second sequence such that the variance is 0.01. Also scale the values of the third sequence such that the variance of the sequence is 0.001.

(*b*) Use Oja's SGA algorithm and estimate all the principal eigenvectors of the covariance matrix associated with the stochastic data generated in part (*a*). Assume random initial synaptic weights. This process could require several training epochs, that is, several presentations of the 1,000 stochastic vectors. For this approach to train the neural network, the initial learning rate parameter can be recomputed for every training epoch, with the synaptic weights from the previous training epoch being used as the initial weights for the next training epoch.

(*c*) Repeat part (*b*) but use Sanger's GHA. Compare the results that you obtain, especially the amount of training time necessary for convergence.

9.8. In Section 9.3.2 the Karhunen-Oja symmetric subspace learning rule was derived from a mean-square error criterion given in (9.38). The resulting discrete-time vector-matrix learning rule is given in (9.45). The mean-square error criterion can be modified to include a positive definite symmetric weighting matrix $\boldsymbol{S} \in \Re^{n \times n}$, that is, $L(\boldsymbol{W}) = \frac{1}{2}\boldsymbol{e}^T\boldsymbol{S}\boldsymbol{e}$.

(*a*) Using $L(\boldsymbol{W}) = \frac{1}{2}\boldsymbol{e}^T\boldsymbol{S}\boldsymbol{e}$ as the error cost function, derive the discrete-time weighted symmetric subspace learning rule in vector-matrix form.

(*b*) From the results derived in part (*a*), let $\boldsymbol{S} = \boldsymbol{I}$, and with the proper assumptions show what results is the learning rule given in (9.45). Therefore, the weighted symmetric subspace learning rule is the general case.

9.9. As stated in Section 9.3.6, standard linear PCA can arise as an optimal solution to the maximization of linearly transformed variances $\mathrm{E}\{[\boldsymbol{w}_i^T \boldsymbol{x}]^2\}$, for $i = 1, 2, \ldots, m$, or the outputs of a linear network under the orthogonality constraints, that is, $\boldsymbol{W}\boldsymbol{W}^T = \boldsymbol{I} \in \Re^{m \times m}$. Derive the discrete-time learning rule for extracting the first principal eigenvector $\boldsymbol{w}_1$ using the maximization of variance approach. The learning rule that you derive should be identical to the result given in (9.25), which was obtained using the minimization of representation error formulation of the cost function.
Hint: Formulate the cost function (to be maximized) as $L(\boldsymbol{w}_1) = \sigma_{y_1}^2 - \lambda_1(\boldsymbol{w}_1^T \boldsymbol{w}_1 - 1)$, where $y_1 = \boldsymbol{w}_1^T \boldsymbol{x}$, $\sigma_{y_1}^2 = \boldsymbol{w}_1^T \boldsymbol{C}_x \boldsymbol{w}_1$, and $\boldsymbol{C}_x = \mathrm{E}[\boldsymbol{x}\boldsymbol{x}^T]$. This formulation of the cost function is based on the Lagrange multiplier method (cf. Sect. A.6.2) for constrained optimization problems, where the Lagrange multiplier is the associated first principal eigenvalue λ_1 and the constraint is $\boldsymbol{w}_1^T \boldsymbol{w}_1 = 1$. Also, since the first principal eigenvector is being estimated using the learning rule, only instantaneous values of the input are considered.

9.10. Design an APEX neural network to implement the two learning rules for estimating both the feedforward and lateral weights. Use the APEX algorithm summarized in Section 9.3.5. Using a fixed learning rate parameter, test your neural network on the data generated in Problem 9.7(*a*). As with Oja's SGA and Sanger's GHA methods, this neural network also typically requires repeated presentations of the training input vectors. Compare your results to the results obtained in Problem 9.7. What conclusions can be drawn?

9.11. It can be shown that Sanger's GHA method and Kung and Diamantaras's APEX algorithm produce the same stable points after the networks are trained. Prove this.

9.12. (*a*) In Section 9.5 the PLSR1 calibration algorithm is presented. Write a computer program, preferably in MATLAB, to implement the calibration process (skip the first step in the algorithm for mean-centering and variance-scaling of the data). The number of PLSR factors should be a variable in your program.
(*b*) Write two additional computer programs to implement the two PLSR1 prediction algorithms given in Section 9.5, that is, both *prediction method 1* and *prediction method 2*.

9.13. Many times there are several components of interest that reside in measurement data collected from a particular process. It was shown in Example 9.4, for the synthetic near-infrared data, that the single-component PLSR1 algorithm, realized as a neural process (PLSNET), could extract the information content with respect to the *single* target values (reference concentrations) and yield much greater predictive performance than the CLS method. When it is desired to use PLSR for *multiple* components of interest (several sets of target values), the PLSR1 algorithm cannot be used directly. *Multiple-component* PLSR methods have been developed [7, 10, 13]; however, typically a sacrifice in performance can be expected when using these methods to develop a single model that can predict multiple components. As an alternative, the PLSR1 algorithm can be used, but applied separately for each component (i.e., different sets of target values). In other words, for every component of interest for which there are target values (reference values) available, the PLSR1 algorithm can be used independently, generating a separate set of weight loading vectors, loading vectors, and regression coefficients for each successive component of interest.

(*a*) Design a neural network for PLSR (i.e., PLSNET). That is, develop PLSNET-C for the calibration phase, and PLSNET-P to use for prediction. Write MATLAB functions for both.

(*b*) Write a computer program that can perform CLS.

(*c*) On the McGraw-Hill web site, download the data designated for this problem at http://www.mhhe.com/engcs/electrical/ham. You will find two sets of data: (1) the same data that were used in Example 9.4 (this is designated as P9_13_Data_Set_1), that is, the training and test data, and (2) another data set (P9_13_Data_Set_2) that consists of an independent-variable block matrix for training with an associated dependent-variable block with two sets of target values, and another equivalent set of data for testing. The measurement data in the second data set are the same as in the first data set, except an additional component of interest is present; thus, there is a second column of target values in the training and test data sets.

(*d*) Using the PLSNET developed in part (*a*), replicate the results shown in Example 9.4, using the first data set that only contains the single component in the data. You will generate one set of weight loading vectors, loading vectors, and regression coefficients. Plot the regression results for predictions on the test data, and compute the SEP (which should be identical to the value given in Example 9.4). The PLSR1 calibration and prediction computer programs from Problem 9.12 can be used to verify the number of PLSR factors that are necessary to retain.

(*e*) Using your computer program for CLS from part (*b*), develop the CLS calibration model and generate the predictions on the test data. Plot the regression results for the predictions on a separate plot, and compute the SEP.

(*f*) Using the second set of data that contains two components of interest and PLSNET, select the proper number of PLS factors and generate the appropriate *two* sets of weight loading vectors, loading vectors, and regression coefficients. *Two* calibration models will be developed. The PLSR1 calibration and prediction computer programs from Problem 9.12 can again be useful to verify the number of PLSR factors that are necessary to retain. Generate the predictions for both components, using the test data and the appropriate weight loading vectors, loading vectors, and regression coefficients associated with the particular component of interest. Plot the regression results for both components separately. Calculate the SEP for both components.

(*g*) Using your computer program for CLS from part (*b*), develop two CLS calibration models for the two different components, and generate the predictions on the two sets of test data. Plot the regression results for the predictions on separate plots, and compute the SEP for both components.

(*h*) Compare the performance results obtained in parts (*f*) and (*g*). What can you conclude?

(*i*) Compare the predictive performance for the first component obtained in part (*f*) using PLSNET to the results obtained in (*d*). What can you conclude?

REFERENCES

1. D. F. Morrison, *Multivariate Statistical Methods*, 3rd ed., New York: McGraw-Hill, 1990.
2. S. Haykin, *Adaptive Filter Theory*, 2nd ed., Englewood Cliffs, NJ: Prentice-Hall, 1991.
3. I. T. Jolliffe, *Principal Component Analysis*, New York: Springer-Verlag, 1986.
4. N. Draper and H. Smith, *Applied Regression Analysis*, 2nd ed., New York: Wiley-Interscience, 1981.
5. H. Hotelling, "Analysis of a Complex of Statistical Variables into Principal Components," *Journal of Educational Psychology*, vol. 24, 1933, pp. 498–520.
6. K. V. Mardia, J. T. Kent, and J. M. Bibby, *Multivariate Analysis*, New York: Academic, 1979.
7. H. Martens and T. Naes, *Multivariate Calibration*, New York: Wiley, 1989.
8. B. R. Kowalski, ed., *Chemometrics—Mathematics and Statistics in Chemistry*, NATO ASI Series, Series C: Mathematical and Physical Sciences, vol. 138, Boston: D. Reidel, 1984.
9. F. M. Ham and I. Kostanic, "Partial Least-Squares: Theoretical Issues and Engineering Applications in Signal Processing," *Journal of Mathematical Problems in Engineering*, vol. 2, no. 1, 1996, pp. 63–93.
10. P. Geladi and B.R. Kowalski, "Partial Least-Squares Regression: A Tutorial," *Analytica Chimica Acta,* no. 185, 1986, pp. 1–17.
11. K. R. Beebe and B. R. Kowalski, "An Introduction to Multivariate Calibration and Analysis," *Analytical Chemistry*, vol. 59, no. 417, 1987, pp. 1007A–17A.
12. D.M. Haaland and E. V. Thomas, "Partial Least-Squares Methods for Spectral Analyses. 1. Relation to Other Quantitative Calibration Methods and the Extraction of Qualitative Information," *Analytical Chemistry*, vol. 60, no. 11, 1988, pp. 1193–202.
13. A. Lorber, L. E. Wagen, and B. R. Kowalski, "A Theoretical Foundation for the PLS Algorithm," *Journal of Chemometrics*, vol. 1, 1987, pp. 19–31.
14. S. D. Brown, "Chemometrics," *Analytical Chemistry*, vol. 62, 1990, pp. 84R–101R.
15. R. C. Gonzalez and R. E. Woods, *Digital Image Processing*, Reading, MA: Addison-Wesley, 1992.
16. A. D. Whalen, *Detection of Signal in Noise*, New York: Academic, 1971.
17. G. E. P. Box, G. M. Jenkins, and G. C. Reinsel, *Time Series Analysis—Forcasting and Control*, 3rd ed., Englewood Cliffs, NJ: Prentice-Hall, 1994.
18. E. Oja, "A Simplified Neuron Model as a Principal Component Analyzer," *Journal of Mathematical Biology*, vol. 15, 1982, pp. 267–73.
19. P. Baldi and K. Hornik, "Neural Networks and Principal Component Analysis: Learning from Examples Without Local Minima," *Neural Networks*, vol. 2, 1989, pp. 53–8.
20. E. Oja and J. Karhunen, "On Stochastic Approximation of the Eigenvectors and Eigenvalues of the Expectation of a Random Matrix," *Journal of Mathematical Analysis and Applications*, vol. 104, 1985, pp. 69–84.
21. T. D. Sanger, "Optimal Unsupervised Learning in a Single-Layer Linear Feedforward Neural Network," *Neural Networks*, vol. 2, 1989, pp. 459–73.
22. P. Földiak, "Forming Sparse Representations by Local Anti-Hebbian Learning," *Biological Cybernetics*, vol. 64, 1990, pp. 165–70.

23. P. Földiak, "Adaptive Network for Optimal Linear Feature Extraction," in *Proceedings of the International Joint Conference on Neural Networks*, Washington, DC, vol. 1, 1989, pp. 401–5.
24. F. M. Silva and L. B. Almeida, "A Distributed Decorrelation Algorithm," in *Neural Networks Advances and Applications*, ed. E. Gelenbe, Amsterdam: North-Holland, 1991, pp. 145–63.
25. S. Y. Kung, K. I. Diamantaras, and J. S. Taur, "Neural Networks for Extracting Pure/Constrained/Oriented Principal Components," in *SVD and Signal Processing*, ed. R. J. Vaccaro, Amsterdam: Elsevier Science, 1991, pp. 57–81.
26. E. Oja, "Principal Components, Minor Components and Linear Neural Networks," *Neural Networks*, vol. 5, 1992, pp. 927–35.
27. P. F. Baldi and K. Hornik, "Learning in Linear Neural Networks: A Survey," *IEEE Transactions on Neural Networks*, vol. 6, 1995, pp. 837–58.
28. L.-H. Chen and S. Chang, "An Adaptive Learning Algorithm for Principal Component Analysis," *IEEE Transactions on Neural Networks*, vol. 6, 1995, pp. 1255–63.
29. E. Oja, "Neural Networks, Principal Components, and Subspaces," *International Journal of Neural Systems*, vol. 1, 1989, pp. 61–8.
30. S. Y. Kung, K. I. Diamantaras, and J. S. Taur, "Adaptive Principal Component Extraction (APEX) and Applications," *IEEE Transactions on Signal Processing*, vol. 42, 1994, pp. 1202–17.
31. J. Rubner and P. Tavan, "A Self-Organizing Network for Principal Component Analysis," *Europhysics Letters*, vol. 10, no. 7, 1989, pp. 693–8.
32. R. Lenz and M. Österberg, "Computing the Karhunen-Loève Expansion with a Parallel, Unsupervised Filter System," *Neural Computation*, vol. 4, 1992, pp. 382–92.
33. R. H. White, "Competitive Hebbian Learning: Algorithm and Demonstrations," *Neural Networks*, vol. 5, 1992, pp. 261–75.
34. J. Karhunen and J. Joutsensalo, "Tracking of Sinusoidal Frequencies by Neural Network Learning Algorithms," in *Proceedings of IEEE International Conference on Acoustics, Speech, and Signal Processing*, Toronto, Canada, 1991, pp. 69–84.
35. K. Hornik and C.-M. Kuan, "Convergence Analysis of Local Feature Extraction Algorithms," *Neural Networks*, vol. 5, 1992, pp. 229–40.
36. T. D. Sanger, "An Optimality Principle for Unsupervised Learning," in *Advances in Neural Information Processing Systems*, ed. D.S. Touretzky, Palo Alto, CA: Morgan Kaufmann, 1989, pp. 11–19.
37. A. Krogh and J. A. Hertz, "Hebbian Learning of Principal Components," in *Parallel Processing in Neural Systems and Computers*, eds. R. Eckmiller, G. Hartmann, and G. Hauske, Amsterdam: North-Holland, 1990, pp. 183–6.
38. A. Cichocki and R. Unbehauen, *Neural Networks for Optimization and Signal Processing*, New York: Wiley, 1993.
39. J. Hertz, A. Krogh, and R. G. Palmer, *Introduction to the Theory of Neural Computation*, Redwood City , CA: Addison-Wesley, 1991.
40. S. Haykin, *Neural Networks—A Comprehensive Foundation*, New York: Macmillan, 1994.
41. J. Karhunen and J. Joutsensalo, "Generalizations of Principal Component Analysis, Optimization Problems, and Neural Networks," *Neural Networks*, vol. 8, 1995, pp. 549–62.
42. E. Oja and J. Karhunen, "Nonlinear PCA: Algorithms and Applications," Technical Report: A18 (Otaniemi 1993), Helsinki University of Technology (Laboratory of Computer and Information Sciences), Espoo, Finland, 1993.

43. J. Karhunen and J. Joutsensalo, "Representation and Separation of Signals Using Nonlinear PCA Type Learning," *Neural Networks*, vol. 7, 1994, pp. 113–27.
44. J. Joutsensalo, "Nonlinear Data Compression and Representation by Combining Self-Organizing Map and Subspace Rule," *Proceedings of IEEE International Conference on Neural Networks '94*, Orlando, FL, vol. 2, 1994, pp. 637–40.
45. F. Palmieri, "Hebbian Learning and Self-Association in Nonlinear Neural Networks," *Proceedings of IEEE International Conference on Neural Networks '94*, Orlando, FL, vol. 2, 1994, pp. 1258–63.
46. A. Sudjianto and M. H. Hassoun, "Nonlinear Hebbian Rule: A Statistical Interpretation," *Proceedings of IEEE International Conference on Neural Networks '94*, Orlando, FL, vol. 2, 1994, pp. 1247–52.
47. L. Xu, "Theories for Unsupervised Learning: PCA and Its Nonlinear Extensions," *Proceedings of IEEE International Conference on Neural Networks '94*, Orlando, FL, vol. 2, 1994, pp. 1252–7.
48. J. Karhunen, "Optimization Criteria and Nonlinear PCA Neural Networks," *Proceedings of IEEE International Conference on Neural Networks '94*, Orlando, FL, vol. 2, 1994, pp. 1241–6.
49. S. Y. Kung, *Digital Neural Networks*, Englewood Cliffs, NJ: Prentice-Hall, 1993.
50. R. Linsker, "Designing a Sensory Processing System: What Can Be Learned from Principal Component Analysis?" in the *Proceedings of the International Joint Conference on Neural Networks*, Washington, DC, vol. 2, 1990, pp. 291–7.
51. H. Köhnel and P. Taven, "The Anti-Hebbian Rule Derived from Information Theory," in *Parallel Processing in Neural Systems and Computers*, eds. R. Eckmiller, G. Hartmann, and G. Hauske, Amsterdam: North-Holland, 1990, pp. 187–90.
52. A. Carlson, "Anti-Hebbian Learning in a Non-Linear Neural Network," *Biological Cybernetics*, vol. 64, 1990, pp. 171–6.
53. L. Xu, E. Oja, and C. Y. Suen, "Modified Hebbian Learning for Curve and Surface Fitting," *Neural Networks*, vol. 5, 1992, pp. 441–57.
54. S. Becker, "Unsupervised Learning Procedures for Neural Networks," *International Journal of Neural Systems*, vol. 2, 1991, pp. 17–33.
55. J. A. Sirat, A Fast Neural Algorithm for Principal Component Analysis and Singular Value Decomposition," *International Journal of Neural Systems*, vol. 2, 1991, pp. 147–55.
56. Y. Chauvin, "Constrained Hebbian Learning: Gradient Descent to Global Minima in an *N*-Dimesional Landscape," *International Journal of Neural Systems*, vol. 2, 1991, pp. 35–46.
57. E. Oja, H. Ogawa, and J. Wangviwattana, "Principal Components Analysis by Homogeneous Neural Networks, Part I: The Weighted Subspace Criterion," *IEICE Transactions on Information and Systems* (Japan), vol. E75-D, 1992, pp. 366–75.
58. E. Oja, H. Ogawa, and J. Wangviwattana, "Principal Components Analysis by Homogeneous Neural Networks, Part II: Analysis and Extensions of the Learning Rule," *IEICE Transactions on Information and Systems* (Japan), vol. E75-D, 1992, pp. 376–82.
59. L. Xu, A. Krzyzak, and E. Oja, "Neural-Net Method for Dual Subspace Pattern Recognition," in *Proceedings of the International Joint Conference on Neural Networks*, Seattle, WA, vol. 2, 1991, pp. 379–84.
60. J. Karhunen and J. Joutsensalo, "Learning of Sinusoidal Frequencies by Nonlinear Constrained Hebbian Algorithms," in *Neural Networks for Signal*

Processing II, eds. S. Y. Kung, F. Fallside, J. A. Sorenson, and C. A. Kamm, New York: IEEE Press, 1992, pp. 39–48.

61. J. Joutsensalo and J. Karhunen, "A Nonlinear Extension of the Generalized Hebbian Learning," *Neural Processing Letters*, vol. 2, 1995, pp. 5–8.
62. J. Karhunen and J. Joutsensalo, "Frequency Estimation by a Hebbian Subspace Learning Algorithm," in *Artificial Neural Networks*, eds. T. Kohonen et al., Amsterdam: North-Holland, 1991, pp. 1637–40.
63. J. Karhunen and J. Joutsensalo, "Nonlinear Hebbian Algorithms for Sinusoidal Frequency Estimation," in *Artificial Neural Networks*, eds. I. Aleksander and J. Taylor, Amsterdam: North-Holland, 1992, pp. 1199–202.
64. S. Amari, "Mathematical Theory of Neural Learning," *New Generation Computing*, vol. 8, 1991, pp. 281–94.
65. D. G. Luenberger, *Linear and Nonlinear Programming*, 2nd ed., Reading, MA: Addison-Wesley, 1984.
66. S. Y. Kung, "Adaptive Principal Component Analysis Via an Orthogonal Learning Network," in *Proceedings of the International Symposium on Circuits and Systems*, New Orleans, 1990, pp. 719–22.
67. K. I. Diamantaras, "Principal Component Learning Networks and Applications," *Ph.D. dissertation*, Princeton, NJ: Princeton University, 1992.
68. S. Bannour and M. R. Azimi-Sadjadi, "Principal Component Extraction Using Recursive Least Squares Method," in *Proceedings of the International Joint Conference on Neural Networks*, Singapore, 1991, pp. 2110–5.
69. S. Y. Kung and K. I. Diamantaras, "A Neural Network Learning Algorithm for Adaptive Principal Component Extraction (APEX)," in *Proceedings of the International Conference on Acoustics, Speech, and Signal Processing*, Albuquerque, NM, 1990, vol. 2, pp. 861–4.
70. J. Karhunen and J. Joutsensalo, "Learning of Robust Principal Component Subspace," in *Proceedings of the International Joint Conference On Neural Networks*, Nagoya, Japan, October 1993, pp. 2409–12.
71. J. Karhunen and J. Joutsensalo, "Nonlinear Generalizations of Principal Component Learning Algorithms," in *Proceedings of the International Joint Conference On Neural Networks*, Nagoya, Japan, October 1993, pp. 2599–602.
72. A. Cichocki and R. Unbehauen, "Robust Estimation of Principal Components by Using Neural Network Learning Algorithms," *Electronic Letters*, vol. 29, 1993, pp. 1869–70.
73. L. Xu and A. Yuille, "Self-Organizing Rules for Robust Principal Component Analysis," in *Advances in Neural Information Processing Systems*, vol. 5, eds. S. J. Hanson, J. D. Cowan, and C. L. Giles, San Mateo, CA: Morgan Kaufmann, 1993, pp. 467–74.
74. J. Joutsensalo, "Conventional and Neural Methods for Estimating Principal Component Type Subspaces with Application to Sinusoidal Frequency Estimation," *Ph.D. dissertation*, Helsinki, Finland: Helsinki, University of Technology, 1994.
75. J. Karhunen and J. Joutsensalo, "Representation and Separation of Signals Using Nonlinear PCA Type Learning," Technical Report A17 (Otaniemi 1993), Helsinki University of Technology (Laboratory of Computer and Information Sciences), Espoo, Finland, 1993.
76. L. Xu, "Least MSE Reconstruction for Self-Organization," *Proceedings of the International Joint Conference On Neural Networks*, Part 2, Singapore, November 1991, pp. 2368–73.
77. L. E. Russo, "An Outer Product Neural Network for Extracting Principal Components from a Time Series," in *Neural Networks for Signal Processing*

(Proceedings of the 1991 IEEE Workshop), eds. B. H. Juang, S. Y. Kung, and C. A. Kamm, New York: IEEE Press, 1991, pp. 161–70.

78. F. L. Luo, R. Unbehauen, and A. Cichocki, "A Minor Component Analysis," *Neural Networks*, vol. 10, 1997, pp. 291–7.
79. K. Matsuoka and M. Kawamoto, "A Neural Network that Self-Organizes to Perform Three Operations Related to Principal Component Analysis," *Neural Networks*, vol. 7, 1994, pp. 753–65.
80. F. M. Ham and I. Kim, "Extension of the Generalized Hebbian Algorithm for Principal Component Extraction," in *Proceedings of Applications and Science of Neural Networks, Fuzzy Systems and Evolutionary Computation*, eds. J. C. Bezdek, B. Bosacchi, and D. B. Fogel, San Diego: SPIE, July 19–24, 1998, vol. 3455, pp. 274–85.
81. K. I. Diamantaras and S. Y. Kung, *Principal Component Neural Networks—Theory and Applications*, New York: Wiley-Interscience, 1996.
82. F. Peper and H. Noda, "A Symmetric Linear Neural Network that Learns Principal Components and Their Variances," *IEEE Transactions on Neural Networks*, vol. 7, 1996, pp. 1042–6.
83. W. Skarbek, A. Cichocki and W. Kasprzak, "Principal Subspace Analysis for Incomplete Image Data in One Learning Epoch," *Neural Network World*, vol. 6, 1996, pp. 375–81.
84. F. L. Luo, R. Unbehauen and Y. D. Li, "A Principal Component Analysis Algorithm with Invariant Norm," *Neurocomputing*, vol. 8, 1995, pp. 213–21.
85. K. I. Diamantaras and S. Y. Kung, "Cross-Correlation Neural Network Models," *IEEE Transactions on Signal Processing*, vol. 42, 1994, pp. 3218–23.
86. D. Obradovic and G. Deco, "An Information Theory Based Learning Paradigm for Linear Feature Extraction," *Neurocomputing*, vol. 12, 1996, pp. 203–21.
87. M. D. Plumbley, "Lyapunov Functions for Convergence of Principal Component Algorithms," *Neural Networks,* vol. 8, 1995, pp. 11–23.
88. J. Dehaene, M. Moonen, and J. Vandewalle, "An Improved Stochastic Gradient Algorithm for Principal Component Analysis and Subspace Tracking," *IEEE Transactions on Signal Processing,* vol. 45, 1997, pp. 2582–6.
89. W. Y. Yan, U. Helmke, and J. B. Moore, " Global Analysis of Oja's Flow for Neural Networks," *IEEE Transactions on Neural Networks*, vol. 5, 1994, pp. 674–83.
90. Q. Zhang and Y. W. Leung, "Energy Function for the One Unit Oja Algorithm," *IEEE Transactions on Neural Networks*, vol. 6, 1995, pp. 1291–3.
91. T. Chen, Y. Hua and W. Y. Yan, "Global Convergence of Oja's Subspace Algorithm for Principal Component Extraction," *IEEE Transactions on Neural Networks*, vol. 9, 1998, pp. 58–67.
92. S. Bannour and M. R. Azami-Sadjadi, "Principal Component Extraction Using Recursive Least Squares Learning," *IEEE Transactions on Neural Networks*, vol. 6, 1995, pp. 457–69.
93. F. Peper and H. Noda, "A Symmetrical Linear Neural Network that Learns Principal Components and Their Variances, *IEEE Transactions on Neural Networks*, vol. 7, 1996, pp. 1042–7.
94. V. Solo and X. Kong, "Performance Analysis of Adaptive Eigenanalysis Algorithms," *IEEE Transactions on Signal Processing*, vol. 46, 1998, pp. 636–46.
95. C. Chatterjee, V. P. Roychowdhury, and E. K. P. Chong, "On Relative Convergence Properties of Principal Component Analysis Algorithms," *IEEE Transactions on Neural Networks*, vol. 9, 1998, pp. 319–29.
96. G. H. Golub and C. F. Van Loan, *Matrix Computations*, 3rd ed., Baltimore, MD: Johns Hopkins University Press, 1996.

97. J. H. Wilkinson, *The Algebraic Eigenvalue Problem*, London: Oxford University Press, 1965.
98. K. Pearson, "On Lines and Planes of Closest Fit to Systems of Points in Space," *Philosophical Magazine*, ser. 6, vol. 2, 1901, pp. 559–72.
99. L. E. Russo and E. C. Real, "Image Compression Using an Outer Product Neural Network," in *Proceedings of IEEE International Conference on Acoustics, Speech, and Signal Processing*, vol. 2, 1992, pp. 377–80.
100. G. W. Cottrell and P. Munro, "Principal Components Analysis of Images via Back Propagation," in *SPIE Visual Communications and Image Processing*, vol. 1001, 1988, pp. 1070–7.
101. E. Oja and T. Kohonen, "A Subspace Learning Algorithm as a Formalism for Pattern Recognition and Neural Networks," in *Proceedings of IEEE Neural Network Conference*, vol. 1, 1988, pp. 270–84.
102. T. Y. Young and T. W. Calvert, *Classification, Estimation, and Pattern Recognition*, New York: American Elsevier, 1974.
103. P. A. Devijver and J. Kittler, *Pattern Recognition, A Statistical Approach*, Englewood Cliffs, NJ: Prentice-Hall, 1982.
104. A. K. Jain, *Fundamentals of Digital Image Processing*, Englewood Cliffs, NJ: Prentice-Hall, 1989.
105. L. Xu and A. Yuille, "Robust PCA Learning Rules Based on Statistical Physics Approach," *Proceedings of the International Joint Conference on Neural Networks*, Part 1, Baltimore, MD, June 1992, pp. 812–17.
106. M. A. Kramer, "Nonlinear Principal Component Analysis Using Autoassociative Neural Networks," *AIChE Journal*, vol. 37, 1991, pp. 233–43.
107. B. Porat, *Digital Processing of Random Signals*, Englewood Cliffs, NJ: Prentice-Hall, 1994.
108. B. Picinbono, *Random Signals and Systems*, Englewood Cliffs, NJ: Prentice-Hall, 1993.
109. C. W. Therrien, *Discrete Random Signals and Statistical Signal Processing*, Englewood Cliffs, NJ: Prentice-Hall, 1992.
110. C. L. Nikias and J. M. Mendel, "Signal Processing with Higher-Order Spectra," *IEEE Signal Processing Magazine*, vol. 10, 1993, pp. 10–37.
111. L. Parra, G. Deco, and S. Miesbach, "Redundancy Reduction with Information-Preserving Nonlinear Maps," *Network*, vol. 1, 1995, pp. 61–72.
112. P. Comon, "Independent Component Analysis: A New Concept?" *Signal Processing*, vol. 36, 1994, pp. 287–314.
113. E. Oja, "The Nonlinear PCA Learning Rule in Independent Component Analysis," *Neurocomputing*, vol. 17, 1997, pp. 25–45.
114. J. Karhunen, E. Oja, L. Wang, R. Vigario, and J. Joutsensalo, "A Class of Neural Networks for Independent Component Analysis," *IEEE Transactions on Neural Networks*, vol. 8, 1997, pp. 486–504.
115. L. Parra, "Symplectic Nonlinear Independent Component Analysis," in *Advances in Neural Information Processing Systems*, vol. 8, eds. D. S. Touretzky, M. C. Mozer, and M. E. Hasselmo, Cambridge, MA: MIT Press, 1996, pp. 437–43.
116. C. N. Banwell, *Fundamentals of Molecular Spectroscopy*, New York: McGraw-Hill, 1966.
117. B. P. Straughtan and S. Walker, eds., *Spectroscopy*, vol. 2, New York: Wiley, 1976.
118. P. Williams and K. Norris, eds., *Near-Infrared Technology*, St. Paul, MN: American Association of Cereal Chemists, 1987.

119. M. Avram and G. H. Mateescu, *Infrared Spectroscopy*, New York: Wiley-Interscience, 1972.
120. E. R. Malinowski, "Determination of the Number of Factors and the Experimental Error in a Data Matrix," *Analytical Chemistry*, vol. 49, 1977, pp. 612-7.
121. E. R. Malinowski, *Factor Analysis in Chemistry*, 2nd. ed., New York: Wiley, 1991.
122. H. Wold, "Soft Modelling: The Basic Design and Some Extensions," *Systems Under Direct Observation, Causality-Structure-Prediction,* eds. K. G. Joreskog and H. Wold, Amsterdam: North-Holland, 1981.
123. H. Wold, in *Food Research and Drug Analysis,* eds. H. Martens and H. Russwurm, London: Applied Science Publ., 1983.
124. E. R. Malinowski, "Theory of Error in Factor Analysis," *Analytical Chemistry*, vol. 49, 1977, pp. 606–12.
125. D. M. Haaland, "Classical Versus Inverse Least-Squares Methods in Quantitative Spectral Analyses," *Spectroscopy*, vol. 2, 1987, pp. 56–7.
126. F. M. Ham and I. Kostanic, "A Neural Network Architecture for Partial Least-Squares Regression (PLSNET) with Supervised Adaptive Modular Hebbian Learning," *Neural, Parallel, Scientific Computations*, vol. 6, 1998, pp. 35–72.
127. F. M. Ham and I. Kostanic, "A Partial Least-Squares Regression Neural NETwork (PLSNET) with Supervised Adaptive Modular Learning," in *Applications and Science of Artificial Neural Networks II*, eds. S. K. Rogers and D. W. Ruck, Proceedings of SPIE, vol. 2760, 1996, pp. 139–50.
128. T. R. Holcomb and M. Morari, "PLS/Neural Networks," *Computers in Chemical Engineering*, vol. 16, 1992, pp. 393–411.
129. S. J. Qin and T. J. McAvoy, "Nonlinear PLS Modeling Using Neural Networks," *Computers in Chemical Engineering*, vol. 16, 1992, pp. 379–91.
130. E. C. Malthouse, "Nonlinear Partial Least Squares", *Ph.D. dissertation*, Evanston, IL: Northwestern University, 1995.
131. F. M. Ham, I. Kostanic, G. M. Cohen, and B. R. Gooch, "Determination of Glucose Concentrations in an Aqueous Matrix from NIR Spectra Using Optimal Time-Domain Filtering and Partial Least-Squares Regression," *IEEE Transactions on Biomedical Engineering*, vol. 44, 1997, pp. 475–85.
132. F. M. Ham, G. M. Cohen, I. Kostanic, and B. R. Gooch, "Multivariate Determination of Glucose Concentrations from Optimally Filtered Frequency-Warped NIR Spectra of Human Blood Serum," *Journal of Physiological Measurement*, vol. 17, 1996, pp. 1–20.
133. F. M. Ham and T. M. McDowall, "Robust Learning in a Partial Least-Squares Neural Network," *Journal of Nonlinear Analysis, Theory, Methods & Applications*, Proceedings of the Second World Congress of Nonlinear Analysts 1996, July 10–17, Athens Greece, vol. 30, 1997, pp. 2903–14.
134. T. M. McDowall, *A Robust Partial Least-Squares Neural Network, Ph.D. dissertation*, Melbourne: Florida Institute of Technology, May 1997.
135. T. M. McDowall and F. M. Ham, "Robust Partial Least-Squares Regression: A Modular Neural Network Approach," in *Applications and Science of Artificial Neural Networks III*, ed. S. Rogers, Proceedings of SPIE, vol. 3077, 1997, pp. 344–55.
136. T. M. McDowall and F. M. Ham, "Robust Learning in Generalized Partial Least-Squares Regression Modular Neural Network", *Neural Parallel and Scientific Computations*, vol. 6, 1998, pp. 391–415.
137. F. M. Ham and T. M. McDowall, "Inverse Model Formulation of Partial Least-Squares Regression: A Robust Neural Network Approach," (Invited Paper), in

Applications and Science of Computational Intelligence, eds. S. K. Rogers, D. B. Fogel, J. C. Bezdek, and B. Bosacchi, Proceedings of SPIE, vol. 3390, 1998, pp. 36–47.

138. C. Wang, H.-C. Wu, and J. C. Principe, "A Cost Function for Robust Estimation of PCA" in *Applications and Science of Artificial Neural Networks II,* eds. S.K. Rogers and D. W. Ruck, Orlando, FL, Proceedings of SPIE, vol. 2760, 1996, pp. 120–7.
139. P. J. Huber, *Robust Statistics*, New York: Wiley, 1981.
140. F. R. Hampel, P. J. Rousseeuw, E. M. Ronchetti, and W. A. Stahel, *Robust Statistics—The Approach Based on Influence Functions*, New York: Wiley, 1986.
141. D. S. Chen and R. C. Jain, "A Robust Back Propagation Learning Algorithm for Function Approximation," *IEEE Transactions on Neural Networks*, vol. 5, 1994, pp. 467–79.

CHAPTER 10

Identification, Control, and Estimation Using Neural Networks

10.1 INTRODUCTION

This chapter covers applications of various neural networks for signal processing, identification, classification, control, and estimation problems. In particular, parametric identification of linear and nonlinear systems is discussed using autoregressive moving-average (ARMA) and nonlinear ARMA (NARMA) models, respectively. The control of nonlinear systems using neural networks is also presented followed by the blind source separation problem, that is, the application of independent-component analysis (ICA) to separate unknown source signals. Spectrum estimation using partial least-squares regression is presented next. The final section of this chapter presents additional case studies using neural networks to solve two important problems.

10.2 LINEAR SYSTEM REPRESENTATION

We are concerned here with the representation of linear time-invariant discrete-time dynamical systems. We consider two basic approaches to describe dynamical systems: the input/output (or transfer function) representation and the state-space representation. We will restrict our discussions to the single-input, single-output (SISO) case. Because of extremely complex physical phenomena, in certain situations the laws of science are not adequate to sufficiently describe the dynamical system (plant) of interest. However, a model of the system can be constructed from experimental data. That is, the plant can be excited by an appropriate input and its response can be measured. From this input/output experimental data, a model can be constructed by estimating the unknown plant parameters. This is called *system identification* [1–4]. We are interested in developing models for *parametric identification* as opposed to identification of *non-parametric models* [1–4].

In the case of the input/output (transfer function) representation of linear time-invariant discrete-time dynamical systems, the form of the strictly proper, rational transfer function is assumed to be

$$H(z) = \frac{Y(z)}{U(z)} = \frac{b_1 z^{n-1} + b_2 z^{n-2} + b_3 z^{n-3} + \cdots + b_n}{z^n + a_1 z^{n-1} + a_2 z^{n-2} + \cdots + a_n} \tag{10.1}$$

where z is a complex variable. For the state-space representation of a dynamical system, the phase variable canonical form, or the controllable canonical form,[†] (cf. Sect. A.2.12), is assumed. The state equation is written as

$$\dot{\boldsymbol{x}}(t) = \begin{bmatrix} 0 & 1 & 0 & 0 & \cdots & 0 \\ 0 & 0 & 1 & 0 & \cdots & 0 \\ \cdots & \cdots & \cdots & \cdots & \cdots & \cdots \\ 0 & 0 & 0 & \cdots & 1 & 0 \\ 0 & 0 & 0 & \cdots & 0 & 1 \\ -a_n & -a_{n-1} & -a_{n-2} & \cdots & -a_2 & -a_1 \end{bmatrix} \boldsymbol{x}(t) + \begin{bmatrix} 0 \\ 0 \\ 0 \\ \vdots \\ 0 \\ 1 \end{bmatrix} u(t) \tag{10.2}$$

and the output (measurement) equation is given by

$$y(t) = [b_n, b_{n-1}, b_{n-2}, \cdots, b_2, b_1]\boldsymbol{x}(t) \tag{10.3}$$

The objective of parametric identification is to estimate the parameter vector ($\boldsymbol{\theta}$) associated with the system of interest, that is, the system parameters shown in either (10.1) or (10.2) and (10.3)

$$\boldsymbol{\theta} = [a_1, a_2, \cdots, a_n, b_1, b_2, \cdots, b_n]^T \tag{10.4}$$

10.3 AUTOREGRESSIVE MOVING-AVERAGE MODELS

From the transfer function in (10.1), a time-domain difference equation can be written as

$$\begin{aligned} & y(k) + a_1 y(k-1) + a_2 y(k-2) + \cdots + a_n y(k-n) \\ & - b_1 u(k-1) - b_2 u(k-2) - b_3 u(k-3) - \cdots - b_n u(k-n) \\ & = \varepsilon(k; a_1, a_2, \ldots, a_n, b_1, b_2, b_3, \ldots, b_n) \end{aligned} \tag{10.5}$$

where $\varepsilon(k; a_1, a_2 \ldots, a_n, b_1, b_2, b_3, \ldots, b_n)$ is an error term which is zero when the parameter vector

$$\boldsymbol{\theta} = [a_1, a_2, \cdots, a_n, b_1, b_2, \cdots, b_n]^T \tag{10.6}$$

[†]The observable canonical form [5] can also be assumed (this is the dual of the controllable canonical form).

contains the actual or true plant parameters [2], and k is the discrete-time index. Equation (10.5) can be written as

$$y(k) = \boldsymbol{\phi}^T(k)\boldsymbol{\theta} + \varepsilon(k; \boldsymbol{\theta}) \tag{10.7}$$

where

$$\begin{aligned}\boldsymbol{\phi}^T(k) = [&-y(k-1), -y(k-2), \cdots \\ &- y(k-n), u(k-1), u(k-2), u(k-3), \cdots, u(k-n)]\end{aligned} \tag{10.8}$$

For the discrete-time index taken as $k = n, n+1, n+2, \ldots, N$, where n is the *assumed* system dimension and N is the total number of samples used from the data set $\{\boldsymbol{u}(k), \boldsymbol{y}(k)\}$, a set of equations resulting from (10.7) can be written as

$$\boldsymbol{y}(N) = \boldsymbol{\Phi}(N)\boldsymbol{\theta} + \boldsymbol{\varepsilon}(N,\boldsymbol{\theta}) \tag{10.9}$$

where

$$\boldsymbol{y}^T(N) = [y(n), y(n+1), y(n+2), \cdots, y(N)] \tag{10.10}$$

$$\boldsymbol{\Phi}^T(N) = [\boldsymbol{\phi}(n), \boldsymbol{\phi}(n+1), \boldsymbol{\phi}(n+2), \cdots, \boldsymbol{\phi}(N)] \tag{10.11}$$

$$\boldsymbol{\varepsilon}^T(N; \boldsymbol{\theta}) = [\varepsilon(n; \boldsymbol{\theta}), \boldsymbol{\varepsilon}(n+1; \boldsymbol{\theta}), \boldsymbol{\varepsilon}(n+2; \boldsymbol{\theta}), \cdots, \boldsymbol{\varepsilon}(N; \boldsymbol{\theta})] \tag{10.12}$$

and

$$\begin{aligned} \boldsymbol{\Phi}(N) &\in \Re^{(N-n+1)\times 2n} && \text{ARMA data matrix} \\ \boldsymbol{y}(k) &\in \Re^{N-n+1} && \text{output vector} \\ \boldsymbol{\varepsilon}(N, \boldsymbol{\theta}) &\in \Re^{N-n+1} && \text{error vector} \\ \boldsymbol{\theta} &\in \Re^{2n} && \text{system parameter vector} \end{aligned} \tag{10.13}$$

10.4 IDENTIFICATION OF LINEAR SYSTEMS WITH ARMA MODELS

Equation (10.9) can be solved for the parameter vector $\boldsymbol{\theta}$ by using the classical least-squares (CLS) approach (cf. Sect. 9.4). First, a performance measure is defined as

$$J(\boldsymbol{\theta}) = \|\boldsymbol{\varepsilon}(N,\boldsymbol{\theta})\|_2^2 \tag{10.14}$$

where $\boldsymbol{\varepsilon} = \boldsymbol{y} - \boldsymbol{\Phi\theta}$ from (10.7); then this is minimized with respect to $\boldsymbol{\theta}$, that is,

$$\min_{\boldsymbol{\theta}\in\Re^{2N}} J(\boldsymbol{\theta}) = \min_{\boldsymbol{\theta}\in\Re^{2N}} \|\boldsymbol{\varepsilon}(N,\boldsymbol{\theta})\|_2^2 \tag{10.15}$$

where the total number of samples used from the data set $\{\boldsymbol{u}(k), \boldsymbol{y}(k)\}$ denoted by N, is taken as

$$N > 2n \tag{10.16}$$

Minimizing $\|\boldsymbol{\varepsilon}(N,\boldsymbol{\theta})\|_2^2$ with respect to $\boldsymbol{\theta}$ gives the CLS result as

$$\hat{\boldsymbol{\theta}}_{\text{CLS}} = (\boldsymbol{\Phi}^T\boldsymbol{\Phi})^{-1}\boldsymbol{\Phi}^T\boldsymbol{y} \tag{10.17}$$

where the parameter vector $\hat{\boldsymbol{\theta}}_{\text{CLS}}$ is given in (10.6).

The natural question to ask at this point is, Does a unique solution exist? That is, is $\hat{\boldsymbol{\theta}}_{\text{CLS}}$ in (10.17) a unique solution? The answer to this question depends on (1) how the parameter vector $\boldsymbol{\theta}$ is selected and (2) what type of input signal $\{u(k)\}$ is used to excite the system, resulting in the output sequence $\{y(k)\}$. To obtain a *unique* set of parameters, a canonical form for the model or an ARMA model must be selected. The parameter vector $\boldsymbol{\theta}$ is said to be *identifiable* if it has the property that one and only one value of $\boldsymbol{\theta}$ makes $J(\boldsymbol{\theta})$ a minimum. If two parameter vectors $\boldsymbol{\theta}_1$ and $\boldsymbol{\theta}_2$ have the property that $J(\boldsymbol{\theta}_1) = J(\boldsymbol{\theta}_2)$, then they are said to the *equivalent* [2]. The issue of selecting a proper input signal is typically addressed by stating that the input must be "persistently exciting of order n" [1, 2]. Simply put, the input must be capable of sufficiently "exciting" the system that enough information exists at the output of the system to identify its characteristics (or to be able to properly estimate the parameter vector $\boldsymbol{\theta}$). An input signal is persistently exciting of order n if its discrete spectrum is nonzero at least n points over the range $0 \leq \omega < \omega_s/2$, where ω_s is the sampling frequency. In the CLS solution in (10.17), the matrix $\boldsymbol{\Phi}^T\boldsymbol{\Phi} \in \Re^{2n \times 2n}$ must be full-rank; that is, $\rho(\boldsymbol{\Phi}^T\boldsymbol{\Phi}) = 2n$. In Franklin et al. [2], it is stated that an input is persistently exciting of order n if the lower right $(n \times n)$-matrix component (partition) of $\boldsymbol{\Phi}^T\boldsymbol{\Phi}$ [which depends only on the input $\{u(k)\}$] is nonsingular. As an illustration, assume $N = 6$, $n = 3$, and the input is a constant, that is, $u(k) = c$. This gives

$$\boldsymbol{\Phi}^T\boldsymbol{\Phi}$$
$$= \begin{bmatrix}
\sum_{i=1}^{4} y(i+1)y(i+1) & \sum_{i=1}^{4} y(i+1)y(i) & \sum_{i=1}^{4} y(i+1)y(i-1) & -c\sum_{i=1}^{4} y(i+1) & -c\sum_{i=1}^{4} y(i+1) & -c\sum_{i=1}^{4} y(i+1) \\
\sum_{i=1}^{4} y(i)y(i+1) & \sum_{i=1}^{4} y(i)y(i) & \sum_{i=1}^{4} y(i)y(i-1) & -c\sum_{i=1}^{4} y(i) & -c\sum_{i=1}^{4} y(i) & -c\sum_{i=1}^{4} y(i) \\
\sum_{i=1}^{4} y(i-1)y(i+1) & \sum_{i=1}^{4} y(i-1)y(i) & \sum_{i=1}^{4} y(i-1)y(i-1) & -c\sum_{i=1}^{4} y(i-1) & -c\sum_{i=1}^{4} y(i-1) & -c\sum_{i=1}^{4} y(i-1) \\
-c\sum_{i=1}^{4} y(i+1) & -c\sum_{i=1}^{4} y(i) & -c\sum_{i=1}^{4} y(i-1) & 4c^2 & 4c^2 & 4c^2 \\
-c\sum_{i=1}^{4} y(i+1) & -c\sum_{i=1}^{4} y(i) & -c\sum_{i=1}^{4} y(i-1) & 4c^2 & 4c^2 & 4c^2 \\
-c\sum_{i=1}^{4} y(i+1) & -c\sum_{i=1}^{4} y(i) & -c\sum_{i=1}^{4} y(i-1) & 4c^2 & 4c^2 & 4c^2
\end{bmatrix} \tag{10.18}$$

From (10.18) we see that the lower right (3×3)-matrix partition is *singular*. Therefore, a constant input $u(k) = c$ is not persistently exciting of order n, and minimizing $\|\boldsymbol{\varepsilon}(N,\boldsymbol{\theta})\|_2^2$ with respect to $\boldsymbol{\theta}$ does not yield a unique solution.

Two very important input signals that can be used to excite a dynamical system for system identification are Gaussian white noise and a chirp signal [6]. There are other signals that can be used; however, the white noise and chirp signals will typically provide enough excitation to properly stimulate the system for the purpose of identification. Figure 10.1(a) shows a typical chirp signal, and Figure 10.1(b) shows the associated magnitude of its fast Fourier transform (FFT). In Figure 10.1(b) we see that the magnitude response encompasses a large frequency band; thus, there is good assurance that the excitatory input signal will be persistently exciting of order n.

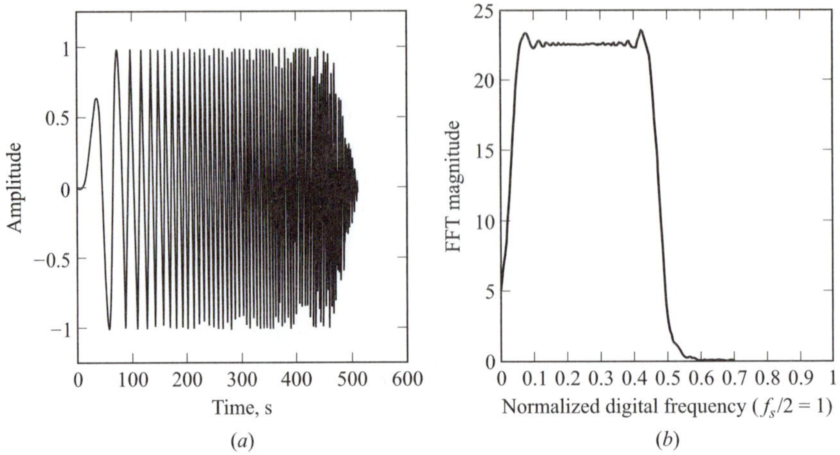

Figure 10.1 (*a*) Chirp signal; (*b*) FFT magnitude of the chirp signal (unity normalized frequency corresponds to the half-sampling frequency point, that is, $f_s/2 = 1$).

Another important point that needs to be discussed is selection of the system dimension, or the order of the system n. This is one of the more challenging problems associated with parametric system identification. The order of the system is an unknown quantity and must be initially selected and then possibly adjusted according to the estimation results that are obtained. Therefore, selecting the order of the system n can be an iterative process. However, physical insight can often yield which range of model orders should be considered. The particular application can also offer knowledge about the model order range to consider. Several methods have been devised to facilitate selection of n. Three of these methods, based on preliminary data analysis, are as follows [1]:

1. Examining the spectral estimate of the system transfer function. This method involves developing a nonparametric estimate of the transfer function from the input/output record, that is, $\{\boldsymbol{u}(k), \boldsymbol{y}(k)\}$.
2. Testing the rank of sample covariance matrices. Sample covariance matrices can be estimated from data record $\{\boldsymbol{u}(k), \boldsymbol{y}(k)\}$, by assuming an ARMA model. Using $\boldsymbol{\phi}(k)$ given in (10.8), a sample covariance matrix can be written as

$$\mathbf{C}_{\hat{n}}(N) = \frac{1}{N}\sum_{k=1}^{N} \boldsymbol{\phi}_{\hat{n}}(k)\boldsymbol{\phi}_{\hat{n}}^T(k) = \frac{1}{N}\boldsymbol{\Phi}_{\hat{n}}^T(N)\boldsymbol{\Phi}_{\hat{n}}(N) \tag{10.19}$$

where $\hat{n}$ is the *estimate* of the *actual* (or *true*) system dimension $\bar{n}$. The sample covariance matrix in (10.19) will be nonsingular for $\hat{n} \leq \bar{n}$ [provided that $u(k)$ is persistently exciting] and singular for $\hat{n} \geq \bar{n}+1$. Obviously $|\boldsymbol{C}_{\hat{n}}(N)| = \det[\boldsymbol{C}_{\hat{n}}(N)]$ could be used to test the quality of the model order estimate $\hat{n}$.

3. Correlating variables. The order determination problem can be viewed as deciding whether to include one more variable in the structure of the model, that is, adding another term $y(k-n-1)$ in (10.5) [or the additive effect of a possible disturbance variable $v(k)$] and deciding whether the addition of this variable has anything to contribute when explicating the output $y(k)$. This can be measured by the correlation between $y(k)$ and $v(k)$. However, there can be a relationship between $y(k)$ and $v(k)$; therefore, the correlation should be measured between $y(k)$ and $\varepsilon(k; \hat{\boldsymbol{\theta}}_N)$, where $\varepsilon(k; \hat{\boldsymbol{\theta}}_N) = y(k) - \hat{y}(k; \hat{\boldsymbol{\theta}}_N)$ constitute the residuals. This is referred to as canonical correlation or partial correlation [7].

There is another method that can be used to determine the system order. This method is partial least-squares regression (PLSR) (cf. Sect. 9.5), or PLSNET (cf. Sect. 9.6). PLSR or PLSNET can be used to estimate the parameter vector, and at the same time the order of the system n can be determined from the factor analysis. This is the subject of the next section.

To conclude this section, note that many other methods can be used to estimate the system parameter vector $\boldsymbol{\theta}$ [1–4]. One very important method involves recursive weighted least squares (RWLS) [1, 2]. The result in (10.17) is referred to as a *batch* solution because all the available data from the input/output record $\{\boldsymbol{u}(k), \boldsymbol{y}(k)\}$ are used simultaneously to estimate $\boldsymbol{\theta}$. The two main advantages of the RWLS method are that (1) RWLS is an iterative algorithm that *adaptively* estimates the parameter vector and (2) because of the weighting function $w(k) = a\gamma^{N+1-k}$ in the recursive algorithm, past data (observations) can be weighted less heavily than current observations, so the weighting function acts as a filter.

10.5 PARAMETRIC SYSTEM IDENTIFICATION OF LINEAR SYSTEMS USING PLSNET

Partial least squares has been applied to the parametric system identification problem [8], as well as PLSNET [9] (cf. Sect. 9.6), using an ARMA model approach. It is necessary to first develop *training* and *test* sets from the input/output data $\{\boldsymbol{u}(k), \boldsymbol{y}(k)\}$, that is,

$$\text{Training data set} = \{\boldsymbol{\Phi}_{\text{train}}(N), \boldsymbol{y}_{\text{train}}(\text{N})\} \tag{10.20}$$

$$\text{Test data set} = \{\boldsymbol{\Phi}_{\text{test}}(N), \boldsymbol{y}_{\text{test}}(\text{N})\} \tag{10.21}$$

where the training and test data sets are taken from different portions of the collected data $\{\boldsymbol{u}(k),\boldsymbol{y}(k)\}$. When the two data sets in (10.20) and (10.21) are generated, it is assumed that the order of the system is *overspecified*; that is, if $\bar{n}$ is considered the *actual* dimension of the system, the *assumed* system dimension is $n > \bar{n}$. The objective of using PLSNET for the parametric system identification problem is to estimate both the system dimension $\hat{n}$, and the parameter vector $\hat{\boldsymbol{\theta}}$.

To select the optimal number of PLSR factors, an independent validation method is used (cf. Sect. 9.4). In the neural network implementation of PLSR, the data matrix (or independent-variable block) is given by

$$A = \boldsymbol{\Phi}(N) \tag{10.22}$$

and the observation vector, target values (or dependent-variable block), is

$$\boldsymbol{c} = \boldsymbol{y}(N) \tag{10.23}$$

By using the training and test data sets in (10.20) and (10.21), respectively, the factor analysis is carried out as in (9.111). When the optimal number of factors is determined, this will give the estimate of the system dimension as

$$\hat{n} = \frac{h^o}{2} \tag{10.24}$$

where h^o is the optimal number of factors determined from the factor analysis [10]. The PLSR factor analysis will yield the system order for the *minimal realization* [5], that is, a system model that is always controllable and observable (cf. Sect. A.2.12). This will be the case when the *first* minimum is selected on the plot of the SEP versus the number of PLSR factors. After the optimal number of factors is determined, a new set of data can be formed for $N > 2\hat{n}$

$$\text{Final data set} = \{\boldsymbol{\Phi}_f(N; \hat{n}), \boldsymbol{y}_f(N; \hat{n})\} \tag{10.25}$$

from the collected data $\{\boldsymbol{u}(k), \boldsymbol{y}(k)\}$. The number of samples used to generate $\boldsymbol{\Phi}_f(N; \hat{n})$ and $\boldsymbol{y}_f(N; \hat{n})$ does not necessarily have to be the same as that used to develop $\{\boldsymbol{\Phi}_{\text{train}}(N), \boldsymbol{y}_{\text{train}}(\text{N})\}$ and $\{\boldsymbol{\Phi}_{\text{test}}(N), \boldsymbol{y}_{\text{test}}(\text{N})\}$ in (10.20) and (10.21), respectively, but must be $N > 2\hat{n}$. An estimate of the parameter vector $\hat{\boldsymbol{\theta}}$ can be obtained by using prediction method 1, given in Section 9.5. From PLSNET-C (cf. Sect. 9.6), by using the final data set in (10.25), the PLSR weight loading vectors $\{\hat{\boldsymbol{w}}_1, \hat{\boldsymbol{w}}_2, \ldots, \hat{\boldsymbol{w}}_{h^o}\}$, the loading vectors $\{\hat{\boldsymbol{b}}_1, \hat{\boldsymbol{b}}_2, , \hat{\boldsymbol{b}}_{h^o}\}$, and the regression coefficients $\{\hat{v}_1, \hat{v}_2, \ldots, \hat{v}_{h^o}\}$ are obtained. Using this information, the following matrices can be formed

$$\hat{\boldsymbol{W}}^T = [\hat{\boldsymbol{w}}_1, \hat{\boldsymbol{w}}_2, \ldots, \hat{\boldsymbol{w}}_{h^o}] \tag{10.26}$$

$$\hat{\boldsymbol{B}}^T = [\hat{\boldsymbol{b}}_1, \boldsymbol{b}_2, \ldots, \hat{\boldsymbol{b}}_{h^o}] \tag{10.27}$$

$$\hat{\boldsymbol{v}}^T = [\hat{v}_1, \hat{v}_2, \ldots, \hat{v}_{h^o}] \tag{10.28}$$

where $\hat{\boldsymbol{W}} \in \Re^{h^o \times h^o}$, $\hat{\boldsymbol{B}} \in \Re^{h^o \times h^o}$, $\hat{\boldsymbol{v}} \in \Re^{h^o \times 1}$, and $h^o = 2\hat{n}$. From (10.26), (10.27), and (10.28), the optimal PLSR calibration model (final calibration coefficients) $\hat{\boldsymbol{b}}_{f\text{PLSR}}$ can be formed as

$$\hat{\boldsymbol{\theta}} = \hat{\boldsymbol{b}}_{f\text{PLSR}} = \hat{\boldsymbol{W}}^T(\hat{\boldsymbol{B}}\hat{\boldsymbol{W}}^T)^{-1}\hat{\boldsymbol{v}} \tag{10.29}$$

Therefore, the parameter vector estimate $\hat{\boldsymbol{\theta}}$ is the PLSR calibration model $\hat{\boldsymbol{b}}_{f\text{PLSR}}$. As always, the PLSR calibration model can also be used for prediction. In this case, the final calibration model would be predicting the output (or response) of the system $\{y(k)\}$. To evaluate the performance of the developed parametric model, another set of data could be generated as in (10.25), using a different portion of the data record $\{\boldsymbol{u}(k), \boldsymbol{y}(k)\}$ that was not used to develop $\hat{\boldsymbol{\theta}} = \hat{\boldsymbol{b}}_{f\text{PLSR}}$, that is,

$$\underset{\text{(for testing)}}{\text{Final data set}} = \{\boldsymbol{\Phi}_{f\text{test}}(N;\hat{n}), \boldsymbol{y}_{f\text{test}}(N;\hat{n})\} \tag{10.30}$$

where $N > 2\hat{n}$. Using $\boldsymbol{\Phi}_{f\text{test}}(N;\hat{n})$ and (10.29), an estimate of $\boldsymbol{y}_{f\text{test}}$ can be determined from

$$\hat{\boldsymbol{y}}_{f\text{test}} = \boldsymbol{\Phi}_{f\text{test}}\hat{\boldsymbol{\theta}} = \boldsymbol{\Phi}_{f\text{test}}\hat{\boldsymbol{b}}_{f\text{PLSR}} \tag{10.31}$$

which can be compared to the actual output $\boldsymbol{y}_{f\text{test}}$.

EXAMPLE 10.1. This example illustrates how PLSNET can be used for parametric system identification. The *actual* discrete-time system is given by the second-order transfer function

$$H(z) = \frac{z - 0.1}{z^2 - 0.5999z + 0.05} \tag{10.32}$$

with the sampling period $T_s = 2\pi/1{,}000\,\text{s}$ and $\bar{n} = 2$ (actual system dimension). Therefore, the actual parameter vector is

$$\boldsymbol{\theta}^T = \left[-0.5999, 0.05, 1, -0.1\right] \tag{10.33}$$

The simulated empirical input/output data $\{\boldsymbol{u}(k), \boldsymbol{y}(k)\}$ are generated in MATLAB using zero-mean, unity-variance Gaussian white noise as the input, and the MATLAB function **`dlsim`** to generate the output data using the parameters from (10.33).

The input and output sequences have 1,024 samples each; however, the first 100 samples in each are used to generate the training data, as shown in (10.20), and the next 100 samples are used to form the test set, as shown in (10.21) with the system (over-) specified as $n = 2$. The ARMA data matrix is given in (10.11). Using the training and test data, a straightforward independent-validation factor analysis reveals that the system dimension should be selected as $\hat{n} = 1$ (see Figure 10.2), because the optimal number of factors from Figure 10.2 is shown to be $h^{\text{o}} = 2$. This indicates that the original second-order system is a *reducible* system;

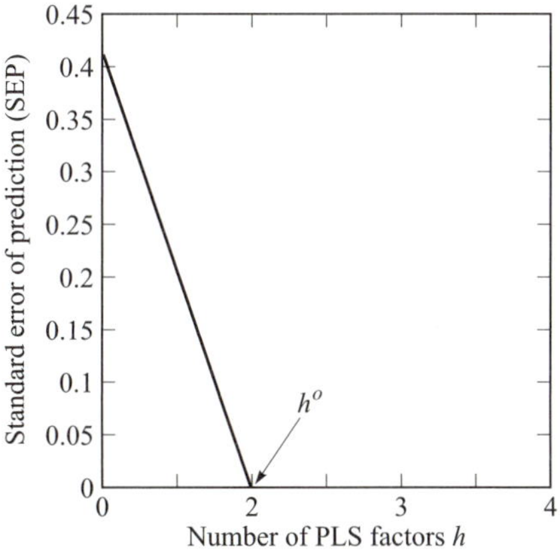

Figure 10.2 Selection of the optimal number of PLSR factors from the PLSR1 results; h^{o} is selected as 2; that is, the estimate of the system order (dimension) is $\hat{n} = 1$ (for the minimal realization of the system).

thus, there must be a pole/zero cancellation in its transfer function. In fact, the original system poles are at

$$\text{poles}_{\text{2nd-order system}} = [0.1, 0.4999] \tag{10.34}$$

and the zero is at

$$\text{zero}_{\text{2nd-order system}} = [0.1] \tag{10.35}$$

therefore, the pole and zero at 0.1 could be canceled. If this pole/zero pair were canceled, the resulting system would be an irreducible (controllable and observable) system.

To find the *minimal realization* system parameters, the final data set (using $\hat{n} = 1$ for the system dimension) must first be generated, as in (10.25). Using PLSNET-C with the final data set, the PLS weight loading vectors, the loading vectors, and the regression coefficients, that is, $\{\hat{\boldsymbol{w}}_h, \hat{\boldsymbol{b}}_h, \hat{v}_h\}$, for $h = 1, 2$, are computed after 3,000 training epochs. The learning rate parameters for PLSNET-C were chosen as

$$\mu_w = \frac{0.1}{\sum_{j=1}^{N-n+1=100} y_{fj}^2} \tag{10.36a}$$

$$\mu_b = \mu_v = (0.05)\mu_w \tag{10.36b}$$

From $\{\hat{\boldsymbol{w}}_h, \hat{\boldsymbol{b}}_h, \hat{v}_h\}$ the three matrices $\hat{\boldsymbol{W}}, \hat{\boldsymbol{B}}, \hat{\boldsymbol{v}}$, given in (10.26) through (10.28) are formed, and the parameter vector for the reduced-order (minimal realization) system is given by

$$\hat{\boldsymbol{\theta}} = \hat{\boldsymbol{b}}_{f\text{PLSR}} = \hat{\boldsymbol{W}}^T(\hat{\boldsymbol{B}}\hat{\boldsymbol{W}}^T)^{-1}\hat{\boldsymbol{v}} = \begin{bmatrix} -0.4999 \\ 1.0 \end{bmatrix} \tag{10.37}$$

which is expected from (10.34) and (10.35). Therefore, the transfer function for the minimal realization system is given by

$$H_{mr}(z) = \frac{1}{z - 0.4999} \tag{10.38}$$

As a simple check using the PLSNET-C results $\{\hat{\boldsymbol{w}}_h, \hat{\boldsymbol{b}}_h, \hat{v}_h\}$, Figure 10.3 shows a plot of the first 50 time samples of the system output for the original (actual) system and the system response using PLSNET-P for the first-order system. From the figure it is obvious that the two responses are essentially identical. The same results were obtained by taking the approach given in (10.31), that is, using the computed parameter vector given in (10.37).

It is interesting to compare the weight loading vectors, the loading vectors, and the regression coefficients using PLSNET-C and the PLSR1 calibration algorithm presented in Section 9.5. The weight loading vectors, the loading vectors, and the regression coefficients using PLSNET-C are given by

$$\hat{\boldsymbol{W}}_{\text{PLSNET-C}} = [\hat{\boldsymbol{w}}_1, \hat{\boldsymbol{w}}_2]_{\text{PLSNET-C}} = \begin{bmatrix} -0.4815 & 0.9431 \\ 0.8764 & 0.3324 \end{bmatrix}$$

$$\hat{\boldsymbol{B}}_{\text{PLSNET-C}} = [\hat{\boldsymbol{b}}_1, \hat{\boldsymbol{b}}_2]_{\text{PLSNET-C}} = \begin{bmatrix} -0.5331 & 0.8900 \\ 0.8471 & 0.4948 \end{bmatrix}$$

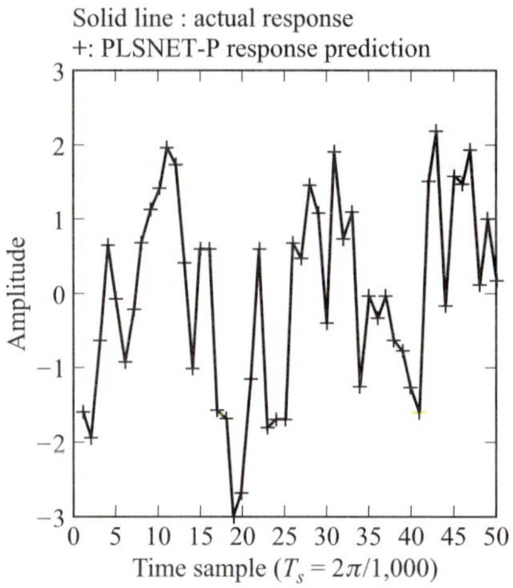

Figure 10.3 Output response of the actual system (solid line) versus the reduced-order (minimal realization) system using PLSNET-P, whose parameters were estimated using PLSNET-C (plotted as discrete samples).

$$\hat{\boldsymbol{v}}_{\text{PLSNET-C}} = \begin{bmatrix} \hat{v}_1 \\ \hat{v}_2 \end{bmatrix}_{\text{PLSNET-C}} = \begin{bmatrix} 1.1136 \\ 0.0500 \end{bmatrix}$$

and the weight loading vectors, the loading vectors, and the regression coefficients using the PLSR1 calibration algorithm are given by

$$\hat{\boldsymbol{W}}_{\text{PLSR1}} = [\,\hat{\boldsymbol{w}}_1, \hat{\boldsymbol{w}}_2\,]_{\text{PLSR1}} = \begin{bmatrix} -0.4854 & 0.8743 \\ 0.8743 & 0.4854 \end{bmatrix}$$

$$\hat{\boldsymbol{B}}_{\text{PLSR1}} = [\,\hat{\boldsymbol{b}}_1, \hat{\boldsymbol{b}}_2\,]_{\text{PLSR1}} = \begin{bmatrix} -0.5329 & 0.8743 \\ 0.8479 & 0.4854 \end{bmatrix}$$

$$\hat{\boldsymbol{v}}_{\text{PLSR1}} = \begin{bmatrix} \hat{v}_1 \\ \hat{v}_2 \end{bmatrix}_{\text{PLSR1}} = \begin{bmatrix} 1.1143 \\ 0.0483 \end{bmatrix}$$

As seen from comparing the values in the arrays, the extracted PLS information is almost identical in both methods.

10.6 NONLINEAR SYSTEM REPRESENTATION

In general, any physical system can be viewed as an operator performing some type of mapping between two spaces [11]. Conceptually, this is illustrated in Figure 10.4. The space containing the domain of the system is commonly referred as to as *input space* U. Similarly, the space containing the results of the system mapping is called the *output space* Y. In most practical applications, the spaces U and Y are vector spaces, subsets of $\Re^{m\times 1}$ and $\Re^{n\times 1}$ respectively. Both the inputs and outputs of the system can be functions of the time

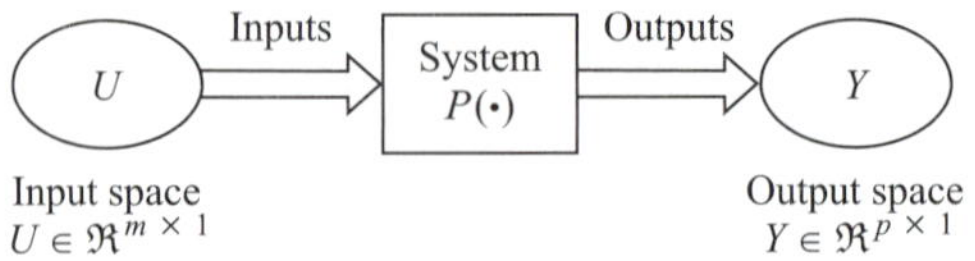

Figure 10.4 A simple model of a dynamical system.

variable, and if that is the case, we consider the system to be *dynamical*. In other words, at any time instant t the dynamic system in Figure 10.4 receives an input $\boldsymbol{u}(t) \subset \mathsf{U}$ and produces an output $\boldsymbol{y}(t) \subset \mathsf{Y}$. For many dynamical systems the output depends not only on the present value of the input variable, but on the past values of the input and output as well. These systems are commonly referred to as systems with *memory*.

Accurate system representation is a fundamental problem in any discipline involving the studying of the system dynamics. In essence, this problem is concerned with finding a convenient way to model the behavior of the system using certain mathematical tools. The efficiency of the model can be judged from various standpoints including accuracy, simplicity, computational considerations, and physical validity. The development of accurate, yet mathematically tractable, models is a complicated science in itself. Fortunately, in practice, we are rarely interested in modeling all possible aspects of the system's behavior, and this allows for development of relatively simple models that are accurate with respect to practical applications.

Based on the nature of the mapping between vector spaces U and Y, systems can be divided into two broad categories, *linear* and *nonlinear*. If the mapping in Figure 10.4 satisfies

$$\mathsf{P}\{A\boldsymbol{u}_1(t) + B\boldsymbol{u}_2(t)\} = A\mathsf{P}\{\boldsymbol{u}_1(t)\} + B\mathsf{P}\{\boldsymbol{u}_2(t)\} \tag{10.39}$$

for every $A, B \in \Re$ and $\boldsymbol{u}_1(t), \boldsymbol{u}_2(t) \subset \mathsf{U}$, the system is said to be linear. On the other hand, if (10.39) does not hold, the system is nonlinear. Linear system theory is a well-developed discipline with many applications. It is based on classical results from linear algebra, complex variable theory, theory of linear operators, and many other areas of mathematics and science as well. Nonetheless, representation, design, identification, and control of *nonlinear systems* are still challenging tasks and are still the subject of extensive research. In this section we present two popular methods for nonlinear dynamical system representation.

10.6.1 Nonlinear Input-State-Output Representation

A wide class of nonlinear dynamical systems can be represented according to

$$\frac{\mathrm{d}\boldsymbol{x}(t)}{\mathrm{d}t} = \dot{\boldsymbol{x}}(t) = f[\boldsymbol{x}(t), \boldsymbol{u}(t)] \tag{10.40}$$

$$\boldsymbol{y}(t) = g[\boldsymbol{x}(t), \boldsymbol{u}(t)] \tag{10.41}$$

where

$x(t) = [x_1(t), x_2(t), \ldots, x_n(t)]^T \in \Re^{n\times 1}$ state vector
$u(t) = [u_1(t), u_2(t), \ldots, u_m(t)]^T \in \Re^{m\times 1}$ input vector
$y(t) = [y_1(t), y_2(t), \ldots, y_p(t)]^T \in \Re^{p\times 1}$ output vector
t = continuous-time variable

The functions $f(\bullet)$ and $g(\bullet)$ represent nonlinear mappings defined by $f : \Re^{(n+m)\times 1} \to \Re^{n\times 1}$ and $g : \Re^{(n+m)\times 1} \to \Re^{p\times 1}$. If the function $g[x(t), u(t)]$ contains $u(t)$ explicitly, we say that system has *direct links* between input and output. For the vast majority of the systems, this is not the case, and very often $g(\bullet)$ is assumed to be only a function of the system states.

Equations (10.40) and (10.41) are generally referred to as *input-state-output* models of nonlinear systems. Obviously, the representation given in (10.40) and (10.41) can be used to describe *continuous* dynamical systems. If the system is discrete in time, the differential equation in (10.40) needs to be transformed to its difference-equation counterpart. Therefore, in the case of a discrete-time system, the input-state-output representation has the form

$$x(k+1) = f[x(k), u(k)] \quad (10.42a)$$

$$y(k) = g[x(k), u(k)] \quad (10.42b)$$

If the mappings $f(\bullet)$ and $g(\bullet)$ are linear, (10.42*a*) and (10.42*b*) simplify and transform to the familiar form of the input-state-output representation of a linear system, given as

Continuous case

$$\dot{x}(t) = A(t)x(t) + B(t)u(t) \quad (10.43)$$

$$y(t) = C(t)x(t) + D(t)u(t) \quad (10.44)$$

Discrete case

$$x(k+1) = A(k)x(k) + B(k)u(k) \quad (10.45)$$

$$y(k) = C(k)x(k) + D(k)u(k) \quad (10.46)$$

where $A \in \Re^{n\times n}$, $B \in \Re^{n\times m}$, $C \in \Re^{p\times n}$, and $D \in \Re^{p\times m}$. In the simplest case when the system is time-invariant, A, B, C, and D are real constant matrices (cf. Sect. A.2.12).

10.6.2 Nonlinear ARMA

Development of the nonlinear system representation in the input-state-output form requires substantial knowledge of the physical processes within the system. An obvious insight that must be obtained in order to formulate the discrete-time state representation in (10.42*a*) and (10.42*b*) is knowledge of the system states. Although in many cases such knowledge is available, having a nonlinear system representation derived only from inputs and outputs of the system itself is of great benefit. Leontaritis and Billings [12] introduced an alternative way of modeling nonlinear dynamical systems: *n*onlinear *a*uto *r*egressive *m*oving *a*verage with e*x*ogenous inputs (NARMAX). A NARMAX model can be viewed as a natural extension of the highly successful ARMAX model for the case of nonlinear dynamic systems. To clarify the point, consider an ARMAX model for a linear dynamic system given as

$$\boldsymbol{y}(k) = \sum_{i=1}^{n_y} \boldsymbol{\theta}_{yi}\boldsymbol{y}(k-i) + \sum_{i=1}^{n_u} \boldsymbol{\theta}_{ui}\boldsymbol{u}(k-i) + \sum_{i=1}^{n_e} \boldsymbol{\theta}_{ei}\boldsymbol{e}(k-i) + \boldsymbol{e}(k) \qquad (10.47)$$

where $\boldsymbol{y}(k) \in \Re^{p\times 1}$, $\boldsymbol{u}(k) \in \Re^{m\times 1}$, and $\boldsymbol{e}(k) \in \Re^{p\times 1}$ are vectors of the linear system outputs, inputs, and measurement or approximation error at the discrete-time instance k, respectively. Also, $\boldsymbol{\theta}_{yi} \in \Re^{r\times p}$, $\boldsymbol{\theta}_{ui} \in \Re^{p\times m}$, and $\boldsymbol{\theta}_{ei} \in \Re^{p\times p}$. If we assume that the system is noise-free, the ARMAX model can be reduced to a much simpler ARMA model discussed in Sections 10.3 and 10.4.

Under some mild assumptions it can be shown that a wide range of nonlinear systems can be represented by the following nonlinear discrete-time difference equation [12]

$$\begin{aligned}\boldsymbol{y}(k) = f[&\boldsymbol{y}(k-1), \ldots, \boldsymbol{y}(k-n_y), \boldsymbol{u}(k-1), \ldots, \boldsymbol{u}(k-n_u),\\ &\boldsymbol{e}(k-1), \ldots, \boldsymbol{e}(k-n_e)] + \boldsymbol{e}(k)\end{aligned} \qquad (10.48)$$

where $\boldsymbol{y}(k) \in \Re^{p\times 1}$, $\boldsymbol{u}(k) \in \Re^{m\times 1}$, and $\boldsymbol{e}(k) \in \Re^{p\times 1}$ have the same interpretation as in (10.47), and $f(\bullet)$ is a nonlinear mapping defined as $f : \Re^{(n_y \cdot p + n_u \cdot m + n_e \cdot p)\times 1} \rightarrow \Re^{p\times 1}$. The matrix equation in (10.48) can be decomposed into p scalar equations as

$$\begin{aligned}y_i(k)& \\ = f_i[&y_1(k-1), \ldots, y_1(k-n_y), y_2(k-1), \ldots, y_2(k-n_y), \ldots, y_p(k-1), \ldots, y_p(k-n_y),\\ &u_1(k-1), \ldots, u_1(k-n_u), u_2(k-1), \ldots, u_2(k-n_u), \ldots, u_m(k-1), \ldots, u_m(k-n_u),\\ &e_1(k-1), \ldots, e_1(k-n_e), e_2(k-1), \ldots, e_2(k-n_e), \ldots, e_p(k-1), \ldots, e_p(k-n_e)]\\ &+ e_i(k)\end{aligned} \qquad (10.49)$$

where $i = 1, 2, \ldots, p$. In a special, but very important, case when the outputs of the system are assumed to be error-free, a NARMAX model can be reduced to a simplified NARX (*n*onlinear *a*uto *r*egressive with e*x*ogenous input) version which can be rewritten as

$$\begin{aligned}\boldsymbol{y}(k) = f[&\boldsymbol{y}(k-1), \boldsymbol{y}(k-2), \ldots, \boldsymbol{y}(k-n_y), \boldsymbol{u}(k-1), \boldsymbol{u}(k-2), \ldots,\\ &\boldsymbol{u}(k-n_u)] + \boldsymbol{e}(k)\end{aligned} \qquad (10.50)$$

or in scalar form

$$\begin{aligned}y_i(k)& \\ = f_i[&y_1(k-1), \ldots, y_1(k-n_y), y_2(k-1), \ldots, y_2(k-n_y), \ldots, y_p(k-1), \ldots, y_p(k-n_y),\\ &u_1(k-1), \ldots, u_1(k-n_u), u_2(k-1), \ldots, u_2(k-n_u), \ldots, u_m(k-1), \ldots, u_m(k-n_u)]\\ &+ e_i(k)\end{aligned} \qquad (10.51)$$

As shown later, it is the NARX form of the NARMAX model that is of fundamental importance for nonlinear system identification.

The NARX form given in (10.51) is very general and encompasses a wide class of nonlinear dynamic systems. However, in practice, the nonlinear function $f(\bullet)$ is rarely known, and therefore, its approximation needs to be developed through a system identification process. One way to approach the problem is based on the well-known Stone-Weierstrass theorem [13] which states that any function can be approximated to any required degree of accuracy by using polynomials. After expansion of (10.51) into an lth-order polynomial, we have

$$\begin{aligned} y_i(k) &= \theta_0^{(i)} + \sum_{i_1=1}^{n} \theta_{i_1}^{(i)} x_{i_1}(k) + \sum_{i_1=1}^{n}\sum_{i_2=i_1}^{n} \theta_{i_1 i_2}^{(i)} x_{i_1}(k) x_{i_2}(k) \\ &\quad + \sum_{i_1=1}^{n}\sum_{i_2=i_1}^{n}\cdots\sum_{i_l=i_{l-1}}^{n} \theta_{i_1 i_2 \ldots i_l}^{(i)} x_{i_1}(k) x_{i_2}(k) \cdots x_{i_l}(k) + e_i(k) \end{aligned} \tag{10.52}$$

where $n = pn_y + mn_u$,

$$x_1(k) = y_1(k-1),\;\; x_2(k) = y_1(k-2),\;\; \ldots\;,\;\; x_{pn_y}(k) = y_p(k-n_y)$$

and

$$x_{pn_y+1}(k) = u_1(k-1),\;\; x_{pn_y+2}(k) = u_1(k-2),\;\; \ldots\;,\;\; x_n(k) = u_1(k-n_u)$$

The polynomial expansion performed in (10.52) transforms the problem of determining the unknown function in (10.51) into a problem of parameter estimation. As an illustration of the above process, consider the following example.

EXAMPLE 10.2. According to (10.50) a NARX model of a SISO system can be formulated as

$$y(k) = f[y(k-1), y(k-2), u(k-1)] \tag{10.53}$$

We want to develop an expression for a polynomial approximation of the nonlinear function in (10.53), assuming that it can be accurately approximated by a polynomial of degree 2. Therefore, according to (10.52) we have

$$\begin{aligned} y(k) &= \theta_0 + \theta_1 y(k-1) + \theta_2 y(k-2) + \theta_3 u(k-1) \\ &\quad + \theta_4 y^2(k-1) + \theta_5 y(k-1)y(k-2) + \theta_6 y(k-1)u(k-1) \\ &\quad + \theta_7 y^2(k-2) + \theta_8 y(k-2)u(k-1) + \theta_9 u^2(k-1) \end{aligned} \tag{10.54}$$

An obvious problem with the NARX model can be seen from the simple nonlinear system in Example 10.2. Since the structure of the nonlinearity $f(\bullet)$ is assumed to be unknown, all polynomial terms need to be included in the analysis. In the general case, this means that for an lth-degree polynomial approximation of the nonlinear system having p outputs and m inputs with maximal lags for outputs and inputs given by n_y and n_u, we need to estimate a total of

$$N = p[1 + n + n(n-1) + \cdots + n(n-1)(n-2)\cdots(n-l+1)] = p\sum_{j=0}^{l} j!\binom{n}{j} \tag{10.55}$$

terms. In (10.55) $n = mn_u + pn_y$. As an example, a fourth-degree polynomial approximation of the SISO system having maximal lags, given by $n_u = n_y = 4$, needs to have

$$N = \sum_{j=0}^{4} j! \binom{8}{j} = 2{,}081 \tag{10.56}$$

coefficients estimated. Furthermore, except for highly nonlinear systems, most of the coefficients are going to be very small, which means that most of the terms in (10.52) will be redundant and should be removed. Having a large number of coefficients does not pose a problem solely from the standpoint of system identification. Most of the time, the identification process is the first step in the design of nonlinear and adaptive controllers. Having a system representation with a large number of coefficients presents an immense challenge from the standpoint of practical implementation. Several schemes have been proposed to circumvent this problem. Two of them are presented now.

Linear regression with orthogonal decomposition

A careful examination of (10.52) reveals that it can be arranged in a simpler form given by

$$\boldsymbol{x}^T(k)\boldsymbol{\theta}^{(i)} = y_i(k) \tag{10.57}$$

where

$$\boldsymbol{x}(k) = \left[1, x_1(k), \cdots, x_n(k), x_1^2(k), \cdots, x_n^2(k), x_1^l(k), \cdots, x_n^l(k)\right]^T \tag{10.58}$$

and

$$\boldsymbol{\theta}^{(i)} = \left[\theta_0^{(i)}, \theta_1^{(i)}, \cdots, \theta_n^{(i)}, \theta_{11}^{(i)}, \cdots, \theta_{nn}^{(i)}, \cdots, \theta^{(i)}_{\underbrace{11\cdots1}_{l}}, \cdots, \theta^{(i)}_{\underbrace{nn\cdots n}_{l}}\right]^T \tag{10.59}$$

Considering M consecutive time lags, a system of equations can be formed as

$$\left[\boldsymbol{x}(k), \boldsymbol{x}(k-1), \cdots, \boldsymbol{x}(k-M)\right]^T \boldsymbol{\theta}^{(i)} = \left[y_i(k), y_i(k-1), \cdots, y_i(k-M)\right] \tag{10.60}$$

Equation (10.60) can be written more compactly as

$$\boldsymbol{P}\boldsymbol{\theta}^{(i)} = \boldsymbol{Y}_i \tag{10.61}$$

where $\boldsymbol{P} \in \Re^{M+1 \times N}$, $\boldsymbol{\theta}^{(i)} \in \Re^{N \times 1}$, $\boldsymbol{Y}_i \in \Re^{M+1 \times 1}$. And N is the total number of system parameters to be estimated for each output and is given by

$$N^{(i)} = \sum_{j=0}^{l} j! \binom{n}{j} \tag{10.62}$$

and

$$n = mn_u + pn_y \tag{10.63}$$

Therefore, the problem of parameter estimation for the polynomial expansion of the NARX model transforms to a linear regression problem given in (10.61). In [14], Billings et al. examine a reliable regression technique that allows the extraction of the parameters of the polynomial approximation of the NARX model. A slightly different algorithm that uses a projection is presented in Reference [15]. An excellent paper by Chen et al. [16] provides a comprehensive survey of the orthogonal least-squares methods and their application to NARX nonlinear system modeling.

Neural network approach

Using neural networks to develop an approximation to the nonlinear mapping of the NARX model provides a viable alternative to the parameter estimation of its polynomial expansion. Any class of neural networks that is capable of nonlinear mapping approximations can be used to *learn* the function $f(\bullet)$. In References [11, 17, 18] it is demonstrated that the use of multilayer perceptrons and recurrent neural networks can accomplish this, and in References [19, 20] radial basis function neural networks are used.

As a final remark, Narendra and Parthasarathy [11] show that a significant benefit in nonlinear system modeling can be drawn from a combination of NARMAX and ARMAX modeling, provided that there exists some prior information about the nonlinearities present in the system. As specified in Reference [11], a nonlinear system can be identified as a member of one of the following model types.

Model 1. The output of the system depends nonlinearly on n_u past values of the input and linearly on n_y past values of the output, and it is given by

$$\boldsymbol{y}(k) = \sum_{i=1}^{n_y} \alpha_i \boldsymbol{y}(k-i) + f_u\left[\boldsymbol{u}(k-1), \boldsymbol{u}(k-1), \cdots, \boldsymbol{u}(k-n_u)\right] + \boldsymbol{e}(k) \tag{10.64}$$

where $\alpha_i \in \Re$ and $f_u : \Re^{m \times n_u} \rightarrow \Re^{p \times 1}$.

Model 2. The output of the system depends linearly on n_u past values of the input and nonlinearly on n_y past values of the output, and it is given by

$$\boldsymbol{y}(k) = f_y\left[\boldsymbol{y}(k-1), \boldsymbol{y}(k-2), \cdots, \boldsymbol{y}(k-n_y)\right] + \sum_{i=1}^{n_u} \boldsymbol{B}_i \boldsymbol{u}(k-i) + \boldsymbol{e}(k) \tag{10.65}$$

where $\boldsymbol{B}_i \in \Re^{p \times m}$ and $f_y : \Re^{p \times n_y} \rightarrow \Re^{p \times 1}$.

Model 3. The output of the system depends nonlinearly on past values of both inputs and outputs, but the nonlinear mappings are separable, that is,

$$\begin{aligned} y(k) = f_y\left[y(k-1), y(k-2), \ldots, y(k-n_y)\right] \\ + f_u\left[u(k-1), u(k-2), \ldots, u(k-n_u)\right] + e(k) \end{aligned} \tag{10.66}$$

where $f_y : \Re^{p\times n_y} \rightarrow \Re^{p\times 1}$ and $f_u : \Re^{m\times n_u} \rightarrow \Re^{p\times 1}$.

Model 4. The dependence between the output and past values of the inputs and outputs is expressed in the form of a nonseparable nonlinearity. Thus, the output is given by

$$y(k) = f\left[y(k-1), \cdots, y(k-n_y), u(k-1), \cdots, u(k-n_u)\right] + e(k) \tag{10.67}$$

where $f : \Re^{(n_y p + n_u m)\times 1} \rightarrow \Re^{p\times 1}$. Obviously, model 4 is the general case of the NARX model discussed in the previous sections, and models 1 through 3 are derived from model 4, assuming a special form of the nonlinear function $f(\bullet)$.

10.7 IDENTIFICATION AND CONTROL OF NONLINEAR DYNAMICAL SYSTEMS

Identification and control are two fundamental tasks of mathematical system theory. While there exists an extensive set of mathematical tools developed for linear systems, the identification and control of nonlinear systems are still challenging tasks. A lack of general theoretical results establishing global controllability, observability, stability, etc., forces us to analyze nonlinear identification and control problems on a case-by-case basis. Recently, considerable effort has been invested in the use of neural networks for nonlinear control and identification [11, 17–21]. Both practical and theoretical results establish the use of neural control as one of the most promising areas of neural network applications.

In this section we present some simple methods for identification and control of nonlinear systems using neural networks. First we treat the problem of system identification, and then we present some methods for nonlinear control design. For the sake of simplicity the discussion is limited to single-input, single-output (SISO) systems. Extension of the methodology presented here to multiple-input, multiple-output (MIMO) systems is straightforward.

10.7.1 Identification of Nonlinear Systems

The general problem of nonlinear system identification is presented in Figure 10.5. Here the parameters of the identification model are estimated adaptively so that the difference between the actual plant output and the output produced by the model is minimized. Ideally, the identification process should be capable of producing an accurate model of the nonlinear system without any prior knowledge of the system dynamics. In essence, this would imply an identification process with both the architecture and parameters of the

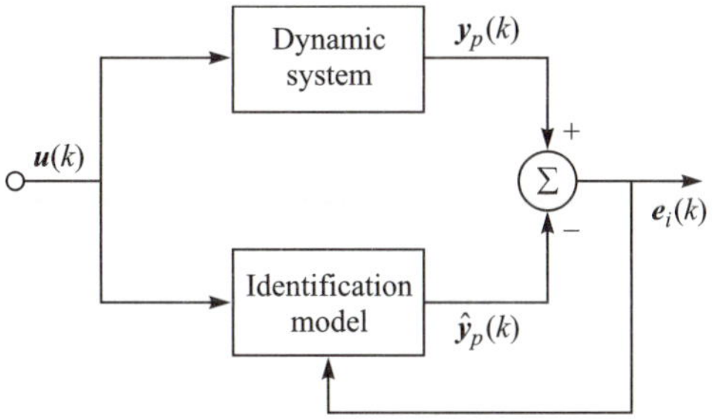

Figure 10.5 General model of the system identification process.

model being adjustable. However, even if this is the case, success in developing the model can be guaranteed only for the input sequences that were used in the identification process. If a particular input sequence was not used in the identification process, the plant and its model may or may not produce the outputs within the desired degree of accuracy.

One of the most useful approaches to nonlinear system identification is based on NARMAX modeling. Two different practical realizations of the identification scheme shown in Figure 10.5, that use the NARMAX modeling approach, are presented in Figure 10.6(*a*) and (*b*). As can be seen from these figures, the two configurations are very similar. In fact, the structure of the nonlinear identification system is identical, and the only difference is in the interconnection between the plant and the *identifier*. The structure shown in Figure 10.6(*a*) is commonly referred to as a *parallel* configuration, and the configuration in Figure 10.6(*b*) is known as a *parallel-series* configuration.

Parallel configuration

An important assumption built into both identification models shown in Figure 10.6(*a*) and (*b*) is bounded-input bounded-output (BIBO) stability of the nonlinear plant for all input sequences used in the identification process. The parallel configuration shown in Figure 10.6(*a*) uses past values of the identifier as delayed versions of the output in the NARMAX model. Referring to (10.48), we see that parallel configuration implements the most general form of the NARMAX model. Since there is a prediction error associated with every output sample that is fed back to the NARMAX model, even if the plant is BIBO-stable, there is no guarantee that the identification model is going to be stable. Exact conditions for the parallel identification process to be stable are still unknown, even for the case of linear systems [11], and for that reason this configuration is seldom used in practical applications.

Parallel-series configuration

Unlike the parallel configuration, in the parallel-series configuration the past values of the actual plant outputs are used in the identification process. Since the plant is assumed to be BIBO-stable, all signals used in the identification process are bounded as well. Furthermore, use of the actual outputs eliminates the measurement error, and the NARMAX model reduces to a

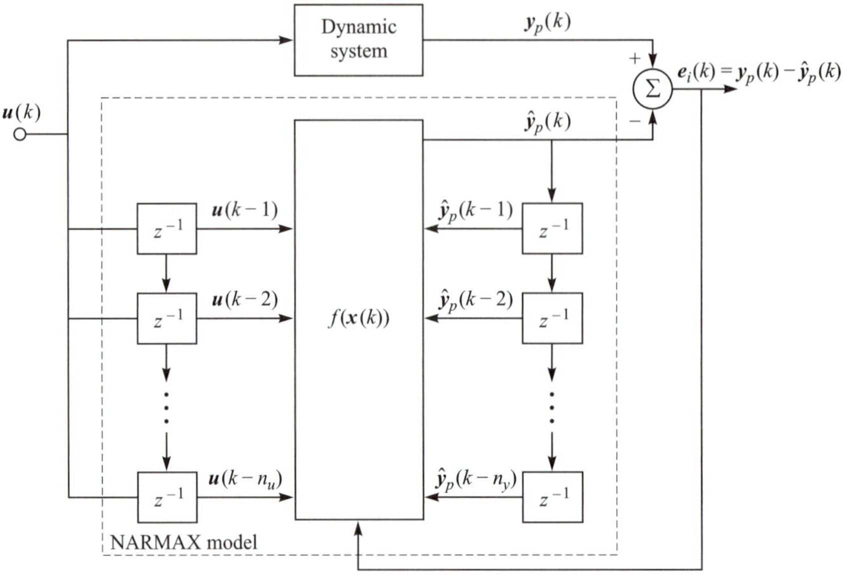

$$\boldsymbol{x}(k) = [\hat{y}_p(k-1), \cdots, \hat{y}_p(k-n_y), \boldsymbol{u}(k-1), \cdots, \boldsymbol{u}(k-n_u)]^T$$

(*a*)

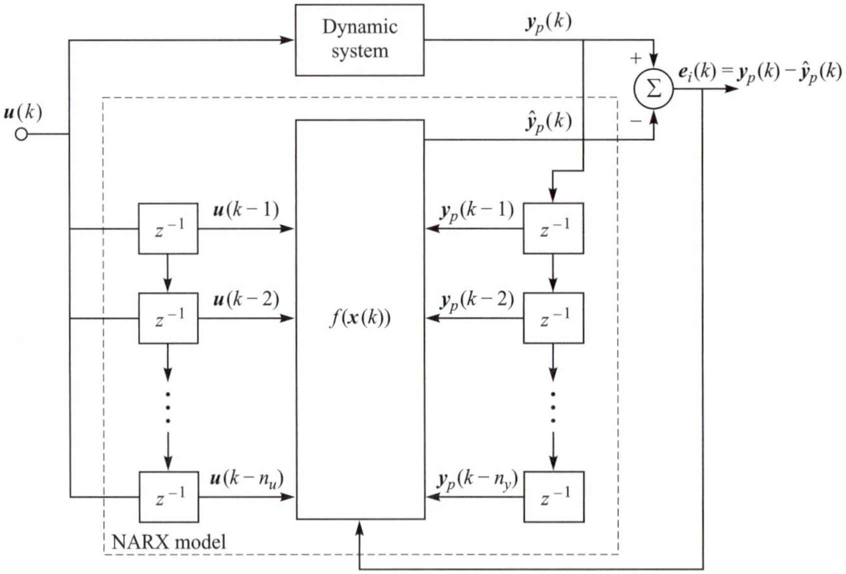

$$\boldsymbol{x}(k) = [y_p(k-1), \cdots, y_p(k-n_y), \boldsymbol{u}(k-1), \cdots, \boldsymbol{u}(k-n_u)]^T$$

(*b*)

Figure 10.6 (*a*) Parallel configuration for system identification; (*b*) parallel-series configuration for system identification.

more manageable NARX model (discussed in the previous section). In light of the above considerations, the parallel-series configuration becomes the configuration of choice for nonlinear system identification.

The most important part of the parallel-series identifier in Figure 10.6(*b*) is the nonlinear mapping. For the case of SISO system, the mapping is given as

$$y(k) = f\left[y(k-1), \cdots, y(k-n_y), u(k-1), \cdots, u(k-n_u)\right] \tag{10.68}$$

Universal approximation properties of some neural network architectures make them prime candidates for a practical realization of the mapping in (10.68). Various network architectures are used to perform the mapping task in (10.68) [11–15, 18–20]. The choice of the neural network type is largely determined by the nature of the identification problem. Here we limit our discussion to the two most popular network architectures, the multilayer perceptron neural network (MLP NN) and the radial basis function neural network (RBF NN).

Identification of nonlinear systems using MLP NN

Use of an MLP NN to approximate the nonlinear mapping of the parallel-series NARX identification model is shown in Figure 10.7. The learning algorithm that is used to train the network is commonly a version of backpropa-

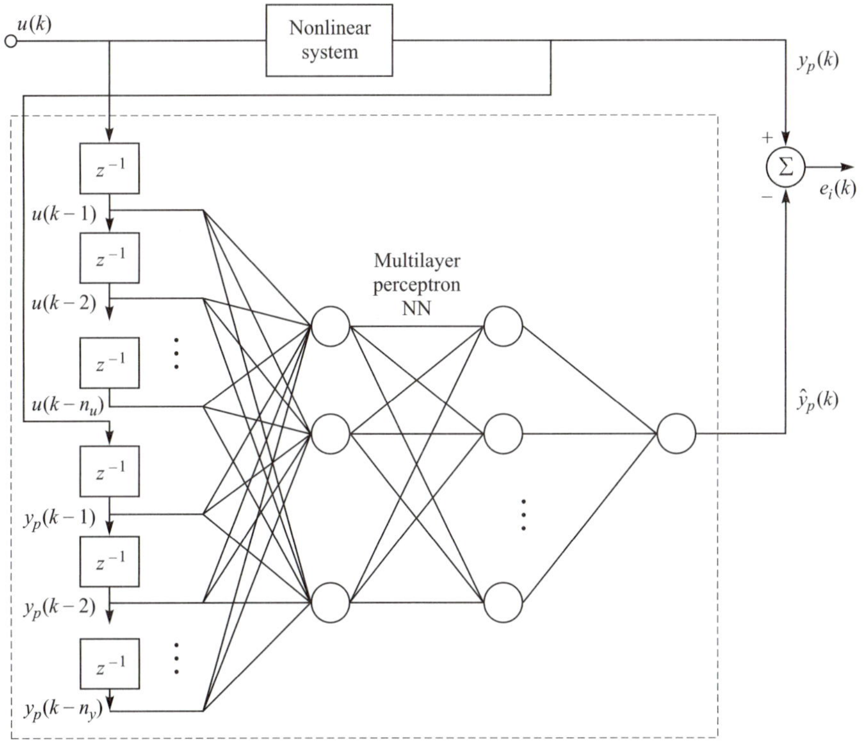

Figure 10.7 Implementation of the parallel-series NARX system identification using a multilayer perceptron neural network.

gation. Two different methods for adaptation of the network weights are very common. The first one is referred to as *pattern learning* [18], and according to this methodology, the weights of the network are updated after each presentation of the input. The second method is an alternative to pattern learning and is called *batch learning*. In this method a sequence of inputs is presented before the network weights are updated. Both of these methods, and tradeoffs between them, are discussed in Chapter 3 (cf. Sect. 3.3.4). Further discussion of these two learning methodologies can be found in References [18, 21]. From a practical standpoint, simulations have demonstrated comparable performances between the two learning rules for small values of the learning rate parameter [18]. With an increase in the learning rate parameter, the batch method shows nonuniform convergence, while the performance of the pattern learning method remains relatively consistent. The biggest advantage of using the MLP NN for approximation of the mapping in (10.68) resides in its simplicity and the fact that it is well suited for online implementation. On the other hand, backpropagation learning creates problems related to the convergence speed of the identification process. In general, any of the methods for accelerating backpropagation learning presented in Chapter 3 may be used to increase the convergence speed. However, this is usually achieved at the expense of increased computational complexity of the learning process.

EXAMPLE 10.3. Consider a system identification problem for a SISO nonlinear system given by the following input-state-output representation:

State equations:

$$x_1(k+1) = \frac{x_2(k)}{x_1(k)+4} \tag{10.69}$$

$$x_2(k+1) = \tanh\{x_1(k) + [1 + x_2(k)]u(k)\} \tag{10.70}$$

Output equation:

$$y(k) = 2x_2(k) \tag{10.71}$$

The parallel-series configuration shown in Figure 10.7 is used to accomplish the identification task. The NARX model has the form given by

$$y(k) = f[u(k-1), \cdots, u(k-4), y(k-1), \cdots, y(k-4)] \tag{10.72}$$

and the MLP NN is designed with two hidden layers and 10 neurons in each layer. During the identification process a uniformly distributed noise sequence in the interval $[-0.5, 0.5]$ is used to provide an input to the system. The neural network identifier is trained using backpropagation. Figure 10.8(a) and (b) shows the output of the system and the NARX model when tested with a square waveform input. The peak amplitude of the square wave input used to generate Figure 10.8(a) is 0.2, and as can be seen, the NARX model accurately approximates the behavior of the nonlinear system. This is to be expected since the magnitude of the waveform is within the range of the random noise used for training of the network. Figure 10.8(b) shows the output of the system and the NARX model when the peak amplitude of the square wave input is 2. As can be seen, there is a substantial difference between the output of the actual nonlinear system and its model. This is due to the amplitude of the input waveform being outside the range of the input patterns used in the development of the NARX model. Therefore, in

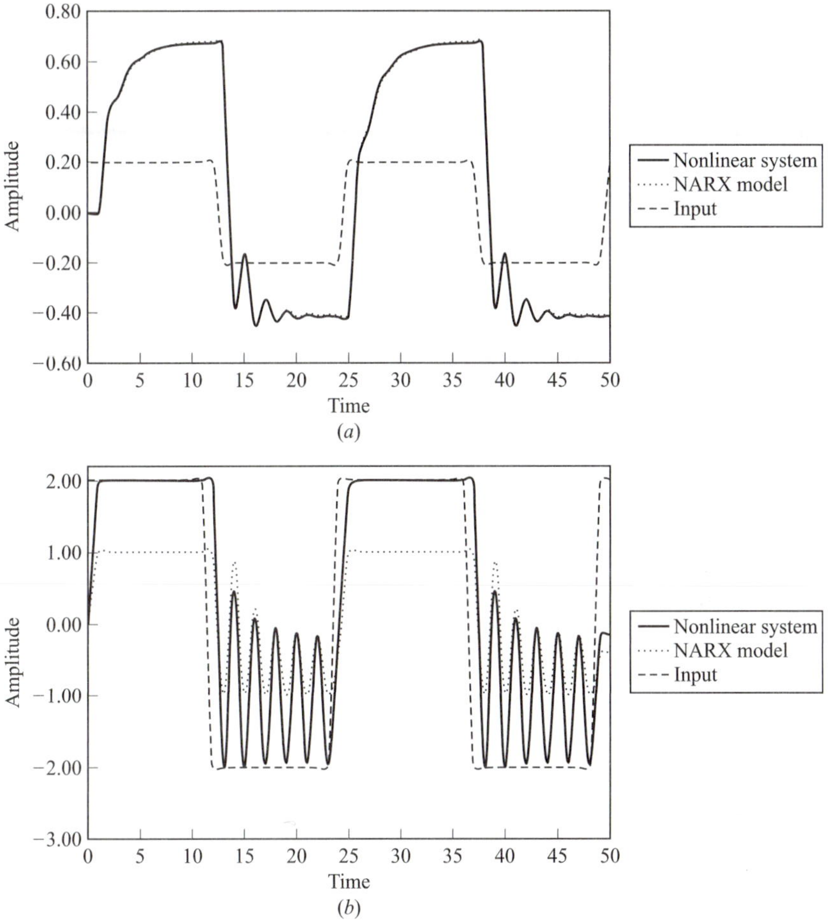

Figure 10.8 (*a*) Simulation of the system identification process in Example 10.2. Response to the square wave input with a peak amplitude of 0.2. (*b*) Response to the square wave input with the peak amplitude of 2.

the nonlinear system identification process, the model exhibits local characteristics, and in general we cannot rely on its accuracy outside of the range of inputs used in the development of the model.

Identification of nonlinear systems using an RBF NN

A depiction of the nonlinear plant identification process using an RBF NN is shown in Figure 10.9. As can be seen, the architecture of the identifier is very similar to the one shown in Figure 10.7 for the case of the MLP NN. Training of the RBF NN in Figure 10.9 can be accomplished in several different ways. For situations when online identification is of greatest importance, the stochastic gradient method suggested in Chapter 3 (cf. Sect. 3.6.2) can be used to determine all three sets of RBF NN parameters: weights, centers, and spread parameters. Although straightforward, this approach

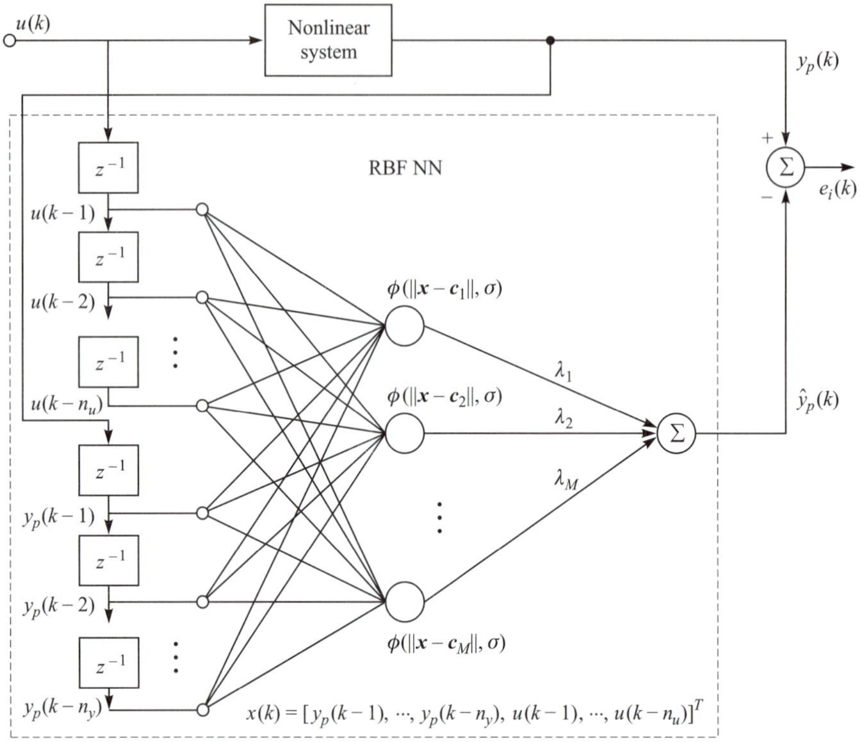

Figure 10.9 Implementation of the parallel-series NARX system identification approach using a radial basis function neural network.

does not utilize the most attractive feature of the RBF NN, its linear "in-parameter" structure. In Section 3.6.1 we demonstrated that once the centers and spread parameters of the network are chosen, the weights of the network can be determined as a solution to the linear regression problem, that is, in "closed form." Furthermore, orthogonal least-squares (OLS) method (cf. Sect. 3.6.3) can be used to select the centers in a judicious manner so the RBF NN identifier is relatively small. The price that is paid for avoiding the use of the stochastic gradient method is that neither the method of fixed centers nor OLS regression is suitable for online identification. In addition, the computational complexity of these methods is significantly higher than that of the stochastic gradient approach.

EXAMPLE 10.4. Consider the same dynamic system given in Example 10.3. This time we use an RBF NN to accomplish the task of the nonlinear mapping in the NARX identifier. Figure 10.10(*a*) and (*b*) shows the results of the system identification process using an RBF NN with 30 neurons in the hidden layer, Gaussian radial basis function, and the spread parameter set to 1. The network was trained using the OLS forward regression method (cf. Sect. 3.6.3). As can be seen from Figure 10.10(*a*), the NARX model predicts the output of the nonlinear system reasonably well when the peak amplitude of the square wave is 0.2. From Figure 10.10(*b*) we see that the NARX model using an RBF NN is incapable of modeling

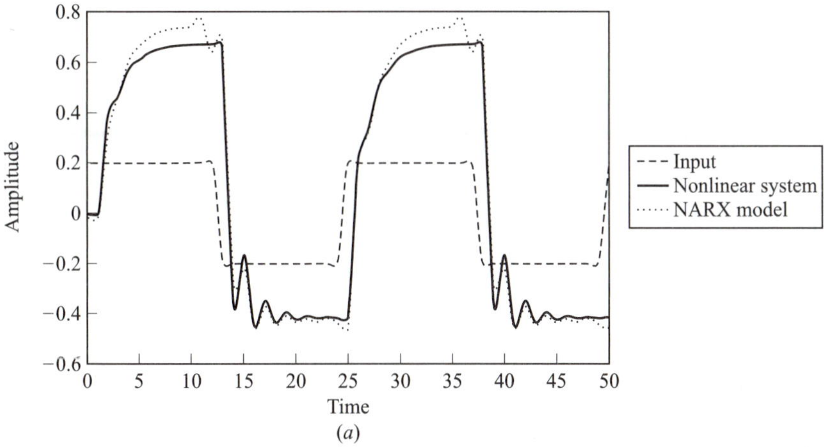

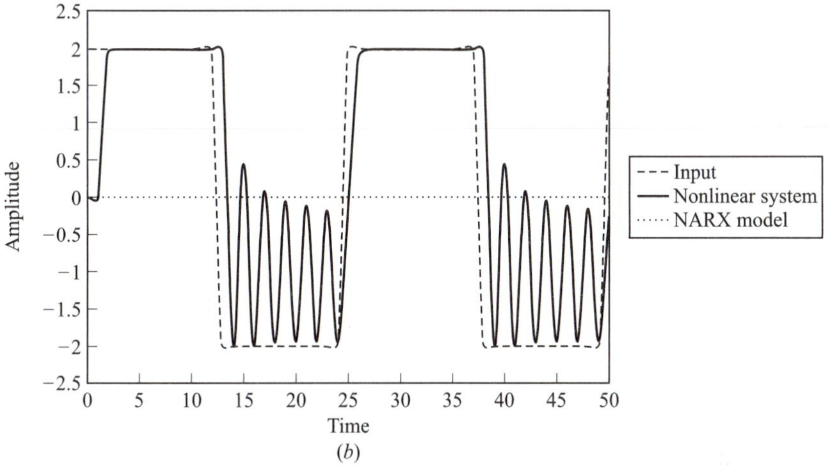

Figure 10.10 (*a*) Simulation of the system identification process in Example 10.3. Response to the square wave input with the peak amplitude 0.2. (*b*) Response to the square wave input with the peak amplitude of 2.

the nonlinear system's behavior for an input peak amplitude of 2. Although this can be explained using the same reasoning given in Example 10.3, we see that the two networks operate in an entirely different manner. While the MLP NN in Example 10.3 is driven into saturation, the RBF NN produces essentially a zero output when the input magnitude is 2, because this is outside the RBF NN's mapping domain.

As a final note, the form used for identification of the nonlinear system corresponds to model 4 described in Section 10.6. This is the general form of the NARX model, and it can be used to model the largest group of nonlinear dynamical systems. However, in many practical cases sufficient knowledge of the nonlinear system's dynamical structure may justify use of one of the simpler forms of the NARX model.

10.7.2 Nonlinear Control

The main objective of nonlinear system control is to maintain the output of a dynamical system within specified limits. Let us consider the situation depicted in Figure 10.11. The nonlinear plant is subjected to an external input $r(t)$. Suppose that the desired output of the plant is known and that it is given in the form of a time function $y_d(t)$. Then the objective of the nonlinear control system is to produce the plant input signal $u(t)$ so that the difference between the actual plant output $y_p(t)$ and the desired output $y_d(t)$ is kept within some specified limit. Depending on the nature of the desired output, we can distinguish between two types of control problems:

1. If $y_d(t) = \text{constant}$, the problem is commonly referred to as a *regulation* problem.
2. If $y_d(t)$ is some nonconstant function of time, the control problem is referred to as a *tracking* problem.

In many cases when the dynamics of the plant is not known in advance, the design of the controller must be carried out in some adaptive fashion. The control algorithm must utilize both the inputs and the outputs of the nonlinear plant, and based on their behavior must modify the internal parameters of the controller and the input signal $u(t)$ so that the control objective is achieved.

Model reference adaptive control

For the adaptive control system shown in Figure 10.11, we have assumed that the desired output of the nonlinear plant is given. Very often the desired behavior of the controlled nonlinear plant is specified in terms of a reference model. The use of the reference model is illustrated in Figure 10.12. This model is a plant of a known dynamical structure that provides the desired output to a given input signal $r(t)$. The objective of the control design is to minimize the difference between the output of the reference model and the nonlinear plant output for a specified range of input values. In general, the reference model can be selected as either linear or nonlinear. However, from the controller design point of view, there is definite interest in linear reference models since they allow the use of powerful tools from linear system theory.

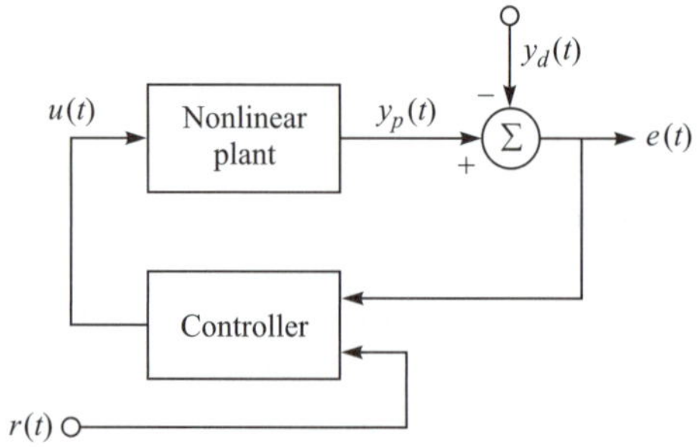

Figure 10.11 General nonlinear control problem.

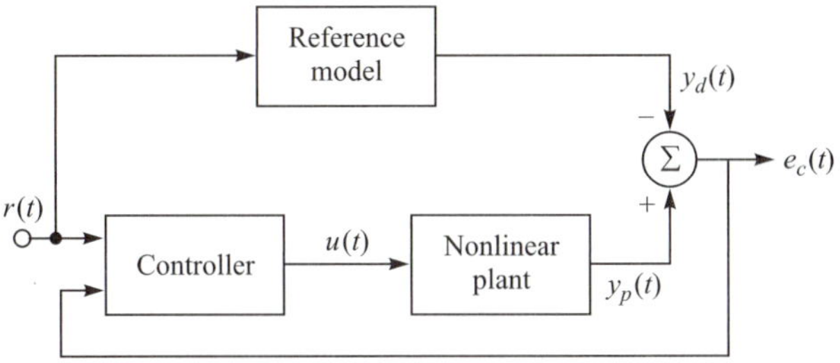

Figure 10.12 *M*odel *r*eference *a*daptive *c*ontrol (MRAC) block diagram.

Direct and indirect control

Traditionally, two different approaches have been used in the adaptive control of dynamical systems: *direct* and *indirect* control. An example of a system implementing direct control is shown in Figure 10.12. The difference between the desired and the actual output of the nonlinear plant is used to adjust the parameters of the controller itself in a direct manner. On the other hand, in the case of indirect control, the difference between the reference model output and the actual output of the plant is used in a system identification process. The design of the controller is performed on the basis of the plant model developed through the identification process. The block diagram showing the process of indirect control is presented in Figure 10.13. At first glance, the structure of the direct adaptive controller in Figure 10.12 may look more appealing since there is no system identification process. However, general methods for adjusting the parameters of the controller based on the input $r(t)$ and output error $e_c(t)$ do not exist. This is attributed to the unknown plant dynamics that resides between the error output and the output of the controller. If indirect control is used, the error between the nonlinear plant and the reference model can be backpropagated through the known dynamics of the plant's model and used in updating the controller's parameters.

Direct and indirect control are strongly motivated by the great success that these methods have shown for linear systems. A detailed treatment of

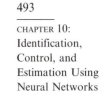

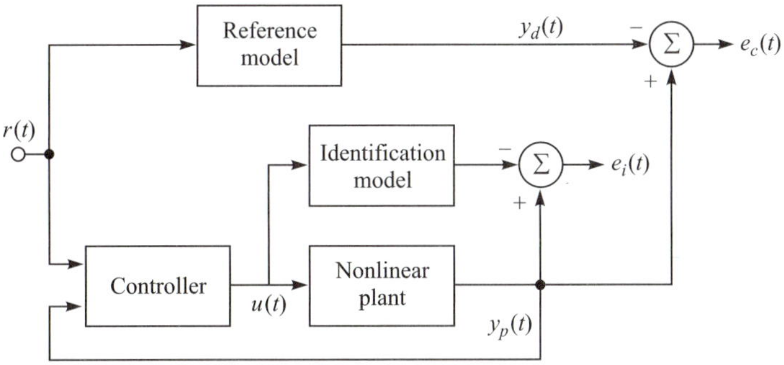

Figure 10.13 Block diagram of the indirect adaptive control approach.

both methods can be found in Reference [22]. The structure of the identification model and the controller design is essentially the same. The fundamental difference between the two lies in the replacement of the linear gains in the identification model with the nonlinear neural network mappings.

Neural networks for indirect control

As previously mentioned, direct adaptive control of nonlinear systems is a very difficult task, and at present for this type of control there are no comprehensive methods [11]. Training algorithms for most of the neural networks used in nonlinear system control involve backpropagation of the error signal representing the difference between the actual and desired system response. Currently, there is no general theory to support the backpropagation of the error signal through the unknown dynamics of the nonlinear plant. Until such theory is available, the predominant method for adaptive control of nonlinear systems is indirect control.

The first step in indirect control is nonlinear system identification. A common approach is to identify the nonlinear system as one of the four nonlinear system model types described in Section 10.6. Once the identification model tracks the behavior of the nonlinear system with sufficient accuracy, it can be used in the MRAC controller training process. Because the dynamics of the neural network system identification model are known, the error signal can be backpropagated and used to adjust the weights of the neural controller. The block diagram of an indirect neural network controller is shown in Figure 10.14.

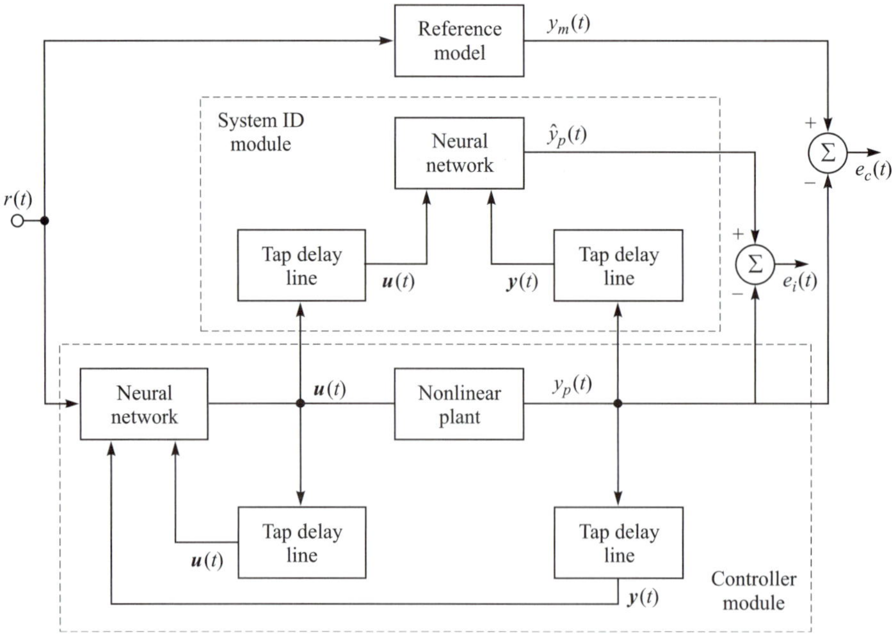

Figure 10.14 Block diagram of an indirect controller using a neural network.

Common types of neural networks used for indirect control shown in Figure 10.14 are the MLP NN or RBF NN. In general, any neural network architecture that performs a nonlinear mapping can be used. Narendra and Parthasarathy [11] discuss the use of recurrent networks (cf. Chap. 5) as well.

Guaranteeing the stability of the control process shown in Figure 10.14 is a challenging task. Since the plant is nonlinear, notions of controllability and observability for linear systems do not apply, and the design of the controller must be evaluated carefully and in many cases on a trial-and-error basis. The choice of the neural network in Figure 10.14, as well as the methodology used in the controller design, depends on our knowledge of the plant behavior. For example, if the plant is stable in the BIBO sense, system identification can be performed offline, using the method discussed in Section 10.7.

If the knowledge of the plant's dynamics is limited, both identification and control must be conducted online, and identification and control must proceed simultaneously. In these circumstances the entire structure represents a complicated nonlinear system, and the individual stability of the components does not guarantee the overall system stability. Special attention must be paid to the rate of the weight adjustments for the networks in system identification and the controller module. In general, the simulations reported in Reference [11] indicate that for stable and efficient online control, the identification process must be sufficiently accurate. This can be accomplished in two ways:

1. The rate of the weight updates for the system identification network is chosen to be greater than the update rate for the controller network.
2. The update of the weights in the controller network is not performed before the magnitude of the identification error falls below a specified value.

Which of the two methods is more appropriate is case-dependent.

Let us now review two examples of indirect nonlinear control that use the neural network-based system identification and controller structure approach. The purpose of the examples is to illustrate the process of designing a controller as well as demonstrating some commonly encountered problems.

EXAMPLE 10.5. MRAC Control of Model 2 Nonlinear Dynamical System. Consider a discrete nonlinear system specified by the following difference equation

$$y_p(k+1) = \frac{y_p(k) + y_p(k-1)}{1 + y_p^2(k) + 2y_p^2(k-1)} + u(k) \tag{10.73}$$

The input/output system representation corresponds to model 2 discussed in Section 10.6. In the process of controller design we assume that the output equation has the form

$$y_p(k+1) = f\big(y_p(k), y_p(k-1)\big) + u(k) \tag{10.74}$$

where the function $f(\bullet)$ is unknown. A reference model specifying the desired behavior of the system is given by

$$y_m(k+1) = 0.5y_m(k) + 0.3y_m(k-1) + r(k) \tag{10.75}$$

where $r(k)$ is the reference input. As shown, the reference model is a stable linear dynamical system. The error between the model and the output of the plant at any arbitrary time instant k is given by

$$e_c(k+1) = y_p(k+1) - y_m(k+1) \tag{10.76}$$

Substituting (10.74) and (10.75) into (10.76), we can write

$$e_c(k+1) = f\big(y_p(k), y_p(k-1)\big) + u(k) - 0.5y_m(k) - 0.3y_m(k-1) - r(k) \tag{10.77}$$

The objective of the controller design is to minimize the error given in (10.77). Setting the right-hand side of (10.77) equal to zero and solving for the control input, we obtain

$$u(k) = 0.5y_m(k) + 0.3y_m(k) + r(k) - f\big(y_p(k), y_p(k-1)\big) \tag{10.78}$$

However, since the nonlinear function $f(\bullet)$ is unknown, a neural network approximation has to be used. The control input to the plant will be generated according to

$$u(k) = 0.5y_m(k) + 0.3y_m(k-1) + r(k) - N\big(y_p(k), y_p(k-1)\big) \tag{10.79}$$

Equation (10.79) represents the mapping performed by the nonlinear controller. Substituting (10.79) into (10.74), the output of the controlled plant at time sample k is given by

$$\begin{aligned} y_p(k+1) = f\big(y_p(k), y_p(k+1)\big) - N\big(y_p(k), y_p(k-1)\big) \\ + 0.5y_p(k) + 0.3y_p(k-1) + r(k) \end{aligned} \tag{10.80}$$

From (10.80) we see that the difference between the output of the nonlinear plant and the output of the reference model depends directly on the accuracy of the neural network's approximation.

The first step in the indirect controller design is the system identification process. Assuming the plant is BIBO-stable, we perform system identification in an off-line manner. From (10.74) it follows that

$$f_y\big(y_p(k), y_p(k-1)\big) = y_p(k+1) - u(k) \tag{10.81}$$

By providing inputs to the plant and recording its outputs we can use (10.81) and generate the input patterns for training the neural network. In short, the neural network is to be trained to perform the mapping

$$\big[y_p(k), y_p(k-1)\big]^T \rightarrow y_p(k+1) - u(k) \tag{10.82}$$

In this example we train an RBF NN with 48 neurons in one hidden layer and a spread parameter $\sigma = 1$. The inputs are drawn as random numbers from a normal distribution having a standard deviation of 2. Once the system is properly identified, (10.79) can be used as a nonlinear controller. Figure 10.15(a) shows the test input signal

$$y_t(t) = \frac{1}{2}[\sin(10\pi t) + \sin(25\pi t + 0.5)] \tag{10.83}$$

Figure 10.15(b) shows a comparison between the response of the reference model and the controlled nonlinear plant to the signal in (10.83). As can be seen, the difference between the two is negligible.

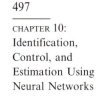

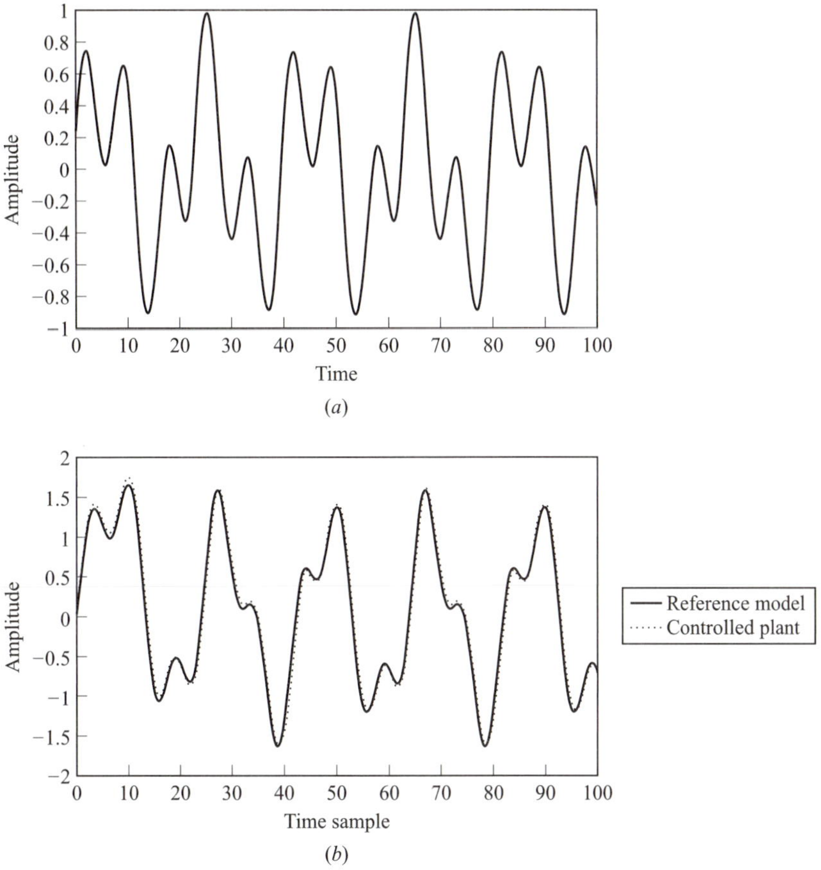

Figure 10.15 (*a*) Test signal given in (10.83); (*b*) responses to the test signal.

EXAMPLE 10.6. MRAC Control of Model 4. Consider a nonlinear plant given by the following difference equation

$$y_p(k+1) = \frac{1 + y_p(k) + y_p(k-1)}{1 + y_p^2(k) + 2y_p^2(k-1)} \tan^{-1}\left(u^3(k)\right) \tag{10.84}$$

Let us assume that we need to design a neural network controller so that the overall behavior of the controlled system can be described by the reference model equation, given as

$$y_m(k+1) = 0.5y_m(k) + 0.3y_m(k-1) + r(k) \tag{10.85}$$

The structure of the neural network controller is given in Figure 10.16.

In the first stage of the controller design we need to perform system identification; that is, the network $\mathsf{NN_I}$ needs to be trained to model the behavior of the nonlinear plant. The NARX model corresponding to the plant can be written as

$$y_p(k+1) = f\left[u(k), y_p(k), y_p(k-1)\right] = f(\boldsymbol{x}) \tag{10.86}$$

where $\boldsymbol{x} = \left[u(k), y_p(k), y_p(k-1)\right]^T$. Using the methods described in Section 10.7.1, an MLP NN with two hidden layers and 30 neurons in each layer is

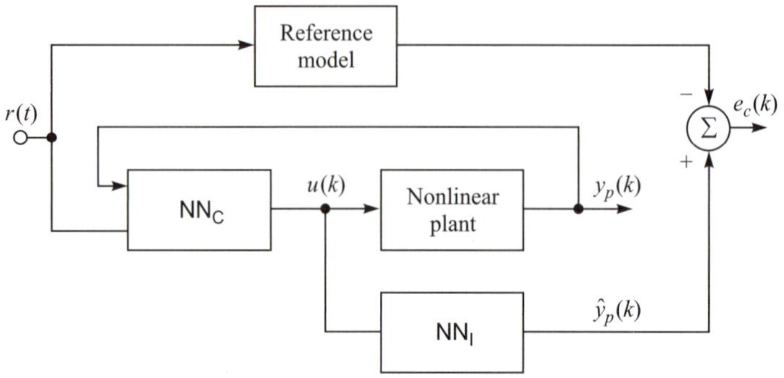

Figure 10.16 Structure of the neural network controller for Example 10.6.

used to perform the identification task. The network is trained using the back-propagation algorithm where the input sequence is selected as uniformly distributed noise in the interval [-4, 4].

Once the plant is identified, the task of the controller design reduces to training a neural network $\mathsf{NN_C}$. The NARX mapping performed by $\mathsf{NN_C}$ is given as

$$u(k) = g\big[r(k), y_p(k), y_p(k-1)\big] \approx g\big[r(k), \hat{y}_p(k), \hat{y}_p(k-1)\big] \tag{10.87}$$

or

$$u(k) = \mathsf{NN_C}(\boldsymbol{y}) \tag{10.88}$$

where $\boldsymbol{y} = \big[r(k), \hat{y}_p(k), \hat{y}_p(k-1)\big]^T$. In this example the $\mathsf{NN_C}$ is selected as an RBF neural network with 1,000 centers $\boldsymbol{c}_i = [c_{i1}, c_{i2}, c_{i3}] \in \Re^{3\times 1}$, and $c_i \in [-1, 1]$. The mapping performed by $\mathsf{NN_C}$ is of the form

$$\mathsf{NN_C}(\boldsymbol{y}) = \sum_{i=1}^{N_c} \lambda_i \phi(\|\boldsymbol{y} - \boldsymbol{c}_i\|) \tag{10.89}$$

where

$$\phi(v) = \exp\left(-\frac{v^2}{\sigma^2}\right) \tag{10.90}$$

and λ_i are the network's weights, and σ is the spread parameter selected as $\sigma = 0.2$.

The error between the controlled plant and the reference model given in (10.85) is

$$e_c(k+1) = y_p(k+1) - y_m(k+1) \approx \hat{y}_p(k+1) - y_m(k+1) \tag{10.91}$$

A cost function can be formed as

$$J(k+1) = \frac{1}{2}e_c^2(k+1) = \frac{1}{2}\big[\hat{y}_p(k+1) - y_m(k+1)\big]^2 \tag{10.92}$$

The network $\mathsf{NN_C}$ is trained to minimize the function in (10.92). Using the steepest descent approach, the weights can be adjusted according to

$$\lambda_i(k+1) = \lambda_i(k) - \mu \left.\frac{\partial J(k+1)}{\partial \lambda_i}\right|_{\lambda_i=\lambda_i(k)} \tag{10.93}$$

The gradient in (10.93) can be evaluated as

$$\begin{aligned} \frac{\partial J(k+1)}{\partial \lambda_i} &= \frac{\partial}{\partial \lambda_i}\left\{\frac{1}{2}\left[\hat{y}_p(k+1) - y_m(k+1)\right]^2\right\} \\ &= e_c(k+1)\frac{\partial \hat{y}(k+1)}{\partial \lambda_i} \\ &= e_c(k+1)\frac{\partial \mathsf{NN_I}}{\partial u(k)}\frac{\partial u(k)}{\partial \lambda_i} \end{aligned} \tag{10.94}$$

Using (10.88) and (10.89), we have

$$\frac{\partial u(k)}{\partial \lambda_i} = \phi(\|\mathbf{y} - \mathbf{c}_i\|) \tag{10.95}$$

The derivative of the network output with respect to the signal $u(k)$ can be approximated as

$$\frac{\partial \mathsf{NN_I}}{\partial u(k)} \approx \frac{\mathsf{NN_I}(u(k)+\varepsilon) - \mathsf{NN_I}(u(k))}{\varepsilon} \tag{10.96}$$

where ε is selected as a number much smaller than the learning rate parameter.

Finally, using (10.94), (10.95), and (10.96), we can formulate a learning rule for adjusting the weights of the controller RBF neural network $\mathsf{NN_C}$ as

$$\lambda_i(k+1) = \lambda_i(k) - \mu e_c(k+1)\frac{\mathsf{NN_I}(u(k)+\varepsilon) - \mathsf{NN_I}(u(k))}{\varepsilon}\phi(\|\mathbf{y} - \mathbf{c}_i\|) \tag{10.97}$$

Equation (10.97) is used to determine the weights of the controller network. Figure 10.17 shows the comparison between the output of the reference model and the controlled nonlinear plant given the test signal

$$s(n) = \frac{1}{4}[\sin(10\pi n) + \sin(20\pi n)] \tag{10.98}$$

As can be seen, the difference is relatively small.

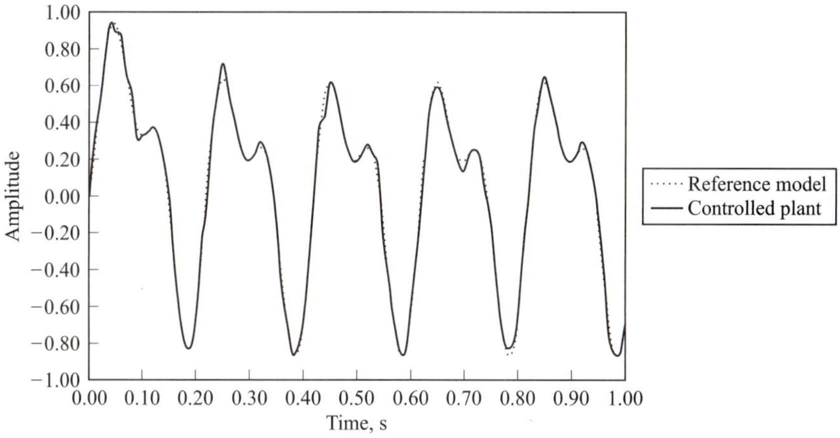

Figure 10.17 Comparison between the response of controlled plant and reference model to the test signal given in (10.98).

10.8 INDEPENDENT-COMPONENT ANALYSIS: BLIND SEPARATION OF UNKNOWN SOURCE SIGNALS

The main objectives of this section will be to present a neural network approach for performing independent-component analysis (ICA), and then a fast fixed-point algorithm for ICA. The basic idea of ICA is presented first. Several examples are also shown that illustrate the utility of ICA for signal processing and image processing problems.

10.8.1 Overview of Independent-Component Analysis

ICA can be thought of as an extension of principal component analysis (PCA) (cf. Sect. 9.2). It is used primarily to separate unknown source signals from their linear mixtures, or the blind source signal separation problem [23–26]. ICA can also be used for feature extraction. The characteristics of the transmission channel do not have to be known to separate the source signals from a set of noisy observable (measured) signals. Blind source separation techniques can be applied to array processing, medical signal processing, communications, speech processing, image processing, and many other areas. Generally speaking, there are two types of blind source separation problems: those that involve *instantaneous* mixtures and those involving *convolutive* mixtures. We will only be concerned with problems involving instantaneous mixtures.

The main difference between PCA and ICA is that instead of the uncorrelatedness property associated with standard PCA, in ICA the coefficients of the linear expansion of the data vectors must be mutually independent, or as independent as possible. What this means is that higher-order statistics [27, 28] must be used to determine the ICA expansion. In standard PCA, second-order statistics provide only decorrelation. Higher-order statistics are useful when dealing with non-Gaussian processes, non-minimum-phase problems, colored noise, or even nonlinear processes [27]. Therefore, it is not surprising that in the neural network implementation of ICA nonlinearities must be used in the learning phase, even though the final input/output mapping is linear [29].

Independent-component analysis basics

We assume that there exist q *zero-mean*, wide-sense stationary source signals $s_1(k), s_2(k), \ldots, s_q(k)$ for $k = 1, 2, \ldots$ (the discrete-time index or for images the pixels) that are scalar-valued and mutually statistically independent for each sample value k. The independence condition (cf. Sect. A.7.1) can be formally defined by stating that the joint probability density of the source signals is equal to the product of the marginal probability densities of the individual signals, that is,

$$\wp[s_1(k), s_2(k), \ldots, s_q(k)] = \wp[s_1(k)]\wp[s_2(k)]\ldots\wp[s_q(k)] = \prod_{i=1}^{q} \wp[s_i(k)] \quad (10.99)$$

The individual source signals are also assumed to be unknown (unobservable); however, we do have access to a set of h noisy linear mixtures of the unknown signals $x_1(k), x_2(k), \ldots, x_h(k)$. These measured signals are given by

$$x_j(k) = \sum_{i=1}^{q} s_i(k) a_{ij} + n_j(k) \tag{10.100}$$

for $j = 1, 2, \ldots, h$, the elements a_{ij} are assumed to be not known, and $n_j(k)$ is additive measurement noise. We can now define the vectors $\boldsymbol{x}(k) = [x_1(k),\ x_2(k), \cdots, x_h(k)]^T$, $\boldsymbol{x}(k) \in \Re^{hx1}$, $\mathbf{s}(k) = [s_1(k),\ s_2(k), \cdots, s_q(k)]^T$, $\mathbf{s}(k) \in \Re^{qx1}$ (the *source vector* consisting of the q independent components), and $\boldsymbol{A} = [\mathbf{a}_1, \mathbf{a}_2, \cdots, \mathbf{a}_q]$, $\boldsymbol{A} \in \Re^{hxq}$ (the *mixing matrix*), where the column vectors of $\boldsymbol{A}$ are the *basis vectors* of the ICA expansion. Equation (10.100) can now be written in vector-matrix form as

$$\boldsymbol{x}(k) = \mathbf{A}\mathbf{s}(k) + \boldsymbol{n}(k) = \sum_{i=1}^{q} s_i(k)\boldsymbol{a}_i + \boldsymbol{n}(k) \tag{10.101}$$

referred to as the ICA expansion. We will assume the mixing matrix $\boldsymbol{A}$ contains at least as many rows as columns ($h \geq q$), and it has full column rank, that is, $\rho(\boldsymbol{A}) = q$ (i.e., the mixtures of the source signals are all different).

Ambiguities associated with ICA

There are several ambiguities associated with the use of ICA:

1. The amplitudes of the separated signals (independent components) cannot be determined. This is due to both $\boldsymbol{s}$ and $\boldsymbol{A}$ being unknown; see (10.101). Any scalar multiplier in one of the sources can always be canceled by dividing the corresponding column of $\boldsymbol{A}$ (i.e., $\boldsymbol{a}_i$) by the same scalar.
2. There is also a sign ambiguity associated with the separated signals; that is, the independent components can be multiplied by -1 without affecting the model.
3. The order of the independent components cannot be determined. Again this is due to both $\boldsymbol{s}$ and $\boldsymbol{A}$ being unknown. Any independent component can be defined as the "first" one. To show this, let $\boldsymbol{P} > 0$ ($\boldsymbol{P} \in \Re^{qxq}$) be a permutation matrix. Then the ICA model can be written as (for the noise-free case)

$$\boldsymbol{x} = \boldsymbol{A}\boldsymbol{P}^{-1}\boldsymbol{P}\boldsymbol{s} = (\boldsymbol{A}\boldsymbol{P}^{-1})(\boldsymbol{P}\boldsymbol{s}) = \tilde{\boldsymbol{A}}\tilde{\boldsymbol{s}} \tag{10.102}$$

where $\tilde{\boldsymbol{A}}$ is the "new" unknown mixing matrix, and the vector $\tilde{\boldsymbol{s}}$ contains the reordered independent-component variables.

10.8.2 Independent-Component Analysis Using Neural Networks

Our discussions here pertaining to the neural network approach for blind source separation using ICA follows the presentation by Karhunen et al. [29]. Figure 10.18 shows the basic neural network architecture design to perform the separation of source signals (i.e., estimate the independent compo-

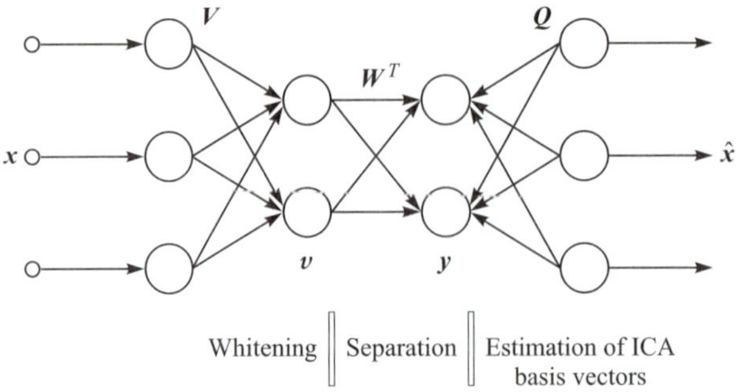

Figure 10.18 The ICA network. The three layers perform whitening, separation, and estimation of the basis vectors. The weight matrices that are necessary to determine are $\boldsymbol{V}$, $\boldsymbol{W}^T$, and $\boldsymbol{Q}$. (*Source:* Adapted with permission from Karhunen et al. [29], p. 490, © 1997 IEEE.)

nents) and to estimate the basis vectors of the ICA expansion [i.e., estimate the column vectors of the mixing matrix $\boldsymbol{A}$ in (10.101)].

Prewhitening process

The whitening process that precedes the separation step (i.e., prewhitening) is a critical procedure. This process normalizes the variances of the observed signals to unity. In general, separation algorithms that use prewhitened inputs often have better stability properties and converge faster. However, whitening the data can make the separation problem more difficult if the mixing matrix $\boldsymbol{A}$ is ill-conditioned or if some of the source signals are relatively *weak* compared to the other signals [30, 31]. The input vectors $\boldsymbol{x}(k)$ are whitened by applying the transformation

$$\boldsymbol{v}(k) = \boldsymbol{V}\boldsymbol{x}(k) \tag{10.103}$$

where $\boldsymbol{v}(k)$ is the kth whitened vector and $\boldsymbol{V}$ is the whitening matrix. The whitening matrix can be determined in two ways: by using a batch approach or by neural learning. For the batch approach, if PCA is used to determine the whitening matrix, it is given as

$$\boldsymbol{V} = \boldsymbol{D}^{-1/2}\boldsymbol{E}^T \tag{10.104}$$

where $\boldsymbol{V} \in \Re^{qxh}$, $\boldsymbol{D}= \mathrm{diag}\left[\lambda_1, \lambda_2, \ldots, \lambda_q\right] \in \Re^{qxq}$, and $\boldsymbol{E} = \left[\boldsymbol{c}_1, \boldsymbol{c}_2, \ldots, \boldsymbol{c}_q\right] \in \Re^{hxq}$, with λ_i the ith-largest eigenvalue of the covariance matrix $\boldsymbol{C}_x = \mathrm{E}\{\boldsymbol{x}(k)\boldsymbol{x}^T(k)\} \in \Re^{hxh}$, and $\boldsymbol{c}_i$ for $i = 1, 2, \ldots, q$ are associated (principal) eigenvectors. Therefore, the transformation in (10.103) actually consists of two steps, that is, compression and whitening. The compression step consists of selecting the proper value for q (the number of source signals). Therefore, the PCA described above for the whitening can also be used to select (i.e., estimate) the number of source signals q to be recovered (or the number of independent components) if the noise term $\boldsymbol{n}(k)$ in (10.101) is assumed to be zero-mean Gaussian white noise with covariance matrix $\mathrm{E}\{\boldsymbol{n}(k)\boldsymbol{n}^T(k)\} = \sigma^2\boldsymbol{I}_h$. In the noise covariance matrix, σ^2 is the conjoint variance of the components of the noise vector $\boldsymbol{n}(k)$. The noise vector is assumed to be uncorrelated with the sources $s_i(k)$, for $i = 1, 2, \ldots, q$. Given these assumptions, the covariance matrix of the data vectors $\boldsymbol{x}(k)$ is given by

$$\mathrm{E}\{\boldsymbol{x}(k)\boldsymbol{x}^T(k)\} = \sum_{i=1}^{q} \mathrm{E}\{\boldsymbol{s}_i^2(k)\}\boldsymbol{a}_i\boldsymbol{a}_i^T + \sigma^2\boldsymbol{I}_h \tag{10.105}$$

The q largest eigenvalues of the covariance matrix in (10.104), that is, $\lambda_1, \lambda_2, \ldots, \lambda_q$, are some linear combination of the source signal powers $\mathrm{E}\{\boldsymbol{s}_i^2(k)\}$ added to the noise power σ^2. Therefore, the remaining $h - q$ eigenvalues correspond to only noise (theoretically these eigenvalues are equal to σ^2). The q largest *signal* eigenvalues will bc distinctly larger than the remaining *noise* eigenvalues if the signal-to-noise ratio is large enough. In practice, the eigenvalues of the input covariance matrix are determined from the time average of the covariance matrix over the available data vectors, given by

$$\boldsymbol{C}_x \cong \frac{1}{N}\sum_{i=1}^{N} \boldsymbol{x}_i(k)\boldsymbol{x}_i^T(k) \tag{10.106}$$

where N is the total number of input vectors.

A stochastic approximation algorithm to learn the whitening matrix is given by

$$\boldsymbol{V}(k+1) = \boldsymbol{V}(k) - \mu(k)[\boldsymbol{v}(k)\boldsymbol{v}^T(k) - \boldsymbol{I}]\boldsymbol{V}(k) \tag{10.107}$$

where it is recommended to adjust the learning rate parameter according to (9.48), that is,

$$\mu(k) = \frac{1}{\dfrac{\gamma}{\mu(k-1)} + \|\boldsymbol{v}(k)\|_2^2} \qquad 0 < \gamma \leq 1.0 \tag{10.108}$$

where γ is the forgetting factor. When the whitening (orthogonal) transformation $\boldsymbol{V}$ is applied to the inputs as in (10.103), the resulting whitened outputs $\boldsymbol{v}(k)$ will possess the whiteness condition, that is,

$$\mathrm{E}\{\boldsymbol{v}(k)\boldsymbol{v}^T(k)\} = \boldsymbol{I}_q \tag{10.109}$$

where $\boldsymbol{v}(k)$ is defined in (10.103).

Separation process

The separation process can be carried out by using many different methods [26, 30, 32]. Approximating contrast functions, maximized by separating matrices, have been developed [26]. However, contrast functions typically require extensive batch computations using estimated higher-order statistics of the data and lead to very complicated adaptive separation algorithms. As we will see, it is sufficient to use the kurtosis (fourth-order cumulant) of the data. In Section 10.8.3 we present an alternative to the neural learning approach presented here that converges much faster. Another class of separation methods involves using neural networks to perform the separation of the source signals [31]. In Figure 10.18, the second stage of the architecture is responsible for the separation of the whitened signals $\boldsymbol{v}$. The linear separation transformation is given by

$$\boldsymbol{y}(k) = \boldsymbol{W}^T \boldsymbol{v}(k) \tag{10.110}$$

where $\boldsymbol{W} \in \Re^{qxq}(\boldsymbol{W}^T\boldsymbol{W} = \boldsymbol{I}_q)$ is the separation matrix. Thus the separated signals are the outputs of the second stage, that is, $\hat{\boldsymbol{s}}(k) = \boldsymbol{y}(k)$. An interesting observation is that once the source signal $\boldsymbol{s}(k)$ has been estimated, this means that the pseudoinverse of $\boldsymbol{A}$, that is, $\boldsymbol{A}^+$, must have been "blindly" determined as well [refer to (10.101)].

One very straightforward neural learning method to determine the separation matrix is based on the nonlinear PCA subspace learning rule [33–37] (cf. Sect. 9.3.6) given by

$$\boldsymbol{W}(k+1) = \boldsymbol{W}(k) + \mu(k)\{\boldsymbol{v}(k) - \boldsymbol{W}(k)g[\boldsymbol{y}(k)]\}g[\boldsymbol{y}^T(k)] \tag{10.111}$$

where $\boldsymbol{v}(k)$ is the prewhitened input vector given in (10.103), and the function $g(\bullet)$ is a suitably chosen nonlinear function usually selected to be odd in order to ensure both stability and signal separation. It is recommended that the learning rate parameter $\mu(k)$ be adjusted according to the adaptive scheme given in (10.108), with $\boldsymbol{v}(k)$ replaced by $\boldsymbol{y}(k)$. Also, for good convergence, it is best to select the initial weight matrix $\boldsymbol{W}(0)$ to have as columns a set of orthonormal vectors. Typically, the nonlinear function $g(\bullet)$ is chosen as

$$g(t) = \beta \tanh(t/\beta) \tag{10.112}$$

where $g(t) = \mathrm{d}f(t)/\mathrm{d}t$ and $f(t) = \beta^2 \ln[\cosh(t/\beta)]$, the logistic function (cf. Sect. 9.7). This is not an arbitrary choice for the nonlinearity in the learning rule of (10.111). It is motivated by the fact that when determining the ICA expansion *higher-order statistics* are needed. This can be seen by observing another neural learning rule to perform separation of unknown signals. This learning rule is called the *bigradient algorithm* [29, 36, 37] given by

$$\boldsymbol{W}(k+1) = \boldsymbol{W}(k) + \mu(k)\boldsymbol{v}(k)g[\boldsymbol{y}^T(k)] + \gamma(k)\boldsymbol{W}(k)[\boldsymbol{I} - \boldsymbol{W}^T(k)\boldsymbol{W}(k)] \tag{10.113}$$

where $\gamma(k)$ is another *gain* parameter, typically about 0.5 or 1. This is a stochastic gradient algorithm that maximizes or minimizes the performance criterion

$$J(\boldsymbol{W}) = \sum_{i=1}^{q} \mathrm{E}\{f(y_i)\} \tag{10.114}$$

under the constraint that the weight matrix $\boldsymbol{W}$ be orthonormal. The orthonormal constraint in (10.114) is realized in the learning rule in (10.113) in an additive manner. With the appropriate function $f(\bullet)$ in (10.114), the performance criterion would involve the sum of the fourth-order statistics (fourth-order cumulants) of the outputs, that is, the *kurtosis* [27]. Therefore, the criterion would be either *minimized* for sources with a *negative kurtosis* (platykurtic) or *maximized* for sources with a *positive kurtosis* (leptokurtic). Source signals that have a *negative kurtosis* are often called *sub-Gaussian* signals, and sources that have a *positive kurtosis* are referred to as *super-Gaussian* signals. In (10.114) the expectation operator would be dropped because we only consider instantaneous values. We now write the logistic function $f(t) = \ln[\cosh(t)]$ (for $\beta = 1$) in terms of a Taylor series expansion

$$f(t) = \ln[\cosh(t)] = \frac{t^2}{2} - \frac{t^4}{12} + \frac{t^6}{45} - \cdots \tag{10.115}$$

The second-order term $t^2/2$ is on the average constant due to the whitening. The nonlinearity would then be given by $g(t) = df(t)/dt = \tanh(t) = t - t^3/3 + 2t^5/15 - \cdots$, and the cubic term will be dominating (an odd function) if the data are prewhitened.

Estimation of the ICA basis vectors

This is the last stage in Figure 10.18. Two methods are presented here to estimate the ICA basis vectors, or the column vectors of the mixing matrix $\boldsymbol{A}$ in (10.101). The first method is a batch approach where the estimate of $\boldsymbol{A}$, that is, $\hat{\boldsymbol{A}}$, is given by

$$\hat{\boldsymbol{A}} = \boldsymbol{E}\boldsymbol{D}^{1/2}\boldsymbol{W} \tag{10.116}$$

where $\boldsymbol{D}$ is the eigenvalue matrix shown in (10.104), $\boldsymbol{E}$ has columns that are the associated eigenvectors shown in (10.104), and $\boldsymbol{W}$ is the separation matrix. The second method is a neural approach for estimating the ICA basis vectors. From Figure 10.18, the last stage gives an estimate of the observed data as

$$\hat{\boldsymbol{x}} = \boldsymbol{Q}\boldsymbol{y} \tag{10.117}$$

Comparing (10.117) with (10.101) for $\boldsymbol{n} = \boldsymbol{0}$ (i.e., $\boldsymbol{x} = \boldsymbol{A}\boldsymbol{s}$), we see that $\boldsymbol{Q} = \hat{\boldsymbol{A}}$ since $\boldsymbol{y} = \hat{\boldsymbol{s}}$. Therefore, the columns of the $\boldsymbol{Q}$ matrix are estimates of the columns of $\boldsymbol{A}$, the ICA basis vectors. A neural learning algorithm can be derived from a representation error performance measure given by

$$J(\boldsymbol{Q}) = \frac{1}{2}\|\boldsymbol{x} - \hat{\boldsymbol{x}}\|_2^2 = \frac{1}{2}\|\boldsymbol{x} - \boldsymbol{Q}\boldsymbol{y}\|_2^2 \tag{10.118}$$

Taking a steepest descent approach given by $\boldsymbol{Q}(k+1) = \boldsymbol{Q}(k) - \mu\nabla_{\boldsymbol{Q}}J(\boldsymbol{Q})$, the neural learning rule for estimating the ICA basis vectors is

$$\boldsymbol{Q}(k+1) = \boldsymbol{Q}(k) + \mu(k)[\boldsymbol{x}(k) - \boldsymbol{Q}(k)\boldsymbol{y}(k)]\boldsymbol{y}^T(k) \tag{10.119}$$

where $\mu > 0$ is the learning rate parameter that can be adapted during learning using (10.108) with $\boldsymbol{v}(k)$ replaced by $\boldsymbol{Q}(k)\boldsymbol{y}(k)$.

EXAMPLE 10.7. The first example involves separation of three sinusoidal signals with frequencies $f_1 = 500$ Hz, $f_2 = 600$ Hz, and $f_3 = 1000$ Hz that are sampled at $f_s = 10$ kHz. The original signals are shown in Figure 10.19. The mixing matrix used is the same one used by Karhunen et al. [29] in their first example,

$$\boldsymbol{A} = \begin{bmatrix} 0.0891 & 0.3906 & -0.3408 \\ -0.8909 & -0.6509 & 0.8519 \\ 0.4454 & 0.6509 & -0.3976 \end{bmatrix} \tag{10.120}$$

Therefore, there will be three "observed" signals generated from $\boldsymbol{x}(k) = \boldsymbol{A}\boldsymbol{s}(k)$, for $k = 1, 2, \ldots, 100$, and they are shown in Figure 10.20. This represents a set of three instantaneous mixtures. The observations are first prewhitened

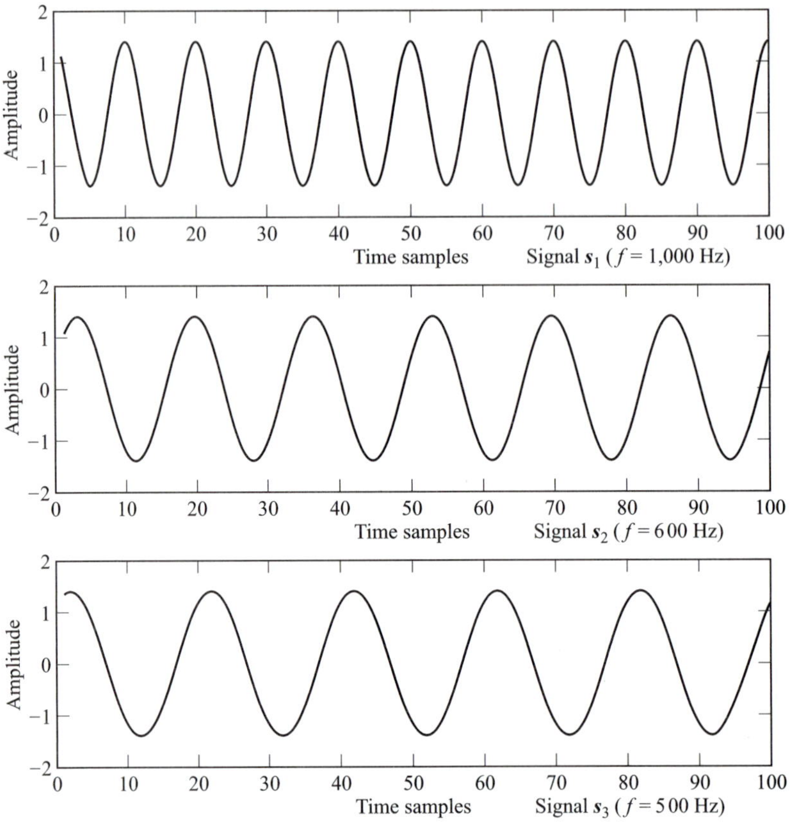

Figure 10.19 Three original sinusoidal source signals.

using the batch whitening process given in (10.103). In this example no compression is necessary; therefore, $h = q = 3$. The nonlinear PCA subspace learning rule in (10.111) is used to perform the separation. The forgetting factor used in (10.108) is $\gamma = 0.9$, $\beta = 1$ [for the nonlinearity in (10.112)], and 100 training epochs were required for convergence. A set of random initial weights was selected from a Gaussian distribution with zero mean and unity variance; then the columns of the weight matrix were orthonormalized. The separated signals are shown in Figure 10.21. Because we know the correct answer, the correlation coefficient can be computed for each of the separated signals with respect to the known (actual) source signals. These correlation coefficients are shown in Figure 10.21. Note that the correlations of the separated signals with respect to the actual source signals are almost perfect. The negative correlation coefficient indicates that a 180° phase shift has occurred in the output of the ICA separation process. Note that the ordering of the output signals is different from that of the original signals shown in Figure 10.19.

For this problem we also want to estimate the ICA basis vectors [i.e., the column vectors of the mixing matrix $\boldsymbol{A}$ given in (10.120)]. Using the batch approach given in (10.116), we see the estimate of $\boldsymbol{A}$ is

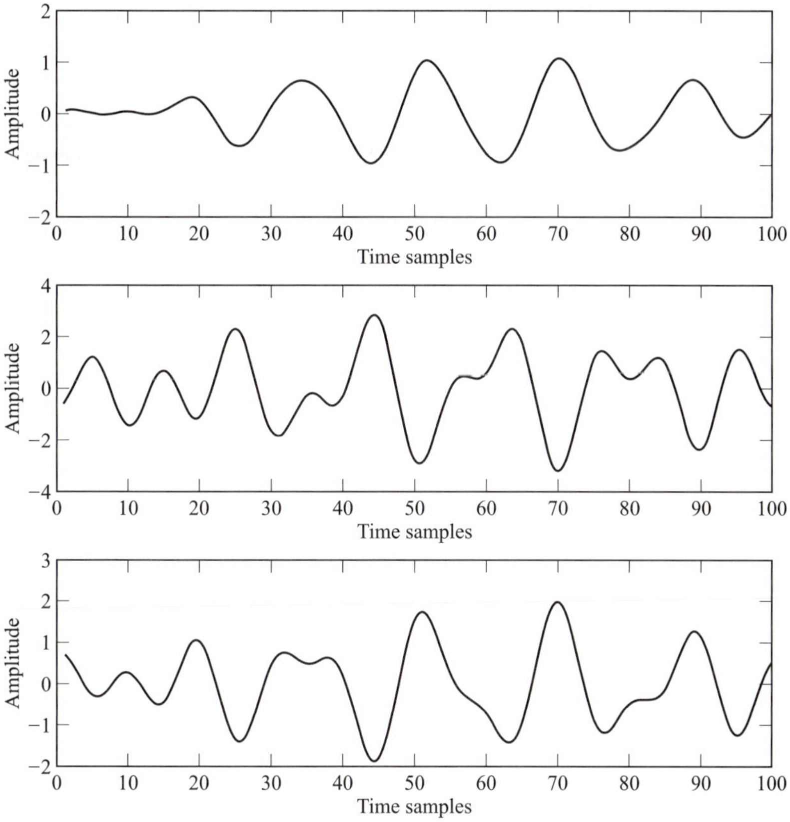

Figure 10.20 "Observed" mixed signals.

$$\hat{A} = \begin{bmatrix} 0.1101 & 0.3478 & 0.3807 \\ -0.9372 & -0.8376 & -0.6109 \\ 0.4749 & 0.3985 & 0.6323 \end{bmatrix} \tag{10.121}$$

Comparing these results with the actual mixing matrix in (10.120), we see that the estimates of the $\boldsymbol{A}$ column vectors are not exact; however, they are relatively close. Note that the ordering of the column vectors of $\hat{\boldsymbol{A}}$ is different from that of the original mixing matrix. The neural learning approach is used next to estimate the basis vectors of the ICA. A set of random weights is selected for the initial weight matrix $\boldsymbol{Q}$, and $\gamma = 0.9$ for the forgetting factor in the adaptive learning rate parameter in (10.108). After only five training epochs, the network converged, and the resulting estimate of the mixing matrix is given by

$$\hat{A} = \begin{bmatrix} 0.1095 & 0.3461 & 0.3788 \\ -0.9323 & -0.8333 & -0.6079 \\ 0.4724 & 0.3964 & 0.6292 \end{bmatrix} \tag{10.122}$$

Again, the estimate of the mixing in (10.122) is not exact; however, it is relatively close to the actual mixing matrix shown in (10.120).

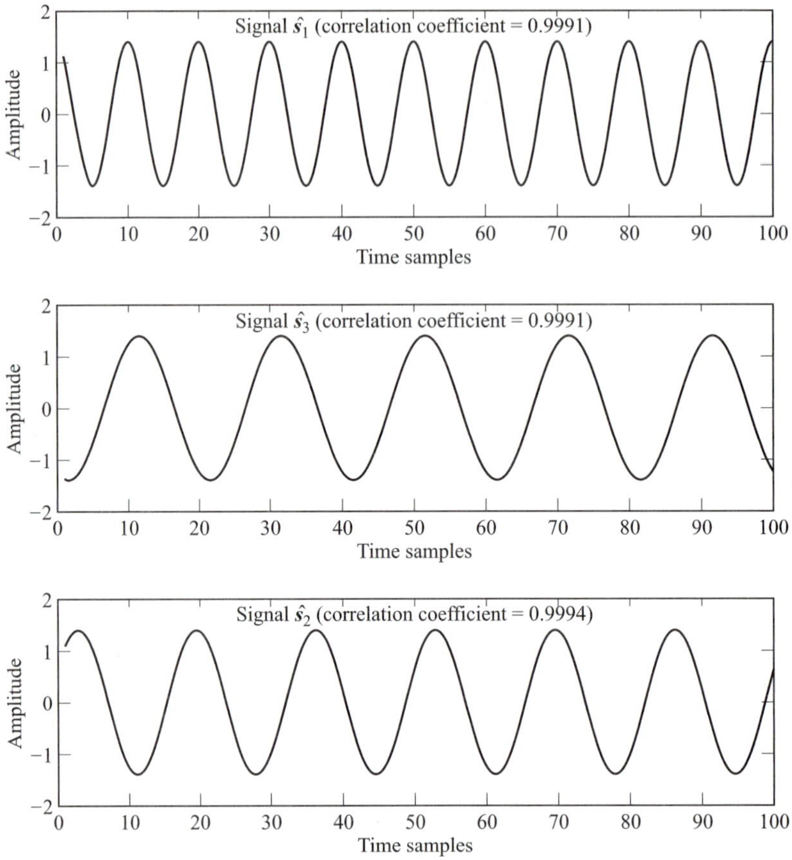

Figure 10.21 Separated sinusoidal source signals using the nonlinear PCA subspace learning rule.

EXAMPLE 10.8. This example involves separating three different instantaneously mixed infrasonic signals [38–41], see Figure 10.22(a). These infrasonic signals were recorded as three separate events and are shown in Figure 10.22(b). The three events are (1) an infrasonic signal from a volcanic eruption [40] at Galunggung, Java; (2) a mountain-associated wave [41] originating from New Zealand; and (3) an internal atmospheric gravity infrasound wave [38]. Gravity infrasound waves are due to temperature inversions occurring in the atmosphere. All these events were recorded at Windless Bight, Antarctica, between 1981 and 1983 using a large four-sensor infrasonic array (F-array). Figure 10.23 shows the geometry of the F-array used to collect the data. The infrasonic signals were sampled with a nominal sampling frequency of 1 Hz. Infrasonic waves are subaudible acoustic waves [38] typically in the frequency range $0.01 < f < 10\,\text{Hz}$, and they can result from many natural and artificial phenomena; see Figure 10.22(a). Infrasound sensors typically have a bandwidth of 0.01 to 10 Hz. Infrasonic waves can result from volcanic eruptions, mountain-associated waves, auroral waves, earthquakes, meteors, avalanches, severe weather, quarry blasting, high-speed aircraft, gravity waves, microbaroms, and nuclear explosions [42]. It has been discovered that elephants (and possibly other animals) communicate using infrasound [43]. To

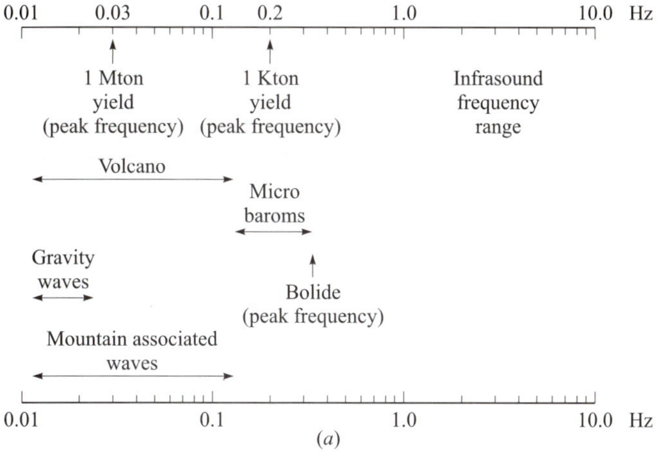

(*a*)

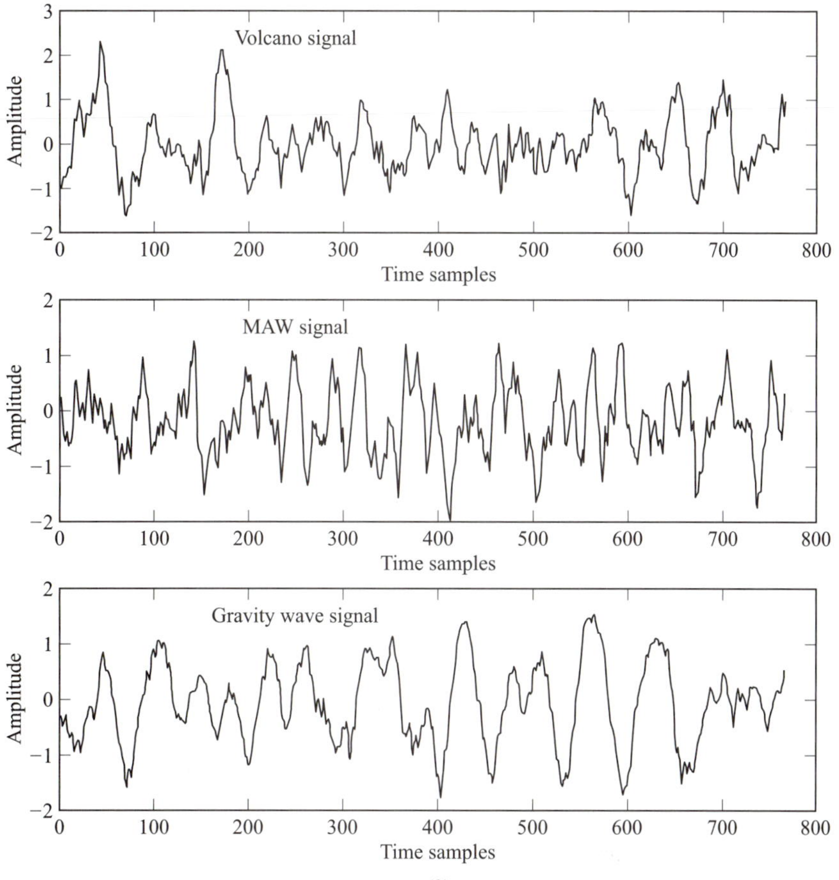

(*b*)

Figure 10.22 (*a*) Infrasound frequency range; (*b*) three original infrasonic source signals.

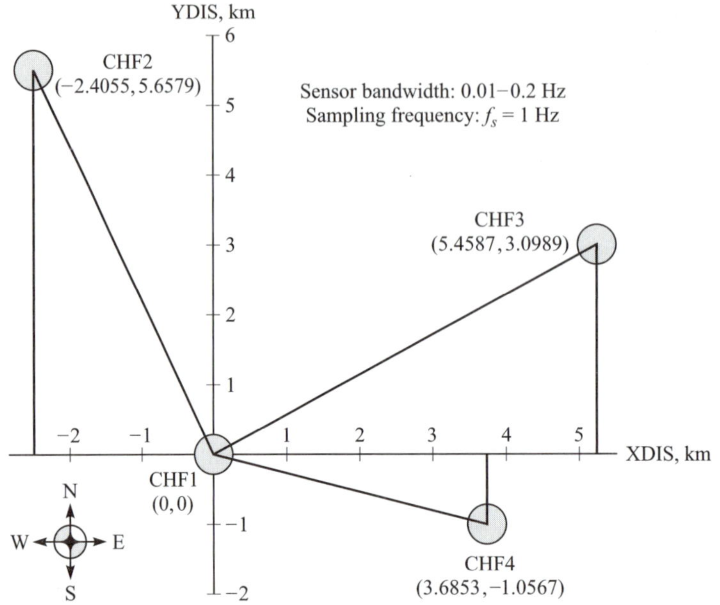

Figure 10.23 Windless Bight, Antarctica, infrasound sensor F-array configuration.

ensure compliance with the Comprehensive Nuclear Test Ban Treaty's (CTBT) ban on nuclear explosions, an International Monitoring System (IMS) will have a global infrasonic, seismic, hydroacoustic, and radionuclide data collection system [42]. For the infrasonic network, experts now recommend a frequency range of 0.02 to 5.0 Hz for the sensors.

The three signals are *artificially* mixed using a random mixing matrix given by

$$\boldsymbol{A} = \begin{bmatrix} 0.3050 & 0.9708 & 0.4983 \\ 0.8744 & 0.9901 & 0.2140 \\ 0.0150 & 0.7889 & 0.6435 \\ 0.7680 & 0.4387 & 0.3200 \end{bmatrix} \tag{10.123}$$

The numbers in (10.123) are uniformly distributed random numbers chosen from a uniform distribution on the interval [0, 1]. Therefore, four observed mixed signals are generated from $\boldsymbol{x}(k) = \boldsymbol{A}\boldsymbol{s}(k)$, for $k = 1, 2, \ldots, 768$, and are shown in Figure 10.24. These signals have had their mean values removed. The eigenvalues of the covariance matrix of the observed data, given by (10.106), are $\lambda_1 = 2.1346$, $\lambda_2 = 0.1976$, $\lambda_3 = 0.0434$, and $\lambda_4 = -3.7772 \times 10^{-16}$. The fourth eigenvalue is considerably smaller than the first three. Therefore, only the first three largest ones need to be retained, and from (10.104) the whitening matrix $\boldsymbol{V} \in \Re^{3\times 4}$ provides both whitening of the observed data and compression. So $h = 4$ and $q = 3$ (the number of source signals to be recovered). As with the previous example, the nonlinear PCA subspace learning rule in (10.111) is used to perform the separation. A set of random initial weights was selected from a Gaussian distribution with zero mean and unity variance; then the columns of the weight matrix were orthonormalized. The forgetting factor used in (10.108) is $\gamma = 0.9$, $\beta = 1$ [for the nonlinearity in (10.111)], and 250 training epochs were required for

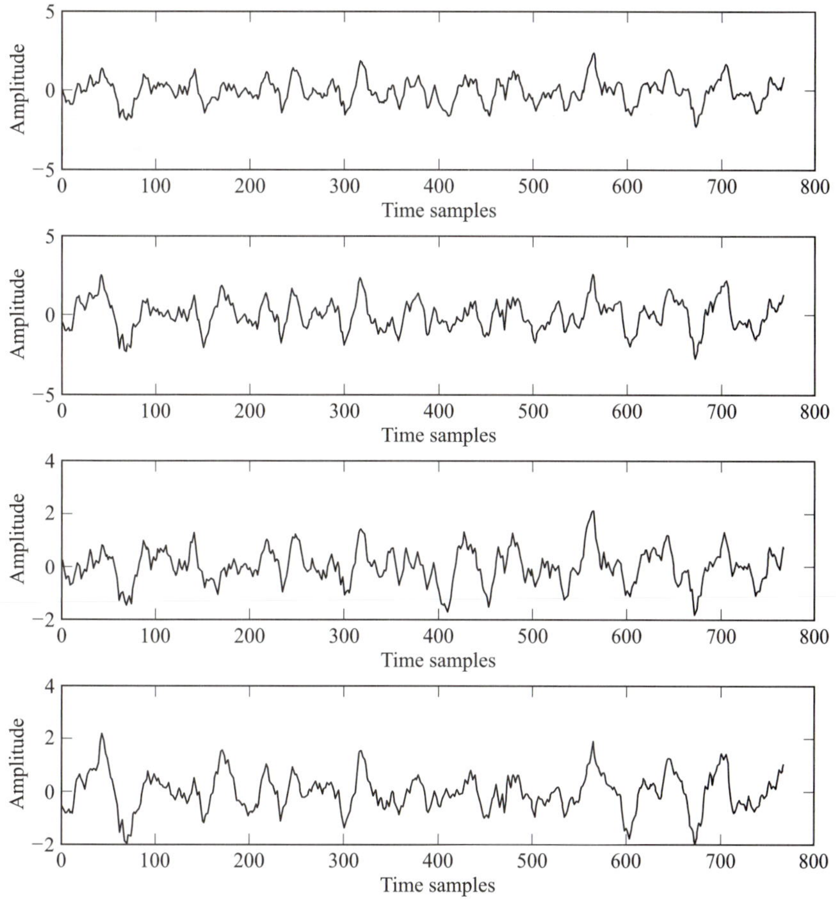

Figure 10.24 "Observed" mixed infrasound signals.

convergence. The separated signals are shown in Figure 10.25. Because we know what the source signals are, the correlation coefficient can be computed for each of the separated signals with respect to the known (actual) source signals. These correlation coefficients are shown in Figure 10.25. The correlations of the separated signals with respect to the actual source signals are almost perfect. The negative correlation coefficient indicates that a sign change has occurred in the output of the ICA separation process. Unlike in the previous problem, the output ordering of the separated signals is the same as that of the original source signals shown in Figure 10.22(*b*). ICA has also been applied to signal separation of the four-channel array signals for a single event, that is, a volcanic eruption [44]. The infrasonic signals used in the ICA process are convolutive mixtures. The results were very revealing, exposing a "hidden" microbarom signal in the dominant volcano infrasound signature.

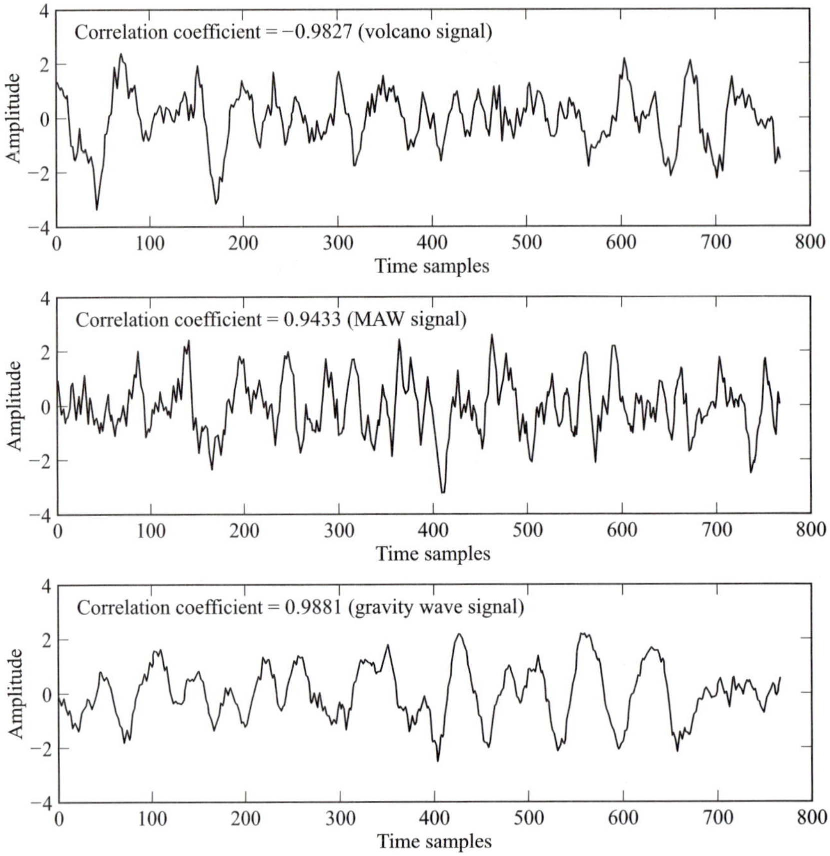

Figure 10.25 Separated infrasonic source signals using the nonlinear PCA subspace learning rule.

10.8.3 Fast Fixed-Point Algorithm for ICA

At least one drawback of using the nonlinear PCA subspace learning rule given in (10.111) for ICA to perform blind separation of source signals is its relatively slow convergence. The fast fixed-point algorithm (FFPA) for ICA [45, 32] converges rapidly to the most accurate solution allowed by the data structure. The FFPA does not depend on any user-specified parameters, and it can compute all non-Gaussian independent components regardless of their probability distributions (actually one signal can be Gaussian). The convergence speed is cubic, and compared to gradient-based algorithms, the FFPA can be 10 to 100 times faster [45].

We will view the problem in a slightly different manner from the presentation in Section 10.8.2. The basic linear relationship for the ICA problem is taken to be

$$\boldsymbol{x} = \boldsymbol{A}\boldsymbol{s} \tag{10.124}$$

which is the same relationship given in (10.101), except that the noise term is not considered and the time dependence is dropped. It is assumed that $\boldsymbol{s} \in \Re^{q \times 1}$ (the vector of q independent components) is zero-mean and unit-variance and has elements that are mutually statistically independent; the observed vector $\boldsymbol{x} \in \Re^{h \times 1}$ has h measured variables where $h \geq q$; and $\boldsymbol{A} \in \Re^{h \times q}$ (the mixing matrix) is assumed to be full rank. As explained in Section 10.8.2, prewhitening (or sphering) the measured data in $\boldsymbol{x}$ can often improve the stability and convergence properties of certain ICA algorithms. A transformation $\boldsymbol{V}$ can be found using standard PCA methods, such that the observed data are linearly transformed to a vector

$$\boldsymbol{v} = \boldsymbol{V}\boldsymbol{x} \tag{10.125}$$

The elements of vector $\boldsymbol{v}$ are mutually uncorrelated and have unit variance. Therefore, the correlation matrix (or covariance matrix since $\boldsymbol{v}$ is zero-mean) is the unit or identity matrix, that is, $\mathrm{E}[\boldsymbol{v}\boldsymbol{v}^T] = \boldsymbol{I}$. As discussed in Section 10.8.2, the dimension of vector $\boldsymbol{v}$ is also reduced to q (the number of independent components) in this process (i.e., data compression). Therefore, the PCA prewhitening process actually serves two purposes: sphering the data and determining the number of independent components.[†] Substituting (10.124) into (10.125), we have

$$\boldsymbol{v} = \boldsymbol{V}\boldsymbol{A}\boldsymbol{s} = \boldsymbol{B}\boldsymbol{s} \tag{10.126}$$

where $\boldsymbol{B} = \boldsymbol{V}\boldsymbol{A}$ is an orthogonal matrix. This can be shown by applying the previous assumptions made, and from (10.126) writing

$$\underbrace{\mathrm{E}[\boldsymbol{v}\boldsymbol{v}^T]}_{\boldsymbol{I}} = \mathrm{E}[\boldsymbol{B}\boldsymbol{s}\boldsymbol{s}^T\boldsymbol{B}^T] = \boldsymbol{B}\underbrace{\mathrm{E}[\boldsymbol{s}\boldsymbol{s}^T]}_{\boldsymbol{I}}\boldsymbol{B}^T = \boldsymbol{B}\boldsymbol{B}^T = \boldsymbol{I} \tag{10.127}$$

Therefore, the problem is reduced to determining an orthogonal matrix $\boldsymbol{B} \in \Re^{q \times q}$ that can be used to perform the separation of the independent signals from (10.126), that is,

$$\hat{\boldsymbol{s}} = \boldsymbol{B}^T\boldsymbol{v} \tag{10.128}$$

Therefore, the pseudoinverse of the mixing matrix $\boldsymbol{A}$ is given by $\boldsymbol{A}^+ = \boldsymbol{B}^T\boldsymbol{V}$.

The FFPA for performing ICA is based on a highly efficient fixed-point iteration scheme for finding the local extrema of the kurtosis of a linear combination of the observed variables. In general, the kurtosis (or fourth-order cumulant) [27, 28] of a zero-mean random variable x is given by

$$\mathrm{kurt}(x) = \mathrm{E}[x^4] - 3(\mathrm{E}[x^2])^2 \tag{10.129}$$

For two independent random variables x_1 and x_2, it is true that $\mathrm{kurt}(x_1 + x_2) = \mathrm{kurt}(x_1) + \mathrm{kurt}(x_2)$. In addition, for the zero-mean random variable x and the scalar α, it is true that $\mathrm{kurt}(\alpha x) = \alpha^4 \mathrm{kurt}(x)$. A linear combination of the prewhitened observed variables can be written as $\boldsymbol{w}^T\boldsymbol{v}$,

[†]A standard PCA approach is not always the best method for computing the *number* of independent components. There are other methods that are better suited in certain cases, for example, an extension of maximum-likelihood estimation known as Akaike's information theoretic criterion (AIC), and from coding theory, minimum description length (MDL) [46].

and this linear combination can be searched such that it possesses maximal or minimal kurtosis, where the weight vector $\boldsymbol{w}$ is bounded, that is, $\|\boldsymbol{w}\|_2 = 1$. The deflation (FFPA) is based on this idea, and each $\boldsymbol{w}_i$ $(i = 1, 2, \ldots, q)$ vector found by this algorithm is a column vector of the orthogonal matrix $\boldsymbol{B}$. From [45], the algorithm for estimating *one* independent component is given by the following:

Fast fixed-point algorithm for ICA (one component)

Step 1. Prewhiten the observed data $\boldsymbol{x}$ to obtain vector $\boldsymbol{v}$.

Step 2. Randomly set the values of the initial weight vector $\boldsymbol{w}(0)$ (note the subscript is dropped here because we are only initially concerned with finding one independent (component), and normalize to unit length, that is,

$$\boldsymbol{w}(0) \leftarrow \frac{\boldsymbol{w}(0)}{\|\boldsymbol{w}(0)\|_2}$$

and set $j = 1$.

Step 3. Let

$$\boldsymbol{w}(j) = \mathrm{E}[\boldsymbol{v}(\boldsymbol{w}^T(j-1)\boldsymbol{v})^3] - 3\boldsymbol{w}(j-1)$$

The expectation operator can be estimated using a relatively large number of $\boldsymbol{v}$ vectors.

Step 4. Normalize $\boldsymbol{w}(j)$ to unit length:

$$\boldsymbol{w}(j) \leftarrow \frac{\boldsymbol{w}(j)}{\|\boldsymbol{w}(j)\|_2}$$

Step 5.† If $|\boldsymbol{w}^T(j)\boldsymbol{w}(j-1)|$ is not close to 1, then let $j \leftarrow j+1$ and go back to step 3. Otherwise, output vector $\boldsymbol{w}(j)$.

Step 6. Using $\boldsymbol{w}(j)$, one of the separated source signals is given by

$$s(k) = \boldsymbol{w}^T(j)\boldsymbol{v}(k) \qquad k = 1, 2, \ldots$$

To estimate q independent components, we run the algorithm given above q times. However, to ensure that different independent components are estimated each time, an orthogonalizing projection is included inside the recursive loop given above. The basic idea is that estimation of the independent components can be carried out one by one if the "current" solution found $\boldsymbol{w}(j)$ is projected on the space that is orthogonal to the columns of matrix $\boldsymbol{B}$ previously found. So we will define $\tilde{\boldsymbol{B}}$ as a matrix whose column vectors are the previously computed column vectors of $\boldsymbol{B}$. The projection operation is added to the beginning of step 4 above; that is, step 4 now becomes

Step 4. Let $\boldsymbol{w}(j) \leftarrow \boldsymbol{w}(j) - \tilde{\boldsymbol{B}}\tilde{\boldsymbol{B}}^T\boldsymbol{w}(j)$, then

$$\boldsymbol{w}(j) \leftarrow \frac{\boldsymbol{w}(j)}{\|\boldsymbol{w}(j)\|_2}$$

The initial random vector should also be projected in this manner before starting the iterations.

---

†An alternate stopping criterion is based on checking

$$\min\{\|\boldsymbol{w}(j) - \boldsymbol{w}(j-1)\|_2, \|\boldsymbol{w}(j) + \boldsymbol{w}(j-1)\|_2\} < \varepsilon$$

where $\varepsilon = 10^{-4}$ is a reasonable value.

EXAMPLE 10.9. In this example we use the deflation fast fixed-point algorithm for ICA given above and apply it to separating mixed digital images. The original images are shown in Figure 10.26. Each digital image has 243×351 (= 85,293) pixels, and each 6-bit pixel has a 64 gray-level range. The computed kurtosis for each image was negative with the exception of image (*a*) in Figure 10.26; it had a positive kurtosis. Figure 10.26(*a*) is artificially generated mosaic tile, Figure 10.26(*d*) is uniformly distributed noise, Figure 10.26(*e*) is a binary periodic checker pattern, and the remaining figures are natural scenes. These six images were artificially mixed using a nonorthogonal, full-rank, 6×6, mixing matrix $\boldsymbol{A}$. This matrix was generated in MATLAB from uniformly distributed random numbers in the range $[0, 1]$. Specifically, if we define images (*a*) through (*f*) in Figure 10.26 to be the arrays $\boldsymbol{P}_1, \boldsymbol{P}_2, \ldots, \boldsymbol{P}_6$, respectively, we can generate a set of "observed signals," carrying out the following procedure using the ICA expansion in (10.124):

Generate each source signal:
$$\begin{aligned} \boldsymbol{s}_1 &= \text{vec}(\boldsymbol{P}_1) & \boldsymbol{s}_2 &= \text{vec}(\boldsymbol{P}_2) \\ \boldsymbol{s}_3 &= \text{vec}(\boldsymbol{P}_3) & \boldsymbol{s}_4 &= \text{vec}(\boldsymbol{P}_4) \\ \boldsymbol{s}_5 &= \text{vec}(\boldsymbol{P}_5) & \boldsymbol{s}_6 &= \text{vec}(\boldsymbol{P}_6) \end{aligned} \tag{10.130}$$

Form source signal matrix:
$$\boldsymbol{S} = [\boldsymbol{s}_1, \boldsymbol{s}_2, \boldsymbol{s}_3, \boldsymbol{s}_4, \boldsymbol{s}_5, \boldsymbol{s}_6] \tag{10.131}$$

where $\boldsymbol{S} \in \Re^{85,293 \times 6}$.

Form matrix of observed signals:
$$\boldsymbol{X} = \boldsymbol{A}\boldsymbol{S}^T \tag{10.132}$$

where $\boldsymbol{X} \in \Re^{6 \times 85,293}$. Therefore, matrix $\boldsymbol{X}$ has *rows* that are each mixed image in vector form. Each mixed image is now extracted from the rows of $\boldsymbol{X}$ by effectively undoing the vec operation on the respective rows, assuming the appropriate column length which is 243 (pixels) (this is the number of rows of the original images in Figure 10.26). Figure 10.27 shows all six mixed images. The mean was next removed from each image, and the results prewhitened. The prewhitened images are shown in Figure 10.28.

A MATLAB m-file function was written to implement the deflation FFPA discussed above with the stopping parameter set at $\varepsilon = 10^{-4}$ (the alternate stopping criterion was used). Using the six prewhitened inputs $\boldsymbol{v}_1, \boldsymbol{v}_2, \ldots, \boldsymbol{v}_6$, convergence was achieved relatively fast to extract the six independent-component (IC) images. For each of the six extracted independent components, Figure 10.29 shows the convergence profile. Figure 10.30 shows the extracted independent-component images. There are several issues to discuss regarding these results. First, because amplitude information of the source signals cannot be preserved, each extracted independent-component image shown in Figure 10.30 had the output gray-level range rescaled to a 64 gray-level range. Second, note that the output images are ordered differently from the original source images shown in Figure 10.26. Third, note that in Figure 10.30(*a*), (*d*), and (*f*), these are "negatives" compared to the original images in Figure 10.26(*e*), (*b*), and (*c*), respectively. This again is one of the ambiguities of ICA (as explained in Sect. 10.8.1). Figure 10.31 shows these three images that have been "inverted" to eliminate the sign ambiguity. Fourth, in Figure 10.30(*c*) and (*b*), these are not exact replicas of the original source signals given in Figure 10.26(*f*) and (*b*), respectively. This can be attributed, at least in part, to the fact that the mutual independence of original source images in Figure 10.26 was not tested. In conclusion, after inverting the three negatives in Figure 10.30, the final independent component images show very good separation results using the deflation FFPA. Figure 10.32 shows the MATLAB function **`ffpica.m`** used to perform the deflation FFPA. The non-

Figure 10.26 Six original source digital images used in Example 10.9. [For (b), (c), and (f),]

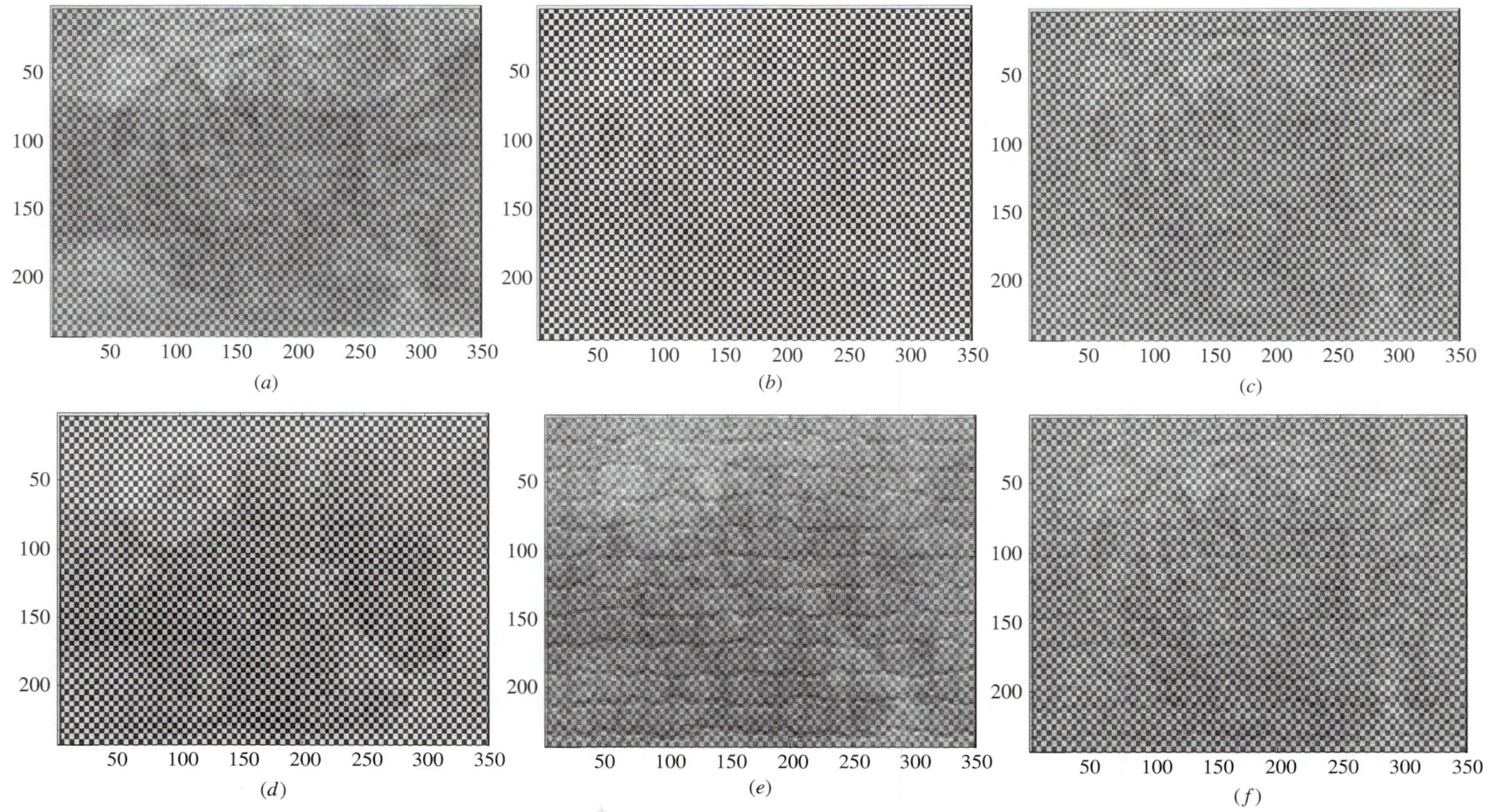

Figure 10.27 Mixed images, where the transpose of the vec of each image constitutes one observed signal, that is, $\boldsymbol{x}_i$. These images are linear mixtures of the source images shown in Figure 10.26.

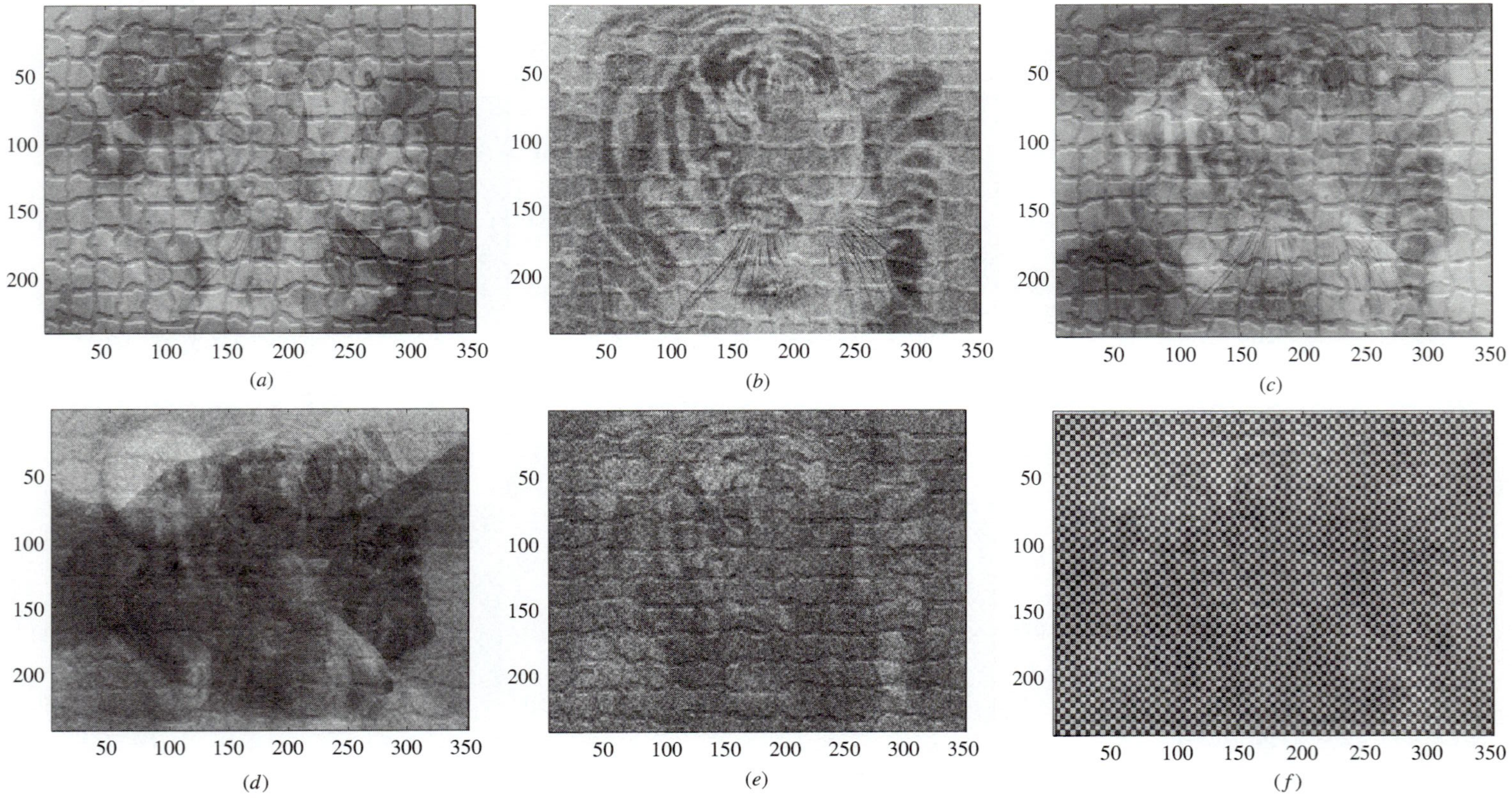

Figure 10.28 PCA prewhitened images.

CHAPTER 10:
Identification,
Control, and
Estimation Using
Neural Networks

IC No.	Symbol	No. of iterations (j)
(1)	○	10
(2)	×	6
(3)	+	6
(4)	*	5
(5)	□	7
(6)	▽	1

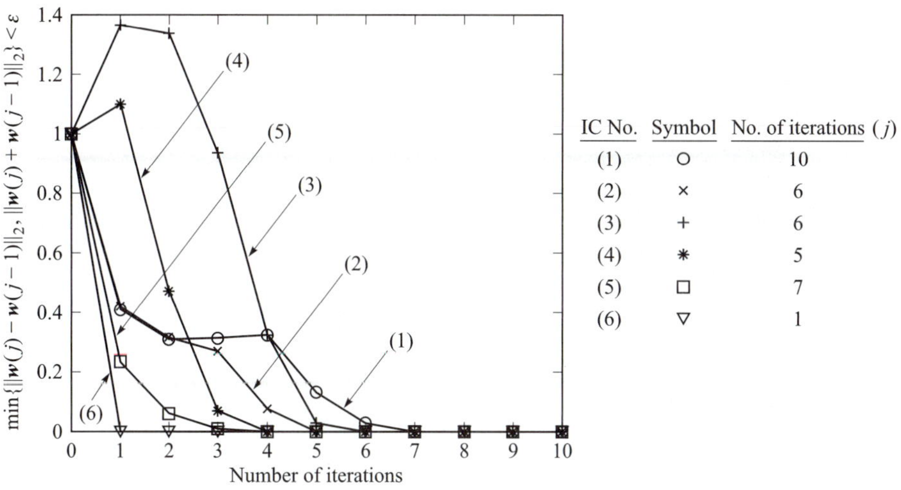

Figure 10.29 Convergence profiles for each independent component extracted using the deflation FFPA.

linearity that is necessary to carry out the separation in the FFPA is a cubic nonlinearity. This can be seen in Figure 10.32 (the fifth line from the bottom of the code). In the program, the maximum number of iterations for any one independent component is set to **max_iteration=1000**. Also note in Figure 10.32, this function performs the necessary prewhitening of the data. No "compression" of the data is necessary since $q = h$; that is, the number of independent components is the same as the number of observed signals. The deflation FFPA program shown in Figure 10.32 would have to be modified if it were necessary to estimate the number of independent components.

10.9
SPECTRUM ESTIMATION OF SINUSOIDS IN ADDITIVE NOISE

Power spectrum estimation of a given stochastic signal is one of the fundamental signal processing problems. If the duration of the signal record is relatively long, it is well known that conventional Fourier analysis can provide an accurate picture of the signal's spectral content. In other words, a signal's frequency representation can be easily obtained from its discrete Fourier transform as long as the record of the signal is sufficiently long. However, in many practical applications there is a need to estimate the spectrum of signals that have a short time duration. In such circumstances, Fourier analysis becomes inaccurate and we are forced to search for alternatives. The problem of estimating the spectrum of short-time duration signals occurs naturally if the signals are nonstationary, or if there is a need for fast spectrum estimates and accumulation of long signal sequences is prohibitive.

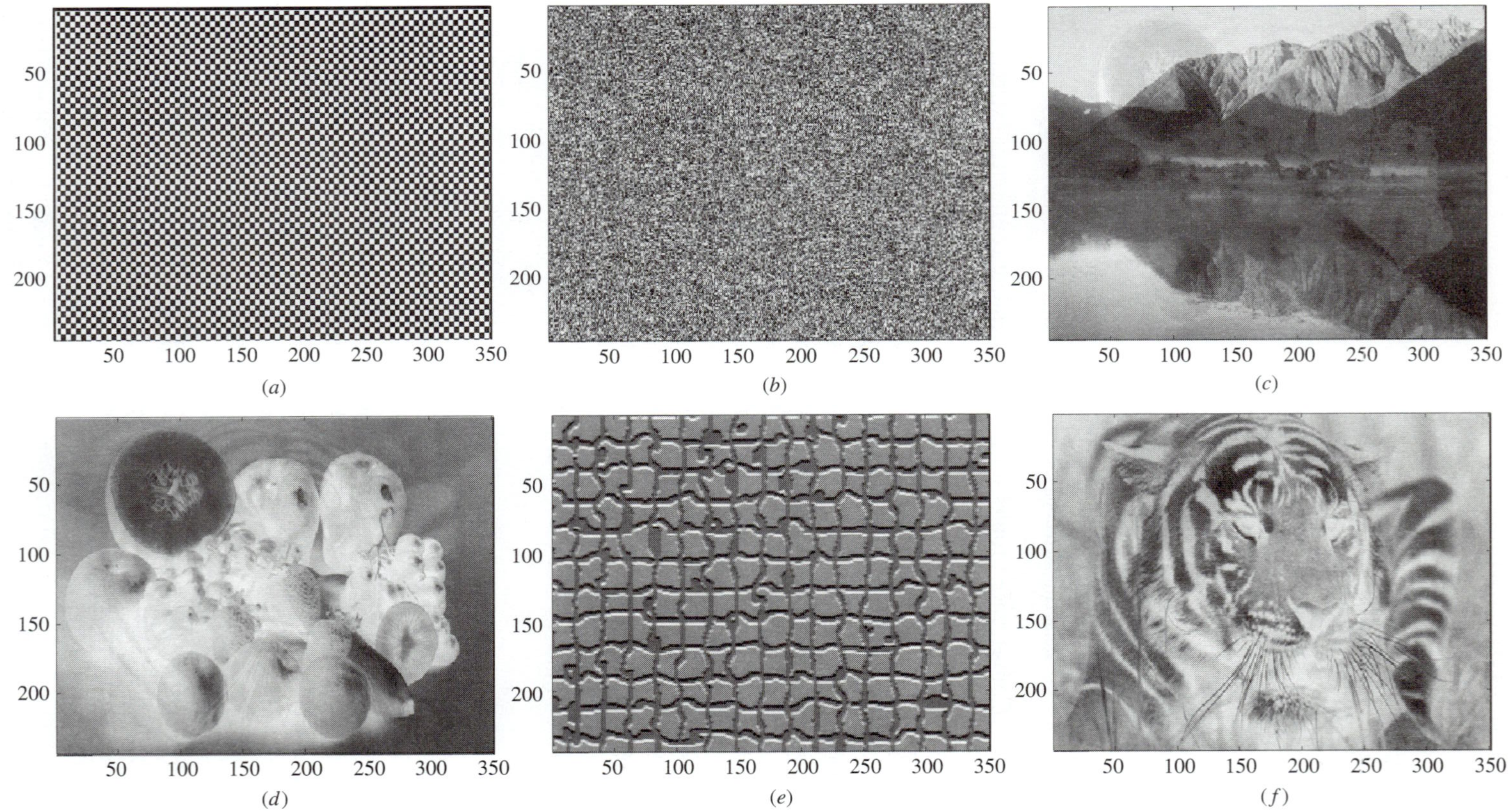

Figure 10.30 Extracted independent component images. These images were obtained using the deflation FFPA. [For (*c*), (*d*), and (*f*),]

Figure 10.31 Separated images shown in Figure 10.30(*a*), (*d*), and (*f*) that have been inverted. This was necessary because of the ICA sign ambiguity.

```
function [output] = ffpica(mixed_signal, max_iteration, epsilon)

% Fast Fixed-Point Independent Component Analysis
% (Deflation Method)
X = mixed_signal;
[num_IC, num_sample] = size(X);

% Remove the mean of row vectors
meanX = mean (X')';
X = X - meanX * ones (1,size (X, 2));

% PCA prewhitening of the data
covX = cov(X', 1); % covariance matrix of X
[E, D] = eig(covX);
whitening_matrix = inv (sqrt (D)) * E';
whitened_X = whitening matrix * X; % whitened signals

% Calculate the ICA using the fixed point algorithm
B = zeros (num_IC);
for i = 1:num_IC
   w = rand(num_IC, 1) - 0.5; % Initialize the weight vector
   w = w - B * B' * w;
   w = w / norm(w);
   w_old = zeros(size(w));
   for j = 1 : max_iteration
      w = w - B * B' * w;
      w = w / norm(w);
      if norm(w - w_old) < epsilon | norm(w + w_old) < epsilon

         B(:, i) = w;
         W(i,:) = w' * whitening_matrix;
         break;
       end
       w_old = w;
       u = whitened_X' * w;
       w = (whitened_X * (u .^ 3)) / num_sample - 3 * w;
       w = w / norm(w);
    end
end
output = W * mixed_signal;
```

Figure 10.32 MATLAB function **ffpica.m** to perform the deflation FFPA.

In this section we examine the application of the partial least-squares regression (cf. Sect. 9.5) or PLSNET (cf. Sect. 9.6) algorithm in a specific spectrum estimation problem, estimation of sinusoids in additive noise. Although this problem is of somewhat limited scope, it is frequently analyzed due to its practical importance. It can be shown that there is an equivalence between the frequency estimation of the sinusoids in additive noise and the problem of estimating the direction of arrival (DOA) of planar electromagnetic waves [47, 48]. The DOA problem is a fundamental problem of radar signal processing, and as such, it has been studied extensively [49, 50]. Although very interesting, a thorough analysis of the two problems is outside the scope of this book and in this section we limit ourselves to the presentation of one of the methods for spectrum estimation that can be used to provide a viable alternative to classical Fourier analysis.

10.9.1 Problem Statement

Consider a signal that consists of K sinusoidal components corrupted by additive noise. Let us assume that the signal is sampled with the sampling frequency f_s and that a total of N samples are available for processing. The ith sample of the signal can be represented as

$$u(i) = \sum_{k=1}^{K} a_k \cos\left(2\pi \frac{f_k}{f_s} i + \theta_k\right) + \vartheta(i) \tag{10.133}$$

where a_k, f_k, and θ_k are unknown amplitudes, frequencies, and phases of sinusoidal components, respectively, and $\vartheta(i)$ is the component attributed to the presence of additive noise. The task of spectrum estimation is to determine the unknown parameters of the sinusoids from the samples of the composite signal given in (10.133). If the number of samples is large, individual sinusoidal components become mutually orthogonal. In such circumstances an accurate estimate of the power spectrum can be obtained from the Fourier transform of the signal

$$S(\omega) = |U(\omega)|^2 = \left|\sum_{i=1}^{N} u(i)e^{-j\omega i}\right|^2 \tag{10.134}$$

where $S(\omega)$ is the estimate of the power spectrum and ω is the normalized frequency. However, if the number of samples is small, the power spectrum estimate provided in (10.134) does not provide sufficient resolution, and we need to search for alternative methods. Also, note that (10.134) represents a general approach to the problem, and it does not take advantage of the special form of the signal given in (10.133).

Although it may not be apparent from (10.133), the estimation of the unknown sinusoidal parameters can be reduced to the problem of estimation of their frequencies. Once the frequencies are known, the amplitudes and phases can be easily determined. To see this, let us rewrite (10.133) as

$$\begin{aligned} u(i) &= \sum_{k=1}^{K} a_k \frac{e^{j[2\pi(f_k/f_s)i+\theta_k]} + e^{-j[2\pi(f_k/f_s)i+\theta_k]}}{2} + \vartheta(i) \\ &= \sum_{k=1}^{K} \left(\frac{1}{2}a_k e^{j\theta_k}\right)e^{j2\pi(f_k/f_s)i} + \left(\frac{1}{2}a_k e^{-j\theta_k}\right)e^{-j2\pi(f_k/f_s)i} + \vartheta(i) \end{aligned} \tag{10.135}$$

or

$$u(i) = \sum_{k=1}^{K} \left(A_k e^{j2\pi(f_k/f_s)i} + A_k^* e^{-j2\pi(f_k/f_s)i}\right) + \vartheta(i) \tag{10.136}$$

If the frequencies of the sinusoidal signals are known, (10.136) can be written for each available sample. In vector-matrix form this becomes

$$\begin{bmatrix} u(1) \\ u(2) \\ \vdots \\ u(N) \end{bmatrix} = \begin{bmatrix} e^{j2\pi(f_1/f_s)} & \cdots & e^{j2\pi(f_K/f_s)} & e^{-j2\pi(f_1/f_s)} & \cdots & e^{-j2\pi(f_K/f_s)} \\ e^{j2\pi(f_1/f_s)2} & \cdots & e^{j2\pi(f_K/f_s)2} & e^{-j2\pi(f_1/f_s)2} & \cdots & e^{-j2\pi(f_K/f_s)2} \\ \vdots & & & & & \vdots \\ e^{j2\pi(f_1/f_s)N} & \cdots & e^{j2\pi(f_K/f_s)N} & e^{-j2\pi(f_1 f_s)N} & \cdots & e^{-j2\pi(f_K f_s)N} \end{bmatrix} \begin{bmatrix} A_1 \\ \vdots \\ A_K \\ A_1^* \\ \vdots \\ A_K^* \end{bmatrix} + \begin{bmatrix} \vartheta(1) \\ \vartheta(2) \\ \vdots \\ \vartheta(N) \end{bmatrix} \tag{10.137}$$

or in a more compact form

$$U = \boldsymbol{\Phi a} + \boldsymbol{\upsilon} \tag{10.138}$$

where $U \in \Re^{N\times 1}$, $\boldsymbol{\Phi} \in C^{N\times 2K}$, $\boldsymbol{a} \in C^{2K\times 1}$, and $\boldsymbol{\upsilon} \in \Re^{N\times 1}$. Assuming that $N \geq 2K$, the set of coefficients $\{A_k\}$ can be obtained as a solution of the system of linear equations in (10.138). For example, the least-squares solution of the system in (10.138) is given as (cf. Sect. 8.3)

$$\boldsymbol{a} = \left(\boldsymbol{\Phi}^H \boldsymbol{\Phi}\right)^{-1} \boldsymbol{\Phi}^H U \tag{10.139}$$

The solution in (10.139) assumes that matrix $\boldsymbol{\Phi}$ has full rank, that is,

$$\text{rank}(\boldsymbol{\Phi}) = 2K \tag{10.140}$$

To see that this is always the case, let us define

$$r_k = e^{j2\pi(f_k/f_s)} \tag{10.141}$$

and

$$r_{-k} = e^{-j2\pi(f_k/f_s)} \tag{10.142}$$

Now matrix $\boldsymbol{\Phi}$ can be rewritten as

$$\boldsymbol{\Phi} = \begin{bmatrix} r_1 & \cdots & r_K & r_{-1} & \cdots & r_{-K} \\ r_1^2 & \cdots & r_K^2 & r_{-1}^2 & \cdots & r_{-K}^2 \\ \cdots & \cdots & \cdots & \cdots & \cdots & \cdots \\ r_1^N & \cdots & r_K^N & r_{-1}^N & \cdots & r_{-K}^N \end{bmatrix} \tag{10.143}$$

From (10.143) we see that $\boldsymbol{\Phi}$ takes the form of a Vandermonde matrix (cf. Sect. A.2.19), and since $r_i \neq r_j$ for every i and j, it has to be full-rank.

10.9.2 PLSR Solution to Frequency Estimation

Assuming that the sinusoidal amplitudes and phases can be estimated according to (10.139), we turn our attention back to the problem of estimating the frequencies. To derive the PLSR solution to the problem, we consider the filter structure presented in Figure 10.33. The filter is performing the function of one-step forward prediction of the time series $\{u(i)\}$ given in (10.133). The output of the filter is the difference between the value of the signal at the ith sample and its predicted value obtained from M previous samples, that is,

$$e(i) = u(i) - \hat{u}(i) = u(i) - \sum_{m=1}^{M} u(i-m)w_m \qquad (10.144)$$

Taking the z transform of (10.144), we have

$$E(z) = U(z)\left[1 - \sum_{m=1}^{M} w_m z^{-m}\right] = \frac{U(z)}{z^M}\left[z^M - \sum_{m=1}^{M} w_m z^{M-m}\right] \qquad (10.145)$$

or

$$E(z) = U(z)H(z) \qquad (10.146)$$

where $H(z)$ represents the overall transfer function of the prediction filter in Figure 10.33. For the moment assume that $\vartheta(i) = 0$, for $i = 1, 2, \ldots, N$, and that we are able to determine the coefficients of the filter so that the error between the predictions $\hat{u}(i)$ and signal values $u(i)$ is zero. In this case we have

$$U(z)H(z) \equiv 0 \qquad (10.147)$$

or

$$U\left(e^{j\omega}\right)H\left(e^{j\omega}\right) = 0, \qquad \forall\omega \qquad (10.148)$$

Since the input signal is the sum of sinusoids, its spectrum consists of discrete components, as shown in Figure 10.34.

In order for the output of the filter to be zero for all i, the zeros of the filter's transfer function have to be located exactly at the frequencies of the sinusoids. Therefore, the problem of frequency estimation reduces to the problem of finding the zeros of the one-step prediction filter transfer function. So far we have assumed that the system is noiseless. In reality, the positions of the

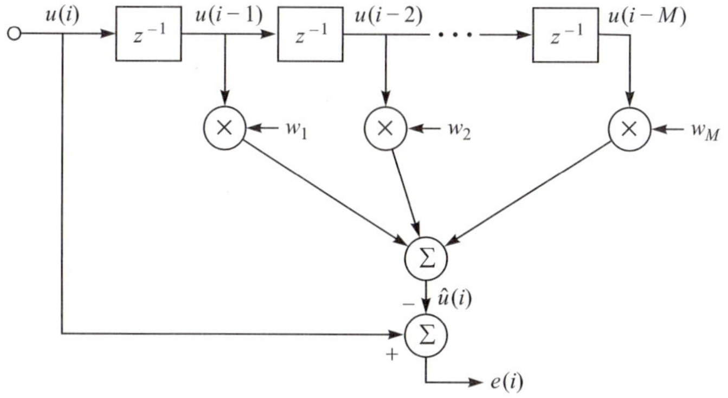

Figure 10.33 Transversal filter performing one-step prediction of a time series.

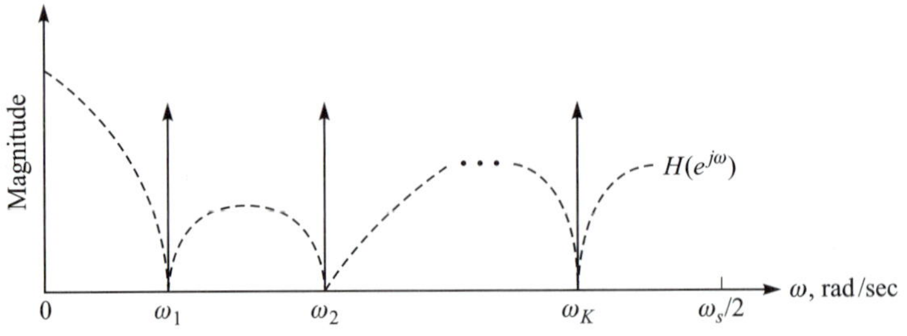

Figure 10.34 Transfer function of one-step prediction filter.

zeros will be slightly perturbed due to the presence of the additive noise. However, unless there are impulses in the noise power spectrum we expect the perturbations to be relatively small.

Equation (10.147) shows that to estimate the sinusoidal frequencies, we need to find the tap weights of the prediction filter. Rewriting (10.144) for $i = M+1, M+2, \ldots, N$, we have

$$u(M+1) = \sum_{m=1}^{M} u(M+1-m)w_m + e(M+1) \tag{10.149}$$

$$u(M+2) = \sum_{m=1}^{M} u(M+2-m)w_m + e(M+2) \tag{10.150}$$

$$\vdots$$

$$u(N) = \sum_{m=1}^{M} u(N-m)w_m + e(N) \tag{10.151}$$

or in vector-matrix form we have

$$\boldsymbol{c} = \boldsymbol{A}\boldsymbol{w} + \boldsymbol{e} \tag{10.152}$$

where:

$$\boldsymbol{c} = \left[u(M+1), u(M+2), \cdots, u(N)\right]^T \tag{10.153}$$

$$\boldsymbol{A} = \begin{bmatrix} u(M) & u(M-1) & \cdots & u(1) \\ u(M+1) & u(M) & \cdots & u(2) \\ \cdots & \cdots & \cdots & \cdots \\ u(N-1) & u(N-2) & \cdots & u(N-M) \end{bmatrix} \tag{10.154}$$

$$\boldsymbol{w} = \left[w_1, w_2, \ldots, w_M\right]^T \tag{10.155}$$

and

$$\boldsymbol{e} = \left[e(M+1), e(M+2), \ldots, e(N)\right]^T \tag{10.156}$$

Estimating the tap weights in (10.152) is the same problem that we dealt with in Chapter 9 (cf. Sects. 9.5 and 9.6). The solution of the tap weights using the PLSR method is given as

$$w = \hat{W}^T \left(\hat{B} \hat{W}^T \right)^{-1} \hat{v} \tag{10.157}$$

where $\hat{W}$, $\hat{B}$ and $\hat{v}$ are appropriate matrices of the PLSR model.†

Once the tap weights of the prediction filter are found, the frequencies of the sinusoids can be found as zeros of the filter's transfer function. However, as previously mentioned, due to the presence of additive noise the position of the zeros may be perturbed. For this reason it is customary to define a quasi-spectrum function as

$$S_e(\omega) = \frac{1}{\left|H\left(e^{j\omega}\right)\right|^2} = \frac{1}{\left|1 - \sum_{i=1}^{M} w_i e^{j\omega}\right|^2} \tag{10.158}$$

The frequencies of the sinusoids can be located as the values of ω for which the function in (10.149) has peaks.

EXAMPLE 10.10. Consider an input signal given as the sum of three sinusoidal components

$$u(i) = \cos[2\pi(0.3)\, i] + \cos\left[2\pi(0.35)\, i + \frac{\pi}{4}\right] + \cos\left[2\pi(0.1)\, i + \frac{\pi}{3}\right] + \vartheta(i) \tag{10.159}$$

where $\vartheta(i)$ is the additive noise component. For this example we assume that the noise is Gaussian with zero mean and a standard deviation of 0.4. This yields a signal-to-noise ratio (SNR) of approximately

$$\frac{S}{N} = 10 \log \left(\frac{\frac{1}{2} + \frac{1}{2} + \frac{1}{2}}{0.4^2} \right) = 9.7\,\text{dB} \tag{10.160}$$

A total of 30 samples of the signal are assumed available for signal processing. The structure of the one-step prediction filter is shown in Figure 10.33, with 10 tap-delay elements. Figure 10.35(a) shows the position of the transfer function zeros obtained in 20 trials of the PLSR spectrum estimation algorithm. The weights of the filter were estimated using 10 PLSR factors. The zeros of the filter's transfer function can be partitioned into two sets. The elements of the first set are zeros grouped in the vicinity of the unit circle. Those zeros correspond to the frequencies of the sinusoidal components for the signal given in (10.159). Since there are three sinusoidal components, the plot in Figure 10.35(a) shows three clusters of filter zeros in each half of the unit circle. The second set consists of zeros that are randomly positioned within the unit circle. The positions of these zeros are determined by the actual samples of the additive noise, and hence, they vary from trial to trial. An example of a quasi-spectrum plot obtained in one of the trials is shown in Figure 10.35(b). We see that the peaks of the quasi-spectrum function occur at the frequencies of the individual sinusoidal components.

Dimension of the PLSR model

An important part of PLS regression modeling is deciding on the number of factors. Intuitively, with the introduction of additional PLSR factors we would expect the one-step prediction filter to become more accurate.

†Note that PLSNET (cf. Sect. 9.6) can also be used to compute $\{\hat{W}, \hat{B}, \hat{v}\}$.

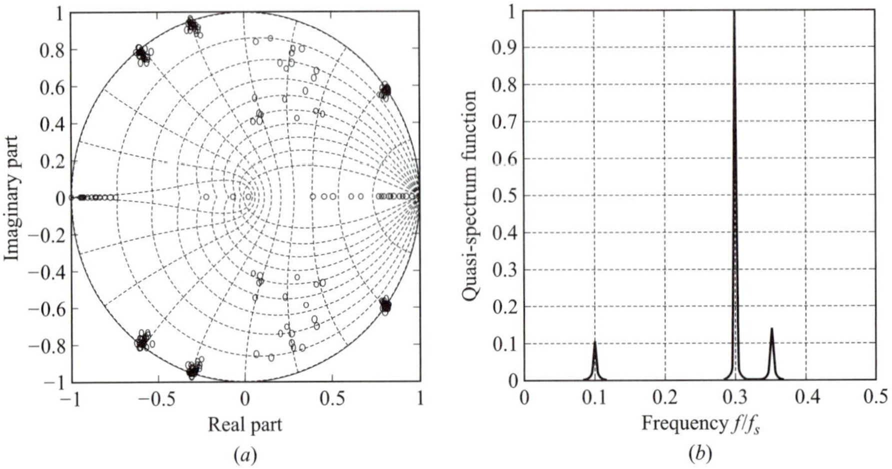

Figure 10.35 Spectrum estimation using the PLS regression. (*a*) Position of the transfer function zeros of the one-step prediction filter in Example 10.10. The results are obtained from 20 independent trials. (*b*) An example of a normalized quasi-spectrum plot obtained in one of the trials. The normalization is performed so that the maximum value is 1.

However, the increase in the prediction accuracy may be a result of data overfitting. Data overfitting can cause spurious peaks in the estimation of the signal's spectrum. Figure 10.36 shows the clustering of the transfer function zeros for the estimation problem given in Example 10.10 for different numbers of PLSR factors. When the number of PLSR factors is smaller than the number of individual sinusoidal components, the estimation algorithm does not provide sufficient spectral resolution. For example, in the case of one PLSR factor we see that there is only one cluster of zeros positioned in the vicinity of the unit circle. A similar situation is obtained for two PLSR factors. In the case of three PLSR factors we notice three well-formed clusters that correspond to the frequencies of the three sinusoidal components. Further addition of PLSR factors tends to increase the resolution of the algorithm. However, this may result in data overfitting, and clear evidence of this can be seen in Figure 36(*d*) where we have used all 10 PLSR factors. In this case, we notice that some of the zeros that are not part of the three clusters may end up in the vicinity of or even on the unit circle. In the quasi-spectrum this would result in spurious spectral peaks that could be interpreted as additional components in the signal's spectrum.

From our discussion of PLSR in Sect. 9.5, we know that if all regression factors are used, the PLS becomes equivalent to classical least-squares estimation. The example results shown in Figure 10.36 demonstrates that the reduction of the number of factors increases the robustness of the estimation compared to its CLS counterpart. This is a well-known technique used in many spectrum estimation algorithms that are commonly referred to as *eigenanalysis algorithms for spectrum estimation*. More details on this group of algorithms and many other spectrum estimation techniques can be found in [47–50].

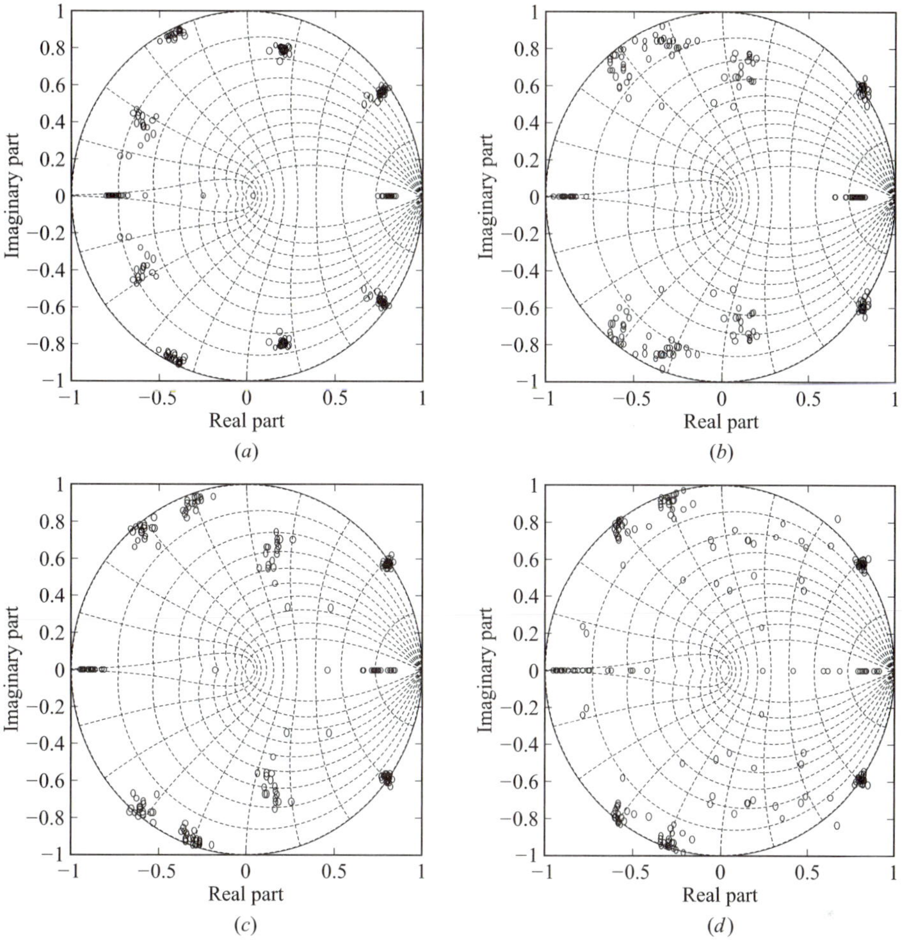

Figure 10.36 The zero positions of a one-step predictor transfer function as a function of the number of PLS factors. (*a*) One PLS factor, (*b*) two PLS factors, (*c*) three PLS factors, (*d*) 10 PLS factors.

10.10 OTHER CASE STUDIES

Highlighted in this section are two additional applications of neural networks. These examples illustrate two important problems that can be solved by a neurocomputing approach.

10.10.1 Estimation of Glucose Concentrations from Synthetic NIR Data

This example involves the use of synthetic near-infrared (NIR) data that emulate different concentrations of glucose in a simple aqueous matrix [51, 52]. The data used are similar to those in Example 9.4 in Sec. 9.6 (see Figure

9.20). The only difference in the data for this example is that the additive noise is not the same as it was used previously (i.e., a different Monte Carlo run to generate the data). For a detailed explanation of how the data are generated and what is represented in the data, see Example 9.4. Figures 10.37 and 10.38 show the MATLAB m-files that were used to generate the data.

The purpose of this example is to compare the performances of (1) classical least-squares regression methods (cf. Sect. 9.4), (2) a standard multilayer perceptron neural network trained by backpropagation (BP) (cf. Sect. 3.3.1), and (3) the partial least-squares neural network (cf. Sect. 9.6). The details of each approach are as follows.

CLS regression

The CLS calibration model $\hat{\boldsymbol{b}}_{f\,\mathrm{CLS}}$ was generated according to (9.98), that is,

$$\hat{\boldsymbol{b}}_{f\,\mathrm{CLS}} = (\boldsymbol{A}_{\mathrm{train}}^T \boldsymbol{A}_{\mathrm{train}})^{-1} \boldsymbol{A}_{\mathrm{train}}^T \boldsymbol{c}_{\mathrm{train}} \tag{10.161}$$

```
%
% Generates Synthetic Near-Infrared (NIR) Data
%

% Spectrum of the component of interest (could be NIR spectrum of
  glucose)
        Spi = .6*gaussd(30,100,15) + .3*gaussd(50,100,70);
% GAUSSD generates a Gaussian distribution
% Spectrum of obscuring component (NIR spectrum of water)
        Spo = .8*gaussd(10,100,20) + .6*gaussd(20,100,80);
        A = zeros(200,100);
        for i = 1:200
            A(i,:) = i*Spi;
        end
% Concentrations (could be glucose concentrations)
        p = ones(100,1);
        C1 = 1000*atan(.0001*A*p);
% Addition of the zero-mean Gaussian noise
        An = A + randn(200,100);
% Addition of the obscuring component
        An1 = zeros(200,100);
        for i =1:200
             An1(i:) = An(i,:) + (1000+30*randn)*Spo;
        end
% Form the Training and Test Data
% Training Spectra (each row is a NIR spectrum)
        TRAIN = An1 (1:2:200,:);
% Training Concentrations (target values)
        TRAINC = C1(1:2:200,:);
% Test Spectra (each row is a NIR spectrum)
        TEST = An1(2:2:200,:);
% Test Concentrations (target values)
        TESTC = C1(2:2:200,:);
clear p i A An An1 C1 Spi Spo
```

Figure 10.37 MATLAB m-file to generate synthetic NIR data. The **gaussd** function is shown in Figure 10.38.

```
function out=gaussd(d,np,c)

% out=gaussed(d,np,c)
% generates a Gaussian distribution
% out = output distribution
% d   = distance from center
% np  = number of points out contains
% c   = peak center

d = d/2;
for k = 1:np
   out(k) = exp(-((k-c)/d)^2);
end
```

Figure 10.38 MATLAB function to generate a Gaussian curve.

Figure 10.39 shows 25 of the 100 training spectra used to develop the CLS calibration model. Using this model and the test data, another 100 spectra (which look similar to the data shown in Fig. 10.39), the predicted (estimated) glucose concentrations are given by

$$\hat{c}_{\text{test CLS}} = \boldsymbol{A}_{\text{test}} \hat{\boldsymbol{b}}_{f\,\text{CLS}} \tag{10.162}$$

The performance of the calibration model given in (10.161) is assessed by computing the standard error of prediction. Figure 10.40 shows the CLS prediction results, where the CLS method yielded SEP = 67.1 mg/dL. This

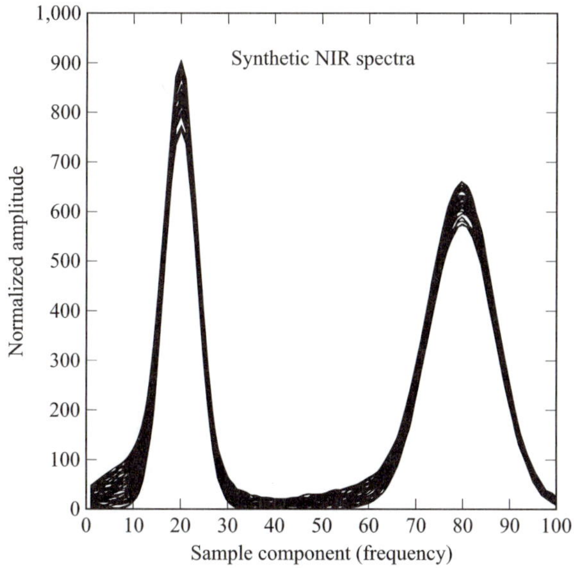

Figure 10.39 Training data for all three methods. This represents 25 percent of the total amount of training data (i.e., 25 spectra).

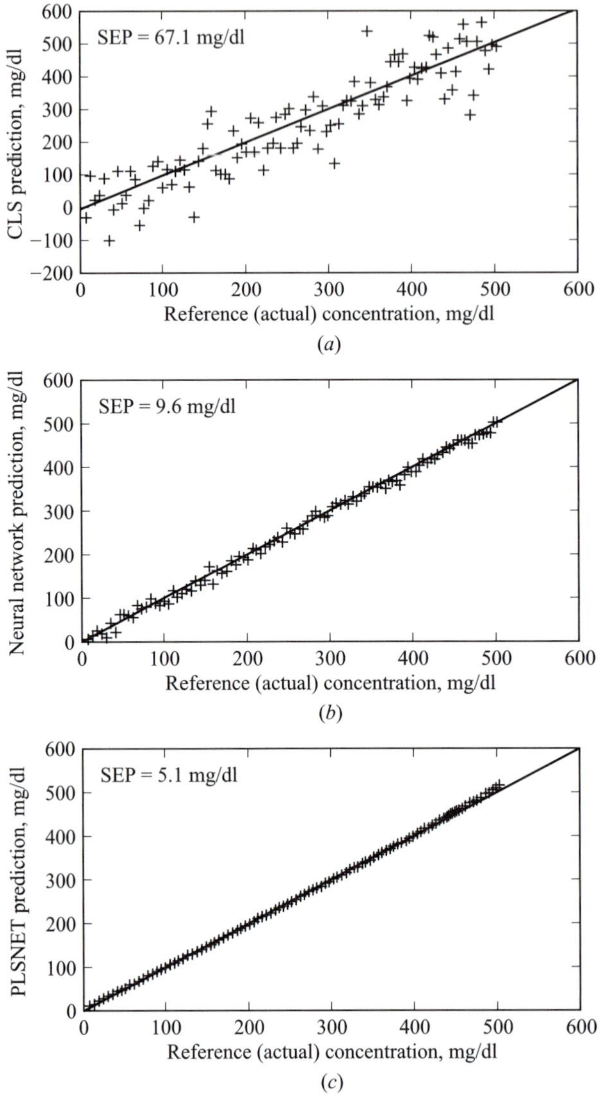

Figure 10.40 (*a*) Glucose concentration prediction using CLS. (*b*) Glucose concentration predictions using an MLP NN trained by BP (100/30/1-layer network). (*c*) Glucose prediction results using PLSNET (number of factors retained is 2).

is a very large error in comparison to what would be a realistically acceptable value for a noninvasive glucose monitoring system [51].

MLP NN trained by BP

The network architecture used for the MLP NN is shown in Figure 10.41. The MATLAB neural network toolbox (version 2) was used to carry out the training of the MLP NN. The initial synaptic weights and biases were set using the function **`initff`** with a **`tansig`** activation function for all three layers.

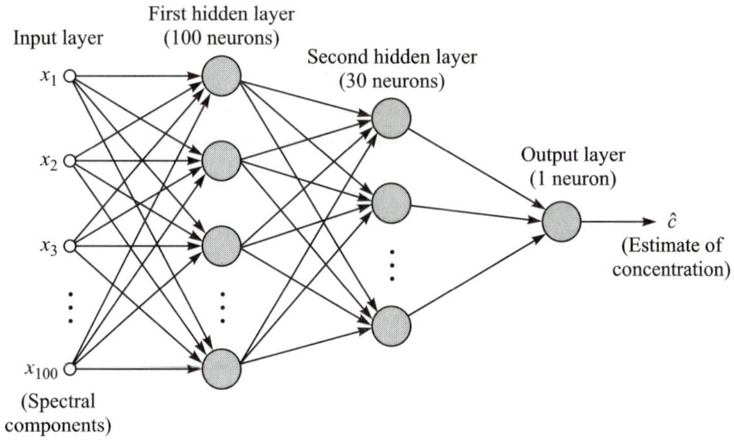

Figure 10.41 Three-layer MLP NN trained by BP.

The **tansig** activation function in MATLAB is the *hyperbolic tangent sigmoid* activation function referred to in Section 2.3. The function **trainbpx** was used to train the network with **tansig** activation functions for all three layers. This MLP NN training function is available in the neural network toolbox and adaptively adjusts the learning rate parameter. The **trainbpx** function was modified to randomize the inputs (100 total) for each training epoch. The initial learning rate parameter was set to 10^{-4}, the momentum was set to 0.9, and the sum-squared error (SSE) goal was set to 10^{-2}. The training data (both the input data, i.e., the 100 spectra, and target values, i.e., the glucose concentrations) were scaled according to the maximum spectral amplitude in the set of 100 spectra (**max_input** = 925.3427). This is necessary because of the activation function limits. The error goal was reached in 2,920 training epochs. The MATLAB neural network toolbox function **simff** was used after training to predict the test concentrations from the test spectra. An important final step after the predictions are obtained is to scale all the values by the maximum spectral amplitude value, that is, **max_input** = 925.3427, that was used to scale the training data. The SEP yielded using the MLP NN was SEP = 9.6 mg/dL. Figure 10.40(*b*) shows the MLP NN prediction results. Comparing Figures 10.40(*a*) and (*b*), we see a dramatic difference in the regression results. The MLP NN is effectively able to "learn" the nature of the "glucose spectral characteristics" from the input spectra (recall, the glucose spectra is "buried" in the signal, see Figure 10.39) presented to it, and correlate the various glucose concentrations associated with the spectral amplitude information in the spectra. The question is: Can we improve on these results?

PLSNET

PLSNET-C (cf. Sect. 9.6) was used next to extract the first two factors of the training data (of the possible 100). The factor analysis used in Example 9.4 already established that two factors are optimal. Figure 9.16 shows the neural network architecture for just two stages (factors). The initial weights were set to normally distributed random numbers with zero mean and unity variance. The learning rate parameters μ_w, μ_b, and μ_v were set as

$$\mu_w = \frac{0.1}{\sum_{j=1}^{100} c_{\text{train}\,j}^2} \tag{10.163}$$

and

$$\mu_b = \mu_v = 0.05\mu_w \tag{10.164}$$

After 10,000 training epochs, the network converged to the weight loading vectors shown in Figure 10.42. These results look similar to those shown in Figure 9.20(*e*). The first weight loading vector has characteristics similar to the water spectrum shown in Figure 9.20(*a*). PLSNET-P was used, along with the extracted weight loading vectors $\{\hat{\boldsymbol{w}}_1, \hat{\boldsymbol{w}}_2\}$, the loading vectors $\{\hat{\boldsymbol{b}}_1, \hat{\boldsymbol{b}}_2\}$, and the regression coefficients $\{\hat{v}_1, \hat{v}_2\}$, to predict the test concentrations from the test spectra. The SEP yielded using PLSNET-P was SEP = 5.1 mg/dL, and Figure 10.40(*c*) shows the prediction results versus the actual concentrations. Comparison all three methods clearly shows that PLSNET can better predict the concentrations than either CLS regression or the MLP NN trained by BP.

10.10.2 Event Classification Using Infrasonic Data

In this example infrasonic data are processed to classify physical events such as volcanic activity and mountain-associated waves (cf. Sect. 10.8.2). Figure 10.43(*a*) shows a set of four-channel signal recordings for the Galunggung,

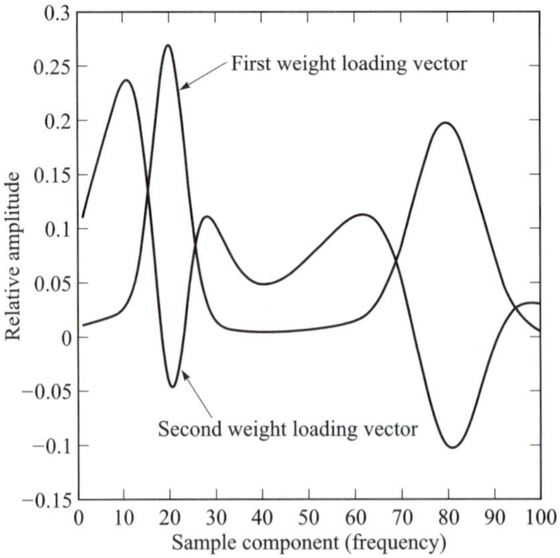

Figure 10.42 First two weight loading vectors extracted by PLSNET-C for the training data.

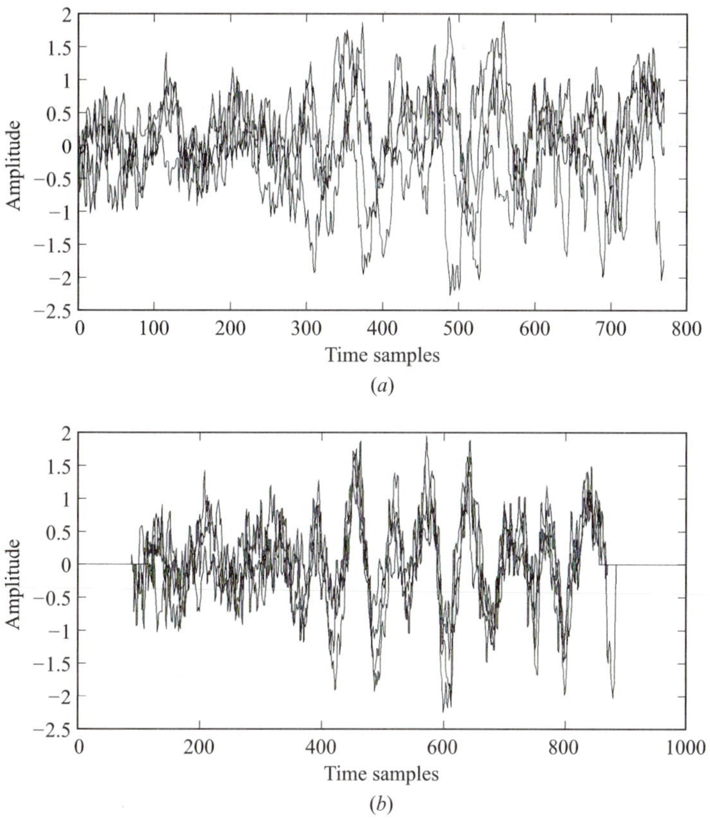

Figure 10.43 (*a*) Set of four-channel signals from the Windless Bight, Antarctica, infrasound sensor before beamforming for the Galunggung, Java, volcanic eruption in 1982. (*b*) Same four signals after beamforming.

Java, volcanic eruption in 1982 (before beamforming). These signals were recorded at Windless Bight, Antarctica, with an infrasound sensor array (F-array). The infrasonic signals were sampled with a nominal sampling frequency of 1 Hz. Figure 10.43(*b*) shows the same set of signals after beamforming. The beamforming results in a set of signals that are adjusted to compensate for the signal time lag over the distances between the sensors in the array (see Figure 10.23). Compensating for the time lags results in the four signals from the different channels to be aligned to an arbitrary, yet common, time-synchronized reference point (this is CHF 1 in Figure 10.23). The raw time-domain signals were "padded" with an appropriate number of zero-amplitude time samples before application of the beamforming. This will ensure that that no information is lost during the adjustment of the signals.

A set of 152 infrasonic signals were used to train and test an MLP NN trained by BP [53–55]. These 152 signals constitute of a set of 28 volcanic events (VOL) and 10 mountain-associated waves (MAW) [56, 57]. For the volcanic events, 6 are for the eruption of the El Chichon, Mexico, volcano, and 22 are for the volcano eruption in Galunggung, Java. Both sets of infrasonic volcano data were recorded in 1982. The source for the MAW recorded at Windless Bight, Antarctica, in 1983 corresponds to azimuth band associated with mountains in New Zealand. For each of the two different events (VOL and MAW) there are four signals each from the F-array sensors. The 152 were divided into a training data set [76 signals: 56 VOL (12 El Chichon and 44 Galunggung) and 20 MAW signals] and a test data set using the other half of the data. When the data are divideded up for training and testing, the four-channel signals are kept together.

The "raw" beamformed infrasonic data, for example, shown in Figure 10.43(*b*), are preprocessed using methods that are common in speech recognition systems [58]. Although the two phenomena are quite different, there are characteristics that are similar. For example, even though there are obviously differences in the channel or propagation media and the frequency ranges of interest, there are similarities in some of the sources. For example, the obvious similarity is between a volcanic eruption and a human voice. The following preprocessing steps were carried out on the raw signals to develop feature vectors that can be used for training and testing.

Data preprocessing steps

1. Remove (subtract) the mean value from the signal.
2. Hamming window is applied to the signal.
3. Compute the power spectral density (PSD) of the signal.
4. Mel-frequency scaling is applied to the PSD [59]. This is an adaptation of the cepstrum [47] of a signal used for speech signals, specifically utilized in speech recognition applications. Given the PSD from the previous step $\boldsymbol{S}(k)$ (where k is a discrete frequency), it is modified according to

$$\boldsymbol{S}_m(k) = \alpha \ln [\beta \boldsymbol{S}(k)]$$

   where $\alpha = 1125$ and $\beta = 0.0016$. These are standard (empirically determined) values used for mel-scaling speech data.
5. Given $\boldsymbol{S}_m(k)$ from the previous step, the discrete inverse cosine transform [47] is taken to yield

$$\boldsymbol{x}_m(n) = \frac{1}{N} \sum_{k=0}^{N-1} \boldsymbol{S}_m(k) \cos\left(\frac{2\pi kn}{N}\right) \qquad \text{for } n = 0, 1, 2, \ldots, N-1$$

   where N is the total number of time-domain samples and $\boldsymbol{x}_m \in \Re^{1 \times n}$.
6. Take the derivative (using the difference operator) of the sequence $\boldsymbol{x}_m(n)$, that is, $\boldsymbol{x}'_m(n)$.
7. Concatenate the derivative sequence $\boldsymbol{x}'_m(n)$ with the cepstral sequence $\boldsymbol{x}_m(n)$ to form the augmented sequence

$$\boldsymbol{x}^a_m = [\boldsymbol{x}'_m(n) | \boldsymbol{x}_m(n)]$$

8. Take the absolute value of the elements in the sequence $\boldsymbol{x}^a_m$ to get

$$\boldsymbol{x}^a_{m,\text{abs}} = |\boldsymbol{x}^a_m|$$

9. Next take the $\ln(x^a_{m,\text{abs}})$ from the previous step to get

$$x^a_{m,\text{abs, ln}} = \ln\left(x^a_{m,\text{abs}}\right)$$

10. Finally, scale the entire data set with respect to the largest amplitude.

Using partial least-squares regression (cf. Sect. 9.5), a preliminary analysis was performed on the feature space to determine the best combination of mel-frequency cepstral coefficients and the associated derivative coefficients. The best combination was 15 cepstral derivatives along with 25 cepstral coefficients. Figure 10.44(a) shows selected representative feature vectors for the two volcanic eruptions, and Figure 10.44(b) shows selected representative feature vectors for the MAW.

After all the signals are preprocessed, training and tests sets are formed, that is, $\{A_{\text{train}}, c_{\text{train}}\}$ and $\{A_{\text{test}}, c_{\text{test}}\}$. Each of the *rows* of A_{train} and A_{test} contains the 40-element feature vectors. The matrices c_{train} and c_{test} consist of the appropriate target vector for the particular event that the neural network will attempt to classify. The binary target vectors are established as

$$[1, 0]^T = \text{volcano} \qquad \text{and} \qquad [0, 1]^T = \text{MAW}$$

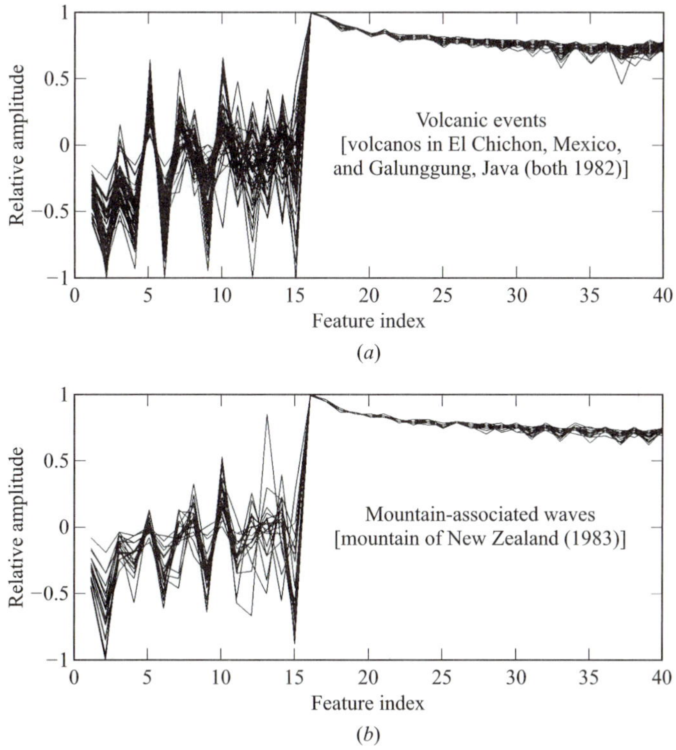

Figure 10.44 (a) Typical volcano infrasonic feature vectors. (b) Typical MAW infrasonic feature vectors. Both sets of data have feature vectors consisting of 15 cepstral derivatives and 25 cepstral coefficients.

Three different event classifiers are used and their performances compared: multilayer perceptron neural network trained by backpropagation, PLSR, and a radial basis function neural network.

The MLP NN is a 40/80/2-layer network with binary sigmoid (log-sigmoid) activation functions. The presentation of the input training feature vectors was randomized every training epoch. The error goal (sum-squared error) was set to SSE = 0.1, the initial learning rate parameter was set to 0.0001, and the momentum parameter was set to 0.9. The MATLAB neural network toolbox function **`trainbpx`** [60] was used for training the MLP NN; however, it was modified to allow randomization of the inputs. The initial weights and biases in the MLP NN were set using the MATLAB neural network toolbox function **`initff`**, and the function **`simuff`** was used to simulate the network for testing.

Because binary target vectors were used and not scalars, the PLSR algorithm presented in Sect. 9.5 cannot be used. A more general form of the PLSR algorithm was used that allows for multiple-component targets. The **`pls`** function in the MATLAB chemometrics toolbox [61] was used along with the function **`plspred`** for the testing phase. In addition, the function **`plspress`** was used to evaluate the performance of the PLSR classifier. This MATLAB function computes the predicted residual error sum of squares (PRESS) which can be thought of as a generalization of the standard error of prediction presented in Sect. 9.4. The PRESS must now be used since multiple-

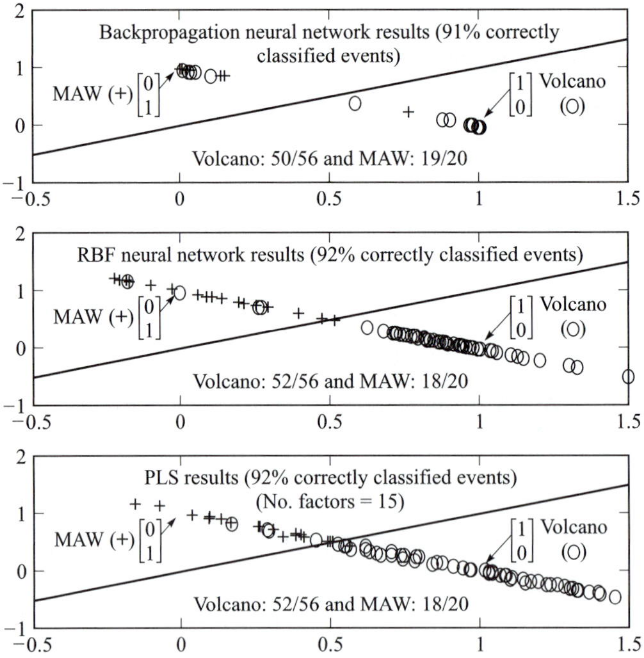

Figure 10.45 Classification results using MLP NN trained by BP, RBF NN, and PLSR.

component targets are being used instead of scalars. The PRESS is a weighted sum of squares of the prediction errors for all target values. A total of 15 PLS factors was determined to be the optimal number to retain for prediction.

The RBF neural network consists of 54 hidden-layer neurons and Gaussian radial basis functions with spread = 0.775, and the SSE goal was set to SSE = 0.1. The MATLAB neural network toolbox function **solverb** was used to design the network, and **simurb** was used for the testing phase.

Figure 10.45 shows the results using the three classifiers. The percentage of correctly classified events is essentially the same for all three methods. However, the MLP NN trained by BP appears to be more robust than in the other two methods; that is, the clusters formed by this network are much "tighter" than those formed by the RBF NN and PLSR.

PROBLEMS

10.1. Assume a second-order discrete-time system whose z transform transfer function is given by

$$H(z) = \frac{Y(z)}{U(z)} = \frac{z-0.1}{z^2-0.8999+0.08}$$

with the sampling period $T_s = 2\pi/1{,}000$ s, and the actual (true) system dimension is given by $\bar{n} = 2$. Therefore, the actual parameter vector is given by $\boldsymbol{\theta} = [-0.8999, 0.08, 1, -0.1]^T$.

(*a*) Generate the simulated system data $\{\boldsymbol{u}(k), \boldsymbol{y}(k)\}$, using the MATLAB function **dlsim**. Specifically, assume the system input sequence $\boldsymbol{u}(k)$ (1,024 samples) to be zero-mean unity-variance Gaussian noise (i.e., 1,024 samples from a Gaussian distribution with zero mean and unity variance).

(*b*) From the data generated in part (*a*), form the ARMA data matrices for the training and test data using the first 100 samples for the training data $\{\boldsymbol{\Phi}_{\text{train}}(N), \boldsymbol{y}_{\text{train}}(N)\}$ and the next 100 samples for the test data $\{\boldsymbol{\Phi}_{\text{test}}(N), \boldsymbol{y}_{\text{test}}(N)\}$. Assume the system to be (over-) specified dimensionally as $n = 2$.

(*c*) Use PLSR (cf. Sect. 9.5) to determine the optimal number of factors. That is, using the training and test data from part (*b*), perform an independent-validation factor analysis to reveal the "true" system dimension. Use the PLSR1 calibration algorithm and the PLSR1 prediction algorithm (method 2) given in Sect. 9.5.

(*d*) Generate the final set of test data, that is, Final data set $= \{\boldsymbol{\Phi}_f(N; \hat{n}), \boldsymbol{y}_f(N; \hat{n})\}$, where $\hat{n}$ is the estimate of the system dimension determined from your results in part (*c*). From this final data set determine the parameter vector using the PLSR1 prediction algorithm (method 1).

(*e*) Plot the original system output response, using the training data. On the same graph plot the discrete-time samples from the PLSR1 prediction algorithm (method 2) results using the training data, and using the system dimension that you determined in part (*c*).

(*f*) Repeat all the above parts in this problem, using PLSNET-P instead of the PLSR1 calibration algorithm to determine the weight loading vectors, the loading vectors, and the regression coefficients. Determine the parameter vector, and compare these results to your previous parameter vector.

(*g*) If you found that the system is *reducible*, determine the parameter vector from the original set of test data generated in part (*b*), using only the PLSR1 calibration algorithm and the PLSR1 prediction algorithm (method 1).

10.2. Repeat all parts in Problem 10.1 with the transfer function given as

$$H(z) = \frac{Y(z)}{U(z)} = \frac{1}{z^2 - z + 0.5}$$

and in part (*b*) assume the system to be (over-) specified dimensionally as $n = 4$.

10.3. Consider a nonlinear system given by the following difference equation

$$y_p(k) = 0.2y_p(k) + 0.5y_p(k-1) + y_p(k-2) + f[u(k)]$$

where

$$f(u) = 4\sin(4u) + \frac{1}{1+u^2}$$

As can be seen, the system's dynamics correspond to model 1 described in Section 10.6.2.

(*a*) Demonstrate that the system is stable and that any bounded input sequence results in a bounded output sequence (BIBO stability criterion).
(*b*) Assuming that the function $f(u)$ is unknown, design a neural network identifier using a series-parallel model of the form:
$$\hat{y}(k+1) = 0.2y_p(k) + 0.5y_p(k-1) + y_p(k-2) + N[u(k)]$$
where $N[u(k)]$ is the mapping performed by the single-input single-output RBF NN. Experiment with a different number of neurons in the hidden layer. In the process of identification use uniformly distributed random inputs in the interval $[-2, 2]$. Test the behavior of the model for inputs outside the interval used for model training, and comment on the success of the model.
(*c*) Repeat part (*b*), using MLP NN.

10.4. Consider an example of a nonlinear communication channel with multipath signal propagation. Assume that the input/output equation characterizing the channel can be approximated as

$$y(k) = 2\arctan\{0.5[x(k) - 0.1x(k-1) + 0.4x(k-3)]\} + v(k)$$

where $v(k)$ is the white noise introduced by the channel. An example of a nonlinear equalizer that can be used to compensate for the channel nonlinearities is shown in Figure 10.46.

(*a*) Generate a random bipolar input sequence, and using the equation above for the nonlinear channel, generate the corresponding outputs. Assume that $v(k)$ is white Gaussian noise having zero mean and standard deviation of $\sigma = 0.1$.
(*b*) Use the equalizer structure shown in Figure 10.46, and train the neural network to perform the equalization process. Use either an RBF or an MLP NN to accomplish this task. Experiment with different structures and sizes for the network. If the equalizer is to work properly in the environment where the number and delay of the multipaths change over time, which neural network architecture is preferred?

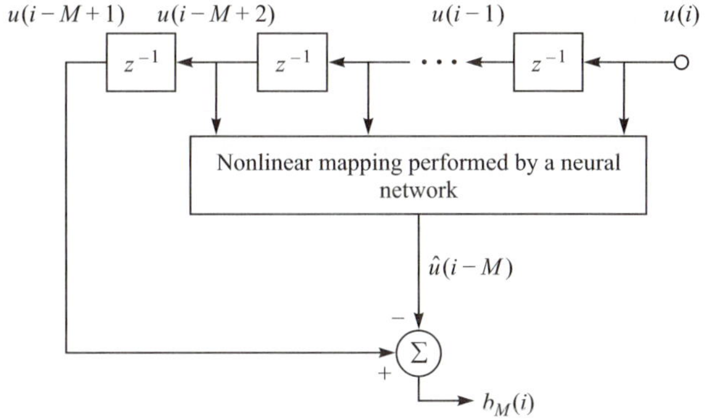

Figure 10.46 Nonlinear equalization using a neural network.

(*c*) What is the minimum number of delay taps that needs to be used in the equalizer structure? Justify the answer.

(*d*) If the data bit rate is 100 kbits/s and the maximum multipath delay expected is 50 μs, estimate the number of taps in the nonlinear equalizer necessary to accomplish the equalization task.

10.5. In this problem we want to blindly separate two signals. The first signal is a zero-mean square wave, and the second signal is uniformly distributed noise in the interval [-1, 1]. These signals can be generated with MATLAB code given as

```
% Generate Two Signals
t=0:0.001:0.199;
% Deterministic Square Wave
s1=square(2*pi*30*t);
% Impulse Noise in the Interval [-1, 1]
s2=2*(rand(1,200)-0.5*ones(1,200));
% Plot Figures
subplot(2,1,1), plot(t,s1), axis([0 0.2 -1.5 1.5])
title('Square Wave'), xlabel('Time (sec)')
ylabel('Amplitude')
subplot(2,1,2), plot (t,s2), axis([0 0.2 -1.5 1.5])
title('Impulse Noise'), xlabel('Time (sec)')
ylabel'Amplitude')
```

The signals **s1** and **s2** are the source signals. Using the mixing matrix

$$A = \begin{bmatrix} 0.2258 & 0.3686 \\ -0.1013 & 0.1264 \\ -0.1416 & -0.2588 \\ -0.2147 & 0.4781 \end{bmatrix}$$

generate four *observable* signals as $X = AS$, where $X \in \Re^{4\times 200}$ (each *row* is an observable signal), and $S =$ [**s1**; **s2**]. Assume that the only information available is the four observable signals. Perform the following steps:

(*a*) Prewhiten the observable data, using the batch approach given in (10.116). This will necessitate performing PCA on the observable data to determine the number of source signals (perform this step even though this information is obvious). Make sure that the mean is removed from the four observable signals in $\boldsymbol{X}$, and that the output of the whitening process is properly scaled.

(*b*) Using the results from part (*a*), determine the separation weight matrix $\boldsymbol{W}$ using the learning rule given in (10.111), where the nonlinear function $g(\bullet)$ is the derivative of the logistic function. Be sure to adjust the learning rate parameter according to (10.108) with $\boldsymbol{v}(k)$ replaced by $\boldsymbol{y}(k)$. Use the following weight matrix to initialize your network:

$$\boldsymbol{W}(0) = \begin{bmatrix} 0.9762 & -0.2171 \\ 0.2171 & 0.9762 \end{bmatrix}$$

(*c*) The independent components from part (*b*), that is, $\boldsymbol{y} = \boldsymbol{W}^T\boldsymbol{v}$, should be original source signals ($\boldsymbol{y} = \hat{\boldsymbol{s}}$). Compute the correlation coefficients between the two separated signals and the two original source signals (you will compute four numbers). Two of these correlation coefficients should be much larger than the other two.

10.6. Repeat Problem 10.5; however, the idea is to now separate four different source signals: three deterministic signals (square wave, sawtooth signal, and a sinusoidal wave) and a signal from uniformly distributed noise in the interval $[-1, 1]$. These signals can be generated with MATLAB code given as

```
% Generate Four Signals
t = 0:0.001:0.199;
s1 = square(2*pi*30*t);
% Sine Wave
s2=sin(2*pi*45*t);
% Sawtooth Waveform
s3=sawtooth(2*pi*50*t);
% Impulse Noise in the Interval [-1,1]
s4=2*(rand(1,200)-0.5*ones(1,200));
% Plot Figures
subplot(4,1,1), plot(t,s1), axis([0 0.2 -1.5 1.5])
title('Square Wave'), xlabel('Time (Sec)')
ylabel('Amplitude')
subplot(4,1,2), plot(t,s2), axis([0 0.2 -1.5 1.5])
title('Sine Wave'), xlabel('Time (Sec)')
ylabel('Amplitude')
subplot(4,1,3), plot(t,s3), axis([0 0.2 -1.5 1.5])
title('Sawtooth Waveform'), xlabel('Time (Sec)')
ylabel('Amplitude')
subplot(4,1,4), plot(t,s4), axis([0 0.2 -1.5 1.5])
title('Impulse Noise'), xlabel('Time (Sec)')
ylabel('Amplitude')
```

The signals **s1**, **s2**, **s3**, and **s4** are the source signals. Using the mixing matrix

$$A = \begin{bmatrix} 0.4501 & -0.4815 & -0.3237 & -0.1471 \\ -0.2689 & 0.3214 & -0.0943 & 0.3132 \\ 0.1068 & -0.0553 & 0.4355 & -0.4901 \\ -0.0140 & 0.1154 & 0.4169 & -0.3611 \\ 0.3913 & 0.2919 & -0.0897 & -0.2972 \\ 0.2621 & 0.4218 & 0.3936 & -0.3013 \\ -0.0435 & 0.2382 & -0.4421 & 0.1038 \end{bmatrix}$$

generate seven *observable* signals as $\boldsymbol{X} = \boldsymbol{AS}$, where $\boldsymbol{X} \in \Re^{7\times 200}$ (each *row* is an observable signal), and $\boldsymbol{S} = [$**s1**; **s2**; **s3**; **s4**$]$. Assume that the only information available is the four observable signals. Use the following weight matrix to initialize your network:

$$\boldsymbol{W}(0) = \begin{bmatrix} 0.5628 & -0.0889 & -0.7278 & -0.3816 \\ -0.1329 & 0.1824 & -0.5475 & 0.8058 \\ 0.6855 & 0.6051 & 0.3456 & 0.2109 \\ 0.4424 & -0.7698 & 0.2259 & 0.4008 \end{bmatrix}$$

10.7. Apply the fast fixed-point algorithm for ICA presented in Section 10.8.3 to Problem 10.6. Compare the results you obtain with those from the neural network approach, that is, the (Karhunen-Oja) nonlinear PCA subspace learning rule given in (10.111). How do the two approaches compare for convergence speed and accuracy of results? What can you conclude?

10.8. Design an adaptive network, that is, a single neuron processor, for estimating the parameters $\{\alpha_1, \alpha_2, \alpha_3\}$ and $\{\beta_1, \beta_2, \beta_3\}$ of a periodic signal $y(t)$ that is corrupted by noise $e(t)$, that is, $z(t) = y(t) + e(t)$. Therefore, $z(t)$ is the observable signal, or the observable noisy version of the signal

$$y(t) = \alpha_1 \sin(\omega t) + \beta_1 \cos(\omega t) + \alpha_2 \sin(5\omega t) + \beta_2 \cos(5\omega t) + \alpha_3 \sin(7\omega t) + \beta_3 \cos(7\omega t)$$

where $\omega = 2\pi f\big|_{f=1\text{Hz}}$. Assume the noise to be zero-mean Gaussian white noise with the variance $\sigma^2 = 0.1$. When simulating the function, assume the sampling frequency $f_s = 1{,}000$ Hz, and generate 1,025 data points. Test the performance of your adaptive network with the parameters $\alpha_1 = \beta_1 = 1$, $\alpha_2 = 0.5$, $\beta_2 = 1/\alpha_2$, $\alpha_3 = 0.25$, and $\beta_3 = 1/\alpha_3$.

10.9. In the frequency estimation problem presented in Section 10.9 we used PLS regression to solve for the tap weights of the linear one-step prediction filter—see Equations (10.157) and (10.158). In Chapter 9 we discussed the principal component regression method that can be used to solve the same problem (cf. Sect. 9.4).

(*a*) Formulate the frequency estimation problem using PCR.
(*b*) Write a MATLAB routine implementing your PCR-based spectrum estimation algorithm. Make the number of PCR factors one of the input parameters to the routine.
(*c*) Use the spectrum estimation problem presented in Example 10.10 to test your routine.
(*d*) Compare your results to the results obtained using PLSR.

10.10. In frequency estimation problems, the length of the one-step prediction filter plays an important role. Consider a signal given as a sum of four sinusoids

$$u(i) = \cos[2\pi(0.1i)] + \cos\left[2\pi(0.2i) + \frac{\pi}{4}\right] + \cos[2\pi(0.38i)] + \cos[2\pi(0.4i)] + \vartheta(i)$$

where $\vartheta(i)$ is the part of signal attributed to additive noise. Assume that the noise is Gaussian with zero mean and standard deviation of 0.35. Assume that $N = 40$ samples of the signal are available for processing.

(*a*) What is the minimum length of the prediction filter that can be used for this problem?

(*b*) Write a MATLAB routine implementing the classical least-squares method to solve for the one step predictor filter coefficients—see Equations (9.98) and (9.99). Allow for the filter order, that is, the number of taps in the delay line, to be one of the input parameters.

(*c*) Use your CLS routine to perform sinusoid frequency estimation for the signal given above. Experiment with different orders of the prediction filter.

(*d*) Lang and McClellan [62] suggest that the order of the filter equal $M = N/3$. On the other hand, Tuffs and Kumaresan [63] have experimentally determined the value of $M = 3N/4$. Compare the results of your experiments with the values suggested by these researchers.

10.11. In our presentation of the spectrum estimation problem we have used a one-step prediction method. This approach has led to the regression equations given in (10.152). Often, the problem can be formulated using the forward-backward linear predictor shown in Figure 10.47. In this case the regression model can be formulated in accordance with

Forward prediction error:

$$f_M(i) = u(i) - \sum_{k=1}^{M} w_k u(i-k)$$

Backward prediction error:

$$b_M(i) = u(i-M) - \sum_{k=1}^{M} w_k u(i-M+k)$$

Setting the optimization criterion as:

$$J(\boldsymbol{w}) = \sum_{i=M+1}^{N} \left[|f_M(i)|^2 + |b_M(i)|^2 \right]$$

we obtain the regression model in the form

$$\boldsymbol{c} = \boldsymbol{A}\boldsymbol{w} + \boldsymbol{e}$$

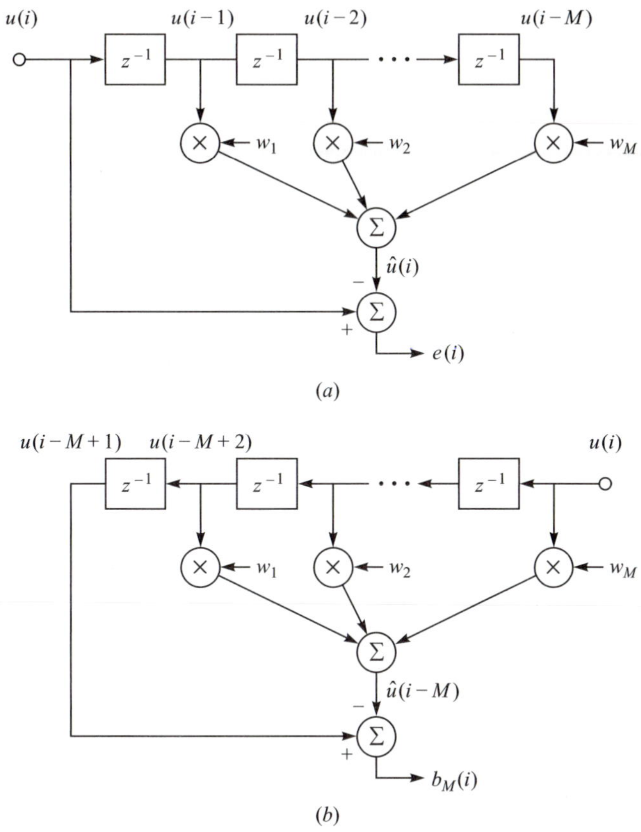

Figure 10.47 (*a*) One-step forward predictor filter; (*b*) one-step backward prediction filter.

where:

$$A = \begin{bmatrix} u(M) & u(M-1) & \cdots & u(1) \\ u(M+1) & u(M) & \cdots & u(2) \\ \vdots & \vdots & & \vdots \\ u(N-1) & u(N-2) & \cdots & u(N-M) \\ \vdots & \vdots & & \vdots \\ u(2) & u(3) & \cdots & u(M+1) \\ u(3) & u(4) & \cdots & u(M+2) \\ \vdots & \vdots & & \vdots \\ u(N-M+1) & u(N-M+2) & \cdots & u(N) \end{bmatrix}$$

$$\boldsymbol{c} = \left[u(M+1), u(M+2), \cdots, u(N) | u(1), u(2), \cdots, u(N-1) \right]^T$$

and

$$\boldsymbol{e} = \left[f_M(M+1), \cdots, f_M(N) | b_M(M+1), \cdots, b_M(N) \right]^T$$

Once the coefficients of the filter are found, (10.158) can be used to provide the estimate of the spectrum.

(*a*) Using the signal $u(i)$ defined in Problem 10.10, formulate the forward-backward prediction problem.

(*b*) Solve the problem using the PLS regression technique.

(*c*) Experimentally determine the optimal length of the forward-backward prediction filter.

REFERENCES

1. L. Ljung, *System Identification: Theory for the User*, Englewood Cliffs, NJ: Prentice-Hall, 1987.
2. G. F. Franklin, J. D. Powell, and M. L. Workman, 2nd ed., *Digital Control of Dynamical Systems*, 2nd ed., Reading, MA: Addison-Wesley, 1990.
3. T. Söderström and P. Stoica, *System Identification*, New York: Prentice-Hall, 1989.
4. J.-N. Juang, *Applied System Identification*, Englewood Cliffs, NJ: Prentice-Hall, 1994.
5. C.-T. Chen, *Linear System Theory and Design*, 2nd ed., New York: Oxford University Press, 1999.
6. P. L. Bogler, *Radar Principles with Applications to Tracking Systems*, New York: Wiley, 1990.
7. N. R. Draper and H. Smith, *Applied Regression Analysis*, 2nd ed., New York: Wiley, 1981.
8. A. M. Qabazard, "System Identification Using Partial Least Squares," Ph.D. Dissertation, Melbourne, FL: Florida Institute of Technology, 1995.
9. F. M. Ham and I. Kostanic, "A Partial Least-Squares Regression Neural NETwork (PLSNET) with Supervised Adaptive Modular Learning," *Applications and Science of Artificial Neural Networks II*, eds. S. K. Rogers and D. W. Ruck, Proceedings of SPIE, vol. 2760, 1996, pp. 139–50.
10. F. M. Ham and I. Kostanic, "Partial Least-Squares: Theoretical Issues and Engineering Applications in Signal Processing," *Mathematical Problems in Engineering*, vol. 2, 1995, pp. 63–93.
11. K. S. Narendra, and K. Parthasarathy, "Identification and Control of Dynamical Systems Using Neural Networks," *IEEE Transactions on Neural Networks*, vol. 1, no. 1, 1990, pp. 4-27.
12. I. J. Leontaritis and S. A. Billings, "Input-Output Parametric Models for Non-Linear Systems. Part I: Deterministic Non-Linear Systems, Part II: Stochastic Non-Linear Systems," *International Journal of Control*, vol. 41, 1987, pp. 303–44.
13. H. L Royden, *Real Analysis*, New York: Macmillan, 1964.
14. S. A. Billings, S. Chen and M. J. Korenberg, "Identification of MIMO Non-linear Systems Using Forward-Regression Orthogonal Estimator," *International Journal of Control*, vol. 49, 1989, pp. 2157–89.
15. A. Desrochers and S. Mohseni, "On Determing the Structure of Non-linear Systems," *International Journal of Control*, vol. 40, 1984, pp. 923–938.
16. S. Chen, S. A. Billings, and W. Luo, "Orthogonal Least-Squares Methods and Their Application to Non-Linear System Identification," *International Journal of Control*, vol. 50, 1989, pp. 1873–86.
17. K. S. Narendra and K. Parthasarathy, "Gradient Methods for Optimization of Dynamic Systems Containing Neural Networks," *IEEE Transactions on Neural Networks,* vol. 2, no. 2, 1991, pp. 252–62.

18. S. Z. Quin, H. T Su, and T. J. McAvoy, "Comparison of Four Neural Net Learning Methods for Dynamic System Identification," *IEEE Transactions on Neural Networks*, vol. 3, no. 1, 1992, pp. 122–30.
19. S. Chen, S. A. Billings, C. F. N. Cowan, and P. M. Grant, "Practical Identification of NARMAX Models Using Radial Basis Functions," *International Journal of Control*, vol. 52, no. 6, 1990, pp. 1327–50.
20. S. Chen, S. A. Billings, C. F. N. Cowan and P. M. Grant, "Non-linear System Identification Using Radial Basis Functions," *International Journal of Systems Science,* vol. 21, no. 12, 1990, pp. 2513–39.
21. P. J. Werbos, "Generalization of Backpropagation with Application to Recurrent Gas Market Model," *Neural Networks*, vol. 1, no. 4, 1988, pp. 270–80.
22. K. S. Narendra and A. M. Annaswamy, *Stable Adaptive Systems*, Englewood Cliffs, NJ: Prentice Hall, 1989.
23. C. Jutten and J. Herault, "Independent Component Analysis (INCA) versus Principal," in *Signal Processing IV: Theories and Applications*, Proc. *EUSIPCO*-88, eds. J. Lacoume, et al., Amsterdam, The Netherlands: Elsevier, 1988, pp. 643–6.
24. C. Jutten and J. Herault, "Blind Separation of Sources, Part I: An Adaptive Algorithm Based on Neuromimetic Archicture," *Signal Processing*, vol. 24, 1991, pp. 1–10.
25. P. Comon, "Independent Component Analysis," in *Proceedings of the International Signal Processing Workshop on Higher-Order Statistics,* Chamrousse, France, 1991, pp. 111–20. [Republished in *Higher-Order Statistics*, ed. J. L. Lacoume, Amsterdam, The Netherlands: Elsevier, 1992, pp. 29–38.]
26. P. Comon, "Independent Component Analysis—A New Concept?" *Signal Processing*, vol. 36, 1994, pp. 287–314.
27. J. M. Mendel, "Tutorial on Higher-Order Statistics (Spectra) in Signal Processing and System Theory: Theoretical Results and Applications," *Proceedings of the IEEE,* vol. 79, 1991, pp. 278–305.
28. C. L. Nikias and J. M. Mendel, "Signal Processing with Higher-Order Spectra," *IEEE Signal Processing Magazine*., vol. 10, 1993, pp. 10–37.
29. J. Karhunen, E. Oja, L. Wang, R. Vigario, and J. Joutsensalo, "A Class of Neural Networks for Independent Component Analysis," *IEEE Transactions on Neural Networks*, vol. 8, 1997, pp. 486–504.
30. J.-F. Carduso and B. H. Laheld, "Equivariant Adaptive Source Separation," *IEEE Transactions on Signal Processing*, vol. 44, 1996, pp. 3017–30.
31. J. Karhunen, "Neural Approaches to Independent Component Analysis and Source Separation," In *Proceedings of 4th European Symposium on Artificial Neural Networks, ESANN'96*, Bruges, Belgium, April 1996, pp. 249–66.
32. A. Hyvärinen, "Fast and Robust Fixed-Point Algorithms for Independent Component Analysis," *IEEE Transactions on Neural Networks*, vol. 10, 1999, pp. 626–34.
33. E. Oja, H. Ogawa, and J. Wangviwattana, "Learning in Nonlinear Constrained Networks," in *Artificial Neural Networks Proceedings of ICANN-91*, eds. T. Kohonen et al., Amsterdam, The Netherlands: North-Holland, 1991, pp. 385–90.
34. J. Karhunen and J. Joutsensalo, "Representation and Separation of Signals Using Nonlinear PCA Type Learning," *Neural Networks*, vol. 7, 1994, pp. 113–27.
35. J. Karhunen and J. Joutsensalo, "Generalizations of Principal Component Analysis, Optimization Problems, and Neural Networks," *Neural Networks*, vol. 8, 1995, pp. 549–62.
36. L. Wang and E. Oja, "A Unified Neural Bigradient Algorithm for Robust PCA and MCA," International Journal of Neural Systems., vol. 7, 1996, pp. 53–67.

37. L. Wang, J. Karhunen, and E. Oja, "A Bigradient Optimization Approach for Robust PCA, MCA, and Source Separation," In *Proceedings of 1995 IEEE International Conference on Neural Networks,* Perth, Australia, November 1995, pp. 1684–9.
38. V. N. Valentina, *Microseismic and Infrasonic Waves*, Research Reports in Physics, New York: Springer Verlag, 1992.
39. A. D. Pierce, *Acoustics: An Introduction to Its Physical Principles and Applications*, Swickley, PA, Acoustical Society of America, 1989.
40. C. R. Wilson and R. B. Forbes, "Infrasonic Waves from Alaskan Volcanic Eruptions," *Journal of Geophysical Research*. vol. 74, 1969, pp. 1812–36.
41. A. J. Bedard, "Infrasound Originating near Mountain Regions in Colorado," *Journal of Applied Meteorology*, vol. 17, 1978, p. 1014.
42. National Research Council, *Comprehensive Nuclear Test Ban Treaty Monitoring*, Washington: National Academy Press, 1997.
43. K. B. Payne, *Silent Thunder: In the Presence of Elephants*, New York: Simon & Schuster, 1998.
44. F. M. Ham, N. A. Faour, and J. C. Wheeler, "Infrasound Signal Separation Using Independent Component Analysis," in *Proceedings of the 21st Seismic Research Symposium: Technologies for Monitoring the Comprehensive Nuclear-Test-Ban Treaty*, Las Vegas, NV, Sept. 21–24, 1999, vol. 2, pp. 133–40.
45. A. Hyvärinen and E. Oja, "A Fast Fixed-Point Algorithm for Independent Component Analysis," *Neural Computation*, vol. 9, 1997, pp. 1483–92.
46. G. Deco and D. Obradovic, *An Information-Theoretic Approach to Neural Computing*, New York: Springer-Verlag, 1996.
47. J. G. Proakis and D. G. Manolakis, *Digital Signal Processing: Principles, Algorithms, and Applications*, 3rd ed., Upper Saddle River, NJ: Prentice-Hall, 1996.
48. S. Haykin, *Adaptive Filter Theory*, 2nd ed., Upper Saddle River, NJ: Prentice-Hall, 1991.
49. *Proceedings of the IEEE—Special Issue on Spectral Estimation,* vol. 70. no. 9, 1982.
50. K. Hamid and M. Viberg, "Two Decades of Array Signal Processing Research, the Parametric Approach," *IEEE Signal Processing Magazine*, vol. 13, no. 4, 1996, pp. 67–94.
51. F. M. Ham, I. Kostanic, G. M. Cohen, and B. R. Gooch, "Determination of Glucose Concentrations in an Aqueous Matrix from NIR Spectra Using Optimal Time-Domain Filtering and Partial Least-Squares Regression," *IEEE Transactions on Biomedical Engineering*, vol. 44, 1997, pp. 475–85.
52. F. M. Ham and I. Kostanic, "Partial Least-Squares: Theoretical Issues and Engineering Applications in Signal Processing," *Journal of Mathematical Problems in Engineering*, vol. 2, 1996, pp. 63–93.
53. F. M. Ham, T. A. Leeney, H. M. Canady, and J. C. Wheeler, "Discrimination of Volcano Activity and Mountain Associated Waves Using Infrasonic Data and a Backpropagation Neural Network," *SPIE Conference on Applications and Science of Computational Intelligence II*, eds. K. L. Priddy, P. E. Keller, D. B. Fogel, and J.C. Bezdek, vol. 3722, April 1999, pp. 344–56.
54. F. M. Ham, T. A. Leeney, H. M. Canady, and J. C. Wheeler, "Volcano Event Classification Using Infrasonic Data and a Backpropagation Neural Network" (abstract), *EOS Transactions (Supplement), AGU,* vol. 79, no. 45, 1998, p. F620.
55. F. M. Ham, T. A. Leeney, H. M. Canady, and J. C. Wheeler, "An Infrasonic Event Neural Network Classifier," *International Joint Conference on Neural Networks*, Washington, DC, July 10–16, 1999, vol. 6, pp. 3768–73.

56. C. R. Wilson, J. V. Olson, and R. Richards, "Library of Typical Infrasonic Signals," Report Prepared for ENSCO, Inc., Melbourne, FL (Subcontract no. 269343-2360.009), vols. 1–4, 1996.
57. K. D. Hutchenson, "Acquisition of Historical Infrasonic Data," Final Technical Report, Contract no. F08650-95-D-0033, ARS-97-012, ENSCO, Inc., Melbourne, FL, 1997.
58. R. J. Mammone, X. Zhang, and R. P. Ramachandran, "Robust Speaker Recognition: A Feature-Based Approach," *IEEE Signal Processing Magazine*, vol. 13, 1996, pp. 58–71.
59. S. B. Davis and P. Mermelstein, "Comparison of Parametric Representations for Monosyllabic Word Recognition in Continuously Spoken Sentences," *IEEE Transactions on Acoustics, Speech, and Signal Processing*, vol. 28, 1980, pp. 357–66.
60. H. Demuth and M. Beale, *Neural Network Toolbox—For Use with MATLAB*, Natick, MA: MathWorks, 1995.
61. R. Kramer, *Chemometrics Toolbox—For Use with MATLAB*, Natick, MA: MathWorks, 1993.
62. S. W. Lang and J. H. McClellan, "Frequency Estimation with Maximum Entropy Spectral Estimation," *IEEE Transactions on Acoustics, Speech, and Signal Processing,* vol. ASSP-28, 1980, pp. 716–24.
63. D. W. Tuffs and R. Kumaresan, "Estimation of Frequencies of Multiple Sinusoids: Making Linear Perform Like Maximum Likelihood," *Proceedings of IEEE*, vol. 70, 1982, pp. 975–89.

APPENDIX A

Mathematical Foundation for Neurocomputing

A.1 INTRODUCTION

To properly analyze and design artificial neural networks, a solid mathematical foundation is necessary. This appendix serves two purposes. It presents some of the necessary mathematical results in the areas of linear algebra, nonlinear programming, general system theory, and stochastic processes in order to study neural networks; and it also presents selected mathematical definitions and establishes notations that are frequently used throughout the book. We intend to present the necessary mathematical background material for neurocomputing and not to be concerned with a rigorous "theorem, proof" presentation of the mathematics. The reader who must refer to the material presented in this appendix should come away with a working knowledge of the mathematical skills necessary to study neural networks.

A.2 LINEAR ALGEBRA

A.2.1 Fields and Vector Spaces

Fields

DEFINITION A.1. A *field* $\mathscr{F}$ consists of a set of elements and two operations called addition "+" and multiplication "•". The two operations are defined over $\mathscr{F}$ such that they satisfy the following conditions:

1. For every pair of elements α and β in $\mathscr{F}$ there corresponds an element $\alpha + \beta$ in $\mathscr{F}$, called the *sum* of α and β, and an element $\alpha \cdot \beta$ (or $\alpha\beta$) in $\mathscr{F}$, called the *product* of α and β.

2. Addition and multiplication are both commutative; that is, for any α and β in $\mathscr{F}$,

$$\alpha+\beta=\beta+\alpha \qquad \text{and} \qquad \alpha\bullet\beta=\beta\bullet\alpha$$

3. Addition and multiplication are both associative; that is, for any α, β, and γ in $\mathscr{F}$,

$$(\alpha+\beta)+\gamma=\alpha+(\beta+\gamma) \qquad \text{and} \qquad (\alpha\bullet\beta)\bullet\gamma=\alpha\bullet(\beta\bullet\gamma)$$

4. Multiplication is distributive with respect to addition; that is, for any α, β, γ in $\mathscr{F}$,

$$\alpha\bullet(\beta+\gamma)=(\alpha\bullet\beta)+(\alpha\bullet\gamma)$$

5. Contained in $\mathscr{F}$ is an element denoted by 0 and another element denoted by 1, such that $\alpha+0=\alpha$ and $1\bullet\alpha=\alpha$, for every α in $\mathscr{F}$.
6. For every α in $\mathscr{F}$, there is an element β in $\mathscr{F}$ such that $\alpha+\beta=0$. The element β is called the *additive inverse*.
7. For every α in $\mathscr{F}$ that is not the 0 element, there is an element γ in $\mathscr{F}$ such that $\alpha\bullet\gamma=1$. The element γ is called the *multiplicative inverse*.

The set of all real numbers with addition and multiplication constitutes a field, that is, $\Re$. The set of complex numbers with addition and multiplication is also a field, that is, $\mathscr{C}$. For the set of complex numbers, the multiplicative inverse of a complex number $\sigma=\alpha+j\beta$ (where $j=\sqrt{-1}$) can be written as $1/\sigma=\bar{\sigma}/|\sigma|^2$, where $\bar{\sigma}=\alpha-j\beta$ (complex conjugate of σ) and $|\sigma|=\sqrt{\alpha^2+\beta^2}$ (magnitude, or modulus, of σ). The angle (or argument) associated with the complex number σ is given by $\angle\sigma=\arg(\sigma)=\tan^{-1}(\beta/\alpha)$.

The set {0, 1} does not form a field with the usual definition of addition and multiplication. This is simple to see because $1+1=2$ is not in the set {0, 1}. However, with the operations defined as

$$0+0=0 \quad 1+0=1 \quad 1+1=0 \quad 0\bullet1=0 \quad 0\bullet0=0 \quad \text{and} \quad 1\bullet1=1$$

then the set {0, 1} is a field (i.e., the set {0, 1} with the defined addition and multiplication operations satisfies the seven conditions listed above for a field). This field is called the field of *binary numbers*.

The set of all 2×2 matrices of the form

$$\begin{bmatrix} w & -z \\ z & w \end{bmatrix}$$

where w and z are arbitrary *real* numbers, along with the standard definitions of matrix addition and multiplication, forms a field. In this case, the elements 0 and 1 of the field are

$$\begin{bmatrix} 0 & 0 \\ 0 & 0 \end{bmatrix} \text{ null matrix} \qquad \text{and} \qquad \begin{bmatrix} 1 & 0 \\ 0 & 1 \end{bmatrix} \text{ identity matrix}$$

respectively. However, the set of all 2×2 matrices does not form a field; for example, in some cases a multiplicative matrix inverse may not exist.

The set of positive real numbers does not form a field because an additive inverse does not exist. The set of integers does not form a field because a multiplicative inverse does not exist. A set of elements that satisfies all the properties in Definition A.1 except property 7 is considered a *ring*. The set of polynomials also does not form a field because a multiplicative inverse does not exist.

Vector spaces

A vector space has a simple geometric interpretation. For example, in an ordinary two-dimensional geometric plane, if our defined reference point is the origin, then every point in the plane is considered a vector. That is, all "arrows" (vectors) drawn from the origin to any point in the plane have a specified direction and magnitude. Each of the vectors can undergo contraction or expansion, and any two vectors can be added. However, two vectors cannot be multiplied (i.e., vector multiplication is not defined). This plane is called a *vector space* (or *linear space*, or *linear vector space*). A vector space is always defined over a specified field for scalar multiplication (vector contraction or expansion) and vector addition.

DEFINITION A.2. A vector (linear) space over a field $\mathscr{F}$ is designated as $(\mathscr{X}, \mathscr{F})$ and contains the set of elements $\mathscr{X}$, called vectors (of arbitrary length), a field $\mathscr{F}$, and two operations, that is, *scalar multiplication* and *vector addition*. These two operations are defined over $\mathscr{X}$ and $\mathscr{F}$ such that they satisfy the following conditions:

1. For every pair of vectors $\boldsymbol{x}_1$ and $\boldsymbol{x}_2$ in the set of vectors $\mathscr{X}$, there corresponds a vector $\boldsymbol{x}_1 + \boldsymbol{x}_2$ also in $\mathscr{X}$, called the *sum* of $\boldsymbol{x}_1$ and $\boldsymbol{x}_2$.
2. Vector addition is commutative; that is, for any $\boldsymbol{x}_1$, $\boldsymbol{x}_2$ in $\mathscr{X}$,
$$\boldsymbol{x}_1 + \boldsymbol{x}_2 = \boldsymbol{x}_2 + \boldsymbol{x}_1$$
3. Vector addition is associative; that is, for $\boldsymbol{x}_1$, $\boldsymbol{x}_2$, and $\boldsymbol{x}_3$ in $\mathscr{X}$,
$$(\boldsymbol{x}_1 + \boldsymbol{x}_2) + \boldsymbol{x}_3 = \boldsymbol{x}_1 + (\boldsymbol{x}_2 + \boldsymbol{x}_3)$$
4. The set of vectors $\mathscr{X}$ contains a vector denoted by $\mathbf{0}$, such that $\mathbf{0} + \boldsymbol{x} = \boldsymbol{x}$ for every $\boldsymbol{x}$ in $\mathscr{X}$. The vector $\mathbf{0}$ is called the zero vector, or the origin in the vector space.
5. For every $\boldsymbol{x}$ in $\mathscr{X}$, there is a vector $\boldsymbol{y} = -\boldsymbol{x}$ in $\mathscr{X}$ such that $\boldsymbol{x} + \boldsymbol{y} = \mathbf{0}$.
6. For every α in $\mathscr{F}$ and every $\boldsymbol{x}$ in $\mathscr{X}$, there corresponds a vector $\alpha\boldsymbol{x}$ in $\mathscr{X}$ called the *scalar product* of α and $\boldsymbol{x}$.
7. Scalar multiplication is associative; that is, for any α, β in $\mathscr{F}$ and any $\boldsymbol{x}$ in $\mathscr{X}$,
$$\alpha(\beta\boldsymbol{x}) = (\alpha\beta)\boldsymbol{x}$$
8. Scalar multiplication is distributive with respect to vector addition; that is, for any α in $\mathscr{F}$ and any $\boldsymbol{x}_1$, $\boldsymbol{x}_2$ in $\mathscr{X}$,
$$\alpha(\boldsymbol{x}_1 + \boldsymbol{x}_2) = \alpha\boldsymbol{x}_1 + \alpha\boldsymbol{x}_2$$
9. Scalar multiplication is distributive with respect to scalar addition; that is, for any α, β in $\mathscr{F}$ and any $\boldsymbol{x}$ in $\mathscr{X}$,
$$(\alpha + \beta)\boldsymbol{x} = \alpha\boldsymbol{x} + \beta\boldsymbol{x}$$
10. For any $\boldsymbol{x}$ in $\mathscr{X}$, $1\boldsymbol{x} = \boldsymbol{x}$, where 1 is the unity element 1 in $\mathscr{F}$.

In general, a field forms a vector space over itself. That is, the elements contained in a vector within the set of vectors are elements of a field; and if the field over which the vector space is defined (for scalar multiplication and vector addition) is the same field, then all 10 conditions in Definition A.2

are satisfied. Two very important vector spaces are $(\Re,\Re)$and $(\mathscr{C},\mathscr{C})$. It is obvious that $(\mathscr{C},\Re)$ is a vector space but $(\Re,\mathscr{C})$ is not because scalar multiplication will not yield, in general, vectors whose elements are in the field of real numbers. Also $(\Re(s),\Re(s))$ and $(\Re(s),\Re)$ are vector spaces, where $\Re(s)$ denotes the field of rational functions with real coefficients and independent variable s. However, $(\Re,\Re(s))$ is not a vector space. Also, the set of real-valued piecewise continuous functions defined over the interval $(-\infty,\infty)$ forms a vector space over the field of real numbers. Addition and multiplication are defined in the usual manner. This type of vector space is more specifically called a *function space*.

We typically define the dimension of the vector space, that is, the length (or dimension) of the vectors. For example, $(\Re^n,\Re)$ denotes a vector space over the field of real numbers where the vectors have length n; therefore, the vector space is an *n-dimensional real vector space*. And $(\mathscr{C}^n,\mathscr{C})$ is the *n-dimensional complex vector space*, and $(\Re^n(s),\Re(s))$ is the *n-dimensional rational vector space*. In this textbook we typically simplify the designation of a vector space by indicating $\mathscr{F}^n$ as the n-dimensional vector space over the field $\mathscr{F}$. We will be even more specific and distinguish between column and row vectors in a vector space. For example, $\Re^{n\times 1}$ is an n-dimensional real vector space; however, the vectors are considered *column vectors*, and $\Re^{1\times n}$ is an n-dimensional real vector space with row vectors.

Another important vector space is based on the set of polynomials $\Re_n\{s\}$, that is, polynomials (with real coefficients) in s of degree n or less. And $(\Re_n\{s\},\Re)$ is a vector space where vector addition is defined as

$$\sum_{i=1}^{n}\alpha_i s^i+\sum_{i=1}^{n}\beta_i s^i=\sum_{i=1}^{n}(\alpha_i+\beta_i)s^i$$

and scalar multiplication is defined as

$$\gamma\left(\sum_{i=1}^{n}\alpha_i s^i\right)=\sum_{i=1}^{n}\gamma\alpha_i s^i$$

for $\alpha_i,\beta_i,\gamma\in\Re$.

DEFINITION A.3. Let $(\mathscr{X},\mathscr{F})$ be a vector space, and also let $\mathscr{Y}$ be a subset of $\mathscr{X}$. Then $(\mathscr{Y},\mathscr{F})$ is a *subspace* of $(\mathscr{X},\mathscr{F})$ if under the operations of the vector space $(\mathscr{X},\mathscr{F})$, $\mathscr{Y}$ forms a vector space over the field $\mathscr{F}$.

A.2.2 Matrix Representations and Operations

Matrices and vectors

A matrix is an array of elements in which the elements are typically numerical values; however, the entries in the array can be functions. In its most general form, a matrix is rectangular; special cases include square matrices, vectors (row and column vectors), and scalars. Let $A\in\Re^{n\times m}$ denote all rectangular matrices with real entries (numbers) and with dimension n (number of rows in the array) by m (number of columns in the array) that is,

$$A = \begin{bmatrix} a_{11} & a_{12} & \cdots & a_{1m} \\ a_{21} & a_{22} & \cdots & a_{2m} \\ \cdots & \cdots & \cdots & \cdots \\ a_{n1} & a_{n2} & \cdots & a_{nm} \end{bmatrix}$$

The matrix dimension is simply written as $n \times m$. And $\boldsymbol{A} \in \mathscr{C}^{n\times m}$ denotes all rectangular matrices with complex elements. Matrix $\boldsymbol{A}$ can also be written as $\boldsymbol{A}=[a_{ij}]_{n\times m}$; therefore, a_{ij} is the element in the matrix located in the ith row and jth column. Special cases in $\boldsymbol{A} \in \Re^{n\times m}$ include these: (1) if $m = n$, $\boldsymbol{A} \in \Re^{n\times n}$ denotes all square matrices with real elements; (2) if $m = 1$, $\boldsymbol{x} \in \Re^{nx1}$ denotes all column vectors with n real elements; and (3) $\boldsymbol{y} \in \Re^{1\times n}$ denotes all row vectors with n real elements. If we denoted the columns of the matrix $\boldsymbol{A} \in \Re^{n\times m}$ as $\boldsymbol{a}_j$ for $j = 1, 2, \ldots, m$, the matrix $\boldsymbol{A}$ can be written as $\boldsymbol{A} = [\boldsymbol{a}_1, \boldsymbol{a}_2, \ldots, \boldsymbol{a}_m]$.

Matrix (vector) addition and subtraction

Given $\boldsymbol{A} \in \Re^{n\times m}$ and $\boldsymbol{B} \in \Re^{n\times m}$, then

$$\boldsymbol{C} = \boldsymbol{A} \pm \boldsymbol{B} \qquad \text{where } \boldsymbol{C} \in \Re^{n\times m}$$

Therefore, the two matrices must have the same number of rows and columns to be conformable for addition or subtraction. If $\boldsymbol{x} \in \Re^{n\times 1}$and $\boldsymbol{y} \in \Re^{n\times 1}$, then adding or subtracting two column vectors results in

$$\boldsymbol{z} = \boldsymbol{x} \pm \boldsymbol{y} \qquad \text{where } \boldsymbol{z} \in \Re^{n\times 1}$$

For two row vectors that have the same length, the result of their sum and difference is obvious.

Matrix multiplication

Given $\boldsymbol{A} \in \Re^{n\times m}$ with elements a_{ij} and $\boldsymbol{B} \in \Re^{m\times p}$ with elements b_{jk}, multiplying $\boldsymbol{A}$ by $\boldsymbol{B}$ gives

$$\boldsymbol{C} = \boldsymbol{AB} \in \Re^{nxp} \qquad \text{with elements } c_{ik} \quad (\Re^{n\times p} \leftarrow \Re^{n\times m} \times \Re^{m\times p})$$

Each element in $\boldsymbol{C}$ can be written as

$$c_{ik} = \sum_{j=1}^{m} a_{ij} b_{jk} \qquad \text{or} \qquad \boldsymbol{AB} = \left[\sum_{j=1}^{m} a_{ij} b_{jk}\right]_{n\times p}$$

From this result we can see that the number of *rows* of matrix $\boldsymbol{B}$ must be equal to the number of *columns* of $\boldsymbol{A}$, if $\boldsymbol{A}$ and $\boldsymbol{B}$ are to be conformable for matrix multiplication. Thus, $\boldsymbol{BA}$ does not make any sense because $\boldsymbol{B}$ has p columns and $\boldsymbol{A}$ has n rows, which obviously do not match. Therefore, in general, matrix multiplication does not possess the commutative property. However, matrix multiplication does possess associative and distributive properties. A special case of matrix multiplication involves multiplying a matrix by a vector; that is, given $\boldsymbol{A} \in \Re^{n\times m}$ and $\boldsymbol{x} \in \Re^{m\times 1}$

$$\boldsymbol{b} = \boldsymbol{Ax} \in \Re^{n\times 1} \qquad \text{with elements } b_i \ (\Re^{n\times 1} \leftarrow \Re^{n\times m} \times \Re^{m\times 1})$$

Transpose

The transpose of a matrix $A \in \Re^{n\times m}$ is denoted as A^T. To *transpose* a matrix $A = [a_{ij}]_{n\times m}$, the rows and columns of the matrix are interchanged, that is,

$$A^T = [a_{ij}]^T_{n\times m} = [a_{ji}]_{m\times n} \in \Re^{m\times n} \tag{A.1}$$

The following are useful properties of the transpose:

1. $(A^T)^T = A$
2. Given $A \in \Re^{n\times m}$ and $B \in \Re^{m\times p}$, $(AB)^T = B^T A^T \in \Re^{p\times n}$
3. $(A + B)^T = A^T + B^T$

Diagonal, symmetric, and identity matrices

A *diagonal* matrix is a square matrix with only diagonal elements; that is, all other elements (off-diagonal elements) are zero. For example, given $A \in \Re^{n\times n}$ as a diagonal matrix, we will write A as

$$A = \text{diag}[a_{11}, a_{22}, \ldots, a_{nn}]$$

where

$$A = \begin{bmatrix} a_{11} & 0 & \cdots & 0 \\ 0 & a_{22} & \cdots & 0 \\ 0 & \cdots & \ddots & 0 \\ 0 & \cdots & 0 & a_{nn} \end{bmatrix}$$

If $A^T = A$, then A is said to be a *symmetric* matrix. Obviously, any diagonal matrix is symmetric. A diagonal matrix with only unity elements on the diagonal is called an *identity* matrix. For example, an $n \times n$ identity matrix I_n has n unity elements on the diagonal and the off-diagonal elements are zeros, that is,

$$I_n = \begin{bmatrix} 1 & & & \mathbf{0} \\ & 1 & & \\ & & \ddots & \\ \mathbf{0} & & & 1 \end{bmatrix} = \text{diag}[1, 1, \ldots, 1] = [e_1, e_2, \ldots, e_n] \tag{A.2}$$

where the jth column of I_n is e_j; for example, if $j = 2$, $e_2 = [0, 1, 0, \ldots, 0]^T$.

A.2.3 Inner and Outer Products

Assume two n-dimensional column vectors $x \in \Re^{n\times 1}$ and $y \in \Re^{n\times 1}$. The inner product of these two vectors is given by

$$\langle x, y\rangle = x^T y = y^T x = \langle y, x\rangle = \sum_{i=1}^{n} y_i x_i \tag{A.3}$$

If x and y have complex elements, that is, $x \in \mathscr{C}^{n\times 1}$ and $y \in \mathscr{C}^{n\times 1}$, the inner product of x with y is given by

$$\langle \boldsymbol{x}, \boldsymbol{y} \rangle = \bar{\boldsymbol{x}}^T \boldsymbol{y} = \boldsymbol{x}^* \boldsymbol{y} = \sum_{i=1}^{n} \bar{x}_i y_i \tag{A.4}$$

where $\boldsymbol{x}^* = \bar{\boldsymbol{x}}^T$ is the complex conjugate transpose of vector $\boldsymbol{x}$. If we now assume two n-dimensional column vectors that have time-varying elements $\boldsymbol{x}(t)$ and $\boldsymbol{y}(t)$, then on the vector space of continuous real functions in the interval $t_1 < t < t_2$, the inner product of $\boldsymbol{x}$ with $\boldsymbol{y}$ is given by

$$\begin{aligned} \langle \boldsymbol{x}(t), \boldsymbol{y}(t) \rangle &= \frac{1}{t_2 - t_1} \int_{t_1}^{t_2} \boldsymbol{x}^T(t) \boldsymbol{y}(t)\, dt \\ &= \frac{1}{t_2 - t_1} \int_{t_1}^{t_2} \left[\sum_{i=1}^{n} x_i(t) y_i(t) \right] dt \end{aligned} \tag{A.5}$$

The *outer product* of two vectors $\boldsymbol{x} \in \Re^{n \times 1}$ and $\boldsymbol{y} \in \Re^{n \times 1}$ results in a *rank* 1 (see Sect. A.2.5) matrix that has dimension $n \times n$, that is,

$$\boldsymbol{A} = \boldsymbol{x}\boldsymbol{y}^T = \begin{bmatrix} x_1 \\ x_2 \\ \vdots \\ x_n \end{bmatrix} \begin{bmatrix} y_1 & y_2 & \cdots & y_n \end{bmatrix} = \begin{bmatrix} x_1 y_1 & x_1 y_2 & \cdots & x_1 y_n \\ x_2 y_1 & x_2 y_2 & \cdots & x_2 y_n \\ \cdots & \cdots & \cdots & \cdots \\ x_n y_1 & x_n y_2 & \cdots & x_n y_n \end{bmatrix} \tag{A.6}$$

A.2.4 Linear Independence of Vectors

DEFINITION A.4. Let $\boldsymbol{a}_1, \boldsymbol{a}_2, \ldots, \boldsymbol{a}_m$ be a set of vectors in $\Re^{n \times 1}$, and let $\alpha_1, \alpha_2, \ldots, \alpha_m$ be a set of scalars in $\Re$. We can form a *linear combination* of the vectors with the scalars as

$$\boldsymbol{b} = \alpha_1 \boldsymbol{a}_1 + \alpha_2 \boldsymbol{a}_2 + \cdots + \alpha_m \boldsymbol{a}_m = \sum_{i=1}^{m} \alpha_i \boldsymbol{a}_i$$

If the linear combination equals zero, that is, $\boldsymbol{b} = \boldsymbol{0}$, for only $\alpha_i = 0$ $(i = 1, 2, \ldots, m)$, then the set of vectors $\{\boldsymbol{a}_i\}$ is said to be *linearly independent*. However, if there are any $\alpha_i \neq 0 \ni \boldsymbol{b} = \boldsymbol{0}$, then the set of vectors $\{\boldsymbol{a}_i\}$ is said to be *linearly dependent*.

DEFINITION A.4.1. Let $\boldsymbol{a}_1, \boldsymbol{a}_2, \ldots, \boldsymbol{a}_m$ be a set of vectors in $\Re^{n \times 1}$, and let $\alpha_1, \alpha_2, \ldots, \alpha_m$ be a set of scalars in $\Re$. The set of all linear combinations of $\boldsymbol{a}_1, \boldsymbol{a}_2, \ldots, \boldsymbol{a}_m \in \Re^{n \times 1}$ is a subspace called the *span* of $\boldsymbol{a}_1, \boldsymbol{a}_2, \ldots, \boldsymbol{a}_m$ denoted by

$$\text{span}\{\boldsymbol{a}_1, \boldsymbol{a}_2, \ldots, \boldsymbol{a}_m\} \triangleq \{\boldsymbol{a} = \alpha_1 \boldsymbol{a}_1 + \alpha_2 \boldsymbol{a}_2 + \cdots + \alpha_m \boldsymbol{a}_m : \underset{1 \le i \le m}{\alpha_i} \in \Re\}$$

If we form an $n \times m$ matrix from the set of vectors $\{\boldsymbol{a}_i\}$, that is, $\boldsymbol{A} = [\boldsymbol{a}_1, \boldsymbol{a}_2, \ldots, \boldsymbol{a}_m] \in \Re^{n \times m}$, then the columns of matrix $\boldsymbol{A}$ are linearly independent if and only if $\boldsymbol{A}^T \boldsymbol{A}$ is *nonsingular* (cf. Sect. A.2.7), or equivalently $\boldsymbol{A}^T \boldsymbol{A}$ has *full rank* (cf. Sect. A.2.5). The rows of $\boldsymbol{A} \in \Re^{n \times m}$ are linearly independent if and only if $\boldsymbol{A}\boldsymbol{A}^T$ is nonsingular (or $\boldsymbol{A}\boldsymbol{A}^T$ has full rank).

A.2.5 Rank of a Matrix and Linear Independence

DEFINITION A.5. The *rank* of a matrix $\boldsymbol{A}$ is the *maximum* number of linearly independent columns, or the *maximum* number of linearly independent rows. The rank of a matrix is denoted by $\rho(\boldsymbol{A})$. A matrix $\boldsymbol{A} \in \Re^{n \times m}$ has *full rank* if $\rho(\boldsymbol{A}) = \min\{n, m\}$. The rank of a matrix can also be defined as the dimension of the largest nonsingular (cf. Sect. A.2.7) minor (square matrix) contained in $\boldsymbol{A}$.

Given $\boldsymbol{A} \in \Re^{n \times m}$, for $\min\{n, m\} = m$, and $\rho(\boldsymbol{A}) < m$, then matrix $\boldsymbol{A}$ is said to be *rank-deficient*. The rank of a matrix is sometimes referred to as the *underlying dimensionality* of the matrix. Some useful properties involving the rank of a matrix $\boldsymbol{A} \in \Re^{n \times m}$ are as follows:

1. $\rho(\boldsymbol{A}^T) = \rho(\boldsymbol{A})$
2. $\rho(\boldsymbol{A}^T\boldsymbol{A}) = \rho(\boldsymbol{A})$
3. $\rho(\boldsymbol{A}\boldsymbol{A}^T) = \rho(\boldsymbol{A})$
4. Given $\boldsymbol{A} \in \Re^{n \times m}$ and $\boldsymbol{B} \in \Re^{m \times p}$, then $\rho(\boldsymbol{A}) + \rho(\boldsymbol{B}) \le m + \rho(\boldsymbol{AB})$
5. Given $\boldsymbol{A} \in \Re^{n \times m}$ and $\boldsymbol{B} \in \Re^{m \times p}$, then

$$\rho(\boldsymbol{A}) + \rho(\boldsymbol{B}) - m \le \rho(\boldsymbol{AB}) \le \min\{\rho(\boldsymbol{A}), \rho(\boldsymbol{B})\} \qquad \text{Sylvester's inequality}$$

A.2.6 Matrix Definiteness

DEFINITION A.6. A symmetric matrix $\boldsymbol{A} \in \Re^{n \times n}$ is said to be *positive definite* if $\boldsymbol{x}^T\boldsymbol{A}\boldsymbol{x} > 0$, $\forall \boldsymbol{x} \in \Re^{n \times 1}$ (except $\boldsymbol{x} = \boldsymbol{0}$). Matrix $\boldsymbol{A}$ is said to be *negative definite* if $\boldsymbol{x}^T\boldsymbol{A}\boldsymbol{x} < 0$, *positive semidefinite* (or *nonnegative definite*) if $\boldsymbol{x}^T\boldsymbol{A}\boldsymbol{x} \ge 0$, and *negative semidefinite* (or *nonpositive definite*) if $\boldsymbol{x}^T\boldsymbol{A}\boldsymbol{x} \le 0$.

For symmetric $\boldsymbol{A} \in \Re^{n \times n}$ (i.e., $\boldsymbol{A}^T = \boldsymbol{A}$), we can also say if $\boldsymbol{A}$ is positive definite, then the *eigenvalues* (cf. Sect. A.2.9) of $\boldsymbol{A}$ are all real and positive; if $\boldsymbol{A}$ is negative definite, then the eigenvalues of $\boldsymbol{A}$ are all real and negative; if $\boldsymbol{A}$ is positive semidefinite, then some eigenvalues of $\boldsymbol{A}$ can be zero (but not all), and the rest must be real and positive; and if $\boldsymbol{A}$ is negative semidefinite, then some eigenvalues of $\boldsymbol{A}$ can be zero (but not all), and the rest must be real and negative. If a symmetric matrix $\boldsymbol{A} \in \Re^{n \times n}$ has both positive and negative real eigenvalues, $\boldsymbol{A}$ is said to be *indefinite*. Matrices $\boldsymbol{A}^T\boldsymbol{A}$ and $\boldsymbol{A}\boldsymbol{A}^T$ are positive semidefinite for any $\boldsymbol{A}$.

Given symmetric $\boldsymbol{A} \in \Re^{n \times n}$, for simplicity we will write

1. $\boldsymbol{A} > 0$, positive definite $\boldsymbol{A}$.
2. $\boldsymbol{A} < 0$, negative definite $\boldsymbol{A}$.
3. $\boldsymbol{A} \ge 0$, positive semidefinite (nonnegative definite) $\boldsymbol{A}$.
4. $\boldsymbol{A} \le 0$, negative semidefinite (nonpositive definite) $\boldsymbol{A}$.

A.2.7 Inverse and Pseudoinverse of a Matrix

Inverse of a matrix

Given the matrix $\boldsymbol{A} \in \Re^{n \times n}$ with $\rho(\boldsymbol{A}) = n$, then $\boldsymbol{A}$ is said to have an inverse, or $\boldsymbol{A}$ is *nonsingular*, or the columns (rows) of $\boldsymbol{A}$ are linearly independent. The inverse of $\boldsymbol{A}$ is denoted as $\boldsymbol{A}^{-1}$, and $\boldsymbol{A}\boldsymbol{A}^{-1} = \boldsymbol{A}^{-1}\boldsymbol{A} = \boldsymbol{I}_n$. If $\rho(\boldsymbol{A}) < n$, then $\boldsymbol{A}$ is rank-deficient and $\boldsymbol{A}$ is said to be *singular*. As we can see, the inverse of a matrix is a generalization of the familiar notion of the reciprocal of a scalar (multiplicative inverse), which was discussed in Section A.2.1 in defining a field.

DEFINITION A.7. The inverse of the nonsingular matrix $\boldsymbol{A} \in \Re^{n\times n}$ is given as

$$\boldsymbol{A}^{-1} = \frac{\text{adj}(\boldsymbol{A})}{|\boldsymbol{A}|} = \frac{[\text{cof}(\boldsymbol{A})]^T}{|\boldsymbol{A}|} \tag{A.7}$$

where adj $\triangleq$ adjoint, cof $\triangleq$ cofactor, and $|\boldsymbol{A}|$ denotes the determinant of $\boldsymbol{A}$ (explained below). The determinant of a matrix $\boldsymbol{A} \in \Re^{n\times n}$ can also be denoted as $\det(\boldsymbol{A})$. If $|\boldsymbol{A}| = 0$, then $\boldsymbol{A}$ is singular (i.e., an inverse does not exist for $\boldsymbol{A}$). We can also say that $\boldsymbol{A}$ is rank-deficient, that is, $\rho(\boldsymbol{A}) < n$, if $\boldsymbol{A}$ has linearly dependent rows (or columns) or if $\boldsymbol{A}$ has at least one zero eigenvalue.

These are some useful properties involving matrix inversion (given $\boldsymbol{A} \in \Re^{n\times n}$, $\boldsymbol{B} \in \Re^{n\times n}$, $\boldsymbol{C} \in \Re^{m\times m}$, $\boldsymbol{u} \in \Re^{n\times 1}$, $\boldsymbol{v} \in \Re^{n\times 1}$):

1. $(\boldsymbol{A}^{-1})^{-1} = \boldsymbol{A}$
2. $(\boldsymbol{A}\boldsymbol{B})^{-1} = \boldsymbol{B}^{-1}\boldsymbol{A}^{-1}$
3. $(\boldsymbol{A}^T)^{-1} = (\boldsymbol{A}^{-1})^T = \boldsymbol{A}^{-T}$
4. $(\boldsymbol{A} + \boldsymbol{u}\boldsymbol{v}^T)^{-1} = \boldsymbol{A}^{-1} - \dfrac{(\boldsymbol{A}^{-1}\boldsymbol{u})(\boldsymbol{v}^T\boldsymbol{A}^{-1})}{1 + \boldsymbol{v}^T\boldsymbol{A}^{-1}\boldsymbol{u}}$
5. $(\boldsymbol{C} + \boldsymbol{D}\boldsymbol{B}\boldsymbol{E})^{-1} = \boldsymbol{C}^{-1} - \boldsymbol{C}^{-1}\boldsymbol{D}(\boldsymbol{E}\boldsymbol{C}^{-1}\boldsymbol{D} + \boldsymbol{B}^{-1})^{-1}\boldsymbol{E}\boldsymbol{C}^{-1}$
6. $(\boldsymbol{C} - \boldsymbol{D}\boldsymbol{B}^{-1}\boldsymbol{E})^{-1} = \boldsymbol{C}^{-1} + \boldsymbol{C}^{-1}\boldsymbol{D}(\boldsymbol{B} - \boldsymbol{E}\boldsymbol{C}^{-1}\boldsymbol{D})^{-1}\boldsymbol{E}\boldsymbol{C}^{-1}$

The determinant of a matrix $\boldsymbol{A} \in \Re^{n\times n}$ plays an important role in many different areas. We saw above how the inverse of a matrix can be defined in terms of the matrix determinant. The determinant of a 2×2 matrix $\boldsymbol{A} \in \Re^{2\times 2}$ is computed as

$$|\boldsymbol{A}| = \begin{vmatrix} a_{11} & a_{12} \\ a_{21} & a_{22} \end{vmatrix} = a_{11}a_{22} - a_{12}a_{21}$$

and the determinant of a 3×3 matrix $\boldsymbol{A} \in \Re^{3\times 3}$ is computed as

$$|\boldsymbol{A}| = \begin{vmatrix} a_{11} & a_{12} & a_{13} \\ a_{21} & a_{22} & a_{23} \\ a_{31} & a_{32} & a_{33} \end{vmatrix} = a_{11}\begin{vmatrix} a_{22} & a_{23} \\ a_{32} & a_{33} \end{vmatrix} - a_{12}\begin{vmatrix} a_{21} & a_{23} \\ a_{31} & a_{33} \end{vmatrix} + a_{13}\begin{vmatrix} a_{21} & a_{22} \\ a_{31} & a_{32} \end{vmatrix}$$

Given $\boldsymbol{A} \in \Re^{n\times n}$ and $\boldsymbol{B} \in \Re^{n\times n}$, these are some properties of determinants:

1. $|\boldsymbol{A}| = 0$, if all elements in any one row (or column) of $\boldsymbol{A}$ are zero.
2. $|\boldsymbol{A}^T| = |\boldsymbol{A}|$
3. $|\boldsymbol{A}\boldsymbol{B}| = |\boldsymbol{B}\boldsymbol{A}| = |\boldsymbol{A}||\boldsymbol{B}|$

4. $$\begin{vmatrix} a_{11} & 0 & 0 & \cdots & 0 \\ a_{21} & a_{22} & 0 & \cdots & 0 \\ a_{31} & a_{32} & a_{33} & \cdots & 0 \\ \cdots & \cdots & \cdots & \cdots & \cdots \\ a_{n1} & a_{n2} & a_{n3} & \cdots & a_{nn} \end{vmatrix} = \begin{vmatrix} a_{11} & a_{12} & a_{13} & \cdots & a_{1n} \\ 0 & a_{22} & a_{23} & \cdots & a_{2n} \\ 0 & 0 & a_{33} & \cdots & a_{3n} \\ \cdots & \cdots & \cdots & \cdots & \cdots \\ 0 & 0 & 0 & \cdots & a_{nn} \end{vmatrix}$$
$$= a_{11}a_{22}a_{33}\cdots a_{nn} = \prod_{i=1}^{n} a_{ii}$$

5. If a matrix $\boldsymbol{B}$ is formed from $\boldsymbol{A}$ by interchanging two rows or two columns of $\boldsymbol{A}$, then $|\boldsymbol{A}| = -|\boldsymbol{B}|$.
6. If a matrix $\boldsymbol{B}$ is formed from $\boldsymbol{A}$ by multiplying every element of a row or column of $\boldsymbol{A}$ by a scalar $k \in \Re$, then $|\boldsymbol{A}| = (1/k)|\boldsymbol{B}|$.
7. $|k\boldsymbol{A}| = k^n|\boldsymbol{A}|$
8. If a matrix $\boldsymbol{B}$ is formed from $\boldsymbol{A}$ by adding a constant times one row (or column) of $\boldsymbol{A}$ to another row (or column) of $\boldsymbol{A}$, then $|\boldsymbol{A}| = |\boldsymbol{B}|$.
9. If two rows (or columns) of $\boldsymbol{A}$ are equal, then $|\boldsymbol{A}| = 0$ [i.e., if any rows or columns of $\boldsymbol{A}$ are linearly dependent, or $\rho(\boldsymbol{A}) < n$, then $|\boldsymbol{A}| = 0$].
10. If $\lambda_1, \lambda_2, \ldots, \lambda_n$ are the eigenvalues of $\boldsymbol{A} \in \Re^{n\times n}$ (cf. Sect. A.2.9), then $|\boldsymbol{A}| = \prod_{i=1}^{n} \lambda_i$; therefore, if any eigenvalue of $\boldsymbol{A}$ is zero, then $|\boldsymbol{A}| = 0$ and $\boldsymbol{A}$ is *singular*.
11. If $\rho(\boldsymbol{A}) = n$ (or all columns or rows are linearly independent, or $|\boldsymbol{A}| \neq 0$, or $\boldsymbol{A}$ is nonsingular, or $\boldsymbol{A}$ has an inverse), then $|\boldsymbol{A}^{-1}| = 1/|\boldsymbol{A}|$.
12. Given $\boldsymbol{A} \in \Re^{n\times n}$, $\boldsymbol{B} \in \Re^{n\times m}$, $\boldsymbol{C} \in \Re^{m\times n}$, and $\boldsymbol{D} \in \Re^{m\times m}$, if $\boldsymbol{A}$ and $\boldsymbol{D}$ are invertible, then $\det(\boldsymbol{A})\det(\boldsymbol{D} - \boldsymbol{C}\boldsymbol{A}^{-1}\boldsymbol{B}) = \det(\boldsymbol{D})\det(\boldsymbol{A} - \boldsymbol{B}\boldsymbol{D}^{-1}\boldsymbol{C})$.

Pseudoinverse of a matrix

For a square matrix that is singular, or a rectangular matrix, a generalized inverse can be computed. This can be very useful for solving a system of simultaneous linear algebraic equations of the form $\boldsymbol{Ax} = \boldsymbol{b}$, where $\boldsymbol{A} \in \Re^{m\times n}$, $\boldsymbol{x} \in \Re^{n\times 1}$, and $\boldsymbol{b} \in \Re^{m\times 1}$. The Moore-Penrose generalized inverse (or pseudoinverse) of $\boldsymbol{A}$ is denoted by $\boldsymbol{A}^+$, and has the following properties:

1. $\boldsymbol{A}^+ = \left(\boldsymbol{A}^T\boldsymbol{A}\right)^{-1}\boldsymbol{A}^T$
2. $\boldsymbol{A}^+\boldsymbol{A}\boldsymbol{A}^+ = \boldsymbol{A}^+$
3. $\boldsymbol{A}\boldsymbol{A}^+\boldsymbol{A} = \boldsymbol{A}$
4. $(\boldsymbol{A}\boldsymbol{A}^+)^T = \boldsymbol{A}\boldsymbol{A}^+$
5. $(\boldsymbol{A}^+\boldsymbol{A})^T = \boldsymbol{A}^+\boldsymbol{A}$

If $m = n$, then $\boldsymbol{A}$ is a square matrix and $\boldsymbol{A}^+ = \boldsymbol{A}^{-1}$, if $\rho(\boldsymbol{A}) = n$. However, if we assume $m > n$ (overdetermined case) and define an error vector $\boldsymbol{e} \in \Re^{m\times 1}$, such that, $\boldsymbol{e} = \boldsymbol{Ax} - \boldsymbol{b}$, and also define an error function given by

$$\mathscr{E}(\boldsymbol{x}) = \frac{1}{2}\|\boldsymbol{e}\|_2^2 = \frac{1}{2}\|\boldsymbol{Ax} - \boldsymbol{b}\|_2^2 = \frac{1}{2}(\boldsymbol{Ax} - \boldsymbol{b})^T(\boldsymbol{Ax} - \boldsymbol{b}) \tag{A.8}$$

then minimizing $\mathscr{E}(\boldsymbol{x})$ with respect to the $\boldsymbol{x}$ (i.e., $\partial\mathscr{E}(\boldsymbol{x})/\partial\boldsymbol{x} = \boldsymbol{0}$) leads to the *normal equations*

$$\boldsymbol{A}^T\boldsymbol{Ax} - \boldsymbol{A}^T\boldsymbol{b} = \boldsymbol{0} \tag{A.9}$$

Solving the normal equations for the *least-squares* solution $\boldsymbol{x}^*$ to $\boldsymbol{Ax} = \boldsymbol{b}$ gives

$$\boldsymbol{x}^* = (\boldsymbol{A}^T\boldsymbol{A})^{-1}\boldsymbol{A}^T\boldsymbol{b} \tag{A.10}$$

if $\boldsymbol{A}^T\boldsymbol{A}$ is nonsingular [that is, $\rho(\boldsymbol{A}) = n$]. We define the pseudoinverse, that is, the Moore-Penrose generalized inverse, as $\boldsymbol{A}^+ = (\boldsymbol{A}^T\boldsymbol{A})^{-1}\boldsymbol{A}^T$, therefore, $\boldsymbol{x}^* = \boldsymbol{A}^+\boldsymbol{b}$. If $m < n$ (underdetermined case), then the pseudoinverse is given by $\boldsymbol{A}^+ = \boldsymbol{A}^T(\boldsymbol{A}\boldsymbol{A}^T)^{-1}$, if $\boldsymbol{A}\boldsymbol{A}^T$ is nonsingular [i.e., $\rho(\boldsymbol{A}) = m$]. These are some other useful properties associated with pseudoinverse matrices:

1. $(\alpha\boldsymbol{A})^+ = \alpha^{-1}\boldsymbol{A}^+ \qquad \text{if } \alpha \neq 0$
2. $(\boldsymbol{A}^+)^+ = \boldsymbol{A}$
3. $(\boldsymbol{A}^+)^T = (\boldsymbol{A}^T)^+$
4. $\boldsymbol{A}^+ = (\boldsymbol{A}^T\boldsymbol{A})^+\boldsymbol{A}^T = \boldsymbol{A}^T(\boldsymbol{A}\boldsymbol{A}^T)^+$
5. $\boldsymbol{A}\boldsymbol{A}^T(\boldsymbol{A}^+)^T = \boldsymbol{A}$
6. $\boldsymbol{A}^+\boldsymbol{A}\boldsymbol{A}^T = \boldsymbol{A}^T$
7. $(\boldsymbol{A}^+)^T\boldsymbol{A}^T\boldsymbol{A} = \boldsymbol{A}$
8. $\boldsymbol{A}^T\boldsymbol{A}\boldsymbol{A}^+ = \boldsymbol{A}^T$
9. $\rho(\boldsymbol{A}^+) = \rho(\boldsymbol{A}) = \rho(\boldsymbol{A}^T)$

One of the problems with defining the pseudoinverse of a matrix in this manner is that the matrix must be full-rank, that is, $\rho(\boldsymbol{A}) = n$ when $m \geq n$ and $\rho(\boldsymbol{A}) = m$ when $m < n$. This problem can be overcome by defining the pseudoinverse of a matrix through its singular-value decomposition (SVD) (cf. Sect. A.2.14).

A.2.8 Orthogonal and Unitary Matrices, and Conjugate Vectors

Orthogonal and unitary matrices

Assume a set of nonzero vectors $\{\boldsymbol{q}_1, \boldsymbol{q}_2, \ldots, \boldsymbol{q}_n\}$ where $\boldsymbol{q}_i \in \Re^{n\times 1}$ for $i = 1, 2, \ldots, n$. The vectors in the set $\{\boldsymbol{q}_i\}$, for $i = 1, 2, \ldots, n$, are considered *orthogonal* if

$$\boldsymbol{q}_i^T\boldsymbol{q}_j = 0 \qquad \text{for} \quad i \neq j$$

If the set of vectors is *orthonormal*, then

$$\boldsymbol{q}_i^T\boldsymbol{q}_j = \left\{\begin{matrix} 0 & \text{for } i \neq j \\ 1 & \text{for } i = j \end{matrix}\right\} = \delta_{ij} \tag{A.11}$$

where δ_{ij} is the *Kronecker delta*. If we define a square matrix $\boldsymbol{Q} \triangleq \{\boldsymbol{q}_1, \boldsymbol{q}_2, \ldots, \boldsymbol{q}_n\}$, then $\boldsymbol{Q} \in \Re^{n\times n}$ is called an *orthogonal matrix* and $\boldsymbol{Q}^T\boldsymbol{Q} = \boldsymbol{Q}\boldsymbol{Q}^T = \boldsymbol{I}_n$. The complex counterpart of an orthogonal matrix is called a *unitary* matrix. Therefore, if $\boldsymbol{Q} \in \mathscr{C}^{n\times n}$ is unitary, then $\boldsymbol{Q}^*\boldsymbol{Q} = \boldsymbol{I}_n$, where $\boldsymbol{Q}^*$ is the complex conjugate transpose of $\boldsymbol{Q}$, that is, $\boldsymbol{Q}^* = \bar{\boldsymbol{Q}}^T$. The complex conjugate transpose of $\boldsymbol{Q}$ is also written as $\boldsymbol{Q}^H$, the *Hermitian transpose* of $\boldsymbol{Q}$. A matrix $\boldsymbol{Q} \in \mathscr{C}^{n\times n}$ is *normal* if $\boldsymbol{Q}\boldsymbol{Q}^H = \boldsymbol{Q}^H\boldsymbol{Q}$ [1].

An important property of orthogonal matrices is that they do not have any effect on inner products. For example, given $\boldsymbol{Q} \in \Re^{n\times n}$ orthogonal and $\boldsymbol{x}, \boldsymbol{y} \in \Re^{n\times 1}$, then

$$\langle Qx, Qy \rangle = (Qx)^T Qy = x^T Q^T Qy = x^T y = \langle x, y \rangle$$

and

$$\| Qx \|_2 = [(Qx)^T Qx]^{1/2} = (x^T Q^T Qx)^{1/2} = (x^T x)^{1/2} = \|x\|_2 = \langle x, x \rangle^{1/2}$$

Conjugate vectors

DEFINITION A.8. Given a symmetric matrix $Q \in \Re^{n \times n}$ and two vectors $d_1 \in \Re^{n \times 1}$ and $d_2 \in \Re^{n \times 1}$, d_1 and d_2 are said to be *conjugate* with respect to Q (or Q-orthogonal) if

$$d_1^T Q d_2 = 0$$

A set of (nonzero) vectors $\{d_0, d_1, \ldots, d_{n-1}\}$, where $d_i \in \Re^{n \times 1}$, for $i = 1, 2, \ldots, n$, is said to be a Q-orthogonal set if

$$d_i^T Q d_j = 0 \qquad \text{for } i \neq j \tag{A.12}$$

From this definition we see that if $Q = I_n$ then the concept of conjugacy is equivalent to the notion of orthogonality. Now if $Q \in \Re^{n \times n}$ is positive definite ($Q > 0$) and the set of nonzero vectors $\{d_0, d_1, \ldots, d_{n-1}\}$ is Q-orthogonal, then these vectors are linearly independent.

A.2.9 Eigenvalues and Eigenvectors

Given a matrix $A \in \Re^{n \times n}$, for a scalar λ and a nonzero vector v, if

$$Av = \lambda v \tag{A.13}$$

then λ is an *eigenvalue* of A and v is the associated *eigenvector*. Considering all eigenvalues and eigenvectors of A, the standard eigenvalue problem is given by

$$(\lambda_i I - A) v_i = \mathbf{0} \qquad \text{or} \qquad (A - \lambda_i I) v_i = \mathbf{0} \qquad \text{for } i = 1, 2, \ldots, n \tag{A.14}$$

This system of equations has a solution if and only if

$$|\lambda I - A| = 0 \tag{A.15}$$

which is referred to as the *characteristic equation* of A. The roots of the polynomial $|\lambda I - A|$ are the eigenvalues; that is, $\{\lambda_i\}$ for $i = 1, 2, \ldots, n$, and for each eigenvalue the associated eigenvectors can be found from $(\lambda_i I - A) v_i = \mathbf{0}$, provided the eigenvalues are *distinct*. The set of eigenvectors associated with the nonzero eigenvalues of A is always *linearly independent*.

In general, the eigenvalues can be distinct or nondistinct, and they can be real or complex. However, because A is considered a real matrix, if any eigenvalues are complex, they must exist in complex conjugate pairs. The set of eigenvalues of a matrix is sometimes referred to as the *spectrum* of A, denoted by $\sigma(A)$, and a specific eigenvalue is denoted by $\lambda(A)$, that is, $\lambda_i(A)$ for $i = 1, 2, \ldots, n$. For A with distinct eigenvalues, if we form a matrix $V = [v_1, v_2, \ldots, v_n] \in \Re^{n \times n}$ whose columns are the associated eigenvectors, then if we form

$$V^{-1}AV = \Lambda \tag{A.16}$$

the matrix $\Lambda = \text{diag}[\lambda_1, \lambda_2, \ldots, \lambda_n]$ has the eigenvalues of A on the diagonal, and A is said to be *diagonalized.* The matrix V is referred to as a *similarity transformation* (cf. Sect. A.2.10). For a nonsingular matrix $A \in \Re^{n \times n}$, that is, $\rho(A) = n$, all eigenvalues will be nonzero, and from (A.16) it is easy to see that the eigenvalues of A^{-1} will be the reciprocals of the eigenvalues of A, that is, $\Lambda^{-1} = \text{diag}[1/\lambda_1, 1/\lambda_2, \cdots, 1/\lambda_n]$.

For *nondistinct* eigenvalues of $A \in \Re^{n \times n}$, A may or may not be diagonalizable. For simplicity, but without loss of generality, we will assume that there exists only one set of eigenvalues of A that are repeated. The multiplicity m is the number of repeated eigenvalues of A; therefore, $m \leq n$, and if $m < n$, then the remaining $n - m$ eigenvalues are considered distinct. The *nullity* of A, that is, $\nu(A)$, is given by

$$\nu(A) = n - \rho(A) \tag{A.17}$$

We will assume an eigenvalue of A with multiplicity m, that is, λ_k (for $k = 1, 2, \ldots, m$), or $\lambda_1 = \lambda_2 = \cdots = \lambda_m = \lambda$. If $\nu(\lambda I - A) = m$, then there exist m linearly independent eigenvectors associated with the m repeated eigenvalues. This is sometimes referred to as *full degeneracy*. The m linearly independent eigenvectors associated with the eigenvalues $\lambda_k = \lambda$, for $k = 1, 2, \ldots, m$, can be found from the nonzero columns of the matrix

$$\frac{1}{m-1} \left\{ \frac{d^{m-1}}{d\lambda^{m-1}} [\text{adj}(\lambda I - A)] \right\} \Bigg|_{\lambda=\lambda_k}$$

The remaining $n - m$ eigenvectors associated with the remaining distinct eigenvalues can be found from $(\lambda_j I - A)v_j = \mathbf{0}$ for $j = m + 1, m + 2, \ldots, n$. We can form a similarity transformation V from the entire set of n linearly independent eigenvectors (the first m columns of V are the eigenvectors associated with the eigenvalue with multiplicity m, and the remaining $n - m$ columns are eigenvectors associated with the distinct eigenvalues). Using the transformation V, we can form

$$V^{-1}AV = \text{diag}[\lambda, \lambda, \cdots, \lambda, \lambda_{m+1}, \lambda_{m+2}, \cdots, \lambda_n] \tag{A.18}$$

and we see that A can still be diagonalized. However, if $\nu(\lambda I - A) = 1$ (this sometimes is referred to as *simple degeneracy*), then there is only one eigenvector associated with $\lambda_1 = \lambda_2 = \cdots = \lambda_m = \lambda$, regardless of the multiplicity m. The one eigenvector can be found from the standard expression $(\lambda I - A)v = \mathbf{0}$. However, there are $m-1$ additional vectors associated with the repeated eigenvalue with multiplicity m, called *generalized eigenvectors* for which all m vectors are linearly independent. The $m-1$ generalized eigenvectors can be found from

$$
\begin{aligned}
(\lambda \boldsymbol{I} - \boldsymbol{A})\boldsymbol{v}_2 &= -\boldsymbol{v}_1 \\
(\lambda \boldsymbol{I} - \boldsymbol{A})\boldsymbol{v}_3 &= -\boldsymbol{v}_2 \\
(\lambda \boldsymbol{I} - \boldsymbol{A})\boldsymbol{v}_4 &= -\boldsymbol{v}_3 \\
&\vdots \\
(\lambda \boldsymbol{I} - \boldsymbol{A})\boldsymbol{v}_{m-1} &= -\boldsymbol{v}_{m-2} \\
(\lambda \boldsymbol{I} - \boldsymbol{A})\boldsymbol{v}_m &= -\boldsymbol{v}_{m-1}
\end{aligned} \tag{A.19}
$$

where $\boldsymbol{v}_1$ is determined by solving the standard expression $(\lambda \boldsymbol{I} - \boldsymbol{A})\boldsymbol{v}_1 = \boldsymbol{0}$. The remaining $n-m$ eigenvectors associated with the remaining distinct eigenvalues can be found from $(\lambda_j \boldsymbol{I} - \boldsymbol{A})\boldsymbol{v}_j = \boldsymbol{0}$ for $j = m+1, m+2, \ldots, n$. We can form a similarity transformation $\boldsymbol{V}$ from the entire set of n linearly independent eigenvectors (the first m columns of $\boldsymbol{V}$ are the generalized eigenvectors associated with the eigenvalue with multiplicity m, and the remaining $n-m$ columns are eigenvectors associated with the distinct eigenvalues). Using the transformation $\boldsymbol{V}$, we can form

$$
\boldsymbol{V}^{-1}\boldsymbol{A}\boldsymbol{V} = \begin{bmatrix}
\lambda & 1 & 0 & 0 & \cdots & 0 & 0 & \cdots & 0 \\
0 & \lambda & 1 & 0 & 0 & 0 & 0 & \cdots & 0 \\
0 & 0 & \lambda & 1 & 0 & 0 & 0 & \cdots & 0 \\
 & \vdots & & \ddots & 1 & & & \vdots & \\
0 & 0 & \cdots & 0 & \lambda & 0 & 0 & \cdots & 0 \\
0 & 0 & 0 & \cdots & 0 & \lambda_{m+1} & 0 & \cdots & 0 \\
0 & 0 & 0 & \cdots & 0 & 0 & \lambda_{m+2} & 0 & 0 \\
 & & \vdots & & & 0 & 0 & \ddots & \\
0 & 0 & 0 & \cdots & 0 & 0 & \cdots & 0 & \lambda_n
\end{bmatrix} \tag{A.20}
$$

In Section A.2.11 we discuss the issues relating to $1 < \nu(\lambda_k \boldsymbol{I} - \boldsymbol{A}) < m$, that is, when the nullity of $(\lambda_k \boldsymbol{I} - \boldsymbol{A})$ is between 1 (simple degeneracy) and the multiplicity m (full degeneracy).

The *trace* of a square matrix is defined as the sum of its diagonal elements. For example, the trace of $\boldsymbol{A} \in \Re^{n \times n}$ is given by

$$
\text{trace}(\boldsymbol{A}) = \text{tr}(\boldsymbol{A}) = \sum_{i=1}^{n} a_{ii}
$$

Some of the properties of the trace are summarized as follows:

1. $\text{trace}(\boldsymbol{A}) = \text{trace}(\boldsymbol{A}^T)$
2. Given $\boldsymbol{A} \in \Re^{n \times m}$ and $\boldsymbol{B} \in \Re^{m \times n}$, then

$$
\begin{aligned}
&\text{trace}(\boldsymbol{AB}) = \text{trace}(\boldsymbol{BA}) = \text{trace}(\boldsymbol{A}^T\boldsymbol{B}^T) = \text{trace}(\boldsymbol{B}^T\boldsymbol{A}^T) \\
&\text{trace}(\boldsymbol{AA}^T) = \text{trace}(\boldsymbol{A}^T\boldsymbol{A})
\end{aligned}
$$

3. Given $\boldsymbol{A}$, $\boldsymbol{B} \in \Re^{n \times n}$ and α, $\beta \in \Re$, then

$$
\text{trace}(\alpha\boldsymbol{A} + \beta\boldsymbol{B}) = \alpha\,\text{trace}(\boldsymbol{A}) + \beta\,\text{trace}(\boldsymbol{B})
$$

4. $\text{trace}(\boldsymbol{A}) = \sum_{i=1}^{n} a_{ii} = \sum_{i=1}^{n} \lambda_i$, where λ_i for $i = 1, 2, \ldots, n$ are the eigenvalues of $\boldsymbol{A}$.

If we let $\Delta(\lambda) = |\lambda \boldsymbol{I} - \boldsymbol{A}| = \lambda^n + \alpha_1 \lambda^{n-1} + \alpha_2 \lambda^{n-2} + \cdots + \alpha_{n-1}\lambda + \alpha_n$ be the *characteristic polynomial* of $\boldsymbol{A} \in \Re^{n\times n}$, then

$$\Delta(\boldsymbol{A}) = \boldsymbol{A}^n + \alpha_1 \boldsymbol{A}^{n-1} + \alpha_2 \boldsymbol{A}^{n-2} + \cdots + \alpha_{n-1}\boldsymbol{A} + \alpha_n \boldsymbol{I} = \boldsymbol{0}$$

In other words, every square matrix satisfies its own characteristic equation. This is known as the *Cayley-Hamilton theorem.*

These are some additional properties of eigenvalues and eigenvectors:

1. If $\boldsymbol{x}$ is an eigenvector of $\boldsymbol{A}$ corresponding to the eigenvalue λ and $\boldsymbol{A}$ is invertible, then $\boldsymbol{x}$ is an eigenvector of $\boldsymbol{A}^{-1}$ corresponding to its eigenvalue $1/\lambda$.
2. If $\boldsymbol{x}$ is an eigenvector of $\boldsymbol{A}$, then $k\boldsymbol{x}$ (where k is a nonzero constant) is also an eigenvector of $\boldsymbol{A}$, where $\boldsymbol{x}$ and $k\boldsymbol{x}$ both correspond to the same eigenvalue.
3. A matrix and its transpose have the same eigenvalues.
4. The eigenvalues of a lower and upper triangular matrix are the main-diagonal elements.
5. If $\boldsymbol{x}$ is an eigenvector of $\boldsymbol{A}$ corresponding to the eigenvalue λ, then $\boldsymbol{x}$ is an eigenvector of $\boldsymbol{A} - \alpha \boldsymbol{I}$ corresponding to the eigenvalue $\lambda - \alpha$ for any scalar α.

A.2.10 Similarity Transformations

Given the matrices $\boldsymbol{A} \in \Re^{n\times n}$ and $\boldsymbol{P} \in \Re^{n\times n}$, where it is assumed $\rho(\boldsymbol{P}) = n$,

$$\bar{\boldsymbol{A}} = \boldsymbol{P}^{-1}\boldsymbol{A}\boldsymbol{P}$$

Then $\boldsymbol{A}$ and $\bar{\boldsymbol{A}}$ have the same eigenvalues (i.e., $\boldsymbol{A}$ and $\bar{\boldsymbol{A}}$ are similar matrices), or equivalently they have the same characteristic equations, that is, $\Delta(\lambda) = |\lambda \boldsymbol{I} - \boldsymbol{A}| = 0$ and $\bar{\Delta}(\lambda) = \left|\lambda \boldsymbol{I} - \bar{\boldsymbol{A}}\right| = 0$. This can be proved by the following

$$\bar{\Delta}(\lambda) = \left|\lambda \boldsymbol{I} - \bar{\boldsymbol{A}}\right| = \left|\lambda \boldsymbol{I} - \boldsymbol{P}^{-1}\boldsymbol{A}\boldsymbol{P}\right| = \left|\boldsymbol{P}^{-1}\lambda \boldsymbol{I}\boldsymbol{P} - \boldsymbol{P}^{-1}\boldsymbol{A}\boldsymbol{P}\right| = \boldsymbol{P}^{-1}|\lambda \boldsymbol{I} - \boldsymbol{A}|\boldsymbol{P} = 0$$

and now premultiplying $\bar{\Delta}(\lambda) = \boldsymbol{P}^{-1}|\lambda \boldsymbol{I} - \boldsymbol{A}|\boldsymbol{P} = 0$ by $\boldsymbol{P}$ and postmultiplying by $\boldsymbol{P}^{-1}$ give

$$\underbrace{\boldsymbol{P}\boldsymbol{P}^{-1}}_{\boldsymbol{I}}|\lambda \boldsymbol{I} - \boldsymbol{A}|\underbrace{\boldsymbol{P}\boldsymbol{P}^{-1}}_{\boldsymbol{I}} = 0 \quad \Rightarrow \quad \Delta(\lambda) = |\lambda \boldsymbol{I} - \boldsymbol{A}| = 0 \quad \Rightarrow \quad \bar{\Delta}(\lambda) = \Delta(\lambda)$$

In Section A.2.9 one very important similarity transformation was presented, that is, $\boldsymbol{V} = [\boldsymbol{v}_1, \boldsymbol{v}_2, \ldots, \boldsymbol{v}_n]$, where $\boldsymbol{v}_i$ for $i = 1, 2, \ldots, n$ are the eigenvectors of $\boldsymbol{A} \in \Re^{n\times n}$ when $\boldsymbol{A}$ has distinct eigenvalues. The result is shown in (A.16) where $\boldsymbol{\Lambda} = \text{diag}[\lambda_1, \lambda_2, \ldots, \lambda_n]$, that is, the similarity transform that diagonalizes $\boldsymbol{A}$. If $\boldsymbol{A} \in \Re^{n\times n}$ is symmetric, $\boldsymbol{A}^T = \boldsymbol{A}$, then the similarity transformation $\boldsymbol{P}$ has the property $\boldsymbol{P}^{-1} = \boldsymbol{P}^T$ and for the case when $\boldsymbol{P} = \boldsymbol{V} = [\boldsymbol{v}_1, \boldsymbol{v}_2, \ldots, \boldsymbol{v}_n]$ (the eigenvectors of $\boldsymbol{A}$), then $\boldsymbol{V}^T\boldsymbol{A}\boldsymbol{V} = \boldsymbol{\Lambda} = \text{diag}[\lambda_1, \lambda_2, \ldots, \lambda_n]$. In this case, $\boldsymbol{A}$ is called *normal*, and $\boldsymbol{V}$ is *orthogonal* (cf. Sect. A.2.8). Also, using the orthogonal similarity transform $\boldsymbol{V}$, the matrix $\boldsymbol{A}$ can be written as

$$A = V\Lambda V^T = \sum_{i=1}^{n} \lambda_i v_i v_i^T \tag{A.21}$$

called the eigenvalue (spectral) decomposition (EVD) of A. For A symmetric, the minimum and maximum eigenvalues satisfy

$$\lambda_{\min}(A) = \min_{v \neq 0} \frac{v^T A v}{v^T v} \qquad \text{and} \qquad \lambda_{\max}(A) = \max_{v \neq 0} \frac{v^T A v}{v^T v} \tag{A.22}$$

There are many other similarity transformations that result in A transformed to special forms. In the next section we discuss one such transformation that leads to the Jordan canonical form.

Two matrices are called similar if they are related through a similarity transformation. If $A, B \in \Re^{n \times n}$ are *similar*, then

1. $|A| = |B|$ ________
2. $\text{trace}(A) = [\text{trace}(\bar{A})]$ (where $\bar{A}$ is the complex conjugate of A).
3. $A^k = B^k$ is similar for $k \geq 1$.
4. $(A^k)^T = (B^k)^T$ is similar for $k \geq 1$.
5. A is invertible $\Leftrightarrow$ B is invertible $\Leftrightarrow$ A^{-k} and B^{-k} are similar for $k \geq 1$.
6. A is involutory, skew involutory, idempotent, tripotent, and nilpotent $\Leftrightarrow$ B is involutory, skew involutory, idempotent, tripotent, and nilpotent (cf. Sect. A.2.18).

A.2.11 Jordan Canonical Form

Every matrix which maps the n-dimensional complex vector space into itself, that is, A: $(\mathscr{C}^n, \mathscr{C}) \to (\mathscr{C}^n, \mathscr{C})$, has a *Jordan canonical form* (or *Jordan form representation*). The matrix $\Lambda = \text{diag}[\lambda_1, \lambda_2, \ldots, \lambda_n]$ in (A.16), which results from the similarity transformation $V = [v_1, v_2, \ldots, v_n]$, where v_i for $i = 1, 2, \ldots, n$ are the eigenvectors of $A \in \Re^{n \times n}$, when A has distinct eigenvalues, is in Jordan canonical form. In this case, Λ has n Jordan blocks of *order* 1. This is actually a special case of the Jordan canonical form for a matrix.

The Jordan canonical form is typically thought of as a *block diagonal* matrix, as shown in (A.20). In this situation, the matrix $A \in \Re^{n \times n}$ had an eigenvalue λ that was repeated m times with $\nu(\lambda I - A) = 1$ (i.e., the nullity equal to 1, or simple degeneracy), and the remaining $n - m$ eigenvalues were considered distinct. Therefore, the Jordan canonical form of A, that is, $A_J = V^{-1}AV$ in (A.20), has one Jordan block (the first block) of order m, associated with the eigenvalue λ with multiplicity m; and the remaining $n - m$ Jordan blocks are of order 1 and are associated with the distinct eigenvalues. In this case, the nullity $\nu(\lambda I - A) = 1$ indicated that there is one Jordan block associated with the eigenvalue λ with multiplicity m.

For the full-degeneracy case, that is, when $\nu(\lambda I - A) = m$ (the multiplicity), the Jordan canonical form $A_J = V^{-1}AV = \text{diag}[\lambda, \lambda, \ldots, \lambda, \lambda_{m+1}, \lambda_{m+2}, \ldots, \lambda_n]$ in (A.18) was a diagonal matrix; that is, A was diagonalizable. For this case, the nullity $\nu(\lambda I - A) = m$ indicated that there are m Jordan blocks associated with the eigenvalue λ with multiplicity m.

Therefore, given a matrix $A \in \Re^{n \times n}$ with an eigenvalue λ with multiplicity m and the remaining $n - m$ eigenvalues considered to be distinct, the nullity $\nu(\lambda I - A) = q$ will indicate the number of Jordan blocks associated with the multiple eigenvalue. If $q = 1$, we said this is simple degeneracy, and for $q = m$ this is full degeneracy. When $1 < q < m$, more information is usually needed to determine the structure of the Jordan blocks [2]. For example, assume $A \in \Re^{4 \times 4}$, and the eigenvalues are given as $\lambda_1 = \lambda_2 = \lambda_3 = \lambda_4 = \lambda$, and $m = 4$. If $\nu(\lambda I - A) = 2$, this indicates that the Jordan canonical form of A has two Jordan blocks; however, this could be two Jordan blocks of order 2, or one Jordan block of order 3 and one Jordan block of order 1, that is,

$$A_J = \begin{bmatrix} \lambda & 1 & 0 & 0 \\ 0 & \lambda & 0 & 0 \\ 0 & 0 & \lambda & 1 \\ 0 & 0 & 0 & \lambda \end{bmatrix} \tag{A.23}$$

or

$$A_J = \begin{bmatrix} \lambda & 1 & 0 & 0 \\ 0 & \lambda & 1 & 0 \\ 0 & 0 & \lambda & 0 \\ 0 & 0 & 0 & \lambda \end{bmatrix} \tag{A.24}$$

or

$$A_J = \begin{bmatrix} \lambda & 0 & 0 & 0 \\ 0 & \lambda & 1 & 0 \\ 0 & 0 & \lambda & 1 \\ 0 & 0 & 0 & \lambda \end{bmatrix} \tag{A.25}$$

which is simply (A.24) with the two Jordan blocks reordered. A Jordan form representation of a matrix A: $(\mathscr{C}^n, \mathscr{C}) \to (\mathscr{C}^n, \mathscr{C})$ is unique up to the ordering of the Jordan blocks. If it is desired to compute the *generalized eigenvectors* (cf. Sect. A.2.9) of matrix A, the Jordan canonical form can always be found from $V^{-1}AV = A_J$.

In Section A.2.9 we saw that matrices that have distinct eigenvalues can be diagonalized. These matrices are referred to as being *nondefective*. Matrices that do not have a complete set of eigenvectors are considered to be *defective*. Therefore, for matrices with repeated eigenvalues that cannot be diagonalized, a Jordan canonical form matrix can be found (i.e., a block diagonal matrix), these matrices are referred to as being *defective*.

A.2.12 State-Space Description of Dynamical Systems

State-space models of dynamical systems are very important in control system design and analysis, signal processing, and many other areas. The state-space model for a time-invariant, linear, continuous-time system is given by

$$\dot{\boldsymbol{x}}(t) = \boldsymbol{A}\boldsymbol{x}(t)+\boldsymbol{B}\boldsymbol{u}(t) \qquad \text{where } \dot{\boldsymbol{x}}(t) \triangleq \frac{\mathrm{d}\boldsymbol{x}(t)}{\mathrm{d}t} \tag{A.26}$$

and

$$\boldsymbol{y}(t) = \boldsymbol{C}\boldsymbol{x}(t)+\boldsymbol{D}\boldsymbol{u}(t) \tag{A.27}$$

where (A.26) is the *state equation* and (A.27) is the system *output equation*, and together the two equations are frequently referred to as the *dynamical equation* for a system. In (A.26) the state vector

$$\boldsymbol{x}(t) = \left[x_1(t), x_2(t), \cdots, x_n(t)\right]^T \in \Re^{n\times 1} \tag{A.28}$$

contains the state variables of the system that describe the *internal* behavior of the system, and

$$\boldsymbol{u}(t) = \left[u_1(t), u_2(t), \cdots, u_p(t)\right]^T \in \Re^{p\times 1} \tag{A.29}$$

is a vector of system (control) inputs, $\boldsymbol{A} \in \Re^{n\times n}$, and $\boldsymbol{B} \in \Re^{nxp}$. The output equation in (A.27) is often referred to as the measurement equation or the observations, where

$$\boldsymbol{y}(t) = \left[y_1(t), y_2(t), \cdots, y_q(t)\right]^T \in \Re^{q\times 1} \tag{A.30}$$

$\boldsymbol{C} \in \Re^{q\times n}$ and $\boldsymbol{D} \in \Re^{q\times p}$. For a set of initial conditions $\boldsymbol{x}(t_0) = \boldsymbol{x}_0$ and a defined input to the system $\boldsymbol{u}(t)$, (A.26) can be solved for $\boldsymbol{x}(t)$, and this in turn can be substituted into the output equation in (A.27) to determine the system output $\boldsymbol{y}(t)$. The solution to the state equation in (A.26) is given as

$$\boldsymbol{x}(t) = \boldsymbol{\phi}(t-t_0)\boldsymbol{x}_0 + \int_{t_0}^{t} \boldsymbol{\phi}(t-\tau)\boldsymbol{B}\boldsymbol{u}(\tau)\mathrm{d}\tau \qquad t \geq t_0 \tag{A.31}$$

In (A.31), $\boldsymbol{\phi}(t-t_0)$ is referred to as the *state transition matrix*. For a linear time-invariant system, the state transition matrix can be written as

$$\boldsymbol{\phi}(t-t_0) = e^{\boldsymbol{A}(t-t_0)} \tag{A.32}$$

that is, a *matrix exponential* function. The state transition matrix can be written using the *Laplace transform*; that is, given $f(t)$ for $t \geq 0$, the Laplace transform of $f(t)$ is given by

$$F(s) = \mathscr{L}\{f(t)\} = \int_0^{\infty} f(t)e^{-st}\,\mathrm{d}t \tag{A.33}$$

The *inverse Laplace transform* is given by

$$f(t) = \mathscr{L}^{-1}\{F(s)\} = \frac{1}{2\pi j}\int_{c-j\infty}^{c+j\infty} F(s)e^{st}\,\mathrm{d}s$$

where the integral is evaluated along the path $s = c + j\omega$ in the complex plane from $c - j\infty$ to $c + j\infty$, where c is any real number for which the path $s = c + j\omega$ lies in the region of convergence of $F(s)$ [3]. Therefore,

$$\boldsymbol{\phi}(t-t_0) = \mathscr{L}^{-1}\{\boldsymbol{\Phi}(s)\}\big|_{t\leftarrow t-t_0} \tag{A.34}$$

where

$$\mathbf{\Phi}(s) = (s\boldsymbol{I} - \boldsymbol{A})^{-1} \tag{A.35}$$

is referred to as the *resolvent matrix*. Substituting (A.31) into the output equation in (A.27) gives

$$\boldsymbol{y}(t) = \underbrace{\boldsymbol{C}\boldsymbol{\phi}(t - t_0)\boldsymbol{x}_0}_{\text{natural response}} + \underbrace{\boldsymbol{C}\int_{t_0}^{t} \boldsymbol{\phi}(t - \tau)\boldsymbol{B}\boldsymbol{u}(\tau)\,d\tau + \boldsymbol{D}\boldsymbol{u}(t)}_{\text{driven response}} \qquad t \geq t_0 \tag{A.36}$$

If the initial-condition vector is set equal to zero in (A.36), that is, $\boldsymbol{x}_0 = \mathbf{0}$, the output can be written as

$$\begin{aligned}\boldsymbol{y}(t) &= \int_{t_0}^{t} \boldsymbol{C}\boldsymbol{\phi}(t - \tau)\boldsymbol{B}\boldsymbol{u}(\tau)\,\mathrm{d}\tau + \boldsymbol{D}\boldsymbol{u}(t) \\ &= \int_{t_0}^{t} [\boldsymbol{C}\boldsymbol{\phi}(t - \tau)\boldsymbol{B} + \boldsymbol{D}\delta(t - \tau)]\boldsymbol{u}(\tau)\,\mathrm{d}\tau \qquad t \geq t_0\end{aligned} \tag{A.37}$$

where

$$\boldsymbol{h}(t - \tau) = \begin{cases} \boldsymbol{C}\boldsymbol{\phi}(t - \tau)\boldsymbol{B} + \boldsymbol{D}\delta(t - \tau) & \text{for } t \geq \tau \\ \mathbf{0} & \text{for } t < \tau \end{cases} \tag{A.38}$$

and $\boldsymbol{h}(t - \tau)$ is the system *impulse response matrix*. Now the output can be written in terms of the system impulse response matrix

$$\boldsymbol{y}(t) = \int_{t_0}^{t} \boldsymbol{h}(t - \tau)\boldsymbol{u}(\tau)\,d\tau \qquad t \geq t_0 \tag{A.39}$$

which shows that for zero initial conditions, the system output is the convolution of the impulse response of the system with the input to the system. If we take the Laplace transform on both sides of (A.39), we obtain

$$\boldsymbol{Y}(s) = \boldsymbol{H}(s)\boldsymbol{U}(s) \tag{A.40}$$

where

$$\boldsymbol{H}(s) = \begin{bmatrix} H_{11}(s) & H_{12}(s) & \cdots & H_{1p}(s) \\ H_{21}(s) & H_{22}(s) & \cdots & H_{2p}(s) \\ \cdots & \cdots & \cdots & \cdots \\ H_{q1}(s) & H_{q2}(s) & \cdots & H_{qp}(s) \end{bmatrix} \tag{A.41}$$

is the *transfer function matrix* of the system. Therefore, the transfer function matrix of the system is the Laplace transform of the impulse response matrix, that is,

$$\boldsymbol{H}(s) = \boldsymbol{C}(s\boldsymbol{I} - \boldsymbol{A})^{-1}\boldsymbol{B} + \boldsymbol{D} = \frac{1}{\det(s\boldsymbol{I} - \boldsymbol{A})}\boldsymbol{C}[\operatorname{cof}(s\boldsymbol{I} - \boldsymbol{A})]^T\boldsymbol{B} + \boldsymbol{D} \tag{A.42}$$

DEFINITION A.9. The *characteristic polynomial* of a proper, rational matrix $\boldsymbol{H}(s)$ is defined to be the least common denominator of all minors of $\boldsymbol{H}(s)$. The *degree* of $\boldsymbol{H}(s)$ (also called the McMillan degree or the Smith-McMillan degree), denoted by deg[$\boldsymbol{H}(s)$], is defined as the degree of the characteristic polynomial of $\boldsymbol{H}(s)$.

If $\boldsymbol{H}(s)$ is a scalar, that is, $H(s)$, the characteristic polynomial reduces to the denominator of $H(s)$. To determine the characteristic polynomial and the degree of a transfer function matrix given as

$$\boldsymbol{H}(s) = \begin{bmatrix} \frac{2s+1}{s+2} & \frac{4s}{s+2} \\ \frac{s+1}{s+2} & \frac{2s+1}{s+2} \end{bmatrix}$$

we first determine the minors of order 1. They are the entries of the matrix, that is,

$$\frac{2s+1}{s+2} \qquad \frac{4s}{s+2} \qquad \frac{s+1}{s+2} \qquad \text{and} \qquad \frac{2s+1}{s+2}$$

The minor of order 2 for $\boldsymbol{H}(s)$ is $1/(s+2)^2$; therefore, the characteristic polynomial of $\boldsymbol{H}(s)$ is $\Delta(s) = (s+2)^2$, and the degree of $\boldsymbol{H}(s)$ is deg[$\boldsymbol{H}(s)$] = 2.

Given the state-space model of a time-invariant, linear, continuous-time system, as described in (A.26) and (A.27), we can state the following:

1. The dynamical system is (asymptotically) stable if all the eigenvalues of $\boldsymbol{A}$ have real parts that are strictly negative. Also, for a single-input, single-output (SISO) system, the system is (asymptotically) stable if all the *poles* of the system lie strictly in the left-half s plane. The poles of a system are the roots of the denominator polynomial (characteristic polynomial) $D(s)$ of the transfer function $Y(s)/U(s) = H(s) = N(s)/D(s)$ [the roots of the numerator polynomial $N(s)$ are called the system *zeros*]. If $\boldsymbol{A}$ has any (nonrepeated) eigenvalues with zero real parts and all others have negative real parts, the system is said to be *stable in the sense of Lyapunov* (or *marginally stable*).
2. The dynamical system is *controllable* if the $\{\boldsymbol{A}, \boldsymbol{B}\}$ pair is controllable, or
$$\rho(\boldsymbol{S}) = \rho\left[\boldsymbol{B}, \boldsymbol{AB}, \boldsymbol{A}^2\boldsymbol{B}, \cdots, \boldsymbol{A}^{(n-1)}\boldsymbol{B}\right] = n \tag{A.43}$$
where $\boldsymbol{S} = [\boldsymbol{B}, \boldsymbol{AB}, \boldsymbol{A}^2\boldsymbol{B}, \cdots, \boldsymbol{A}^{(n-1)}\boldsymbol{B}] \in \Re^{n\times np}$ is the *controllability matrix*.
3. The dynamical system is *observable* if the $\{\boldsymbol{A}, \boldsymbol{C}\}$ pair is observable, or
$$\rho(\boldsymbol{L}) = \rho\left[\boldsymbol{C}^T, \boldsymbol{A}^T\boldsymbol{C}^T, (\boldsymbol{A}^2)^T\boldsymbol{C}^T, \cdots, (\boldsymbol{A}^{(n-1)})^T\boldsymbol{C}^T\right]^T = n \tag{A.44}$$
where $\boldsymbol{L} = [\boldsymbol{C}^T, \boldsymbol{A}^T\boldsymbol{C}^T, (\boldsymbol{A}^2)^T\boldsymbol{C}^T, \cdots, (\boldsymbol{A}^{(n-1)})^T\boldsymbol{C}^T]^T \in \Re^{n\times nq}$ is the *observability matrix*.
4. If (A.26) and (A.27) describe an *open-loop system*, closed-loop feedback control can be applied to the system using *state-variable feedback control*, where the control law is given by
$$\boldsymbol{u}(t) = -\boldsymbol{Kx}(t) + \boldsymbol{r}(t) \tag{A.45}$$

where $\boldsymbol{K} \in \Re^{p \times n}$ is the closed-loop system *gain matrix* and $\boldsymbol{r}(t) \in \Re^{p \times 1}$ is the *system reference command input*. If $\boldsymbol{r}(t) \neq \boldsymbol{0}$, the closed-loop system is a *tracking system*; however, if $\boldsymbol{r}(t) = \boldsymbol{0}$, the closed-loop system is a *regulator*. State-variable feedback control can be used for *pole-placement* design. Substituting (A.45) into (A.26) gives the closed-loop system state equation

$$\dot{\boldsymbol{x}}(t) = (\boldsymbol{A} - \boldsymbol{BK})\boldsymbol{x}(t) + \boldsymbol{Br}(t) \tag{A.46}$$

The design objective is to determine the gain matrix $\boldsymbol{K}$ such that the closed-loop system eigenvalues (poles) are at the *desired* locations, where

$$|\lambda \boldsymbol{I} - \boldsymbol{A} + \boldsymbol{BK}| = 0 \tag{A.47}$$

is the closed-loop characteristic equation. For a single-input system, if the system is controllable, the poles of the closed-loop system can be placed arbitrarily and the gain matrix is unique.

5. If we assume a SISO system, that is, $\dot{\boldsymbol{x}}(t) = \boldsymbol{Ax}(t) + \boldsymbol{B}u(t)$ and $y(t) = \boldsymbol{Cx}(t) + \boldsymbol{D}u(t)$, and $\boldsymbol{A} \in \Re^{n \times n}$ (the system *plant*) has distinct eigenvalues, then we form a similarity (*equivalence*) transformation from the eigenvectors of $\boldsymbol{A}$, that is, $\boldsymbol{P} = [\boldsymbol{p}_1, \boldsymbol{p}_2, \ldots, \boldsymbol{p}_n] \in \Re^{n \times n}$ (we referred to this as the matrix $\boldsymbol{V}$ previously) [recall: $\rho(\boldsymbol{V}) = n$], and using $\boldsymbol{P}$, we define a new state vector $\bar{\boldsymbol{x}}(t) = \boldsymbol{P}^{-1}\boldsymbol{x}(t)$. Substituting $\boldsymbol{x}(t) = \boldsymbol{P}\bar{\boldsymbol{x}}(t)$ into $\dot{\boldsymbol{x}}(t) = \boldsymbol{A}x(t) + \boldsymbol{B}$ $u(t)$ gives an *equivalent* state equation

$$\dot{\bar{\boldsymbol{x}}}(t) = \underbrace{\boldsymbol{P}^{-1}\boldsymbol{AP}}_{\boldsymbol{\Lambda}}\bar{\boldsymbol{x}}(t) + \underbrace{\boldsymbol{P}^{-1}\boldsymbol{B}}_{\boldsymbol{\Gamma}}u(t) = \boldsymbol{\Lambda}\bar{\boldsymbol{x}}(t) + \boldsymbol{\Gamma}u(t) \tag{A.48}$$

and the equivalent output equation is given by

$$y(t) = \underbrace{\boldsymbol{CP}}_{\bar{\boldsymbol{C}}}\bar{\boldsymbol{x}}(t) + \boldsymbol{D}u(t) = \bar{\boldsymbol{C}}\bar{\boldsymbol{x}}(t) + \boldsymbol{D}u(t) \tag{A.49}$$

In (A.48), $\boldsymbol{\Lambda} = \boldsymbol{P}^{-1}\boldsymbol{AP}$ is a diagonal matrix with the eigenvalues of $\boldsymbol{A}$ on the diagonal (cf. Sect. A.2.9); therefore, the system dynamics are completely decoupled. Note that the two equivalent systems have the same eigenvalues, and the input and output are the same for the two systems.

6. If we assume a SISO system as in item 5, with the $\{\boldsymbol{A}, \boldsymbol{B}\}$ pair controllable, then there exists a similarity transformation given by

$$\bar{\boldsymbol{x}}(t) = \boldsymbol{Qx}(t) = \begin{bmatrix} \tilde{\boldsymbol{Q}} \\ \tilde{\boldsymbol{Q}}\boldsymbol{A} \\ \vdots \\ \tilde{\boldsymbol{Q}}\boldsymbol{A}^{n-1} \end{bmatrix} \boldsymbol{x}(t) \tag{A.50}$$

where

$$\tilde{\boldsymbol{Q}} = \begin{bmatrix} 0 & \cdots & 0 & 1 \end{bmatrix} \left[\boldsymbol{B}, \boldsymbol{AB}, \boldsymbol{A}^2\boldsymbol{B}, \cdots, \boldsymbol{A}^{(n-1)}\boldsymbol{B}\right]^{-1} \tag{A.51}$$

and $\boldsymbol{S} = \left[\boldsymbol{B}, \boldsymbol{AB}, \boldsymbol{A}^2\boldsymbol{B}, \cdots, \boldsymbol{A}^{(n-1)}\boldsymbol{B}\right]$ is the system controllability matrix. The equivalent system that results from using the transformation in (A.50) is given by

$$\dot{\bar{x}}(t) = \underbrace{QAQ^{-1}}_{\bar{A}}\,\bar{x}(t) + \underbrace{QB}_{\bar{B}}\,u(t) \tag{A.52}$$

$$y(t) = \underbrace{CQ^{-1}}_{\bar{C}}\,\bar{x}(t) + \underbrace{D}_{\bar{D}}\,u(t) \tag{A.53}$$

where

$$\bar{A} = QAQ^{-1} = \begin{bmatrix} 0 & 1 & 0 & 0 & \cdots & 0 \\ 0 & 0 & 1 & 0 & \cdots & 0 \\ \cdots & \cdots & \cdots & \cdots & \cdots & \cdots \\ 0 & 0 & 0 & \cdots & 1 & 0 \\ 0 & 0 & 0 & \cdots & 0 & 1 \\ -\alpha_n & -\alpha_{n-1} & -\alpha_{n-2} & \cdots & -\alpha_2 & -\alpha_1 \end{bmatrix} \tag{A.54}$$

$$\bar{B} = QB = [0, \cdots, 0, 1]^T \tag{A.55}$$

$$\bar{C} = Q^{-1}C = [\bar{c}_1, \bar{c}_2, \cdots, \bar{c}_n] \tag{A.56}$$

$$\bar{D} = D \tag{A.57}$$

From (A.52) and (A.53), the equivalent system given by

$$\dot{\bar{x}}(t) = \bar{A}\bar{x}(t) + \bar{B}u(t) \tag{A.58}$$

$$y(t) = \bar{C}\bar{x}(t) + \bar{D}u(t) \tag{A.59}$$

is known as the *phase variable canonical form* of (A.26) and (A.27), for a SISO system, or the *controllable canonical form*. The matrix in (A.54) is referred to as the *companion matrix* (or *Frobenius matrix*). The characteristic equation of $\bar{A}$ in (A.54) is given by

$$\Delta(\lambda) = |\lambda I - \bar{A}| = \lambda^n + \alpha_1\lambda^{n-1} + \alpha_2\lambda^{n-2} + \cdots + \alpha_{n-1}\lambda + \alpha_n = 0 \tag{A.60}$$

7. The dual canonical form of the controllable canonical form in item 6 is the *observable canonical form*, written as

$$\dot{\breve{x}}(t) = \breve{A}\breve{x}(t) + \breve{B}u(t) \tag{A.61}$$

$$y(t) = \breve{C}\breve{x}(t) + \breve{D}u(t) \tag{A.62}$$

where $\breve{A} = \bar{A}^T$, $\breve{B} = \bar{C}^T$ [from (A.56)], $\breve{C} = \bar{B}^T$ [from (A.55)], and $\breve{D} = D$.

A.2.13 Vector and Matrix Norms

The concept of a *vector norm* is a generalization of the notion of length. If we assume any vector $x \in \Re^{n\times 1}$ (or $x \in \mathscr{C}^{n\times 1}$) and any scalar $\alpha \in \Re$ (or $\alpha \in \mathscr{C}$), then any real-valued function of x, denoted by $\|x\|$, can be defined as a vector norm if it has the following properties:

1. $\|x\| \geq 0$, and $\|x\| = 0 \Leftrightarrow x = \mathbf{0}$

2. $\|\alpha \boldsymbol{x}\| = |\alpha| \|\boldsymbol{x}\|$
3. $\|\boldsymbol{x}_1 + \boldsymbol{x}_2\| \leq \|\boldsymbol{x}_1\| + \|\boldsymbol{x}_2\|$. (This is known as the *triangle inequality*.)

The L_p-norm (or p norm) of a vector $\boldsymbol{x} = [x_1, x_2, \ldots, x_n]^T$ is defined as

$$\|\boldsymbol{x}\|_p \triangleq \left[\sum_{i=1}^{n} |x_i|^p \right]^{1/p} \tag{A.63}$$

where $p \in \Re^+$. However, p usually takes on positive *integer* values, $1, 2, \ldots, \infty$, and the corresponding norms are referred to as the $1, 2, \ldots$, infinity norms, respectively. A classic result using the L_p-norm is the *Holder inequality*, given as

$$|\langle \boldsymbol{x}, \boldsymbol{y} \rangle| = |\boldsymbol{x}^T \boldsymbol{y}| \leq \|\boldsymbol{x}\|_p \|\boldsymbol{y}\|_q \qquad \frac{1}{p} + \frac{1}{q} = 1 \tag{A.64}$$

for $\boldsymbol{x}, \boldsymbol{y} \in \Re^{n \times 1}$. The *1-norm* or the L_1-norm (or absolute value norm) from (A.63) is defined as

$$\|\boldsymbol{x}\|_1 \triangleq \sum_{i=1}^{n} |x_i| \tag{A.65}$$

The *2-norm* or L_2-norm (or *Euclidean norm*) from (A.63) is defined as

$$\|\boldsymbol{x}\|_2 \triangleq \left[\sum_{i=1}^{n} x_i^2 \right]^{1/2} = (\boldsymbol{x}^T \boldsymbol{x})^{1/2} = \langle \boldsymbol{x}, \boldsymbol{x} \rangle^{1/2} \tag{A.66}$$

Using the Euclidean norm, the *Cauchy-Schwartz inequality* is given by

$$|\langle \boldsymbol{x}, \boldsymbol{y} \rangle| = |\boldsymbol{x}^T \boldsymbol{y}| \leq \|\boldsymbol{x}\|_2 \|\boldsymbol{y}\|_2 \tag{A.67}$$

for $\boldsymbol{x}, \boldsymbol{y} \in \Re^{n \times 1}$, which is a special case of the Holder inequality given in (A.64). The *infinity norm* or L_∞*-norm* (or *Chebyshev norm*) is defined as

$$\|\boldsymbol{x}\|_\infty = \max(|x_1|, |x_2|, \ldots, |x_n|) \tag{A.68}$$

The *negative-infinity norm* or $L_{-\infty}$*-norm* is defined as

$$\|\boldsymbol{x}\|_{-\infty} = \min(|x_1|, |x_2|, \ldots, |x_n|) \tag{A.69}$$

The *inner-product-generated norm* is defined as

$$\|\boldsymbol{x}\|_W = \langle \boldsymbol{x}, \boldsymbol{x} \rangle_W^{1/2} = [(\boldsymbol{W}\boldsymbol{x}) \bullet (\overline{\boldsymbol{W}\boldsymbol{x}})]^{1/2} = [(\boldsymbol{W}\boldsymbol{x})^T (\overline{\boldsymbol{W}\boldsymbol{x}})]^{1/2} = [\boldsymbol{x}^T \boldsymbol{W}^T \bar{\boldsymbol{W}} \bar{\boldsymbol{x}}]^{1/2} \tag{A.70}$$

A related norm is the *weighted Euclidean norm*, given by

$$\|\boldsymbol{x}\|_{2-Q} = (\boldsymbol{x}^T \boldsymbol{Q} \boldsymbol{x})^{1/2} \tag{A.71}$$

where $\boldsymbol{Q} \in \Re^{n \times n}$, $\boldsymbol{Q}^T = \boldsymbol{Q}$, and $\boldsymbol{Q} > 0$.

A *unit vector* is a vector whose norm is unity. A non-zero vector is *normalized* by dividing every element of the vector by its norm; therefore, normalized vectors are unit vectors. A set of vectors that is orthogonal in which each has unit length is called *orthonormal*. The *distance* between two vectors $\boldsymbol{x}, \boldsymbol{y} \in \Re^{n\times 1}$ is $\|\boldsymbol{x}-\boldsymbol{y}\|$ (which depends on the type of norm selected).

A norm for a square matrix $\boldsymbol{A} \in \Re^{n\times n}(\boldsymbol{A} \in \mathscr{C}^{n\times n})$, denoted by $\|\boldsymbol{A}\|$, is a real-valued function that must satisfy the following conditions:

1. $\|\boldsymbol{A}\| \geq 0$
2. $\|\boldsymbol{A}\| = 0 \Leftrightarrow \boldsymbol{A} = \boldsymbol{0}$
3. $\|\alpha\boldsymbol{A}\| = |\alpha|\|\boldsymbol{A}\|$, for any scalar α.
4. Given $\boldsymbol{A}, \boldsymbol{B} \in \Re^{n\times n}$ $(\boldsymbol{A}, \boldsymbol{B} \in \mathscr{C}^{n\times n})$,

$$\|\boldsymbol{A}+\boldsymbol{B}\| \leq \|\boldsymbol{A}\| + \|\boldsymbol{B}\| \qquad \text{triangle inequality}$$

5. Given $\boldsymbol{A}, \boldsymbol{B} \in \Re^{n\times n}$ $(\boldsymbol{A}, \boldsymbol{B} \in \mathscr{C}^{n\times n})$,

$$\|\boldsymbol{A}\boldsymbol{B}\| \leq \|\boldsymbol{A}\|\|\boldsymbol{B}\| \qquad \text{consistency condition}$$

One important norm is the *Frobenius norm* of a matrix $\boldsymbol{A} \in \Re^{n\times n}(\boldsymbol{A} \in \mathscr{C}^{n\times n})$ $(\boldsymbol{A} = [a_{ij}]$, for $i, j = 1, 2, \ldots, n)$, given by

$$\|\boldsymbol{A}\|_F = \left(\sum_{i=1}^{n}\sum_{j=1}^{n}|a_{ij}|^2\right)^{1/2} \tag{A.72}$$

It can be shown that the Frobenius norm can be computed as the square root of the sum of the squares of the nonzero *singular values* (cf. Sect. A.2.14) of $\boldsymbol{A}$.

Induced norms of matrices are induced by vector norms. Each vector p norm induces (or generates) the matrix p norm

$$\|\boldsymbol{A}\|_p = \sup_{\boldsymbol{x}\neq\boldsymbol{0}} \frac{\|\boldsymbol{A}\boldsymbol{x}\|_p}{\|\boldsymbol{x}\|_p} = \sup_{\|\boldsymbol{x}\|_p=1} \|\boldsymbol{A}\boldsymbol{x}\|_p \tag{A.73}$$

where "sup" means *supremum*, the least upper bound of $\|\boldsymbol{A}\boldsymbol{x}\|_p$. From (A.73) it follows that

$$\|\boldsymbol{A}\boldsymbol{x}\|_p \leq \|\boldsymbol{A}\|_p\|\boldsymbol{x}\|_p \tag{A.74}$$

If a vector norm is *compatible* with a matrix norm, then (A.74) must hold. Induced norms are always compatible with vector norms that generated them. Compatibility is not restricted to induced norms. For example, the Frobenius norm is compatible with the Euclidean vector norm, even though the Euclidean norm did not induce the Frobenius matrix norm.

Some useful induced norms are as follows:

1. The L_1-*matrix* norm (which is induced by the L_1-*vector* norm)

$$\|\boldsymbol{A}\|_1 = \max_{j=1,2,\ldots,n}\left(\sum_{i=1}^{n}|a_{ij}|\right) \tag{A.75}$$

   This is the largest *column* sum of absolute values.

2. The L_∞ matrix norm (which is induced by the L_∞-*vector* norm):

$$\|\boldsymbol{A}\|_\infty = \max_{i=1,2,\cdots,n}\left(\sum_{j=1}^{n}|a_{ij}|\right) \tag{A.76}$$

This is the largest *row* sum of absolute values.

3. The *spectral* norm (which is induced by the *Euclidean* norm)

$$\|A\|_2 = [\lambda_{\max}(A^* A)]^{1/2} \tag{A.77}$$

which is the square root of the largest eigenvalue of $A^* A = \bar{A}^T A$. This can be computed numerically as the largest *singular value* (cf. Sect. A.2.14) of A.

The *spectral radius* of a square matrix $A \in \Re^{n \times n} (A \in \mathscr{C}^{n \times n})$ is denoted by $\sigma_r(A)$ and is the largest absolute value of any eigenvalue of A. Therefore, given the eigenvalues of A as λ_i for $i = 1, 2, \ldots, n$, the spectral radius of A is

$$\sigma_r(A) = \max_{1 \le i \le n} |\lambda_i| \tag{A.78}$$

For any matrix norm

$$\sigma_r(A) \le \|A\| \tag{A.79}$$

which provides bounds on the eigenvalues of A. An equivalent expression for the spectral radius of a matrix A can be written as

$$\sigma_r(A) = \lim_{m \to \infty} \|A^m\|^{1/m} \tag{A.80}$$

Gerschgorin's theorem [1] can provide a means of estimating the spectral radius of a matrix A.

A.2.14 Singular-Value Decomposition

We assume $A \in \Re^{m \times n}$, and the real orthogonal matrices

$$U = [u_1, u_2, \ldots, u_m] \in \Re^{m \times m} \tag{A.81}$$

$$V = [v_1, v_2, \ldots, v_n] \in \Re^{n \times n} \tag{A.82}$$

exist, such that we can write

$$U^T A V = \text{pseudodiag}[\sigma_1, \sigma_2, \ldots, \sigma_p] = S \tag{A.83}$$

where $S \in \Re^{m \times n}$,

$$p = \min\{m, n\} \tag{A.84}$$

and

$$\sigma_1 \ge \sigma_2 \ge \cdots \ge \sigma_p \ge 0 \tag{A.85}$$

Because the matrices U and V are orthogonal, (A.83) can be written as

$$A = USV^T = \sum_{i=1}^{r} \sigma_i u_i v_i^T \tag{A.86}$$

and is known as the *singular-value decomposition* of matrix A, where $\rho(A) = r$ (the *index* of the smallest singular value). If

$$\sigma_1 \ge \sigma_2 \ge \cdots \ge \sigma_r > \sigma_{r+1} = \sigma_{r+2} = \cdots = \sigma_p = 0 \tag{A.87}$$

then we can define

$$U_r \triangleq U(:, 1 : r) \tag{A.88}$$

$$V_r \triangleq V(:, 1 : r) \tag{A.89}$$

$$S_r \triangleq S(1 : r, 1 : r) \tag{A.90}$$

where $U_r \in \Re^{m \times r}$, $V_r \in \Re^{n \times r}$, and $S_r \in \Re^{r \times r}$. The notation used in (A.88), (A.89), and (A.90) is derived from MATLAB's method of indexing arrays; for example, in (A.88), $U(:, 1 : r)$ means to take all rows, thus the first ":" and "$1{:}r$" means to take columns 1 through r. Therefore, (A.86) can be written as

$$A = U_r S_r V_r^T = \sum_{i=1}^{r} \sigma_i u_i v_i^T \tag{A.91}$$

The relationship between S_r and S can be observed directly from Figure A.1.

It is now straightforward to understand how MATLAB uses the SVD of a matrix to numerically compute its rank. That is, the number of nonzero singular values of a matrix $A \in \Re^{m \times n}$ is its *rank*. From (A.86), two expressions can be written

$$Av_i = \sigma_i u_i \tag{A.92}$$

$$A^T u_i = \sigma_i v_i \tag{A.93}$$

where $i = 1, 2, \ldots, \min\{m, n\}$, $u_i \in \Re^{m \times 1}$ (left singular vectors), and $v_i \in \Re^{n \times 1}$ (right singular vectors). This shows the similarity to the standard eigenvalue problem (cf. Sect. A.2.9).

A special case for the SVD arises when $A \in \Re^{n \times n}$, $A^T = A$, and $A \geq 0$. In this case the singular values are the eigenvalues of A, that is,

$$S = \begin{bmatrix} \lambda_1 & & & \mathbf{0} \\ & \lambda_2 & & \\ & & \ddots & \\ \mathbf{0} & & & \lambda_n \end{bmatrix} = \mathrm{diag}[\lambda_1, \lambda_2, \ldots, \lambda_n] = \Lambda \in \Re^{n \times n} \tag{A.94}$$

where $\lambda_1 \geq \lambda_2 \geq \cdots \geq \lambda_n \geq 0$ are the *real* (because A is symmetric) eigenvalues of A corresponding to the *orthogonal* (because A is symmetric) set of eigenvectors v_i, for $i = 1, 2, \ldots, n$. In addition, $U = V$; that is, the right and left singular vectors are the same. Without loss of generality, we will assume $\rho(A) = n$. Given the SVD of A, that is, $A = USV^T$, we can form

$$U^T A V = S = \left[\begin{array}{c|c} S_r & \mathbf{0}_{12} \\ \hline \mathbf{0}_{21} & \mathbf{0}_{22} \end{array} \right]$$

where: $\mathbf{0}_{12} \in \Re^{r \times n - r}$ (null matrix)
$\mathbf{0}_{21} \in \Re^{m - r \times r}$ (null matrix)
$\mathbf{0}_{22} \in \Re^{m - r \times n - r}$ (null matrix)
$[\rho(A) = r]$

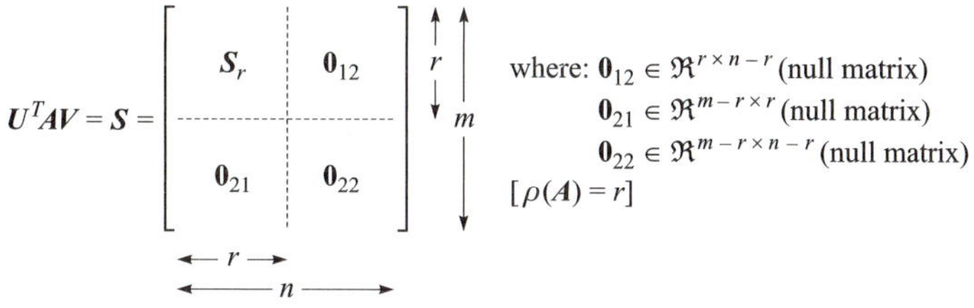

Figure A.1 Relationship between S_r and S in the SVD of a matrix $A \in \Re^{m \times n}$.

$$A^T A = (VS^T U^T)(USV^T) = VS^T \underbrace{U^T U}_{I} SV^T = V \underbrace{S^T S}_{S^2} V^T = VS^2 V^T \tag{A.95}$$

However, from (A.94), $S^2 = \Lambda^2$, and (A.95) can be written as

$$A^T A = V\Lambda^2 V^T = AA^T \qquad \text{because } U = V \tag{A.96}$$

and this is the eigenvalue decomposition of $A^T A$ (or AA^T) (cf. Sect. A.2.10). It is now straightforward to understand how the SVD of a matrix can be used to compute its spectral norm (cf. Sect. A.2.13). That is, the largest singular value of $A^T A$ is λ_1^2, which gives the spectral norm of A as $\sqrt{\lambda_1^2} = \lambda_1$. However, this can be determined from the SVD of A; that is, the largest singular value of A is $\sigma_1 = \lambda_1$ [see (A.94)].

The Frobenius norm [see Sect. A.2.13, Equation (A.72)] of a matrix A can also be determined from its SVD. Another form of the Frobenius norm is given as

$$\|A\|_F = [\text{trace}(A^T A)]^{1/2} = [\text{trace}(AA^T)]^{1/2} \tag{A.97}$$

If we write A in terms of its SVD, that is, $A = USV^T$, then (A.97) can be written as

$$\begin{aligned} \|A\|_F &= [\text{trace}(A^T A)]^{1/2} = [\text{trace}(AA^T)]^{1/2} \\ &= [\text{trace}(S^T S)]^{1/2} = [\text{trace}(SS^T)]^{1/2} \\ &= \left(\text{trace}\begin{bmatrix} \sigma_1^2 & & & \mathbf{0} \\ & \sigma_2^2 & & \\ & & \ddots & \\ \mathbf{0} & & & \sigma_p^2 \end{bmatrix}\right)^{1/2} = \left(\sum_{i=1}^{p} \sigma_i^2\right)^{1/2} \end{aligned} \tag{A.98}$$

(recall: $p = \min\{m, n\}$).

Let's revisit the matrix pseudoinverse problem in Section A.2.7, in relation to solving the system of simultaneous linear algebraic equations $Ax = b$, for $A \in \Re^{m \times n}$, $x \in \Re^{n \times 1}$, and $b \in \Re^{m \times 1}$. If we write A in terms of its SVD from (A.91), that is, $A = U_r S_r V_r^T$, where $r = \rho(A)$, the error function in (A.8) can now be written as

$$\mathscr{E}(x) = \frac{1}{2}\|Ax - b\|_2^2 = \frac{1}{2}x^T V_r S_r^T S_r V_r^T x - x^T V_r S_r^T U_r^T b + \frac{1}{2}b^T b \tag{A.99}$$

Taking the partial derivative of $\mathscr{E}(x)$ in (A.99), with respect to the vector x, and setting the result equal to zero give

$$V_r S_r^T S_r V_r^T x - V_r S_r^T U_r^T b = \mathbf{0} \tag{A.100}$$

Premultiplying both sides of (A.100) by V_r^T gives $S_r^T S_r V_r^T x = S_r^T U_r^T b$, and premultiplying both sides of this result by $(S_r^T S_r)^{-1}$ gives

$$V_r^T x = (S_r^T S_r)^{-1} S_r^T U_r^T b = S_r^{-1} U_r^T b \tag{A.101}$$

Now the least-squares solution x^* to $Ax = b$ can be obtained from (A.101) by premultiplying both sides by V_r

$$x^* = V_r S_r^{-1} U_r^T b \tag{A.102}$$

and the pseudoinverse of A from (A.102) is given by

$$A^+ = V_r S_r^{-1} U_r^T \tag{A.103}$$

However, the pseudoinverse of A in (A.103), written in terms of the SVD of A, can also be written as

$$A^+ = V S^+ U^T \tag{A.104}$$

where

$$S^+ = \begin{bmatrix} S_r^{-1} & 0 \\ 0 & 0 \end{bmatrix}^T = \begin{bmatrix} S_r^{-1} & 0_{r\times m-r} \\ 0_{n-r\times r} & 0_{n-r\times m-r} \end{bmatrix} \in \Re^{n\times m}$$

Therefore, S^+ is simply computed by taking the reciprocal of the *nonzero* singular values (then the transpose of the matrix is taken), and the linear least-squares solution of $Ax = b$ in (A.102) can be written using (A.104) as

$$x^* = V S^+ U^T b = A^+ b \tag{A.105}$$

A.2.15 Matrix Condition Number

The *condition number* of a matrix $A \in \Re^{m\times n}$ is defined as

$$\text{cond}_p(A) \triangleq \|A\|_p \|A^+\|_p \tag{A.106}$$

where p can be associated with any matrix norm and A^+ is the pseudoinverse of A. Matrices with small condition numbers are referred to as *well-conditioned* matrices, and those with large condition numbers are referred to as *ill-conditioned* matrices. In many applications the *L_2-norm condition number* of a matrix $A \in \Re^{m\times n}$ is used. Therefore, the condition number can be written as

$$\text{cond}_2(A) = \|A\|_2 \|A^+\|_2 = \frac{\sigma_1}{\sigma_p} \tag{A.107}$$

that is, the ratio of the largest singular value of A to its smallest one, where $p = \min\{m, n\}$. This is the condition number that MATLAB yields when the function **cond** is used to compute the condition number of a matrix.

The condition number of a matrix arises naturally in analyzing the sensitivity of the solution of a system of simultaneous linear algebraic equations $Ax = b$. For simplicity, we assume $A \in \Re^{n\times n}$ and $\rho(A) = n$. If there exists an error in b due to roundoff (or measurement) error, that is, $b + \Delta b$, then the solution will be changed, that is, $x + \Delta x$. Therefore, we can write the system of equations as $A(x + \Delta x) = (b + \Delta b)$; subtracting $Ax = b$ gives

$$A\,\Delta x = \Delta b \tag{A.108}$$

Solving for $\Delta \boldsymbol{x}$ gives

$$\Delta \boldsymbol{x} = \boldsymbol{A}^{-1} \Delta \boldsymbol{b} \tag{A.109}$$

Using the property in (A.74) for matrix norms, we can write

$$\|\Delta \boldsymbol{x}\| \leq \|\boldsymbol{A}^{-1}\| \|\Delta \boldsymbol{b}\| \tag{A.110}$$

Dividing both sides of (A.110) by $\|\boldsymbol{x}\|$ and multiplying and dividing the right-hand side by $\|\boldsymbol{b}\|$ give

$$\frac{\|\Delta \boldsymbol{x}\|}{\|\boldsymbol{x}\|} \leq \frac{\|\boldsymbol{b}\| \left\|\boldsymbol{A}^{-1}\right\|}{\|\boldsymbol{x}\|} \frac{\|\Delta \boldsymbol{b}\|}{\|\boldsymbol{b}\|} \tag{A.111}$$

However, $\|\boldsymbol{b}\| = \|\boldsymbol{A}\boldsymbol{x}\|$; therefore, (A.111) can be written as

$$\frac{\|\Delta \boldsymbol{x}\|}{\|\boldsymbol{x}\|} \leq \left[\frac{\|\boldsymbol{A}\boldsymbol{x}\|}{\|\boldsymbol{x}\|}\right] \left\|\boldsymbol{A}^{-1}\right\| \frac{\|\Delta \boldsymbol{b}\|}{\|\boldsymbol{b}\|} \tag{A.112}$$

Moreover, using (A.74), we can also write $\|\boldsymbol{A}\| \, \|\boldsymbol{x}\| \geq \|\boldsymbol{A}x\|$, and dividing both sides by $\|\boldsymbol{x}\|$ gives $\|\boldsymbol{A}\| \geq \|\boldsymbol{A}x\|/\|\boldsymbol{x}\|$. Using this result, (A.112) can be written as

$$\frac{\|\Delta \boldsymbol{x}\|}{\|\boldsymbol{x}\|} \leq \underbrace{\|\boldsymbol{A}\| \left\|\boldsymbol{A}^{-1}\right\|}_{\text{cond}(\boldsymbol{A})} \frac{\|\Delta \boldsymbol{b}\|}{\|\boldsymbol{b}\|} \tag{A.113}$$

or

$$\frac{\|\Delta \boldsymbol{x}\|}{\|\boldsymbol{x}\|} \leq \text{cond}(\boldsymbol{A}) \frac{\|\Delta \boldsymbol{b}\|}{\|\boldsymbol{b}\|} \tag{A.114}$$

Therefore, in (A.114) the *relative solution error* $\|\Delta \boldsymbol{x}\|/\|\boldsymbol{x}\|$ is less than (or equal to) the condition number of $\boldsymbol{A}$, that is, cond($\boldsymbol{A}$), times the *relative problem error*, where the right-hand side provides an upper bound. If the problem error is taken as $\Delta \boldsymbol{A}$ (i.e., there exists an error in $\boldsymbol{A}$ instead of $\boldsymbol{b}$), then (A.114) changes to

$$\frac{\|\Delta \boldsymbol{x}\|}{\|\boldsymbol{x} + \Delta \boldsymbol{x}\|} \leq \text{cond}(\boldsymbol{A}) \frac{\|\Delta \boldsymbol{A}\|}{\|\boldsymbol{A}\|} \tag{A.115}$$

If there exist perturbations in both $\boldsymbol{A}$ and $\boldsymbol{b}$, that is, $\Delta \boldsymbol{A}$ and $\Delta \boldsymbol{b}$, respectively, then it can be shown that the inequality is given by

$$\frac{\|\Delta \boldsymbol{x}\|}{\|\boldsymbol{x}\|} \leq \text{cond}(\boldsymbol{A}) \left(\frac{\|\Delta \boldsymbol{b}\|}{\|\boldsymbol{b}\|} + \frac{\|\Delta \boldsymbol{A}\|}{\|\boldsymbol{A}\|} \right) \tag{A.116}$$

A.2.16 Partitioned Matrix Operations

There are many matrix operations that can be simplified by properly partitioning the matrix (or matrices). Some useful matrix operations for partitioned matrices are given below.

Matrix determinant

Given $\boldsymbol{A} \in \Re^{n \times n}$, $\boldsymbol{B} \in \Re^{n \times m}$, $\boldsymbol{C} \in \Re^{m \times n}$, $\boldsymbol{D} \in \Re^{m \times m}$, and $\boldsymbol{M} \in \Re^{(n+m) \times (n+m)}$ partitioned as

$$\boldsymbol{M} = \begin{bmatrix} \boldsymbol{A} & \boldsymbol{B} \\ \boldsymbol{C} & \boldsymbol{D} \end{bmatrix} \tag{A.117}$$

then if $\det(\boldsymbol{A}) \neq 0$,

$$\det(\boldsymbol{M}) = \det(\boldsymbol{A}) \det(\boldsymbol{D} - \boldsymbol{C}\boldsymbol{A}^{-1}\boldsymbol{B}) \tag{A.118}$$

Given $\boldsymbol{A} \in \Re^{m \times m}$, $\boldsymbol{B} \in \Re^{m \times p}$, $\boldsymbol{C} \in \Re^{p \times m}$, $\boldsymbol{D} \in \Re^{pxp}$, and $\boldsymbol{M} \in \Re^{(m+p) \times (m+p)}$ partitioned as

$$\boldsymbol{M} = \begin{bmatrix} \boldsymbol{A} & \boldsymbol{B} \\ \boldsymbol{C} & \boldsymbol{D} \end{bmatrix} \tag{A.119}$$

if $\det(\boldsymbol{D}) \neq 0$, then

$$\det(\boldsymbol{M}) = \det(\boldsymbol{D}) \det(\boldsymbol{A} - \boldsymbol{B}\boldsymbol{D}^{-1}\boldsymbol{C}) \tag{A.120}$$

For $\boldsymbol{A}$ and $\boldsymbol{B}$ square matrices,

$$\det\begin{bmatrix} \boldsymbol{A} & \boldsymbol{B} \\ \boldsymbol{0} & \boldsymbol{C} \end{bmatrix} = \det(\boldsymbol{A}) \det(\boldsymbol{C}) \qquad \text{or} \qquad \det\begin{bmatrix} \boldsymbol{A} & \boldsymbol{0} \\ \boldsymbol{B} & \boldsymbol{C} \end{bmatrix} = \det(\boldsymbol{A}) \det(\boldsymbol{C}) \tag{A.121}$$

Given $\boldsymbol{B} \in \Re^{n \times m}$ and $\boldsymbol{C} \in \Re^{m \times n}$, then

$$\det\begin{bmatrix} \boldsymbol{I}_n & \boldsymbol{B} \\ -\boldsymbol{C} & \boldsymbol{I}_m \end{bmatrix} = \det(\boldsymbol{I}_m - \boldsymbol{C}\boldsymbol{B}) = \det(\boldsymbol{I}_n - \boldsymbol{B}\boldsymbol{C}) \tag{A.122}$$

Inverse of partitioned matrices

Given $\boldsymbol{B} \in \Re^{n \times m}$ and $\boldsymbol{C} \in \Re^{m \times n}$, then

$$\begin{bmatrix} \boldsymbol{I}_n & \boldsymbol{B}_{n \times m} \\ \boldsymbol{0}_{m \times n} & \boldsymbol{I}_m \end{bmatrix}^{-1} = \begin{bmatrix} \boldsymbol{I}_n & -\boldsymbol{B}_{n \times m} \\ \boldsymbol{0}_{m \times n} & \boldsymbol{I}_m \end{bmatrix} \in \Re^{(n+m) \times (n+m)} \tag{A.123}$$

and

$$\begin{bmatrix} \boldsymbol{I}_n & \boldsymbol{0}_{n \times m} \\ \boldsymbol{C}_{m \times n} & \boldsymbol{I}_m \end{bmatrix}^{-1} = \begin{bmatrix} \boldsymbol{I}_n & \boldsymbol{0}_{n \times m} \\ -\boldsymbol{C}_{m \times n} & \boldsymbol{I}_m \end{bmatrix} \in \Re^{(n+m) \times (n+m)} \tag{A.124}$$

Given $\boldsymbol{A}, \boldsymbol{B} \in \Re^{n \times n}$ and $\rho(\boldsymbol{A}) = \rho(\boldsymbol{B}) = n$, then

$$\begin{bmatrix} \boldsymbol{A} & \boldsymbol{0} \\ \boldsymbol{P} & \boldsymbol{B} \end{bmatrix}^{-1} = \begin{bmatrix} \boldsymbol{A}^{-1} & \boldsymbol{0} \\ -\boldsymbol{B}^{-1}\boldsymbol{P}\boldsymbol{A}^{-1} & \boldsymbol{B}^{-1} \end{bmatrix} \tag{A.125}$$

and

$$\begin{bmatrix} \boldsymbol{A} & \boldsymbol{Q} \\ \boldsymbol{0} & \boldsymbol{B} \end{bmatrix}^{-1} = \begin{bmatrix} \boldsymbol{A}^{-1} & -\boldsymbol{A}^{-1}\boldsymbol{Q}\boldsymbol{B}^{-1} \\ \boldsymbol{0} & \boldsymbol{B}^{-1} \end{bmatrix} \tag{A.126}$$

Given $\boldsymbol{A} \in \Re^{n\times n}$, $\boldsymbol{B} \in \Re^{n\times m}$, $\boldsymbol{C} \in \Re^{m\times n}$, $\boldsymbol{D} \in \Re^{m\times m}$, then

$$\begin{bmatrix} \boldsymbol{A} & \boldsymbol{B} \\ \boldsymbol{C} & \boldsymbol{D} \end{bmatrix}^{-1} = \begin{bmatrix} \boldsymbol{A}^{-1} + \boldsymbol{E}_A \Delta_A^{-1} \boldsymbol{F}_A & -\boldsymbol{E}_A \Delta_A^{-1} \\ -\Delta_A^{-1} \boldsymbol{F}_A & \Delta_A^{-1} \end{bmatrix} \tag{A.127}$$

if $\Delta_A = \boldsymbol{D} - \boldsymbol{C}\boldsymbol{A}^{-1}\boldsymbol{B}$ (known as the *Schur complement* of $\boldsymbol{A}$) and $\boldsymbol{A}$ are invertible, where $\boldsymbol{E}_A = \boldsymbol{A}^{-1}\boldsymbol{B}$ and $\boldsymbol{F}_A = \boldsymbol{C}\boldsymbol{A}^{-1}$. Also, if $\boldsymbol{D}$ and $\Delta_D = \boldsymbol{A} - \boldsymbol{B}\boldsymbol{D}^{-1}\boldsymbol{C}$ (known as the *Schur complement* of $\boldsymbol{D}$) are invertible, then

$$\begin{bmatrix} \boldsymbol{A} & \boldsymbol{B} \\ \boldsymbol{C} & \boldsymbol{D} \end{bmatrix}^{-1} = \begin{bmatrix} \Delta_D^{-1} & -\Delta_D^{-1} \boldsymbol{F}_D \\ -\boldsymbol{E}_D \Delta_D^{-1} & \boldsymbol{E}_D \Delta_D^{-1} \boldsymbol{F}_D + \boldsymbol{D}^{-1} \end{bmatrix} \tag{A.128}$$

where $\boldsymbol{E}_D = \boldsymbol{D}^{-1}\boldsymbol{C}$ and $\boldsymbol{F}_D = \boldsymbol{B}\boldsymbol{D}^{-1}$. Moreover, if $\boldsymbol{A}$, $\Delta_A = \boldsymbol{D} - \boldsymbol{C}\boldsymbol{A}^{-1}\boldsymbol{B}$, and $\Delta_D = \boldsymbol{A} - \boldsymbol{B}\boldsymbol{D}^{-1}\boldsymbol{C}$ are invertible, then

$$\Delta_D^{-1} = \boldsymbol{A}^{-1} + \boldsymbol{E}_A \Delta_A^{-1} \boldsymbol{F}_A \tag{A.129}$$

Rank of partitioned matrices

Given $\boldsymbol{A} \in \Re^{n\times m}$, $\boldsymbol{B} \in \Re^{p\times q}$, and $\boldsymbol{C} \in \Re^{p\times m}$, then

$$\rho(\boldsymbol{A}) + \rho(\boldsymbol{B}) = \rho\begin{bmatrix} \boldsymbol{A} & \boldsymbol{0} \\ \boldsymbol{0} & \boldsymbol{B} \end{bmatrix} \le \rho\begin{bmatrix} \boldsymbol{A} & \boldsymbol{0} \\ \boldsymbol{C} & \boldsymbol{B} \end{bmatrix} \tag{A.130}$$

and

$$\rho(\boldsymbol{A}) + \rho(\boldsymbol{B}) = \rho\begin{bmatrix} \boldsymbol{0} & \boldsymbol{A} \\ \boldsymbol{B} & \boldsymbol{0} \end{bmatrix} \le \rho\begin{bmatrix} \boldsymbol{0} & \boldsymbol{A} \\ \boldsymbol{B} & \boldsymbol{C} \end{bmatrix} \tag{A.131}$$

Given $\boldsymbol{A} \in \Re^{n\times n}$, $\boldsymbol{B} \in \Re^{n\times m}$, $\boldsymbol{C} \in \Re^{p\times n}$, and $\boldsymbol{D} \in \Re^{p\times m}$, if $\boldsymbol{A}$ is invertible, then

$$\begin{bmatrix} \boldsymbol{A} & \boldsymbol{B} \\ \boldsymbol{C} & \boldsymbol{D} \end{bmatrix} = \begin{bmatrix} \boldsymbol{I} & \boldsymbol{0} \\ \boldsymbol{C}\boldsymbol{A}^{-1} & \boldsymbol{I} \end{bmatrix} \begin{bmatrix} \boldsymbol{A} & \boldsymbol{0} \\ \boldsymbol{0} & \boldsymbol{D} - \boldsymbol{C}\boldsymbol{A}^{-1}\boldsymbol{B} \end{bmatrix} \begin{bmatrix} \boldsymbol{I} & \boldsymbol{A}^{-1}\boldsymbol{B} \\ \boldsymbol{0} & \boldsymbol{I} \end{bmatrix} \tag{A.132}$$

and

$$\rho\begin{bmatrix} \boldsymbol{A} & \boldsymbol{B} \\ \boldsymbol{C} & \boldsymbol{D} \end{bmatrix} = n + \rho(\boldsymbol{D} - \boldsymbol{C}\boldsymbol{A}^{-1}\boldsymbol{B}) \tag{A.133}$$

If $p = m$, see (A.118).

Given $\boldsymbol{A} \in \Re^{n\times m}$, $\boldsymbol{B} \in \Re^{n\times p}$, $\boldsymbol{C} \in \Re^{p\times m}$, and $\boldsymbol{D} \in \Re^{p\times p}$, if $\boldsymbol{D}$ is invertible, then

$$\begin{bmatrix} \boldsymbol{A} & \boldsymbol{B} \\ \boldsymbol{C} & \boldsymbol{D} \end{bmatrix} = \begin{bmatrix} \boldsymbol{I} & \boldsymbol{B}\boldsymbol{D}^{-1} \\ \boldsymbol{0} & \boldsymbol{I} \end{bmatrix} \begin{bmatrix} \boldsymbol{A} - \boldsymbol{B}\boldsymbol{D}^{-1}\boldsymbol{C} & \boldsymbol{0} \\ \boldsymbol{0} & \boldsymbol{D} \end{bmatrix} \begin{bmatrix} \boldsymbol{I} & \boldsymbol{0} \\ \boldsymbol{D}^{-1}\boldsymbol{C} & \boldsymbol{I} \end{bmatrix} \tag{A.134}$$

and

$$\rho\begin{bmatrix} \boldsymbol{A} & \boldsymbol{B} \\ \boldsymbol{C} & \boldsymbol{D} \end{bmatrix} = p + \rho(\boldsymbol{A} - \boldsymbol{B}\boldsymbol{D}^{-1}\boldsymbol{C}) \tag{A.135}$$

If $n = m$, see (A.120).

Given $\boldsymbol{A} \in \Re^{n\times m}$ and $\boldsymbol{B} \in \Re^{p\times m}$, then

$$\rho\begin{bmatrix} \boldsymbol{A} \\ \boldsymbol{B} \end{bmatrix} \leq \rho(\boldsymbol{A}) + \rho(\boldsymbol{B}) \tag{A.136}$$

and if $\boldsymbol{A} \in \Re^{n\times m}$ and $\boldsymbol{B} \in \Re^{nxp}$

$$\rho[\boldsymbol{A}, \boldsymbol{B}] \leq \rho(\boldsymbol{A}) + \rho(\boldsymbol{B}) \tag{A.137}$$

A.2.17 The Kronecker product and sum

The Kronecker product and sum [4, 5] are useful for problems in system theory. Given the matrices $\boldsymbol{A} \in \Re^{p\times q}$ and $\boldsymbol{B} \in \Re^{m\times n}$, the *Kronecker product* is defined as

$$\boldsymbol{A} \otimes \boldsymbol{B} \triangleq \begin{bmatrix} a_{11}\boldsymbol{B} & a_{12}\boldsymbol{B} & \cdots & a_{1q}\boldsymbol{B} \\ a_{21}\boldsymbol{B} & a_{22}\boldsymbol{B} & \cdots & a_{2q}\boldsymbol{B} \\ \cdots & \cdots & \cdots & \cdots \\ a_{p1}\boldsymbol{B} & a_{p2}\boldsymbol{B} & \cdots & a_{pq}\boldsymbol{B} \end{bmatrix} = \boldsymbol{C} \in \Re^{pm\times qn} \tag{A.138}$$

Given the matrices $\boldsymbol{N} \in \Re^{n\times n}$ and $\boldsymbol{M} \in \Re^{m\times m}$, the *Kronecker sum* is defined as

$$\boldsymbol{N} \oplus \boldsymbol{M} \triangleq \boldsymbol{N} \otimes \boldsymbol{I}_m + \boldsymbol{I}_n \otimes \boldsymbol{M} = \boldsymbol{P} \in \Re^{nm\times nm} \tag{A.139}$$

The Kronecker sum essentially allows matrices that are normally nonconformable for addition to be "summed." Assuming $\boldsymbol{A} \in \Re^{p\times q}$, a very important vector-valued function of a matrix is defined as [4, 5]

$$\operatorname{vec}(\boldsymbol{A}) \triangleq \Big[\underbrace{(a_{11}, a_{21}, \cdots, a_{p1})}_{\text{1st column of } \boldsymbol{A}}, \underbrace{(a_{12}, a_{22}, \cdots, a_{p2})}_{\text{2nd column of } \boldsymbol{A}}, \cdots, \underbrace{(a_{1q}, a_{2q}, \cdots, a_{pq})}_{q\text{th column of } \boldsymbol{A}}\Big]^T \tag{A.140}$$

where $\operatorname{vec}(\boldsymbol{A}) \in \Re^{pq\times 1}$. The kth *row* of matrix $\boldsymbol{A}$ will be written as $\boldsymbol{A}_{k:}$, and the kth *column* will be denoted by $\boldsymbol{A}_{:k}$. Therefore, (A.140) can be written as

$$\operatorname{vec}(\boldsymbol{A}) \triangleq \begin{bmatrix} \boldsymbol{A}_{:1} \\ \boldsymbol{A}_{:2} \\ \vdots \\ \boldsymbol{A}_{:q} \end{bmatrix} \in \Re^{pq\times 1} \tag{A.141}$$

The vec operator above can be modified to only select the major diagonal elements of a square matrix; that is, this vector-valued function of a square matrix $\boldsymbol{N} \in \Re^{n\times n}$ is defined as

$$\operatorname{vecd}(\boldsymbol{N}) \triangleq [n_{11}, n_{22}, \cdots, n_{nn}]^T \in \Re^{n\times 1} \tag{A.142}$$

Also, given $\boldsymbol{C} \in \Re^{p\times q}$ with $\boldsymbol{D} \in \Re^{m\times q}$ (note that the matrices have the same number of columns), the *Khatri-Rao product* is defined as

$$D \odot C \triangleq \left[D_{:1} \otimes C_{:1}, D_{:2} \otimes C_{:2}, \cdots, D_{:q} \otimes C_{:q} \right] \in \Re^{mp \times q} \tag{A.143}$$

Some useful properties involving the Kronecker product, Kronecker sum, the vec operator and the Khatri-Rao product are as follows:

1. $(A \otimes B)^T = (A^T \otimes B^T)$
2. $(A \otimes B) \otimes C = A \otimes (B \otimes C)$
3. $(A \otimes B)(D \otimes G) = AD \otimes BG$
4. $(\alpha A) \otimes B = A \otimes (\alpha B)$, where $\alpha \in \Re$
5. $(A + H) \otimes (B + R) = A \otimes B + A \otimes R + H \otimes B + H \otimes R$
6. $(A + H) \otimes B = A \otimes B + H \otimes B$
7. $A \otimes (B + R) = A \otimes B + A \otimes R$
8. $(N \otimes M)^{-1} = N^{-1} \otimes M^{-1}$, if N and M are invertible
9. $\det(N \otimes M) = (\det N)^m (\det M)^n = \det(M \otimes N)$
10. $\text{trace}(N \otimes M) = \text{trace}(N)\text{trace}(M)$
11. $(I_m \otimes N)(M \otimes I_n) = (M \otimes I_n)(I_m \otimes N)$
12. $f(N \otimes I_m) = f(N) \otimes I_m$, where $f(\bullet)$ is an analytic function
13. $f(I_m \otimes N) = I_m \otimes f(N)$
14. $e^{N \otimes M} = e^N \otimes e^M$
15. $(I_p \otimes z)A = A \otimes z$
16. $A(I_q \otimes z^T) = A \otimes z^T$
17. $\text{vec}(A + H) = \text{vec}(A) + \text{vec}(H)$
18. $\text{vec}(ADB) = (B^T \otimes A)\text{vec}(D)$
19. $\text{vec}(AD) = (I_s \otimes A)\,\text{vec}(D) = (D^T \otimes I_p)\,\text{vec}(A) = (D^T \otimes A)\,\text{vec}(I_q)$
20. $\text{vec}(AD) = \sum_{k=1}^{q} D_{:k}^T \otimes A_{:k}$
21. $A^{[2]} = A \otimes A$ (the *Kronecker square*)
22. $A^{[k+1]} = A \otimes A^{[k]}$
23. $(AD)^{[k]} = A^{[k]} D^{[k]}$
24. $M \otimes A = (U_1 \otimes U_2)(S_1 \otimes S_2)(V_1 \otimes V_2)^T$, where M and A have the SVDs $M = U_1 S_1 V_1^T$ and $A = U_2 S_2 V_2^T$
25. $\text{trace}(ADW) = [\text{vec}(A^T)]^T (I_p \otimes D)\,\text{vec}(W)$
26. $\text{trace}(A^T H) = [\text{vec}(A)]^T \text{vec}(H)$
27. $A \odot (D^T \odot F^T) = (A \odot D^T) \odot F^T$
28. $(A \otimes B)(F \odot G) = AF \odot BG$
29. $\text{vec}(AVD) = (D^T \odot A)\,\text{vecd}(V)$, providing $V \in \Re^{q \times q}$ is diagonal
30. $N \oplus N = A \otimes I_n + I_n \otimes A$
31. If λ_i, for $i = 1, 2, \cdots, n$ are the eigenvalues of N with associated eigenvectors z_i and β_j, for $j = 1, 2, \ldots, m$ are the eigenvalues of M with associated eigenvectors y_j, then we can write

$$\boldsymbol{\lambda}(N \otimes M) = \boldsymbol{\lambda}(N) \otimes \boldsymbol{\lambda}(M)$$

when N and M are nonsingular. That is, the eigenvalues of the Kronecker product of two nonsingular matrices are given by the Kronecker product of the eigenvalues [contained in the vectors $\boldsymbol{\lambda}(N)$ and $\boldsymbol{\lambda}(M)$] of the individual matrices, where the vector dimensions are $\boldsymbol{\lambda}(N \otimes M)_{nm \times 1}$, $\boldsymbol{\lambda}(N)_{n \times 1}$, and $\boldsymbol{\lambda}(M)_{m \times 1}$. An alternate means of expressing this is to say that the eigenvalues of $N \otimes M$ are the mn numbers $\lambda_i \beta_j$, for $i = 1, 2, \ldots, n$ and $j = 1, 2, \ldots, m$.

32. The eigenvalues of $N \oplus M = (N \otimes I_m) + (I_n \otimes M)$ are the mn numbers $\lambda_i + \beta_j$, for $i = 1, 2, \ldots, n$ and $j = 1, 2, \ldots, m$ (see property 31).
33. The eigenvectors of $N \otimes M$ and $N \oplus M$ correspond to the eigenvalues $\lambda_i \beta_j$ and $\lambda_i + \beta_j$, respectively, and are $z_i \otimes y_j$, $i = 1, 2, \ldots, n$ and $j = 1, 2, \ldots, m$ (see property 31).

The dimensions of the matrices and vectors used in the properties above are:

$A \in \Re^{p \times q}$	$D \in \Re^{q \times s}$	$H \in \Re^{p \times q}$	$R \in \Re^{s \times t}$
$B \in \Re^{s \times t}$	$F \in \Re^{q \times u}$	$M \in \Re^{m \times m}$	$W \in \Re^{s \times p}$
$C \in \Re^{r \times 1}$	$G \in \Re^{t \times u}$	$N \in \Re^{n \times n}$	$y \in \Re^{m \times 1}$
			$z \in \Re^{n \times 1}$

The well-known Lyapunov equation [6] is given as

$$AX + XA^T = -C \tag{A.144}$$

where $A \in \Re^{n \times n}$, $C \in \Re^{n \times n}(C^T = C$ and $C \geq 0)$, and the solution $X \in \Re^{n \times n}(X^T = X$ and $X \geq 0)$. This equation cannot be solved directly for X using standard methods from linear algebra. However, if we first form the matrix vector on both sides of (A.144)

$$\text{vec}(AX + XA^T) = -\text{vec}(C) \tag{A.145}$$

and use property 19 and (A.139), we can write (A.145) as

$$(A \oplus A)\,\text{vec}(X) = -\text{vec}(C) \tag{A.146}$$

To solve for vec(X), $A \oplus A$ must be invertible, that is,

$$\text{vec}(X) = -(A \oplus A)^{-1}\,\text{vec}(C) \tag{A.147}$$

therefore, $A \oplus A$ cannot have any zero eigenvalues. Therefore, using property 32, we can conclude that the necessary and sufficient condition for the existence of a unique solution for (A.144) is $\lambda_i(A) + \lambda_j(A) \neq 0$ for $i, j = 1, 2, \ldots, n$. Therefore, if A is asymptotically stable (cf. Sect. A.2.12), a unique solution exists for (A.144); however, if A is marginally stable, a unique solution does not exist. It also follows that if A is unstable and possesses pairs of complex conjugate (or strictly real) eigenvalues forming symmetry about the $j\omega$ axis in the complex plane, then (A.144) will not have a unique solution. Following the same approach above, it is simple to show that if $A \in \mathscr{C}^{n \times n}$ and $C \in \mathscr{C}^{n \times n}$, (A.144) will have a unique solution $X \in \mathscr{C}^{n \times n}$ if $\lambda_i(A) + \bar{\lambda}_j(A) \neq 0$ for $i, j = 1, 2, \ldots, n$ (where $\bar{\lambda}$ is the complex conjugate of λ).

A.2.18 Summary of Important Properties for Real and Complex Square Matrices

The following is a summary of some of the important properties for real and complex square matrices.

Real matrices

If $A \in \Re^{n\times n}(A = [a_{ij}],$ for $i, j = 1, 2, \ldots, n)$, then:

1. A is symmetric $\Leftrightarrow$ $A^T = A$
2. A is skew symmetric $\Leftrightarrow$ $A^T = -A$
3. A is symmetric $\Leftrightarrow$ $a_{ij} = 0, \forall i \neq j$
4. A is normal $\Leftrightarrow$ $AA^T = A^T A$
5. A is positive definite $\Leftrightarrow$ $A^T = A$ and $x^T Ax > 0$, for $x \in \Re^{n\times 1}$ and $x \neq 0$
6. A is positive semidefinite (nonnegative definite) $\Leftrightarrow$ $A^T = A$ and $x^T Ax \geq 0$, $x \in \Re^{n\times 1}$ and $x \neq 0$
7. A is orthogonal $\Leftrightarrow$ $A^T A = AA^T = I$
8. A is involutary $\Leftrightarrow$ $A^2 = I$
9. A is skew involutary $\Leftrightarrow$ $A^2 = -I$
10. A is idempotent $\Leftrightarrow$ $A^2 = A$
11. A is tripotent $\Leftrightarrow$ $A^3 = A$
12. A is nilpotent $\Leftrightarrow$ $A^k = 0$, for some $k > 0$
13. A is upper triangular $\Leftrightarrow$ $a_{ij} = 0 \ \forall i > j$
14. A is lower triangular $\Leftrightarrow$ $a_{ij} = 0 \ \forall i < j$

If $A \in \Re^{2n\times 2n}$ and $J_n \in \Re^{2n\times 2n}$ is defined by

$$J_n \triangleq \begin{bmatrix} \mathbf{0} & I_n \\ -I_n & \mathbf{0} \end{bmatrix}$$

then

1. A is Hamiltonian $\Leftrightarrow$ $J^{-1}A^T J = -A \Leftrightarrow (JA)^T = JA$
2. A is symplectic $\Leftrightarrow$ $J^{-1}A^T J = A^{-1} \Leftrightarrow A^T JA = J$

If $A, B \in \Re^{n\times n}$ are *congruent*, then

1. $\rho(A) = \rho(B)$
2. A^T and B^T are congruent
3. A is invertible $\Leftrightarrow$ B is invertible $\Rightarrow$ A^{-1} and B^{-1} are congruent
4. A is symmetric, skew symmetric, positive definite, positive semidefinite $\Leftrightarrow$ B is symmetric, skew symmetric, positive definite, positive semidefinite

If A, $B \in \Re^{n\times n}$ are *orthogonally similar*, then A is symmetric, skew symmetric, positive definite, positive semidefinite, normal, orthogonal, involutary, skew involutary, idempotent, tripotent, and nilpotent if and only if B is symmetric, skew symmetric, positive definite, positive semidefinite, normal, orthogonal, involutary, skew involutary, idempotent, tripotent, and nilpotent.

Complex matrices

If $A \in \mathscr{C}^{n\times n}$, then

1. A is normal $\Leftrightarrow$ $AA^* = A^* A$
2. A is Hermitian $\Leftrightarrow$ $A^* = A$
3. A is skew Hermitian $\Leftrightarrow$ $A^* = -A$
4. A is positive definite $\Leftrightarrow$ $A^* = A$ and $x^* Ax > 0$, for $x \in \mathscr{C}^{n\times 1}$
5. A is positive semidefinite $\Leftrightarrow$ $A^* = A$ and $x^* Ax \geq 0$, for $x \in \mathscr{C}^{n\times 1}$
6. A is unitary $\Leftrightarrow$ $A^* A = AA^* = I$

Circulant matrices

A *circulant matrix* is a square matrix in which every row, beginning with the second, can be obtained from the preceding row by shifting each element in the row one column to the right, with the first element in the first row finally residing as the last element in the last row. Every circulant matrix has the general form

$$A = \begin{bmatrix} a_1 & a_2 & a_3 & a_4 & \cdots & a_n \\ a_n & a_1 & a_2 & a_3 & \cdots & a_{n-1} \\ a_{n-1} & a_n & a_1 & a_2 & \cdots & a_{n-2} \\ a_{n-2} & a_{n-1} & a_n & a_1 & \cdots & a_{n-3} \\ \cdots & \cdots & \cdots & \cdots & \cdots & \cdots \\ a_2 & a_3 & a_4 & a_5 & \cdots & a_1 \end{bmatrix} \tag{A.148}$$

Properties of circulant matrices are as follows:

1. A circulant matrix $A_{n\times n}$ has eigenvalues given by
$$\lambda_i = a_1 + a_2 r_i + a_3 r_i^2 + \cdots + a_n r_i^{n-1}, \qquad \text{for } i = 1, 2, \ldots, n$$
where $[a_1, a_2, a_3, \cdots, a_n]$ is the first row of A and r_i is one of the distinct solutions of $r^n = 1$. The corresponding eigenvectors are given by $v_i = [1, r_i, r_i^2, \cdots, r_i^{n-1}]^T$.
2. If A and B are circulant matrices of the same order and α and β are any two scalars, then $\alpha A + \beta B$ is also a circulant matrix.
3. If a circulant matrix is nonsingular, then its inverse is a circulant matrix.
4. The product of two circulant matrices A and B of the same order is also a circulant matrix, and the product is commutative, that is, $AB = BA$.

Band matrices

A square matrix $A_{n\times n}$ with elements $[a_{ij}]$ for $i, j = 1, 2, \ldots, n$ is a *band matrix* of width $2\kappa + 1$ if $a_{ij} = 0$ when $|i - j| > \kappa$ for some nonnegative integer $0 \le \kappa \le n-1$. Band matrices have the general form shown in Figure A.2. All nonzero elements in a band matrix are positioned on the main diagonal and all other κ superdiagonals and subdiagonals. A band matrix with $\kappa = 0$ is a (strictly) diagonal matrix. The sums, products, and transposes of $n \times n$ dimensional band matrices with a width of $2\kappa+1$ are also band matrices of the same width.

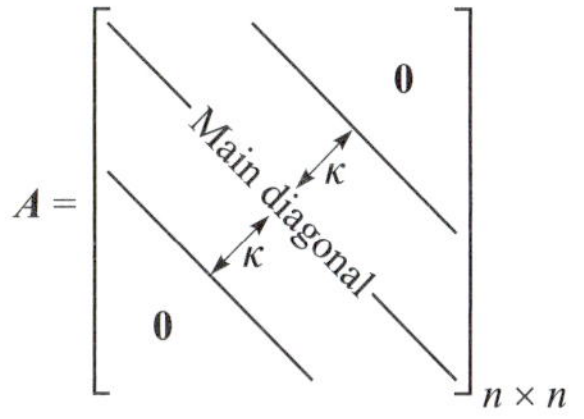

Figure A.2 General form of a band matrix.

A *Toeplitz matrix* is a band matrix in which each diagonal consists of identical elements; however, different diagonals may contain different elements. Every nonzero circulant matrix is a Toeplitz matrix of full width. An example of a Toeplitz matrix that is also a circulant matrix is

$$\boldsymbol{A} = \begin{bmatrix} 1 & 2 & 3 & 4 & 5 \\ 5 & 1 & 2 & 3 & 4 \\ 4 & 5 & 1 & 2 & 3 \\ 3 & 4 & 5 & 1 & 2 \\ 2 & 3 & 4 & 5 & 1 \end{bmatrix}$$

Tridiagonal Matrices

A *tridiagonal matrix* is a band matrix that has width 3 ($\kappa = 1$). Therefore, the only nonzero elements are on the main diagonal, the superdiagonal, and the subdiagonal (all other elements in the matrix are zero). If we assume a tridiagonal Toeplitz matrix of order $n \times n$, that is, $\boldsymbol{T}_{n \times n}$, with elements on the main diagonal given by α, elements on the superdiagonal given by β, and elements on the subdiagonal given by γ, the eigenvalues of $\boldsymbol{T}$ are given by

$$\lambda_k = \alpha + 2\sqrt{\beta\gamma}\cos\left(\frac{k\pi}{n+1}\right) \tag{A.149}$$

for $k = 1, 2, \ldots, n$.

Hessenberg form

A square matrix is in *Hessenberg form* if all elements below the subdiagonal are zero. Every real square matrix $\boldsymbol{A} \in \Re^{n \times n}$ is congruent to a matrix in Hessenberg form. An example of a matrix in Hessenberg form is

$$\boldsymbol{A} = \begin{bmatrix} 2 & -4 & 0 & 0 & 0 \\ 3 & 2 & -4 & 0 & 0 \\ 0 & 3 & 2 & -4 & 0 \\ 0 & 0 & 3 & 2 & -4 \\ 0 & 0 & 0 & 3 & 2 \end{bmatrix}$$

This matrix is also a band matrix of width 3 ($\kappa = 1$), a tridiagonal matrix, and a Toeplitz matrix.

Hilbert matrices

A *Hilbert matrix* is a square symmetric matrix $\boldsymbol{H} \in \Re^{n \times n}(\boldsymbol{H} = [h_{ij}]$, $i, j = 1, 2, \ldots, n)$, with elements given by $h_{ij} = 1/(i + j - 1)$. This is a classic example of a badly conditioned matrix. The Hilbert matrix is a special case of a *Cauchy matrix* [7].

Hankel matrices

A *Hankel matrix* $\boldsymbol{H} = [h_{ij}]$, for $i, j = 1, 2, \ldots, n$, is symmetric and has identical elements on the *antidiagonals*. In its basic form, the elements of a Hankel matrix are completely determined by a defined vector $\boldsymbol{v} = [v_k]$ for $k = 1, 2, \ldots, n$, that is,

$$h_{ij} = v_{i+j-1} \tag{A.150}$$

The elements of $\boldsymbol{H}$ are set to zero for the indexing of the vector $\boldsymbol{v}$ greater than its length (i.e., $> n$) in (A.150).

Consider a proper rational transfer function (cf. Sect. A.2.12)

$$H(s) = \frac{\beta_0 s^n + \beta_1 s^{n-1} + \beta_2 s^{n-2} + \cdots + \beta_{n-2} s^2 + \beta_{n-1} s^1 + \beta_n}{s^n + \alpha_1 s^{n-1} + \alpha_2 s^{n-2} + \cdots + \alpha_{n-2} s^2 + \alpha_{n-1} s^1 + \alpha_n} \tag{A.151}$$

expanded into an infinite power series in the variable s

$$H(s) = h(0) + h(1)s^{-1} + h(2)s^{-2} + h(3)s^{-3} + \cdots \tag{A.152}$$

The coefficients in (A.152) $h(i)$, for $i = 0, 1, 2, 3, \ldots$, called the *Markov parameters*, can be written recursively as

$$\begin{aligned}
h(0) &= \beta_0 \\
h(1) &= -\alpha_1 h(0) + \beta_1 \\
h(2) &= -\alpha_1 h(1) - \alpha_2 h(0) + \beta_2 \\
&\vdots \\
h(n) &= -\alpha_1 h(n-1) - \alpha_2 h(n-2) - \cdots - \alpha_n h(0) + \beta_n \\
h(n+i) &= -\alpha_1 h(n+i-1) - \alpha_2 h(n+i-2) - \cdots - \alpha_n h(i)
\end{aligned} \tag{A.153}$$

for $i = 1, 2, 3, \ldots$. The matrix $\boldsymbol{H}(\alpha, \beta)$ (Hankel matrix of order $\alpha \times \beta$), can be formed as

$$\boldsymbol{H}(\alpha, \beta) = \begin{bmatrix} h(1) & h(2) & h(3) & \cdots & h(\beta) \\ h(2) & h(3) & h(4) & \cdots & h(\beta+1) \\ h(3) & h(4) & h(5) & \cdots & h(\beta+2) \\ \cdots & \cdots & \cdots & \cdots & \cdots \\ h(\alpha) & h(\alpha+1) & h(\alpha+2) & \cdots & h(\alpha+\beta-1) \end{bmatrix} \tag{A.154}$$

from the Markov parameters $h(i)$, for $i = 1, 2, 3, \ldots$ [note that $h(0)$ is not directly involved in forming $\boldsymbol{H}(\alpha, \beta)$]. The Hankel matrix in (A.154) is used in control theory to find irreducible realizations of the transfer function given in (A.151).

Vandermonde matrices

A *Vandermonde matrix* is a square matrix $\boldsymbol{V}_{n \times n}$, and it has as its columns the eigenvectors of a companion or Frobenius matrix (cf. Sect. A.2.12). The general form of a Vandermonde matrix is

$$\boldsymbol{V} = \begin{bmatrix} 1 & 1 & 1 & \cdots & 1 \\ \lambda_1 & \lambda_2 & \lambda_3 & \cdots & \lambda_n \\ \lambda_1^2 & \lambda_2^2 & \lambda_3^2 & \cdots & \lambda_n^2 \\ \cdots & \cdots & \cdots & \cdots & \cdots \\ \lambda_1^{n-1} & \lambda_2^{n-1} & \lambda_3^{n-1} & \cdots & \lambda_n^{n-1} \end{bmatrix} \tag{A.155}$$

where λ_i for $i = 1, 2, \ldots, n$ are the n eigenvalues of the companion matrix. The determinant of the Vandermonde matrix is given by

$$\det(V) = (\lambda_2 - \lambda_1)(\lambda_3 - \lambda_2)(\lambda_3 - \lambda_1)(\lambda_4 - \lambda_3)(\lambda_4 - \lambda_2)(\lambda_4 - \lambda_1)\cdots(\lambda_n - \lambda_1) \tag{A.156}$$

If the eigenvalues of the companion matrix are distinct, we see from (A.156) that the determinant of the Vandermonde matrix will be nonzero.

Hadamard matrices

A *Hadamard matrix* [8] is a square matrix $\boldsymbol{H} \in \Re^{n \times n}(\boldsymbol{H} = [h_{ij}], i, j = 1, 2, \ldots, n)$, with elements $h_{ij} = \pm 1$, and for which the rows of $\boldsymbol{H}$ are mutually orthogonal. Hadamard matrices can exist for only certain n. Specifically, a necessary condition for their existence is if $n > 2$, then n must be a multiple of 4. For Hadamard matrices we have $\boldsymbol{H}^T\boldsymbol{H} = \boldsymbol{H}\boldsymbol{H}^T = n\boldsymbol{I}$; therefore, it follows that $\boldsymbol{H}^{-1} = n^{-1}\boldsymbol{H}^T$.

Inertia matrices

A square matrix $\boldsymbol{A}$ is *congruent* (*Hermitian congruent*, or *conjunctive*) to a square matrix $\boldsymbol{B}$ of the same dimension if there exists a nonsingular matrix $\boldsymbol{Q}$ such that

$$\boldsymbol{A} = \boldsymbol{Q}\boldsymbol{B}\boldsymbol{Q}^T (\boldsymbol{A} = \boldsymbol{Q}\boldsymbol{B}\boldsymbol{Q}^H) \tag{A.157}$$

Every $n \times n$ Hermitian matrix with rank r is congruent to a unique matrix in the partitioned form

$$\boldsymbol{U} = \begin{bmatrix} \boldsymbol{I}_p & \boldsymbol{0} & \boldsymbol{0} \\ \boldsymbol{0} & \boldsymbol{I}_q & \boldsymbol{0} \\ \boldsymbol{0} & \boldsymbol{0} & \boldsymbol{0} \end{bmatrix} \tag{A.158}$$

known as an *inertia matrix*.

Sylvester's law of inertia: Two Hermitian matrices are congruent if and only if they are congruent to the same inertia matrix, and then they both have p positive eigenvalues, q negative eigenvalues, and $\gamma = n - p - q$ zero eigenvalues (p is the *index* of $\boldsymbol{A}$ and $s = p - q$ is the *signature* of $\boldsymbol{A}$).

Pascal matrices

Pascal matrices are positive definite symmetric test matrices that have integer entries and are made up from Pascal's triangle. A Pascal matrix has the interesting property that its inverse also has integer entries. The elements of a Pascal matrix $\boldsymbol{P}_n \in \Re^{n \times n}$ are defined by

$$p_{ij} = \frac{(i+j-2)!}{(i-1)!(j-1)!} = \binom{i+j-2}{j-1} \tag{A.159}$$

In MATLAB, the function **`pascal(n)`** will generate an n-dimensional (square) Pascal matrix whose elements are defined by (A.159). The function **`pascal(n,1)`** in MATLAB will generate a scaled and transposed Cholesky factor $\boldsymbol{S}$ (lower triangular matrix) that is involutary, that is, it is its own inverse ($\boldsymbol{S}^2 = \boldsymbol{I}$).

Kahan matrices

A *Kahan matrix* is an upper triangular matrix $\boldsymbol{K}_n(\theta) \in \Re^{n\times n}$ of order n with the parameter θ and is defined by

$$\boldsymbol{K}_n(\theta) = \text{diag}[1, s, s^2, \cdots, s^{n-1}] \begin{bmatrix} 1 & c & c & \cdots & c \\ 0 & 1 & c & \cdots & c \\ 0 & 0 & 1 & \cdots & c \\ \cdots & \cdots & \cdots & \cdots & \cdots \\ 0 & 0 & 0 & \cdots & 1 \end{bmatrix} \tag{A.160}$$

where $s = \sin(\theta)$ and $c = \cos(\theta)$. For different-order matrices and the angle θ, various matrices of varying *condition* can be created. In general, the smaller the angle θ, the more ill-conditioned the matrix becomes.

MathWorks, Inc., has a "Test Matrix Toolbox" available by anonymous ftp; the URL is as follows: ftp://ftp.mathworks.com/pub/contrib/linalg/testmatrix. The manual is also available at the same location as testmatrix.ps. The toolbox contains a collection of MATLAB m-files with test matrices, routines for visualizing matrices, and other routines that provide useful additions to the existing MATLAB functions.

A.3 PRINCIPLES OF MULTIVARIABLE ANALYSIS

A.3.1 Sets and Functions

Sets

Consider σ to be a member or element of the set Σ, that is, $\sigma \in \Sigma$. If an element does not belong to the set Σ, we write $\sigma \notin \Sigma$. We typically write the elements of a set inside a set of brackets, for example, $\Sigma = \{-1, 0, 1\}$, or the set of positive integers $\Omega = \{1, 2, 3, \ldots, n, \ldots\}$. We can also define a set of elements by a certain property; for example, we can say that set A is the set of all elements σ in Σ which have the property P, that is,

$$A = \{\sigma \in \Sigma : P(\sigma)\} \tag{A.161}$$

or this can also be written more compactly as

$$A = \{\sigma : P(\sigma)\} \tag{A.162}$$

when the set Σ is implied. Although we typically think of a set as having some members, there is one (and only one) set that has no members, called the *empty set* $\emptyset$. If each σ in the set A is in set B, that is, $\sigma \in A \Rightarrow \sigma \in B$, then we say that A is a *subset* of B (or A is contained in B) and we write $A \subset B$. If $A \subset B$ and $B \subset A$, then $B = A$. Moreover, for any set A, each member of the empty set $\emptyset$ (for which there are none) is a member of A, and we can write $\emptyset \subset A$. Therefore, the empty set is a subset of every set. A set C that has elements which belong to either a set A or a set B is called the *union* of the two sets and is written as

$$A \cup B = C = \{\sigma : \sigma \in A \vee \sigma \in B\} \tag{A.163}$$

If A and B are subsets of C, we can define their *intersection* to be the set of all elements which belong to both A and B, written as

$$A \cap B = C = \{\sigma : \sigma \in A \wedge \sigma \in B\} \tag{A.164}$$

If A is a subset of B, then the *complement* of A, that is, $\tilde{A}$ (relative to B), is defined as the set of elements *not* in A. This is written as

$$\tilde{A} = \{\sigma \in B : \sigma \notin A\} \tag{A.165}$$

Then it follows that $\tilde{\tilde{A}} = A$, $A \cup \tilde{A} = B$, and $A \cap \tilde{A} = \emptyset$. De Morgan's laws are given by

$$A \,\tilde{\cup}\, B = \tilde{A} \cap \tilde{B} \qquad \text{and} \qquad A \,\tilde{\cap}\, B = \tilde{A} \cup \tilde{B} \tag{A.166}$$

If A is a subset of the set of real numbers B, that is, $A \subset B$, then the smallest element of the set B which is greater than or equal to every element in A is called the *least upper bound* or the *supremum* (sup) of A. We write supremum of A as

$$\sup A \qquad \text{or} \qquad \sup_{\sigma \in A} \sigma \qquad \text{or} \qquad \sup\{\sigma : \sigma \in A\} \tag{A.167}$$

Conversely, the *greatest lower bound* of a set of real numbers A, also called the *infimum* (inf) of A, is the largest number among the lower bounds of A, written as

$$\inf A \qquad \text{or} \qquad \inf_{\sigma \in A} \sigma \qquad \text{or} \qquad \inf\{\sigma : \sigma \in A\} \tag{A.168}$$

Note that $\inf_{\sigma \in A} \sigma = -\sup_{\sigma \in A} -\sigma$.

Functions

A function is essentially a restricted assignment rule; that is, a function f from (or on) a set X to (or into) a set Y is the rule which assigns to each $x \in X$ a unique element $f(x) \in Y$. The collection $\mathfrak{G}$ of ordered pairs $\langle x, f(x) \rangle$ in $X \times Y$ (the Cartesian product) is called the *graph* of the function f. Many times a function is defined to be its graph. Moreover, the term *mapping* is often used as a synonym for *function*. We can express that f is a function on X into Y as

$$f : X \to Y$$

The set X is called the *domain* of f, and the set of values taken by f is called the *range* of f. The range of a function is typically smaller than Y; however, if the range of f is Y, then it is said that f is a function *onto* Y (or f is *surjective*).

If we assume $A \subset X$, then the *image* under the function f of A is defined as the set of elements in Y such that $y = f(x)$ for some $x \in A$. This image is denoted by $f[A]$. Thus the range of f is $f[X]$, and f is onto Y if and only if $Y = f[X]$. Now if $B \subset Y$, we can define the *inverse image* $f^{-1}[B]$ of B to be the set of those $x \in X$ for which $f(x)$ is in B. Note that f is onto Y if and only if the inverse image of each nonempty subset of Y is nonempty. A function $f: X \to Y$ is called *one-to-one*, or *injective*, or *univalent*, if $f(x_1) = f(x_2)$ only when $x_1 = x_2$. If a function is one-to-one from X onto Y, this function is called a one-to-one correspondence (or *bijective*) between X and Y. In this case there is a function $g: Y \to X$ such that $g(f(x)) = x$ and $f(g(y)) = y$, $\forall x$ and y. The function g is called the *inverse* of the function f and is denoted by f^{-1}. If $f: X \to Y$ and $g: Y \to Z$, we will define a new function $h: X \to Z$ such that we can write $h(x) = g(f(x))$. The function h is referred to as the *composition* of g with f and is denoted by $g \circ f$.

Functions of several variables are important in the study of neurocomputing. If we assume a vector $\boldsymbol{x} \in \Re^{n \times 1}$ (where $\boldsymbol{x} = [x_1, x_2, \cdots, x_n]^T$), the minimization of a function $f(\boldsymbol{x})$ over a set of specified vectors $\mathscr{X}$ can be written as (cf. Sect. A.5)

$$\min_{\boldsymbol{x} \in \mathscr{X}} f(\boldsymbol{x})$$

Let us now define a convex set and then a convex function.

DEFINITION A.10. A set Σ is *convex* if, for any elements $\boldsymbol{x}, \boldsymbol{y} \in \Sigma$,

$$\beta \boldsymbol{x} + (1 - \beta)\boldsymbol{y} \in \Sigma \qquad \forall\ 0 \le \beta \le 1$$

Definition A.10 simply says that if $\boldsymbol{x}$ and $\boldsymbol{y}$ are in Σ, then a line segment connecting $\boldsymbol{x}$ and $\boldsymbol{y}$ is also in Σ. Every set that is defined by a system of linear constraints is a convex set. Figure A.3 shows examples of convex and nonconvex sets.

DEFINITION A.11. A function f defined on a convex set Σ is a *convex function* if for every $\boldsymbol{x}, \boldsymbol{y} \in \Sigma$ and every β $(0 \le \beta \le 1)$ we can write

$$f(\beta \boldsymbol{x} + (1 - \beta)\boldsymbol{y}) \le \beta f(\boldsymbol{x}) + (1 - \beta) f(\boldsymbol{y})$$

Also for every $\beta (0 < \beta < 1)$and $\boldsymbol{x} \ne \boldsymbol{y}$, function f is considered to be a *strictly convex function* if

$$f(\beta \boldsymbol{x} + (1 - \beta)\boldsymbol{y}) < \beta f(\boldsymbol{x}) + (1 - \beta) f(\boldsymbol{y})$$

An example of a convex function is shown in Figure A.4. Geometrically, this means that a function is convex if a line that is drawn connecting any two points on its graph lies strictly above the graph (and nowhere below the graph). Intuitively, the graph of the function is bowl-shaped. Note that if a function is convex, then the negative of the function is *concave*.

Let $\boldsymbol{F}(\boldsymbol{x})$ be an m-dimensional vector of functions of the n-dimensional vector $\boldsymbol{x}$, that is, $\boldsymbol{F}(\boldsymbol{x}) = [f_1(\boldsymbol{x}), f_2(\boldsymbol{x}), \cdots, f_m(\boldsymbol{x})]^T$. Then $\boldsymbol{F}$ is *Lipschitz continuous* on an open set† Σ if for some constant β

$$\|\boldsymbol{F}(\boldsymbol{x}) - \boldsymbol{F}(\boldsymbol{y})\| \le \beta \|\boldsymbol{x} - \boldsymbol{y}\| \qquad \forall \boldsymbol{x}, \boldsymbol{y} \in \Sigma \tag{A.169}$$

†An open set here is considered to be an open sphere with center $\boldsymbol{x}$ and radius $\varepsilon > 0$, that is, $S(\boldsymbol{x}, \varepsilon)$. Therefore, (A.169) can also be written as $\|\boldsymbol{F}(\boldsymbol{x}) - \boldsymbol{F}(\boldsymbol{y})\| \le \beta \|\boldsymbol{x} - \boldsymbol{y}\|$, $\forall \boldsymbol{y} \in S(\boldsymbol{x}, \varepsilon)$. The function $\boldsymbol{F}$ can also be referred to as being *locally Lipschitz continuous* of rank β at $\boldsymbol{x}$.

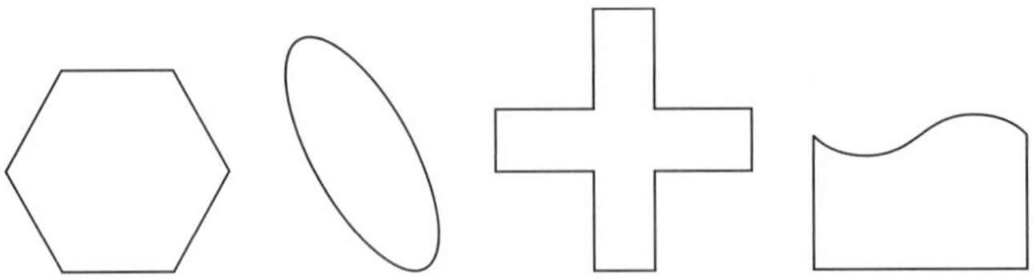

Figure A.3 (a), (b) Examples of convex sets. (c), (d) Examples of nonconvex sets.

A.3.2 Quadratic Forms

A *quadratic form* in $\boldsymbol{x} \in \Re^{n\times 1}$ (where $\boldsymbol{x} = [x_1, x_2, \cdots, x_n]^T$) can be written as

$$q = \sum_{i=1}^{n}\sum_{j=1}^{n} p_{ij}x_i x_j \tag{A.170}$$

where p_{ij}, for $i, j = 1, 2, \ldots, n$, are real coefficients and $q \in \Re$. The expression in (A.170) can be written compactly in vector-matrix form as

$$q = \boldsymbol{x}^T \boldsymbol{P}\boldsymbol{x} \tag{A.171}$$

where $\boldsymbol{P} \in \Re^{n\times n}(\boldsymbol{P} = [p_{ij}])$ ($\boldsymbol{P}$ is usually assumed to be symmetric, $\boldsymbol{P}^T = \boldsymbol{P}$). The quadratic expression in (A.171) can also be written as

$$q = \text{trace}(\boldsymbol{P}\boldsymbol{x}\boldsymbol{x}^T) = \text{trace}(\boldsymbol{x}\boldsymbol{x}^T\boldsymbol{P}) \tag{A.172}$$

For $\boldsymbol{x} \in \mathscr{C}^{n\times 1}$ and $\boldsymbol{P} \in \mathscr{C}^{n\times n}$, the complex quadratic form is given by

$$q = \boldsymbol{x}^H \boldsymbol{P}\boldsymbol{x} \tag{A.173}$$

where $\boldsymbol{P}$ is assumed to be Hermitian, $\boldsymbol{P}^H = \boldsymbol{P}$. The quadratic expression can also be written as $q = \langle \boldsymbol{P}\boldsymbol{x}, \boldsymbol{x}\rangle$ (cf. Sect. A.2.3), called the Euclidean inner

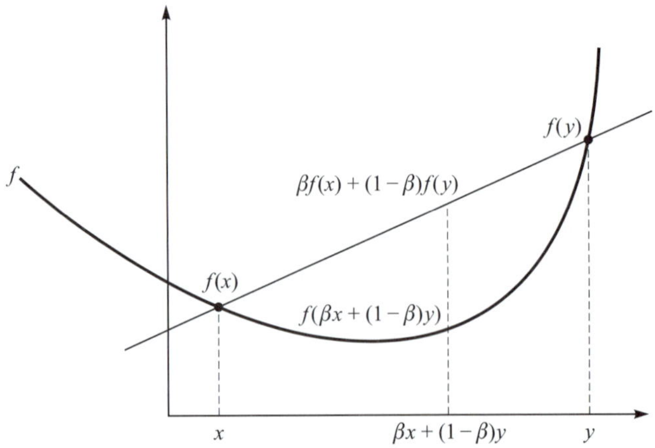

Figure A.4 Example of a convex function.

product, and will be a real quantity if $\boldsymbol{P}^H = \boldsymbol{P}$. Refer to Section A.2.6 on determining the definiteness of a symmetric matrix with respect to a quadratic form. A special case of the quadratic form is for $\boldsymbol{P}$ diagonal. In this case, there are no cross-product terms, that is, $p_{ij} = 0, \forall i \neq j$.

A.3.3 Chain Rule

If $q(x) = u(x)v(x)$, then

$$\frac{\mathrm{d}}{\mathrm{d}x} q(x) = \frac{\mathrm{d}}{\mathrm{d}x}(uv) = u\frac{\mathrm{d}v}{\mathrm{d}x} + v\frac{\mathrm{d}u}{\mathrm{d}x} \tag{A.174}$$

If $q(x) = (u(x)/v(x))$ $(v \neq 0)$, then

$$\frac{\mathrm{d}}{\mathrm{d}x} q(x) = \frac{\mathrm{d}}{\mathrm{d}x}\left(\frac{u}{v}\right) = \frac{v\left(\frac{\mathrm{d}u}{\mathrm{d}x}\right) - u\left(\frac{\mathrm{d}v}{\mathrm{d}x}\right)}{v^2} \tag{A.175}$$

If $q(x) = (c/u(x))$ $(c \in \Re)$, then

$$\frac{\mathrm{d}}{\mathrm{d}x} q(x) = \frac{\mathrm{d}}{\mathrm{d}x}\left(\frac{c}{u}\right) = c\frac{\mathrm{d}}{\mathrm{d}x}\left(\frac{1}{u}\right) = -\frac{c}{u^2}\frac{\mathrm{d}}{\mathrm{d}x}(u) \tag{A.176}$$

If $q(x) = u^n(x)$, then

$$\frac{\mathrm{d}}{\mathrm{d}x} q(x) = \frac{\mathrm{d}}{\mathrm{d}x}(u^n) = nu^{n-1}\frac{\mathrm{d}}{\mathrm{d}x}(u) \tag{A.177}$$

Assume $y = f(u)$ and $u = g(x)$. Therefore, we can write y as a function of a function, $y = f(g(x))$. Now if y is a differentiable function of u, and if u is a differentiable function of x, then $y = f(g(x))$ is a differentiable function of x and can be written as

$$\frac{\mathrm{d}y}{\mathrm{d}x} = \frac{\mathrm{d}y}{\mathrm{d}u}\frac{\mathrm{d}u}{\mathrm{d}x} \tag{A.178}$$

If $z = f(x, y)$ is a continuous function of the variables x and y with continuous partial derivatives $\partial z/\partial x$ and $\partial z/\partial y$, and if x and y are differentiable functions of the variable t, $x = g(t)$, and $y = h(t)$, then z is also a function of the variable t and $\mathrm{d}z/\mathrm{d}t$ (called the *total derivative* of z with respect to t) is given by

$$\frac{\mathrm{d}z}{\mathrm{d}t} = \frac{\partial z}{\partial x}\frac{\mathrm{d}x}{\mathrm{d}t} + \frac{\partial z}{\partial y}\frac{\mathrm{d}y}{\mathrm{d}t} \tag{A.179}$$

If $z = f(x, y)$ is a continuous function of variables x and y with continuous partial derivatives $\partial z/\partial x$ and $\partial z/\partial y$, and if x and y are continuous functions $x = g(r, s)$ and $y = h(r, s)$ of the independent variables r and s, then z is a function of r and s with

$$\frac{\partial z}{\partial r} = \frac{\partial z}{\partial x}\frac{\partial x}{\partial r} + \frac{\partial z}{\partial y}\frac{\partial y}{\partial r} \quad \text{and} \quad \frac{\partial z}{\partial s} = \frac{\partial z}{\partial x}\frac{\partial x}{\partial s} + \frac{\partial z}{\partial y}\frac{\partial y}{\partial s} \tag{A.180}$$

Suppose that

$$f(\boldsymbol{x}) = g(\boldsymbol{x})h(\boldsymbol{x}) \tag{A.181}$$

where both g and h are continuously differentiable scalar functions of the vector $\boldsymbol{x} \in \Re^{n\times 1}(\boldsymbol{x} = [x_1, x_2, \cdots, x_n]^T)$. Then

$$\nabla_x f(\boldsymbol{x}) = \nabla_x g(\boldsymbol{x}) h(\boldsymbol{x}) + \nabla_x h(\boldsymbol{x}) g(\boldsymbol{x}) \tag{A.182}$$

where

$$\nabla_x f(\boldsymbol{x}) \stackrel{\Delta}{=} \frac{\partial f(\boldsymbol{x})}{\partial \boldsymbol{x}} = \left[\frac{\partial f}{\partial x_1}, \frac{\partial f}{\partial x_2}, \cdots, \frac{\partial f}{\partial x_n}\right]^T \tag{A.183}$$

is the *gradient* of f with respect to $\boldsymbol{x}$.

A.3.4 Matrix Calculus

A.3.4.1 Differentiation of scalar functions with respect to a vector

Assume $\boldsymbol{x}, \boldsymbol{y} \in \Re^{n\times 1}$, $\boldsymbol{P} \in \Re^{n\times n}$, and $\boldsymbol{A} \in \Re^{m\times n}$. Then

1. $\dfrac{\partial}{\partial \boldsymbol{x}}(\boldsymbol{x}^T \boldsymbol{y}) = \boldsymbol{y}$
2. $\dfrac{\partial}{\partial \boldsymbol{x}}(\boldsymbol{y}^T \boldsymbol{x}) = \boldsymbol{y}$
3. $\dfrac{\partial}{\partial \boldsymbol{x}}(\boldsymbol{x}^T \boldsymbol{x}) = 2\boldsymbol{x}$
4. $\dfrac{\partial}{\partial \boldsymbol{x}}(\boldsymbol{P}\boldsymbol{x}) = \boldsymbol{P}^T$
5. $\dfrac{\partial}{\partial \boldsymbol{x}}(\boldsymbol{x}^T \boldsymbol{P}\boldsymbol{y}) = \boldsymbol{P}\boldsymbol{y}$
6. $\dfrac{\partial}{\partial \boldsymbol{x}}(\boldsymbol{y}^T \boldsymbol{P}\boldsymbol{x}) = (\boldsymbol{y}^T \boldsymbol{P})^T = \boldsymbol{P}^T \boldsymbol{y}$
7. $\dfrac{\partial}{\partial \boldsymbol{x}}(\boldsymbol{x}^T \boldsymbol{P}\boldsymbol{y}) = \dfrac{\partial}{\partial \boldsymbol{x}}(\boldsymbol{y}^T \boldsymbol{P}\boldsymbol{x}) = \boldsymbol{P}\boldsymbol{y}$, for $\boldsymbol{P}^T = \boldsymbol{P}$
8. $\dfrac{\partial}{\partial \boldsymbol{x}}(\boldsymbol{x}^T \boldsymbol{P}\boldsymbol{x}) = \boldsymbol{P}\boldsymbol{x} + \boldsymbol{P}^T \boldsymbol{x}$
9. $\dfrac{\partial}{\partial \boldsymbol{x}}(\boldsymbol{x}^T \boldsymbol{P}\boldsymbol{x}) = 2\boldsymbol{P}\boldsymbol{x}$, for $\boldsymbol{P}^T = \boldsymbol{P}$
10. $\dfrac{\partial}{\partial \boldsymbol{x}}(\boldsymbol{x} - \boldsymbol{y})^T \boldsymbol{P}(\boldsymbol{x} - \boldsymbol{y}) = 2\boldsymbol{P}(\boldsymbol{x} - \boldsymbol{y})$, for $\boldsymbol{P}^T = \boldsymbol{P}$
11. $\dfrac{\partial}{\partial \boldsymbol{x}^T}(\boldsymbol{A}\boldsymbol{x}) = \boldsymbol{A}$
12. $\dfrac{\partial}{\partial \boldsymbol{x}}\|\boldsymbol{x}\|_2 = \dfrac{\boldsymbol{x}}{\|\boldsymbol{x}\|_2}$
13. $\dfrac{\partial}{\partial \boldsymbol{x}}(\boldsymbol{x}^T \boldsymbol{P}\boldsymbol{x}) = \boldsymbol{P}\boldsymbol{x} + \text{vec}(\boldsymbol{x}^T \boldsymbol{P}) = \boldsymbol{P}\boldsymbol{x} + \boldsymbol{P}^T \boldsymbol{x}$
14. $\dfrac{\partial}{\partial \boldsymbol{x}}(\boldsymbol{A}\boldsymbol{x}) = (\boldsymbol{I}_n \otimes \boldsymbol{A})\,\text{vec}(\boldsymbol{I}_n) = \text{vec}(\boldsymbol{A})$

15. $\dfrac{\partial}{\partial \boldsymbol{x}^T}(\boldsymbol{x} \otimes \boldsymbol{x}) = \dfrac{\partial}{\partial \boldsymbol{x}^T}(\boldsymbol{x} \odot \boldsymbol{x}) = \boldsymbol{I}_n \otimes \boldsymbol{x} + \boldsymbol{x} \otimes \boldsymbol{I}_n$

16. $\dfrac{\partial}{\partial \boldsymbol{x}}(\boldsymbol{x}^T \odot \boldsymbol{A}) = \dfrac{\partial}{\partial \boldsymbol{x}}(\boldsymbol{A} \odot \boldsymbol{x}^T) = \boldsymbol{I}_n \odot \boldsymbol{A}$

A.3.4.2 Differentiation of scalar functions with respect to a matrix

For matrices $\boldsymbol{A}$, $\boldsymbol{B}$, and $\boldsymbol{C}$ of appropriate dimensions, we can write

1. $\dfrac{\partial}{\partial \boldsymbol{A}} \text{trace}(\boldsymbol{A}) = \boldsymbol{I}$
2. $\dfrac{\partial}{\partial \boldsymbol{A}} \text{trace}(\boldsymbol{BAC}) = \boldsymbol{B}^T\boldsymbol{C}^T$
3. $\dfrac{\partial}{\partial \boldsymbol{A}} \text{trace}(\boldsymbol{BA}^T\boldsymbol{C}) = \boldsymbol{CB}$
4. $\dfrac{\partial}{\partial \boldsymbol{A}} \text{trace}(\boldsymbol{ABA}^T) = \boldsymbol{AB}^T + \boldsymbol{AB}$
5. $\dfrac{\partial}{\partial \boldsymbol{A}} \text{trace}(\boldsymbol{ABA}) = \boldsymbol{A}^T\boldsymbol{B}^T + \boldsymbol{B}^T\boldsymbol{A}^T$
6. $\dfrac{\partial}{\partial \boldsymbol{A}} \text{trace}(\boldsymbol{BACA}) = \boldsymbol{B}^T\boldsymbol{A}^T\boldsymbol{C}^T + \boldsymbol{C}^T\boldsymbol{A}^T\boldsymbol{B}^T$
7. $\dfrac{\partial}{\partial \boldsymbol{A}} \text{trace}(\boldsymbol{BACA}^T) = \boldsymbol{B}^T\boldsymbol{AC}^T + \boldsymbol{BAC}$
8. $\dfrac{\partial}{\partial \boldsymbol{A}} \text{trace}(\boldsymbol{A}^T\boldsymbol{A}) = 2\boldsymbol{A}$
9. $\dfrac{\partial}{\partial \boldsymbol{A}} \text{trace}(\boldsymbol{BA}^T\boldsymbol{AC}) = \boldsymbol{ACB} + \boldsymbol{AB}^T\boldsymbol{C}^T$
10. $\dfrac{\partial}{\partial \boldsymbol{A}} \text{trace}(\boldsymbol{BAA}^T\boldsymbol{C}) = \boldsymbol{B}^T\boldsymbol{C}^T\boldsymbol{A} + \boldsymbol{CBA}$
11. $\dfrac{\partial}{\partial \boldsymbol{A}} \text{trace}(\boldsymbol{BA}^T\boldsymbol{AB}^T) = 2\boldsymbol{AB}^T\boldsymbol{B}$
12. $\dfrac{\partial}{\partial \boldsymbol{A}} \text{trace}(\boldsymbol{B}^T\boldsymbol{AA}^T\boldsymbol{B}) = 2\boldsymbol{BB}^T\boldsymbol{A}$
13. $$\dfrac{\partial}{\partial \boldsymbol{A}} \text{trace}\{\boldsymbol{B}(\boldsymbol{A}^T\boldsymbol{A})^2\boldsymbol{B}^T\} = \dfrac{\partial}{\partial \boldsymbol{A}} \text{trace}(\boldsymbol{BA}^T\boldsymbol{AA}^T\boldsymbol{AB}^T) = 2\boldsymbol{AA}^T\boldsymbol{AB}^T\boldsymbol{B} + 2\boldsymbol{AB}^T\boldsymbol{BA}^T\boldsymbol{A}$$
14. $\dfrac{\partial}{\partial \boldsymbol{A}} \text{trace}(e^{\boldsymbol{A}}) = e^{\boldsymbol{A}}$
15. $\dfrac{\partial}{\partial \boldsymbol{A}} |\boldsymbol{BAC}| = |\boldsymbol{BAC}|(\boldsymbol{A}^{-1})^T$
16. $\dfrac{\partial}{\partial \boldsymbol{A}} \text{trace}(\boldsymbol{A}^k) = k(\boldsymbol{A}^{k-1})^T$

17. $\dfrac{\partial}{\partial A}\text{trace}(BA^k) = \left(\sum_{i=0}^{k-1} A^i B A^{k-i-1}\right)^T$

18. $\dfrac{\partial}{\partial A}\text{trace}(BA^{-1}C) = -(A^{-1}CBA^{-1})^T$

19. $\dfrac{\partial}{\partial A}\log|A| = (A^T)^{-1}$

20. $\dfrac{\partial}{\partial A}|A^T| = \dfrac{\partial}{\partial A}|A| = |A|(A^T)^{-1}$

21. $\dfrac{\partial}{\partial A}|A^k| = k|A^k|(A^T)^{-1}$

We conclude this section by addressing the derivative of a matrix $A(\tau) \in \Re^{m\times n}$ [or $A(\tau) \in \mathscr{C}^{m\times n}$] with respect to a scalar $\tau \in \Re$ (or $\tau \in \mathscr{C}$) written as

$$\frac{dA(\tau)}{d\tau} \triangleq \left[\frac{da_{ij}(\tau)}{d\tau}\right]$$

where $a_{ij}(\tau)$, for $i = 1, 2, \ldots, m$ and $j = 1, 2, , n$, are the elements of $A(\tau)$.

A.3.5 Hessian Matrix

If $f(x)$ is a real scalar function of the vector $x \in \Re^{n\times 1}$, in Section A.3.3 we defined the *gradient* of $f(x)$ with respect to x, denoted by $\nabla_x f(x) = \partial f(x)/\partial x$, for $f\colon \Re^n \to \Re$. This assumes that $f(x)$ is of class C^1, that is, $f(x) \in C^1$. If $f(x)$ is of class C^2, the Hessian matrix $f(x)$ with respect to x is defined as

$$\nabla_x^2 f(x) = \frac{\partial^2 f(x)}{\partial x\,\partial x} = \begin{bmatrix} \dfrac{\partial^2 f(x)}{\partial x_1\,\partial x_1} & \dfrac{\partial^2 f(x)}{\partial x_1\,\partial x_2} & \dfrac{\partial^2 f(x)}{\partial x_1\,\partial x_3} & \cdots & \dfrac{\partial^2 f(x)}{\partial x_1\,\partial x_n} \\ \dfrac{\partial^2 f(x)}{\partial x_2\,\partial x_1} & \dfrac{\partial^2 f(x)}{\partial x_2\,\partial x_2} & \dfrac{\partial^2 f(x)}{\partial x_2\,\partial x_3} & \cdots & \dfrac{\partial^2 f(x)}{\partial x_2\,\partial x_n} \\ \dfrac{\partial^2 f(x)}{\partial x_3\,\partial x_1} & \dfrac{\partial^2 f(x)}{\partial x_3\,\partial x_2} & \dfrac{\partial^2 f(x)}{\partial x_3\,\partial x_3} & \cdots & \dfrac{\partial^2 f(x)}{\partial x_3\,\partial x_n} \\ \cdots & \cdots & \cdots & \cdots & \cdots \\ \dfrac{\partial^2 f(x)}{\partial x_n\,\partial x_1} & \dfrac{\partial^2 f(x)}{\partial x_n\,\partial x_2} & \dfrac{\partial^2 f(x)}{\partial x_n\,\partial x_3} & \cdots & \dfrac{\partial^2 f(x)}{\partial x_n\,\partial x_n} \end{bmatrix} = \left[\frac{\partial^2 f(x)}{\partial x_i\,\partial x_j}\right] \tag{A.184}$$

where $i, j = 1, 2, \ldots, n$. The Hessian matrix is symmetric, that is

$$\left[\frac{\partial^2 f(x)}{\partial x_i\,\partial x_j}\right] = \left[\frac{\partial^2 f(x)}{\partial x_j\,\partial x_i}\right]$$

The Hessian matrix (or simply the Hessian) of $f(x)$ with respect to x is also denoted as $\nabla_x^2 f(x) = H(x)$.

A3.6 Jacobian Matrix

If $\boldsymbol{f}(\boldsymbol{x})$ is a real vector function of the vector $\boldsymbol{x} \in \Re^{n\times 1}$ where $\boldsymbol{f}: \Re^n \to \Re^m$, that is,

$$\boldsymbol{f}(\boldsymbol{x}) = [f_1(\boldsymbol{x}), f_2(\boldsymbol{x}), \cdots, f_m(\boldsymbol{x})]^T \tag{A.185}$$

the first-order derivative of $\boldsymbol{f}(\boldsymbol{x})$ involves differentiating each component function $f_i(\boldsymbol{x})$ for $i = 1, 2, \ldots, m$ individually as

$$\nabla\boldsymbol{f}(\boldsymbol{x}) = [\nabla f_1(\boldsymbol{x}), f_2(\boldsymbol{x}), \cdots, \nabla f_m(\boldsymbol{x})] \tag{A.186}$$

where

$$\nabla f_i(\boldsymbol{x}) = \left[\frac{\partial f_i}{\partial x_1}, \frac{\partial f_i}{\partial x_2}, \cdots, \frac{\partial f_i}{\partial x_n}\right]^T \tag{A.187}$$

The individual gradients $\nabla f_1(\boldsymbol{x})$, $\nabla f_2(\boldsymbol{x}), \ldots, \nabla f_m(\boldsymbol{x})$ make up the columns of the $n \times m$ matrix $\nabla\boldsymbol{f}(\boldsymbol{x})$; and its transpose, the $m \times n$ matrix $\nabla^T\boldsymbol{f}(\boldsymbol{x}) = \boldsymbol{J}(\boldsymbol{x})$, is called the *Jacobian matrix*. Therefore, the elements of the Jacobian matrix can be written as

$$\boldsymbol{J}(\boldsymbol{x}) = \left[\left.\frac{\partial f_i(\boldsymbol{x})}{\partial x_j}\right|_{ij}\right] \in \Re^{m\times n} \tag{A.188}$$

for $i = 1, 2, \ldots, m$ and $j = 1, 2, \ldots, n$. More specifically, the Jacobian matrix can be written as

$$\boldsymbol{J}(\boldsymbol{x}) = \begin{bmatrix} \nabla^T f_1(\boldsymbol{x}) \\ \nabla^T f_2(\boldsymbol{x}) \\ \vdots \\ \nabla^T f_m(\boldsymbol{x}) \end{bmatrix} = \begin{bmatrix} \frac{\partial f_1}{\partial x_1} & \frac{\partial f_1}{\partial x_2} & \cdots & \frac{\partial f_1}{\partial x_n} \\ \frac{\partial f_2}{\partial x_1} & \frac{\partial f_2}{\partial x_2} & \cdots & \frac{\partial f_2}{\partial x_n} \\ \cdots & \cdots & \cdots & \cdots \\ \frac{\partial f_m}{\partial x_1} & \frac{\partial f_m}{\partial x_2} & \cdots & \frac{\partial f_m}{\partial x_n} \end{bmatrix} = \frac{\partial \boldsymbol{f}(\boldsymbol{x})}{\partial \boldsymbol{x}} \tag{A.189}$$

Assume two real vector functions $\boldsymbol{f}(\boldsymbol{x})$ and $\boldsymbol{g}(\boldsymbol{x})$, for $\boldsymbol{x} \in \Re^{n\times 1}$, where $\boldsymbol{f}(\boldsymbol{x}) = [f_1(\boldsymbol{x}), f_2(\boldsymbol{x}), \cdots, f_m(\boldsymbol{x})]^T$ and $\boldsymbol{g}(\boldsymbol{x}) = [g_1(\boldsymbol{x}), g_2(\boldsymbol{x}), \cdots, g_m(\boldsymbol{x})]^T$. We can write (using the appropriate chain rule)

$$\frac{\partial}{\partial \boldsymbol{x}}(\boldsymbol{f}^T\boldsymbol{g}) = \left(\frac{\partial \boldsymbol{f}}{\partial \boldsymbol{x}}\right)^T \boldsymbol{g} + \left(\frac{\partial \boldsymbol{g}}{\partial \boldsymbol{x}}\right)^T \boldsymbol{f} \tag{A.190}$$

where $\partial\boldsymbol{f}/\partial\boldsymbol{x} = \nabla^T\boldsymbol{f}(\boldsymbol{x}) = \boldsymbol{J}_f(\boldsymbol{x})$ is the Jacobian of $\boldsymbol{f}$ with respect to $\boldsymbol{x}$, and $\partial\boldsymbol{g}/\partial\boldsymbol{x} = \nabla^T\boldsymbol{g}(\boldsymbol{x}) = \boldsymbol{J}_g(\boldsymbol{x})$ is the Jacobian of $\boldsymbol{g}$ with respect to $\boldsymbol{x}$. The result in (A.190) is an n-dimensional column vector.

A.3.7 Taylor Series Expansion

Given the real-valued scalar function $f(\boldsymbol{x})$ of the real vector $\boldsymbol{x} \in \Re^{n\times 1}$, we can write the multivariable *Taylor series expansion* of $f(\boldsymbol{x})$ about the point $\boldsymbol{x}_k$ as

$$
\begin{aligned}
f(\boldsymbol{x}) &= f(\boldsymbol{x}_k + \Delta\boldsymbol{x}) \\
&= f(\boldsymbol{x})|_{\boldsymbol{x}=\boldsymbol{x}_k} + \Delta\boldsymbol{x}^T\, \nabla f(\boldsymbol{x})|_{\boldsymbol{x}=\boldsymbol{x}_k} + \frac{1}{2}\, \Delta\boldsymbol{x}^T\, \nabla^2 f(\boldsymbol{x})\, \Delta\boldsymbol{x}|_{\boldsymbol{x}=\boldsymbol{x}_k} \\
&\quad + \text{higher-order terms}
\end{aligned} \tag{A.191}
$$

where $\nabla f(\boldsymbol{x})|_{\boldsymbol{x}=\boldsymbol{x}_k} = \nabla f(\boldsymbol{x}_k)$ is the gradient of $f(\boldsymbol{x})$ with respect to $\boldsymbol{x}$ evaluated at $\boldsymbol{x}_k$, and $\nabla^2 f(\boldsymbol{x})|_{\boldsymbol{x}=\boldsymbol{x}_k} = \nabla^2 f(\boldsymbol{x}_k)$ is the Hessian of $f(\boldsymbol{x})$ with respect to $\boldsymbol{x}$ evaluated at $\boldsymbol{x}_k$. The partial derivative of $f(\boldsymbol{x})$ in (A.191) with respect to $\boldsymbol{x}$ (neglecting the higher-order terms and taking into account that the Hessian matrix is symmetric) is given by

$$
\frac{\partial f(\boldsymbol{x})}{\partial \boldsymbol{x}} \cong \nabla f(\boldsymbol{x}_k) + \nabla^2 f(\boldsymbol{x}_k)\, \Delta\boldsymbol{x} \tag{A.192}
$$

at $\boldsymbol{x} = \boldsymbol{x}_k + \Delta\boldsymbol{x}$.

We will let $f(\boldsymbol{x})$, for $\boldsymbol{x} \in \Re^{n\times 1}$, be a real-valued scalar function of class C^1. Taylor's theorem (or the mean-value theorem) is given as

$$
f(\boldsymbol{x}) = f(\boldsymbol{x}_k + \Delta\boldsymbol{x}) = f(\boldsymbol{x}_k) + \Delta\boldsymbol{x}^T\, \nabla f(\boldsymbol{x}_k + \beta\, \Delta\boldsymbol{x}) \tag{A.193}
$$

where β is a scalar ($0 \le \beta \le 1$). Moreover, if $f(\boldsymbol{x})$ is of class C^2, then there exists a scalar β ($0 \le \beta \le 1$) such that

$$
f(\boldsymbol{x}) = f(\boldsymbol{x}_k + \Delta\boldsymbol{x}) = f(\boldsymbol{x}_k) + \Delta\boldsymbol{x}^T\, \nabla f(\boldsymbol{x}_k) + \frac{1}{2}\, \Delta\boldsymbol{x}^T\, \nabla^2 f(\boldsymbol{x}_k + \beta\, \Delta\boldsymbol{x})\, \Delta\boldsymbol{x} \tag{A.194}
$$

A.4 LYAPUNOV'S DIRECT METHOD

If we assume a homogeneous linear time-invariant system described by (cf. Sect. A2.12)

$$
\dot{\boldsymbol{x}}(t) = \boldsymbol{A}\boldsymbol{x}(t) \tag{A.195}
$$

where $\boldsymbol{x} \in \Re^{n\times 1}$ and $\boldsymbol{A} \in \Re^{n\times n}$, we can define a Lyapunov function as follows.

DEFINITION A.12. A time-invariant *Lyapunov function*, denoted by $V(\boldsymbol{x})$, is any scalar function of the state vector $\boldsymbol{x}$ that satisfies the following conditions $\forall t \ge t_0$ and for all $\boldsymbol{x}$ in the neighborhood of the origin:

1. Function $V(\boldsymbol{x})$ and its first partial derivatives with respect to the arguments $x_1, x_2, \ldots, x_n$, that is, $\boldsymbol{x} = [x_1, x_2, \cdots, x_n]^T$, exist and are continuous.
2. $V(\boldsymbol{0}) = 0$
3. $V(\boldsymbol{x}) > 0$ for $\boldsymbol{x} \ne \boldsymbol{0}$ (positive definite).

An equilibrium state $\boldsymbol{x}_e$ of the system in (A.195) is *asymptotically stable*, or equivalently matrix $\boldsymbol{A}$ has eigenvalues with strictly negative real parts, if

$$
\dot{V}(\boldsymbol{x}) < 0 \qquad \text{negative definite} \tag{A.196}
$$

where $\dot{V}(\boldsymbol{x})$ is the total derivative (cf. Sect. A.3.3) of the Lyapunov function

$$\dot{V}(\boldsymbol{x}) = \frac{\partial V}{\partial x_1}\frac{\mathrm{d}x_1}{\mathrm{d}t} + \frac{\partial V}{\partial x_2}\frac{\mathrm{d}x_2}{\mathrm{d}t} + \cdots + \frac{\partial V}{\partial x_n}\frac{\mathrm{d}x_n}{\mathrm{d}t} = \underbrace{\left[\frac{\partial V}{\partial x_1}, \frac{\partial V}{\partial x_2}, \cdots, \frac{\partial V}{\partial x_n}\right]}_{\nabla_x^T V(\boldsymbol{x})} \underbrace{\begin{bmatrix} \frac{\mathrm{d}x_1}{\mathrm{d}t} \\ \frac{\mathrm{d}x_2}{\mathrm{d}t} \\ \vdots \\ \frac{\mathrm{d}x_n}{\mathrm{d}t} \end{bmatrix}}_{\frac{\mathrm{d}\boldsymbol{x}}{\mathrm{d}t}}$$

$$= \nabla_x^T V(\boldsymbol{x}) \frac{\mathrm{d}\boldsymbol{x}}{\mathrm{d}t} \tag{A.197}$$

In the case of a time-invariant system, $\boldsymbol{x}_e = \boldsymbol{0}$ is always an equilibrium state of $\dot{\boldsymbol{x}}(t) = \boldsymbol{A}\boldsymbol{x}(t)$, where the equilibrium state is a solution of $\dot{\boldsymbol{x}} = \boldsymbol{0}$ (or $\boldsymbol{A}\boldsymbol{x} = \boldsymbol{0}$) and has the property $\boldsymbol{x}_e = \mathrm{e}^{\boldsymbol{A}t}\boldsymbol{x}_e \forall t \geq 0$. If

$$\dot{V}(\boldsymbol{x}) \leq 0 \qquad \text{negative semidefinite} \tag{A.198}$$

then the equilibrium state is considered to be *stable in the sense of Lyapunov* (*stability isL*). An equivalent statement can be made regarding stability *isL* in relation to the eigenvalues of $\boldsymbol{A}$ in (A.195); that is, $\boldsymbol{A}$ can possess some eigenvalues with zero real parts providing they are not repeated eigenvalues. A Lyapunov function is often referred to as an *energy function*. However, many times a Lyapunov function may not have any physical significance in relation to, or represent the energy of, a dynamical system.

For this linear time-invariant case, a quadratic form can be taken for the Lyapunov function as

$$V(\boldsymbol{x}) = \boldsymbol{x}^T\boldsymbol{P}\boldsymbol{x} \tag{A.199}$$

where $\boldsymbol{P} \in \Re^{n\times n}$ and $\boldsymbol{P}^T = \boldsymbol{P}$, and the derivative of $V(\boldsymbol{x})$ is written as

$$\dot{V}(\boldsymbol{x}) = \dot{\boldsymbol{x}}^T\boldsymbol{P}\boldsymbol{x} + \boldsymbol{x}^T\boldsymbol{P}\dot{\boldsymbol{x}} \tag{A.200}$$

using the appropriate chain rule. Equation (A.200) can be rewritten using (A.195) as

$$\dot{V}(\boldsymbol{x}) = \boldsymbol{x}^T\boldsymbol{A}^T\boldsymbol{P}\boldsymbol{x} + \boldsymbol{x}^T\boldsymbol{P}\boldsymbol{A}\boldsymbol{x} = \boldsymbol{x}^T\left(\boldsymbol{A}^T\boldsymbol{P} + \boldsymbol{P}\boldsymbol{A}\right)\boldsymbol{x} \tag{A.201}$$

where it is required that $\dot{V}(\boldsymbol{x}) < 0$ (negative definite) in order for an equilibrium state of the system to be asymptotically stable. From (A.201) we will let

$$\boldsymbol{A}^T\boldsymbol{P} + \boldsymbol{P}\boldsymbol{A} = -\boldsymbol{Q} \tag{A.202}$$

and now in order for an equilibrium state of the system to be asymptotically stable, $\boldsymbol{Q} > 0$; that is, $\boldsymbol{Q}$ must be positive definite and symmetric ($\boldsymbol{Q}^T = \boldsymbol{Q}$). Therefore, a standard test to determine if a system whose homogeneous state equation is written as $\dot{\boldsymbol{x}}(t) = \boldsymbol{A}\boldsymbol{x}(t)$ has an asymptotically stable equilibrium state or not, is to first select a positive definite matrix $\boldsymbol{Q}$ and solve (A.202) (called a *Lyapunov equation*) for $\boldsymbol{P}$. Then if $\boldsymbol{P} > 0$ (i.e., positive definite), the quadratic function in (A.199) is a Lyapunov function and the equilibrium state of the system is asymptotically stable.

For the more general case

$$\frac{\mathrm{d}\boldsymbol{x}(t)}{\mathrm{d}t} = \boldsymbol{f}(\boldsymbol{x}, t) \tag{A.203}$$

where we assume nonlinear time-varying dynamics, a time-varying Lyapunov function is defined as follows:

DEFINITION A.13. A time-varying Lyapunov function, denoted by $V(\boldsymbol{x}, t)$, is any scalar function of the state vector $\boldsymbol{x} \in \Re^{n\times 1} (\boldsymbol{x} = [x_1, x_2, \cdots, x_n]^T)$ and time t that obeys the following conditions for $t \geq t_0$ and for all $\boldsymbol{x}$ in the neighborhood of the origin (the equilibrium state is the zero state):

1. Function $V(\boldsymbol{x}, t)$ and its first partial derivatives with respect to the state variables $x_1, x_2, \cdots, x_n$ and t exist and are all continuous.
2. $V(\boldsymbol{0}, t) = 0$
3. $V(\boldsymbol{x}, t) \geq \alpha\|\boldsymbol{x}\| > 0$ for $\boldsymbol{x} \neq \boldsymbol{0}$ and $t \geq t_0$, where $\alpha(0) = 0$ [$\alpha(t)$ is a continuously nonincreasing scalar function of t].

If a time-varying Lyapunov function can be found for the system $\dot{\boldsymbol{x}}(t) = \boldsymbol{f}(\boldsymbol{x}, t)$, where $\boldsymbol{f}(\boldsymbol{0}, t) = \boldsymbol{0}$, and $\dot{V}(\boldsymbol{x}, t) < 0$ for $\boldsymbol{x} \neq \boldsymbol{0}$, then the state $\boldsymbol{x} = \boldsymbol{0}$ is asymptotically stable.

Consider a linear time-invariant digital system described by the difference equation

$$\boldsymbol{x}(k+1) = \boldsymbol{A}\boldsymbol{x}(k) \tag{A.204}$$

where $\boldsymbol{x} \in \Re^{n\times 1}$ and $\boldsymbol{A} \in \Re^{n\times n}$. If we define our Lyapunov function as $V(\boldsymbol{x}) = \boldsymbol{x}^T\boldsymbol{P}\boldsymbol{x}$, (where $\boldsymbol{P} \in \Re^{n\times n}$, $\boldsymbol{P} > 0$, and $\boldsymbol{P}^T = \boldsymbol{P}$) and a *difference operator* operating on $V(\boldsymbol{x})$ as

$$\Delta V[\boldsymbol{x}(k)] = V[\boldsymbol{x}(k+1)] - V[\boldsymbol{x}(k)] \tag{A.205}$$

then (A.205) can be written as

$$\Delta V[\boldsymbol{x}(k)] = \boldsymbol{x}^T(k+1)\boldsymbol{P}\boldsymbol{x}(k+1) - \boldsymbol{x}^T(k)\boldsymbol{P}\boldsymbol{x}(k) \tag{A.206}$$

Substituting (A.204) into (A.206), we obtain

$$\begin{aligned}\Delta V[\boldsymbol{x}(k)] &= \boldsymbol{x}^T(k)\boldsymbol{A}^T\boldsymbol{P}\boldsymbol{A}\boldsymbol{x}(k) - \boldsymbol{x}^T(k)\boldsymbol{P}\boldsymbol{x}(k)\\ &= \boldsymbol{x}^T(k)[\boldsymbol{A}^T\boldsymbol{P}\boldsymbol{A} - \boldsymbol{P}]\boldsymbol{x}(k)\end{aligned} \tag{A.207}$$

The equilibrium state $\boldsymbol{x}_e = \boldsymbol{0}$ is asymptotically stable if and only if $\boldsymbol{A}^T\boldsymbol{P}\boldsymbol{A} - \boldsymbol{P} < 0$ (negative definite) in (A.207). That is, if we define a matrix $\boldsymbol{Q} \in \Re^{n\times n}$, $\boldsymbol{Q} > 0$, $\boldsymbol{Q}^T = \boldsymbol{Q}$, such that

$$\boldsymbol{A}^T\boldsymbol{P}\boldsymbol{A} - \boldsymbol{P} = -\boldsymbol{Q} \tag{A.208}$$

and if the solution $\boldsymbol{P}$ to (A.208) is positive definite, then from (A.207) $\Delta V[\boldsymbol{x}(k)] = -\boldsymbol{x}^T(k)\boldsymbol{Q}\boldsymbol{x}(k) < 0$, and the equilibrium state $\boldsymbol{x}_e = \boldsymbol{0}$ of (A.204) is asymptotically stable.

A.5 UNCONSTRAINED OPTIMIZATION METHODS

A.5.1 Necessary and Sufficient Conditions for an Extremum

An unconstrained optimization problem can be stated as follows: Find a vector $\boldsymbol{x} \in \Re^{n\times 1}$ that minimizes the real-valued scalar function

$$\mathscr{E} = \mathscr{E}(\boldsymbol{x})$$

called the *cost function*, or the *energy* (or *objective*) *function*. Therefore, we can state the unconstrained optimization problem as

$$\underset{\boldsymbol{x}}{\text{minimize}}\, \mathscr{E}(\boldsymbol{x}) \tag{A.209}$$

where there are no constraints imposed on the elements of the design vector $\boldsymbol{x} = [x_1, x_2, \cdots, x_n]^T$. Without loss of generality, an equivalent statement for the unconstrained optimization problem would be to maximize the negative of the same cost function

$$\underset{\boldsymbol{x}}{\text{maximize}} -\mathscr{E}(\boldsymbol{x})$$

We will consider $\boldsymbol{x}^*$as the *global minimizer* of $\mathscr{E}(\boldsymbol{x})$ if

$$\mathscr{E}(\boldsymbol{x}^*) \le \mathscr{E}(\boldsymbol{x}) \qquad \forall \boldsymbol{x} \in \Re^{n\times 1} \tag{A.210}$$

and if $\mathscr{E}(\boldsymbol{x}^*) < \mathscr{E}(\boldsymbol{x})$, $\forall \boldsymbol{x} \in \Re^{n\times 1}$, then $\boldsymbol{x}^*$ is considered to be a *strict global minimizer*. Not all functions have a finite global minimizer. Moreover, even though a global minimizer may exist for a function, this does not ensure that the function will have a strict global minimizer. Ideally, we would be pleased to find a global minimizer for any function that we might encounter; however, this is unrealistic. Many useful optimization methods are based on information about the particular function at a specific point, and thus, this information is valid within a small neighborhood of the specific point. If no other information is available or additional assumptions are made about the problem, it is not possible to guarantee that a global solution has been found (or one exists). One very important exception to this occurs when the function $\mathscr{E}$ is convex (which is true for linear programming problems). If we were considering problems with constraints, the set of feasible points Σ would be defined by a set of constraints, and this set would also have to be convex. Note that for problems without constraints, the set Σ would be $\Re^n$.

If the global minimizer cannot be found for a particular function, then the next best thing would be to find a solution that is better than any of the surrounding points. Therefore, we would like to find a *local minimizer* of the function $\mathscr{E}$, that is, a point that satisfies

$$\mathscr{E}(\boldsymbol{x}^*) \le \mathscr{E}(\boldsymbol{x}) \qquad \forall \boldsymbol{x} \in \Re^{n\times 1} \ni \|\boldsymbol{x} - \boldsymbol{x}^*\| < \varepsilon \tag{A.211}$$

where ε is some small positive number that may be dependent on $\boldsymbol{x}^*$. If the point $\boldsymbol{x}^*$ (local solution) is a *strict local minimizer*, then

$$\mathscr{E}(\boldsymbol{x}^*) < \mathscr{E}(\boldsymbol{x}) \qquad \forall \boldsymbol{x} \in \Re^{n\times 1} \ni \boldsymbol{x} \ne \boldsymbol{x}^* \text{ and } \|\boldsymbol{x} - \boldsymbol{x}^*\| < \varepsilon \tag{A.212}$$

Strict local minimizers can be identified in many important cases by computing the first and second derivatives of $\mathscr{E}$ with respect to $\boldsymbol{x}$ evaluated at $\boldsymbol{x} = \boldsymbol{x}^*$. Therefore, strict local minimizers can be revealed using algorithms that compute the first and second derivatives of the problem cost function $\mathscr{E}$. Many algorithms only allow a *stationary point* to be determined, especially those that compute only the first derivative. For the unconstrained case, a stationary point is a point where the first derivatives of $\mathscr{E}$ are equal to zero. Even though the global solution may be desired for a particular problem, many times a local solution is satisfactory if it produces an acceptable reduction in the cost (objective) function.

Assuming the first and second derivatives of $\mathscr{E}(\boldsymbol{x})$ exist, a vector $\boldsymbol{x}^*$ is a strict local minimizer of $\mathscr{E}(\boldsymbol{x})$ if

$$\nabla_{\boldsymbol{x}}\mathscr{E}(\boldsymbol{x})|_{\boldsymbol{x}=\boldsymbol{x}^*} = \frac{\partial}{\partial \boldsymbol{x}}\mathscr{E}(\boldsymbol{x})|_{\boldsymbol{x}=\boldsymbol{x}^*} = \boldsymbol{0}$$

(i.e., the gradient of $\mathscr{E}(\boldsymbol{x})$ with respect to $\boldsymbol{x}$ evaluated at the point $\boldsymbol{x}^*$ equals zero), and the Hessian matrix $\nabla^2\mathscr{E}(\boldsymbol{x})$ evaluated at the point $\boldsymbol{x}^*$ is positive definite, that is, $\boldsymbol{x}^T\nabla^2\mathscr{E}(\boldsymbol{x})|_{\boldsymbol{x}=\boldsymbol{x}^*}\boldsymbol{x} > 0$, $\forall \boldsymbol{x} \in \Re^{n\times 1}$ (except $\boldsymbol{x} \neq \boldsymbol{0}$). Recall that the Hessian matrix is always symmetric (cf. Sect. A.3.5). Therefore, informally, the necessary and sufficient conditions for a point $\boldsymbol{x}^*$ to be a strict local minimizer is $\mathscr{E}(\boldsymbol{x}^*) < \mathscr{E}(\boldsymbol{x})$ for all $\boldsymbol{x}$ in $0 < \|\boldsymbol{x} - \boldsymbol{x}^*\| < \varepsilon$ $(\varepsilon > 0)$, if $\nabla_{\boldsymbol{x}}\mathscr{E}(\boldsymbol{x}^*) = 0$ and the Hessian matrix $\nabla^2\mathscr{E}(\boldsymbol{x}^*)$ is symmetric and positive definite.

A.5.2 Steepest Descent

The method of steepest descent (often referred to as the gradient method) is one of the oldest and most widely used numerical optimization techniques for minimizing a function of several variables. It is typically the benchmark technique to which other methods are compared. Many other methods are variations of steepest descent developed in an attempt to improve the convergence properties of the new algorithms. If we assume $\mathscr{E}$ (an energy function) is a function of several variables $\boldsymbol{x} = \left[x_1, x_2, \cdots, x_n\right]^T$, where $\boldsymbol{x} \in \Re^{n\times 1}$, and has continuous partial derivatives on $\Re^n$, then the gradient of $\mathscr{E}$ with respect to the vector $\boldsymbol{x}$ is given by $\nabla_{\boldsymbol{x}}\mathscr{E}(\boldsymbol{x}) \in \Re^{n\times 1}$ and sometimes is written as $\boldsymbol{g} \triangleq \nabla_{\boldsymbol{x}}\mathscr{E}(\boldsymbol{x})$. The method of steepest descent can be defined for the discrete-time case as

$$\boldsymbol{x}_{k+1} = \boldsymbol{x}_k - \alpha_k \boldsymbol{g}_k \tag{A.213}$$

where k is the discrete-time index, $\boldsymbol{x}_k = \boldsymbol{x}(k)$, $\boldsymbol{g}_k \triangleq \nabla_{\boldsymbol{x}}\mathscr{E}(\boldsymbol{x}_k)$, and α_k is a nonnegative scalar that minimizes $\mathscr{E}(\boldsymbol{x}_k - \alpha \boldsymbol{g}_k)$. Steepest descent searches from the point $\boldsymbol{x}_k$ along the direction of the "negative gradient" $-\boldsymbol{g}_k$ to a *minimum* point on this line, where this minimum point is taken as $\boldsymbol{x}_{k+1}$. Figure A.5 illustrates the steepest descent iterative process.

The discrete-time steepest descent method is a special case of the general class of dynamic gradient systems for unconstrained optimization. As previously mentioned, many gradient descent methods have their roots in the method of steepest descent (and Newton's methods). These methods trans-

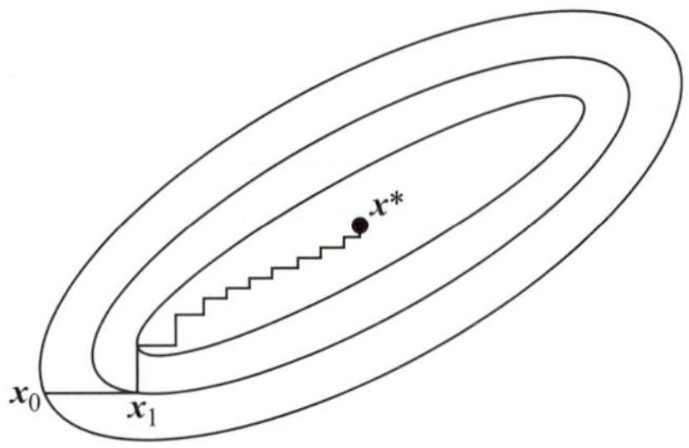

Figure A.5 Steepest descent iterative process.

form the optimization (minimization) problem to an associated system of first-order differential equations written as

$$\frac{\mathrm{d}\boldsymbol{x}(t)}{\mathrm{d}t} = -\boldsymbol{\mu}(\boldsymbol{x}, t)\nabla_{\boldsymbol{x}}\mathscr{E}(\boldsymbol{x}) \tag{A.214}$$

with the initial conditions $\boldsymbol{x}(t_0 = 0) = \boldsymbol{x}_0$, where $\boldsymbol{\mu}(\boldsymbol{x}, t)$ is a symmetric positive definite matrix referred to as the learning matrix. To find the vector $\boldsymbol{x}^*$ that minimizes the cost function $\mathscr{E}(\boldsymbol{x})$, the system of ordinary differential equations written in vector-matrix form in (A.214) (along with the initial conditions) must be solved. The first concern in finding $\boldsymbol{x}^*$ is that of stability. The stability issue for the system of differential equations in (A.214) can be resolved by taking the derivative of the cost or energy (Lyapunov) function with respect to time

$$\frac{\mathrm{d}\mathscr{E}}{\mathrm{d}t} = \frac{\partial\mathscr{E}}{\partial\boldsymbol{x}}\frac{\mathrm{d}\boldsymbol{x}}{\mathrm{d}t} = \nabla_{\boldsymbol{x}}^T\mathscr{E}(\boldsymbol{x})\frac{\mathrm{d}\boldsymbol{x}}{\mathrm{d}t} = -\nabla_{\boldsymbol{x}}^T\mathscr{E}(\boldsymbol{x})\boldsymbol{\mu}(\boldsymbol{x}, t)\nabla_{\boldsymbol{x}}\mathscr{E}(\boldsymbol{x}) < 0 \tag{A.215}$$

(cf. Sect. A.4). Therefore, according to the quadratic expression in (A.215), the condition for stability is that the learning matrix must be symmetric and *positive definite*. This guarantees the energy function $\mathscr{E}(\boldsymbol{x})$ will decrease with time and will converge to a stable local minimum (equilibrium point) as $t \to \infty$. In other words, the local minimum of the energy function is given by the solution trajectory of the gradient system

$$\boldsymbol{x}^* = \lim_{t\to\infty} \boldsymbol{x}(t)$$

The entries in the learning matrix in (A.214) will dictate the speed of convergence to the minimum.

Various choices of (or algorithms for developing) the learning matrix entries lead to different gradient methods. The simplest form arises when the learning matrix is taken as the identity matrix multiplied by a scalar μ (the learning rate parameter). From (A.214), the resulting system of differential equations is given as

$$\frac{\mathrm{d}\boldsymbol{x}(t)}{\mathrm{d}t} = -\mu\nabla_{\boldsymbol{x}}\mathscr{E}(\boldsymbol{x}) = -\mu\boldsymbol{g}(\boldsymbol{x}) \tag{A.216}$$

with the initial conditions $\boldsymbol{x}(0) = \boldsymbol{x}_0$. As time progresses, the trajectory $\boldsymbol{x}(t)$ moves in a direction that has the sharpest rate of decrease, called the direction of steepest descent. In (A.216), the continuous-time form of the learning rule is guaranteed to converge according to this steepest descent if the learning rate parameter is a positive number ($\mu > 0$). In discrete-time form, the steepest descent continuous-time learning rule in (A.216) becomes

$$\underbrace{\frac{\boldsymbol{x}_{(k+1)T_s} - \boldsymbol{x}_{kT_s}}{T_s}}_{\approx \frac{\mathrm{d}\boldsymbol{x}(t)}{\mathrm{d}t}} = -\tilde{\mu}\nabla_{\boldsymbol{x}}\mathscr{E}(\boldsymbol{x}_k) = -\tilde{\mu}\boldsymbol{g}_k$$

or

$$\boldsymbol{x}_{k+1} = \boldsymbol{x}_k - \mu\boldsymbol{g}_k \tag{A.217}$$

where $\boldsymbol{x}_k = \boldsymbol{x}_{kT_s}$, T_s is the sampling period, $\mu = T_s\tilde{\mu}$, and $\boldsymbol{x}(0) = \boldsymbol{x}_0$ is the initial-condition vector. Convergence (stability) is guaranteed for (A.217) if $0 \le \mu \le \mu_{\max}$. The learning rate parameter (or integration step size) can also change in time μ_k, for example, adaptively change according to an algorithm similar to the annealing schedules discussed in Section 2.5.1 for the LMS algorithm.

A.5.3 Newton's Methods

Newton's methods involve a local approximation of the energy (or objective) function $\mathscr{E}(\boldsymbol{x})$, to be minimized, in the form of a quadratic function. This quadratic approximation to $\mathscr{E}(\boldsymbol{x})$ is local about the current point $\boldsymbol{x}_k$ and is minimized exactly. In the scalar case, our objective is to minimize $\mathscr{E}(x)$ with respect to x when at the point x_k, where a measurement is made, we know $\{\mathscr{E}(x_k),\mathscr{E}'(x_k),\mathscr{E}''(x_k)\}$ [e.g., where $\mathscr{E}'(x_k)$ is the first derivative of $\mathscr{E}(x_k)$ with respect to x]. Then it is possible to construct a quadratic function q, where this function at the point x_k agrees with $\mathscr{E}$ up to the second derivative, that is,

$$q(x) = \mathscr{E}(x_k) + \mathscr{E}'(x_k)(x - x_k) + \frac{1}{2}\mathscr{E}''(x_k)(x - x_k)^2 \tag{A.218}$$

An *estimate* of the minimum point x_{k+1} of $\mathscr{E}$ can be calculated by finding the point where the derivative of q vanishes; that is, from (A.218) we can write

$$\dot{q}(x) = \frac{\mathrm{d}q(x)}{\mathrm{d}x} = \mathscr{E}'(x_k) + \mathscr{E}''(x_k)(x - x_k)\big|_{x=x_{k+1}} = 0$$

or

$$\mathscr{E}'(x_k) + \mathscr{E}''(x_k)x_{k+1} - \mathscr{E}''(x_k)x_k = 0$$

and solving for x_{k+1} gives the *classical Newton method* result

$$x_{k+1} = x_k - \frac{\mathscr{E}'(x_k)}{\mathscr{E}''(x_k)} \tag{A.219}$$

It is interesting to note that the result in (A.219) does not depend on the value $\mathscr{E}(x_k)$. In general, Newton's method can be viewed as a technique for iteratively solving equations of the form $y(x) = 0$, and when applied to minimization problems, we set $y(x) = \mathscr{E}'(x)$. Therefore, Newton's method can be written as

$$x_{k+1} = x_k - \frac{y(x_k)}{y'(x_k)} \tag{A.220}$$

for iteratively determining the roots of $y(x) = 0$. This is illustrated in Figure A.6.

For the multiple-variable case, that is, $\boldsymbol{x} = [x_1, x_2, \cdots, x_n]^T$, the scalar function $\mathscr{E}(\boldsymbol{x})$ is approximated locally by a quadratic function, and this approximate function is then minimized exactly. Therefore, near the point $\boldsymbol{x}_k$ we can approximate the function $\mathscr{E}$ by the truncated Taylor series expansion given as

$$\mathscr{E}(\boldsymbol{x}) \cong \mathscr{E}(\boldsymbol{x}_k) + \nabla_{\boldsymbol{x}}^T \mathscr{E}(\boldsymbol{x}_k)(\boldsymbol{x} - \boldsymbol{x}_k) + \frac{1}{2}(\boldsymbol{x} - \boldsymbol{x}_k)^T \boldsymbol{H}_k(\boldsymbol{x} - \boldsymbol{x}_k) \tag{A.221}$$

where $\nabla_{\boldsymbol{x}}\mathscr{E} \in \Re^{n\times 1}$ [and $\boldsymbol{g}_k = \nabla_{\boldsymbol{x}}\mathscr{E}(\boldsymbol{x}_k)$] is the gradient of $\mathscr{E}$ with respect to the vector $\boldsymbol{x}$ and $\boldsymbol{H} = \nabla^2\mathscr{E} \in \Re^{n\times n}$[and $\boldsymbol{H}_k = \boldsymbol{H}(\boldsymbol{x}_k) = \nabla^2\mathscr{E}(\boldsymbol{x}_k)$] is the Hessian matrix (cf. Sect. A.3.5). Now the point $\boldsymbol{x}_{k+1}$ that minimizes the truncated Taylor series expansion in (A.221) must satisfy

$$\nabla_{\boldsymbol{x}}\mathscr{E}(\boldsymbol{x})\big|_{\boldsymbol{x}=\boldsymbol{x}_{k+1}} = 0$$

and is given by

$$\boldsymbol{x}_{k+1} = \boldsymbol{x}_k - \boldsymbol{H}_k^{-1}\nabla_{\boldsymbol{x}}\mathscr{E}(\boldsymbol{x}_k) \tag{A.222}$$

or

$$\boldsymbol{x}_{k+1} = \boldsymbol{x}_k - \boldsymbol{H}_k^{-1}\boldsymbol{g}_k \tag{A.223}$$

From (A.222) or (A.223) we see that the second-order sufficiency conditions require that the Hessian matrix be positive definite at the minimum point $\boldsymbol{x} = \boldsymbol{x}^*$. In the case where the objective function is purely quadratic, that is,

$$\mathscr{E}(\boldsymbol{x}) = \frac{1}{2}\boldsymbol{x}^T\boldsymbol{Q}\boldsymbol{x} + \boldsymbol{b}^T\boldsymbol{x} + a \tag{A.224}$$

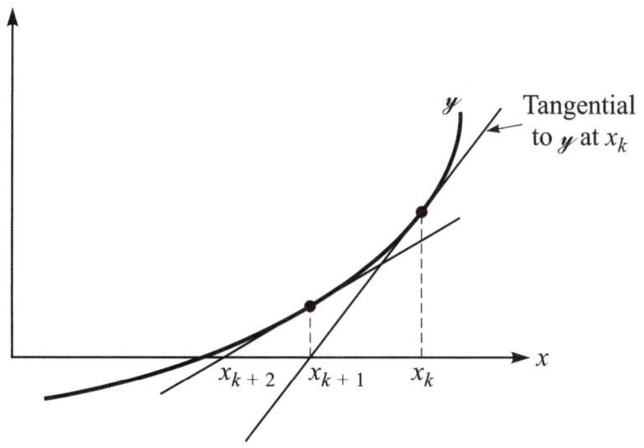

Figure A.6 Newton's method used to iteratively determine the roots of a function y.

the gradient is given by $\nabla_x \mathscr{E}(\boldsymbol{x}) = \boldsymbol{Qx} + \boldsymbol{b}$ and the Hessian matrix is constant, that is, $\nabla_x^2 \mathscr{E}(\boldsymbol{x}) = \boldsymbol{H} = \boldsymbol{Q}$. Therefore, comparing these results with the classical Newton's method given in (A.222), we see that starting at any initial point $\boldsymbol{x}_0$, the minimum of the quadratic function will be arrived at in just one step.

A.5.4 Modified Newton and Quasi-Newton Methods

Modified Newton methods

Under ideal circumstances the convergence rate of Newton's method is quadratic, whereas steepest descent (the simplest Newton method) has linear convergence. However, at points that are remote from the "solution," modifications to the Hessian may be necessary to ensure positive definiteness of this matrix and to ensure descent. The first modification usually involves introducing a *search parameter* α such that,

$$\boldsymbol{x}_{k+1} = \boldsymbol{x}_k - \alpha_k(\boldsymbol{H}_k^{-1}\boldsymbol{g}_k) \tag{A.225}$$

where $\alpha_k > 0$ is selected to minimize the objective function $\mathscr{E}$. Near the solution we expect $\alpha_k \cong 1$; however, this parameter can guard against the possibility of an *increasing* objective function due to nonquadratic terms in the actual objective function. Equation (A.225) is referred to as a *limited-step Newton formula.*

The second modification to the basic Newton method involves perturbing the Hessian matrix in the event it becomes ill-conditioned, that is, if it tends to a singular condition. This can be accomplished many ways. One method is to perform an *LDU decomposition* (cf. Sect. 7.3) on the Hessian matrix $\boldsymbol{H}$, that is, $\boldsymbol{H} = \boldsymbol{LDU} = \boldsymbol{LDL}^T$, since $\boldsymbol{H}$ is symmetric. The diagonal elements of $\boldsymbol{D} \in \Re^{n \times n}$ indicate the definiteness of $\boldsymbol{H}$, and if $\boldsymbol{H}$ is tending to a singular matrix, at least one of the diagonal elements of $\boldsymbol{D}$ will approach zero as k increases. This zero (or almost zero or nonpositive) element (or elements) of the $\boldsymbol{D}$ matrix can be replaced with a small positive number that is large enough to ensure that the modified Hessian matrix $\boldsymbol{H}$ is positive definite with an improved condition number. This can be done for each iteration by carrying out the following procedure:

Modified Newton method

Step 1. Select an initial solution vector $\boldsymbol{x}_0$ and a convergence tolerance ε.
Step 2. For $k = 0, 1, 2, \ldots$, compute $\boldsymbol{g}_k = \nabla_x \mathscr{E}[\boldsymbol{x}_k]$ (where $\mathscr{E}$ is the objective function). If $\|\boldsymbol{g}_k\| < \varepsilon$, then stop.
Step 3. Compute $\boldsymbol{H} = \boldsymbol{LDL}^T$.
Step 4. Modify the diagonal elements of $\boldsymbol{D}$ as necessary, $\boldsymbol{D} \leftarrow \boldsymbol{D}_m$.
Step 5. Compute the search direction from

$$(\boldsymbol{LD}_m\boldsymbol{L}^T)\boldsymbol{d}_k = -\boldsymbol{g}_k$$

Step 6. Perform a line search to determine

$$\boldsymbol{x}_{k+1} = \boldsymbol{x}_k + \alpha_k \boldsymbol{d}_k$$

the new solution estimate, where α_k is selected such that

$$\min_{\alpha \geq 0} \mathscr{E}(\boldsymbol{x}_k + \alpha \boldsymbol{d}_k)$$

Then go to step 2.

Alternatively, a diagonal matrix can be added to $\boldsymbol{H}$; that is, replace $\boldsymbol{H}$ by $\varepsilon_k \boldsymbol{I} + \boldsymbol{H}_k$, where ε_k is the smallest nonnegative constant such that the matrix $\varepsilon_k \boldsymbol{I} + \boldsymbol{H}_k$ has eigenvalues greater than or equal to $\delta > 0$ (where δ is determined according to a reasonable condition number for $\boldsymbol{H}$). Then define the *direction vector*

$$\boldsymbol{d}_k = -(\varepsilon_k \boldsymbol{I} + \boldsymbol{H}_k)\boldsymbol{g}_k \tag{A.226}$$

and iterate according to

$$\boldsymbol{x}_{k+1} = \boldsymbol{x}_k + \alpha_k \boldsymbol{d}_k \tag{A.227}$$

where α_k minimizes $\mathscr{E}(\boldsymbol{x}_k - \alpha \boldsymbol{d}_k)$, for $\alpha \geq 0$. We can readily see from (A.226) and (A.227) that the method of steepest descent is a special case of Newton's method. Specifically, in (A.226), if $\varepsilon_k \boldsymbol{I} + \boldsymbol{H}_k = \boldsymbol{I}_n$ every iteration, then the direction vector is always taken as the negative of the gradient of $\mathscr{E}$; that is, $\boldsymbol{d}_k = -\boldsymbol{g}_k$. And the resulting iterative expression in (A.227) reduces to (A.213) for steepest descent. Moreover, adding $\varepsilon_k \boldsymbol{I}$ to the Hessian matrix $\boldsymbol{H}_k$ is equivalent to the previous modified Newton method using the *LDU* decomposition.

Quasi-Newton methods

Many times only the gradient of the objective function $\mathscr{E}$ is available, and not the Hessian matrix. In these cases, so-called quasi-Newton methods (also called variable-metric methods) can be used. The basic idea behind any quasi-Newton method is that the inverse of the Hessian matrix is approximated each step of the gradient descent optimization procedure in lieu of the actual inverse as in (A.223). One of the most popular quasi-Newton methods today is the Broyden-Fletcher-Goldfarb-Shanno (BFGS) algorithm. The details of the algorithm are given below [9].

Broyden-Fletcher-Goldfarb-Shanno algorithm

Step 1. Select an initial solution vector $\boldsymbol{x}_0$ and an initial Hessian approximation $\boldsymbol{B}_0$ (this could be $\boldsymbol{B}_0 = \boldsymbol{I}$).

Step 2. For $k = 0, 1, 2, \ldots,$ if $\boldsymbol{x}_k$ is optimal (in some sense), then stop.

Step 3. Else compute the gradient of the objective function $\mathscr{E}$, that is, $\boldsymbol{g}_k = \nabla_{\boldsymbol{x}} \mathscr{E}(\boldsymbol{x}_k)$, then solve $\boldsymbol{B}_k \boldsymbol{d}_k = -\boldsymbol{g}_k$ for $\boldsymbol{d}_k$.

Step 4. Perform a line search to determine $\boldsymbol{x}_{k+1} = \boldsymbol{x}_k + \alpha_k \boldsymbol{d}_k$, where α_k is selected such that $\min_{\alpha \geq 0} \mathscr{E}(\boldsymbol{x}_k + \alpha \boldsymbol{d}_k)$.

Step 5. Compute $\boldsymbol{\delta}_k = \boldsymbol{x}_{k+1} - \boldsymbol{x}_k$ and $\boldsymbol{y}_k = \boldsymbol{g}_{k+1} - \boldsymbol{g}_k$.

Step 6. Compute

$$\boldsymbol{B}_{k+1} = \boldsymbol{B}_k - \frac{(\boldsymbol{B}_k \boldsymbol{\delta}_k)(\boldsymbol{B}_k \boldsymbol{\delta}_k)^T}{\boldsymbol{\delta}_k^T \boldsymbol{B}_k \boldsymbol{\delta}_k} + \frac{\boldsymbol{y}_k \boldsymbol{y}_k^T}{\boldsymbol{y}_k^T \boldsymbol{\delta}_k}$$

where $\boldsymbol{B}_k$ is the current estimate of the Hessian matrix $\nabla_{\boldsymbol{x}}^2 \mathscr{E}(\boldsymbol{x}_k)$.

Step 7. Go to step 2.

Step 6 above used to update the estimate of the Hessian matrix is considered a *rank-2* formula [9]. Rank-2 update formulas ensure the Hessian matrix approximations are both symmetric and positive definite.

A.5.5 Conjugate Gradient Method

The conjugate gradient method was originally developed to solve

$$Qx = b \tag{A.228}$$

for $x \in \Re^{n\times 1}$ given $Q \in \Re^{n\times n}(Q^T = Q \text{ and } Q > 0)$ and $b \in \Re^{n\times 1}$. Solving (A.228) is equivalent to minimizing the scalar function [10]

$$\mathscr{E}(x) = \frac{1}{2}x^T Qx - x^T b \tag{A.229}$$

In the conjugate gradient method a set of direction vectors $\{d_0, d_1, \ldots, d_{n-1}\}$ is generated that are *conjugate* with respect to the matrix Q (cf. Sect. A.2.8, Definition A.8), that is, $d_i^T Qd_j = 0$ for $i \neq j$. A conjugate direction vector is generated at the kth iteration of the iterative process by adding to the calculated current negative gradient vector of the objective function a linear combination of the previous direction vectors. The advantages of the conjugate gradient method are that (1) a very simple formula is used to determine the new direction vector, (2) this makes the conjugate gradient method only slightly more complicated that than the method of steepest descent, and (3) because the direction vectors are based on the calculated gradients, the process makes good uniform progress toward the solution at every step. Note that this is inconsequential for the purely quadratic case; however, it is important for generalizations of the conjugate gradient method to nonquadratic problems. The conjugate gradient algorithm is summarized as follows:

Conjugate gradient algorithm

Step 1. Start with any $x_0 \in \Re^{n\times 1}$. Define the initial direction vector as

$$d_0 = -g_0 = -\nabla_x \mathscr{E}(x_k)|_{k=0} = b - Qx_0$$

Step 2. $\alpha_k = -\dfrac{g_k^T d_k}{d_k^T Qd_k}$ where $g_k = Qx_k - b$

Step 3. $x_{k+1} = x_k + \alpha_k d_k$

Step 4. $d_{k+1} = -g_{k+1} + \beta_k d_k$ where $\beta_k = \dfrac{g_{k+1}^T Qd_k}{d_k^T Qd_k}$

An alternate form for β_k is given by

$$\beta_k = \frac{g_{k+1}^T g_{k+1}}{g_k^T g_k}$$

Step 5. Go to step 2.

This algorithm converges in a finite number of steps, and the quadratic problem convergence is achieved in no more than n steps. Note that in the conjugate gradient algorithm the first step is identical to that in the steepest descent method. In step 2 of the algorithm shown above, the parameter α_k is given. This parameter can be determined from $\alpha_k = \min_{\alpha \geq 0} \mathscr{E}(x_k + \alpha d_k)$. That is, for the purely quadratic case $\mathscr{E}(x) = {}^1\!/_2\, x^T Qx - x^T b$ from (A.229). Therefore,

$$\begin{aligned}\mathscr{E}(\boldsymbol{x}_k + \alpha \boldsymbol{d}_k) &= \frac{1}{2}(\boldsymbol{x}_k + \alpha \boldsymbol{d}_k)^T \boldsymbol{Q}(\boldsymbol{x}_k + \alpha \boldsymbol{d}_k) - (\boldsymbol{x}_k + \alpha \boldsymbol{d}_k)^T \boldsymbol{b} \\ &= \frac{1}{2}(\boldsymbol{x}_k^T \boldsymbol{Q} \boldsymbol{x}_k + 2\alpha \boldsymbol{d}_k^T \boldsymbol{Q} \boldsymbol{x}_k + \alpha^2 \boldsymbol{d}_k^T \boldsymbol{Q} \boldsymbol{d}_k) - \boldsymbol{x}_k^T \boldsymbol{b} - \alpha \boldsymbol{d}_k^T \boldsymbol{b}\end{aligned} \tag{A.230}$$

Computing the gradient of (A.230) with respect to the parameter α and setting the result equal to zero gives

$$\begin{aligned}\frac{\partial \mathscr{E}(\boldsymbol{x}_k + \alpha \boldsymbol{d}_k)}{\partial \alpha} &= \boldsymbol{d}_k^T \boldsymbol{Q} \boldsymbol{x}_k - \boldsymbol{d}_k^T \boldsymbol{b} + \alpha \boldsymbol{d}_k^T \boldsymbol{Q} \boldsymbol{d}_k = \boldsymbol{d}_k^T (\underbrace{\boldsymbol{Q} \boldsymbol{x}_k - \boldsymbol{b}}_{\boldsymbol{g}_k}) + \alpha \boldsymbol{d}_k^T \boldsymbol{Q} \boldsymbol{d}_k \\ &= \boldsymbol{d}_k^T \boldsymbol{g}_k + \alpha \boldsymbol{d}_k^T \boldsymbol{Q} \boldsymbol{d}_k = \boldsymbol{g}_k^T \boldsymbol{d}_k + \alpha \boldsymbol{d}_k^T \boldsymbol{Q} \boldsymbol{d}_k = 0\end{aligned} \tag{A.231}$$

Solving for $\alpha = \alpha_k$ in (A.231) gives

$$\alpha_k = -\frac{\boldsymbol{g}_k^T \boldsymbol{d}_k}{\boldsymbol{d}_k^T \boldsymbol{Q} \boldsymbol{d}_k} \tag{A.232}$$

which is the expression for α_k shown above in step 2 of the conjugate gradient algorithm.

Now let us extend these results to nonquadratic problems. It is assumed that near the solution point the problem is approximately quadratic. There are several approaches for this; however, we present only one approach, based on *line search methods*, and two variations of the algorithm.

Fletcher-Reeves conjugate gradient algorithm (with restart)

The objective is to minimize $\mathscr{E}(\boldsymbol{x})$, where $\boldsymbol{x} \in \Re^{n \times 1}$ and $\mathscr{E}$ is not necessarily a quadratic function.

Fletcher-Reeves conjugate gradient algorithm (with restart)

Step 1. Set $\boldsymbol{x}_0$.
Step 2. Compute

$$\boldsymbol{g}_0 = \nabla_{\boldsymbol{x}} \mathscr{E}(\boldsymbol{x}_0) = \left. \frac{\partial \mathscr{E}(\boldsymbol{x})}{\partial \boldsymbol{x}} \right|_{\boldsymbol{x}=\boldsymbol{x}_0}$$

Step 3. Set $\boldsymbol{d}_0 = -\boldsymbol{g}_0$.
Step 4. Compute

$$\boldsymbol{x}_{k+1} = \boldsymbol{x}_k + \alpha_k \boldsymbol{d}_k \qquad \text{where } \alpha_k = \min_{\alpha \geq 0} \mathscr{E}(\boldsymbol{x}_k + \alpha \boldsymbol{d}_k).$$

Step 5. Compute $\boldsymbol{g}_{k+1} = \nabla_{\boldsymbol{x}} \mathscr{E}(\boldsymbol{x}_{k+1})$.
Step 6. Compute

$$\boldsymbol{d}_{k+1} = -\boldsymbol{g}_{k+1} + \beta_k \boldsymbol{d}_k \qquad \text{where } \beta_k = \frac{\boldsymbol{g}_{k+1}^T \boldsymbol{g}_{k+1}}{\boldsymbol{g}_k^T \boldsymbol{g}_k}$$

Steps 4 through 6 are carried out for $k = 0, 1, \ldots, n-1$.
Step 7. Replace $\boldsymbol{x}_0$ by $\boldsymbol{x}_n$ and go to step 1.
Step 8. Continue until convergence is achieved. Termination criterion could be $\|\boldsymbol{d}_k\| < \varepsilon$ (where ε is an appropriate predetermined small number).

Computing β_k as shown in step 6 above is known as using the *Fletcher-Reeves* formula. Two other alternative methods for computing β_k are as follows:

Polak-Ribiere method Hestenes-Stiefel method

$$\beta_k = \frac{(\mathbf{g}_{k+1} - \mathbf{g}_k)^T \mathbf{g}_{k+1}}{\mathbf{g}_k^T \mathbf{g}_k} \qquad \beta_k = \frac{(\mathbf{g}_{k+1} - \mathbf{g}_k)^T \mathbf{g}_{k+1}}{\mathbf{d}_k^T (\mathbf{g}_{k+1} - \mathbf{g}_k)}$$

The conjugate gradient method is intermediate between the steepest descent and quasi-Newton methods. The *restart* feature in the above algorithm (step 7) is important for the cases when the objective function is not quadratic. The Fletcher-Reeves conjugate gradient algorithm is restarted by a search in the *steepest descent* direction after each n iterations (or when a nondescent search direction is generated). The pure steepest descent step taken every n steps serves as a "spacer step." The restart feature of the algorithm is important for global convergence, because in general one cannot guarantee that the directions $\mathbf{d}_k$ generated are *descent* directions. Note that in step 4 above, α_k must be derived from the particular objective function, which can be difficult in some cases.

A.6 CONSTRAINED NONLINEAR PROGRAMMING

A.6.1 Kuhn-Tucker Conditions

The Kuhn-Tucker conditions are the necessary conditions for optimization problems with *inequality constraints*. We will let $\mathbf{x}^*$ be a local minimizer (minimum point) of the scalar function $\mathscr{E}$ subject to the constraints $g_j(\mathbf{x}) \geq 0$, where $j = 1, 2, \ldots, m$, and $\mathbf{x} \in \Re^{n\times 1}$. Therefore, it is desired to

$$\text{Minimize} \quad \mathscr{E}(\mathbf{x}) \tag{A.233}$$

$$\text{subject to} \quad g_j(\mathbf{x}) \geq 0, \qquad \text{for} \quad j = 1, 2, \ldots, m \tag{A.234}$$

We can now construct a *Lagrangian function* [9–11] given by

$$\mathscr{L}(\mathbf{x}, \boldsymbol{\lambda}) = \mathscr{E}(\mathbf{x}) - \sum_{j=1}^{m} \lambda_j g_j(\mathbf{x}) = \mathscr{E}(\mathbf{x}) - \boldsymbol{\lambda}^T \mathbf{g}(\mathbf{x}) \tag{A.235}$$

where $\mathbf{g} \in \Re^{mx1}$ is a vector of constraints, and $\boldsymbol{\lambda} \in \Re^{mx1}$ is a vector of *Lagrange multipliers*. Again $\mathbf{x}^*$ is considered a local minimum point for the problem in (A.223) and (A.234), and it is also considered a *regular point*† for the constraints. Then there exists a vector of Lagrange multipliers $\boldsymbol{\lambda}^*$ such that the following conditions are satisfied (called the *Kuhn-Tucker conditions*):

† The local minimum point $\mathbf{x}^*$ is said to be a *regular point* of the constraints if the gradient vectors $\nabla_{\mathbf{x}} g_j(\mathbf{x}^*)$, $j \in J$, are linearly independent, where J is a set of indices corresponding to active inequality constraints at $\mathbf{x}^*$, that is, $J = \{j\colon 1 \leq j \leq m,\ g_j(\mathbf{x}^*) = 0\}$ (i.e., all indices corresponding to the equality constraints at $\mathbf{x}^*$).

1. $\nabla_x \mathscr{L}(\boldsymbol{x}^*, \boldsymbol{\lambda}^*) = \mathbf{0}$, that is

$$\nabla_x \mathscr{E}(\boldsymbol{x}^*) - \sum_{j=1}^{m} \lambda_j^* \nabla_x g_j(\boldsymbol{x}^*) = \frac{\partial \mathscr{E}(\boldsymbol{x}^*)}{\partial x_i} - \sum_{j=1}^{m} \lambda_j^* \frac{\partial g_j(\boldsymbol{x}^*)}{\partial x_i} = \mathbf{0} \tag{A.236}$$

for $i = 1, 2, \ldots, n$

2. $$\lambda_j^* g_j(\boldsymbol{x}^*) = 0 \qquad j = 1, 2, \ldots, m \tag{A.237}$$

3. $$\lambda_j^* \geq 0 \qquad j = 1, 2, \ldots, m \tag{A.238}$$

4. $$g_j(\boldsymbol{x}^*) \geq 0 \qquad j = 1, 2, \ldots, m \tag{A.239}$$

It is assumed in (A.236) through (A.239) that $\mathscr{E}(\boldsymbol{x})$ and $g_j(\boldsymbol{x})$ have continuous first-order partial derivatives. A point $\boldsymbol{x} \in \Sigma \subset \Re^{n \times 1}$ that satisfies all the constraints is called *feasible*. If the set of feasible points Σ is nonempty, the optimization problem is called *consistent*. A feasible point $\boldsymbol{x}^*$ is a local minimizer if it is a local minimizer of the scalar function $\mathscr{E}(\boldsymbol{x})$ on the set of feasible points Σ. The condition in (A.237), that is, $\boldsymbol{\lambda}^{*T} \boldsymbol{g}(\boldsymbol{x}^*) = 0$, is referred to as the *complementary slackness condition*. Since the vectors $\boldsymbol{\lambda}^*$ and $\boldsymbol{g}(\boldsymbol{x}^*)$ are both nonnegative, it implies that $\lambda_j^* g_j(\boldsymbol{x}^*) = 0$ for each j. What this means is that either a constraint is inactive or its associated Lagrange multiplier is zero. Specifically, any inactive constraint has a Lagrange multiplier equal to zero. If the Lagrange multipliers associated with the active constraints are all positive, then *strict complementarity* exists; otherwise, if a Lagrange multiplier corresponding to an active constraint is zero, the constraint is considered to be *degenerate* [9].

A.6.2 Lagrange Multiplier Methods

Here we want to address constrained optimization problems, and we consider the following nonlinear programming problem:

Minimize	$\mathscr{E}(\boldsymbol{x})$		(A.240*a*)
Subject to	$g_j(\boldsymbol{x}) \leq c_j$	for $j = 1, 2, \ldots, m_1$	(A.240*b*)
	$g_j(\boldsymbol{x}) \geq c_j$	for $j = m_1 + 1, \ldots, m_2$ $(m_1 \leq m_2)$	(A.240*c*)
	$g_j(\boldsymbol{x}) = c_j$	for $j = m_2 + 1, \ldots, m$ $(m_2 \leq m)$	(A.240*d*)

where $\boldsymbol{x} \in \Re^{n \times 1}$, and it is assumed that $\mathscr{E}(\boldsymbol{x})$ and $g_j(\boldsymbol{x})$, for $j = 1, 2, \ldots, m$, have continuous first-order partial derivatives. For this nonlinear programming problem we can define the Lagrangian function as

$$\mathscr{L}(\boldsymbol{x}, \boldsymbol{\lambda}) = \mathscr{E}(\boldsymbol{x}) - \sum_{j=1}^{m} \lambda_j [g_j(\boldsymbol{x}) - c_j] \tag{A.241}$$

If we assume $\boldsymbol{x}^*$ is a regular point and a local minimizer for this problem, then there exists at least one nonzero vector $\boldsymbol{\lambda}^* \in \Re^{m \times 1}$ such that

1. $$\lambda_j^* \leq 0 \qquad j = 1, 2, \ldots, m_1 \tag{A.242a}$$

2. $$\lambda_j^* \geq 0 \qquad j = m_1 + 1, \ldots, m_2 \tag{A.242b}$$

3. $$\lambda_j^* \text{ of arbitrary sign} \qquad \text{for } j = m_2 + 1, \cdots m \tag{A.242c}$$

4. $$\lambda_j^* = 0 \qquad \text{for } j \in J_0 \tag{A.242d}$$

where J_0 is the set of indices j from $j = 1, 2, \ldots, m_2$ for which the inequalities are satisfied at $\boldsymbol{x}^*$ as strict inequalities,

5. $$\lambda_j^*[g_j(\boldsymbol{x}^*) - c_j)] = 0 \qquad \text{for } j = 1, 2, \ldots, m_2 \tag{A.242e}$$

6. $$\nabla_{\boldsymbol{x}}\mathscr{E}(\boldsymbol{x}^*) - \sum_{j=1}^{m} \lambda_j^* \nabla_{\boldsymbol{x}} g_j(\boldsymbol{x}^*) = \mathbf{0} \tag{A.242f}$$

For the special case of minimization of the objective function $\mathscr{E}(\boldsymbol{x})$ subject to only equality constraints $g_j(\boldsymbol{x}) = c_j$ for $j = 1, 2, \cdots, m$, the conditions given above reduce to a single relationship

$$\nabla_{\boldsymbol{x}}\mathscr{E}(\boldsymbol{x}^*) - \sum_{j=1}^{m} \lambda_j^* \nabla_{\boldsymbol{x}} g_j(\boldsymbol{x}^*) = \mathbf{0} \tag{A.243}$$

and the other conditions are superfluous.

An example of using the Lagrange multiplier approach with a single equality constraint is to determine the *largest* volume that can be fitted into an ellipsoid

$$\frac{x^2}{a^2} + \frac{y^2}{b^2} + \frac{z^2}{c^2} = 1 \tag{A.244}$$

It is assumed that each edge of the box is parallel to a rectangular coordinate axis, and each of the eight corners of the box will lie on the ellipsoid. We let the corner in the first octant have the coordinates (x, y, z); therefore, the dimensions of the box are $2x$, $2y$, and $2z$, and the volume is $V = 8xyz$. It is desired to determine the maximum V subject to the constraint in (A.244); therefore, a Lagrangian function can be formed as

$$\mathscr{L}(x, y, z, \lambda) = 8xyz - \lambda\left(\frac{x^2}{a^2} + \frac{y^2}{b^2} + \frac{z^2}{c^2} - 1\right) \tag{A.245}$$

The necessary conditions from (A.243) are

$$\frac{\partial \mathscr{L}}{\partial x} = 8yx - 2\lambda\frac{x}{a^2} = 0 \tag{A.246a}$$

$$\frac{\partial \mathscr{L}}{\partial y} = 8xz - 2\lambda\frac{y}{b^2} = 0 \tag{A.246b}$$

$$\frac{\partial \mathscr{L}}{\partial z} = 8xy - 2\lambda\frac{z}{c^2} = 0 \tag{A.246c}$$

Dividing each equation in (A.246a) through (A.246c) by 2; multiplying (A.246a) by x, (A.246b) by y, and (A.246c) by z; and adding give

$$12xyz - \lambda \underbrace{\left(\frac{x^2}{a^2} + \frac{y^2}{b^2} + \frac{z^2}{c^2}\right)}_{1} = 0 \Rightarrow \lambda = 12xyz \tag{A.247}$$

Now substituting the result in (A.247) into the necessary conditions in (A.246*a*) through (A.246*c*) gives

$$yz(a^2 - 3x^2) = 0 \qquad xz(b^2 - 3y^2) = 0 \qquad xy(c^2 - 3z^2) = 0 \tag{A.248}$$

Since the *maximum* volume is desired, from the three expressions in (A.248) we want the positive values of x, y, and z. Thus,

$$x = \frac{a}{\sqrt{3}} \qquad y = \frac{b}{\sqrt{3}} \qquad z = \frac{c}{\sqrt{3}} \tag{A.249}$$

and

$$\lambda = 12xyz = \frac{4abc}{\sqrt{3}} \tag{A.250}$$

Therefore, the box of maximum volume that can fit into the ellipsoid in (A.244) has dimensions $(2a/\sqrt{3}, 2b/\sqrt{3}, 2c/\sqrt{3})$, and the maximum volume is

$$V_{\max} = \frac{8abc}{3\sqrt{3}} \tag{A.251}$$

A.7 RANDOM VARIABLES AND STOCHASTIC PROCESSES

A.7.1 Random Variables

Any presentation of random variables begins with a discussion of *probability*. There are several ways to define probability. The four most common are (1) axiomatic (measure), (2) relative frequency (Von Mises), (3) a priori definition as a ratio of favorable to total number of alternatives (the classical approach), and (4) a measure of belief (inductive reasoning). However, the two definitions that are the most useful are the relative-frequency approach and the axiomatic approach. The relative-frequency approach attempts to attach some physical significance to the concept of probability. Thus, this approach makes it possible to relate probabilistic concepts to the real world. The axiomatic approach treats the probability of an event as a number that satisfies certain postulates, but is otherwise undefined. The number does not necessarily have to relate to anything in the real world and is not relevant in developing the mathematical structure that emerges from these postulates.

Relative-frequency approach

The relative-frequency approach to probability is closely related to the frequency of occurrence of a particular *event*. An event is an incident that may or may not happen. For example, tossing a coin can result in either a head or a tail, and each is an event. The idea of an *experiment* and the *outcomes* of that particular experiment are important in understanding this concept more precisely. Tossing a coin, throwing a die, drawing a card, observing a voltage greater than zero (less than zero)—all are examples of an experiment. In each experiment listed, the number of outcomes is finite. This is the special case of discrete probability. If the experiment involves observing a voltage in a continuous range of possible values, then there are an infinite number of outcomes (this is an example of continuous probability). In the case of the coin toss experiment, we expect, for a relatively large number of trials, that heads will occur one-half of the time and tails the other half of the time. Therefore, we would assign a probability of $^1/_2$ to each of the two events. Generally, if an experiment is performed N times and we expect event A to occur N_A times, then we assume the probability of event A, that is, $\Pr(A)$, to be

$$\Pr(A) = \frac{N_A}{N} \tag{A.252}$$

The number N_A is not the actual number of times that event A occurs in N trials, but only the number we assume occurs based on our intuition about the experiment. If a certain experiment has possible outcomes $A, B, C, \ldots, M$ (only one outcome can occur on any given trial), the possible events are said to be *mutually exclusive*. If event A is expected to occur N_A times out of N trials, event B is expected to occur N_b times, and so on, then

$$N_A + N_B + N_C + \cdots + N_M = N \tag{A.253}$$

and dividing both sides by N gives

$$\frac{N_A}{N} + \frac{N_B}{N} + \frac{N_C}{N} + \cdots + \frac{N_M}{N} = 1 \tag{A.254}$$

From (A.252), we can write (A.254) as

$$\Pr(A) + \Pr(B) + \Pr(C) + \cdots + \Pr(M) = 1 \tag{A.255}$$

Four important statements can be summarized as follows:

1. $0 \leq \Pr(A) \leq 1$
2. $\Pr(A) + \Pr(B) + \Pr(C) + \cdots + \Pr(M) = 1$, assuming a complete set of mutually exclusive events.
3. An *impossible* event is represented by $\Pr(A) = 0$.
4. A *certain* event is represented by $\Pr(A) = 1$.

If there is an interest in more than one event at a time, then one must consider *joint probability*. For example, $\Pr(A, B)$ expresses the probability of the joint occurrence of events A and B. The joint probability does not necessarily equal the product of the individual (*marginal*) probabilities; that is, $\Pr(A, B, \ldots, M) \neq \Pr(A)\Pr(B) \cdots \Pr(M)$.

Another important type of probability is *conditional probability*. For example, $\Pr(A|B)$ represents the probability of one event A given that another event B has occurred. In general, we can write

$$\Pr(A, B) = \Pr(A|B)\Pr(B) = \Pr(B|A)\Pr(A) \tag{A.256}$$

Two random events are statistically independent if and only if

$$\Pr(A, B) = \Pr(A)\Pr(B) \tag{A.257}$$

Axiomatic approach

This approach relates probability theory to the concepts in set theory. A *probability space* ($\mathscr{S}$) is defined whose elements are all the possible outcomes from an experiment. Then for each event a number is assigned referred to as the *probability* of the event. The various subsets of $\mathscr{S}$ can be identified with the events. We denote the event as A and the probability of this event as $\Pr(A)$. This assigned number is chosen in a manner to satisfy the following three conditions (or axioms):

1. $\Pr(A) \geq 0$ (A.258)
2. $\Pr(\mathscr{S}) = 1$ (A.259)
3. If $AB = \emptyset$, then $\Pr(A + B) = \Pr(A) + \Pr(B)$ (A.260)
   where AB is the *product* or *intersection* and $\emptyset$ is the *null set* or *empty set* (cf. Sect. A.3.1).

The entire body of probability can be deduced from these three axioms. Some important corollaries can be deduced from these axioms:

- Since $\mathscr{S}\emptyset = \emptyset$ and $\mathscr{S} + \emptyset = \emptyset$ (where $\mathscr{S} + \emptyset$ is the sum or union (cf. Sect. A.3.1), using (A.260), it follows that

$$\Pr(\mathscr{S} + \emptyset) = \Pr(\mathscr{S}) \qquad \text{therefore} \qquad \Pr(\emptyset) = 0 \tag{A.261}$$

- Since $A\tilde{A} = \emptyset$, where $\tilde{A}$ is the *complement* of A (cf. Sect. A.3.1) and $A + \tilde{A} = \mathscr{S}$, again it follows from (A.260) that

$$\Pr(A + \tilde{A}) = \Pr(A) + \Pr(\tilde{A}) = \Pr(\mathscr{S}) = 1 \tag{A.262}$$

- From (A.262) and (A.258), it follows that

$$\Pr(A) = 1 - \Pr(\tilde{A}) \leq 1 \tag{A.263}$$

  Therefore, the probability of an event must be a number between 0 and 1.
- If A and B are not mutually exclusive

$$\Pr(A + B) = \Pr(A) + \Pr(B) - \Pr(AB) \leq \Pr(A) + \Pr(B) \tag{A.264}$$

An important *conditional probability* property is

$$\Pr(A|B) = \frac{\Pr(AB)}{\Pr(B)} \qquad \Pr(B) > 0 \tag{A.265}$$

where $\Pr(AB)$ is the probability of event AB.

Independence

Two events A and B are independent if and only if

$$\Pr(AB) = \Pr(A)\Pr(B) \tag{A.266}$$

The concept of a *random variable* can be summarized in the following definition.

DEFINITION A.14. A real *random variable* X is a real function whose domain is the sample space $\mathscr{S}$ (i.e., $\mathscr{S} = \{\alpha\}$, the set of all possible outcomes of a random experiment) and

1. The set $\{X \leq x\}$ is an event for any real number $x \in \Re$.
2. The probability of events $\{X = +\infty\}$ and $\{X = -\infty\}$ equals zero, that is, $\Pr(X = +\infty) = \Pr(X = -\infty) = 0$.

Therefore, from Definition A.14, a random variable is simply a real-valued function defined over the sample space $\mathscr{S}$; or it can be thought of as a numerical description of the outcome of a random experiment. When the outcome of the random experiment is α, the random variable X has a value denoted as $X(\alpha)$. If a random variable can assume any value within a specified range (possible infinite), then it is considered a continuous random variable. Discrete random variables can only assume one of a countable set of values; however, they can be treated by exactly the same methods as continuous random variables. In what follows, we will be concerned with only continuous random variables, unless stated otherwise.

A.7.2 Probability Distribution Functions

Continuous random variables can be considered within the framework of probability concepts by defining the events to be associated with the probability space through the *probability distribution function* [12]. We let X be a random variable and x be any allowed value of this random variable. Then the probability distribution function is defined to be the probability of the event that the observed random variable X is less than or equal to the allowed value x, that is,

$$\mathscr{P}_X(x) = \Pr(X \leq x)^{\dagger} \tag{A.267}$$

A probability distribution function is a probability; therefore, it must satisfy the same properties as probabilities given in Section A.7.1. However, this function is also a function of x (the possible values of the random variable X) and thus must generally be defined for all values of x. The properties of probability distribution functions are

1. $0 \leq \mathscr{P}_X(x) \leq 1 \qquad -\infty < x < \infty$
2. $\mathscr{P}_X(-\infty) = 0 \qquad \mathscr{P}_X(\infty) = 1$
3. $\mathscr{P}_X$ is nondecreasing as x increases.
4. $\Pr(x_1 < X \leq x_2) = \mathscr{P}_X(x_2) - \mathscr{P}_X(x_1)$

The probability distribution function can also be used to express the probability of the event that the observed random variable X is greater than but

---

†The symbol $\mathscr{P}_X(x)$, denoting the probability distribution function, is used in lieu of $F_X(x)$, typically found in the mathematics literature. Similarly, $\mathscr{p}_X(x)$ will be used for the probability density function instead of $f_X(x)$.

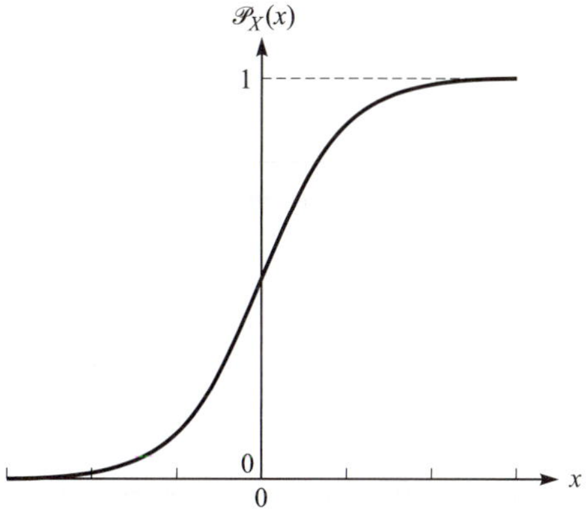

Figure A.7 Typical probability distribution function.

not equal to x. This event is simply the complement of the event having probability $\mathscr{P}_X(x)$, that is,

$$\Pr(X > x) = 1 - \mathscr{P}_X(x) \qquad (A.268)$$

A typical probability distribution function is shown in Figure A.7.

A.7.3 Probability Density Functions

The probability density function is a more convenient form for the probability model for a single random variable. The (marginal) probability density function is the derivative of the probability distribution function (when the derivative exists), that is,

$$p_X(x) = \frac{d\mathscr{P}_X(x)}{dx} \qquad (A.269)$$

The general properties of probability density functions are

1. $p_X(x) \geq 0 \qquad -\infty < x < \infty$
2. $\int_{-\infty}^{\infty} p_X(x)\, dx = 1$
3. $\mathscr{P}_X(u) = \int_{-\infty}^{x} p_X(u)\, du$
4. $\int_{x_1}^{x_2} p_X(x)\, dx = \Pr(x_1 < X \leq x_2)$

Figure A.8 shows a typical probability density function.

A.7.4 Expectation, Mean Values, and Moments

The concept of finding average values for time functions is familiar to engineers and scientists. Time averages are also important for random functions of time; however, they have no meaning for a single random variable (which is defined as the value of the time function at a single instant of time). In the case

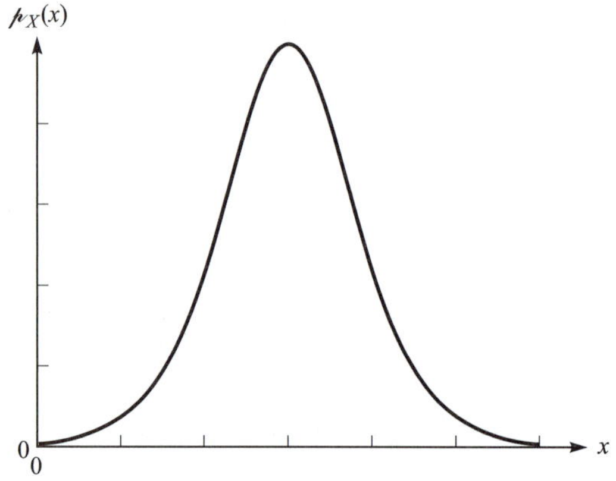

Figure A.8 Probability density function (associated with the probability distribution function in Figure A.7).

of a random variable, it is necessary to find the average value by integrating over the range of possible values that the random variable may assume. This operation is referred to as *ensemble averaging*, and the result is called the *mean value*. The mean value of a random variable X is given by

$$\mathrm{E}[X] = \bar{X} = \int_{-\infty}^{\infty} x p(x)\, \mathrm{d}x \tag{A.270}$$

where $p(x) = p_X(x)$ (the probability density of the random variable—the subscript X will be omitted), and $\mathrm{E}[X]$ is read as the *expectation* of the random variable X (or the *expected value* of X). The expected value of a function of x, that is, $f(x)$, can be determined in a similar manner

$$\mathrm{E}[f(X)] = \int_{-\infty}^{\infty} f(x) p(x)\, \mathrm{d}x \tag{A.271}$$

A function of particular importance is $f(x) = x^n$. This function leads to the general moments of a random variable, thus

$$\mathrm{E}[X^n] = \overline{X^n} = \int_{-\infty}^{\infty} x^n p(x)\, \mathrm{d}x \tag{A.272}$$

When $n = 1$ in (A.272), this leads to the mean value, as previously discussed; and when $n = 2$, this leads to the *mean-square value*

$$\mathrm{E}[X^2] = \overline{X^2} = \int_{-\infty}^{\infty} x^2 p(x)\, \mathrm{d}x \tag{A.273}$$

The *central moments* of a random variable are also important and are defined as the moments of the difference between a random variable and its mean value. Therefore, the nth central moment is defined as

$$\mathrm{E}[(X - \bar{X})^n] = \overline{(X - \bar{X})^n} = \int_{-\infty}^{\infty} (x - \bar{X})^n p(x)\, \mathrm{d}x \tag{A.274}$$

From (A.274) we see that the first central moment ($n = 1$) is zero. The second central moment ($n = 2$) has a special name, the *variance* (σ^2), and from (A.274) it is written as

$$\sigma^2 = \mathrm{E}[(X - \bar{X})^2] = \overline{(X - \bar{X})^2} = \int_{-\infty}^{\infty} (x - \bar{X})^2 p(x)\,\mathrm{d}x \qquad \text{(A.275)}$$

The variance can also be written as

$$\begin{aligned}\sigma^2 &= \mathrm{E}[(X - \bar{X})^2] = \mathrm{E}[X^2 - 2X\bar{X} + \bar{X}^2] \\ &= \mathrm{E}[X^2] - 2\mathrm{E}[X]\bar{X} + \bar{X}^2 = \overline{X^2} - 2\overline{X}\,\overline{X} + \bar{X}^2 = \overline{X^2} - \bar{X}^2\end{aligned} \qquad \text{(A.276)}$$

Therefore, from (A.276) we see that the variance of a random variable can be expressed as the difference between its mean-square value and the square of its mean value. Also, the square root of the variance σ is known as the *standard deviation*.

Of particular interest is the *Gaussian* (or *normal*) density function. The mathematical expression for the Gaussian density function is

$$p(x) = \frac{1}{\sqrt{2\pi}\sigma} \exp\left[\frac{-(x - \bar{X})^2}{2\sigma^2}\right] \qquad -\infty < x < \infty \qquad \text{(A.277)}$$

where $\bar{X}$ is the mean and σ^2 is the variance. Figure A.9 shows the probability density function for a Gaussian random variable.

A *uniform probability density function* can be written as

$$p(x) = \begin{cases} \dfrac{1}{x_2 - x_1} & x_1 < x \le x_2 \\ 0 & \forall \text{ other } x \end{cases}$$

and is shown in Figure A.10.

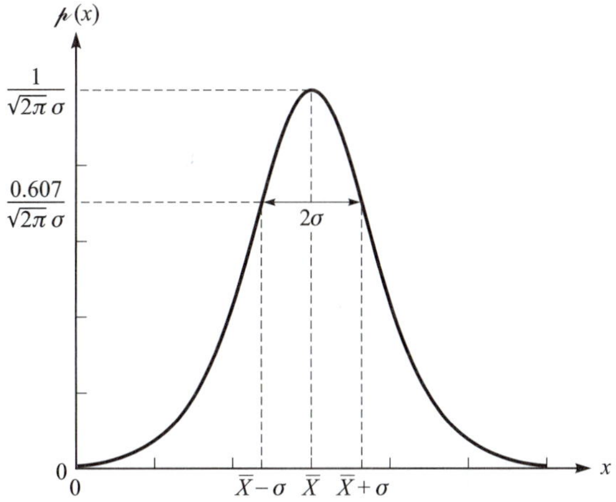

Figure A.9 Probability density function for a Gaussian random variable.

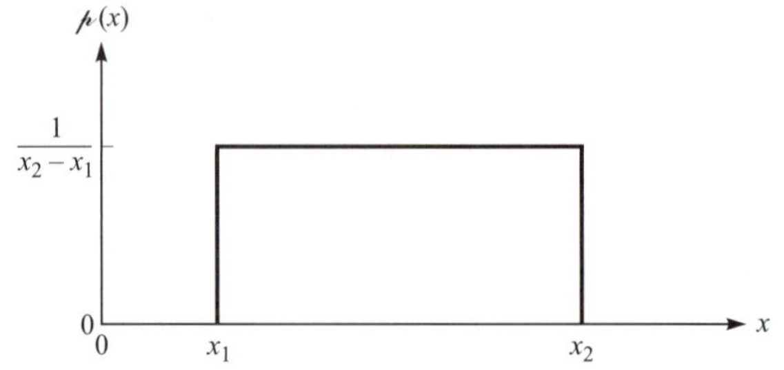

Figure A.10 Uniform probability density function.

The mean value can be computed using (A.270) as

$$\bar{X} = \mathrm{E}[X] = \int_{-\infty}^{\infty} x p(x)\,\mathrm{d}x = \int_{x_1}^{x_2} x \frac{1}{x_2 - x_1}\,\mathrm{d}x = \frac{1}{x_2 - x_1} \frac{x^2}{2}\Big|_{x_1}^{x_2}$$
$$= \frac{1}{x_2 - x_1} \frac{x_2^2 - x_1^2}{2} = \frac{x_1 + x_2}{2} \tag{A.278}$$

and the mean-square value can be computed using (A.273) as

$$\overline{X^2} = \mathrm{E}[X^2] = \int_{-\infty}^{\infty} x^2 p(x)\,\mathrm{d}x = \int_{x_1}^{x_2} x^2 \frac{1}{x_2 - x_1}\,\mathrm{d}x = \frac{1}{x_2 - x_1} \frac{x^3}{3}\Big|_{x_1}^{x_2}$$
$$= \frac{1}{x_2 - x_1} \frac{x_2^3 - x_1^3}{3} = \frac{1}{3}(x_2^2 + x_1 x_2 + x_1^2) \tag{A.279}$$

The variance is given, by using (A.276), (A.278), and (A.279), by

$$\sigma^2 = \overline{X^2} - \bar{X}^2 = \frac{(x_2 - x_1)^2}{12} \tag{A.280}$$

A.7.5 Stochastic Processes

Assume a random experiment with specified outcomes α forms a sample space $\mathcal{S}$, where certain subsets of $\mathcal{S}$ are called events and the probabilities of these events are specified. To every outcome α we can assign (according to a defined rule) a time function $X(t, \alpha)$. For each α, $X(t, \alpha)$ forms a family of functions, and this family is called a *stochastic* (or *random*) *process*. Therefore, a stochasic process is a function of two variables t (time) and α (random experiment outcome), and four different situations can be represented by $X(t, \alpha)$:

1. A family of time functions (t and α variables).
2. A random variable (t fixed and α variable).
3. A single time function (t variable and α fixed), called a *realization* or *sample path* of the stochastic process.
4. A single number (t and α fixed).

Typically, the notation used to depict a stochastic process is $X(t)$, thus omitting the dependence on α (this is normally understood from the context).

For a specific t, $X(t)$ is a random variable, and the distribution function of this random variable in general will depend on t and is given by

$$\mathscr{P}(x; t) = \Pr\{X(t) \leq x\} \tag{A.281}$$

Given two real numbers x and t, the function $\mathscr{P}(x; t)$ equals the probability of the event $\{X(t) \leq x\}$ that consists of all outcomes α such that at a specified time t the functions $X(t)$ of the process do not exceed the given number x. The function $\mathscr{P}(x; t)$ is called the *first-order distribution* of process $X(t)$. The corresponding density function can be obtained by differentiating the distribution function with respect to x

$$p(x; t) = \frac{\partial \mathscr{P}(x; t)}{\partial x} \tag{A.282}$$

At two different times t_1 and t_2 we can consider the random variables $X(t_1)$ and $X(t_2)$. Their *joint distribution* depends, in general, on t_1 and t_2 and can be written as

$$\mathscr{P}(x_1, x_2; t_1, t_2) = \mathscr{P}\{X(t_1) \leq x_1, X(t_2) \leq x_2\} \tag{A.283}$$

The function $\mathscr{P}(x_1, x_2; t_1, t_2)$ is called the *second-order distribution* of process $X(t)$. The corresponding density function is given by

$$p(x_1, x_2; t_1, t_2) = \frac{\partial^2 \mathscr{P}(x_1, x_2; t_1, t_2)}{\partial x_1 \, \partial x_2} \tag{A.284}$$

Also,

$$p(x_1; t_1) = \int_{-\infty}^{\infty} p(x_1, x_2; t_1, t_2) \, dx_2 \tag{A.285}$$

and

$$\mathscr{P}(x_1, \infty; t_1, t_2) = \mathscr{P}(x_1; t_1) \tag{A.286}$$

In addition, the conditional density is given by

$$p(x_1, t_1 | X_2(t_2) = x_2) = \frac{p(x_1, x_2; t_1, t_2)}{p(x_2; t_2)} \tag{A.287}$$

Stationary, nonstationary, and wide-sense-stationary processes

If all marginal and joint density functions of a particular process do not depend upon the choice of the time origin, the process is said to be *stationary*. Therefore, all the corresponding mean values and moments are constants and do not depend on time. If any of the probability density functions change with the choice of the time origin, the process is *nonstationary*. Therefore, one or more of the mean values or moments will also depend on time.

In a strict sense, stationary processes do not physically exist; however, there are many physical situations in which the process does not appreciably change during the time period over which it is being observed. Therefore, a more relaxed requirement would be that the mean of a process is a constant, and its autocorrelation depends only on the time difference $t_2 - t_1$; then the process is said to be *wide-sense-stationary* (wss). Typically, there is no need to distinguish between a stationary process and a wide-sense-stationary one.

Ergodic and nonergodic stochastic processes

There are some stationary stochastic processes that possess the property that almost every member of the ensemble exhibits the same statistical behavior that the entire ensemble possesses. In these cases, it is possible to determine this statistical behavior by analyzing only one typical sample function, and these processes are said to be *ergodic*. For ergodic processes, the mean values and moments can be determined by time averages as well as ensemble averages. For example, the nth general moment can be found from

$$\mathrm{E}[X^n] = \overline{X^n} = \int_{-\infty}^{\infty} x^n \not{p}(x)\,\mathrm{d}x = \lim_{T\to\infty} \frac{1}{2T}\int_{-T}^{T} X^n(t)\,\mathrm{d}t \tag{A.288}$$

A process that does not possess the property shown in (A.288) is said to be *nonergodic*.

Mean, autocorrelation, and autocovariance functions

The *mean* $m_X(t)$ of a stochastic process $X(t)$ is the expected value of $X(t)$, that is,

$$m_X(t) = \mathrm{E}[X(t)] = \int_{-\infty}^{\infty} x \not{p}(x; t)\,\mathrm{d}x \tag{A.289}$$

If $X(t)$ is a sample function from a stochastic process, and we assume two times t_1 and t_2, we can consider the random variables $X(t_1) = X_1$ and $X(t_2) = X_2$. Then the *autocorrelation* function $R_X(t_1, t_2)$ is defined as

$$R_X(t_1, t_2) = \mathrm{E}[X_1 X_2] = \int_{-\infty}^{\infty}\int_{-\infty}^{\infty} x_1 x_2 \not{p}(x_1, x_2; t_1, t_2)\,\mathrm{d}x_1\,\mathrm{d}x_2 \tag{A.290}$$

This definition is applicable for both stationary and nonstationary stochastic processes. However, we are typically interested in stationary processes; therefore, (A.290) can be simplified. For a wide-sense-stationary process, all ensemble averages are independent of the time origin, and accordingly

$$R_X(t_1, t_2) = R_X(t_1 + T, t_2 + T) = \mathrm{E}[X(t_1 + T)X(t_2 + T)] \tag{A.291}$$

Since this equation is independent of the choice of the time origin, we can let $T = -t_1$ and (A.291) can be written as

$$R_X(t_1, t_2) = R_X(0, t_2 - t_1) = \mathrm{E}[X(0)X(t_2 - t_1)] \tag{A.292}$$

and because this expression depends on only the time difference $t_2 - t_1$, we let $\tau = t_2 - t_1$ and write (A.292) as

$$R_X(\tau) = R_X(t_2 - t_1) = \mathrm{E}[X(t_1)X(t_1 + \tau)] \tag{A.293}$$

From (A.293), we see that the autocorrelation function depends only on τ and not on the value of t_1. Because the autocorrelation function does not depend on the particular time t_1 at which the ensemble averages are taken for a wide-sense-stationary process, we can write (A.293) as

$$R_X(\tau) = \mathrm{E}[X(t)X(t+\tau)] \tag{A.294}$$

A *time autocorrelation* function can also be defined for a particular sample function as

$$\mathscr{R}_x(\tau) = \lim_{T\to\infty} \frac{1}{2T}\int_{-T}^{T} x(t)x(t+\tau)\,\mathrm{d}t \tag{A.295}$$

For the special case of an ergodic process, $\mathscr{R}_x(\tau)$ is the same for every $x(t)$ and is equal to $R_X(\tau)$, that is,

$$\mathscr{R}_x(\tau) = R_X(\tau) \qquad \text{for an ergodic process} \tag{A.296}$$

The general properties of autocorrelation functions for stationary processes are as follows:

1. $R_X(0) = \overline{X^2}$. The mean-square value of a stochastic process can be obtained by setting $\tau = 0$ in the autocorrelation function.
2. $R_X(\tau) = R_X(-\tau)$. The autocorrelation function is an even function of τ.
3. $|R_X(\tau)| \le R_X(0)$. The largest value of the autocorrelation function will always occur at $\tau = 0$.
4. If the stochastic process $X(t)$ has a mean value, then $R_X(\tau)$ will have a constant component.
5. If $X(t)$ has a periodic component, then $R_X(\tau)$ will also have a periodic component with the same period.
6. If $\{X(t)\}$ is ergodic and zero-mean and has no periodic component, then

   $$\lim_{|\tau|\to\infty} R_X(\tau) = 0$$

   As τ becomes large, the random variables tend to become statistically independent, because the effect of past values tends to "die out" as time progresses.

The *autocovariance* function of a stochastic process $X(t)$ is the covariance of the random variables $X(t_1)$ and $X(t_2)$, that is

$$C_X(t_1, t_2) = \mathrm{E}\{[X(t_1) - m_X(t_1)][X(t_2) - m_X(t_2)]\} \tag{A.297}$$

It is straightforward to show that the autocovariance function can be written as

$$C_X(t_1, t_2) = R_X(t_1, t_2) - m_X(t_1)m_X(t_2) \tag{A.298}$$

From (A.298) it is easy to see that if $X(t)$ is zero-mean, the autocorrelation and autocovariance functions are the same for the process.

The variance of $X(t)$ is given by

$$\begin{aligned}\sigma^2_{X(t)} &= \mathrm{E}\{[X(t) - m_X(t)]^2\} = C_X(t, t)\\ &= R_X(t, t) - m^2_X(t)\end{aligned} \tag{A.299}$$

The *correlation coefficient* of $X(t)$ is defined as the correlation coefficient of $X(t_1)$ and $X(t_2)$

$$\rho_X(t_1, t_2) = \frac{C_X(t_1, t_2)}{\sqrt{C_X(t_1, t_1)}\sqrt{C_X(t_2, t_2)}} \tag{A.300}$$

The correlation coefficient is a measure of the extent to which a random variable can be predicted as a linear function of another.

A.7.6 Vector Stochastic Processes

Suppose that $x_1(t), x_2(t), \ldots, x_n(t)$ are n scalar stochastic processes (which are possibly mutually dependent). Then

$$\boldsymbol{x}(t) = [x_1(t), x_2(t), \ldots, x_n(t)]^T \tag{A.301}$$

is called a *vector stochastic process*. The *mean* of the vector stochastic process is given by

$$\boldsymbol{m}_x(t) = \mathrm{E}[\boldsymbol{x}(t)] \tag{A.302}$$

the *correlation matrix* is

$$\boldsymbol{R}_x(t_1, t_2) = \mathrm{E}\{\boldsymbol{x}(t_1)\boldsymbol{x}^T(t_2)\} \tag{A.303}$$

and the *covariance matrix* is given by

$$\boldsymbol{C}_x(t_1, t_2) = \mathrm{E}\{[\boldsymbol{x}(t_1) - \boldsymbol{m}_x(t_1)][\boldsymbol{x}(t_2) - \boldsymbol{m}_x(t_2)]^T\} \tag{A.304}$$

If $\boldsymbol{x}(t)$ is a wide-sense-stationary stochastic process, the mean $\boldsymbol{m}_x$ in (A.302) is constant, the correlation matrix $\boldsymbol{R}_x(t, t)$ in (A.303) is finite for all t, and the covariance matrix in (A.304) depends only on $t_2 - t_1$, that is, $\boldsymbol{C}_x(t_2 - t_1)$.

A Gaussian (normal) vector stochastic process $\boldsymbol{x}$ is a vector stochastic process in which the set of n random variables has a joint probability distribution that is Gaussian. The corresponding probability density function for $\boldsymbol{x}$ is then given as

$$\not{p}_x(\boldsymbol{\xi}) = \frac{1}{(2\pi)^{n/2}[\det(\boldsymbol{C}_x)]^{1/2}} \exp\left[\frac{-(\boldsymbol{\xi} - \boldsymbol{m}_x)^T \boldsymbol{C}_x^{-1}(\boldsymbol{\xi} - \boldsymbol{m}_x)}{2}\right] \tag{A.305}$$

where $\boldsymbol{\xi}$ is the allowed value of $\boldsymbol{x}$, and it is assumed that the inverse of the covariance matrix $\boldsymbol{C}_x$ exists.

A.7.7 Power Spectral Density Functions and Matrices

The Fourier transform of the autocorrelation function of a wide-sense-stationary stochastic process is called the *power spectral density* function and is denoted as $S_x(\omega)$. Assuming a wss stochastic process $x(t)$, the power spectral density function is given by

$$S_x(\omega) = \int_{-\infty}^{\infty} R_x(\tau)e^{-j\omega\tau}\, d\tau \tag{A.306}$$

The mean-square value of the stochastic process can be found from the power spectral density function as

$$\overline{x^2} = \frac{1}{2\pi}\int_{-\infty}^{\infty} S_x(\omega)\,\mathrm{d}\omega \tag{A.307}$$

If $\boldsymbol{x}(t)$ is a vector-valued wss stochastic process as in (A.301), with a correlation matrix $\boldsymbol{R}_x(\tau)$, the power spectral density matrix is written as

$$\boldsymbol{S}_x(\omega) = \int_{-\infty}^{\infty} \boldsymbol{R}_x(\tau)e^{-j\omega\tau}\,\mathrm{d}\tau \tag{A.308}$$

The general properties of power spectral density matrices are as follows:

1. $\boldsymbol{S}_x(-\omega) = \boldsymbol{S}_x^T(\omega), \forall\omega$
2. $\boldsymbol{S}_x^*(\omega) = \boldsymbol{S}_x(\omega), \forall\omega$ (where the asterisk denotes the complex conjugate transpose)
3. $\boldsymbol{S}_x(\omega) \geq 0, \forall\omega$ [i.e., $\boldsymbol{S}_x(\omega)$ is positive semidefinite, or nonnegative definite]

Exponentially correlated noise example

We will consider a scalar wss stochastic process $x(t)$ with an autocorrelation function given as

$$R_x(\tau) = \sigma^2 e^{-|\tau|/\theta} \qquad \theta > 0 \tag{A.309}$$

The power spectral density function can be found by taking the Fourier transform of $R_x(\tau)$ as shown in (A.306):

$$\begin{aligned} S_x(\omega) &= \sigma^2\int_{-\infty}^{\infty} e^{-|\tau|/\theta}e^{-j\omega\tau}\,\mathrm{d}\tau \\ &= \sigma^2\int_{-\infty}^{0} e^{\tau/\theta}e^{-j\omega\tau}\,\mathrm{d}\tau + \sigma^2\int_{0}^{\infty} e^{-\tau/\theta}e^{-j\omega\tau}\,\mathrm{d}\tau \\ &= \frac{2\sigma^2\theta}{1+\omega^2\theta^2} \qquad \theta > 0 \end{aligned} \tag{A.310}$$

A.7.8 Linear Systems Driven by White Noise and Spectral Factorization

Linear systems driven by white noise

Consider the linear time-invariant system described by

$$\begin{aligned} \dot{\boldsymbol{x}}(t) &= \boldsymbol{A}\boldsymbol{x}(t) + \boldsymbol{B}\boldsymbol{w}(t) \\ \boldsymbol{x}(0) &= \boldsymbol{x}_0 \end{aligned} \tag{A.311}$$

where $\boldsymbol{w}(t)$ is *white noise* with constant intensity $\boldsymbol{V}$; that is, $\boldsymbol{w}(t)$ is a wide-sense-stationary process with a power spectral density matrix given by

$$\boldsymbol{S}_w(\omega) = \boldsymbol{V} \tag{A.312}$$

It is assumed that $\boldsymbol{A}$ is asymptotically stable (cf. Sect. A.2.12). The initial-condition vector $\boldsymbol{x}_0$ in (A.311) is a random variable independent of $\boldsymbol{w}(t)$, with mean $\boldsymbol{m}_0$ and $\boldsymbol{Q}_0 = \mathrm{E}[(\boldsymbol{x}_0 - \boldsymbol{m}_0)(\boldsymbol{x}_0 - \boldsymbol{m}_0)^T]$ its variance matrix. Then $\boldsymbol{x}(t)$ has mean

$$\boldsymbol{m}_x(t) = \boldsymbol{\phi}(t)\boldsymbol{m}_0 = e^{\boldsymbol{A}t}\boldsymbol{m}_0 \tag{A.313}$$

where $\boldsymbol{\phi}(t) = e^{\boldsymbol{A}t}$ is the state transition matrix (cf. Sect. A.2.12). The steady-state variance matrix corresponding to $\boldsymbol{x}(t)$ can be found by solving the steady-state algebraic Lyapunov equation

$$\boldsymbol{AQ} + \boldsymbol{QA}^T + \boldsymbol{BVB}^T = \boldsymbol{0} \tag{A.314}$$

for $\boldsymbol{Q}$ [13]. The steady-state variance matrix can also be found from [13]

$$\boldsymbol{Q} = \int_0^\infty e^{\boldsymbol{A}\tau}\boldsymbol{BVB}^T e^{\boldsymbol{A}^T\tau}\,\mathrm{d}\tau \tag{A.315}$$

or from [13]

$$\boldsymbol{Q} = \int_{-\infty}^{\infty} (j\omega\boldsymbol{I} - \boldsymbol{A})^{-1}\boldsymbol{BVB}^T(-j\omega\boldsymbol{I} - \boldsymbol{A}^T)^{-1}\,\mathrm{d}f \tag{A.316}$$

where $\omega = 2\pi f$ rad/s.

Spectral factorization

Consider an asymptotically stable linear time-invariant system with a transfer function matrix $\boldsymbol{H}(s)$, where s is the Laplace variable. If the input to this system is a realization of a wss stochastic process $\boldsymbol{u}(t)$ with power spectral density matrix $\boldsymbol{S}_u(\omega)$, then the output of the system is a realization of a wss stochastic process $\boldsymbol{y}(t)$ with power spectral density matrix given by [13]

$$\boldsymbol{S}_y(\omega) = \boldsymbol{H}(j\omega)\boldsymbol{S}_u(\omega)\boldsymbol{H}^T(-j\omega) \tag{A.317}$$

where $\boldsymbol{H}(j\omega) = \boldsymbol{H}(s)|_{s=j\omega}$ is the sinusoidal steady-state transfer function matrix. For a scalar system with a transfer function $H(s)$, and a scalar wss stochastic input $u(t)$ with a power spectral density function $S_u(\omega)$, the power spectral density function of the output $y(t)$ is given by

$$S_y(\omega) = H(j\omega)H(-j\omega)S_u(\omega) \tag{A.318}$$

In other words, the power spectral density function of the output is proportional to the square of the magnitude of the transfer function. The result in (A.318) can also be expressed in terms of the Laplace transform as

$$S_y(s) = H(s)H(-s)S_u(s) \tag{A.319}$$

This approach requires the use of the two-sided Laplace transform [14].

Many times it is necessary to model a stochastic process, given only the power spectral density of the process. More specifically, given the power spectral density function of a wss stochastic process, what are the characteristics of a *shaping filter*, such that when this shaping filter has its input white noise, the output is a realization of a wss stochastic process with the defined power spectral density function? The answer is actually quite simple, and the method is straightforward and is referred to as *spectral factorization* [14–16]. The method involves decomposing the power spectral density function of the process into its positive and negative time parts. The positive time part is the necessary *shaping filter*, such that when white noise is introduced to it, the response will be a realization of the stochastic process. This can be seen from (A.318). If the input to the shaping filter is considered to be zero-mean, unity-variance white noise, the power spectral density of the output is $S_y(\omega) = H(j\omega)H(-j\omega)$, because the spectral density of the white noise is $S_w(\omega) = 1$; recall that the autocorrelation of the white noise is $R_w(\tau) = \delta(\tau)$. Therefore, $H(j\omega)$ is the desired shaping filter sinusoidal steady-state transfer function that must be *unraveled* (decomposed) from $S_y(\omega)$. The following examples will illustrate the procedure.

In this first example the power spectral density of the process of interest is for exponentially correlated noise given in (A.310), that is,

$$S_y(\omega) = \frac{2\sigma^2\theta}{1+\omega^2\theta^2} \qquad \theta > 0 \tag{A.320}$$

This function can be written (decomposed) as

$$S_y(\omega) = \underbrace{\frac{\sqrt{2\sigma^2\theta}}{1+j\omega\theta}}_{\substack{\text{positive}\\\text{time part}}}\;\underbrace{\frac{\sqrt{2\sigma^2\theta}}{1-j\omega\theta}}_{\substack{\text{negative}\\\text{time part}}} \tag{A.321}$$

Therefore, the sinusoidal steady-state transfer function of the shaping filter, when driven by white noise, will have as its output the desired stochastic process. This transfer function is given as

$$H(j\omega) = \frac{\sqrt{2\sigma^2\theta}}{1+j\omega\theta} \tag{A.322}$$

and in terms of the Laplace transform the transfer function is

$$H(s) = \frac{\sqrt{2\sigma^2\theta}}{1+\theta s} \tag{A.323}$$

The result in (A.323) can be derived directly from (A.319) when the power spectral density of the exponentially correlated noise process is given in terms of the Laplace transform, that is,

$$S_y(s) = \frac{2\sigma^2\theta}{1-\theta^2 s^2} \tag{A.324}$$

In the second example the power spectral density function of the stochastic process is given as (in terms of the Laplace transform)

$$S_y(s) = \frac{-s^2+1}{s^2+64} \tag{A.325}$$

Completing the square on the denominator polynomial yields the set of roots $-2 \pm j2$ and $2 \pm j2$, and the decomposition gives

$$S_y(s) = \underbrace{\frac{s+1}{s^2+4s+8}}_{\substack{\text{positive}\\ \text{time part}}} \underbrace{\frac{-s+1}{s^2-4s+8}}_{\substack{\text{negative}\\ \text{time part}}} \tag{A.326}$$

Therefore, the shaping filter transfer function is given by

$$H(s) = \frac{s+1}{s^2+4s+8} \tag{A.327}$$

A.8 FUZZY SET THEORY

A classical set A (cf. Sect. A.3.1) is defined as a collection of elements or objects $x \in X$ such that each x can either belong to or not belong to the set A, $A \subset X$ (i.e., A is a subset of X). By defining a *characteristic function* (or *membership function*) on each element x in A, a classical set can be represented by a set of ordered pairs $(x, 0)$ or $(x, 1)$, where 1 indicates membership and 0 nonmembership. The characteristic function is also known as an *indicator function*, defined as

$$I_A(x) = \begin{cases} 1 & \text{if } x \in A \\ 0 & \text{if } x \notin A \end{cases} \tag{A.328}$$

The indicator functions of the intersection or union of two sets A and B are easily expressed in terms of the indicator function of a set A and the indicator function of a set B

$$I_{A\cap B}(x) = \min[I_A(x), I_B(x)] \tag{A.329}$$

$$I_{A\cup B}(x) = \max[I_A(x), I_B(x)] \tag{A.330}$$

Similarly, the indicator function for the complement of a set A (i.e., $\tilde{A}$) is given by

$$I_{\tilde{A}}(x) = 1 - I_A(x) \tag{A.331}$$

The condition for A to be a subset of B can be expressed as

$$A \subset B \Leftrightarrow I_A(x) \le I_B(x) \qquad \forall x \in X \tag{A.332}$$

Unlike in the conventional sets as described above, each element of the universal set X is an element of the *fuzzy* A to a *certain degree*. Hence, the characteristic function, denoting the degree of membership of an element in a given fuzzy set, is allowed to have any value between 0 and 1. If X is a collection of objects denoted generically as $\{x\}$, then a fuzzy set A in X is defined as a set of ordered pairs such that

$$A = \{(x, m_A(x)) | x \in X\} \tag{A.333}$$

where $m_A(x)$ is the membership function of x in A, which maps X to the interval [0, 1]. Typically the elements of X are ordered, and set A is specified by listing only the membership function values. When $m_A(x)$ takes on only the values 0 and 1, then A is nonfuzzy (or *crisp*), and $m_A(x)$ is identical to the characteristic function of a nonfuzzy set.

Zadeh [17] defined the fuzzy intersection (MIN) and the fuzzy union (MAX) of fuzzy sets A and B, and the complement of A, by expressions exactly analogous to those for standard sets. In a similar manner, Zadeh proposed defining A to be a (fuzzy) subset of B by using the membership functions for A and B, that is,

$$A \subset B \Leftrightarrow m_A(x) \leq m_B(x) \qquad \forall x \in X \tag{A.334}$$

This is referred to as the *dominated membership function*. For example, if $A = \{0.3, 0.0, 0.7\}$ and $B = \{0.4, 0.7, 0.9\}$, then A is a fuzzy subset of B, but B is not a fuzzy subset of A.

A.9 SELECTED TRIGONOMETRIC IDENTITIES

Pythagorean formulas

$$\sin^2 \alpha + \cos^2 \alpha = 1$$

$$1 + \tan^2 \alpha = \sec^2 \alpha$$

$$1 + \cot^2 \alpha = \csc^2 \alpha$$

Reciprocal formulas

$$\sin \alpha = \frac{1}{\csc \alpha} \qquad \cos \alpha = \frac{1}{\sec \alpha} \qquad \tan \alpha = \frac{1}{\cot \alpha}$$

$$\csc \alpha = \frac{1}{\sin \alpha} \qquad \sec \alpha = \frac{1}{\cos \alpha} \qquad \cot \alpha = \frac{1}{\tan \alpha}$$

Quotient formulas

$$\sin \alpha = \frac{\tan \alpha}{\sec \alpha} \qquad \cos \alpha = \frac{\cot \alpha}{\csc \alpha} \qquad \tan \alpha = \frac{\sin \alpha}{\cos \alpha}$$

$$\csc \alpha = \frac{\sec \alpha}{\tan \alpha} \qquad \sec \alpha = \frac{\csc \alpha}{\cot \alpha} \qquad \cot \alpha = \frac{\cos \alpha}{\sin \alpha}$$

Product formulas

$$\sin\alpha = \tan\alpha\cos\alpha \qquad \cos\alpha = \cot\alpha\sin\alpha \qquad \tan\alpha = \sin\alpha\sec\alpha$$
$$\cot\alpha = \cos\alpha\csc\alpha \qquad \sec\alpha = \csc\alpha\tan\alpha \qquad \csc\alpha = \sec\alpha\cot\alpha$$

$$\sin\alpha\sin\beta = \frac{1}{2}\cos(\alpha-\beta) - \frac{1}{2}\cos(\alpha+\beta)$$
$$\cos\alpha\cos\beta = \frac{1}{2}\cos(\alpha-\beta) + \frac{1}{2}\cos(\alpha+\beta)$$
$$\sin\alpha\cos\beta = \frac{1}{2}\sin(\alpha+\beta) + \frac{1}{2}\sin(\alpha-\beta)$$
$$\cos\alpha\sin\beta = \frac{1}{2}\sin(\alpha+\beta) - \frac{1}{2}\sin(\alpha-\beta)$$

Angle-sum and angle-difference formulas

$$\sin(\alpha+\beta) = \sin\alpha\cos\beta + \cos\alpha\sin\beta$$
$$\sin(\alpha-\beta) = \sin\alpha\cos\beta - \cos\alpha\sin\beta$$
$$\cos(\alpha+\beta) = \cos\alpha\cos\beta - \sin\alpha\sin\beta$$
$$\cos(\alpha-\beta) = \cos\alpha\cos\beta + \sin\alpha\sin\beta$$
$$\tan(\alpha+\beta) = \frac{\tan\alpha + \tan\beta}{1 - \tan\alpha\tan\beta}$$
$$\tan(\alpha-\beta) = \frac{\tan\alpha - \tan\beta}{1 + \tan\alpha\tan\beta}$$
$$\sin(\alpha+\beta)\sin(\alpha-\beta) = \sin^2\alpha - \sin^2\beta = \cos^2\beta - \cos^2\alpha$$
$$\cos(\alpha+\beta)\cos(\alpha-\beta) = \cos^2\alpha - \sin^2\beta = \cos^2\beta - \sin^2\alpha$$

Double-angle formulas

$$\sin 2\alpha = 2\sin\alpha\cos\alpha = \frac{2\tan\alpha}{1+\tan^2\alpha}$$
$$\cos 2\alpha = \cos^2\alpha - \sin^2\alpha = 2\cos^2\alpha - 1 = 1 - 2\sin^2\alpha = \frac{1-\tan^2\alpha}{1+\tan^2\alpha}$$
$$\tan 2\alpha = \frac{2\tan\alpha}{1-\tan^2\alpha} \qquad \cot 2\alpha = \frac{\cot^2\alpha - 1}{2\cot\alpha}$$

Power formulas

$$\sin^2\alpha = \frac{1}{2}(1 - \cos 2\alpha) \qquad \sin^3\alpha = \frac{1}{4}(3\sin\alpha - \sin 3\alpha)$$

$$\sin^4\alpha = \frac{1}{8}(3 - 4\cos 2\alpha + \cos 4\alpha)$$

$$\cos^2\alpha = \frac{1}{2}(1 + \cos 2\alpha) \qquad \cos^3\alpha = \frac{1}{4}(3\cos\alpha + \cos 3\alpha)$$

$$\cos^4\alpha = \frac{1}{8}(3 + 4\cos 2\alpha + \cos 4\alpha)$$

$$\tan^2\alpha = \frac{1 - \cos 2\alpha}{1 + \cos 2\alpha} \qquad \cot^2\alpha = \frac{1 + \cos 2\alpha}{1 - \cos 2\alpha}$$

Half-angle formulas

$$\sin\frac{\alpha}{2} = \pm\sqrt{\frac{1 - \cos\alpha}{2}} \qquad \cos\frac{\alpha}{2} = \pm\sqrt{\frac{1 + \cos\alpha}{2}}$$

$$\tan\frac{\alpha}{2} = \pm\sqrt{\frac{1 - \cos\alpha}{1 + \cos\alpha}} = \frac{1 - \cos\alpha}{\sin\alpha} = \frac{\sin\alpha}{1 + \cos\alpha}$$

$$\cot\frac{\alpha}{2} = \pm\sqrt{\frac{1 + \cos\alpha}{1 - \cos\alpha}} = \frac{1 + \cos\alpha}{\sin\alpha} = \frac{\sin\alpha}{1 - \cos\alpha}$$

Euler's formulas

$$e^{\pm j\alpha} = \cos\alpha \pm j\sin\alpha \qquad j = \sqrt{-1}$$

$$\sin\alpha = \frac{e^{j\alpha} - e^{-j\alpha}}{2j} \qquad \cos\alpha = \frac{e^{j\alpha} + e^{-j\alpha}}{2}$$

$$\tan\alpha = -j\left(\frac{e^{j\alpha} - e^{-j\alpha}}{e^{j\alpha} + e^{-j\alpha}}\right) = -j\left(\frac{e^{j2\alpha} - 1}{e^{j2\alpha} + 1}\right)$$

Function-sum and function-difference formulas

$$\sin\alpha + \sin\beta = 2\sin\frac{1}{2}(\alpha + \beta)\cos\frac{1}{2}(\alpha - \beta)$$

$$\sin\alpha - \sin\beta = 2\cos\frac{1}{2}(\alpha + \beta)\sin\frac{1}{2}(\alpha - \beta)$$

$$\cos\alpha + \cos\beta = 2\cos\frac{1}{2}(\alpha + \beta)\cos\frac{1}{2}(\alpha - \beta)$$

$$\cos\alpha - \cos\beta = -2\sin\frac{1}{2}(\alpha + \beta)\sin\frac{1}{2}(\alpha - \beta)$$

$$\tan\alpha + \tan\beta = \frac{\sin(\alpha + \beta)}{\cos\alpha\cos\beta} \qquad \tan\alpha - \tan\beta = \frac{\sin(\alpha - \beta)}{\cos\alpha\cos\beta}$$

REFERENCES

1. G. H. Golub and C. F. Van Loan, *Matrix Computations*, 3rd ed., Balitmore, MD: Johns Hopkins University Press, 1996.
2. C. T. Chen, *Linear System Theory and Design*, New York: Holt, Rinehart and Winston, 1984.
3. C. V. Churchill, J. W. Brown, and R. F. Verhey, *Complex Variables and Applications*, 3rd ed., New York: McGraw-Hill, 1976.
4. J. W. Brewer,"Kronecker Products and Matrix Calculus in System Theory," *IEEE Transactions on Circuits and Systems*, vol. CAS-25, 1978, pp. 772–81.
5. R. A. Horn and C. R. Johnson, *Topics in Matrix Analysis*, New York: Cambridge University Press, 1991.
6. K. Zhou, J. C. Doyle, and K. Glover, *Robust and Optimal Control*, Upper Saddle River, NJ: Prentice-Hall, 1996.
7. N. J. Higham, *Accuracy and Stability of Numerical Algorithms*, Philadelphia, PA: Society for Industrial and Applied Mathematics, 1996.
8. W. D. Wallis, "Hadamard Matrices," in *Combinatorial and Graph-Theoretical Problems in Linear Algebra*, eds. R. A. Brualdi, S. Friedland, and V. Klee, vol. 50, New York: Springer-Verlag, 1993, pp. 235–43.
9. S. G. Nash and A. Sofer, *Linear and Nonlinear Programming*, New York: McGraw-Hill, 1996.
10. D. G. Luenberger, *Linear and Nonlinear Programming*, 2nd ed., Reading, MA: Addison-Wesley, 1984.
11. M. Avriel, *Nonlinear Programming: Analysis and Methods*, Englewood Cliffs, NJ: Prentice-Hall, 1976.
12. A. Papoulis, *Probability, Random Variables, and Stochastic Processes*, 3rd ed., New York: McGraw-Hill, 1991.
13. H. Kwakernaak and R. Sivan, *Linear Optimal Control Systems*, New York: Wiley-Interscience, 1972.
14. R. G. Brown and P. Y. C. Hwang, *Introduction to Random Signals and Applied Kalman Filtering*, 2nd ed., New York: Wiley, 1992.
15. B. Picinbono, *Random Signals and Systems*, Englewood Cliffs, NJ: Prentice-Hall, 1993.
16. T. K. Moon and W. C. Stirling, *Mathematical Methods and Algorithms*, Upper Saddle River, NJ: Prentice-Hall, 2000.
17. L. A. Zadeh, "Fuzzy Sets," *Information and Control*, vol. 8, 1965, pp. 338–53.

Name Index

Subject Index